CRC
HANDBOOK OF
LABORATORY
SAFETY 5TH EDITION

CRC
HANDBOOK OF
LABORATORY
SAFETY 5TH EDITION

A. KEITH FURR

CRC Press
Boca Raton   London   New York   Washington, D.C.

Library of Congress Cataloging-in-Publication Data

Furr, A. Keith, 1932-
    CRC handbook of laboratory safety / A. Keith Furr. -- 5th ed.
      p.   cm.
    Includes bibliographical references and index.
    ISBN 0-8493-2523-4 (alk. paper)
    1. Chemical laboratories--Safety measures--Handbooks, manuals, etc. I. Title.

QD51 .C73 2000
542′.1′0289—dc211

99-086773
CIP

Visit the CRC Press Web site at www.crcpress.com

No claim to original U.S. Government works
International Standard Book Number 0-8493-2523-4
Library of Congress Card Number 99-086773
Printed in the United States of America   4 5 6 7 8 9 0
Printed on acid-free paper

In rereading the foreword for the 4th edition, I noted the statement "There does not appear to be much pending for the immediate future." I was wrong. There have been significant changes in many areas, such as respiratory protection, and dramatic changes in the Clean Air Act which have made it difficult to use incineration as a means of disposing of hospital, medical, and infectious waste. Perhaps the most significant change however has been in means of communication, with the explosive growth of the Internet. This has placed a tremendous amount of information available to anyone with a computer and a modem. Indeed, there is so much information, one must be careful to select that which is useful and accurate. This resource has greatly influenced preparation of this handbook.

At first glance, one might assume that little has changed in much of this edition. Again, this would not be correct. The same general topics do remain for the most part, but several older and now obsolete articles have been completely removed and replaced, either with new material on the same subject or by completely new material, representing over a hundred pages. Where the material may at first glance look familiar, please look more carefully. Every word on every page has been scrutinized and there are literally hundreds of changes to bring the material up to date or clarify the presentation.

There are new figures illustrating new material, and new tables. Full use of the Internet has been made to make sure the information is as up-to-date as possible as of the end of the summer of 1999. In addition to the usual journal articles as references, most topics now include Internet references which were used and which I believe will be helpful.

I am pleased with this edition and believe it is the most authoritative of those for which I have been responsible. One point I wish to close with and that is, safety in the laboratory is not just a rigid adherence to regulatory standards and guidelines. It must take into account human factors as well, and unfortunately the first part of the old adage "To err is human, to forgive is divine" is all too true. Human nature being what it is, the vast proportion of breakdowns in laboratory safety are due to human error, sometimes due to oversights but also sometimes due to a feeling that it's not important or individuals feel they will not make a mistake. This is reflected throughout the handbook and reflects either my personal experience or observations. I hope that no one is bothered by this intrusion.

I hope you will find the handbook useful as many have been kind enough to tell me they did the previous editions.

A. Keith Furr, Ph.D., was, until his retirement in late 1994, Head of the Department of Environmental Health and Safety at Virginia Polytechnic Institute and State University, Blacksburg, Virginia, and Professor of Nuclear Science and Engineering. He received an A.B. degree, cum laude, from Catawba College in 1954, an M.S. degree from Emory University in 1955, and a Ph.D. from Duke University in 1962. From 1960 until 1971, he was in the Department of Physics at VPI & SU where he attained the rank of Professor. In 1971, he transferred to Engineering as Professor of Nuclear Science and Engineering. In 1975, he established the Environmental Health and Safety Department at the University. A unique feature of this department is that it eventually included a University volunteer rescue squad composed entirely of students. In addition to other assignments, he was Director of the Nuclear Reactor Facility and Head of the Neutron Activation Analysis Laboratory. During the early 1970s, he created an undergraduate program in Radiation Safety and afterward he participated in a broad program in Industrial Safety in the University's Department of Industrial and Systems Engineering. In recent years, he played a leadership role in developing a unique program in correcting indoor air quality problems in the University. He belongs to the Health Physics Society, the Campus Safety Association, the National Safety Council, and the National Fire Protection Association. He has published over 60 articles in professional journals, many in the area of environmental studies, three encyclopedia articles and was editor and principal contributor to the two previous editions of this handbook. After his retirement, he became a member of the advisory board of the Laboratory Safety & Environmental Management Newsletter and Conference. He has contributed numerous articles to the Newsletter. Dr. Furr has been active in working with public bodies to develop programs that respond to environmental emergencies; that address the disposal of hazardous materials, infectious wastes, and other solid wastes; and that are environmentally and economically sound. He was Chair of the Montgomery County Local Emergency Planning Committee and a member of the County Solid Waste and Recycling Committee. He was also Chair of the Blacksburg Telecommunications Committee and of a group of active Internet users called the Blacksburg Electronic Village Seniors.

He and his wife moved to Brooksville, Florida in May, 1998, where most of this handbook was prepared.

John W. Cure, III, Health Physics Consultation, Lynchburg, Virginia

Richard F. Desjardins, M.D., Wilmington, North Carolina

Lawrence G. Doucet, P.E., Doucet & Mainka, **P.C.,** Peekskill, New York

Caldwell N. Dugan, Division of Institutional Resources, National Science Foundation, Washington, D.C.

Frank A. Graf, Jr., Research Scientist, Westinghouse Hanford Company, Richland, Washington

Scott A. Heider, Division of Institutional Resources, National Science Foundation, Washington, D.C.

Harold Horowitz, Division of Institutional Resources, National Science Foundation, Washington, D.C.

Alvin B. Kaufman, Litton Systems Divison, Litton Industries, Woodland Hills, California

Edwin N. Kaufman, Senior Scientist, Douglas Aircraft Co., Woodland Hills, California

David M. Moore, D.V.M., Director, Laboratory Animal Resources, College of Veterinary Medicine, Virginia Polytechnic Institute and State University, Blacksburg, Virginia

Eric W. Spencer, Brown University, Providence, Rhode Island

William L. Sprout, M.D., Medical Consultant, Haskell Laboratory, E.I. du Pont de Nemours & Co., Newark, Delaware

Norman V. Steere, Laboratory Safety and Design Consultant, Minneapolis, Minnesota

M.A. Trevino, M.D., Medical Director, Quimca Fluor, S.A. de C.V., Matamoros, Tamaulipas, Mexico

Paul Woodruff, Environmental Resources Management, Inc., West Chester, Pennsylvania

Note that these are only those individuals specifically named in the text. As noted in the Dedication, the actual number of contributors was much, much greater.

There are many individuals who must be thanked for their assistance with this book. As with the 4th edition, my son Rob Furr who assisted me with solving technical problems. Individuals such as John Cure, Lawrence Doucet, and David Moore who contributed material for the 4th edition that has been carried over to this edition. Many other individuals who gave lectures at meetings and with whom I talked at these meetings, whose insights and actual knowledge found their way into the material. I hope that I've acknowledged all of these at the relevant points in the text, by means of either a mention in the text or as a reference. I may have missed some inadvertently, and if so, I hope that they accept my sincere apologies. One group which I cannot overlook, but whom I cannot specifically name in most cases are those individuals who compiled the numerous standards and guidelines to which I have lavishly referred in this book. There are those who complain about regulatory standards but one must have a mark at which to aim. There are also numerous persons and organizations who have made information available on the Internet and this information has been extremely helpful.

I must single out my wife for this edition for even more thanks than in the previous ones. Not only did she tolerate the chaos and confusion alluded to in the last edition, but we moved during the course of preparing this edition from a home in which we had lived for 30 years. She assumed much of the burden of that move and getting the new home organized. In addition, she read every word of this edition to proofread it and make suggestions to improve its clarity. The publisher also proofread the manuscript and eliminated still more errors. I now more fully appreciate proofreaders! However, I assume full responsibility for any residual errors since I followed up the previous reviews with a final review. I hope there are none but I am sure some may still lurk within the pages.

Finally, I want to thank the purchasers of previous editions for their support. If they had not made the 3rd and 4th editions a success, this one would not have been done.

Chapter 1

INTRODUCTION

Safety and health programs for industrial organizations began to be significant early in the 20th century, and since then, there were slow but steady improvements in the industrial working environment, until, in the early 1970s, the Federal Occupational Health and Safety Act was passed as a national program to establish minimum standards for safety for industrial workers. In order to comply with this Act, corporations in the United States have had to create formal safety programs, and usually, internal safety organizations to comply with the regulations. The more enlightened of these firms, recognizing the value to themselves and their employees of keeping employees safe and well, strongly support these internal units. Smaller firms sometimes have made a lesser commitment proportionally to safety, although there are increasingly fewer exceptions. The standards incorporated in the regulations originally were based on previously developed industrial standards created by various organizations. Some important groups were not covered for a time, including public employees and effectively laboratory research workers. For the latter group, the standards, based on industrial processes, simply did not translate well to a laboratory scale environment. Many industrial organizations proceeded to develop their own safety programs for laboratory personnel that were quite effective. Progressive universities and colleges have created good safety programs, but again the strengths of these programs until relatively recently were in the crafts and maintenance staff. Because of unique features of the academic environment, safety programs affecting research operations typically were less comprehensive than in industry. Until recently, when the OSHA standard for laboratory facilities was enacted, even some larger educational institutions, and a very high percentage of moderate to smaller schools have had minimal safety programs for their scientific employees. In consequence, safety may not have been stressed as much as it should have been in academic research institutions. In some schools, students can still graduate without having received any formal safety training, or in some cases without even being exposed to good safety practices. Since attitudes, once established, are very difficult to change, the attitude that safety is of secondary importance still may be carried over into professional careers for a significant fraction of current laboratory research personnel. Fortunately, this scenario is changing for the better as individuals become more aware of the consequences of not doing things well.

Over a period of years, the regulatory environment has grown to encompass more of the research environment, beginning with the users of radioactive materials in the 1940s and 1950s. If an organization used radioactive materials, it was required to have a radiation safety officer or committee, because these were mandated by the Nuclear Regulatory Commission (NRC), or its antecedents, under their licensing regulations. Approximately half the states have assumed this responsibility as a surrogate for the NRC. In recent years, additional regulations or

1

guidelines have been passed explicitly covering laboratory operations affecting the disposal of chemicals, infectious waste, exposure to human blood, tissue and other body fluids, the use of animals and human subjects in research, and work involving recombinant DNA. Concern about risks to workers and the environment from chemicals led to regulations requiring hazard information to be given to individuals exposed to chemicals, and to local communities, where the potential exposure to hazardous chemical incidents existed. These last regulations covered industrial operations as well as research personnel. Finally, in 1991, OSHA passed a standard specifically designed for the laboratory environment which required operators of research laboratories to provide laboratory employees protection equivalent to that enjoyed by industrial workers covered by the OSHA General Industry Standard. This laboratory oriented standard is a performance standard, which is a virtual necessity in view of the tremendous variation in laboratory operations. Although it provides flexibility, it is no less demanding in the net safety provided laboratory personnel than specific standards. Additional hazard specific standards have also been established. The 1992 Bloodborne Pathogen standard resulting from concerns about exposures to AIDS and hepatitis probably affects more laboratory workers than emergency care providers, the group that most persons immediately think of in the context of accidental exposure to blood, tissue, and other bodily fluids. Infectious waste rules in force in many states affect laboratories performing basic research in the life sciences as well as medical facilities. The Americans With Disabilities Act has many implications for the laboratory environment, although it too was intended to be applicable to employees in all occupations. Part of the difficulty in developing an effective laboratory safety program lies in the nature of laboratories. Research activities conducted within laboratories normally are extremely varied and change frequently. The processes and materials in use may present unidentified problems. Research materials may be being synthesized for the first time or may be being used in novel ways. Flammable solvents are probably the most common class of chemicals in use, and there are ample sources of ignition in most laboratories. Because of the changing needs in the laboratory and the scale of most reactions over short periods, even a well-managed facility tends to accumulate a large and varied inventory of partially used containers of chemicals. In one moderate, not atypical, research laboratory at the author's institution a recent inventory found 1041 different chemicals on the shelves, only a handful of which were currently in use or anticipated to be in use in the near future. Many of these will never be used since research personnel no longer trust their quality. Laboratory equipment is often fabricated or modified within the laboratory or in an instrument shop maintained by the facility. Devices manufactured or modified locally obviously are extremely unlikely to have been tested by any certified safety testing organization. Laboratory facilities that may have been well designed for their intended initial use may easily become wholly inadequate in terms of electrical services, ventilation, or special equipment such as hoods, as programs change or as new occupants move into the space. Subsequent changes are more often made solely to accommodate more activities rather than to improve safety. It is also quite possible that a laboratory was not designed properly initially. Relatively few architectural firms appear to really know how to design laboratories, although this situation is definitely improving. Engineering factors are often the ones sacrificed to achieve other goals. Buildings and spaces originally built to house classrooms in a university are regularly converted to serve expanding research programs. These converted spaces rarely serve well as laboratory structures without substantial and difficult to obtain sums being needed to renovate them. Even with the best of intentions, physical solutions to safety problems that may work well in large-scale industrial processes are often extremely difficult to scale to the laboratory environment and alternatives may be expensive.

Individuals managing laboratories are usually very capable persons who tend to be strongly goal oriented, often to the unintended exclusion of other factors. Scientists designing an experiment may inadvertently neglect some peripheral factors affecting safety because they are not directly applicable to attaining the research goal. They also may not remember that the technicians or students under them may not have the years of experience and training that research directors do. They may not evaluate all of the consequences if some item of equipment were to fail or if personnel make errors. It is true, too, that familiarity all too often can lead to a casual approach to routine procedures. Also, and unfortunately as with any other large group, the research community is not immune to the presence of a few persons who cut corners to serve their own interests. The growing competitiveness in finding funds for research, to publish, and to obtain tenure, as examples, in academic environments must exacerbate this problem. The pressures can be intense. It is probably unwise to depend wholly on the professional expertise of the research scientist or his voluntary use of safety programs. There are too many other factors influencing his priorities. Yet, in the laboratory standard, much of the responsibility for laboratory safety is placed squarely on the shoulders of the laboratory research director. It is essential that procedures be established to inform laboratory managers and laboratory staff of appropriate means to create an effective program and to monitor their compliance with the program. The organizational structure of many research institutions, especially at academic institutions, also contributes to the variability and general weakness in the safety programs, due to fragmentation of responsibility. Strong line organizations found in businesses and factories are often very weak in universities. Individual research programs in an academic laboratory are typically defined solely by the laboratory director, virtually independently of the department head. The typical academic department head may or may not establish the broad areas within the department that will be emphasized, but generally will intervene to a minimal degree in the conduct of research in individual laboratories. Intervention is construed as an infringement of their academic freedom by many scientists. Similar gaps exist between the department head and his dean and so on through successive administrative layers. Thus, carrying out a safety program is a difficult logistical problem in colleges and universities. The situation in commercial laboratories is better, since programs are less often defined solely by the laboratory director. However, even in industry, the laboratory director usually has more autonomy and discretion than a line supervisor in a factory, so that communication and carrying out of safety policies are somewhat more difficult in commercial laboratories also.

None of the difficulties associated with laboratory safety programs cited in the previous paragraphs have disappeared in the past few decades. The complexity and cost of effective laboratory safety programs have increased too, adding to the problem. Competitive pressures have significantly increased both in academia and in commercial facilities. New mandatory regulations and standards have been enacted, so "doing what has to be done" has become a much larger task. The acceptance of the status quo involving health and safety issues has become increasingly unfashionable and unpalatable. Expectations of laboratory workers for their personal well-being have continued to increase. They have become very concerned about the effects of their working environment on their health and safety and they expect something to be done to eliminate perceived problems. The public is less likely now to accept the premise that the scientists "know what they are doing" and may be concerned about releases of dangerous materials into the environment. Acquiring liability followed by litigation initiated by an employee, due to a weak safety program is a possibility that management cannot ignore. Lack of resources for meeting safety issues is not an acceptable defense. A good case can be made that without these higher expectations and possible repercussions, many regulations governing health and safety in the workplace may not have been passed.

Because of the new regulations and new attitudes, it has become necessary for overall management of research institutions to take an increasingly active role in safety programs, although the new lab standard places much of the direct responsibility on the research manager. Organizations can no longer legally choose to have an effective safety program or not. They must provide support to the laboratory directors to aid them in creating their own safety programs and to monitor that the laboratory's programs are being properly done. They have to provide training to employees and take positive steps to encourage employees at all levels to actively support and apply standard safety policies for everyone. Resources have had to be identified to comply with more comprehensive and complex regulations. Safety departments often have had to be enlarged at a time when available funds were likely to be decreasing. However, it cannot be overlooked that individual laboratory directors have been assigned the primary responsibility in managing their operations safely. Therefore, they are more important than ever as key persons in carrying out the institutional safety program. This is not a role to which many of them are accustomed, and many of them are not comfortable in it. Many are concerned about the personal responsibility implicit in these requirements, as they should be. Deliberate lack of compliance with the regulations could carry legal liabilities. Old habits are difficult to shed and, for many, there will be a difficult transition before safety is as high a priority in the laboratory as it should be, even under the impetus of the laboratory standard. It has become a major responsibility of management to assist and guide the laboratory directors and, where necessary, ensure that they do comply.

A critical component of an organization's safety effort is an effective safety department. The primary objective of such a department is to provide guidance to the remainder of the organization, including upper management, not act as a policing agent. Due to the complexity of modern research operations and the regulatory climate, it is the responsibility of the members of the safety and health organization to be the primary source of knowledge of current regulations and their interpretation. They must also be aware of technological advances so that the advice and training, which is their responsibility to provide, will be the best possible in terms of effectiveness and efficiency. A second major responsibility of the safety organization is to monitor the performance of the employees and management of the organization, and unfortunately, this does require policing functions. Management may not realize this function is required, but it is a major responsibility implicit in the OSHA Act. Safety and health departments must be given enforcement authority by the parent organization to properly monitor compliance with safety issues. It is this enforcement function that must be managed most sensitively by safety and health personnel. It is critical that it is done well. If done heavy-handedly, the scientific personnel will be resentful, but if safety personnel are too timid, they will be ignored by both workers and management. Without active and visible support by management, it is very likely that even well structured safety and health programs will not succeed. Operational duties are also now assigned to most safety and health departments, such as disposal of hazardous waste. Finally, in the current litigious climate, the safety and health department must keep superb records to protect workers, themselves, and their organization.

Safety personnel must understand in defining their role that they are not the primary function of the organization of which they are a part. A widget manufacturer makes widgets as its first priority. A university has students to educate and scholarly research to perform. A good safety program, if managed properly, will facilitate attainment of these primary goals, not hinder it. A safety department must act so that this is true and is perceived to be true by all of their clients' constituencies. Safety departments need to establish themselves with an image of a strong emphasis on service. Service as an operating premise does not mean that the safety department is to take the safety burden entirely upon itself. It cannot. The department must provide an appropriate structure for both employees and management to conduct operations safely

Without the support of management, no safety department can succeed. Support entails many factors. Resource support is obviously very important in physically enabling the department to function. If the safety department is not provided with a reasonable level of support, protestations of support by management will seem insincere and hollow. Conversely when management "puts its money where its mouth is," the conclusion will be drawn by the employees that management does see value in the programs affecting their safety and health. There are other ways that management can illustrate support. To whom does the head of the department report? Is it a mid-level manager, or is it a senior person such as a vice president? If confrontations do arise (and these may occur, especially in an academic research laboratory where full professors are accustomed to operating with relatively few constraints), will the safety professional be supported when he is right? Does the safety department participate in making operational decisions or policy decisions? The status of the safety department will be enhanced if the answers to these questions are positive. If the answers are negative, the effectiveness of the organization's safety program will be seriously diminished.

There is a management tool called "strategic planning" which can be used to define the function of any organization and, if used properly, can make the entire organization function better. Management structures come and go, but strategic planning should be adaptable to any type of organization. A first step in doing strategic planning is to define the organization's clients and stakeholders. Most persons take too limited a view of who these are. For a safety and health department, the most obvious "clients" are the employees and management. Other internal units within the organization also can be considered clients or at least, stakeholders, since it frequently requires collaborative efforts to achieve specific goals and if they share common objectives, obviously the performance of both will be enhanced. Other important clients or stakeholders are regulatory agencies, the local community, and the public. Each of these clients will have their own agenda and the perceived needs of each client group should be explicitly identified in terms of whether the safety department's mission meets these needs.

Once the clients have been defined, the objectives and goals need to be defined. Goals are usually more generalized, such as to reduce the number of accidents within the organization, while an objective may be more specific, such as to reduce the number of back injuries by 25% within the next 12-month period. When doing this exercise, it may be found that there may be dozens of goals, and even more objectives. Not all of these will necessarily be equally important.

The strengths and weaknesses of the department should then be determined. Among the strengths would be the skills of departmental personnel, the funds and equipment available to the department, the relationships between the department and its clients, and the internal relationships among departmental personnel. The same things could be weaknesses as well if deficiencies exist. However, it is too easy to blame lack of success on lack of resources. Failure to make optimum use of available resources is a measure of inefficiency, which the strategic planning process can help minimize.

Once the preliminary work is done, the department can define its programs to maximize their effectiveness. The programs need to be ranked. Not all are equally productive or attainable in enhancing safety and health. Note that in each of the preceding few paragraphs, it is the department that is to do all these things, not the organization, nor the department head. Everybody in the department and, desirably, representatives of the clients should be involved. Everyone needs to "buy in" to the process and have a share in the result. The department head and the more experienced persons can and should provide guidance and leadership. However, there is obvious value in fresh ideas and approaches that may be provided by newer personnel. A younger person can be as productive in developing new ideas as persons "set in their ways," but they may need to be encouraged to contribute their ideas. No one individual should dominate the strategic planning process.

The cooperation of all the client groups is important, but that of the organization's employees is essential. Employees must see the actual value of the safety program to themselves. Reasonable persons are not going to seriously advocate doing things unsafely, but many honestly feel that they are doing things as safely as possible and resent being told otherwise. Many conscientiously feel that a formal safety program, with the attendant rules, regulations, and all the accompanying "red tape" is not necessary and is counterproductive. An astute safety professional, with the active support of management, can overcome this attitude. This can be done by education and training, so that people know what is expected of them and why; by actively involving the employees in developing good approaches to achieve compliance with standards; by minimizing the administrative burden on individuals; by demonstrating that the safety program helps the organizations personnel solve their problems; and by making sure that the safety program clearly provides a safer environment. Training in management techniques to enhance a safety processional's "people skills" is a sound investment by management. Training for safety professionals is frequently thought of only in terms of technical skills. In scientific research, typically many employee clients will have advanced degrees, especially those ultimately in charge of a laboratory, while many safety professionals will not have comparable credentials. This can be a hurdle which, with proper training of both parties, can be overcome. A good, meaningful laboratory safety program is possible and, under today's regulations, mandatory. Under the OSHA laboratory safety standard, a research organization must require each laboratory within it to have a satisfactory written "industrial hygiene plan" or, in other words, an effective safety and health program for that facility. According to the standard, the "plan" is the responsibility of the person ultimately in charge of the laboratory with the organization required to monitor the performance of the laboratory compliance with the plan. This emphasis on the individual laboratory manager does not preclude the organization, through its safety and health departments, from providing training and guidance in the preparation of the plans, including prototypes or templates tailored for individual laboratories.

Since so much responsibility is placed on the laboratory manager and the laboratory staff, the goal of this handbook is to help the individual research scientist define the requirements of a successful program. Recommendations usually will be based on current regulations or accepted practices. Sometimes the recommendations will go further where this appears desirable or where the opportunity exists for innovative approaches. No one should be afraid to try an innovative and more cost-effective program to achieve a desirable goal. In the latter case, the differences between the required procedures and the modified approach should be clear. The area of emphasis for much of the material will be the chemistry laboratory, but there will be separate sections for several other important areas. Where material corresponding to the content of this edition of the Handbook was included in the earlier editions of the book, the earlier material will have been thoroughly reviewed and brought up to date, or projected into the future, when changes can be anticipated clearly.

I.   LABORATORY SAFETY AS A COOPERATIVE RESPONSIBILITY

The OSHA Laboratory Safety Standard has significantly changed the ground rules governing laboratory safety since it was introduced some years ago. As noted in the previous section, much of the responsibility for developing and carrying out a laboratory safety program is assigned to the individual ultimately responsible for the laboratory. This standard preempts most, but not quite all, of the general industry standards. It is a performance standard that permits substantial flexibility in how the health and safety of the laboratory employees are to be assured. Laboratories do not exist in a vacuum, although they are treated as individual units for the purposes of the standard. Other agencies and individuals within the organization must share much of the responsibility for achieving a successful program. Compliance with the requirements of the standard will be discussed in detail in later chapters, but the roles of other organizational units will be briefly discussed in this section.

A. Human Resources
The Human Resources Department (sometimes called "Personnel or Employee Relations, etc.") is the first contact that most employees have with an organization. This department should have the responsibility of informing the new employee of the basic policies of the organization in an initial orientation program. It is important that they include the rights and responsibilities of the employee involving safety as well as working hours, benefits, etc.

Employees should receive written documents detailing the general safety policies of the organization, where additional information can be obtained, and to whom the employee can turn to express safety concerns. The initial impression of the importance placed on safety is made at this time, and every effort should be made to ensure that it is positive. For laboratory employees, the role of the central organization in supporting the safety programs of the individual workplaces and monitoring compliance with work place safety procedures needs to be clearly explained. The Safety and Health Department is the logical one to provide the orientation material in this area, and may be asked to provide a presenter.

Detailed safety procedures should not be covered at the initial orientation program point where laboratory safety is concerned, since laboratory activities vary widely from laboratory to laboratory. Information concerning specific safety procedures for a given laboratory is the responsibility of the person in charge of the facility.

The Human Resources Department is also often the department charged with handling workers' compensation claims. Thus, it has a direct concern in reduction of injuries to minimize the cost of such claims. In some organizations, the equivalent department is Risk Management as far as insurance claims are concerned. The Safety and Health Department may also be affiliated with the Risk Management Department. Clearly the Human Resources Department plays a critical role in an effective laboratory safety program.

In many organizations, the group charged with assuring that no discrimination occurs and that everyone is assured of an equal opportunity is also in Human Resources. In others, this group is autonomous or reports through a different chain of command, to avoid any conflict of interest. There are obvious general safety implications for disabled persons, such as both audible and visual fire alarms or safe, unblocked evacuation routes. However, one of the biggest laboratory-related safety issues for the EEO group is that of exposure of younger women to chemicals in the laboratory. Fetal vulnerability is an example of a very sensitive topic where this applies. In some areas, information is sufficiently developed to allow well-defined policies. For example, substantial information is available on absorption of radioactive material by the fetus or the effects of external radiation on fetal growth. As a result, the rights and responsibilities of fertile women have been delineated as an NRC Regulatory Guide for several years and have been incorporated in the latest version of the Federal Title 10, Part 20. On the other hand, information about the effect on the fetus or the reproductive process of most of the thousands of chemicals in use is very limited. In such a situation, women may have legitimate concerns about their exposures and about possible unfair exclusion from some areas of employment. It is the responsibility of the EEO office to provide them with a means of expressing these concerns. EEO officers must work with the employees and other parts of the organization to provide adequate safeguards to protect the rights of women employees. The new Americans With Disabilities Act of 1990 has placed some real teeth into requirements for making appropriate provisions for disabled employees and others. The definition of disabilities is no longer necessarily limited to those with obvious physical dysfunctions. An individual who is demonstrably hypersensitive to certain chemical fumes may also be considered disabled and thus eligible for accommodation under this statute, if these chemicals are part of the work environment. Further, Congress is considering possible legislation on indoor air quality that might be applicable in such a case. It will be necessary for research facilities to work closely with the Human Resource/EEO Departments to ensure that they comply with this politically sensitive issue.

B. Legal Department

There are many occasions for the legal office of an organization to concern itself with laboratory safety. Many of these are straightforward, concerned with contract terms and conditions for sponsored research. Contracts for public academic institutions, for example, usually cannot contain a "hold harmless and indemnification clause" since this in effect waives the protection provided by the immunity to suit claimed for many state agencies. The legal counsel for the organization must find alternative language or an alternative procedure for satisfying this contract condition if it is invoked by the contract language. There are many restrictions built into law or policies for various fund granting agencies, and it is the responsibility of the legal office to ensure that these terms are met. The question of personal liability often arises as well. Laboratory managers often ask about the risk of personal exposure to litigation as a result of the manner in which they enforce the organizations safety program. It is obviously impossible to monitor every action by laboratory staff to assure compliance with safety and still have time to do research. However, what are the boundaries of adequate supervision? Finally, most organizations have grievance and disciplinary procedures for actions that involve alleged violations of safety procedures. The legal counsel will necessarily become involved in these.

Besides these issues, academic institutions are increasingly vulnerable to citations by regulatory agencies, many including substantial fines. The organization's legal representative, with the input of safety and health personnel, should carefully review citations to determine if there is a problem with the language or the application of specific regulations to the activity being cited. It is unlikely that they will be sufficiently knowledgeable about the technical aspects of the basis of the citation to adequately represent the organization or to prepare a response. The safety and health department, the scientist, and the legal counsel must all cooperate in preparing a response, to minimize the level of the citation and to avoid excessive costs or too restrictive abatement agreements. Safety professionals, however, have a responsibility to individuals as well so that they occasionally find themselves in a position where this aspect puts them at odds with the organization, and in such a case, it is especially important that the safety professionals have a good rapport with the legal counsel so that a mutually beneficial solution can be obtained.

There are opportunities to have informal meetings or discussions with most regulatory agencies before making a formal response. An attorney should be a participant in these meetings and may be the appropriate person to represent the organization in negotiations. The final reply should definitely be carefully reviewed by an attorney representing the organization. Inappropriate or imprecise language is the natural breeding ground for litigation. Lawsuits by outside groups, such as citizens concerned about the environment, represent similar legal challenges, again requiring a coordinated response. Many organizations now include a legal representative on their safety committees. Sensitive decisions by these committees should be reviewed by a legal advisor.

Liability insurance—the ability to obtain it or not and the cost of the insurance when obtainable—has become a major issue. The insurance manager has become a key person in processing contracts where safety is implicated. For example, contractors handling hazardous laboratory waste for an organization are typically required to carry substantial amounts of liability insurance, in addition to general liability insurance, to cover their errors in handling the waste. The contracts that waste contractors offer also often contain the phrase "indemnify and hold harmless." Most public facilities cannot sign a contract containing this clause since it, in effect, forfeits some of their constitutional rights. A $5,000,000 policy is a typical requirement, and many small firms cannot afford the insurance. This may limit the organization's ability to provide services at reasonable costs or may limit access to firms providing specialized critical expertise. With the support of the legal counsel, the purchasing department, and the risk manager, occasionally it is reasonable to go unprotected where the risk is judged to be sufficiently small by the technical safety advisor. The insurance manager is also usually eager to work with the safety group and personnel departments to stem the rising tide of workers' compensation claims and other safety-related claims.

C. Purchasing Department

In most organizations, the Purchasing Department processes orders for every item purchased. The cooperation of the Purchasing Department can facilitate the creation of safe laboratory conditions. With their assistance, it is possible to add safety specifications, and sometimes to limit eligibility of vendors of products to ensure that the items ordered will function safely. This may occasionally increase costs, but can significantly enhance safety. A good example of how additional quality specifications can be used to enhance client acceptance involves the choice of chemical splash goggles. Chemical splash goggles that will minimally meet ANSI standards can be purchased very cheaply. However, the least expensive units are typically uncomfortable, are hot, fog quickly, or have a combination of these undesirable qualities. For short-term, very limited wear, they are reasonably satisfactory, but after a short interval, the wearer usually removes them or slides them up on the forehead. In either case, they would not offer the needed protection. With the assistance of the Purchasing Department, additional specifications for antifogging coatings, better air flow than that provided by the typical side-ports, hypoallergenic materials, etc., can be added to the specifications. Limiting the choice to a single brand can still allow bidding for the best price if there are a number of distributors. If the improvement in quality is sufficiently clear, then it is often possible to justify use of specifications even when there is only a single source. Major items such as refrigeration units can be required to be safe for flammable material storage. Fume hoods can be required to be reviewed by the safety department and the organization's engineering staff to ensure that blowers and other features are adequate. Chemical purchases, with new computer technology, can be tracked from the time of order to the point at which they are fully used or disposed of as surplus or waste, if chemical purchases can be ordered and delivered to a single point of receipt. This last possibility will allow compliance with many regulations involving chemicals, which would be otherwise virtually impossible.

The area of chemical purchasing is an especially important one where cooperation of the Safety and Health Department and Purchasing can significantly enhance compliance with safety standards. If an agreement can be reached to establish that all chemical purchases will be made through a single purchasing, receiving and internal distribution facility, this would tremendously facilitate the required chemical material management program for an organization. Failure to establish such an arrangement, on the other hand, makes a chemical management program extremely difficult.

D. Facilities Department

A key issue in the new OSHA laboratory safety standard is for research facilities to be suitable for the purpose for which they are to be used. To many research personnel, this simply means sufficient space with appropriate equipment and services. They do not think of their area in the context of the whole facility or of the many safety-related codes that affect its design. The user's considerations are largely involved with their research program. Research personnel depend on architects and contractors or internal equivalents from planning, physical plant, and maintenance departments to design or renovate their spaces to comply with codes and regulations, if it is not too inconvenient or expensive. They have a point. They are the *raison d'etre* for the space being required. However, current regulations make specific demands on the design and construction of facilities. Unfortunately, other than the fire code requirements, of which most design firms are reasonably well aware, many firms do not take measures to assure that they are current on changes in other regulatory requirements involving safety. With their lack of background in the specific disciplines involved, they may not even know the right questions to ask which would be applicable to a given facility. It is critical that at least four parties are involved in facility design and construction: the users, the architects (both external and internal), contractors, and the safety department. The need to involve safety goes beyond regulations. There are many issues that are covered only by guidelines, and others that have been revealed by experience to be important where even adequate guidelines have not been developed. For example, until quite recently, the needs for ventilation were assumed to be met if sufficient fresh air were made available. The characteristics of the supplied air have been found to be much more critical than previously thought. Low levels of various airborne contaminants, well below levels recommended as acceptable by OSHA PELs or the ACGIH TLVs, have been found to cause problems for exposures over extended periods. Wherever it is feasible, within reason, facilities should meet more than minimum standards, especially if life cycle costs can be reduced . It should also be recognized that it is likely that more stringent regulations can be anticipated.

The physical plant department has the responsibility to provide most of the services that are needed to keep buildings functional including heat, light, utilities, and custodial services, make repairs, and do renovations. However, few laboratory workers really appreciate what they are asking of these persons. Many laboratory workers have had the experience of custodial workers being afraid to enter a laboratory with a radiation sign on the door and make allowances for their fears. However, the same persons expect a mechanic to work on a chemically contaminated exhaust motor to a fume hood without question or to work on the roof while other nearby fume hoods are still emitting exhaust fumes. Maintenance workers are becoming increasingly concerned about their exposure to toxic materials while performing maintenance in laboratory facilities. The risk may or may not be significant, but maintenance staff do not really *know* that it is not. The OSHA hazard communication standard that went into effect on May 25, 1986, requires the facility manager to inform maintenance personnel of potential risks. Maintenance personnel may become even more reluctant to perform laboratory maintenance upon receiving these warnings, unless they can be honestly assured of the lack of risk to themselves or that measures are available to them so that they will be protected from exposure. It may be difficult to document this lack of risk. In their own areas of expertise, maintenance personnel have many skills and considerable knowledge but they are not scientists. The lack of familiarity with the potential risks in the research being conducted in the facility where they are to work may lead to perhaps unreasonable fears. It will require a combination of education, tact, and accommodation in reducing risks by laboratory personnel to assuage these fears so they may continue receiving needed laboratory maintenance services.

In order to comply with the spirit of the law for both the organization's own maintenance personnel and personnel of outside contractors, procedures are needed to assure that maintenance personnel are properly informed of exposure risks. One of the most difficult situations involves work on or near fume hood exhausts. In a research building that can have separate blowers for hoods from many different laboratories, at any given time it is hard to know what is currently being exhausted or what residues are present from times past. Individuals working on fume hood exhaust components have been known to have severe reactions to these residues. A method to control access to such areas must be established, so that steps can be taken to prevent unnecessary exposures to protect the worker and minimize liability to the organization.

E.   Management

This section would not be complete without mentioning the vital role of central management. They establish the atmosphere and provide much of the needed operational resources. A positive attitude on their part in support of an effective laboratory health and safety program is critical. However, the other units must understand that not all problems can be solved efficiently by resources alone, and they cannot measure this support solely by the level of physical support assigned to safety. In one sense, measuring quality of programs by measurement of resources made available is to measure the programs by how inefficient they are. Parkinson's Law, which involves the principle that work will expand to use all resources provided, is certainly applicable to safety programs as with any other type. However, management must provide *sufficient* resources to make it possible to develop and implement a *cost-effective* program. They must also provide the moral support for the program to be managed with full regard toward meeting its goal of striving for a safe and healthy environment for everyone.

F. Organizational Structure

A major and valid justification for OSHA to adopt the laboratory safety standard to supersede the general industry standards and the hazard communication standard for the laboratory environment was that the laboratory environment is radically different from industrial facilities and most other types of occupations. The organizational structure for research institutions is not nearly so structured, especially in the academic environment, and operations are different in size and character. The laboratory standard uses this difference as a criterion as to whether the laboratory standard is to be applied to a program.

A major factor that is important to consider in developing a safety and health program is the way directives are transmitted from one level to another. Laboratory supervisors in universities and in many commercial research institutions do not have the same degree of communication with, or responsibilities to, higher level management as do typical industrial supervisors. They are much more independent and tend to resent "unwarranted" interference. In a university, the term "academic freedom" is often called into use when a researcher feels imposed upon. Scientific personnel especially do not want outsiders to attempt to intervene in their program of research. Indeed, scientists have some justification for being concerned about anyone becoming too familiar with their work, as priority of discovery is an extremely important factor to them. Even a nonspecialist in a field, such as a safety professional, may inadvertently provide some key information to a competitor. A major role for safety personnel, with these factors in mind, is to define the organizational constraints and the importance of following safe practices for the laboratory supervisor, who has been assigned the responsibility to integrate safety concepts within a written health and safety program under the laboratory standard.

Since every laboratory operation differs in detail, safety personnel cannot be expected to prepare a written plan for each facility. For example, a research university is likely to have a thousand or more different laboratories. Attempting to draft a separate plan for each laboratory is far too big a task for most safety departments. It is reasonable, however, to expect a safety professional to design a standardized template and *help* the laboratory manager develop his own plan and written operational procedures. Even this effort will require a major effort for both the manager and the laboratory safety staff of most health and safety departments. When a new regulation or policy needs to be introduced, generally it is not feasible to send it through a chain of command structure. As noted earlier, in a university, deans or department heads have extremely limited control over the academic staff, outside of classroom teaching assignments. In industry, the chain of command is better defined, since the general mission is usually determined by management. This factor typically defines research targets more explicitly, but in industrial facilities engaged in basic research, there is often similarity to the academic environment.

Due to the relative independence of the research investigator, the laboratory environment leads to the need for much self-supervision, especially for the laboratory director or manager, who is typically judged primarily by the quantity and quality of his work. The laboratory staff interacts with the organization primarily through the person ultimately responsible for their employment, so it is this person who legitimately is held responsible for carrying out of the safety plan for the unit. A laboratory represents a considerable investment for academic and other research institutions, so success is critical to the career of a research scientist. This difference may encourage the laboratory scientist to take risks greater than he might otherwise to achieve desired results. This usually is not done maliciously but because scientists are typically strongly goal oriented. In addition, they work under a great deal of stress, so "extraneous" safety factors generally are not first among those considered.

The interface between the safety and health department and laboratories has evolved rapidly in recent years. Not too long ago the role of a safety department was primarily to advise and train employees, mainly those providing support functions. As regulations became more complex, new sophisticated responsibilities became necessary, dealing with health and safety issues more germane to laboratory work. By default, safety and health departments assumed many of these functions. They were delegated the functions and powers of the federal and state agencies whose regulations were applicable to the organization. Often, they were also charged with the responsibility of managing the resources to achieve compliance with the regulations. With the laboratory standard, the pendulum has swung back to a degree, in that the responsibility for implementing the standard has been explicitly assigned to the individual laboratory. However, the authority to ensure that the laboratories fulfill their obligations has been assigned to the organization. The organization, in turn, usually returns the responsibility to the Safety and Health Department. There are any number of possible working arrangements for a safety program. The alternatives should have several common characteristics to be effective. Among these are:

- Assignment of safety responsibility to a senior executive, such as a vice president, with the manager of safety and health programs reporting to this person.
- The formation of a Safety and Health Department, most commonly designated as the Environmental Health and Safety Department (EH&S), staffed by professionals and with sufficient resources and personnel to perform their function. In an academic environment, there may be some advantages for the head (and possibly some senior staff) of the safety department to have academic credentials. Since most laboratory heads at a university have a doctorate and often judge themselves and others by their scholarly activity, an individual with similar credentials may find it easier to gain and hold their respect. In larger, more sophisticated commercial research organizations, the same argument may hold true. Under the OSHA Laboratory Standard, an individual, presumably from the EH&S Department, is required to be identified as the organizational Chemical Hygiene Officer (CHO). This individual clearly should have appropriate credentials in chemistry as well as chemically related safety practices. Most larger research organizations represent potential significant environmental risks for the area in which they are located. A representative of the EH&S Department should be represented on the Local Emergency Planning Committee for a designated local governmental area, required under the provisions of the SARA (Superfund Amendments and Reauthorization Act) Title III, EPA regulations.
- The creation of one or more safety committees at the institutional level. In an industrial situation, one safety committee may suffice, but in a research facility, especially one with a wide variety of programs, it could be desirable for these to be more specialized. Examples could include a radiation safety committee, an institutional biosafety committee, a general laboratory safety committee (possibly separated into chemical and life science areas), a general safety committee, a human subjects review board, a fire and emergency committee, one charged with ameliorating problems for those with disabilities, and an animal care committee. All of these are either mandated or recommended by various standards if the organization has research programs in corresponding areas. Some of these special interest committees will have more authority than others, due to the underlying strength of the applicable regulatory standard. For example, the use of radioactive materials is very strictly regulated by the Nuclear Regulatory Commission or, in agreement states, by the state surrogate agency. Failure by the institution to comply with the regulations and the terms of the institution's license can, and frequently does, result in substantial fines, as well as national publicity. A comprehensive committee, formed of the heads of the special interest committees, may be a useful vehicle to recommend or define broad safety policies to ensure consistency in the various areas of responsibilities. Each committee should have a definite, written charge, as well as definite rules of procedures. An

appropriate person from the EH&S department should serve as an *ex officio* member on each of these committees. A major benefit of an effective committee is that it assumes, along with the safety professional, part of the responsibility for actions, procedures, and policies that could be perceived by the regulated group as being onerous. It partially insulates the safety professional from being held solely responsible for the constraints imposed by the regulations.

- Because of the independence of the various divisions, or schools, in larger organizations, the separate internal administrative units may find it desirable to establish their own safety committees. The unit committees would adapt the overall institutional safety policies which must be followed in the context of their own operations and areas of responsibility. It would probably be desirable to identify an individual as the division chemical hygiene officer to act as liaison with the individual representing the central authority. This position is not required, however, for a subunit (other than the laboratory) of an organization under the OSHA laboratory standard.

- Each department that includes chemical laboratories as a normal function should identify a single individual, equivalent within the department to the organizational chemical hygiene officer. This person would act as liaison with the institutional safety department, the safety committees, and the laboratories within the department. In many organizations, serving in such a position is not especially beneficial to the career of the individual, so some organizations tend to shift such responsibilities onto less productive personnel. This is not appropriate. The assignment should be rotated among active and productive individuals. A laboratory hygiene officer might be appointed as the primary contact on safety issues, other than the laboratory manager, if a laboratory unit is large enough by itself. Often the person ultimately in charge of a laboratory facility spends very little time in the laboratory, leaving day-to-day operations under the direction of a lab supervisor or a senior technician. The senior scientists typically have many other responsibilities to distract them, not the least of which is obtaining funds to support the laboratory.

Selection of suitable members of the committees at the various levels is extremely critical. As just noted, less productive individuals are often given the more demanding and less glamorous committee assignments. Safety committees, if performing as they should, are very demanding of their members. There is also reluctance by many very active and productive persons to accept responsibilities not immediately germane to their own programs. The members should be drawn from active, productive, and respected laboratory personnel, for limited terms if necessary to secure their agreeing to serve. Individuals who would be most affected should play a major role in defining policies. Policies and procedures adopted by such respected, active research scientists will be much more likely to be accepted by their peers than would otherwise be the case. Candidates for a committee position should be encouraged to look at the assignment as an opportunity to positively influence a program that is important to them. Most persons will be flattered to be asked and will accept appointment if this is the approach taken.

Some care must be taken to define "productive" personnel. The most productive individuals are not always the most visible or the most vocal. The actual achievers may be too busy to promote themselves and simply are quietly effective. This latter class should be sought out and the former passed over. A safety committee member must be able to consider issues objectively and be willing to act, even if occasionally their own immediate self-interests might be adversely affected. "Difficult" individuals should not necessarily be excluded. Often, if they are given an opportunity to participate, such persons can become a committee's strongest advocate. The background, credentials, and references of candidates for membership on a safety committee should be checked as carefully as if the candidate were a job applicant.

Many regulatory agencies now require that independent "lay" persons be added to the regulatory committees. These individuals, if they are to act as a "conscience" for the committee or at least present a different point of view, as is really their role, should truly be independent. They should have no possible conflict of interest ties to the organization. Among the affiliations that might disqualify an individual would be a personal relationship (spouse, brother or sister, life-partner, etc.) with a person employed by the organization, or a financial relationship (a retired former employee perhaps, a paid consultant, or one with a significant investment in a commercial firm). They should also be sufficiently technically qualified to understand the area of responsibility assigned to the committee. Unfortunately, these two provisions often make it difficult to obtain suitable lay members.

Nothing will destroy a committee more rapidly than for the members to feel that the work of the committee serves no useful purpose. The committee must be provided with a specific charge which must include clearly defined objectives and expectations. There must be meaningful work for the committee to do at each meeting. The results of their work must be taken seriously and not simply disappear into files. Each meeting should be structured with a definite agenda and definite goals. Much of the responsibility for making a committee work depends upon an effective chair and good staff work. The staff must prepare the working documents and distribute them in sufficient time for review by the members to allow them to be prepared at the time of the meeting. A chair has the responsibility to conduct the meeting fairly, with everyone having enough time to contribute, but the chair must also see that the business is conducted expeditiously. Meetings should be no longer than necessary.

It would be very desirable for chairs of the various safety committees to have administrative experience. They should have experience in managing a budget and managing people so that they can provide guidance to the committee should it take actions that might demand resources that would be hard to obtain or cause personnel problems. They should, of course, be knowledgeable in the area of the committee's responsibility. There are occasions when not all these qualities can be found in a single individual, usually the specialized knowledge or recent active laboratory experience being the missing factors. In such cases, senior professional persons who have had to obtain and manage funded projects would be reasonably satisfactory alternatives to administrators. It would be less desirable for the safety professional to be committee chair because of a possible perception of a conflict of interest. There are circumstances, due to experience and training, when exceptions to this "rule" are appropriate. There also can be logistic factors when it would be helpful for the chair to take a directly active role in a safety area, which could make it acceptable for a safety professional to chair a committee.

In the preceding paragraphs, there has been considerable emphasis placed on defining a functional committee structure, but in a research institution, it is unlikely that the busy professionals making up committees would be willing to meet more than once a quarter, except in emergencies. Therefore, safety committees cannot be assigned the day-to-day task of directly managing a safety program. They are, in effect, somewhat analogous to the legislative branch (and on occasion, the judicial branch) of the safety "government," with the safety department being the administrative branch. Safety department staffs must have the authority to act as necessary between committee meetings, with the knowledge that they will be held accountable for their actions by the committees and their administrative superiors. It is up to the head of the safety department to prepare budgets, administer the programs and personnel assigned to the department, and to provide leadership in the area of safety.

Figure 1.1 is a simplified organizational chart that embodies the concepts discussed in this section. Some duties and layers of responsibility might be combined in a small firm or academic institution, while in very large organizations additional functions or layers might be needed. Note that the safety committees are appointed by a senior executive and have direct access to him.

G. The Safety Department

In older descriptions in the literature, the role of the safety department was seen as largely

TYPICAL SAFETY PROGRAM ORGANIZATION

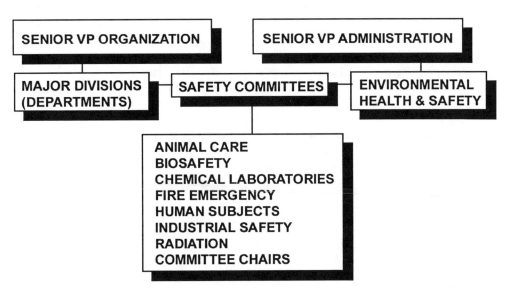

Figure 1.1 In an academic institution, the operations vice president would be the provost and the major divisions would be colleges or schools.

advisory in the scientific work environment. Under the OSHA laboratory safety regulations, each laboratory does have the responsibility for developing and carrying out its own chemical safety plan. However, the central administration of an organization has the responsibility to ensure that the industrial hygiene safety plans for individual laboratories comply with OSHA standards.

Maintaining an awareness of the current status of all environmental health and safety-related regulations, guidelines, and acceptable procedures that affect laboratory operations demands more time and specialized knowledge than a busy laboratory director or most administrators have available. It has become necessary to develop sharply defined specialists, even within the environmental health and safety discipline, because of the growing complexity of the field. This again has led to a strengthening of the role of the safety department because both management and the client groups are often forced to defer to the expertise of the department. Without comparable competence and experience, it could be foolhardy from a standpoint of liability for a laboratory manager or a senior administrator to ignore the opinion of a specialist.

The safety department is often forced to assume another unfamiliar role by directly providing support services, instead of simply monitoring efforts of others, to see that operations are done correctly. The skills, knowledge, and specialized equipment that are needed are not normally available from other support units. A comprehensive safety department needs some laboratory facilities of its own and the trained personnel to use them. However, exclusive use of in-house analyses can lead to concerns of conflicts of interest. Outside analytical services should be used where this is a possibility. A direct consequence of the provision of safety-related services is the need to fund these services. Resources to meet safety needs are typically funneled through the safety department instead of being directly allocated to the research departments. For example, monitoring to see that legal environmental airborne exposure levels of possible contaminants are not exceeded usually requires special equipment and the training to use it, not normally available to individual laboratories. Environmental health and safety departments now typically provide access to medical surveillance programs, training, waste disposal, emergency planning, and a host of other programs to laboratory personnel. The following is a list of common functions of a safety department in a research organization. Not all safety departments would provide all these services, while there are some departments would provide these and more.

1. **Functions Relevant to Laboratories for the Safety Department**
    Evacuation procedures
    Medical
    First aid
    Cardio-pulmonary resuscitation
    Serious injuries, physical and chemical
    Universal precautions/OSHA regulations for bloodborne pathogens contacts
    Chemical, radioactive materials, biological toxins spills and releases
    Corporate institutional generic safety policies/procedures as applicable for:
        Radiation safety
        Chemical laboratory safety
        Biological laboratory safety
        Recombinant DNA laboratory safety
        Animal care facilities
        Human subject research
    Chemical, infectious, radioactive waste procedures
    Personal protective equipment usage and care
    Publication of safety manuals
    Assists departments, individual laboratories with their chemical hygiene plans
    Other OSHA, EPA, NRC required training
    Organizational safety policies and training programs

- **Waste Disposal**
    Prepare bid protocols for waste brokerage firms
    Manage:
        Chemical wastes (including permissible storage, processing, and redistribution)
        Radioactive wastes (including mixed waste)
        Infectious waste
    Contingency planning for emergencies
    Liaison with federal, state, and/or local waste officials

- **Building Safety**
    Building inspections
    Fire system inspections, testing, calibration, and maintenance
    Monitor building evacuation programs
    Interactions with local fire and emergency response groups
    Code review of all renovations and new construction
    Consultation on building design, participates on building committees
    Interaction with on-site construction contractors and architects
    Liaison with fire marshals and building code officials

- **Environmental Protection and Industrial Hygiene**
    Air quality monitoring, testing
    Testing of fume hoods and biological safety cabinets
    Laboratory inspections
    Review laboratory design proposals
    Review research proposals for potential problems

Monitor compliance with Hazard Communication Act for support agencies
Chemical accountability (regulated chemicals, chemical inventories)
Monitor controlled substance licenses
Review and specify personal protective equipment
Review selected equipment purchases for safety features
Maintain laboratory census for safety features
Provide consultation services
Investigate laboratory accidents
Asbestos testing (bulk and air sampling) in existing facilities, equipment
Monitor asbestos removal projects
Release of chemicals to the environment (community right-to-know)
Maintenance of records, database management

- **Medical Program**
  Set criteria for employee participation in medical surveillance
  Initiate medical surveillance/counseling as required
  Provide staff support for occupational physicians
  Maintain all medical records
  Work with personnel department for pre-employment examinations
  Analyze accident records, prepares reports
  Investigate serious accidents
  Act as liaison with local medical emergency groups

- **Radiation Safety**
  Inspect and survey radiation-using facilities
  Review all internal applications for use of ionizing radiation
  Perform hazard analyses for *all* experimental procedures involving usage of ionizing radiation
  Review usage of non-ionizing radiation
  Process all orders for radioactive material
  Receive and check all radioactive material deliveries
  Maintain radioactive material inventory
  Monitor all personnel exposures
  Maintain calibration of all survey instruments
  Maintain all required bioassays
  Manage or oversee all radioactive waste disposal
  Maintain all required records
  Act as liaison with regulatory agencies

- **Organizational Hazard Awareness Program**
  Maintain material safety data sheet files
  Track chemical purchases/employee participation in training programs
  Provide organization's emergency coordination
  Represent organization on SARA, Title III local emergency planning committee
  Responsible for hazard awareness for outside contractors involved with chemical hazards

- **Miscellaneous Responsibilities**
  Maintain awareness of upper administration with status of safety and environmental issues
  Participate in all organizational-wide safety committees
  Provide staff support to safety committees
  Investigate employee complaints

Anticipate potential regulatory actions
Provide sanitation, health evaluations of facilities

Not all of the areas listed above are the sole responsibility of the Safety and Health Department. Many involve or require collaboration with other support departments, with the participation of safety differing in degree. The list reflects a strong direct involvement of safety departments in the affairs of their parent organizations. Because of the increasingly complex knowledge, skills, and resources needed to meet current safety standards, there is really no choice except to centralize the responsibility for dealing with many of these standards. The alternative of making everyone comparably knowledgeable and able to manage their affairs sufficiently safely to avoid potential liability for themselves, as well as be technically proficient, is impractical. The goal is to balance, in the area of safety, as much local autonomy and a sense of personal responsibility as possible among the organization's other employees with the responsibility of management. Safety and health personnel can define the programs needed and in many important areas provide required support services, but they cannot be present all the time to see that everyone acts safely. Safety ultimately must be a local and even individual responsibility.

One of the more difficult parts of developing the Safety Department's role as the central internal agency for these areas of responsibility is to do so without appearing to be intrusive into other department's territory. Safety departments are relatively new departments and as such, are having to fit into an existing corporate or institutional environment, which already had portions of the responsibilities listed assigned to existing departments. Physical plant and/or maintenance departments, for example may feel that many of the facility issues were already being handled very nicely but much of the external regulatory areas in themselves represent new issues of which they are unlikely to have the necessary topical knowledge. The safety department can approach this by offering or suggesting that they will take this new burden on themselves so that the physical plant personnel will not have to develop new expertise, and devote new resources into these areas. In all probability, because of the subject matter involved, the older department may feel relieved that someone else is having to do this. In the author's experience, at least at the worker's level, this was indeed the case.

Despite the sense of the previous few pages identifying broad areas of Safety Department responsibility, Environmental Safety Departments should not be too eager to accept additional responsibilities from other service groups. Responsibility often will be shifted or created without a commensurate transfer or allotment of resources, so that the safety departmental resources, including personnel, physical, and financial resources, may be severely strained. Available resources may be difficult to obtain, especially in times of retrenchment, for most public and private organizations. Care must be taken to set achievable goals and priorities. It is better to do a few things well instead of attempting to do too much and failing to do anything properly. An example involves underground storage tanks. The EPA published regulations in 1988 requiring all underground storage tanks containing petroleum products or other substances that could be hazardous to the environment to be tested and modified, if necessary, to prevent environmental contamination. The owner would be required to clean up the site if environmental contamination were found to have occurred already. It seems logical at first that this "environmental" problem would be the concern of an "environmental" safety department. However, it is really a maintenance problem. This is an example of where the maintenance department might wish to shift responsibility but the role of the environmental health and safety department should be limited to a monitoring of the maintenance department's program. This would ensure that the testing and remediation program would be done properly so that existing problems were corrected and additional problems not created. Other areas which might be questionable would be sanitation, including testing of potable water and oversight of food service cleanliness. Some Safety and Health Departments do have these responsibilities but most do not. If the responsibility were to be transferred to an environmental health and safety department, new staffing and resources devoted to these areas should certainly be provided.

H. Departmental Responsibilities

A department is an entity within an organization with which individuals with more or less similar interests and goals identify themselves and are recognized by the organization as a unit for administrative purposes. A department is the logical unit to establish and administer common safety-related procedures and practices suitable for the laboratory programs in the discipline identified with the department. The leadership of the department provides a natural channel for communication with laboratory personnel and provides a vehicle for enforcing organizational policies or for directing concerns by the department about these policies to higher administrative levels. In the context of the present discussion, the major virtue of a department's leadership is that they are physically present on the scene and represent a recognized source of authority. No safety department is sufficiently large to be present everywhere. Safety personnel must rely on the help of the local departments to implement safety practices and policies. The local departmental leadership must accept the need for a strong safety program and see that a concern for safety is the standard expected of all departmental personnel. Unfortunately, especially in the academic work environment, internal department management structure is relatively weak. All research scientists are expected to obtain their own research funding, hire their own staff, and in most respects, except for teaching assignments, operate almost independently. It is a real challenge for management in such an environment to develop a cohesive safety and health program.

At a minimum, even small departments should designate an individual to be a safety coordinator for the department and the individual laboratories within the department, who may also act as liaison to the institutional or corporate safety committee. Larger departments preferably should establish an internal safety committee with representatives from each major division within the department. Individuals at various levels should be included on the committee, such as laboratory supervisors, laboratory chemical hygiene officers, and technical staff to ensure that all points of view are fairly represented. In an academic department, it might be desirable to include an experienced graduate student. One function of the committee would be to interpret the corporate or institutional policies in the context of the departmental operations and the operations of the individual laboratories. The committee can also advise individuals and make recommendations on improvements for safe operations in laboratories. The support of a committee often can facilitate obtaining needed resources, whereas an individual acting alone might not succeed. Finally, the committee should establish a frequent schedule of inspection of facilities of the laboratories and the support infrastructure for the laboratories within the department. The supervisor responsible for the problem area should be informed if conditions are found which need correction, and a follow-up inspection should be done after a reasonable interval to ensure that the corrections have been made.

Departments also can and should function as a resource center. For example, the Laboratory Safety Standard, under OSHA, requires that certain information be readily available to the employees in the department, e.g., a copy of the organization's industrial hygiene plan and Material Safety Data Sheets. A central hard copy file of these, in a large research organization or university, could easily encompass 5000 to 10,000 records in a massive and difficult-to-manage file. A departmental file would normally be expected to be much smaller and more manageable, although in a large academic chemistry department, the distinction might virtually disappear. Similarly, the department would be a logical group to provide much of the technology-specific training required for the OSHA standard for the department's laboratory employees and graduate students, since as noted previously, the department is normally organized around a common discipline.

The department should also accept some responsibility for the safety of its personnel in terms of resources. There are legitimate questions about the limits of the institutional or organizational responsibility and those of the department or laboratory. It is clear, for example, that such things as utilities, custodial services, and building maintenance are within the purview of the corporation or university. Specialized equipment such as an electron microscope or a highly specialized laboratory environment is equally clearly the responsibility of the local department. However, who is responsible for eyewash stations and deluge showers, fume hoods, glove boxes, chemical waste disposal, respirators, safety glasses, flammable material storage cabinets, and refrigerators, etc.? Some of these would not be needed if it were not for specific research programs or grants. Should not the department or laboratory manager incorporate these costs or at least a portion of them in his own budget? To some degree, the imposition of safety standards and regulations has inculcated, in many persons, a sense that safety is not their own responsibility. Some persons have the attitude that someone besides themselves is responsible for not only the rules, but to provide all the means to achieve compliance with the rules. Awareness that safety is, to a major degree, its own reward seems to have been lost to an extent. As a minimum, it would appear that in soliciting a grant, the investigator and the departmental leadership should bear the responsibility of assuring in advance that the resources are available or attainable, so that the research can be done safely and with appropriate regard for the protection of the environment.

I. Laboratory Responsibilities

Ultimately, the responsibility for being safe and working safely falls to the laboratory personnel themselves. The laboratory supervisor, who may or may not be an active participant in the daily, routine work of the laboratory, still must set the standards of performance expected of everyone within the laboratory. The laboratory supervisor must make it clear by example or by direction that carelessness in safety is no more acceptable than careless and sloppy science. This is not to say that the laboratory must always be immaculate. A busy laboratory is almost always somewhat messy, but adherence to good safety practices should not be allowed to be catch-as-catch can. The supervisor has a responsibility to those under his direction to establish safe work practices and to ensure that the employees are given the opportunity to be informed and fully understand any risks associated with the program of work. No one should be asked to perform an act posing a substantial degree of personal risk, unless that is a normal aspect of their work, such as is expected of firefighters or police. This is not to say that laboratory work must be made totally free of risk because this ostensibly desirable goal is unattainable. It does mean that all reasonable and practical steps that can be done to minimize risks have been taken. The concept of ALARA (As Low As Reasonably Achievable), used in the nuclear field, is a good, pragmatic guideline to follow. Equipment must be maintained in good repair and adequately designed to work properly or it should not be used. Written standard operating procedures are required under the OSHA Laboratory Standard, and individuals must be fully trained in the procedures that apply to their duties. Work should be carefully analyzed to foresee potential accidents or failures. The "what if" criterion should be applied to every procedure used in the facility. Even low probability bets do pay off on occasion, sometimes in a very bad way. Contingency plans must be developed to meet at least the most likely emergencies. To fail to do these things could expose the laboratory director and in turn possibly the department and the university or corporation to charges of willful negligence.

Employees, on the other hand, have an equal share in the responsibility. They must adhere to the safety policies that have been established. They must take the initiative to ensure that they are knowledgeable about good laboratory safety practices, and they must not diverge from these practices because they are too time consuming, too much trouble, or inconvenient at the moment. If they are uncertain of the proper procedure, they must not be reluctant to admit it, but should seek clarification. Often, especially if the laboratory supervisor is not a reasonably active participant in the work in progress, a knowledgeable employee may understand the actual potential risks better than the supervisor. Any employee should offer suggestions to improve safety in the workplace. Unfortunately, this ideal may be a bit unrealistic in that many subordinates are sometimes wary of "making waves" or contradicting their superiors, and not all supervisors are appreciative of suggestions from subordinates. It is this last situation for which the ability to make anonymous complaints, which are certain to be investigated, was incorporated in the Federal OSHA Act and subsequently passed on to the state standards, where states have adopted their own OSHA statutes. Ideally the team of a caring, conscientious laboratory supervisor working in cooperation with competent, intelligent, and imaginative laboratory workers should make it possible to conduct laboratory work with minimal risk.

A "loose" laboratory in the sense of a congenial atmosphere with everyone working as a team is conducive to establishing a workable safety program. It is worthwhile for laboratory management to do things to promote a pleasant atmosphere in the facility, rather than make adherence to good safety practices strictly a disciplinary policy. I once spent a summer at a major governmental research facility, and as I was about to do something which had been a marginally risky common practice at the school where I had done my graduate work, an ordinary laborer tapped me on the shoulder and said "We don't do things that way here, doc." That comment, and the worker's freedom and willingness to make that comment to a professional has remained with me ever since, and provides much of the philosophy underlying this handbook.

REFERENCES

1. **Petersen, D.,** *Techniques of Safety Management,* 2nd ed., McGraw-Hill, New York, 1978.
2. *Prudent Practices for Handling Hazardous Chemicals in Laboratories,* National Academy Press, Washington, D.C., 1981.
3. *Safety in Academic Chemistry Laboratories,* 4th ed., American Chemical Society Washington, D.C., 1985.
4. **Steere, N.V.,** Responsibility for laboratory safety, in *CRC Handbook of Laboratory Safety* 2nd ed., Steere, N.M. (Ed.), CRC Press, Cleveland, OH,1971, 3.
5. **Becker, E.I. and Gatwood, G.T.,** Organization for safety in laboratories, in *CRC Handbook of Laboratory Safety* 2nd ed., Steere, N.V. (Ed.), CRC Press, Cleveland, 1971, 11.
6. **Songer, J.R.,** Laboratory safety program organization, in *Laboratory Safety: Principles and Practices,* Miller, B.M., Gröschel, D.H.M., Richardson, J.H., Vesley, D., Songer, J.H., Housewright, R.D., and Barkley, W.E. (Eds.), American Society for Microbiology Washington, D.C., 1986.
7. **Bilsom, R.E.,** Torts Among the Ivy: Some Aspects of the Civil Liability of Universities, University of Saskatchewan, Saskatoon, Saskatchewan,1986.
8. **Campbell, S.L.,** Expanding environmental duties, in *Occupational Health & Safety,* 61 11, 18, 1992.
9. **Jefferson, E.G.,** Safety is good business, in *Chem. & Eng. News,* November 17, 1986, 3.
10. **Koshland, Jr., D.E.,** The DNA Dragon 1, in *Science,* 237 No. 4821, 1987, 1397.
11. *OSHA Exposure to Hazardous Chemicals in Laboratories,* 29 CFR 1910.1450, 55 FR 3327, January 31, 1990 and FR 7967, March 6, 1990.
12. *OSHA Hazard Communication Standard,* 29 CFR Parts 1910.38, 1910.120, 1910.1200, 1910.1450 and 1910.1500, as amended.
13. *DOT Hazardous Materials, Substances and Waste Regulations,* 49 CFR Parts 171-177, as amended.
14. *EPA Hazardous Waste Regulations,* 40 CFR Parts 260-265, 268, 302, 303, 311, 355, and 370, as amended.
15. *OSHA Occupational Exposure to Human Pathogens,* 29 CFR 1910.1030, 56 FR 64004, December 6, 1991.
16. *EPA Standards for the Tracking and Management of Medical Waste,* 40 CFR 259.
17. *Americans With Disabilities Act of 1990,* Titles I-V, July 26, 1992.
18. *DHEW Guidelines for Research Involving Recombinant DNA Molecules,* FR 51, 16957, May 7, 1986, as amended.

19. *DHHS Protection of Human Subjects,* 45 CFR 46.
20. *NRC Standards for Protection Against Radiation Revision,* 10 CFR Part 20.
21. *Prevalence of Safety and Health Specialists,* Bureau of National Affairs, Occupational Health and Safety Reporter.
22. *Prudent Practices in the Laboratory,* National Research Council, National Academy Press, Washington, D.C. 1995.

Chapter 2

EMERGENCIES

Emergencies are, by definition, not planned. However, planning for emergencies can not only be done, but is an essential component of laboratory safety. This is especially true for the laboratory environment where the *potential* for incidents is much higher than in many other working situations. There are many regulatory standards that now *require* that organizations using chemicals in laboratories and elsewhere, or that produce chemical waste, have formal emergency plans. These plans must cover emergency evacuation and response procedures, emergency equipment to be kept on hand, security, training of personnel handling hazardous chemicals, reference materials, identification of emergency personnel, and access to external resources, including aid agreements with local emergency organizations. For example, every facility with laboratories that come under the OSHA laboratory standard must meet this obligation under Title 29 Part 1910.1450. This includes even relatively small organizations. In addition, OSHA Industry General Standards under 1910.38, 120, and 1200 also provide for emergency planning. The OSHA Bloodborne Pathogen Standard (29 CFR 1910.1030) has provisions for emergency actions in case of an accidental exposure. The RCRA Act considers all organizations that generate more than 100 kg of hazardous waste per month as large generators. Title 40 CFR Part 265.16 and Parts 311 and 355 defines the emergency requirements of RCRA. The Americans With Disabilities Act, Titles II and III, 28 CFR, impacts accessibility for disabled individuals. Each of the regulatory acts will be discussed in more detail in later chapters. However, these regulations simply provide the specifics for a legal mandate to do what every organization handling hazardous material should do anyway.

A realistic appraisal of the circumstances that can lead to emergencies in a laboratory will reveal many foreseeable and controllable problems. Some problems that can be expected to occur might include:

- Fires
- Chemical spills
- Generation of toxic fumes and vapors
- Inhalation, ingestion, or absorption of toxic materials
- Release of compressed toxic, anesthetic, explosive, asphyxiating, and corrosive gases locally or beyond the boundaries of a facility
- Release of radioactive materials
- Release of pathogens and restricted biological materials
- Power failure, involving loss of lights or ventilation
- Electrical shocks
- Explosions, or runaway reactions

- Failure of a facility exhaust system
- Physical injuries to individuals
- Consequences of natural disasters
- A combination of any of these simultaneously

This list is not intended to be complete. Some events are more likely to create immediate and pressing problems than others. It is impossible to anticipate all classes of problems that can occur. Some events are recognized as emergencies more readily while others may not be identified for extended periods of time. Some involve the threat of personal injury, while others impact the environment with little likelihood of immediate injurious effects to individuals. Emergency personnel often mention that many emergencies in which they have been involved were not anticipated and would have been unlikely to have been considered, even by the most careful planning. There is no limit to the variations that human ingenuity and the vagaries of fate can take to modify the factors that control our lives. The inability to foresee all possible emergencies should not inhibit the development of plans to cope with those that *can* be anticipated, or to provide a basic emergency response infra-structure that can be used, even for unanticipated types of emergencies.

The scope of this chapter will be to examine the general principles of emergency preparedness to serve as a guide for preparation of individual, specific action plans, and to provide some useful information to be used in various classes of emergencies. Planning and preparation are necessary to help in identifying and finding the resources needed to support a flexible, effective, and, if needed, rapid response to laboratory emergencies. Injuries, property and environmental damage can be limited if effective emergency procedures already exist and are practiced regularly. Practice is essential to expose deficiencies in the procedures and to familiarize personnel involved with them. Plans that are developed and then filed away are worse than useless. They can provide a false sense of security. In a real emergency, it is essential to *know* immediately what to do or, often, what not to do. Time to read a manual is often not likely to be available.

Before developing the theme suggested in the preceding paragraph, there are some caveats that need to be introduced that are applicable to the contents of this entire chapter. There are advantages in not overreacting. It is easy for well-meaning and knowledgeable individuals to turn a relatively minor event into a major and expensive incident by acting too quickly, without full awareness of the total situation and without consulting other persons involved. No serious worsening of a situation might result in doing absolutely nothing until the situation has been discussed, evaluated, and a plan of action developed. Evacuation, containment, and exclusion of nonessential personnel are the appropriate initial actions in almost every emergency. Unless a situation is clearly deteriorating and shows signs of becoming out of control, a review of the situation and an examination of the response options by emergency and operational personnel is usually desirable. However, this decision is best left to the responsible personnel on the scene.

One other note of caution: no one is expected in the normal course of their work to go to extreme measures, risking their own lives, to cope with an emergency when the risk is certain to be very great. The more responsible action often is to leave the scene when the situation is obviously beyond an individual's capabilities. Doing so makes it possible for emergency response groups to have a competent source of information about the situation when they arrive. It is difficult to do when lives are involved, instead of only property, but there is no point in adding to the loss when the situation is clearly hopeless. It is a judgment call again that can only be made at the time by persons present.

Despite the two cautionary paragraphs immediately preceding, there are steps that individuals and local groups can and should take, when appropriate, to confine and minimize the impact of emergencies. The first few moments of an emergency are frequently the ones that are the most crucial. Actions should be based on training, knowledge, and a due regard for priorities. Protection of life and health should come before protection of property, or reputation, especially the latter. Unfortunately, many persons do not seek help or take inappropriate actions until too late for fear of being blamed for a problem, often allowing the situation to worsen until out of control. Trained and knowledgeable personnel are less likely to make these mistakes.

A. Components of Emergency Preparedness

Emergency preparedness is the responsibility of everyone. Many persons consider this the job of such organizations as fire departments, police departments and rescue squads and do not consider themselves as part of the emergency response. This is not true. Everyone has a role to play and it is the responsibility of emergency planners to define these roles and prepare individuals to carry out their personal responsibilities, even if, in some cases these are limited to alerting others of the problems, evacuating the area and making sure trained groups are notified promptly.

1. Initial Conditions

Basic conditions should exist to ease meeting emergency responsibilities. Some of these conditions should be met before a building is constructed. For example, in the initial planning, the building should have been designed to incorporate safety codes and regulations by the architects, in cooperation with the persons responsible for the programs to be housed in the building. Codes represent minimum requirements which the builder's owners should feel no hesitation in exceeding if it appears needed. Appropriate fixed and movable equipment must be installed or provided, consistent with the concept of a facility that could be operated safely. Code mandated emergency equipment must be available, but decisions must be made about what design features and equipment should be mandatory, what is desirable, and what would be a luxury. Once these decisions are made, leaning, it is hoped, toward the side of enhanced safety, then personnel responsibilities should be considered next. It is necessary to define the role of each person in a facility and to specifically designate which individuals and groups should have the leadership responsibility for emergency planning and emergency response. It is critical that it not be necessary to develop an impromptu plan or seek one buried in a file cabinet.

The emergency plan for a given facility should be a subset of a plan for the entire organization. The infrastructure and planning available to the entire organization can be adapted to the needs of individual needs and individual laboratories. Decisions must be made as to who is responsible for providing emergency response equipment and supplies, and obviously with this decision, the need arises to decide the source of funds. A major decision is to define the type of command structure that will be used and who will be involved. A clearly defined line of authority is needed. The responsibilities of the key individuals and groups must be delineated and boundaries established between local responsibility, institutional responsibility, and outside emergency response agencies. Finally, based on all of the applicable factors, each individual facility can establish a written emergency response plan for itself with specified responses to anticipated classes of emergencies specific to that facility. The organizational plans as well as the plans for smaller units all must be sufficiently flexible to provide responses to unanticipated emergencies.

The following material will elaborate on these points.

2. Facilities, Fixed, and Movable Equipment

Where buildings and facilities already exist prior to developing an emergency plan, it is necessary, of course, to adapt the plan to the existing structure, but if the opportunity arises, there is much that can be done to reduce the severity of later emergencies when designing, building, and equipping a facility. Once built, it is expensive to modify a facility but incorporating safety features in a newly built structure can save substantial costs. For example, renovating existing structures to make them earthquake resistant, unnecessary in many areas while very important in others, is very expensive, but it is possible to do it at much less cost for new buildings where needed. In order to facilitate the design and construction of safe buildings, fire and building codes have been established in most localities that govern new construction and renovations to existing buildings. Generally, under these codes, research laboratories come under the classification of a business use occupancy or occasionally as a hazardous use occupancy where unusually hazardous activities are involved, each of which incorporates different safety requirements. OSHA also has standards in the area of fire safety, as well as ventilation, which must be met. OSHA standards, where applicable*, are consistent in every state, but building codes vary from locality to locality, often depending upon interpretations of a local code official. The requirements for access for the disabled under the ADA clearly affect emergency movements. As a result, fire alarms now require intense strobe lighting devices as well as audible signals. Braille instructions may be required for the blind in parts of a facility. Special chairs may have to be provided for the physically disabled. Places of refuge to which disabled persons can go while awaiting help must be identified. For several other types of risk, special regulations, such as the classification system for recombinant DNA research facilities, also have safety restrictions that must be included in the building design. This latter set of safety restrictions will be reserved to later chapters dealing with these special topics.

Concerns which should be addressed in the designs of laboratory buildings to enhance emergency responses depend upon the classification. For example, if the building is a hazardous use occupancy, most codes will require a sprinkler or other fire suppression systems. If a sprinkler or alarm system is required by a local fire code, then OSHA 1910.37(m&n) requires maintenance and testing. Also for this classification, OSHA will require under 1910.37(f)(2) that the doors swing in the direction of exit travel, yet most building codes have restrictions on doors swinging into corridors to avoid creating obstructions to corridor traffic. In order to satisfy both requirements, doors should be recessed into alcoves inside the laboratory. Even existing facilities may have to be upgraded to meet some code standards.

The size of a building, the number of floors, and the relationship to other structures all enter code decisions affecting safety in emergencies. Addition of equipment to a laboratory, such as a hood, can have serious fire safety implications. Is there adequate makeup air? If not, where can it be obtained? Halls cannot not be used as a plenum or as a supply of makeup air for more than a few hundred ft^3 per minute (cfm) for each laboratory space. Even a small, 4-foot fume hood discharges about 800 cfm, so that one cannot draw the required makeup air in through louvers in the door. Usually, one must go outside for a source of makeup air, but what is the relation of this new inlet air intake to the exhaust system? Toxic fumes could be drawn back into a building. A fume exhaust duct penetrating a floor could allow a fire to spread from one floor to another. Therefore, most codes require fume hood ducts to be enclosed in a fire-rated chase. Because of the expense of constructing a chase, the cost of avoiding worsening the fire separation in a building could preclude installation of the hood, which in turn could preclude using the space for the intended research. One option, to allow future flexibility, is to incorporate external chases as an architectural feature in the design. Energy loss considerations can impact the design of a laboratory. Auxiliary air hoods have been used in the past to reduce the amount of tempered air being "wasted," but there are a number of reasons why this type of hood is less desirable and they are seldom used any more in new construction. In fact, most laboratory designers explicitly prohibit the use of auxiliary air hoods. An alternative is to design a ventilation system for a laboratory to maintain a

---

* About 25 states have adopted their own state OSHA plans which are required to be as stringent as the federal standards; however, public employees in some of these states may not be covered by the OSHA standards.

constant volume of air through a hood while in use, and provide some means of reducing the ventilation requirements for a facility when the hood is not being used. Ventilation will be discussed in much more detail in Chapter III.

The interior arrangements of a laboratory are critical in permitting safe evacuation from the laboratory. The types of accidents listed earlier could pose much more serious risks to individuals should they occur between an individual and the exit from the room. A simple solution for these potential emergencies for larger laboratories is to have two well-separated exits. This is not always possible, especially in smaller laboratories. An alternative would be to evaluate what components of a laboratory are most likely to be involved in an incident and which would increase the hazard if they became involved in an ongoing emergency. These components should be located so that an escape route from the normal work area does not pass by them. Also, portable fire extinguishers, fire blankets, respirators, and other emergency equipment should be located on this same escape route. Eyewash stations and deluge showers should be located close to where injuries are likely to occur, so an individual will not have to move substantial distances while in intense pain or blinded. Aisles should be wide (typically a minimum of 42 to 48 inches), straight, and uncluttered with excess equipment to ease movement in emergencies. A laboratory should have emergency lighting, but many do not. The considerable dangers posed to an individual stumbling around in a pitch dark laboratory should the power fail are obvious. Inexpensive, battery-powered rechargeable units are a potential solution here and are not expensive, even in retrofitting a facility.

Many regulations found in OSHA standards include features that will minimize the scope and impact of an emergency such as a fire. For example, restrictions in 1910.106 on container sizes of flammable liquids and the amounts of these materials that are permitted to be stored outside flammable material storage cabinets are designed to limit the amount of fuel available to a fire and to extend the time before the material could become involved.

Every action should be considered in terms of what would result if the worst happened. In large projects, this is often part of a formal hazard analysis, but this concept should be extended to virtually every decision within a laboratory. For example, a common piece of equipment found in most laboratories is a refrigerator. A refrigeration unit suitable for storing flammables, i.e., containing no internal sources of ignition, costs about two to three times as much as a similar unit designed for home use. It is tempting, especially if money is tight and the immediate need does not require storage of flammables, to save the difference. However, the average lifetime of a refrigeration unit is roughly 15 to 20 years. Who can say what materials research programs will entail over such a long period? If flammable vapors within an ordinary refrigerator should be ignited, a violent explosion is very likely to occur. Employees could be injured or killed and the laboratory, the building, and the product of years of research could be destroyed. Not only would there be immediate problems, but in most cases, replacing laboratory space would be very expensive, currently in the vicinity of $130 to $300 per square foot. Actual construction of replacement space for buildings as complex as most laboratories, from the time of planning to completion of construction, typically takes 4 years or more after the money is obtained.

Many actions are influenced by the costs involved, as in the preceding example. A continuing question involves who should be responsible for paying for safety facilities and equipment. Under the OSHA laboratory standard, the adequacy of a facility to allow work to be done safely is a key condition. There are some straightforward guidelines that can be used:

1. For new construction, safety should be integrated into the building design and the choice of all fixed equipment. The latter should be incorporated in the building furniture and equipment package. This would include major items such as fume hoods, since these are relatively expensive units to retrofit.

2. Certain equipment and operational items common to the entire organization (e.g., fire extinguishers, emergency lighting, deluge showers, eyewash stations, and fire alarm

systems) and maintenance of these items should be just as much an institutional responsibility as provision of utilities.

3. Items which are the result of operations unique to the individual laboratory or operations should be a local responsibility This would include equipment such as flammable material refrigeration units, flammable material storage cabinets (if these are not built in), and specialized safety equipment such as radiation monitors, gas monitors, etc. Some major items which might be included under fixed equipment in new construction might have to be provided by the individual if renovation of a space were to be involved. For example, it might be necessary to construct a shaft to enclose a fume hood duct and to provide a source of additional makeup air for the hood. The expense for personal protective equipment, such as goggles, face masks, respirators, and gloves, should also be provided at either the laboratory or departmental level.

It is unlikely that any individual, whether it is the laboratory supervisor, safety professional, planner, or architect, will alone be sufficiently knowledgeable or have the requisite skills to make appropriate decisions for all of the factors discussed in this section. In addition, every one of these persons will have their own agenda. The inclusion of emergency preparedness features should be explicitly included as one of the charges to the building or project design committee so that these needs can be integrated with function, efficiency, esthetics, and cost.

It was not the intent at this point to elaborate on all the implications of codes as safety issues but, by a few examples, to draw attention to the idea that the root cause of an emergency and the potential for successfully dealing with it could well lie with decisions made years earlier. The point that *was* intended to be made was that laboratory safety and the capability to respond to emergencies does not start and end with teaching good laboratory technique and the adoption of an emergency response plan after beginning operations.

B. Institutional or Corporate Emergency Committee

In most organizations, there are many support groups that have been assigned specific responsibilities in dealing with emergencies which extend beyond those associated only with laboratories. Among these are safety, police or security, maintenance, communications, legal counsel, and media or public relations. Unlike the laboratory supervisor, departmental chair or individual laboratory employee who is primarily concerned with his research or administrative duties, these groups are directly concerned with one or more aspects of emergency response. In larger organizations, fire departments, physicians or medical services, or even more specialized groups may exist in-house. Each of these groups have their own expertise, their own dedicated resources, and their own contacts with outside agencies. Representatives from these agencies will be the ones normally called to the scene of an emergency and will be the ones expected to cope with the situation. This group should form the nucleus of the emergency planning committee but it should also include participation from the remainder of the organization. In the current context, this participation should include comprehensive coverage of the various areas of the corporate or institutional research programs. The committee should have direct access to upper levels of management, and it should also interact closely with safety committees associated with each broad research area, e.g., chemical, radiation, biosafety, and animal care. This committee also needs to coordinate its efforts with non-organizational support groups such as local, state, and federal police authorities, fire departments, rescue units, local emergency planning groups, environmental regulatory agencies such as EPA and local or regional water, air, and waste management agencies, and safety regulatory groups such as OSHA. Note that the emergency committee does not have the responsibility to manage the response to an actual incident. The emergency committee, once formed and its charge

Figure 2.1 A sign such as this placed at each telephone is an effective way to inform people how to notify authorities.

clearly defined, should meet periodically (at least once a year and preferably more often) to review the status of the organization's emergency preparedness, to plan for practice sessions, to review drills that have been conducted, and to investigate and review incidents that have occurred. Reports of these meetings, along with the findings, should be presented to management and to the individual safety committees.

C. Emergency Plan

The initial order of business for the emergency committee is to develop an emergency response plan (ERP). In developing the ERP, the committee should analyze the types of emergencies which could happen, their relative seriousness, and their relative probability of occurrence, in other words, perform an organizational hazard analysis. The emergencies to be considered should specifically include releases of hazardous and toxic chemicals to the environment, as required under SARA, Title III (Superfund Amendments and Reauthorization Act of 1986). Once the classes of emergencies have been defined, each should be analyzed as to the resources, equipment, training, and manpower which would be needed for an adequate response. An integral part of this analysis would be provisional plans for using these resources to respond to potential emergencies. The analysis should include both internal and external resources. Finally, a critical evaluation should be made of the current status of the institutional resources and a recommendation made to correct deficiencies. Based on the preliminary studies, the final plan should be drafted, circulated for review, amended if required, and implemented. The support of management is critical, or this effort would be wasted.

Plans should be developed which would be operative at differing levels. A basic plan should be short and easy to understand and to implement. The simple sign in Figure 2.1 above is effective for most emergencies. The caller is expected to be guided by the person (usually a dispatcher) at the other end of the line for specific guidance for the appropriate response to the immediate problem. The major caveat is that the time to make such a call may not be available prior to evacuation for emergencies representing immediate and worsening emergency situations. Occupants of a facility should be trained to recognize when this condition exists and know how to initiate an evacuation of as large an area as necessary.

1. Laboratory Emergency Plan

Workers in most laboratories normally are intelligent, knowledgeable individuals and can cope with many small emergencies such as a spill of a liter of sulfuric acid or a small fire if they have received appropriate emergency training. Such training is mandatory under the OSHA laboratory standard. A comprehensive laboratory emergency response plan is required under current standards for the risks associated with operations within the facility. The plan needs to include basic information such as risk recognition appropriate to the operations of the facility, means of internal responses to small to moderate emergencies, and evacuation training. All employees in the laboratory must receive instruction on these points at the time of beginning work in the facility, or when any new procedure or operation is introduced posing different risks. In order to identify potential risks, a detailed, thorough hazard analysis needs to have been done, based on the things that *could* go wrong, not just the risks associated with normal operations. Among information which must be included in the plan is where an employee can get not only the laboratory specific plan, but also the organization's overall plan. Another key ingredient of the plan is where safety and health information for the chemicals used in the laboratory, as represented by Material Safety Data Sheets (MSDSs), can be readily provided.

A written emergency plan for an individual laboratory might, in outline, resemble the following:

I.   In bold letters, the basic number to call in the event of an emergency, perhaps 911or possibly an internal number.

II.   A defined line of authority. This should provide the names and home and work telephone numbers of several individuals authorized to make decisions for the facility. They should be persons with direct knowledge of laboratory operations and, at least at the top of the list, persons who can make financial commitments.

III. A list of external persons/groups, with telephone numbers, who can provide emergency assistance relevant to the risks associated with operations. Such a list should include at least the following:

Emergency telephone number- 911, if available in the area
University police or corporate security, if not available through the 911 number
Local government police, if not available through 911 number
Fire department number, if not available through 911 number
Emergency medical care (rescue squad), if not available through 911 number
Nearest Poison Control Center
Nearest hospital
Safety department
Spill control group, if not available through 911 dispatcher or Safety Department
Maintenance department number(s)
Laboratory supervisor business and home telephone number
Secondary laboratory authorities business and home telephone numbers
Departmental or building authority number

IV.   A list of normally required safety procedures appropriate to laboratory operations.

V.   A simplified list of emergency actions to take for most likely emergencies.

VI.   Evacuation instructions, including a map of at least two alternative evacuation routes. The primary route should be identified and normally should be the shortest, most direct means of egress from the facility. A gathering area should be identified to which evacuees would normally go. This is important to allow a "head count" to ensure that everyone did successfully evacuate, and to provide a location where external agencies could come in order to receive information concerning the emergency.

VII. Location of Material Safety Data Sheets and other safety and health reference materials.

VIII. Location of the organization's emergency plan.

IX. Procedures for expanding the emergency response to additional areas of the building and organization when the emergency is a "large" one extending beyond the immediate area. The location of one or more telephones outside of the affected facility but readily accessible should be clearly identified.

Two items need to be placed on or adjacent to the laboratory door to assist emergency responders when lab personnel are not immediately available during an incident: the line of authority, listed in Item II above, and indications of the types of hazards to be found within the laboratory. Some areas have ordinances requiring the use of the National Fire Prevention Association (NFPA) Diamond for the latter purpose, but unfortunately, most laboratories would have at least some material with high-risk ratings in all categories. Pictographic labels identifying classes of hazards within a facility are also used. The best way to alert firefighters would be to have laboratory inventories on a computer database and provision made for emergency response groups to have electronic access to this information. Software is available, although not yet in wide use, which does this.

This plan incorporates some aspects of the Laboratory Industrial Hygiene Plan as required under OSHA, which could be deleted, since the written industrial hygiene plan must be maintained. However, items I, II, III, V, VI, and VII are essential.

The plan just described should be reviewed with each new employee and at least annually for all occupants of a laboratory. An annual practice drill is strongly recommended.

2. Organizational Emergency Plan

There is some overlap between planning for responses to local emergencies in individual laboratories and the response to large-scale emergencies. At the extremes, the distinction is clear. A minor spill or a trash can fire obviously is a minor emergency while a fire that involves an entire building or a major spill where hazardous materials are released into the environment clearly is beyond the capacity of laboratory personnel. Planning needs to provide guidelines to cover the transition between the two levels to ensure that an appropriate response does occur. A comprehensive plan is intended to provide a general infrastructure for all classes of emergencies. Detailed plans are essential for organized emergency groups, but for the use of the general public a basic emergency plan is to evacuate the area or building, and call for emergency help. Often, evacuation will be more than is actually needed, but it is usually a conservative and safe approach. The essential information to enable this can be placed on a single page for a facility. Normally, planning for large-scale emergencies will be the responsibility of the corporate or institutional Emergency Committee, working with internal groups and the Local Emergency Planning Committee (required under SARA Title III) and nearby support agencies.

A basic means of reacting to virtually any emergency for untrained persons would be to place a sign, such as is shown in Figure 2.1, on or near every telephone. In this case, it is up to the individual at the other end of the telephone line, normally a dispatcher, to give verbal directions for subsequent actions. The dispatcher needs to be well trained and provided with a list of individuals and groups whom they would notify of the incident, in an appropriate priority. These individuals, groups, and priorities are defined in the master emergency plan for the organization.

Following is a simplified table of contents for an emergency plan established for an area containing a university, major commercial activities including chemically related industries, transportation sources (highway, rail, and air), and the usual variety of emergency support groups.

1.0   Charge
1.1   Assignment of legal authority and responsibilities
    Charge
    Members of governing body
1.2   Purpose of plan, functional description
1.3   Instruction on how to use the plan
1.4   Initial conditions
    Demographics
    Geographic description
    Natural risks
    Climate
    Time factors
    Local hazard sources
    Utilities
    Local administrative units
    Local emergency units
    Local resources
1.5   Communications
    Notification procedures
    List of agencies/personnel requiring notification
    Telephone lists
        Key personnel and alternates
            Telephone tree
        Emergency assistance numbers
            Local
            Regional
            State
            National
            Commercial
        Regulatory agencies
    Alternative communication options
    Authorized radio coordination procedure
1.6   Incident recognition/response
    Identification of incident
    Response protocol
    Emergency command structure (see Figure 2.2)
        Command center, normal
        At-scene control center
        Emergency coordinator
        On-scene commanders
1.7   Responsibilities of emergency support groups (initial response)
    Fire/rescue/haz-mat teams
    Law enforcement
    Medical
    Communications (public notification/media relations)
    Logistics support
        Transportation
        Public works
        Emergency housing/refuge centers

Administrators (government/corporate/institutional)
Agencies (regional/national/regulatory)
Emergency committee

1.8 Ongoing and completion
Assessment of conditions
Containment
Termination
Recovery
Critique

1.9 Continuing processes
Training
Practice drills
Resource development
Plan review

2.0 Appendices
Incident forms
Mutual aid agreements
Current emergency rosters
Evacuation centers
Hospitals/medical assistance
Social agencies
Emergency equipment lists
Likely incident locations
Cleanup contractors
Experts
Testing laboratories
Maps/overlays
Radio/TV/newspaper contacts
Copies of regulations

All of the groups likely to be involved in the emergency response should possess a copy and be familiar with the organization's emergency response manual. The manual should spell out in detail, but still as simply and as flexibly as possible, the correct response to the classes of emergencies incorporated in the ERR.

It is always the intent of every organization that no emergency will ever occur and for the more unusual situations considered in the ERR, long intervals may pass between incidents. However, it is essential to include provision for periodic review and practice drills in every emergency plan.

a. Emergency Plan Components
A partial list of some of the more common laboratory-related emergency situations was given in Section 2.1. A written response plan should be provided for each of these situations, identifying the likely locations where these classes of problems would be apt to occur, the characteristics of the locations, accessibility, probable means of response, local resources available, contact persons, outside agencies that would need to be notified, and possible refuge areas to which the occupants would evacuate. Important characteristics or questions which need to be addressed would include: is it a multiple story building, what type of construction (combustible or fire resistant), does an alarm and/or sprinkler system exist, are there standpipe connections or hydrants nearby, what is the typical occupancy level at various times of day, are there disabled persons in the building requiring special assistance, are there

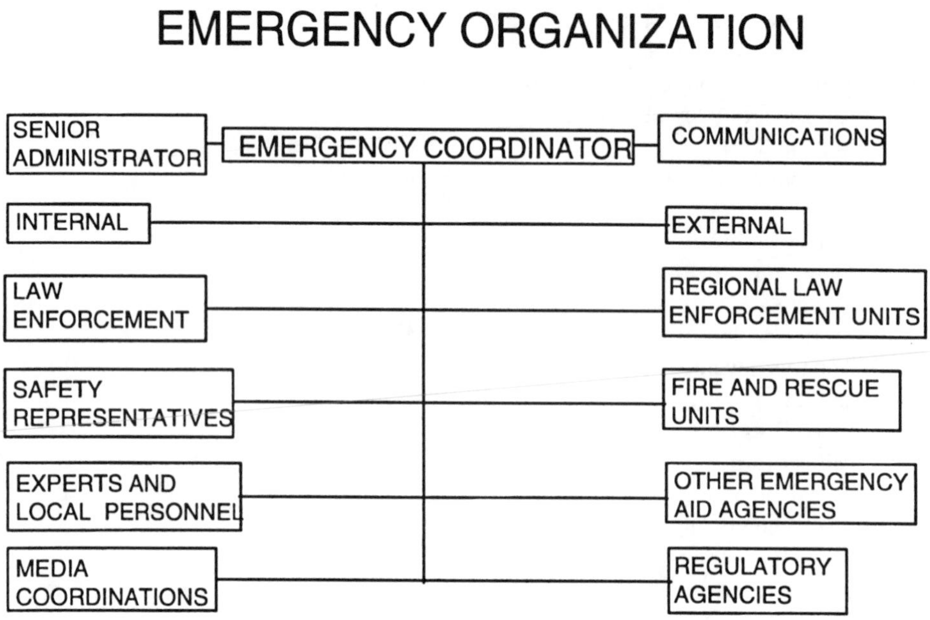

EMERGENCY ORGANIZATION

SENIOR ADMINISTRATOR	EMERGENCY COORDINATOR	COMMUNICATIONS
INTERNAL		EXTERNAL
LAW ENFORCEMENT		REGIONAL LAW ENFORCEMENT UNITS
SAFETY REPRESENTATIVES		FIRE AND RESCUE UNITS
EXPERTS AND LOCAL PERSONNEL		OTHER EMERGENCY AID AGENCIES
MEDIA COORDINATIONS		REGULATORY AGENCIES

Figure 2.2 A typical military-type command structure for responding to a substantial emergency.

hazardous materials in the facility, the kinds and quantities of these materials and what is the potential impact on adjacent structures or areas should hazardous materials be released for various environmental conditions, among other factors. This type of information requires a great deal of time to compile. The compiled information should be placed in a well-organized appendix to the main body of the plan, so that it would not be necessary to wade through what would necessarily be a massive amount of data for larger organizations.

The management structure is critical to controlling emergencies. This needs to be defined in advance. If the organization is sufficiently large, the plan may include managing virtually every aspect internally without utilizing external agencies, unless the scope of the emergency extends beyond the area of the organization's control. In such cases, outside agencies *must* be notified, and they *may* assume partial responsibility for management of the emergency response, but an emergency extending beyond the controlled boundaries will definitely mandate notification of outside agencies. A large organization may have its own fire brigade, police force, safety department, hazardous material response team, rescue squad, and access to experts internally. Most larger corporations and universities have some of these, but typically not all. Smaller firms and colleges might have only a combined security force and a small safety department.

Most emergency plans employ a pseudo-military organization, at least for coordinating the initial response. An individual, with alternates, is identified as the emergency coordinator. If the organization is highly structured, a command center, again with alternates, is identified to which the emergency coordinator and other key individuals will go when an emergency of sufficient scope occurs. This command center should have radio and telephone communication capability, which would be less vulnerable to loss of power and normal communication channels. Radio contact on emergency frequencies should be available to fire and rescue units, nearby hospitals, local and state police, and state emergency response agencies. In large-scale emergencies, even these channels can become overloaded, as will normal telephone lines. Cellular telephone service is an alternative which has become widely available that does not depend upon hard-wired telephone communications. Other advantages of using cellular telephones are that they do not use what may be limited radio channels and are less likely to be overheard by the general public. A chart is shown in Figure 2.2 which reflects this typical command center operation.

The emergency coordinator is a key individual and must be someone who will be accepted as a command figure. The individual ideally should be one to grasp information quickly, be able to integrate it, and come up with appropriate responses. It is critical too that this person be sufficiently flexible mentally that proffered advice is not disregarded out of hand. Since the most often employed emergency response structure is semi-military in nature, a person often designated as the emergency coordinator will be the public safety director. In the context of laboratory emergencies, most public safety managers are likely to have had police training, not scientific training, so having knowledgeable persons present to make technically correct recommendations is very important. These may be from the safety department and/or individuals from the scene of the incident. In addition to the structured internal departments, major resources available at any research-oriented institution are the scientists and technicians who work there. The ones most likely to be helpful for the types of emergencies anticipated in developing the emergency plan should be identified and a master list of their office and home telephone numbers maintained. A copy of the current list should be maintained by the key internal organizations involved in the emergency response plan. A copy of the list should also be personally maintained by the key individuals in these latter organizations, both in their offices and at home. Alternates should always be designated for these key persons, so that backups are available at all times. Radios, cellular telephones, or beeper systems to allow these key persons to be reached when not at their usual locations would be highly desirable.

Organizations having the capability for a response at this level will have some type of security or police force. These individuals are very likely to be the very first "outsiders" arriving on the scene of an emergency and, as such, initiating a first response. Clearly, they need to receive sufficient training to permit them to make an appropriate "first response" evaluation of the incident and set the containment and response mechanisms in process. It is relatively rare, though, that they will have sufficient training to manage the response to technically involved emergencies. Some key personnel among the security or police groups will ideally have been given special "hazardous-materials-incident" training to allow them to initiate or effect an evacuation of affected personnel and provide safety for themselves and for the evacuees, pending further response actions.

In many jurisdictions, the legal responsibility for management of incidents involving hazardous materials has been delegated to the fire department or to specialized hazardous material response teams. When these arrive on the scene, the management responsibility for an incident may shift so that the emergency coordinator, having the ultimate authority, will no longer be a representative of the organization or institution. In such a case, the internal side of the picture would shift to a supportive and/or advisory role. However, in many instances, the fire department, if that is the responsible agency to which authority is delegated, may choose to take substantial advice and guidance from the organization's team or even ask them to continue *de facto* management of the response to the incident. Depending upon the nature of the incident, one or more regulatory agencies may need to be notified promptly. If a significant chemical release is involved which becomes airborne or involves a liquid spill such that hazardous materials escape from the controlled boundaries of a facility, the National Response Center must be notified as well as the local emergency response coordinator (often the sheriff, police chief, or civil disaster coordinator) and state agencies. Other agencies would also be called, as their areas of regulatory concern would become involved. Although these outside regulatory agencies (note the distinction here between regulatory agencies and emergency response agencies) will arrive on the scene, the responsibility for the incident normally remains a local responsibility, unless it truly becomes a massive problem. Written aid agreements need to have been worked out in detail between corporations and institutions with local emergency response organizations.

There are three groups identified in Figure 2.2 that have not been touched upon as yet. No major incident occurs without news media quickly arriving at the scene. Emergency response personnel must not be distracted by these persons, so media contact persons or groups should be established with whom the news representatives may interact. The security or police may need to act to ensure that not only news media but other nonessential persons do not enter the area. In a mature response stage of an emergency, the role of the police will almost certainly have devolved from active management to control of the boundaries of the affected area. The emergency coordinator has to have some resources immediately at his disposal but is unlikely to have access to larger amounts. Typically, when or if these are needed, authorization will have to come from senior administrators with authority to make substantial financial commitments. Finally, communications has been touched upon in terms of contacting agencies, support groups, and the media. The communications team is also responsible to see that all occupants of an area affected by, for example, an airborne plume of a toxic gas, are notified. Time may be critical, so the communications group must have procedures in place to communicate by all reasonable means using radio, TV, roving vehicles equipped with public address systems, and (if time and conditions permit) door to door searches.

A library of reference materials should be maintained for the use of the emergency responders. Following is a short summary of some of the more useful references, many of which are revised frequently. Although these are primarily printed books, today a number of other types of data information sources are becoming widely available for chemical products, primarily as a result of information needs evoked by the OSHA Hazard Communication Standard. An example of these, included in the list, are Material Safety Data Sheets, available directly from the chemical product manufacturer and on the Internet. These are provided when the chemical is first purchased and when significant new information becomes available. Compilations of these are sold as hard bound or looseleaf volumes, on microfiche, or on computer CD-rom disks. The latter contain vast volumes of information on a 4.75 inch plastic disk. Many of these provide quarterly upgrades at reasonable costs. Most government regulatory standards and guides are now directly available on the Internet. There is little reason not to be adequately informed with all of these resources readily available. Many of the information sources listed below are available either directly on the Internet or available through Internet orders. In addition, many of the Internet sites include links to other sites, other than those given below, which provide additional information.

- ACGIH, American Conference of Industrial Hygienists—Threshold Limit Values (TLV) for Chemical and Physical Substances
  1330 Kemper Meadow Drive, Ste. 1600
  Cincinnati, OH 45240
  http://www.acgih.org/
- Chemical Hazards Response Information condensed Guide(CHRIS)
  Available through Federal General Services Administration. See
  http://www.uscg.mil/hq/g-s/g-si/g-sii/dpri-file-
- Department of Transportation Emergency Response Guidebook, DOT Publication NAERG9G (or later version, revised every 3 years)
  http://hazmat.dot.gov/gydebook.htm
  also check
  http://hazmat.dot.gov/ohmforms.htm
- Safe Handling of Compressed Gases in the Laboratory and Plant
  Matheson Gas Products
  PO. Box 85
  East Rutherford, NJ 07073

http://www.mathesongas.com/catalog1.htm
also,
http://www.mathesongas/acorepro.htm
The company also provides MSDS for all their products via the Internet

- List of Certified Poison Control Centers/by state-region
  http://www.medicinenet.com/Art.asp?li=MNI&ag=Y&ArticleKey=869
- Farm Chemicals Handbook
  Meister Publishing Co.
  37733 Euclid Avenue
  Willoughby, OH 44094-5992
  http://www.meisterpro.cm/
- Fire Prevention Guide on Hazardous Materials
  National Fire Protection Association (NFPA)
  1-Batterymarch Park
  P.O. Box 9101
  Quincy, MA 02269-9101
  http://www.nfpa.org/
- First Aid Manual for Chemical Accidents, 2nd Edition
  Lefevre, Marc J.(Editor), Conibear, Shirley (Contributor)
  John Wiley & Sons
  605 Third Avenue
  New York, NY 10158-0012
- Hazardous Materials
  Department of Transportation
  Office of Secretary Transportation
  Washington, DC 20590
  http://hazmat.dot.gov/toc.htm
- Material Safety Data Sheets Master File for Chemicals in Use at the Institution.
  (Available from chemical manufacturer or generic database, often directly on the
  Internet from the manufacturer. Note that there are now a number of commercial
  providers of generic databases, either in hard copy form or in various computer
  formats.) For a free MSDS data base via the Internet, see the following, available
  from Paul Restivo of the University of Kentucky.
- MSDS Data base available from http://www.ilpi.com/msds/index.chtml
- Merck Index
  Merck & Co. Inc.
  Rahway, NJ 07065
- NIOSH/OSHA Pocket Guide to Chemical Hazards, DHHS (NIOSH) Publication No.
  78-2 10
  U.S. Government Printing Office
  Washington, DC 20402
- Physicians' Desk Reference
  Medical Economics Company
  Oradell, NJ 07649
- Prudent Practices for Handling Hazardous Chemicals in Laboratories
  National Academy Press
  2101 Constitution Avenue, NW
  Washington, DC 20418
- Handbook of Chemistry and Physics
  CRC Press LLC
  2000 Corporate Blvd., NW
  Boca Raton, FL 33431

- Laboratory Safety Principles and Practice
  American Society for Microbiology
  1913 I St., N.W
  Washington, DC 20006
- National Health Council,
  1730 M Street, NW, Suite 500
  Washington, DC 20036-4505
  202-785-3910

Internal resources will not always be sufficient to handle an emergency. Therefore, a list of external emergency organizations should be maintained by the organizational emergency groups as well. The following are among those likely to be useful and readily available. Any others that might be useful to you and are available should be identified and added to the list. Currently available telephone numbers are given in some cases. These are subject to change and should be verified before incorporating them in a plan.

- Regional emergency group/coordinator
- Arson and/or bomb squad, if not otherwise identified
- Civil Defense coordinator, if not otherwise identified
- Commercial analytical laboratories
- Commercial environmental emergency response firms
- Law enforcement organizations, e.g., city or county Police Chief or Sheriff, state police, F.B.I.
- Centers for Disease Control, phone no. 404-639-1024 or http://www.cdc.gov/
- CHEMTREC (for chemical and pesticide spills), phone no. 800-424-9300 or http://www.cma.com/
- Compressed Gas Association, phone no. 212-412-9000 or http://www.naturalgas.org/CGA/index.htm
- National Fire Prevention Association, phone no. 617-770-3000 or http://www.nfpa.org/home.html
- National Response Center (USCG and EPA), phone no. 800-424-8802, or http://www.epa.gov:12001/s97is.vts
- Nuclear Regulatory Commission, phone no. 301-492-7000 (also state or regional federal office) or http://nrc.gov/
- Occupational Safety and Health Administration, phone no. 202-245-3045 (also state or regional federal office) also see http://www.osha.gov
- Poison Control Center, phone no. 502-362-2327 also see list of certified poison control centers listed above.

Many of these are sources of information only, and normally do not provide actual assistance for the emergency response. The ones likely to have the capability to do so are the first six. However, the commercial groups listed represent profit-making organizations and the institution or corporation must be willing to pay for their services. Since ultimately the organization (or their insurers) will bear the bulk of the costs for the emergency response, authority must be provided to pay for these services.

b. Emergency Equipment

Another important step in preparing for an emergency is acquiring appropriate equipment, which is kept readily available for use. Some of this should be located in the laboratory area and every laboratory should be furnished with it. Other equipment, because of the cost and relatively rare occasions when it is likely to be needed, should be maintained at a central location. Even the equipment kept centrally needs to be realistically selected. For example, it is neither necessary nor desirable for every organization to maintain an expensive, fully equipped hazardous material emergency response team. Some very large organizations may find them essential but most institutions will not be able to justify the cost.

Some of the emergency equipment needs to be built in, as part of the fixed equipment in the laboratory. Included in this group are the following items:

Eyewash stations—At least one of these, meeting ANSI standard Z358.1-1990, (or preferably the new version- Z358.1-1998) must be placed in an easily accessible location. The travel distance to a unit should be no more than 100 feet according to the standard and travel time should not exceed 10 seconds. According to Andrew Munster, M.D., Secretary of the American Burn Association, "time is critical" and Russell Kilmer of the Polymer Products Division Of the E.I. DuPont Experiment Station in Wilmington, DE, is quoted as saying "Every laboratory in their facility is equipped with an emergency shower or eyewash station to meet their safety requirements...." It is very undesirable for an injured person, possibly blinded by a chemical, to have to find a way to units outside the immediate room, perhaps through a closed door. Proposed standards for disabled individuals have been proposed as ANSI standard 117.1-1992, establishing access clearances and other physical limits. Eyewash stations should be mounted on a plumbed water line, rather than the small squeeze bottles that are sometimes used for the purpose. The squeeze bottles do not contain enough water to be effective. OSHA inspectors are likely to cite a facility in which the bottles represent the only source of water for flushing contaminants from a person's eyes. Where plumbed water lines do not exist, such as in the field, larger self-contained units are available which do provide sufficient water flow for an extended period. Cold water itself can be uncomfortable to the eye, so if possible the eyewash water supply should have a holding tank to ensure that the water is at least near room temperature. In many of the colder areas of the country, tap water may be well below room temperature for several months of the year.

Deluge shower — Eyewash stations and deluge showers ideally should be installed as a unit. The standards cited in the preceding paragraph apply to emergency showers as well. Although the eyes are probably the most critical exposed organs susceptible to damage, chemicals splashed on the face may also splash on the body. A deluge shower should be capable of delivering about a gallon per second with a water pressure of 20 to 50 psi. A common error is to plumb the unit into too small a line incapable of delivering an adequate flow. The water supply should be at least a 1-inch line. Although a floor drain is desirable, it is not essential. One can always mop up afterward. There should be a timed cutoff, however, at about 15 to 20 minutes, after which the unit would need to be reactivated. Cases have occurred where, as an act of vandalism, a deluge shower was activated and rigged so that it would continue to run. In one case, before the problem was discovered, over 30,000 gallons of water flooded the facility. The unit was in the hall outside the laboratory; another argument for placing the units within a lockable room. Care must be taken to ensure that the water from the shower cannot come into contact with electrical wiring, either directly from the shower or by coming into contact with extension cords improperly running across the floor. Again, the units should always be placed in an easily accessible location. Care is essential to maintain clear accessibility. In laboratories, many instances have been noted where limited floor space has resulted in equipment being placed immediately under the showers. The ANSI standards meeting ADA requirements for the disabled cited in the previous section must be maintained.

Fire extinguishers — OSHA requires that every flammable material storage area be equipped with a portable class B fire extinguisher. The standard does not specify the amount of a flammable material which makes a room a storage facility so in effect most laboratories face the need to comply with the standard. The unit should be at least a 12-lb unit and it should not be necessary to travel more than 25 feet to reach it from any point in the laboratory. This specific requirement in the General Industry Standard may be preempted by the OSHA Laboratory Standard, but requirements of that standard provide for emergency response training, which is construed to include training in how to use portable fire extinguishers. If it is intended that employees may attempt to put out small fires and not simply evacuate immediately, then the employees should be trained in the proper use of an extinguisher at the time of employment and receive refresher training annually. Class B extinguishers are, of course, intended for flammable solvents. Other classes of fire exting- uishers are class A, intended for combustible solid materials, such as paper or wood, class C, where electrically live equipment is involved, and class D, where reactive metals, such as sodium, are used. Combination units such as AB or ABC are available, which, although not equally effective for all types of fires, can be used where mixed fuels are involved. More information on fire extinguishers will be found in a later section.

Fire blanket — A fire blanket is a desirable unit to have permanently mounted in a laboratory. The blankets are usually installed in a vertical orientation so that a user need only grasp the handle and roll themselves up in it in order to smother the fire. Some blankets include asbestos in their manufacture; these should not be installed, and existing units should be replaced. The concern is that they could become a source of airborne asbestos fibers, which have known carcinogenic properties. Unfortunately the heavy woolen blankets most often used as alternatives are likely to be stolen. There are fire blankets using fiberglass or special fire-resistant synthetics instead of asbestos or wool available. If a fire situation is a distinct possibility, consideration should be given to providing a woolen blanket saturated with a water-soluble, oil-based gel. This not only protects against fires and aids in escape through an active fire, but can be useful in the emergency treatment of burn victims. These gel blankets have a limited shelf life, are expensive, and are infrequently found in a facility.

Emergency lights — Emergency lighting to enable safe evacuation must be provided by some mechanism. One alternative is to have two sources of commercial power to the lighting circuits in a building. This can be achieved by having a second source external to the building or secondary power sources within the building, but this alternative is defeated in power outages covering a wide area. There are several alternative types of internal power sources including emergency generators; large, uninterruptible power supplies (UPS) to provide power for lights for a substantial area which depends on batteries to provide power for a fairly limited interval; and individual trickle-charged battery-powered lights in individual labora- tories. Generator units require frequent testing under load and thus are a maintenance problem. Uninterruptible power supplies are best suited for maintenance of power to equipment such as computers, where a controlled shutdown is almost essential. The most economical alternative especially in retrofitting an older facility is the individual trickle- charged battery-powered units that come on when the power fails.

First aid kit — One of these needs to be in every laboratory and should be kept in a pre- determined fixed location. They are intended to be used for minor injuries or basic treatment while awaiting more advanced care for major injuries. Access to appropriate emergency medical care is required under OSHA standard 1910.151. Kits should be relatively small units. Packaged units are sold that are adequate for five or six persons. There is little value in having larger units, since in the event of an emergency involving more persons, help definitely will be needed from trained emergency care provider units, including rescue squads and physicians. Present in the kits should be a variety of bandages, adhesive tapes, alcohol swabs, gauze, perhaps some protective creams, and a few cold packs. Special situations could require special items to be available to provide treatment. Items such as iodine, methiolate, and tourniquets are no longer recommended for inclusion in most cases. It is essential that a maintenance program be established to ensure that the kit is always adequately supplied. It is all too easy to use up the supplies without replenishing them.

Fire alarm pull station —The location of the nearest pull station should be familiar to everyone in the laboratory

Special safety equipment —There are many specialized research areas which require special safety items such as explosion-proof wiring, combustible gas monitors, and explosion venting for laboratories working with highly explosive gases. The possibilities are too many to dwell on at this point.

Some emergency equipment need not be built in but should be available. Among these items are the following:

Absorptive material — Probably the most common laboratory accident is a spill from a beaker or a chemical container. The volume is typically fairly small, rarely exceeding more than 4 or 5 liters and usually much less. Of course, there are spills which would require immediate evacuation of the area or even the building, but more frequently the spilled material simply must be contained and cleaned up as quickly as possible. **THIS IS NOT THE RESPONSIBILITY OF THE CUSTODIAL STAFF**. They are not trained to do it properly or safely. Spill kit packages are available commercially to neutralize acids and bases, and to absorb solvents or mercury. Although it is possible to put together similar packages oneself, the commercial packages are convenient to obtain and store. After being used, the materials should be collected and disposed of as hazardous waste.

Personal protective equipment and janitorial supplies — Several miscellaneous items are needed to clean up an area. Among these are plastic and metal buckets, mops, brooms, dust pans, large, heavy-duty polyethylene bags, kraft paper boxes (for broken glass), plastic-coated coveralls, shoe covers, duct tape, and an assortment of gloves. If not kept in an individual laboratory, at least one set should be kept on each hall or floor of a building. Custodians may have some of these materials, but they are not always available to laboratory personnel, especially outside normal working hours when many laboratories are active.

Respirators — Fumes and vapors from many irritating and dangerous materials can be protected against by the use of respirators with appropriate cartridges or filters. If operations are sufficiently standardized so that a standard respirator combination would be effective, they should be kept in an emergency kit. However, cartridge respirators are not intended for protection against materials which are immediately dangerous to life and health (IDLH). Whatever units are provided, laboratory personnel must be trained in the appropriate use of the units and the units must be maintained properly. Respirators should be assigned to specific individuals.

Supplied air escape units — Supplied air units, such as emergency squads might use, are expensive and require a significant level of training to be able to put them on quickly and use them properly. However, small air-supplied units are available at very reasonable prices which only need to be pulled over one's head and activated to provide 5 minutes of air. This is usually sufficient time in which to escape the immediate area of an accident.

Virtually any small to moderate, chemical emergency can be handled with the equipment described above.

A few major items of equipment should be readily available from the safety department, fire department, security force, or perhaps the emergency medical team. Their ready availability is by no means certain, and the institution or corporation should maintain a set of these major items. Many of these items require special training to be used safely.

Oxygen meter — A portable meter should be available to ensure that the oxygen level is above the acceptable limit of 19.5%. It is important to be able to detect oxygen-deficient atmospheres, where the levels are significantly less than the acceptable level.

Combustible gas and toxic fume testing equipment — A number of different types of equipment are sold to test for the presence of toxic fumes. A common type, frequently combined with an oxygen meter, is a device to detect "combustible gases." Specialized units are built to detect other gases such as carbon monoxide and hydrogen sulfide. Very elaborate and, consequently, expensive units, such as portable infrared spectrometers, gas chromatographs, and atomic absorption units, can detect and identify a much greater variety of chemicals, often to very low concentrations. A less expensive alternative is a hand pump, used to pull known quantities of air through detector tubes containing chemicals selected to undergo a color change upon exposure to a specific chemical. All of these can be used to obtain an instantaneous or "grab" reading. Where a longer duration sample is desired, powered pumps can be used to collect samples, and for some chemicals, passive dosimeters can be worn which can be analyzed later in a laboratory. Equipment to meet local needs should be selected. Although sophisticated testing devices are available, there are tens of thousands of possible chemical contaminants. It is essential for emergency personnel to know what to test for to ensure rapid identification. Emergency medical care may be delayed or limited to supportive treatment until positive identification is obtained. Therefore, a list of possible hazardous materials currently in use or stored in significant quantity should be maintained by the laboratory and be available to emergency responders, prior to their entrance into the laboratory.

Supplied air breathing units — These are not to be confused with the escape units previously described and usually will be available from the fire department or, for larger institutions or corporations, from the chemical safety division of the safety department. There are two basic types, one of which provides air from a compressed air tank just as does SCUBA gear. The most common size tank is rated at 30 minutes, which, under conditions of heavy exertion, may last only 20 minutes or less. The second type uses pure oxygen recirculated through a chemical scrubber to extend the life of the supply to 1 hour, or longer for some units. The first of these two types have been available for a longer period and more emergency personnel have been trained to use them. The second does offer a significantly longer working interval. This could be very important. Pairs of either type should be owned so that in the event an individual entering an emergency area is overcome, it would be possible to effect a rescue.

Fire-resistant suits — Special fire-resistant suits are needed to enter burning areas. There are different grades of these which provide varying degrees of protection to fire. Some protect against steam or hot liquids as well. They normally require a self-contained supplied air system to be worn during use.

Chemical-resistant suits — Protection is frequently needed in chemistry incidents for protection against corrosive liquids and vapors. In standardized situations, materials for protective suits can be custom selected for maximum protection for the specific chemicals of concern. Where a variety of chemicals such as acids, bases, and frequently used solvents are involved, a butyl rubber suit is often a reasonable choice. Combination units of chemical and fire resistant entry suits are available.

Clean air supply system — An alternative to self-contained air or oxygen tanks is a compressor system capable of delivering clean air through hoses from outside the area involved in the incident. Persons inside the work area would wear masks connected to the system. Personal air-powered units are available which use small, battery-powered packs to draw local air through a filter and maintain a positive air pressure within the face mask.

High-efficiency particulate and aerosol (HEPA) filtered vacuum cleaner — Ordinary vacuum cleaners, including wet shop vacuums, do not remove very small particulates from the air. They remove larger particles, but the smaller ones pass through the internal container or filter and return to the room. In several instances, this can actually worsen the situation. For example, droplets from a mercury spill can be dispersed back into the air in the form of much smaller droplets and cause the mercury vapor pressure in the air to increase. (mercury vacuums are available which have special design features.) In another actual case, in a carpeted room where large quantities of forms and computer paper were processed, vacuuming with an ordinary vacuum cleaner during normal working hours increased the number of respirable paper dust particles suspended in the air to a level such that several individuals who were allergic to the dust had to be sent out of the area. HEPA filters will remove 0.9997 of all particles from the air which have a diameter of 0.3 microns or greater. They will remove a smaller fraction of particles of smaller sizes, but the smaller particles have difficulty reaching the deep respiratory system, so they are less of a problem.

Radios/cellular phones — Communication between persons entering an accident area and those outside is highly desirable. Emergency groups will have portable radios with frequencies specifically assigned to them. Cellular phones are a recent alternative which provide access through the telephone system to virtually any external resource.

Fire suppressant materials — In addition to water and the usual materials available in portable fire extinguishers, most fire departments now have available foam generators which can saturate a fire area.

Containment materials — In order to prevent the spread of large amounts of liquid chemicals, a supply of diking materials needs to be maintained. Ready access to a supply of bales of straw is a great asset. Straw is cheap, easily handled, and easy to clean up afterward. In the event of a spill reaching a stream, floating booms and skimmers are useful in containing and cleaning up the spill. Booms are not effective for materials more dense than water and not water soluble.

Radiation emergency — Many laboratories use radioactive materials. For emergencies involving these units, in addition to the other emergency equipment, radiation survey instruments must be available or maintained in an emergency kit. The radiation safety office will be able to supply additional units. These should include instruments capable of detecting both low levels of gammas and low energy betas as well as instruments for measuring high levels of contamination. Although low levels are not necessarily dangerous, normally only very restrictive levels of contamination are permissible under established safety limits for most organizations, according to ALARA (As Low As Reasonably Achievable) guidelines.

Miscellaneous clothing — Items needed include a variety of coveralls, including (but not limited to) chemically resistant suits in a range of grades; disposable Tyvek™ coveralls; gloves with different chemical resistances; regular work gloves; Kevlar™, Nomex™, or Zetex™ gloves for hot use; rubber and neoprene boots and shoe covers; head covers; hard hats; chemical splash goggles; safety glasses; and masks.

Miscellaneous tools and paraphernalia — A variety of small tools could be needed, as well as shovels, pickaxes, axes, rope, flares, emergency lights, sawhorses, a bullhorn, a chain saw, a metal cuffing saw, a bolt cutter, and a "laws of life" metal spreader. Special non-sparking tools may be required where sparks may ignite flammable vapors.

Victim protection — In equipping an emergency kit, the emphasis is usually on protecting emergency response personnel. In order to bring a victim out through a fire or chemically dangerous area, blankets, disposable coated Tyvek™ overalls, loose-fitting chemically resistant gloves, and the 5-minute escape air units should be available.

All of the equipment listed in this section must be maintained properly, and a definite maintenance schedule must be established. For example, the integrity of the chemical protective suits must be verified on a 6-month schedule. A maintenance log must be kept in order to confirm that the maintenance program has been done on schedule.

A fire hose is specifically *not* included as a desirable item of emergency equipment that should be available to the usual occupants of a building. Although standards are provided in OSHA for fire brigades, in general, if a fire is sufficiently large to require a fire hose to control it, it is usually too large for anyone except professionals. Building codes frequently require installation of a 1.5 inch emergency hose connection. Often, building officials encourage the owners to request a variance to permit this requirement to be deleted. Many fire departments question the value of such connections or, even if available, whether a hose of this size would be sufficiently useful. Those institutions or corporations that do choose to establish a fire brigade will need to provide training beyond the scope of this book.

c. Basic Emergency Procedures

A list of several common types of emergencies that might occur in a laboratory was given in the introductory section to this chapter. Many of these emergencies, as well as others not mentioned in the list, share common characteristics for the initial response which are important to do *first.* The following material will, for the most part, be in the context of a fire incident, but the recommendations would be the same if a substantial release of a toxic chemical were released and became airborne.

1. Make sure everyone in the immediate vicinity is made aware of the problem. In a busy, active laboratory, an accident can occur in one part of the laboratory and personnel in other areas within the same laboratory could be temporarily unaware of the event. This is especially likely if the space is subdivided or if there are no obvious effects associated with the event, such as a loud sound from an explosion.

2. Confine the emergency if reasonably achievable. Many emergencies can be readily confined if quick action is taken. Small quantities of a spilled chemical can be contained with absorbent materials or toweling by the persons directly involved if the chemical is not immediately dangerous to life and health (IDLH). Individuals should be trained to take these actions, and appropriate containment materials for the materials in use should be conveniently available. In the event that the emergency includes a fire, laboratory personnel, if properly trained, can and should put out small fires with portable fire extinguishers, but a very serious question of judgment is involved. What, precisely is a small fire? One definition is a trash-can size fire, but unless there is a reasonable certainty that the fire can be controlled, then evacuation of the building should be strongly considered and implemented as soon as the situation appears to be deteriorating. Time is likely to be critical if the volume of solvents often available as fuel in a typical laboratory is considered. If more than one person is available, there may be more flexibility. One or more persons may attempt to contain the fire, while others are taking initial steps to evacuate the building. Where it is necessary to evacuate an area larger than a single laboratory, the building's evacuation plan should include measures to ensure that all spaces are checked, including restrooms, janitors closets, etc.

3. Evacuate the building. Whenever the situation is obviously serious, such as a major fire, a moderate-to-large spill of an IDLH material, a rupture of a large gas cylinder, or large spills of ordinarily dangerous materials, such as strong acids, then evacuation procedures for the area or the building must be initiated as soon as possible. Any measures taken in such a case to confine the emergency situation should provide extra time for the evacuation to be carried out safely.

Evacuation is a conservative step and should be implemented whenever any doubt exists of the severity of the situation at hand. It is inconvenient and is disruptive to work activities, but the alternative is far worse if an incident cannot be controlled. The first few minutes of a fire, especially, are very important and any significant delay can make the job of the fire department much more difficult. Once a fire takes hold, it is often very difficult to bring under control. In a laboratory situation, the involvement of the inventory of chemicals can convert a straightforward fire into one which could involve the generation of extremely toxic vapors. Most fire departments are inadequately trained to handle complex chemical fires. Their chemical incident training usually includes situations involving only a single material. Even if the fire is out before they arrive, there are things that the fire department needs to do. They need to check the area to ensure that it is really out. Fire department personnel when they arrive on the scene are usually charged with the legal responsibility for managing and terminating hazardous material incidents. They also need to determine the cause of the fire in order to prepare an accurate incident report. The information they obtain will be needed to determine how to prevent subsequent fires due to the same cause. Where property and personal injuries losses occur, their report will normally be needed by the insurer of the property to determine the amount of cost recovery available.

Normally primary evacuation routes from an area within a building should follow the shortest and most direct route, along corridors designed and constructed to meet standards for exitways. However, since in an emergency any given path may be blocked, one or more alternate secondary routes should be designated. In no instance should an evacuation plan include elevators as part of the evacuation procedure, even for a disabled person. In the case of a fire, elevators should be designed to immediately go to the ground floor and be interlocked to stay there until the danger is over. There are convenient evacuation chairs which a single individual can use to assist disabled persons. One of these should be available and one or more persons designated to provide the required aid.

In any evacuation procedure, standard operating procedures for closing down operations should be included, if there is sufficient time to implement them. Gas should be turned off, along with electric and other types of heaters. Valves on gas cylinders should be turned off, especially if they contain flammable or toxic materials. High voltage equipment should be turned off. Closing sashes on fume hoods may be desirable. Certainly any flammable material storage cabinets should be closed.

Even in the worst situation, there are some simple things which can be done by individuals evacuating the building to confine and minimize the emergency. The highest priority is to protect personnel, so the first thing is to actuate the building alarm, assuming that one exists. If not, then air horns should be used or, failing that, a verbal warning must be issued. Doors to the laboratory should be closed on the way out. Doors between floors should be closed behind those evacuating. Stairwells serve very well as chimneys to carry smoke and fire to upper floors if the doors are not closed. If the building has been built according to code, as briefly discussed earlier in this chapter and covered in much more detail in Chapter 3, then these last two simple steps can significantly retard the spread of a fire or spread of fumes.

If a laboratory is under negative pressure, as most chemical laboratories should be, then the negative pressure will also tend to confine the emergency to a single room. In order to maintain a negative pressure, it may be desirable to leave the sash of a hood open or to leave the hood working, even though there might appear to be concern about the fire spreading through the hood duct. If a hood has been installed properly, the exhaust will be at a negative pressure with respect to the space surrounding the exhaust duct and, as noted earlier, is either going directly outside the building without passing through an intervening floor level or is enclosed in a fire-rated chase. Under either of these conditions, fire being drawn through a hood exhaust should not cause fire to spread to other floors. The door to the room being closed will further reduce the amount of fresh air available to support a fire. If an air exhaust is turned off, any air intake should also be turned off to avoid creating a positive pressure in the room and thus possibly causing extension of the emergency by leakage into corridors.

Evacuation should be done as quickly as possible, but in such a way as to not engender a panic situation. This can best be achieved by having it be a frequently practiced procedure, so that everyone is familiar with the routes. In a corporate situation with a stable personnel complement in the building, drills two or three times a year will quickly accomplish the purpose. In an academic environment, the problem is much more complicated. In most colleges and universities, as many as 8 to 12 classes per day may be held in the same classroom. Classrooms may be assigned by some central authority, not necessarily with regard to the subject matter being taught. This may result, for example, in a professor of economics being assigned a class in a chemistry building for one quarter during a year, and who may not have a class in that building for the remainder of the year. During the course of an academic year, the population in the building may change in a large part every quarter or semester. Because of all these complicating factors, a single drill per academic session could prepare as little as 10% of the population in a building for an actual emergency. Under these circumstances, unusual care should be taken to clearly mark evacuation routes from buildings, and to train those individuals who form the permanent population in the building to take charge during an evacuation. Complicated maps placed at intersections to show evacuation routes are often used, but are difficult to read and interpret in the press of events occurring during a serious emergency situation. A simple but very effective evacuation system is illustrated in Figures 2.3 and 2.4. A distinctive, high contrast, standardized symbol, employed only for marking primary evacuation routes, is placed directly across a corridor from every door opening onto a

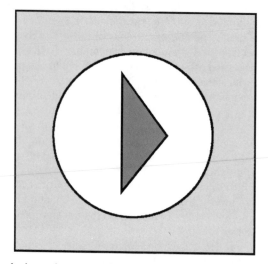

Figure 2.3 A simple sign such as this is easily recognizable as a directional guide.

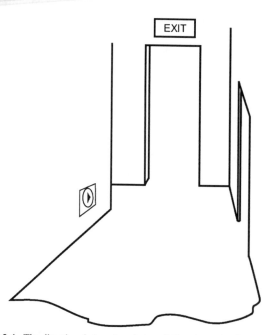

Figure 2.4 The directional sign placed about 2 feet above the floor, opposite doors, at intersections of halls, and about every 30 feet along corridors without doors, provides a guide to exits. If the arrow is phosphorescent, it would be visible even in a power failure.

corridor, at appropriate intervals (30 to 50 feet) along the corridor without doors, and at every branching point along the path of egress. A person totally unfamiliar with a building need only follow the symbols to be conducted to the nearest exit. Smoke tends to rise, so these should be placed a short distance off the floor, so that they would remain visible when signs placed above doors might be obscured. Power can fail, even in buildings equipped with separate emergency power for lights, so if the directional symbol can be made with a phosphorescent paint, it will remain visible for two or more hours, this being ample time to evacuate almost any building. Printing the signs on a fragile substrate which cannot be removed intact will minimize the theft of the signs to be used as decorations in dormitories or residences. This system should be used as a supplement to a code-conforming system of exit lights rather than a substitute. The maps mentioned above are useful when time permits.

A standard part of any emergency evacuation plan should include a previously chosen point of assembly for those evacuating. This should be a location generally upwind from the building being evacuated. Obviously the wind does not always blow from the same direction, so alternative gathering places should be selected. Those individuals most directly involved with the emergency, and presumably the most knowledgeable of the circumstances should be especially certain to remain at the evacuation location and make themselves known to the emergency response groups upon their arrival in order to assist them. It is critical that the emergency responders be aware of the characteristics of the emergency situation which they are facing. There should be a clearly defined line of communication among the persons responsible for the facility. Individuals in authority and with assigned responsibility for the space involved should also go to the assembly point, if not already on the scene, and remain available to assist the emergency personnel in managing the evacuees. Often the police will be among the first emergency groups arriving on the scene and their aid will be invaluable in crowd control. If other senior corporate or institutional officials involve themselves, they should inform the building authorities that they have arrived and they may wish to assume responsibility, although they will typically be less knowledgeable of local circumstances than building and laboratory officials. Procedures should be employed to account for the individuals who are known to have been in the building, and reports of any persons suspected to still be in the building should be made to the authority figures to pass on to the emergency responders. Any decision to allow reentrance to the building approved by the fire department or other emergency group should be disseminated by these persons.

4. Summon aid. Current building codes normally require building alarms to be connected to a central location manned 24 hours per day. Unless the building alarm system is connected to a central station, pulling the alarm will not alert any external agencies. The more sophisticated systems commercially available will pinpoint almost the exact location in a building and will even provide a map showing the best route to the facility for the emergency group. Such systems greatly facilitate the response of an emergency group. Unfortunately, current building codes apply only to new construction, so older buildings may have much less sophisticated systems or none at all. In such cases, using the almost universally available 911 system, would have to suffice. It is imperative that requests for assistance be initiated quickly. However, in a serious, life-threatening emergency, evacuation should not be delayed to call for assistance. Calls can be made from a point outside the area affected by the emergency. Where personnel are available, one individual can be designated to make the appropriate telephone calls while others are engaged in other aspects of the emergency response.

As noted earlier, a key group in responding to most institutional or corporate emergencies is the police department. If they are not the group initially contacted, it is probable that even small departments monitor emergency radio frequencies and will arrive either at nearly the same time as the emergency group summoned or even before. They should have training in a number of key areas, such as how to use fire extinguishers most effectively and how to give first aid and CPR, and should have the capability of independently causing a building to be evacuated. The police should also have at least basic hazardous materials training (usually called "first responder haz-mat" training). This basic training essentially trains them in how to identify a hazardous material incident situation. It does not provide for managing the incident response. It would be desirable if at least a cadre of a police department could receive higher levels of training. Once the fire department, rescue squad, or other emergency group arrives and assumes the responsibility for their duties, the police are needed for crowd control and communications.

In any type of incident in which a spill of a hazardous material has occurred, the standard procedure is to establish a command center outside the periphery of the area affected by the accident and establish a controlled access point for emergency response and, later, decontamination personnel entering the area. All materials and personnel entering and leaving the area should pass through the control point. When the remediation stage is reached, unless there are overriding considerations, decontamination should begin at the periphery and the work program designed to progressively constrict the affected area. Everything collected at the control point, including waste materials, contaminated clothing, and equipment which cannot be cleaned and reclaimed, should be immediately packaged for disposal according to standards applicable to the contaminant. Information and status reports should flow to the command point and overall direction of the response should come from the command center. Emergency response, to be effective, needs central coordination and a clearly defined chain of command.

D. Emergency Procedures for Selected Emergencies
1.   Spills

A chemical spill is probably the most common accident in the laboratory, and in most cases can be cleaned up by laboratory personnel with minimal effort or risk. According to the requirements of the OSHA Hazard Communication Standard, laboratory personnel are required to be trained in the risks associated with the chemicals with which they are working and should know when it is safe to clean up a minor spill. Workers should be especially sure to be familiar with the risks and the corrective actions to be taken in an emergency for chemicals labeled on the container "DANGER" or "WARNING." If personal protective equipment is needed, personnel required to wear it must receive appropriate training in how to use and maintain the equipment.

Paper towels, assuming that the paper would not react with the spilled materials, or absorbent and/or neutralizing materials can be used to clean up minor spills, with the residue being placed in an appropriate chemically resistant container for later disposal. Work surfaces should be nonabsorbent and chemically inert, but in order to avoid possible decontamination of work surfaces, it is often convenient to protect them with plastic-backed absorbent paper. Relatively few materials associated with a cleanup should be placed in ordinary trash receptacles. Most should be disposed of as hazardous waste. Individuals (including students as well as employees, in academic facilities) handling the waste materials should have documented training in the handling of chemical wastes in order to conform with both OSHA and RCRA standards. Chemical containers should not be placed in ordinary trash for disposal. In one actual instance, two different, partially empty chemical containers placed in the ordinary trash in two separate buildings combined in the trash vehicle and resulted in a fire. The fire caused the worker on the truck to be overcome with fumes, requiring emergency medical treatment. The nature of the fumes was unknown until later, when it was possible to retrieve the containers. It should be a general policy that no chemical container should be disposed of in the general trash. If any are disposed of in this manner, they should be triple rinsed and the labels removed. Trash handlers have become very concerned when they see containers with chemical names and chemical hazards on the labels.

Even small spills can often be dangerous if the spilled chemical interacts with the body. Strong acids and bases, as is well known, can cause serious chemical burns to tissue by direct contact. Eyes are especially vulnerable. Chemicals can cause serious injury by ingestion, inhalation, or absorption through the skin. For example, phenol is readily absorbed through the skin, and in relatively small quantities is quite toxic. Vapors from some spilled materials are IDLH by inhalation, even in small quantities. Obviously, work with such materials should be done in a hood where the sash can be lowered should a spill occur. However, if an accident occurs outside a hood involving these very hazardous materials, the area should be evacuated and the door to the laboratory closed and help sought from persons trained and equipped to cope with such dangerous materials. In some instances, if a material is sufficiently volatile to give off enough vapors to be dangerous from a small spill, it is often sufficiently volatile so that it will quickly evaporate. Evacuation will allow time alone to effect a remedy by allowing the vapors to be exhausted through the facility's ventilation system where dilution with the atmosphere should be sufficient to render the vapors harmless.

In most cases, flushing the area of the body affected by a splash of a liquid chemical with copious amounts of low pressure water for 15 to 30 minutes is the best immediate treatment. The best source would be an eyewash station or deluge shower, but in an emergency, if these are not available, any other source of running water (at low pressure) should be used. If the exposure is to the eyes, check for contact lenses and remove them if found and if possible.* Then hold the eyes open while they are being flushed with water. Any clothing or jewelry in the affected area should be removed to ensure thorough cleansing. No neutralizing agents should be employed. If the original exposure was due to a dry chemical, normally the best course would be to brush off loose material and then follow the same course of action.

While washing is taking place, emergency medical help should be summoned, normally by calling 911. Chemical injuries, due to their possible complexity, probably should elicit a response from a crew capable of providing advanced life support-level care. If a severe physical injury has occurred in addition to the chemical exposure, appropriate first aid measures should be taken while waiting for assistance. In order of priority, restoration of breathing and restoration of blood circulation, stopping severe bleeding, and treatment for shock should be done first. These injuries are life threatening. Training in these techniques are available from many sources, such as the Red Cross, the American Heart Association, local rescue squads, and hospitals, usually at minimal or no cost.

Persons involved in the accident or the subsequent treatment of the injured person or persons should remain at the scene until emergency medical aid arrives. It is important that those treating the victim know what chemical was involved. In addition, the persons providing assistance can provide emotional support to the victim. Generally, it is preferable that transport to a hospital be done by the emergency rescue personnel. They are not only trained and qualified to handle many types of medical emergencies, but they will also have communication capability with an emergency medical treatment center. Through this radio contact, they can advise the emergency center physician of the situation and the physician can instruct the emergency team of actions they can initiate immediately. In addition, if special preparations are needed to treat the injured person upon arrival at the emergency center, these can be started during the transport interval.

Some materials, such as mercury, do not appear to pose much of an obvious hazard upon a spill and a cursory clean up may seem to be sufficient. However, mercury can divide into extremely small droplets which can get into cracks and seams in the floor and laboratory

---

* There are two schools of thought on the use of contact lens in the laboratory. Current thinking is that they are permissible, especially if chemical splash goggles are worn over them. Some feel that there is a risk of vapors of tissue corrosive materials finding their way behind the lens by capillary action and do not allow use of contact lens.

Figure 2.5   Accident due to poorly installed and weak shelving.

furniture. Mercury remains in metallic form for a long time after a spill, capable of creating a significant concentration of mercury vapor pressure in a confined, poorly ventilated space. Exposure to these fumes over an extended period can lead to mercury poisoning. After gross visible quantities have been cleaned up by carefully collecting visible drops (preferably with an aspirator), absorbent material specifically intended to absorb mercury should be spread on the floor and left there for several hours. Afterwards, the area of the spill should be vacuumed with a special version of a HEPA filtered vacuum cleaner *adapted for mercury cleanup*. A penknife can be used to check seams in floor tiles and cracks to check if the cleanup has been thoroughly done.

The preceding material on spills assumed that the incident only involved one chemical. Figure 2.5 shows what could have been, but miraculously was not, a major disaster which could have injured several persons. A set of wall shelves put up by laboratory personnel, loaded with a large variety of chemicals, collapsed while no one was working in the area. Here, unlike the incident involving chemicals from containers mixing in a trash truck, several bottles broke with chemicals becoming mixed, no reaction occurred and the damage was limited to the loss of the chemicals. If a vigorous reaction had occurred between the contents of any two of the broken bottles, the resulting heat might well have caused more of the unbroken containers to have ruptured and a major disaster could have resulted. Where multiple chemicals are involved, the same techniques as those used in a simple incident should be applied, with the additional stipulation that unnecessary mixing of chemicals should be carefully avoided.

Spills which result in a substantial release of toxic liquids or airborne vapors such that the release extends beyond the facility boundaries invoke the requirements of the Community Right-To-Know Act. Notification of the local emergency coordinator by the dispatcher would be the first legal step to get the mechanisms moving.

While all of the corrective measures are being taken, the affected area should be secured to ensure that no one is allowed in who is not needed. "Tourists" are not welcome. If necessary, help should be obtained from security or police forces to exclude nonessential persons.

2.   Fire

A second common laboratory emergency involves fire. Laboratory fires stem from many sources, the ubiquitous Bunsen burner, runaway chemical reactions, electrical heating units, failure of temperature controls on equipment left unattended, such as heat baths, stills, etc., overloaded electrical circuits, and other equipment. With a fire, the possibility of the immediate laboratory personnel being qualified and able to cope with the emergency depends very strongly on the size of the fire. As indicated earlier, only if it is clear that the fire can be safely put out with portable extinguishers should a real attempt be made by laboratory personnel to do so. However, trained personnel temporarily can use portable extinguishers for moderate fires which are not gaining ground rapidly to gain time to initiate evacuation procedures.

In order to use an extinguisher effectively, laboratory personnel must receive training in their use. If possible, this training should include hands-on experience. They should be familiar with the different types of extinguishers and the type of fires for which they would be effective.

Class A extinguishers are intended to be used on fires involving solid fuels such as paper, wood, and plastics. Generally a class A extinguisher contains water under pressure. Water acts to cool the fuel during the extinguishing process, which has the advantage that the fuel has to regain kindling temperature once the fire has been put out. The large amount of energy required to convert liquid water into vapor places an added burden on the energy requirement to rekindle the fire in wet fuel. An extinguisher rated IA is intended to be able to put out a fire of 64 square feet if used properly. A typical extinguisher will throw a stream of water up to 30 to 40 feet for approximately 1 minute.

Class B extinguishers, intended for use on petroleum and solvent fires, usually contain carbon dioxide or a dry chemical, such as potassium or sodium bicarbonate. The first of these puts out the fire by removing one of the essential components of a fire, oxygen, by displacing the air in the vicinity of the fire. The second uses a chemical in direct contact with the burning material. Some chemical extinguishers contain materials such as monoammonium phosphate or potassium carbamate, which, even in small sizes, have very impressive ratings for putting out a solvent fire. Chemical extinguishers are messy and can damage electronic equipment. Typical dry chemical or carbon dioxide portable units last on the order of 15 to 30 seconds, and in the case of carbon dioxide units, it is necessary to be within 10 feet of the fire to use them effectively. A third type of unit, no longer being produced, which does not have this latter negative characteristic, contains one of a class of chlorinated fluorocarbons called Halon™. The Montreal Protocol, regulating chlorinated fluorocarbons because of the deteriorating effect of these materials on the atmospheric ozone layer, will eliminate the two major types of Halon™ within a relatively few years. It has not been permitted to produce these materials since January 1, 1994 although existing stocks can continue to be used. For the time being, existing systems will continue to be acceptable, but replenishing units will become increasingly difficult as existing stocks are depleted.

The chlorinated fluorocarbons used are Halon™ 1211 and Halon™ 1301, distinguished chiefly by the fact that the first of these operates at a lower pressure than the second and thus is more common as a portable extinguisher. The following points will apply to the alternative materials now available, which will be described in succeeding paragraphs. Permanently installed systems have tended to be Halon™ 1301. Both types work by interrupting the chemistry of the fire; however, Halon™, being gaseous, can be dissipated easily. Once the air concentration falls below the level at which it is effective, it no longer provides any residual fire protection. One way in which the Halon™ units have been used effectively has been to install them in small storage rooms as ceiling-mounted units. Reasonably priced units were available which went off automatically at temperatures set by fusible links in the heads of the units.

Alternatives to these two types are being sought and hundreds of compounds have been tested and several are now produced commercially. The requirements for the alternatives are

1) comparably effective fire fighting characteristics, 2) low or zero ozone depletion, and 3) low toxicity. The last requirement can be neglected if there is no possibility of human exposure. The compound $CF_3CH_2CF_3$ (FE-36) is a substitute for Halon™ 1211 and CHF_3 (FE-13) is a substitute for Halon 1301™.

Class C extinguishers are intended for electrical fires, which, because of the potential shock hazard, preclude the use of water. Many class B extinguishers are also rated for use on electrical fires. Class D extinguishers are used primarily for reactive metal fires and a few other specialized applications. Due to the extra cost of these units, only those laboratories which actively use reactive metals need to be equipped with class D units.

As has already been noted in several instances, training is required to use a portable extinguisher effectively since the available supply of fire suppression materials last less than 1 minute in most cases. To be most effective, the extinguishing material should be aimed at the base of the fire and worked from the point immediately in front of the extinguisher operator progressively toward the rear of the fire, away from the operator. If more than one person is present, additional extinguishers should be brought to the scene so that as one is used up, another can be quickly brought into use to prevent the fire from regaining vigor. More than one unit at a time can be used, of course. About half of all fires that can be put out with portable extinguishers require only one, but conversely, the other half require more than one.

To be effective, an extinguisher must be full. Units can leak, and unfortunately individuals with juvenile mentalities apparently feel that extinguishers are toys, provided for their amusement. This seems to be an attitude especially prevalent on college and university campuses (most of the problems exist in resident dormitories, but not exclusively so). Therefore, extinguishers in laboratories should be checked frequently by laboratory personnel as well as by fire safety staff. If the unit has a gauge, it should be in the acceptable range. Empty and full weights are indicated on the extinguisher, so weighing will confirm if the unit is full or not. Breakable wire or plastic loops through the handles, which are broken when the unit is used, should be checked to see if they are intact. If a loop is found to be broken, the unit should be checked. Any units found to be discharged should be replaced immediately, preferably as a practical matter within one working day.

Since a hood is where most hazardous laboratory operations should be carried out, a substantial number of laboratory fires occur in them. In the event of a fire in a hood, a simple and often effective procedure to control the fire is to close the sash. This serves two purposes: it isolates the fire from the laboratory and reduces the amount of air available to support combustion. Since a properly installed hood exhausts either directly to the outside or through a fire-rated chase, in many instances a fire in a hood can safely be left to burn itself out, or at least can reasonably be counted upon not to spread while an extinguisher is obtained. If the risk of a fire within a hood is substantial, automatic extinguishers are available that can be mounted within the hood.

In the event a person's clothing catches on fire, it is important not to run because this provides additional air to support the flames. Many authorities recommend that a person aflame should roll on the floor to attempt to smother the flames. In a crowded laboratory there is often a risk of involving solvents and other materials in the fire, however. A deluge shower is an effective way to put out the fire if it is in the immediate area, or, if a fire blanket is available, the fire can be smothered by the person quickly wrapping himself in it. If others are present, they can help smother the flames or they might employ a fire extinguisher to put the fire out. As with any other type of injury or burn, call for emergency medical assistance as quickly as possible. Perform whatever first aid is indicated, if qualified, while waiting for assistance.

3. Explosions

Among many other possibilities, an explosion may result from a runaway chemical reaction, a ruptured high-pressure vessel, reactive metals coming into contact with moisture, degraded ethers set off by friction or shock, or perhaps ignition of confined gases or fumes. Fortunately, explosions are less common in the laboratory than a fire but they still occur too frequently. The use of protective shields and personal protective equipment should be mandatory where the potential is known to be appreciable. Heavy gloves with gauntlets will offer protection to arms and hands. A mask *and* goggles should be used to protect the eyes, face, and throat. When an explosion does occur, in addition to the shock wave and the extreme air pressures which also may occur, flying debris, possibly secondary fires, and spilled chemicals may exacerbate the situation and feed a fire or lead to further reactions. Often there are toxic fumes released which may be the most serious hazard, not only to the persons immediately involved but to others outside the area and to emergency personnel. Initiation of procedures to handle resultant fires and chemical spills are appropriate if the situation is manageable. The most likely physical complications are personal injuries, including injuries to the eye, lacerations, contusions, broken bones, and loss of consciousness. Toxic fumes may cause respiratory injuries, possibly leading to long-lasting, permanent effects, possibly even death. In addition, chemicals may be splashed over the body even more extensively than in a spill, so it may be even more imperative to wash them off. However, it is essential to establish priorities. If breathing is impaired, artificial respiration should be administered, and if heavy bleeding occurs, pressure should be applied to the wound to stop it. These two problems are immediately life threatening. If there is time, and if it appears safe to do so, i.e., it does not appear that the spine has been injured or that other injuries will be worsened by the movement, then injured persons should be removed from the immediate vicinity of the accident. This is partially to protect the rescuer as well as the victim from the effects of chemicals, fumes, and smoke. Basically the same criteria apply as in a fire. Unless it is possible to safely handle the situation with the personnel present, then at least the immediate area should be evacuated, if necessary the building as well, and the fire department and other professional aid summoned. Care should be exercised by the emergency responder that in their efforts to assist injured personnel, that they do not incur injuries to themselves, such as coming into contact with spilled acids.

For most fire departments, a fire or an explosion in a laboratory represents an uncommon occurrence. It would be highly desirable, in the absence of a knowledgeable person immediately on the scene, if information on the contents of the laboratory could be found posted either on the door or close by. Preferably this information should be brief, legible from a distance, and be in a format already familiar to fire personnel. Many localities have attempted to meet these needs by requiring the laboratory to be posted with the NFPA universal hazard diamond in which the degree of danger for reactivity flammability, and health effects are indicated by a numerical rating, with the numerical rating referring to the contents of the laboratory instead of a specific chemical.

An example of an NFPA symbol is shown in Figure 2.6. There are four small diamonds, which together are assembled into a larger one. The four smaller diamonds are blue for health or toxicity, red for flammability, yellow for reactivity, and white for special warnings, such as radiation or carcinogenicity Printed in each segment is a prominent black number showing the degree of hazard involved, ranging from 0 to 4.

The numerical ratings are

0 = according to present data, no known hazard = slight hazard
2 = moderate hazard
3 = severe hazard
4 = extreme hazard

Although this system appears simple, it is difficult to implement meaningfully in practice, since, in a typical laboratory, there may literally be hundreds of chemicals on the shelves.

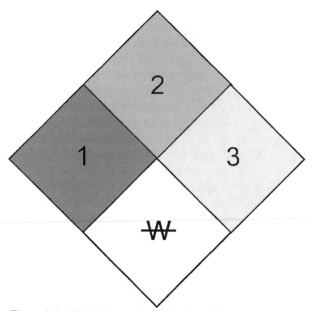

Figure 2.6 NFPA Diamond symbol with arbitrary ratings in the individual diamonds.

How should the rating for the laboratory be established? Should it be determined by the rating of the worst material present for each category or should the rating also depend upon the total amount of each of the chemicals present? For example, if the most flammable chemical present in a laboratory were ether, there would be a substantial difference in risk to firemen responding to a laboratory fire where the amount present was a single 500 milliliter container compared to one in which several 200 liter containers were present. If no allowance is made for the quantity present, both would have the same flammability rating. An alternative would be a subjective rating, combining both the worst-case type of chemical with the amount present to give a rating which in the judgment of the individual doing the rating properly takes into account both factors. The NFPA symbol is best applied to a single container or to an area with a very limited variety of materials present.

Another problem with the use of the NFPA symbol alone is that it may be too concise. Obviously, it does not inform fire personnel of exactly what is present. Under SARA Title III, corporations and institutions are required to provide information to the fire department on the locations and quantities of their hazardous chemical holdings. However, there are some important exceptions, one of the most important of which applies to research laboratories. A hazardous chemical used in a laboratory, under the direction of a competent scientist, even in excess of a reportable quantity (which may range from I to 5000 pounds, depending upon the chemical), need not be reported. Reporting all of the contents of laboratories in a major research facility could overwhelm the ability of a fire department to absorb data. In a major research institution, there may be literally hundreds of laboratories, each with potentially hundreds of different chemicals, with the inventory changing daily. Although a record of the contents would be helpful, even if not completely current, it would be very clumsy to use as a first response tool. In a later chapter, a means of providing computerized data to emergency groups will be discussed.

Possible alternatives that offer the advantage of providing information in a form with which fire departments are familiar would include posting symbols such as those shown in Figure 2.7, or use the DOT placard system, with a space to fill in the approximate amounts of

Figure 2.7 Additional information such as these posted on the outside of laboratory doors can inform emergency responders of possible risks and needed precautions within the facility.

each class present. An advantage of this latter alternative is that every fire and rescue group normally will have an Emergency Response Guide in their vehicles at all times. The response procedures recommended in the guide are very conservative, which is generally desirable.

4. Toxic Air Quality

An uncommon laboratory emergency situation that does need to be mentioned because it often leads to a fatality when it occurs is the danger of entering a space filled with a toxic gas or which is deficient in oxygen. OSHA has issued an updated confined space standard which offers some guidance, although most laboratories would not be anticipated to fall under the provisions of the standard. However, as a result of a fire, a spill of an IDLH substance, a leaking gas cylinder, or an improperly vented experiment releasing toxic fumes, it would be possible for a laboratory to be full of fumes and gases which would be fatal. Even a cylinder full of a nontoxic gas, such as nitrogen, can rupture and displace normal air sufficient to cause asphyxiation. The common practice of riding in an elevator with a 30-L dewar full of liquid nitrogen could prove fatal should the dewar rupture. The volume of the elevator is small, there is no rapid means of escape, and the speed of many freight elevators could mean that it could take far longer to reach the intended floor than most persons could hold their breath. The same concern could exist when riding elevators with full gas cylinders. Not all gases which may be found fairly commonly in use in the laboratory have adequate warning properties. No one should enter a space where this could conceivably be a problem without using a self-contained air breathing apparatus, nor should an individual go in such a space without others being aware of it. There should always be a backup set of self-contained breathing equipment with personnel available, trained, and able to use it to effect a rescue if necessary.

5. Radioactive and Contagious Biological Material Releases

Releases of radioactive material and active contagious biological materials represent two different types of emergencies which cause unusual concern because of the potential danger, perceived by the public, of the problem spreading beyond the immediate scene. In almost every instance, the levels of these two classes of materials used in ordinary laboratories are sufficiently small that the risk to the general public, as well as to properly trained laboratory workers is minimal.

a. Biological Accident

In recent years, the Centers for Disease Control has established a system of classification of laboratories for biological safety defining biological safety levels 1 through 4. Research with organisms posing little or moderate risk, requires only level 1 or 2 facilities, which are essentially open laboratories. Work with organisms, which do pose considerable or sub-stantial risks, requires level 3 or 4 facilities. A characteristic of both level 3 and 4 laboratory facilities is that they are essentially self-contained, with entrance through an anteroom or airlock and with access restricted to authorized personnel. This has greatly limited the possibility of an accident spreading beyond the confines of the facility. The major risks are accidents that cause direct exposures to individuals working in the laboratories. The facilities, especially those intended for higher risk use, are built to allow ease of decontamination to minimize the chances of a continuing source of infection in the event of a spill. Whenever a possibly infectious spill occurs, the immediate emergency procedure is to obtain medical care for the potentially exposed person as quickly as possible and to perform tests to determine if in fact the person involved has received the suspected exposure. Of course, concurrently, care must be taken to contain any spread of the affected area. A baseline medical examination (including a medical history) for each employee at the time of employment, with a serum sample taken for storage at that time, is of great value for comparison at the time of an accident. Because there may be delayed effects, records of any suspected incident need to be maintained indefinitely. As long as contaminated materials removed from the facility are autoclaved or double-bagged followed by incineration, there is little risk to the general public from laboratory research involving biological materials. Recent concerns about the disposal of infectious waste or ''regulated medical waste'' (as is now becoming the acceptable term) have caused a major increase in research into alternative means of rendering these types of waste harmless and *unrecognizable* by the general public. Materials made biologically safe by steam sterilization would still have to be mechanically processed to change their appearance. The concern, of course, is based on the fear that an individual coming into contact with improperly disposed of regulated medical waste could contract a serious disease, specifically AIDS or hepatitis B. Further discussion of these processes will be found in Chapter 4. In addition, the impact of this concern about bloodborne pathogens on emergency responders will be discussed later in this chapter.

Individuals not involved directly in the accident should evacuate the laboratory and the area must be decontaminated by persons wearing proper protective clothing. Only those indi-viduals who have received documented training as required by the OSHA Bloodborne Pathogen standard are allowed to clean up any materials that might be contaminated by human blood, other bodily fluids, mucous, or tissue. It may be necessary to chemically decontaminate the entire exposed space. However, each incident needs to be treated on a case-by-case basis.

If it is necessary to transfer individuals to an emergency facility, all information available should be given to the emergency response personnel and also transmitted to the personnel at the hospital facility. Both of these groups may wish to activate isolation procedures to protect themselves and others.

b. Radiation Incident

Radioactive spills represent another class of accident of special concern. There are circumstances that ameliorate the risk in actual accidents. Although laboratories in which radioactive materials are used are not classified as to the degree of risk as are laboratories using pathogens, they do operate under unusually stringent regulations established by the Nuclear Regulatory Commission (NRC) or an equivalent state agency. The regulations are intended to minimize the amount of material involved in a single incident and to limit the number of persons involved to authorized, trained, and experienced personnel. As a result, an individual involved in a spill generally knows to restrict access to the area of the accident and to avoid spreading the material to uncontaminated areas. Unfortunately, not all researchers exercise the required care, and as a result, there are occasions when radioactive materials may be spread unnecessarily. Every institution licensed to use radioactive materials is required to have a radiation safety program and a radiation safety officer who should be notified immediately in case of an accident. In obtaining the license to use radioactivity, the institution or corporation must demonstrate to the NRC that it has the capability of managing accidents properly In addition, there are requirements governing reports to the NRC, or to the equivalent state agency in an "Agreement" statement, spelled out in Title 10 of the Code of Federal Regulations, Part 20, when an accident occurs. Thus, the response to an emergency involving a release of radioactive material is relatively straightforward. Individuals working with many classes of radioactive materials must wear personal dosimeters (usually a badge containing a material with a known dose response relationship), so that in the event of an incident, their total external exposure can be read from these badges. Nasal swipes can be taken to check for inhaled materials. The clothes and skin of persons in the area and those allowed to leave can be checked with survey meters, which should be present in laboratories using radioactive materials or brought to the scene by radiation safety personnel. Surface contamination within the laboratory and on personnel can be cleaned up with little risk, using proper personal protective equipment to protect those doing it. The protective equipment normally would consist of a cartridge respirator and filter, coveralls of Tyvek™ or a similar material, head and foot covers (these may need to be impregnated with an appropriate plastic material), and "impermeable" gloves (unless chemical solvents are involved, the gloves most commonly used are made of polyethylene). Duct tape is an invaluable asset to seal gaps in the protective clothing around wrists, ankles, and the front opening. If the possibility exists that anyone ingested or inhaled radioactive material, then the individual should undergo further testing. This would include a bioassay for radioactive materials and, possibly, whole body counting at a facility with this capability. Whole body counters are available as mobile units which can be brought to a site should the need be justified. A major advantage of radioactive materials is that instruments exist which can detect radiation from spilled materials to levels well below any defined risk.

A situation in which personal injury is accompanied by a spill of radioactive material onto that person introduces significant complications in the emergency medical response. Radioactive material may have entered the body through a wound, and there is a possibility that both the emergency transport vehicle and the emergency room at the hospital could become contaminated. Due to the small quantities used in most laboratories, the contamination is unlikely to actually be a serious problem, but could be *perceived* as one by emergency medical personnel. In order to reassure them, a radiation safety person should accompany the victim to the emergency center, if possible, and be able to provide information on the nature of the radioactive material, the radiation levels to be expected, and advice on the risks posed by the exposure to the patient and to others. The type of radiation and the chemical or material in which it is present can have a major impact on the actions of the emergency room personnel. Some materials are much worse than others if they have entered the body. As noted above, a bioassay, other specialized tests, and a whole body count of the victim may be needed in order to ascertain that no internal contamination exists.

A sheet of plastic placed between the injured person and the backboard or stretcher and brought up around the person will effectively reduce the amount of contamination of loose material from the patient to the ambulance and the equipment being used, and will serve the same purpose later at the emergency room. If it is felt to be necessary, the emergency personnel can wear particulate masks or respirators to avoid inhalation of any contaminants. Due to the low level of material being used in most laboratories, it is unlikely that emergency personnel will need to be protected from direct radiation from the victim. There have been cases of industrial accidents where this last statement definitely was not true. Emergency equipment used in the course of the emergency response can be readily checked and, if necessary, decontaminated after the patient has been transferred to the emergency room. The patient should be separated from any other occupants of the emergency reception area to avoid any unnecessary exposures, even if they are well within safe limits, again because of the public concern regarding exposures to radioactivity at any level. In the very unusual event that substantial levels of radiation might be involved, the victim should be placed in an isolated room and emergency equipment brought to the room rather than using the normal emergency room. A possible location would be the morgue. In such an incident, it is important to document exposures for everyone involved in the emergency response. Even in low-activity situations, it is good standard practice to survey the interior of the ambulance, the parts of the emergency facility which might have been contaminated, equipment that may have been used, and the emergency personnel involved and make wipe tests for loose contamination. All radiation survey data should be carefully recorded and the records maintained for future use. The records should include estimated dose levels of all personnel participating in the event, based on the proximity to the radiation and the duration of the exposure. There could easily be a need for these data in court at a later time.

6. Multiple Class Emergencies[*]

Emergency response procedures will need to incorporate sufficient flexibility to serve in many nonstandard situations. Unfortunately, one cannot depend upon an accident being of a single type or even limited to one or two complicating factors. Consider the following hypothetical scenario: a laboratory worker puts a beaker containing a volatile solvent, to which a radioactive compound has been added, into an ordinary refrigerator. Due to carelessness, it is not covered tightly. During the next several hours, the concentration of vapors builds up in the confined space and at some point the refrigerator goes through a defrost cycle. The vapor ignites explosively, the refrigerator door is blown off, strikes a worker, and knocks several bottles of chemicals off a shelf. Chemicals from the broken bottles spill onto the floor and onto the injured person. The solvent in the beaker, as well as in several other containers, spills on the floor and ignites. The radioactive material in the beaker and in some of the other containers is spread throughout the laboratory and into adjacent rooms. Although this is posed as a hypothetical situation, it could happen and with the exception of there being no injured person has happened at the author's facility.

In a complicated incident such as the one described, the first priority is preservation of life, even ahead of possible future complications. In the presence of a fire which, in a laboratory containing solvents, always has at least the potential of spreading uncontrollably, evacuation of the injured party should be considered as the first priority, followed by or paralleled by initiating evacuation of the rest of the building. Note that in every case of injury, the comparative risk of further injuring a person by moving them must be compared to the risk of not moving them. Notifying emergency medial services should be done as soon as possible after the removal of the victim to a safe location so treatment of the physical and chemical injuries to the victim can begin. Preliminary steps can be taken prior to the arrival of the emergency medical personnel if done with care not to exacerbate any of the injuries. In the case of the scenario described above, summoning the fire department can take the next priority. Of course, if adequate personnel are available, this step can be taken concurrently with the ones already mentioned. Generally, it is desirable to make these contacts with outside agencies from a place outside of the incident area. Assuming that the fire is manageable, then preliminary steps can be taken for cleanup and decontamination of the spilled chemicals and radioactive material. Unless there appears to be a risk that the contaminated area will spread, perhaps due to runoff of water used in fighting the fire, it is not necessary for these last steps be done in any haste. However, the surrounding area must be cordoned off until measurements and surveys are completed by trained radiation safety and, perhaps, chemical safety personnel. This isolation must be maintained until *a formal* release of the area by the individual in charge, based on the information provided by the safety specialists.

---

[*]   The Editor is indebted to Dr. Richard F. Desjardins, M.D. for his input for this section.

After the incident is over, a review of the causes of the accident and the emergency response should be conducted by the appropriate safety committee or committees. In this case, the laboratory safety committee and the radiation safety committee would probably jointly conduct the review. Basically, there were two root causes of this specific incident. Solvents should not be stored in any container which cannot be tightly sealed, but this would not have caused the explosion if the refrigerator had been suitably designed for storage of flammable materials. These are commercially available, although at a price two to three times more than a unit not designed to be explosion safe against internal flammable vapor releases. Note that the words "explosion proof" are not used here, since this implies that they could operate in an atmosphere of flammable vapors safely. Units meeting this more stringent criteria do exist but at a much higher price.

The subsequent review should consider if anything could have made the incident worse. For example, in the hypothetical accident, the worker could have been alone, although this was not assumed to be the case. In academic research laboratories, research workers, and especially graduate students, tend to work unusual hours as they try to work around their class schedules to meet deadlines imposed by the framework of timetables, deadlines for submission of theses and dissertations, etc. If the injured person had been alone, the potential for a loss of life would have existed.

The situation described in an earlier paragraph illustrates not only that in the real world emergencies can be very complicated, but also illustrates that some emergency responses can wait but others cannot. Components of the emergency that are immediately life threatening must be dealt with promptly, but others, such as cleaning up, can wait to be done carefully and properly after appropriate planning. Any incident also should be treated as a learning opportunity. There were basic operational errors leading to the postulated incident which could be repeated in other laboratories. There were aspects to the incident which would have permitted it to be worse. These should be factored into the emergency plan for the facility if they had not already been considered. If violations of policy had occurred, then the review should point these out and recommend courses of action to prevent future violations. It is not necessary to deliberately embarrass someone but it is important that this concern not conceal true errors which could have been avoided. An emergency plan should not only cover responses to classes of emergencies which have occurred, but should have the capability of reducing the possibility that emergencies will occur.

E. Artificial Respiration, Cardiopulmonary Resuscitation (CPR), and First Aid

In several examples of responses to various emergencies, allusions were made to emergency medical procedures which should be performed. Most of these procedures require prior training. Because of the relatively high probability of accidents in laboratories, it would be desirable if at least a cadre of trained persons was available in every laboratory building.

Both first aid and CPR classes are taught by a number of organizations in almost every community. Among these are the Red Cross, American Heart Association, rescue squads, other volunteer organizations, and many hospitals. Usually, except for a small fee to cover the cost of materials, the classes are free. In addition, labeling regulations and the OSHA Hazardous Communication Standard now require that emergency information be made available on the labels of chemical containers and as part of the training programs. Since in most cases involving a chemical injury the chemical causing the injury will be known, and thus information will be available, the following material on first aid for chemical injuries will be restricted to the case of basic first aid for an injury caused by an unknown chemical. Similarly, since formal class instruction in CPR, which will also cover artificial respiration, is almost always available, the material on CPR will be very basic. CPR should be done only by properly trained individuals, with the training including practice on mannequins. Certification in CPR is easily and readily acquired. It is also important to periodically become recertified, as new concepts and procedures are frequently evolving and presented in the training programs.

In all the following sections, it is assumed that emergency medical assistance will be called for immediately. Emergency medical personnel are trained to begin appropriate treatment upon their arrival. Depending upon the level of training and the availability of telemetry, they normally will have radio contact with a hospital emergency facility or a trauma center and can receive further instruction from a physician while providing immediate care during transit to the treatment center.

The following material is a composite of the information gleaned from a number of different sources. Where sources differed slightly, the more conservative approach was taken, i.e., that approach which appeared to offer the most protection to an injured person, with a second priority being the approach offering the least risk to the individuals providing the assistance. A third criterion was simplicity and the feasibility of performing the procedure with materials likely to be available. It was compiled explicitly in the context of injuries that are likely to occur as a result of laboratory accidents and is not intended to provide a compre-hensive treatment of emergency medical care. It has been reviewed and, where needed, revised by a physician.

Except where mandated by the nature of the problem, such as removal from a toxic atmosphere, or other circumstances immediately dangerous to life and health, no stress is placed on evacuation. Unless there are obvious fractures, there may be injuries to the spine, or broken bones that may puncture vital organs which are not immediately apparent. If it is essential to move the victim, do so very carefully. Use a backboard or as close to an equivalent as possible to keep the body straight, and support the head so that it does not shift. Any inappropriate movement of a fractured neck may damage or even sever the spinal cord and result in paralysis, death, or in a compromising of the patient's airway.

To repeat, before performing any of the more complicated first aid procedures, formal training classes taught by certified instructors should be taken. It is possible for an inexperienced person to cause additional injuries.

1. Artificial Respiration

The lack of oxygen is the most serious problem that might be encountered. If the victim is not breathing or the heart is not beating, then oxygen will not be delivered to the brain. If this condition persists for more than 4 to 6 minutes, it is likely that brain damage will occur. In this first section, it will be assumed that the heart is beating but that the victim is not breathing. This is checked by the lack of motion of the chest.

a. Artificial Respiration, Manual Method

Although mouth-to-mouth or mouth-to-nose artificial respiration is much more effective, an alternative method of artificial respiration will be discussed first. There are occasions when it is not safe to perform direct mouth-to-mouth resuscitation, such as when poisoning by an unknown or dangerous chemical substance is involved, or when the victim has-suffered major facial injuries which make mouth-to-mouth impossible. Since the first of these conditions can be expected to occur in some laboratory accidents, it is good to know that there is an alternative procedure available. The method considered the best alternative is described below.

1. Check the victim's mouth for foreign matter To do this, insert the middle and forefinger into the mouth, inside one cheek and then probe deeply into the mouth to the base of the tongue and the back of the throat, finally sliding your fingers out the opposite side of the mouth. Be aware that a semiconscious patient may bite down on your fingers. It would be wise to insert a folded towel or object that would not break teeth between the teeth while you are doing your examination.

2. Place the victim on his back on a hard surface in a face up position. Problems with aspirating vomitus can be reduced by having the head slightly lower than the trunk of the body. An open airway is essential and can be maintained by placing something, such as a rolled up jacket, under the victim's shoulders to raise them several inches. This will permit the head to drop backwards and tilt the chin up. Turn the head to the side. *Important! Do not do this if there is any suspicion of neck or spinal trauma.*

3. Kneel just behind the victim's head, take the victims wrists, and fold the victim's arms across the lower chest.

4. Lean forward, holding onto the wrists, and use the weight of your upper body to exert steady, even pressure on the victim's chest. Your arms should be approximately straight up when in the forward position. This will cause air to be forced out of the victim's chest. Perform this step in a smooth, flowing motion.

5. As soon as step 4 is completed, take your weight off the victim's chest by straightening up and simultaneously pulling the victim's arms upwards and backwards over his head as far as possible. This will cause air to flow back into the lungs, thereby completing one equivalent breathing cycle.

6. Steps 4 and 5 should be repeated 12 to 15 times per minute to pump air into and out of the victim's lungs. Stop and check frequently that no vomitus or foreign matter has been brought up into the airway If a helper is available, let the helper do this while you continue the pumping process.

7. Continue the procedure until normal breathing is established or until emergency medical personnel arrive. Be ready to recommence the process if breathing difficulties reoccur.

This less effective method was given first since it is human nature to question the need to learn a less effective method if a better one exists, and it has already been covered. However, it is important to limit the number of injured parties as well as to treat those already injured. Since, in the case of an unknown toxic or an especially dangerous substance, the person giving emergency treatment could be exposed to the same material, this procedure will serve in such cases.

b. Artificial Respiration, Mouth-to-Mouth Method

There is a consensus that this is the most effective method of artificial respiration. To be effective, it should be begun as quickly as possible.

1. Check the victim's mouth for foreign matter. Clear out any that is found with your fingers. A cloth such as a towel or handkerchief covering the fingers helps in removal of objects, or even loose solids, which would slide off wet fingers. The towel also will serve to protect the fingers should the patient bite down on them.

2. The victim's air passage must be open. Put your hand under the person's neck and lift (only if no spinal injury is suspected. In that case, pull the lower jaw forward.) Assuming no spinal injury, place your other hand on the victim's forehead and tilt the victim's head back as far as it will reasonably go, essentially straightening the airway. A folded jacket or coat under the shoulders will aid in keeping the head back.

3. Maintain this position, and with the fingers of the hand being used to tilt the head backward, pinch the nostrils closed.

4. Open your mouth widely, take a deep breath, place your mouth firmly around the victim's mouth to get a good seal and blow into the victim's mouth.

5. Try to get a good volume of air into the victim's lungs with each breath.

6. Watch to see the chest rise. When it has expanded, stop blowing and remove your mouth from the victim and check for exhalation by listening for air escaping from the mouth and watching to see the chest fall. This occurs naturally from the inherent elasticity of the chest wall.

7. Steps 4 through 5 constitute one breathing cycle. Assuming it has been successful, i.e.,

there has been no significant resistance to the flow of air during the inflow cycle and air has been exhaled properly, repeat steps 4 through 6 at a rate of 12 to 15 times per minute.

8. If you are not getting a proper exchange of air, check again to see if there is anything obstructing the air passage and make sure you are holding the head properly so as to prevent the tongue from blocking the flow of air.

9. Continue the procedure until normal breathing is established, or until emergency medical personnel arrive. Be ready to recommence the process if breathing difficulties reoccur.

2. Cardiopulmonary Resuscitation

Cardiopulmonary resuscitation combines artificial respiration with techniques to manually and artificially provide blood circulation, and is to be used where the heart is not beating. It requires special training and practice and should not be done except by trained, qualified individuals. As noted earlier, receiving the training is not difficult since it is readily available from several organizations. The material presented here is intended to provide information as to the general procedures, but neither it nor the previous procedures are intended as an instructional guide.

One-person CPR will be the procedure described. In this procedure, the person performing it will provide both artificial breathing and blood circulation. When two persons are available, each person can take the responsibility for one of these functions and can switch from time to time to relieve fatigue.

a.   Initial Steps

1. Check for breathing. Do this by placing your hand under the victim's neck and the other on the forehead. Lift with the hand under the neck and tilt the head back. While doing this, place your ear near the victim's mouth and look toward the chest. If the victim is breathing, you should be able to feel air on your skin as it is being exhaled, you should be able to hear the victim breathe, and you should be able to see the chest rise and fall. Do this for at least 5 seconds.

2. If the victim is not breathing, while holding the victim as in step one and pinching the nostrils closed, open your mouth widely, place it over the victim's mouth, and give two quick full breaths into the victim's mouth.

3. After step 2, repeat step 1 to see if breathing has started. If not, proceed to step 4.

4. Check the victim's pulse with the hand that had been under the victim's neck. Keep the head tilted back with the other hand on the forehead. Check the pulse by sliding the tips of the fingers into the groove on the victim's neck to the side of the Adam's apple nearest you. Again, check for at least 5 seconds.

b.   Formal CPR Procedures

The victim must be on a firm surface. Otherwise, when pressure is applied to the chest, the heart will not be compressed against the backbone as the backbone is pressed into a yielding soft surface. The head of the victim should not be higher than the heart, in order for blood to flow to the brain, as is needed to avoid brain damage. Although the brain averages about 2% of your body weight, it requires 20% of the oxygen you breathe, as well as at least 40 mgm per cent of blood dextrose. Any concentration less will result in unconsciousness and progressive brain damage. Preferably, the feet and legs should be adjusted to be higher than the heart to facilitate blood flowing back to the heart, but this has a lower priority than commencing CPR procedures.

1. Kneel beside the victim, at breast height. Rest on your knees, not your heels.

2. Locate the victim's breastbone. Place the heel of one of your hands on the breastbone so that the lower edge of the hand is about two finger-widths up from the bottom tip of the breastbone. Put your other hand on top of the first hand. Lift your fingers or

otherwise keep from pressing with them. Improper placement of the hands can cause damage during the compression cycle.

3. Place your shoulders directly over the breastbone. Keep your arms straight.
4. To initiate the compression cycle, push straight down, pivoting at the hips.
5. Push firmly and steadily down until the chest has been compressed about 4 to 5 centimeters. Then smoothly relax the pressure until the chest rebounds and is no longer compressed, then start the compression cycle again. The compression-relaxation cycle should be a smooth, continuous process.
6. Continue the chest compression procedure for 15 cycles at the rate of 80 per minute. This should take between 11 and 12 seconds. Then quickly place your mouth over the mouth of the victim and give two quick full breaths. Then return to compressing the chest for another 15 cycles. Be sure to locate your hands properly and compress the chest as in steps 4 and 5.
7. Continue step 6, alternately compressing the chest and providing artificial respiration.
8. Quickly check for a pulse after 1 minute and then every few minutes thereafter. Watch for any signs of recovery.
9. If a pulse is found, then check for breathing. If necessary give artificial respiration only, but check frequently to be sure that the heart is still beating.
10. Continue with whatever portion of the procedure is necessary until the victim is functioning on his own, emergency medical personnel arrive, or it is obvious that efforts will not succeed. A half hour is not unreasonable as a period of actively using CPR.

Thus, to repeat once more, the techniques for artificial respiration and CPR are not difficult but require training and practice. Familiarity with these techniques is likely to be of value at any time.

c. First Aid
i. Severe Bleeding

A person may bleed to death in a very short time from severe or heavy bleeding. Whenever this problem is involved in an accident, it is extremely important to stop it as soon as possible. Arterial bleeding may be frightening, but the muscular artery wall usually contracts to diminish or stop the flow. Venous bleeding is more insidious as it flows steadily The relative absence of muscle in the vein wall does not help to stem the flow. Try to be calm and try to keep the victim calm as well. Bleeding may cause the victim to panic or become overwrought.

1. The most effective treatment is pressure applied directly to the wound over which a sterile dressing has been placed (Figure 2.8).
2. If possible, wash your hands thoroughly both before and after treating a bleeding wound. It would be desirable to wear a latex glove. Apply a sterile dressing if immediately available (a handkerchief for some other cloth to the wound, if not). Then place the palm of the hand directly over the wound and apply pressure. If nothing else is available use your bare hand, but try to find something to use as a dressing as soon as possible.
3. A dressing will help staunch the flow of blood by absorbing the blood and permitting it to clot. Do not remove a dressing if it becomes blood soaked, but leave it in place and apply an additional one on top of the first in order not to disturb any clotting that may have started. Keep pressure on with the hand until you have time to place a pressure bandage over the dressing to keep it in place.
4. Unless there are other injuries, such as a fracture or the possibility of a spinal injury, cases for which the victim should be disturbed as little as possible, the wound should be elevated so that the injured part of the body is higher than the heart. This will reduce the blood pressure to the area of the wound.

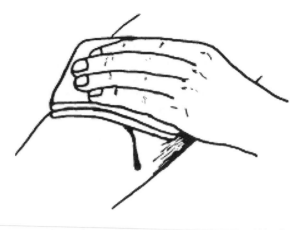

Figure 2.8   Apply pressure directly to a wound to control bleeding.

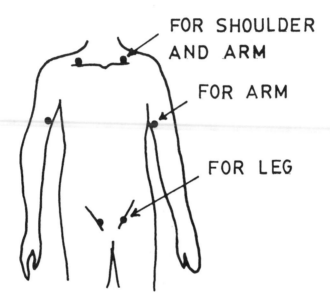

Figure 2.9 Pressure points to control bleeding at the extremities.

5. If bleeding persists and cannot be stopped by direct pressure, putting pressure on the arteries supplying the blood to the area may be needed. In this technique, pressure is applied to the arteries by compressing the artery between the wound and the heart against a bone at the points indicated in Figure 2.9. Since this stops all circulation to points beyond the point of compression, it can cause additional injuries if continued too long. For this reason, it should be discontinued as quickly as possible and the use of direct pressure and elevation to control the bleeding should be resumed, unless this is the only effective technique.

6. As a last resort, since it stops the flow of blood to the limb beyond the point of application, a tourniquet, which should be at least two inches wide, can be applied. An example where the use of a tourniquet might be indicated is to stop bleeding from a severed limb.

ii. Shock

Shock may accompany almost any type of severe injury, exposure to toxic chemicals, a heart attack, loss of blood, burns, or any other severe trauma. It can be recognized from a number of characteristic symptoms: skin cold to the touch (possibly clammy and bluish or pale), weakness, a rapid weak pulse, rapid irregular breath, restlessness, and exhibition of unusual signs of thirst. As the condition worsens, the victim will become unresponsive and the eyes may become widely dilated. The treatment for shock is:

1. The victim should be lying down, although the type of injuries may determine what precisely is the best position. If uncertain, allow the victim to lie flat on his back. Unless it is painful or it makes it harder for the victim to breathe, it will help if the feet are raised 20 to 30 centimeters high.
2. Use blankets to keep the victim from losing body heat, but do not try to add heat.
3. If the victim is conscious and is not vomiting, nor appears likely to do so, then about half a glass of liquid every 15 minutes or so will be helpful. However, do not give fluids if the victim is unconscious or nauseated.

iii. Poisoning by Unknown Chemicals

The rationale in limiting this section to unknown chemicals is that first aid information is readily found when the chemical is known, either immediately on the label of the container or in the MSDS for the chemical (note that current law requires chemical producers to provide MSDSs to those who purchase chemicals from them). An MSDS file should be maintained in every laboratory.

Even if the chemical causing the injury is not known and the victim is unconscious so that no direct information is available, an examination of the circumstances of the accident and an examination of the victim's lips, skin, mouth, and tongue could provide helpful information on whether the victim swallowed a poisonous substance or inhaled a toxic vapor, or whether the injury was due to absorption through the skin. If the abdomen is distended and pressing on it causes pain, the likelihood is that a corrosive or caustic substance has been ingested. Various other symptoms such as nausea, vomiting, or dizziness can occur if the person has ingested or inhaled a toxic substance. If there are blisters or discoloration of the skin, then external exposure is likely. Any information on the nature of the harmful material will be helpful to the emergency personnel or physician who will treat the victim.

iv. Poisoning by Inhalation

1. When poisoning by inhalation is suspected, evacuate the victim to a safe area as soon as possible. If there are fumes still suspected to be present, a rescuer should wear a self-contained respirator. Do not take a chance which might result in a second victim.
2. Check for unusual breath odors if the victim is breathing.
3. Loosen tight clothing around the victim's neck and waist.
4. Maintain an open airway.
5. If the victim is not breathing, perform artificial respiration using the manual method. It is dangerous to the person providing aid to give mouth-to-mouth artificial respiration if the toxic material is not known.

v. Poisoning by Ingestion

1. Examine the lips and mouth to ascertain if the tissues are damaged as a possible indicator that the poison was ingested, although the absence of such signs is not conclusive. Check the mouth and remove any dentures.

2. If the victim is not breathing, perform manual artificial respiration.
3. If the victim becomes conscious, try to get the victim to vomit, unless it is possible that the poisoning is due to strong acids, caustics, petroleum products, or hydrogen peroxide, in which case additional injuries would be caused to the upper throat esophagus, and larynx. Vomiting may be induced by tickling the back of the throat. Lower the head so that the vomit will not reenter the mouth and throat. Dilute the poison in the stomach with water or milk.
4. If the victim has already vomited, collect a sample of the vomit, if possible, for analysis.
5. If convulsions occur, do not restrain the victim, but remove objects with which he might injure himself or orient the victim to avoid his striking fixed, heavy objects.
6. Watch for an obstruction in the victim's mouth. Remove if possible, but do not force fingers or a hard object in between the victim's teeth. If a soft pad can be inserted between the victim's teeth, it will protect the tongue from being bitten. A badly bleeding tongue immensely complicates the patient's problems.
7. Loosen tight clothing, such as a collar, tie, belt, or waistband.
8. If the convulsions cease, turn the victim on his side or face down so that any fluids in the mouth will drain.
9. Treat for shock if the symptoms for shock are noted.

vi. Poisoning by Contact

1. If the chemical got into the victim's eyes, check for and remove any contact lenses. Take the victim immediately to an eyewash station (if one is not available, to a shower or even a sink) and wash the eyes, making sure that the eyelids are held widely open. Wash for at least 15 minutes. If the chemical is caustic rather than acidic, the victim may not feel as much pain and may wish to quit earlier, since an acid causes pain due to the precipitation of a protein complex. An alkali or caustic chemical is more dangerous than an acid as it does not precipitate the protein and continues to penetrate the globe of the eye and may even lead to global rupture. It is imperative that the eye be flushed out thoroughly.
2. Do not use an eye ointment or neutralizing agent.
3. If the chemical only came in contact with the victim's exposed skin, such as the hands, wash thoroughly until the chemical is totally removed.
4. If the chemical was in contact with the clothed portion of the body, remove the contaminated clothes as quickly as possible, protecting your own hands and body, and place the victim under a deluge shower. If the eyes were not affected initially, protect them while washing the contaminated areas. Be careful not to damage the affected skin areas by rubbing too firmly. Let the flowing water rinse the chemical off. A detergent is sometimes used, but be careful not to carry the offending chemical to other parts of the body. Be particularly careful to clean folds, crevices, creases, and groin.
5. If both the eyes and portions of the body were exposed, there should preferably be a combination eyewash and deluge shower unit available. If not, take the victim to the deluge shower and tilt the head back, holding the eyelids widely open, and wash the entire body.

vii. Heat Burns

First Degree (minor): Painful and red. No blisters. Skin elastic. Epidermis only. Minimal swelling.

1. Apply cold water to relieve pain and facilitate healing.
2. Avoid re-exposure, as the already injured skin can be more susceptible to further damage than normal skin.

Second Degree: Severe and painful, but no immediate tissue damage. Pale to red. Weeping blisters, vesicles. Marked swelling. Involves epidermis and dermis.

1. Immerse affected area in cold water to abate the pain.
2. Apply cold, clean cloths to the burned area.
3. Carefully blot dry.
4. Do not break blisters.
5. If legs or arms are involved, keep them elevated with respect to the trunk of the body.

Third Degree: Deep, severe burns, likely tissue damage. White, red, or black and dry and inelastic tissue. No pain, involves full thickness of skin. May involve subcutaneous tissue, muscle, and bone.

1. Do not remove burned clothing from the burned area.
2. Cover the burned area with a thick, sterile dressing or clean cloths.
3. Do not immerse an extensively burned area in cold water, because this could exacerbate the potential for shock and introduce infection. A cold pack may be used on limited areas such as the face.
4. If the hands, feet, or legs are involved, keep them elevated with respect to the trunk of the body.
5. Third-degree burns must be treated by a physician and/or hospital. They may need reconstruction, skin grafting, and prolonged care. Control of infection is mandatory.

REFERENCES

Note that many of the basic references were incorporated in the text as materials needed to plan or facilitate an effective emergency program. The following are additional references used in preparing the material.

1. **Lowery, G.G. and Lowery, R.C.,** *Handbook of Hazard Communications and OSHA Requirements,* Lewis Publishers, Chelsea, MI, 1990.
2. **Lowery, G.G. and Lowery, R.C.,** *Right-to-Know and Emergency Planning,* Lewis Publishers, Chelsea, MI, 1989.
3. **Laughlin, J.W.,** Ed., *Private Fire Protection and Detection,* ISFTA 210 International Fire Training Association, Fire Protection Publications, Oklahoma State University, Stillwater, 1979.
4. ANSI Z358.1-1998, Emergency Eyewash and Shower Equipment, American National Standards Institute, New York, 1981.
5. **Srachta, B.J.,** in *Safety and Health,* National Safety Council, Chicago, 1987, 50.
6. **Steere, N.V.,** Fire, emergency, and rescue procedures, in *CRC Handbook of Laboratory Safety,* Steere, N.V., Ed., CRC Press, Cleveland, OH, 1971, 15.
7. ANSI Z87 1-1979, Practice for Occupational and Educational Eye and Face Protection, American National Standards Institute, New York, 1979.
8. **Schwope, A.D., Costas, P.P., Jackson, J.O., Stull, J.O., and Weitzman, D.J.,** Eds., *Guidelines for the Selection of Chemical Protective Clothing,* 3rd ed., Arthur D. Little, Inc. for U.S. EPA and U.S. Coast Guard, Cambridge, 1987.
9. **McBriarty J.P. and Henry, N.W.,** Eds., *Performance of Protective Clothing:* Fourth Volume, American Society for Testing and Materials, Philadelphia, 1992.
10. ANSI Z88.2-1980 Practices for Respiratory Protection, American National Standards Institute, New York, 1980.

11. **Kairys, C.J.,** Hazmat protection improves with equipment documentation, in *Occupational Health and Safety 56,* No. 12, 20, November 1987.

12. **Still, S. and Still, J.M., Jr.,** Burning issues (charts), Humana Hospital, Augusta, GA.

13. **Schmelzer L.L.,** *Emergency Procedures and Protocols,* Cancer Research Safety Workshop Workbook, Office of Research Safety, National Cancer Institute, Bethesda, MD, 1978, 106.

14. **Gröschel, D.H.M., Dwork, K.G., Wenzel, R.P., and Schiebel, L.W.,** Laboratory accidents with infectious agents, in *Laboratory Safety Principles and Practices,* Miller, B.M., Gröschel, D.H.M., Richardson, J.H., Vesley, D., Songer, J.R., Housewright, R.D., and Barkley, W.E., Eds., American Society of Microbiology, Washington, D.C., 1986, 261.

15. **Edlich, R.E., Levesque, E., Morgan, R.E., Kenney, J.G., Sulboway, K.A., and Thacker, J.G.,** Laboratory personnel as first responders, in *Laboratory Safety Principles and Practices,* Miller, B.M., Gröschel, D.H.M., Richardson, J.H., Vesley, D., Songer, J.R., Housewright, R.D., and Barkley, W.E., Eds., American Society of Microbiology, Washington, D.C., 1986, 279.

16. Emergency first aid guide, appendix 4, in *Laboratory Safety Principles and Practices,* Miller, B.M., Gröschel, D.H.M., Richardson, J.H., Vesley, D., Songer, J.R., Housewright, R.D., and Barkley, W.E., Eds., American Society of Microbiology, Washington, D.C., 1986, 348.

17. *Safety in Academic Chemistry Laboratories,* 5th ed., American Chemical Society Washington, D.C., 1985.

18. *Multimedia Standard First Aid, Student Workbook,* American Red Cross, Washington, D.C., 1981.

19. *Standard First Aid & Personal Safety* 2nd ed., American Red Cross, Washington, D.C., 1979.

20. *Adult CPR Workbook,* American Red Cross, Washington, D.C., 1987.

21. **Hafen, B.Q. and Karren, K.J.,** *First Aid and Emergency Care Workbook,* 3rd ed., Morton Publishing, Englewood, CO, 1984.

22. **Senecab, J.A.,** Halon replacement chemicals: perspectives on the alternatives, in *Fire Technology,* 28(4), 332, November, 1992.

23. **Zurer, P.S.,** Looming ban on production of CFC's, Halon spurs switch to substitutes, *Chemical & Engineering News,* 71(46), 12, November 15, 1993.

24. Health answers available at http://www.healthanswers.com/ -Orbis Broadcast Group, 1110 Sangamon Chicago, IL.

25. *Halon Replacements. Technology and Science,* Andrezej W. Mizolek, Editor, Wing Tsang. ACS symposium Series, Oct. 1997.

Chapter 3

LABORATORY FACILITIES—DESIGN AND EQUIPMENT

I. LABORATORY DESIGN

The design of a laboratory facility depends upon both function and program needs but not strongly upon the discipline involved. Although there are differences among engineering, life sciences and chemistry laboratories, and within the field of chemistry (between laboratories intended for physical chemistry and polymer synthesis, to take two examples), the similarities outweigh the differences except in unusual specialized facilities. Approximately the same amount of space normally is required. Certain utilities are invariably needed. Adequate ventilation is needed to eliminate odors and vapors from the air, which might have the potential to adversely affect the health of the employees, as well as to provide tempered air for comfort. Provision is needed for safely stocking reasonable quantities of chemicals and supplies. As these are used over a period of time, chemical wastes are generated and provisions must be made for temporary storage and disposal of these wastes according to regulatory standards. The laboratories must provide suitable work space for the laboratory workers. Many of these items, as well as others, vary only in degree. Most differences are relatively superficial and are represented primarily by the equipment which each laboratory contains and the selection of research materials used.

Not only are laboratories basically similar, but there is a growing need for "generic" laboratory spaces readily adaptable to different research programs. This is due in part to the manner in which most research is funded today. In industry, laboratory operations are generally goal oriented, i.e., they exist to develop a product, improve a product, or to perform basic research in a field relevant to the company's commercial interests. There is a cost-benefit factor associated with laboratory space which affects the amount of assigned space. In the academic field, research is primarily funded by grants submitted to funding agencies by the faculty. These grants can be from any number of public and private sources, but, with only a moderate number of exceptions, grants are based on submission of a proposal to the funding agency to perform research toward a specific end during a stipulated period of time. At the end of this period, the grant may or may not be renewed; if not, control of the space may be turned over to another investigator. Laboratory space is too limited and too expensive (currently running in the range of $100 to $300 per square foot, dependent upon the complexity of the construction) to be allowed to remain idle. The result has been a trend to design laboratories that are relatively small, typically suitable for no more than two to four persons to work in them simultaneously, with connections to adjacent rooms to permit expansion if needed. Under these circumstances, it will be appropriate in most of this chapter to base the discussion upon a standard module. One potential result of this growing need for flexibility may be an eventual breakdown of the concept of department-owned space for research buildings, i.e., the concept of chemistry or biology buildings. Eventually facilities may be designed toward a given type of use, such as microbiology or polymer chemistry but the users may be assigned suitable space independently of their original departmental affiliation, based, at least in part, on current needs.

Instructional laboratories are an exception in terms of size since they normally are intended for continued basic programs, serving class sizes of 20 or more persons, and so typically are somewhat larger than is needed for research programs. Also, except at advanced levels, the instructional laboratories usually do not conduct experiments or use chemicals having the same degree of risk as do research laboratories. The risk in instructional laboratories is also being reduced by the greater use of smaller quantities of chemicals because of advances in technology, and because of the safety training being routinely provided to the graduate assistant instructors at many schools. However, even in the case of instructional laboratories, many of the basic safety requirements still must be incorporated in the design.

A. Engineering and Architectural Principles

The increasing cost of sophisticated laboratory space dictates a number of design considerations. It is essential that space be used to maximum advantage. Due to the necessity for mechanical services, closets, columns, wall thicknesses, halls, stairs, elevators, and restrooms, the percentage of net assignable space in even a well-designed, efficient building is generally on the order of about 65%. Due to the large number of fume hoods in a typical laboratory building and other ventilation requirements, as well as the increasingly stringent temperature and humidity constraints imposed by laboratory apparatus and computers, heating and ventilation (HVAC) systems are becoming more sophisticated. The engineer must accommodate these needs as well as the need to provide personal comfort, conserve energy, and provide low life-cycle maintenance costs. Stringent new regulatory requirements under the Americans With Disabilities Act to accommodate disabled persons in virtually every program impose costly additional constraints on accessibility and provisions for emergencies. Building designs need to be sufficiently flexible not only to suit different uses based on current technology, but should be sufficiently flexible to adopt technological innovations. For example, provision for installation of additional data, video, and voice lines in excess of earlier needs is almost certainly desirable. Additional electrical capacity should be provided over that meeting current needs. Interaction of the occupants of the building with each other, with outside services, and with other disciplines also mandates a number of design parameters. This latter set of parameters is very dependent upon the specific programs using the building and will require substantial input from the users. Different disciplines perhaps require more variation in provision for the needs of service groups than in the laboratories themselves. Typically all of these design needs must be accommodated within a construction budget, established before the design of the building is in more than a very early conceptual stage, so the design process is a constant series of compromises. It is rare that all of the program desires (as opposed to needs) can be fully satisfied.

To the architect, a very important factor is that the building must meet all the needs in an attractive way. Otherwise, the architect's reputation could be at risk. There is certainly nothing wrong with creating an attractive facility in harmony with its surroundings, as long as this aspect is not achieved at the expense of the basic needs of the users. Generally the most efficient space is a cube, with no more than the minimally required penetrations of the walls and with no embellishments. No one would truly like to see this become the standard, although in the right context, even such a facility could be made very attractive. Buildings should fit into their environment in an aesthetic and congenial manner, but function and use factors should be preeminent in the design.

No mention has been made up to this point of health and safety design factors. They must be incorporated into virtually every other design feature. The location of a building, access to the building, the materials of construction and interior finish, size and quality of doors, width of corridors, length of corridors, number of floors, the number of square feet per floor, selection of equipment, utilities, etc. are impacted by safety and health requirements.

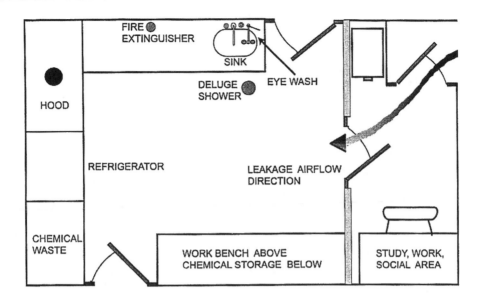

Figure 3.1 Standard Laboratory Module.

Although it would be anticipated that architects and engineers would be thoroughly familiar with applicable safety codes and regulations, experience has shown that this is not necessarily so, especially where they involve safety concepts other than those relating to fire or strength of materials. Even in these areas, the wide range of variability in interpretation of codes often results in a tendency to liberally interpret the codes in favor of increasing the amount of usable space or enhancing the visual aspects of the design. It is surprising how few architectural firms maintain dedicated expertise on their permanent staff in the areas of building code compliance, especially those areas involving health and safety for specialized buildings such as laboratory buildings. Even where such staff personnel are available, there is an inherent problem with a conflict of interest between the code staff and the designers since they are both employed by the same firm with the firm's typical architect owner being strongly design oriented. Of course, the reciprocal is also true. Most safety professionals are not artists, as many architects consider themselves, who can adequately include the aesthetic aspect in their own ideas. The eventual users may not appreciate the sole viewpoints of either of these two groups.

Since relatively few laboratories are built compared to the numbers of other types of buildings, comparatively few firms are really well prepared to design them for maximum safety, especially in terms of environmental air quality and laboratory hazards. For this reason, the eventual owners/users of a planned building should be sure to include persons to work with the architects and contractors. Where this expertise is not available in-house, they should not hesitate to hire appropriate consultants to review the plans and specifications prior to soliciting bids.

Shown in Figure 3.1 above, is a standard laboratory module which forms the basis for much of the material in this chapter. This design, although simply a representative example, does provide a significant number of generally applicable safety features. A slightly larger variation on this design includes a central workbench down the center of the facility, but this represents an obstacle many users prefer not to have. The laboratories on either side can be designed as mirror images of this one and this alternating pattern can be repeated to fill the available space. The two side doors may be operational, as shown here, and provide ready

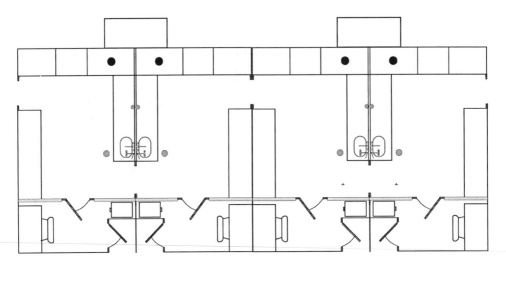

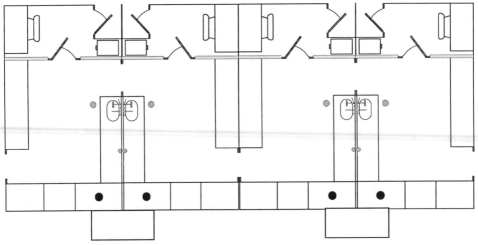

Figure 3.2 Section of a building utilizing the standard module as a recurring element. The single corridor with laboratories on both sides is a very efficient use of a building's space. As shown, all of the laboratories are equipped the same but this can be readily changed by the use of modular casework.

access to adjacent spaces, if needed, for the research program. Most building codes do not require more than a single exit in such a small room unless it is classified by the building code applicable to the facility as a hazardous duty occupancy, so that if access to additional lab modules is not needed, either or both doors can be constructed as breakaway emergency exits or even not constructed initially to allow additional bench or storage space. Where the doors are included, two well separated, readily accessible exits exist from every point within the room, even at the end of a sequence of laboratories.

In this basic 12 foot x 20 foot module, the areas where the likelihood of a violent accident are greatest (within the fume hood) are at the far end of the laboratory, away from the corridor entrance, and are well separated from stored flammable materials and other reagents. The desk area is separated from the work areas by a transparent barrier which, with the door to the laboratory properly closed, isolates the workers, when they are not actively engaged in their research, from both the possible effects of an accident and continuous exposure to the atmospheric pollutants of the laboratory. This latter factor is enhanced by the normal negative atmospheric pressure between the laboratory and the corridor, so that the air in the desk area should be virtually as clean as the corridor air. The transparent barriers also permits the laboratory worker to maintain an awareness of what is transpiring in the work area even when they are not in it. Note that the negative pressure is not such that a major portion of the makeup air is drawn from the corridor. The amount of makeup air from the corridor is limited to about 200 cfm by code requirements. The area at the entrance thus would represent a safe space for employees or students to socialize, study, or even have a drink or snack. The door from the corridor to the laboratory is set into an alcove so that it may open in the direction of exit travel yet not swing into the hall, so as to create an obstruction to traffic in the corridor.

Many of the laboratory's features will be discussed more fully later on but a brief summary of the other safety features which recommend this design will be given here. Possibly the most important is the location of the fume hood which is located in the lowest traffic area in the room and where it would not be necessary to pass by it in the event of an incident requiring evacuation. The hood should be equipped with a velocity sensor which will alert workers if the velocity falls below an acceptable level. The eyewash station and deluge shower are located close to the center of the room such that only a very few steps would be necessary to reach both of them. They can be used simultaneously. There is only a modest amount of chemical storage space, located beneath the work bench. The lack of space strongly encourages maintenance of tight controls on chemical inventories. The fire extinguisher is also located so that it is readily at hand. The makeup air inlet for the room, which is not shown, allows air to be diffused through the ceiling in such a way that it provides minimal disturbance of the air in the vicinity of the fume hood face. If warranted, a relatively inexpensive automatic flooding fire extinguishing system can be provided for the entire room. Similarly to the chemical storage space, the space devoted to the storage of waste chemicals is also modest, encouraging their removal in a timely fashion. A flammable material storage cabinet can take the place of this chemical waste area, with the chemical waste being stored in a small portion of the chemical storage area.

Figure 3.2 provides a simplified illustration of how the modular approach can be integrated into an efficient and safe building design. This figure represents a section of a typical upper floor of a research building. Mechanical services, loading dock and receiving areas, support services, offices, conference rooms, toilets, lounges, and classrooms would be located either on other floors or further along the access corridors. Note that the fume hoods are at a back corner of the laboratory, and immediately outside the building is an external chase to carry the exhaust duct to the roof. The location of the external chase at the juncture of two laboratories allows one chase to serve two laboratories. This external chase solves another problem if all laboratories are not originally equipped with hoods. It would be almost as economical to go back and add a hood in this design as it would be to equip every laboratory with a hood initially.

The internal equipment is shown as the same in each laboratory, but with the exception of equipment dependent on service utilities such as water, and the fume hood exhausts, the internal arrangement is highly flexible. The individual manager can relocate virtually anything else, and with modular casework now available, there would be few restrictions on the arrangements, even in such a small module. Mention has already been made of the ability to add a fume hood later, and flammable material storage in refrigerators or flammable material storage cabinets may or may not be needed. Since the use of laboratories does change over time, the design should provide contingencies for the maximum hazard use, in terms of safety considerations, in the original construction.

The arrangement of laboratories with only a single support corridor, as shown in Figure 3.2 provides an advantageous net to gross square footage. The use of modules, arranged compactly as these naturally permit doing, allows the architect and building owner to achieve

Figure 3.3 The external columns on this laboratory building, located on the
campus of Virginia Polytechnic Institute & State University contain the chases
for the fume hood exhaust ducts, and are located so as to serve two adjacent
laboratories. The exhaust ducts are led to a common plenum and are exhausted
directly upward. The laboratory module in this facility is somewhat larger than
the standard module described in this chapter so as to accommodate windows.
The air intakes for the facility are located to the reader's right and take
advantage of the prevailing winds from that direction.

an efficient building. The external chases lend themselves to an attractive architectural
columnar appearance to the building, otherwise the absence of windows in large segments of
the wall might otherwise appear too austere. An actual building based on this external chase
concept and with modular laboratories is shown in Figure 3.3.

An aspect of the design above, which may not be immediately apparent, is that such a
design is especially appropriate for adding to an older facility which was originally designed
to meet less demanding standards than those of today. The newer component, situated
adjacent to the original structure, and designed to meet current sophisticated research
requirements, can be connected to the older one at appropriate places. By proper construction
and fire separations, it would be possible to treat the old and new components as separate
buildings, even though they are joined, so that it would not be necessary to renovate the older
building to current construction standards. Less demanding operations, such as instructional

laboratories and offices, could remain in the older component, and activities requiring additional and probably more sophisticated services, higher construction standards, etc. would be located in the new area. All of a department's operations would be in the "same" building, which has important logistical and personnel implications, and construction of an entirely new building for a department would be unnecessary This concept is called an "infill" approach and provides some important financial savings, as it can extend the usable life of some older facilities. The methods of joining and maintaining separations between the two components also provide opportunities for architects to express themselves, such as making the less-expensive spaces between the two sections outside of the laboratory facility proper, into attractive communal areas.

REFERENCES

1. **Barker, J.H.,** *Designing for Safer Laboratories,* CDC Laboratory Facilities Planning Committee, Chamblee Facility, 1600 Clifton Rd., Atlanta, GA.
2. Earl Wall and Associates, Basic Program of Space Requirements, Dept. of Chemistry, VPI & SU, Laboratory Layout Studies, Blacksburg, VA, 1980.
3. **Ashbrook, P.C. and Renfrew, M.M.,** Eds., *Safe Laboratories, Principles and Practices for Design and Remodeling,* Lewis Publishers, Chelsea, MI, 1991.
4. *Trends in U.S. Lab Designs for the '90s, Technical Paper No. 90.03,* Hamilton Industries Infobank, New Rivers, WI, May, 1995.
5. **Deluga, G.F.,** Designing a Modern Microbiological/ Biomedical Laboratory: Design Process and Technology: Laboratory Ventilation, Landis & Gyr, Buffalo Grove, IL, July 1996.

B. Building Codes and Regulatory Requirements[*]

There are many codes and standards applicable to building construction. Many of these are incorporated by reference in the OSHA standards. A large number of the codes grew out of a concern for fire safety, and hence this general area is relatively mature. Existing health codes generally address only acute exposures and immediate toxic effects. It has been only relatively recently that concern for long-term systemic effects has been addressed in standards, so there are fewer of them. The current OSHA Laboratory Standard replaces the detailed OSHA standards which were intended primarily for industrial situations and is a performance standard which requires that laboratories prepare and voluntarily comply with an industrial hygiene. The intent of the current standard is to ensure that laboratory employees are provided with at least the equivalent degree of protection as would have been provided by the general industry standards. Since most users of this Handbook may not be familiar with the OSHA General Industry Standards, there will be allusions in the text to these latter requirements as a reference base.

There are specific sources from which a substantial portion of the material in this section will be derived or to which it will be compared. For building codes, the information will be referenced to the BOCA (Building Officials and Code Administrators) code and The Southern Building Code (Southern Building Code Congress International, Inc.). These codes are not used universally and, in fact, differ in detail, but do represent typical codes which, where applicable, provide mandatory standards. Other regional building codes are based on the same

---

[*] The basic material in this chapter concerned with building code requirements was reviewed for the 3rd edition of this handbook by Howard W Summers, former Chief Fire Marshal (retired) for the State of Virginia.

general industrial codes and recommendations of standard-setting organizations, but specific applications of these reference standards in codes for a given area may differ. The material presented here should not be construed as equivalent to either of these codes but instead as being representative of the subject areas under discussion. Standard 45 of the National Fire Protection Association (NFPA), currently under review for revision, is specifically labeled as a laboratory safety standard. It has not been adopted as a formal legal requirement in many localities, but it does provide valuable guidance in certain areas for goals against which both existing and proposed laboratories can be measured. The building codes are primarily concerned with fire and construction safety, with less emphasis on health issues. The materials cited are those most directly affecting the physical safety of building occupants or useful to persons discussing building design with architects and contractors. In addition, there will be other standards, such as the Americans With Disabilities Act (ADA), which will be superimposed on both existing facilities and, especially, new construction, that will also influence the design of laboratories. This last act (ADA) is very broad in its statements, and implementation details may in many cases depend upon litigation. Of course, OSHA also addresses some of the same issues as do the codes but due to the long process involved in modifying the OSHA standards, they tend to lag behind the other sources. The two building codes mentioned are formally revised every three years. Facility designers and users are encouraged to use the more conservative, safety and health wise, of current standards and guidelines.

Standard 45 and the applicable building codes do not always agree, or at least they sometimes lead to different interpretations. The classification of the structure or building in which testing or research laboratories are operated is usually designated under both BOCA and the Southern Code, for example as an educational or business use occupancy, although if the degree of hazard meets a number of specific criteria, a facility may be designated as a hazardous use facility. Under NFPA Standard 45, buildings used for the purpose of instruction by six or more persons are classed as an educational occupancy. The classification is not a trivial question since it evokes a number of different design and construction constraints. A building used primarily for instruction, which might include instructional laboratories, and some testing and research laboratories might be considered primarily an educational occupancy if any research areas were properly separated from the remainder of the building. An educational occupancy is more restrictive than a business occupancy but less so than a hazard use classification. Standard 45 classifies laboratories as class A, B, or C according to the quantities of flammable and combustible liquids contained within them, with A being the most hazardous and C being the least. As discussed later, a system of ratings has been developed in the certain types of facilities for the life sciences to designate laboratories according to four safety levels, with classes 1 and 2 meeting the needs of most laboratory operations, while 3 and 4 are restrictive and very restrictive, respectively. This concept, for consistency, might eventually be considered for laboratories of all types. In a later section of this Chapter, such a proposed classification scheme for chemical laboratories is put forward.

1. Building Classification

For the purposes of this section, the classification of a building will be derived from the two building codes mentioned in the preceding section. The basic classification, therefore, will be either as an educational or business use occupancy. However, since some laboratories and ancillary spaces, such as storerooms, may meet the definitions of High Hazard Use (Group H), the following material will provide some guidance as to whether a given building or facility or part thereof should be considered a Group H occupancy. The standard laboratory module, as shown in Figures 3.1 and 3.2, can meet some of the requirements for high hazard use, e.g., that two or more well-separated exits and the doors swing in the direction of exit travel. The doors from the modules also are set within an alcove so that the door does not swing out into the corridors when opened.

Table 3.1. Exemption Limits in Gallons for Several Classes of Materials
For a Class 2, Hazardous Use Occupancy

Types of Materials	Flammable Liquids 1A	Flammable Liquids 1B	Flammable Liquids 1C	Combustible Liquids II	Combustible Liquids III A	Flammable Oxidizing Cryogenics
Materials not in storage cabinets, building not sprinklered	30	60	120	120	330	45
Materials in storage cabinets or building sprinklered	60	120	240	240	660	45 (in cabinets) 90 (in sprinklered building)
Materials in storage cabinets and building sprinklered	120	180	360	480	1,320	90

The hazard use occupancy group H is divided into 4 levels of hazard, 1-4. Since the OSHA Laboratory Standard does not address manufacturing or pilot process facilities, the following information does not apply to them.

- The highest risk level, listed as H-1 generally is applied to facilities in which activities take place using materials that represent an explosive risk. In the context of laboratories, in addition to materials normally considered explosives, this includes organic peroxides, oxidizers, other highly unstable materials, and pyrophoric materials capable of *detonation,* as opposed to those which do not react as violently. The difference between "detonation" and "deflagration," employed in the description of the second highest level of risk facility is the speed of the reaction process and the speed of propagation of the resultant spread of the affected area. Relatively few laboratory facilities would fall in this category.
- Group H-2 includes facilities using less vigorously reacting materials than those of Group H-1, as well as flammable and combustible liquids, gases, and dusts that are a *deflagration* hazard. Some laboratory facilities could fall in this category if substantial quantities of such materials were involved. However, the OSHA Labora-tory Standard definition would often exclude these facilities.
- Group H-3 facility activities involve materials that represent a physical hazard due to the ability of the materials to support combustion.
- Group H-4 facilities contain materials and involve activities that present health hazards. Laboratories could be found in any of these categories in most major research facilities. Fortunately, however, most laboratories are exempt because they use relatively small quantities of these materials.

Material Safety Data Sheets, which are now required to be provided by distributors and manufacturers of commercial chemicals, give detailed information on the characteristics of all commonly sold laboratory chemicals. The definitions of explosive, flammable, combustible, and various health hazards are consistent with those provided by OSHA in CFR 29, Parts

Table 3.2 Exemption Limits for a Few Critical Classes of Materials Representing Health Hazards For a Class 4, Hazardous Use Occupancy

Types of Materials	Highly Toxic Gases[1,2] (ft^3)	Highly Toxic Solids & Liquids (lbs)
Materials not in storage cabinets, building not sprinklered	0	1
Materials in storage cabinets	20	2
Materials in storage cabinets and building sprinklered	40	4

1. Cabinets here are construed as fume hoods or exhausted gas storage cabinet.
2. Gas cylinders of 20 ft^3 or less stored in gas storage cabinets or fume hoods.

1200, 1450, and 1910, Department of Transportation, CFR 40, Part 173, or other regulatory standards. These are discussed in detail in Chapter 4.

Table 3.1 represents the maximum amount of various classes of materials representing physical hazards allowed in a controlled area, e.g.., laboratories, for a Hazard Class 2 facility. Note that few laboratories will be considered Hazard Class 2 occupancies. Most will be considered Business occupancies, and the limits on flammables in these facilities will be governed by OSHA regulations. The limits for laboratories will be discussed in detail in a later section dedicated to flammable solvents. Similarly, Table 3.2 does the same for materials which represent health risks for a Hazard Class 4. One factor must be borne in mind, no flammable materials may be stored or used in a space that is below grade, i.e., in major part below ground level.

It is possible to have different areas in a building classified differently. If this occurs, then the requirements for each use area shall be met in those areas. Where provisions differ, the requirements providing the greater degree of safety will apply to the entire building, or a complete fire separation must be provided between the two sections. This occurs most frequently when major renovations occur, such as adding a new wing to a building in the infill process or upgrading an area within a building.. Generally the most restrictive height and area restrictions will still apply to the entire building.

2. Types of Construction

There are several classifications of types of construction. Basically without providing complete definitions which may be found in the local building codes (available in most libraries of reasonable size) or, if not, at the office of the local building official, the classifications range from construction materials which are wholly noncombustible (used for buildings where such materials are justified, which includes most laboratory buildings), to intermediate types which may include both combustible and noncombustible, with critical elements still required to be made of noncombustible materials to those in which any materials may be used as long as they meet code acceptable fire resistance. The last of these usually would have height and area restrictions as well which would make it unlikely that laboratory facilities would be of this type of construction. The most fire resistant facilities would have many of their structural elements with fire resistance ratings of 4 or 3 hours. Due to the cost of this level of construction, most laboratory facilities have key structural components with only 2 hour fire resistance ratings, with some other elements having ratings of 1 to 1.5 hours. These lower ratings should not represent an actual decrease in safety for the building's occupants. If the structural components are protected such that the equivalent fire resistance ratings are provided, the fire resistance rating of the component itself can be decreased.

In order to facilitate the use of the following tables, a number of definitions are in order:

- Fire Resistance Rating—The time in hours or fractions thereof that materials or their assemblies will resist fire exposure.
- Fire Separation Assembly—A fire resistance rated assembly designed to restrict the spread of a fire.
- Protected—Construction in which all structural members are constructed or protected in such a manner that the individual unit or the combined assemblage of all such units has the requisite fire resistance rating for its specific use or application.

Walls:

- Bearing Wall— Any wall supporting any additional vertical load in addition to its own weight.
- Fire Wall—A fire resistance-rated wall which is intended to restrict the spread of a fire and which is continuous from the foundation to or through the roof of a building.
- Fire Separation Wall— Similar to a fire wall in that it is intended to restrict the spread of a fire but does not include the requirement of extending from the foundation to the roof of a building.
- Party Wall— A fire wall on an interior lot line used for joint service between two buildings.

As indicated earlier, most laboratory facilities represent a reasonable compromise between safety and cost, generally being of Type II construction for a Business Occupancy. Given in Table 3.3 are the typical required fire ratings for several of the structural components for this construction class.

An important consideration for a building is its size and height. For the type of construction on which the previous three tables are based, a laboratory building would be limited to three stories or 40 feet in height with each story being no more than 14,400 ft^2. There are any number of ways which permit these limits to be exceeded, including building to a higher standard of construction, use of an automatic fire suppression system throughout the building, and other factors depending upon the location of the facility with respect to road access. The question arises however, should such factors be used when viewed in the context of the safety of the occupants? A laboratory building, even though it is designated as a Business occupancy, does represent unique potential safety issues, which are different than many other types of uses found in this classification. Even in a non-laboratory building evacuating perhaps several hundred persons down stairs presents problems. When the source of a fire could involve a bewildering variety of chemicals which might or might not generate fumes much more toxic than the normal smoke fumes, which are usually the major cause of deaths in a fire, should the occupants have to face any more risk than necessary? Where space for construction is a premium, there is a great temptation to at least consider the options available but safety should be given a very high priority.

As just noted above, there are other factors and conditions that may become involved in determining the allowable area, height, etc., in addition to the ones discussed. However, the intent here is not to provide a course in code review, which involves much more sophisticated details than it would be possible to cover in this space, but to provide sufficient basic information for laboratory personnel, to allow them to understand the constraints under which the designer operates. The details of the final design must be negotiated among the architect, contractor, building official, and representatives of the owner. The participation of laboratory

Table 3.3 Fire Ratings in hours for Selected Structural Components For Type II Construction

Load bearing exterior walls	0 or 1
Party and fire walls	2
Interior bearing walls	2
Exit enclosures	2
Exit corridors/ fire partitions	1
Shafts	2
Floors, ceiling assemblies	2
Roofs	1
Beams, girders, trusses (one floor)	2
Columns	2

personnel is essential to define their program needs in the context of what is permissible under the building code and is economically feasible. Code issues are not always clear cut, with much of the actual language subject to interpretation. Also, there are often alternative ways to provide equivalent protection so that requests to code officials for variances, based on this concept, are frequently acceptable.

There will be additional safety issues addressed in many of the following sections where specific design features will be discussed in more detail.

C.   Laboratory Classification

There are no universal safety criteria to classify laboratories which take into account all types of risks. Standard 45 of the National Fire Prevention Association (NFPA) designates chemical laboratories of different degrees of risk, based essentially on fire safety factors, regulating the amount of solvents which each class may contain. The Centers for Disease Control has published and uses a set of guidelines establishing a biological safety level rating system for laboratories in the life sciences and those using animals, based on a number of parameters relating to the infectiousness to humans of the organisms used in the facility. This system parallels an earlier four-level classification scheme developed for those working in recombinant DNA research. Both of these classification schemes are guidelines, not regulations, although they are virtually as effective as standards when funding requests are involved. The military sponsors research involving diseases to which its forces may be exposed and also uses these biological facility classifications. The standards associated with biological organisms are concerned with the potential risk to the public at large as well as to the laboratory workers. The Department of Agriculture regulates the importation, possession, or use of a number of non-indigenous pathogens of domesticated animals. The Drug Enforcement Agency licenses and sets standards for facilities in which controlled substances are employed to ensure that they are used safely and to guard against their loss or theft. The Nuclear Regulatory Commission (NRC) licenses agencies or individuals using radioactive materials to ensure that neither the workers nor the general public are adversely affected by the use of radiation. To obtain an NRC license, one must demonstrate the competence to use the material safely and to be able and willing to meet an extremely detailed set of performance standards. All of these standards have been developed essentially independently and, where a regulatory agency is involved, are administered separately. In many instances laboratory operations will be affected by several sets of regulations. However, even if all of the regulatory standards were imposed simultaneously there would still be many safety factors which would not be included. Thus, it is, at least partially, the responsibility of the institution or corporation to establish additional criteria to properly evaluate the degree of risk in a research program and to assign the program to a space providing the requisite degree of safety.

Working with materials with low risk potentials will obviously be much more tolerant of poor facilities or procedures than using materials involving a high risk, but not totally so. Even a small quantity of a IA flammable solvent such as ether, used in an inadequate facility could lead to a serious accident, while the same quantity, used in a fume hood by a careful worker following sound safety procedures, could be used quite safely. Of course, even the best facilities cannot prevent problems if the personnel using the facilities do not follow good safety practices.

The OSHA Laboratory Standard mandates that performance standards be established in each facility that would ensure that the employees would be as well protected as those working in industrial situations, for which long-established general industry standards apply. This appears to bypass, at least as far as OSHA is concerned, the need for any sort of laboratory classification scheme, leaving the responsibility primarily to the local laboratory or organization. The OSHA Laboratory Standard does not replace the biological guidelines since the OSHA standard does not at this time include pathogens as a possible risk, nor would it supersede radiation safety standards. There is also the difficulty that research programs tend to evolve and could change the level of risk involved over a short period of time. It would be impractical to be continually shifting occupants of space as this occurred. However, the flexibility permitted by the standard laboratory module described earlier in this chapter permits easy and economical changes in a facility to modify the quality of the space for different levels of risk.

Although it is unlikely that a formal system of classifying laboratories according to a comprehensive safety standard is imminent, it surely is incumbent upon an institution or corporation to ensure that research is assigned to space suitably designed and equipped so that research can be performed with a reasonable assurance of safety. If research programs are evaluated properly, it should be possible to assign them to laboratories classified into low, moderate, substantial, and high-risk categories. This type of classification seems to be the simplest and most practical to use and has the further advantage of already being employed in life science laboratories. Before examining the features that might be incorporated into each category which will depend somewhat upon the area of research involved, it might be well to list at least some of the parameters that should be considered in evaluating research programs.

1. Program-Related Factors

Evaluation of programs to permit assignment to the appropriate class of facility should depend upon several factors:

I. Materials
   A. Recognized risks
      1. Flammable
      2. Reactive
      3. Explosive
      4. Acute toxicity
      5. Strongly corrosive, acidic
      6. Known systemic or chronic health effects
         a. Carcinogens
         b. Mutagens, teratogens
         c. Affect reproduction/fertility
         d. Radiation
         e. Pathogens
         f. Affect the respiratory system
         g. Neurotoxic

           h. Known strong allergens
           i.   Sensitizers
           j.   Other known health effects
       7. Physical risks
           a. Electrical
           b. High pressure
           c. Heat and cold
           d. Sound
           e. Non-ionizing radiation/light
           f. Mechanical physical risk factors
           g. Ventilation
       8.   Factors affecting the external environment
    B. Quantities/scale of operations
    C. Procedures
       1.   Standard operating procedures/practices
       2.   Emergency procedures
II. Information/training
    A. Health and safety training
       1.   Documentation of safety and health training for laboratory managers/staff
       2.   Procedures to train new personnel
       3.   Procedures to train all personnel when new materials/new procedures are used
    B. Material Safety Data Sheets available for all chemicals used
    C. Chemical Hygiene Plan in effect
III.   Personnel protection
    A. Exposure monitoring
    B. Personal protective equipment available
    C. Health assurance/medical response program available

The information in Part I above is, in effect, an evaluation of possible negative aspects of the program under consideration, while positive information under each of the items in Parts II and III can be used to offset, to some degree, the needs which must be met by the facility. It is preferable, however, to design-in safety rather than depending upon procedures and administrative rules.

2. Laboratory Class Characteristics
In the following four sections, oriented primarily toward chemistry laboratories, the reader already familiar with laboratory classification guidelines established by the Centers for Disease Control will note that in many respects the recommendations or defining qualities for low, moderate, substantial, and high-risk categories closely parallel those for biosafety levels one through four. It will be noted that this system will involve classifying laboratory facilities by much more than the configuration of bricks and mortar of which they are built, or their contents of a single type or a limited variety of hazardous material, although these aspects will be important. The assumption is also made that for at least the first two levels of risks that a modular facility, not dissimilar to the standard laboratory described at the beginning of this chapter, will form the basis for the facility. Separate major sections in Chapter 5 are devoted to laboratories in the life sciences, animal facilities, and radiation, so reference to topics relevant to those areas will be deferred to those sections.

a. Low-Risk Facility
A low-risk facility is used for work with materials, equipment, or classes of operations, with no known or minimal risk to the workers, the general public, or to the environment. It is possible to work safely with all the necessary materials on open benches. No special protection or enclosures are needed for the equipment or operations. There is a written

laboratory safety plan to which all the employees have access. Laboratory workers have been properly trained in laboratory procedures and are supervised by a trained and knowledgeable person. If there are any potential risks, the employees have been informed of them, how to detect them if they are not immediately obvious, and emergency procedures.

Although the laboratory design requirements are not stringent, features which would be difficult to change, if the utilization should become one which would require a higher classification, should be built to a higher level. Examples of this concept, marked with an asterisk (*), include provisions for easily cleaned and decontaminated floors and laboratory furniture and good ventilation.

Standard Practices

1. Access to the laboratory is limited at the discretion of the laboratory supervisor, as needed.
2. A program exists to ensure that reagents are stored according to compatibility.
3. An annual (or continuous) chemical inventory will be performed and information sent to a central data collection point. Outdated and obsolete chemicals will be disposed of through a centrally managed chemical waste disposal program.
4. The laboratory will be maintained in an orderly fashion.
5. Although it is anticipated that the amount of hazardous chemicals used in a low risk facility will be very limited, all secondary containers containing materials incorporating more than 1.0% of a hazardous component or combination of hazardous components, which will be used more than a single work day, shall be labeled with a label listing the hazardous components (not required under the OSHA Laboratory Standard, but good practice).
6. Any chemical wastes are placed in appropriate and properly identified containers for disposal through a chemical waste disposal program. Broken glass is disposed of in heavy cardboard or kraftboard boxes labeled "broken glass." Any "sharps," as defined under the blood-borne pathogen standard, will be placed in a legal container for disposal as infectious waste. Only ordinary solid, nonhazardous waste may be placed in ordinary trash containers.
7. Eating, drinking, smoking, and application of cosmetics are not permitted in the work area.
8. No food or drink can be placed in refrigeration units used in the laboratory.
9. The telephone numbers of the laboratory supervisor, any alternates, and the department head shall be posted on the outside of the laboratory door or the adjacent wall.

Special Practices

There are no special practices associated with a low-risk laboratory

Special Safety Equipment

1. Any refrigerators or freezers shall be rated as acceptable for "Flammable Material Storage," i.e., be certified as explosion safe, except for ultra-low temperature units.
2. No other special safety equipment is needed.

Laboratory Facilities

1. The floor of the laboratory is designed to be easily cleaned. Seamless floors and curved junctures to walls aid in accomplishing this.*
2. Bench tops should be resistant to the effects of acids, bases, solvents, moderate heat, and should not absorb water. The tops should have few seams or crevices to facilitate cleaning.
3. Furniture should be designed to be sturdy and designed for convenient utilization and modification. Storage spaces should be easily accessible.

4. Aisle spaces should be 40 to 48 inches wide and not constricted to less than 28 inches by any temporary obstacles.
5. Electrical outlets shall be three-wire outlets, with high-quality, low-resistance ground connections. Circuits should be clearly identified to correlate with labels in breaker panels.
6. The laboratory should be supplied with a sink. The plumbing shall be sized to accommodate a deluge shower and eyewash station. With average water pressure, this would normally be a one-inch line or larger.
7. Normal building ventilation is sufficient. However, it is recommended that at least six air changes per hour of 100% fresh air be provided as standard.

b. Moderate-Risk Facility

A moderate-risk facility involves material, practices, and use of equipment such that improper use could pose some danger to the employees, the general public or the environment. Generally, the materials used would have health, reactivity or flammability ratings, according to NFPA Standard 704 of 2 or less. Small quantities of materials with higher ratings might be involved in work being performed in chemical fume hoods or in closed systems. Work with special risks, such as with carcinogens, would not be performed in a moderate-risk facility. Equipment which could pose a physical hazard should have adequate safeguards or interlocks. However, in general, most operations could be safely carried out on an open work bench or without unusual precautions. The amounts of flammables kept in the laboratory meet NFPA standard 45 for Class A laboratories (or less), and when not in use are stored in either a suitable flammable material storage cabinet or other comparable storage unit.

The person responsible for the work being performed in the laboratory is to be a competent scientist. This individual shall develop and implement a safety and health program for the facility that meets the requirements of the OSHA Laboratory Standard. The individual workers are to be fully trained in the laboratory procedures being employed and to have received special training in the risks specifically associated with the materials or work being performed. The workers are to be informed about the means available to them to detect hazardous conditions and the emergency procedures that should be followed, should an incident occur.

Standard Practices

1. Access to the laboratory work area is limited during the periods work is actively in progress, at the discretion of the laboratory supervisor.
2. A program exists to ensure that chemicals are stored properly, according to compatibility. Quantities of chemicals with hazard ratings of 3 or greater are limited to the amount needed for use in a 2-week interval, or in accordance with NFPA standard 45 for flammables, whichever is less.
3. An annual (or continuous) chemical inventory will be performed and sent to a central data collection point, preferably based on a centralized chemical computer management program. Outdated and obsolete chemicals will be disposed of through a centrally managed chemical waste disposal program. Ethers and other materials which degrade to unstable compounds shall be shelf dated for disposal 6 months after being opened (unless a material specific earlier shelf limit is indicated), but no more than 12 months after purchase, even if unopened, unless processed to remove any unstable peroxides that may have formed.
4. A Material Safety Data Sheet file will be maintained for all chemicals purchased for use in the laboratory. The file will be accessible to the employees in the laboratory. This requirement may be met by computer access to a centrally managed MSDS data base. All laboratory workers shall be trained in how to interpret the information in an MSDS.

5. All secondary containers, in which are materials containing more than 10% of a hazardous component or combination of hazardous components, which will be used more than a single work day shall be labeled with a label listing the hazardous components.

6. Any chemical wastes are placed in appropriate and properly identified containers for disposal through a chemical waste disposal program. Broken glass is disposed of in heavy cardboard or kraftboard boxes prominently labeled "broken glass." Any "sharps," as defined under the blood-borne pathogen standard, will be placed in a legal container for disposal as infectious waste. Only ordinary solid, nonhazardous waste may be placed in ordinary trash containers.

7. Ten to twelve air changes per hour of 100% fresh air shall be supplied to the facility. No air shall be recirculated. The ventilation system shall be designed such that the room air balance is maintained at a small negative pressure with respect to the corridors whether the fume hood is on or off.

8. The laboratory will be maintained in an orderly fashion.

9. No food or drink can be placed in refrigeration units used in the laboratory.

10. A placard or other warning device shall be placed on the door or on the wall immediately adjacent to the door identifying the major classes of hazards in the laboratory (See Chapter 2, Figures 2.6 and 2.7).

11. The telephone numbers of the laboratory supervisor, any alternates, and the department head shall be posted on the outside of the laboratory door or the adjacent wall.

Special Practices

1. Work with materials with safety and health ratings of 3 or greater in any category shall be performed in a functioning fume hood.

2. Work with substantial amounts of materials with hazard ratings of 1 or 2 shall be performed in a hood or in an assembly designed to be safe in the event of a worst-case failure.

3. Appropriate personal protective equipment shall be worn in the work area. Because eyes are critical organs very susceptible to chemical injuries or minor explosions, it is strongly recommended that chemical splash goggles be worn whenever the work involved offers any possibility of eye injury. Wearing of contact lenses should follow the safety practices established for the facility, but if an individual must wear them for medical reasons, then that individual should wear chemical splash goggles at all times in the laboratory. A mask may be used to supplement the minimum eye protection.

Special Safety Equipment

1. Any refrigerators or freezers shall be rated as acceptable for "Flammable Material Storage," i.e., be certified as explosion safe, except for ultra-low temperature units.

2. A flammable material storage cabinet, either built-in or free standing, shall be used for the storage of flammable materials.

3. The laboratory shall be equipped with a fume hood.

4. The laboratory shall be equipped with an eyewash station and a deluge shower.

5. The laboratory shall be provided with one or more Class 12 ABC fire extinguishers.

6. A first-aid kit shall be provided and maintained.

7. Any special equipment mandated by the research program shall be provided.

Laboratory Facilities

1. The floor of the laboratory is designed to be easily cleaned. Seamless floors and curved junctures to walls aid in accomplishing this.

2. Bench tops should be resistant to the effects of acids, bases, solvents, and moderate heat, and should not absorb water. To facilitate cleaning, the tops should have few seams or crevices.

3. Furniture should be designed to be sturdy and designed for convenient utilization and modification. Storage spaces should be easily accessible.

4. Aisle spaces should be 40 to 48 inches wide and shall not be constricted to less than 28 inches by any temporary obstacles.

5. Electrical outlets shall be three-wire outlets with high-quality, low-resistance ground connections. Circuits should be clearly identified to correlate with labels in breaker panels.

6. The laboratory shall be supplied with a sink. The trap should be of corrosion-resistant material. The plumbing shall be sized to accommodate the deluge shower and eyewash station. With average water pressure, this would normally be a 1-inch line or larger.

7. Ten to twelve air changes per hour of 100% fresh air shall be supplied to the facility. No air shall be recirculated. The ventilation system shall be designed such that the room air balance is maintained at a small negative pressure with respect to the corridors whether the fume hood is on or off.

8. It is recommended that the facility include a separation of work spaces and desk areas as well as a second exit, as shown in the standard laboratory module, Figure 3.1 (see Chapter 3, Section I. A).

c. Substantial-Risk Facility

For the two lower risk categories, it is possible to be almost completely general since they are specifically intended to be used for only limited risks. However, for both substantial risk and high-risk facilities, the nature of the risk will dictate specific safety-related aspects of the facility. Most of these can be accommodated at the substantial risk level within the standard laboratory module, appropriately modified and equipped.

The use of highly toxic (or having a seriously detrimental health characteristic, such as a potential carcinogen), highly reactive, or highly flammable chemicals or gases would man-date the work being performed within at least a substantial risk facility. If explosives are involved, then the laboratory should be designed with this in mind. Explosion venting may be required in this instance. The location of the facility may be dictated by the need to contain or control the debris or fragments from an explosion. The level of construction may need to be enhanced to make the walls stronger to increase their explosion resistance. The use of toxic or explosive gases may require continuous air monitoring with alarms designed to alert the occupants of levels approaching an action level, which should be no higher than 50% of the level representing either a permissible exposure limit (PEL) or the lower explosive limit (LEL). The alarms must be connected to the building alarm system, which in turn should be connected to a central manned location. Highly flammable materials may require special automatic extinguisher systems, using high-speed fire detectors, such as ultraviolet light sensors coupled with dry chemical or Halon™** comparable fire suppression systems. It may be desirable to have electrical circuits protected by Ground Fault Interruptor (GFI) devices or a readily operable master disconnect switch available. There are, of course, other risks as tabulated in Chapter 3, Section I.C.1., which would require other precautions.

Access to a substantial risk facility should be restricted during operations and at other times to authorized personnel only at the discretion of the laboratory supervisor. The laboratory supervisor shall be a competent scientist, having specific knowledge and training relevant to the risks associated with the program of research in the laboratory. Each person authorized to enter the laboratory shall have received specific safety training appropriate to

---

* Note the discussion in Chapter 2 about the phasing out of the availability of previously popular chlorinated fluorocarbons due to the negative effect these materials have on the earth's ozone layer. In the context of this recommendation, the alternatives described there should be used.

the work and to the materials employed. A formal, written laboratory industrial hygiene plan, including an emergency plan complying with the requirements of the OSHA Laboratory Standard, shall be developed and practiced at least annually. A copy of the emergency plan shall be provided to all agencies, including those outside the immediate facility who would be called upon to respond to an incident. The emergency plan shall include a list of all personnel in the facility with business and home telephone numbers.

Standard Practices

1. Access to the laboratory is limited to authorized personnel only during operations, and to others at times and under such conditions as designated by written rules or as established by the laboratory supervisor.
2. All chemicals must be stored properly according to compatibility. Any chemicals which pose a special hazard or risk shall be limited to the minimum quantities required to meet short-term needs of the research program, and materials not in actual use shall be stored under appropriate, safe conditions. For example, flammables not in use shall be kept in a flammable materials storage cabinet, and excess quantities of explosives should be stored in magazines, away from the immediate facility. Other materials such as drugs or radioactive materials may also require secured storage areas.
3. An annual (or continuous) chemical inventory will be performed and sent to a central data collection point, preferably based on a centralized chemical computer management program. Outdated and obsolete chemicals will be disposed of through a centrally managed chemical waste disposal program. Ethers and other materials which degrade to unstable compounds shall be shelf dated for disposal 6 months after being opened (unless a material specific earlier shelf limit is indicated), but no more than 12 months after purchase, even if unopened, unless processed to remove any unstable peroxides that may have formed.
4. A Material Safety Data Sheet file will be maintained for all chemicals purchased for use in the laboratory. The file will be accessible to the employees in the laboratory. This requirement may be met by computer access to a centrally managed MSDS data base. All laboratory workers shall be trained in how to interpret the information in an MSDS. In some cases, such as experimental compounds being tested, an MSDS may not be available. Any information provided by the manufacturer will be kept in a supplement to the MSDS data base such cases.
5. All secondary containers containing materials having more than 1% of a hazardous component or combination of hazardous components (0.1% for carcinogens), which will be used more than a single work day, shall be labeled with a label listing the hazardous components.
6. Any chemical wastes are placed in appropriate and properly identified containers for disposal through a chemical waste disposal program. Any wastes which pose a special hazard or fall under special regulations and require special handling shall be isolated and a program developed to dispose of them safely and legally. Broken glass is disposed of in heavy cardboard or kraftboard boxes prominently labeled "broken glass." Any "sharps," as defined under the blood-borne pathogen standard, will be placed in a legal container for disposal as infectious waste. Only ordinary solid, nonhazardous waste may be placed in ordinary trash containers.
7. The laboratory will be maintained in an orderly fashion. Any spills or accidents will be promptly cleaned up and the affected area decontaminated or rendered safe, by safety personnel if a major spill or by laboratory personnel if a minor one. Major spills will be reported to the Safety Department.
8. No food or drink can be brought into the operational areas of the laboratory, nor can anyone smoke or apply cosmetics.

9. Any required signs or information posting mandated by any regulatory agency shall be posted on the outside of the door to the entrance to the laboratory. In addition, a placard or other warning device shall be placed on the door or on the wall immediately adjacent to the door identifying any other major classes of hazards in the laboratory (see Section 2.3.4). A sign shall be placed on the door stating in prominent letters, meeting any regulatory standards, "AUTHORIZED ADMISSION ONLY."

10. The telephone numbers of the laboratory supervisor, any alternates, and the department head shall be posted on the outside of the laboratory door or the adjacent wall.

Special Practices

1. Specific policies, depending upon the nature of the hazard, shall be made part of the laboratory industrial hygiene and safety plan and scrupulously followed to minimize the risk to laboratory personnel, the general public, and the environment. Several examples of laboratory practices for various hazards are given below. This list is not intended to be comprehensive, but instead represents some of the more likely special precautions needed for a variety of types of risks.

   - All work with hazardous kinds or quantities of materials shall be performed in a fume hood or in totally enclosed systems. It may be desirable for the hood to be equipped with a permanent internal fire suppression system.
   - Work with explosives shall be limited to the minimum quantities needed. For small quantities used in a hood, an explosion barrier in the hood, with personnel wearing protective eye wear, face masks, and hand protection, may be sufficient protection. For larger quantities, the facility must be specifically designed for the research program.
   - Some gases, such as fluorine, burn with an invisible flame. Apparatus for work with such materials should be placed behind a barrier to protect against an inadvertent introduction of a hand or other part of the body, so as to prevent burns.
   - Systems containing toxic gases that would be immediately dangerous to life and health (IDLH) or gases that could pose an explosive hazard if allowed to escape, especially if they have no sensory warning properties, shall be leak tested prior to use and after any maintenance or modification which could affect the integrity of the system. Where feasible, the gas cylinders may be placed external to the facility and the gases piped into the laboratory to help minimize the quantity of gas available to an incident. Permanently installed gas sensors, capable of detecting levels of gas well below the danger limits may be needed in some cases.
   - Vacuum systems capable of imploding, resulting in substantial quantities of glass shrapnel or flying debris, shall be protected with cages or barriers or, for smaller systems, shall be wrapped in tape.
   - Systems representing other physical hazards, such as high voltage, radiation, intense laser light beams, high pressure, etc., shall be marked with appropriate signs and interlocked so as to prevent inadvertent injuries. The interlocks shall be designed to be fail safe such that no one failure of a component would render the safety interlock system inoperative.

2. Activities in which the attention of the worker is not normally engaged with laboratory operations, such as record maintenance, calculations, discussions, study, relaxation, etc., shall not be performed in the laboratory proper, but shall be performed in an area isolated from the active work area. The segregated desk area of the standard laboratory module is specifically intended to serve this purpose. Depending upon the nature of the hazard, it is usually economically feasible to make at least a portion of the barrier separating the two sections of the laboratory transparent so that continuing operations can be viewed, if necessary.

3. Workers in the laboratory, if they actively use materials for a significant portion of their work week which would pose a significant short- or long-term risk to their health, should participate in a medical surveillance program. Employees shall be provided medical examinations if they work with any material requiring participation in a medical program by OSHA or other regulatory agencies under conditions which do not qualify for an exemption. Employees shall notify the laboratory supervisor as soon as possible of any illness that might be attributable to their work environment. Records shall be maintained of any such incident.

4. No safety feature or interlock of any equipment in the facility shall be disabled without written approval of the laboratory supervisor. Any operations which depend upon the continuing function of a critical piece of safety equipment, such as a fume hood, shall be discontinued should the equipment need to be temporarily removed from service for maintenance. Any such item of equipment out of service shall be clearly indicated with a signed "Out of Service" tag. Only the person originally signing the tag, or a specific, designated alternate, shall be authorized to remove the tag.

5. It shall be mandatory to wear any personal safety equipment required for conducting operations safely in the laboratory.

6. It is recommended that a laboratory safety committee review each new experiment planned for such a facility to determine if the experiment can be carried out safely in the facility. If the risk is such that experiments may affect the environment or the surrounding community, it is recommended that the committee include at least one layperson from the community, not currently affiliated directly or indirectly with the institution or corporation. In this context, "new" is defined as being substantially different in character scope, or scale from any experiment previously approved for the facility.

Special Safety Equipment

1. Any refrigerators or freezers shall be rated as acceptable for "Flammable Material Storage," i.e., be certified as explosion safe, except for ultra-low temperature units.

2. A flammable material storage cabinet, either built-in or free standing, shall be used for the storage of flammable materials.

3. The laboratory shall be equipped with a fume hood. The fume hood should meet any specific safety requirements mandated by the nature of the research program. A discussion of hood design parameters will be found in a later section, but for high hazard use the interior of the hood and the exhaust duct should be chosen for maximum resistance to the reagents used; the blower should either be explosion-proof or, as a minimum, have non-sparking fan blades; the hood should be equipped with a velocity sensor and alarm should the face velocity fall below a "safe" limit; the interior lights should be explosion-proof, and all electrical outlets and controls should be external to the unit. It may be desirable to equip the unit with an internal automatic fire suppression system.

4. The laboratory shall be equipped with an eyewash station and a deluge shower.

5. The laboratory shall be equipped with a fire alarm system connected so as to sound throughout the building (and in a central facility manned 24 hours per day), an appropriate fire suppression system, and be provided with one or more class 12 BC, or larger, fire extinguishers, or class D units if reactive metals are in use.

6. An emergency lighting system shall be provided.

7. A first-aid kit shall be provided and maintained.

8. Any special safety equipment mandated by the research program shall be provided. For example, electrical equipment other than refrigerators may need to be designed to be explosion-safe.

Laboratory Facilities

1. The floor of the laboratory is designed to be easily cleaned. Durable, seamless floors of materials that are substantially impervious to spilled reagents are easily decontaminated, and have curved junctures to walls to aid in accomplishing this.

2. Two well-separated exit doors shall be available to the laboratory which shall swing in the direction of exit travel.

3. Bench tops should be resistant to the effects of acids, bases, solvents, and moderate heat, and should not absorb water. To facilitate cleaning, the tops should have few seams or crevices.

4. Casework should be designed to be sturdy and designed for convenient utilization and modification. Storage spaces should be designed to meet any special requirements and should be easily accessible. It should not be necessary, for example, to stretch to reach any reagent which, if dropped, could represent a safety problem.

5. Aisle spaces should be 40 to 48 inches wide and shall not be constricted to less than 28 inches by any temporary obstacles. The aisles should lead as directly as possible toward a means of egress.

6. The organization of the facility shall be such as to reduce the likelihood of having to pass an originating or secondary hazard to evacuate the facility in the event of an emergency.

7. Electrical outlets shall be three-wire outlets with high-quality, low-resistance ground connections. Circuits should be clearly identified to correlate with labels in breaker panels. Some locations would need to be equipped with ground-fault interrupters (GFIs), such as where electrical connections are near sinks.

8. Laboratories in which the risk of electrical shock is greater than normal may also be equipped with a master "panic" manually operated, electrical disconnect switch, clearly marked and located in a readily accessible location.

9. The laboratory shall be supplied with a sink. The trap shall be of corrosion-resistant material. The plumbing shall be sized to accommodate the deluge shower and eyewash station. With average water pressure, this would normally be a 1-inch line or larger.

10. Ten to twelve air changes per hour of 100% fresh air shall be supplied to the facility. No air shall be recirculated. The ventilation system shall be designed such that the room air balance is maintained at a small negative pressure with respect to the corridors whether the fume hood is on or off. Where toxic and explosive gases and fumes are present, the system is to be designed to be efficient in exhausting these fumes by locating the exhaust intakes either very near the source of fumes or near the floor (except for lighter-than-air or hot gases). Typical air flow patterns are to be such as to draw dangerous fumes away from the normal breathing zones of the laboratory's occupants.

11. The facility shall include a separation of work spaces and desk areas as well as a second exit, equivalent to the arrangement shown in the standard laboratory module, Figure 3.1 (see Chapter 3, Section 3.A).

d. High-Risk Facility

A distinguishing feature of a high-risk facility is that the operations of the laboratory pose an immediate and substantial danger to the occupants, the general public, or the environment if not performed safely in a suitable facility. The users of the facility and those permitted access to it must be limited to those individuals of the highest competence, training, and character. The OSHA required laboratory safety plan must include training specifically tailored to inform the personnel in the facility of the risks to which they are exposed, the mandatory preventive safety procedures which must be followed, and the measures which must be taken in an emergency. Because it is so difficult to guarantee the degree of safety which must be met, a typical academic building would not normally be suitable, nor would most common industrial research facilities, without substantial modifications.

A second distinguishing feature of a high-risk facility is the need for isolation. If, for example, specific exceptions are permitted under the building codes, then a building of use group H (hazard) shall not be located within 200 feet of the nearest wall of buildings of the types most likely to be found in research facilities or isolation obtained by other means. In some cases this is achieved by distance, as above. In other instances, isolation is achieved by building walls and other structural components to a higher than normal level of construction. In cases in which the level of risk is not so much physical, as is basically the concern of most building codes, but involves toxic materials or biologically pathogenic organisms, isolation can be achieved by such devices as airlocks and hermetically sealed doors. Where the risk is biological, isolation may be achieved in part by autoclaving and/or treating and disinfecting all garments, waste, and other items leaving the facility. Personnel may be required to wear self-contained, air-supplied suits while inside the facility or, in extreme cases, conduct all operations inside glove boxes or enclosures using mechanical and electrical manipulating devices. Exhaust air from such a facility may require passing through a flame to kill any active organisms. Where the risk is of this character rather than representing a danger due to fire or explosion, it may be possible to accommodate the facility within a building of generally lower risk level.

It will be noted that the four sections following are similar to those for the substantial risk facility. However, there are some significant differences.

Standard Practices

1. Access to the laboratory is limited to authorized personnel only, except at times and under such conditions as designated by written rules established by the laboratory supervisor and when accompanied by an authorized individual. The doors shall be locked at all times, with a formal key (or equivalent) control program in place.

2. All chemicals must be stored properly, according to compatibility. All chemicals which pose a special hazard or risk shall be limited to the minimum quantities needed for the short-term need of the research program, and materials not in actual use shall be stored under appropriate safe conditions. For example, flammables not in use shall be kept in a flammable material storage cabinet, or excess quantities of explosives shall be stored in magazines, away from the immediate facility. Other materials such as drugs or radioactive materials may also require secured storage areas.

3. An annual (or continuous) chemical inventory will be performed and sent to a central data collection point, preferably based on a centralized chemical computer management program. Outdated and obsolete chemicals will be disposed of through a centrally managed chemical waste disposal program. Ethers and other materials which degrade to unstable compounds shall be shelf dated for disposal 6 months after being opened (unless a material specific earlier shelf limit is indicated), but no more than 12 months after purchase, even if unopened, unless processed to remove any unstable peroxides that may have formed.

4. A Material Safety Data Sheet file will be maintained for all chemicals purchased for use in the laboratory. This requirement may be met by computer access to a centrally managed MSDS data base. In some cases, such as experimental compounds being tested, they are not available. Any information provided by the manufacturer will be kept in such cases. In some instances, such as experimental compounds being tested, these data may not be available. Where equivalent data exist in whole or in part, this information will be made part of the MSDS file. The supplementary MSDS file will be accessible to the employees in the laboratory at all times. All laboratory workers shall be trained in how to interpret the information in an MSDS.

5. All secondary containers containing materials having more than 1% of a hazardous component or combination of hazardous components (0.1% for carcinogens), which will be used more than a single work day, shall be labeled with a label listing the hazardous components.

6. All hazardous wastes are placed in appropriate and properly identified containers for disposal through a hazardous waste disposal program. Any wastes which pose a special hazard, or fall under special regulations and require special handling (such as human blood, tissue, and other bodily fluids regulated under the blood-borne pathogens standard), shall be isolated and a program developed to dispose of them safely and legally. Normal, nontoxic waste shall be disposed of according to standard practices appropriate to such wastes, subject to any restrictions needed to prevent breaching any isolation procedures.

7. The laboratory will be maintained in an orderly fashion. Any spills or accidents will be promptly cleaned up and the affected area decontaminated or rendered safe, by safety personnel if a major spill or by laboratory personnel if a minor one. Major spills will be reported to the Safety Department.

8. No food or drink can be brought into the operational areas of the laboratory, nor can anyone smoke or apply cosmetics.

9. Any required signage or posting mandated by any regulatory agency shall be posted on the outside of the door to the entrance to the laboratory. In addition, a placard or other warning device shall be placed on the door or on the wall immediately adjacent to the door identifying any other major classes of hazards in the laboratory (see Chapter 2, Section C.c). A sign meeting any regulatory standards shall be placed on the door stating in prominent letters, "AUTHORIZED ADMISSION ONLY."

10. The telephone numbers of the laboratory supervisor, any alternates, and the department head shall be posted on the outside of the laboratory door or the adjacent wall.

Special Practices

1. Specific policies, depending upon the nature of the hazard, shall be made part of the OSHA-mandated laboratory safety plan and scrupulously followed to minimize the risk to laboratory personnel, the general public, and the environment. Several examples of laboratory practices for various hazards are given below. This list is not intended to be comprehensive, but instead represents some of the more likely special precautions needed for a variety of types of risks.

   - All work with hazardous kinds or quantities of materials shall be performed in a fume hood or biological safety hood, specifically designed to provide the maximum safety for the hazard involved or in a totally enclosed system. It may be desirable for the hood or enclosed system to be equipped with a permanent internal fire suppression system. If the work involves a material which could be hazardous to the public or to the environment if released, an appropriate filtration system may be provided on the exhaust duct to the hood. If so, then a pressure sensor to measure the pressure drop across the filter would be required to ensure that the filter would be replaced as needed as the static pressure offered increases.

   - Work with explosives shall be limited to the minimum quantities needed. For small quantities used in a hood, an explosion barrier in the hood, with personnel wearing protective eye wear, face masks, and hand protection, may be sufficient protection. Note that most hoods are not designed to provide primary explosion protection. For larger quantities, the facility must be specifically designed for the research program. It is strongly recommended that a formal hazard analysis be completed, following guidelines such as those given in NFPA 49, Appendix C, if explosives are a major factor in designating the facility as a high-risk facility. During periods of maximum risk, occupancy of the facility shall be limited to essential personnel.

- Some gases, such as fluorine, burn with an invisible flame. Apparatus for work with such materials should be placed behind a barrier to protect against an inadvertent introduction of a hand or other part of the body, so as to prevent burns.
- Systems containing toxic gases that would be immediately dangerous to life and health or gases that could pose explosive or health hazard ratings of 3 or 4 (lesser ratings if they provide no physiological warning) if allowed to escape shall be leak tested prior to use and after any maintenance or modification which could affect the integrity of the system. Where feasible, the gas cylinders shall be placed external to the facility and the gases piped into the laboratory to help minimize the quantity of gas available to an incident. As few cylinders as feasible shall be maintained within a given facility, preferably three or less. Permanently installed gas sensors, capable of detecting levels of gas well below the danger limits, may be needed in some cases, such as when escaping gas provides no physiological warning signal.
- Vacuum systems, capable of imploding and resulting in substantial quantities of glass shrapnel or flying debris, shall be protected with cages or barriers, or for smaller systems, shall be wrapped in tape.
- Systems representing other physical hazards, such as high voltage, radiation, intense laser light beams, high pressure, etc., shall be marked with appropriate signs and interlocked so as to prevent inadvertent injuries. The interlocks shall be designed to be fail safe such that no one failure of a component would render the safety interlock system inoperative.

2. Activities in which the attention of the worker is not normally engaged with laboratory operations, such as record maintenance, calculations, discussions, study, relaxation, etc., shall not be performed in the laboratory proper but shall be performed in an area isolated from the active work area. The segregated desk area of the standard laboratory module is specifically intended to serve this purpose. Depending upon the nature of the hazard, it is usually economically feasible to make at least a portion of the upper half of the barrier separating the two sections of the laboratory transparent so that operations can be viewed if necessary.

3. Workers in the laboratory should participate in a medical surveillance program if they actively use materials for a significant portion of their work week which would pose a significant short- or long-term risk to their health. Employees shall be provided medical examinations if they work with any material, such as regulated carcinogens, requiring participation in a medical program by OSHA or another regulatory agency under conditions which do not qualify for an exemption. Employees shall notify the laboratory supervisor as soon as possible of any illness that might be attributable to their work environment. Records shall be maintained of any such incident as defined by the OSHA requirements for maintenance of health records.

4. No safety feature or interlock of any equipment in the facility shall be disabled without written approval of the laboratory supervisor. Any operations which depend upon the continuing function of a critical piece of safety equipment, such as a fume hood, shall be discontinued should the equipment need to be temporarily removed from service for maintenance. Any such item of equipment out of service shall be clearly identified with a signed "Out of Service" tag. Only the person originally signing the tag or a specific, designated alternate shall be authorized to remove the tag.

5. It shall be mandatory to wear any personal safety equipment required for conducting operations safely in the laboratory.

6. It is recommended that a laboratory safety committee review each new experiment planned for such a facility to determine if the experiment can be carried out safely in the facility. If the risk is such that experiments may affect the environment, or the

surrounding community, it is recommended that the committee include at least one layperson from the community, not affiliated directly or indirectly with the institution or corporation. In this context, "new"is defined as being substantially different in character, scope, or scale from any experiment previously approved for the facility.

Special Safety Equipment
1. Any refrigerators or freezers shall be rated as acceptable for "Flammable Material Storage,"i.e., be certified as explosion safe, except for ultra-low temperature units.
2. A flammable material storage cabinet, either built-in or free standing, shall be used for the storage of flammable materials. Any other special storage requirements, such as for locked storage cabinets or safes for drugs or radioactive materials, shall be available and used.
3. If the nature of the research program requires it, the laboratory shall be equipped with a fume hood. The fume hood shall meet any specific safety requirements mandated by the nature of the research program. A discussion of hood design parameters will be found in a later section, but for high hazard use, the interior of the hood and the exhaust duct should be chosen for maximum resistance to the reagents used; the fan should preferably be explosion-proof or, as a minimum, be equipped with nonsparking fan blades; the hood shall be equipped with a velocity sensor and alarm; the interior lights shall be explosion-proof, and all electrical outlets and controls shall be external to the unit. It may be desirable to equip the unit with an internal automatic fire suppression system.
4. The laboratory shall be equipped with an eyewash station and a deluge shower.
5. The laboratory shall be equipped with a fire alarm system connected so as to sound throughout the building (and in a central facility manned 24 hours per day) and an appropriate fire suppression system and be provided with one or more class 12 BC, or larger, fire extinguishers, or class D units if reactive metals are in use.
6. An emergency lighting system shall be provided.
7. A first-aid kit shall be provided and maintained.
8. Any special equipment mandated by the research program shall be provided. For example, electrical equipment other than refrigerators may need to be designed to be explosion-safe.
9. Any special equipment needed to maintain the required isolation for materials in the laboratory shall be provided. Examples are specially labeled waste containers, autoclaves, other decontamination equipment, or disposable clothing.

Laboratory Facilities
1. The floor of the laboratory is designed to be easily cleaned. Durable, seamless floors of materials that are substantially impervious to spilled reagents are easily de-contaminated, and have curved junctures to walls, aid in accomplishing this. The walls are to be similarly painted with a tough, substantially impervious paint (such as epoxy) to facilitate cleaning and decontamination.
2. Two well-separated exit doors shall be available to the laboratory which shall swing in the direction of exit travel.
3. Bench tops should be resistant to the effects of acids, bases, solvents, and moderate heat, and should not absorb water. To facilitate cleaning, the tops should have few seams or crevices. Although not necessarily subjected to the same level of abuse, other surfaces of the furniture should be readily cleaned or decontaminated.
4. Casework should be designed to be sturdy and designed for convenient utilization and modification. Storage spaces should be designed to meet any special requirements and should be easily accessible. It should not be necessary, for example, to stretch to reach any reagent which, if dropped, could represent a safety problem.

5. Aisle spaces should be 40 to 48 inches wide and shall not be constricted to less than 28 inches by any temporary obstacles. The aisles shall lead as directly as possible toward a means of egress.

6. The organization of the facility shall be such as to reduce the likelihood of having to pass an originating or secondary hazard to evacuate the facility in the event of an emergency.

7. Electrical outlets shall be three-wire outlets with high-quality, low-resistance ground connections. Circuits should be clearly identified to correlate with labels in breaker panels. If the nature of the hazard generates potentially explosive or ignitable aerosols, vapors, dusts, or gases, the electrical wiring, lights, and electrical switches shall be explosion-proof. Where connections and switches are near water sources, the circuits should be equipped with ground-fault interrupters (GFIs).

8. Laboratories in which the risk of electrical shock is greater than normal may also be equipped with a master "panic," manually operated electrical disconnect, clearly marked and located in a readily accessible location.

9. The laboratory shall be supplied with a sink. The trap shall be of corrosion-resistant materials. The plumbing shall be sized to accommodate the deluge shower and eyewash station. With average water pressure, this would normally be a 1-inch line or larger.

10. Ten to twelve air changes per hour of 100% fresh air shall be supplied to the facility. Some animal laboratory facilities are designed for 20 air changes per hour. No air shall be recirculated. The ventilation system shall be designed so that the room air balance is maintained at a small, negative pressure with respect to the corridors, whether the fume hood is on or off.* Where toxic and explosive gases and fumes are present, the system is to be designed to be efficient in exhausting these fumes by locating the exhaust intakes either near the source of fumes or near the floor (except for lighter-than-air or hot gases). Typical air flow patterns should draw dangerous fumes away from the normal breathing zones of the laboratory's occupants.

11. The facility shall include a separation of work spaces and desk areas as well as a second exit, equivalent to the arrangement shown in the standard laboratory module, Figure 3.1 (see Chapter 3, Section 3.A) unless the risk is so pronounced as to require complete separation of operational and nonoperational areas.

12. Some high-risk facilities require air locks, changing rooms equipped with showers with "clean" and "dirty" sides, or special equipment to decontaminate materials entering or leaving the facility. The doors to the air locks should be separated by at least 7 feet to prevent both doors from being open simultaneously

D. Access

Much of the present chapter has been spent on details directly concerning the laboratory itself. However, a laboratory is rarely an isolated structure, but is almost always a unit in a larger structure. It often appears that the typical laboratory manager or employee is insufficiently aware of this. If it is necessary to dispose of some equipment, it is often simply placed outside in the hall where it is no longer of concern. The thought that it may reduce the corridor width to well below the required minimum width also probably does not arise. A door swinging into the hall in such a way that it may block the flow of traffic appears similarly unimportant if it preserves some additional floor or wall space within the laboratory. The use of the corridor as a source of make-up air often seems reasonable, yet the possibility of this permitting a fire or toxic fumes to spread from one laboratory to another or to other parts of a building is clear once it has been considered. The natural inclination for

---

* This does not necessarily apply to some biological laboratories or "clean rooms" where a positive pressure is maintained to reduce the likelihood of contamination of the room by external contaminants.

most research personnel is to concentrate one's thoughts on the operations within a laboratory since this is where virtually everything important to them takes place. The ideas presented in the previous sections relating to optimizing safety within the facility are quickly grasped and accepted by most laboratory personnel, but the importance of extending these same concepts beyond the confines of their own laboratory frequently appears to be more difficult to communicate. However, due to the inherent risks in laboratory facilities, it is critical that sufficient, safe means of egress are always available. Except for scale and specific code requirements, most of the principles used in the laboratory to allow safe evacuation extend readily to an entire building.

1. Exitways

An exitway consists of all components of the means of egress leading from the occupied area to the outside of the structure or to a legal place of refuge. The Americans with Disabilities Act (ADA) requirements specifically call for places of refuge as part of new construction where disabled persons can await assistance in an emergency. Included as exitway components are the doors, door hardware, corridors, stairs, ramps, lobbies, and the exit discharge area. The function of the exitway is to provide a rapid, protected way of travel to a final exit from the facility to a street or open area. Elevators are not acceptable as a required means of egress. It is critical that this protected exitway not introduce components that would hamper the free movement of persons using it. Therefore, it should be ample so that overcrowding not occur, contain few obstructions, or unexpected changes in elevation or irregularities, be as direct as possible to the outside, and lead to an outside area sufficiently large and remote from the building so that evacuation to this area would be safe. Building codes are designed to meet these criteria. Remember, that the following sections are only intended to provide an understanding of the intent of the building codes, and the actual application of the codes to a facility must be done by professionals.

a. Required Exits

Any required exitway is required to be maintained available at all times, unless alternate means are approved in advance by a building official,* which will provide equivalent protection. This is probably one of the most common code violations. An extremely serious violation was personally observed by the author while attending a *safety* conference at a major university which provided degrees in safety management. The meetings were held within a large, multistory building containing meeting rooms, dining facilities serving up to 300 persons, and offices. The facility had all but one small, poorly marked, out-of-the-way exit blocked. This condition existed for a period of several weeks during a renovation project, during which full operations continued in the building.

Another common violation of the same type is chaining of exits for protection against theft during low-usage hours. It is common, however, for research buildings to be partially occupied at almost any time. All exits may not be required during periods of low activity but enough legal exits must be available to serve the occupants. It is essential that occupants know which exits remain usable, if some which are normally available are blocked during certain hours. Most of these problems arise because the persons making the decisions to eliminate or reduce the size or number of exits are not personally knowledgeable of the legal requirements and fail to check with those who do.

If sufficient legal exits cannot be maintained during renovations or at other times, the occupancy load must be reduced or perhaps sections of the building, served by the needed exits, should be closed temporarily. At a minimum, each floor of every building with an

---

* A building official in this context is a person or agency specifically authorized to administer and enforce the building code applicable to the building, not a person in charge of a building or facility.

occupancy load up to 500 persons must have at least two legal exits; between 500 and 1000, three exits; and above 1000, at least four exits.

b. Exit Capacity

In designing the needed exits for a facility, it is necessary to consider: (a) the number of occupants in the building; (b) the number that *could be* in the building, if the maximum density of occupants allowed by the building code were present; or (c) the latter number, plus any persons who might have to pass through the building from another space to reach an exit. The exits must be sufficient to accommodate the largest of these three numbers. The maximum floor area allowed per occupant under a typical building code is (space occupied by permanent fixtures is not counted) 100 gross sq. ft. for a laboratory building that does not meet the criteria for a high hazard facility.

The exit capacity from an area must be sufficient for the number of occupants of the space involved. Let us assume that a three story laboratory building can have up to a maximum of 1,000 occupants. If the building does not have a full fire suppression system, the total exit capacity for the stairways leading to at least three exits would be 25 feet. The corridors, doors from the corridors and ramps would have to total almost 17 feet. If the building were to be protected by a full fire suppression system, these could be reduced to just under 17 feet and 12.5 feet, respectively.

The minimum width shall be at least 44 inches for occupant loads greater than 50, or 36 inches for occupant loads of 50 or less.

c. Travel Distance

The characteristics of the routes of egress to an exit are also important, especially in as critical a facility as a laboratory building. Care should be taken, just as within the laboratory, for the distances to be as short and direct as practicable. The location of hazardous areas should be chosen to eliminate or minimize the probability of the direction of travel on a primary or secondary evacuation route being toward a likely hazard during an emergency. The normally allowed maximum travel distances for the business occupancy we have been discussing is 200 feet if the building is not equipped with a fire suppression system and 250 feet if it is so equipped.

As with the standard laboratory module, when a building requires more than one exit, which would almost always be the case for a laboratory research facility, these exits should be as remote as practical from each other. In a facility not served by a fire suppression system, the separations are to be at least half of the maximum diagonal distance of the area served. If there is an approved fire suppression system, this distance can be reduced to one fourth of the diagonal distance. They must also be arranged so that access is available from more than one direction from the area served so that it is unlikely that access from both directions will be blocked in an emergency.

It is acceptable to use an adjacent room or space as a means of egress from a room, as indicated in the standard laboratory module, if the room that provides the path of egress to an exit is not a higher hazard than the original space, and is not subject to locking. Thus, the laboratory modules should be arranged in blocks of comparable level of risk, if this concept is used to provide a second exit from each laboratory.

d. Corridors

In the introduction to this section two points were used as illustration, the first one maintaining the corridors free of obstruction and the second concerning the undesirability of using halls as a source of make-up air. Both of these points are intended to ensure that the corridors remain available for evacuation. A door may not swing into a hall such that it reduces the width of the corridor to less than half the legally required width, nor can the door, when fully open, protrude into the hall more than 7 inches. An obvious implication of the first provision, considering that most laboratory doors are 36 inches wide or wider, when combined with one half of the minimum legal width of 44 inches for a corridor (except for buildings occupied by 50 persons or less), would mean a minimum actual permissible width of 58 inches, unless the doors are recessed into alcoves in the connecting rooms.

Other unnecessary obstacles should be avoided as well, such as low-hanging signs, water fountains, desks, chairs, tables, etc., and similar devices which may protrude into a corridor, or even safety devices such as deluge showers with low hanging chains which could strike a person in the face in a partially dimmed or darkened corridor. The corridors must have a minimum of 80 inches of headroom. Door closers and stops cannot reduce this to less than 78 inches. Between the heights of 27 inches and 80 inches, objects cannot protrude into the corridor by more than 4 inches, approximately the length of a door knob.

If the corridors were to serve as a plenum for return air, they could spread smoke and toxic fumes from the original source to other areas. Further, instead of being a protected exitway, they themselves would represent a danger. In many fires, the majority of those persons that fail to survive often are individuals trapped in smoke-filled corridors and stairs. Laboratories, in general, need to be kept at negative pressure with respect to the corridors, but the 200 or 300 cfm recommended as permissible to enter through an open door, needed to maintain a negative air pressure, normally will not violate the prohibition on the corridors as a plenum. Space above a false ceiling in a corridor can be used as a plenum, if it can be justified for the corridor to not be of rated construction (unlikely for a laboratory building) or if the plenum is separated by fire resistance-rated construction. The use of spaces above false ceilings as plenums should be discouraged for other reasons, however. In recent work involving HVAC systems contaminated with microbiological contaminants, such spaces with slow-moving air have been shown to provide a favorable environment for such organisms to grow. This situation can lead to serious problems for those allergic to biological pollutants.

If a corridor serves as an exit access in the building occupancies which are being considered here, the corridor walls must be of at least 1-hour fire resistance rating. Care must be taken to construct corridor walls that are continuous to the ceiling separation to ensure this rating. Cases have been observed in which the wall was not taken above a suspended ceiling or continued into open service alcoves. Corridor floors should have slip-resistant surfaces.

The eventual point of exit discharge must be to a public way or a courtyard or other open space leading to a public way which is of sufficient width and depth to safely accommodate all of the occupants. On occasion, during renovation or construction projects, the areas outside the exits may not be maintained in such a way as to satisfy this condition. Such situations should be corrected promptly upon discovery. It may be possible to obtain variances from the building official to provide temporary passageways through the affected area.

A fairly common error that tends to creep into older buildings as renovations take place is the creation of dead end corridors. Frequently for other design reasons, more corridors are built in originally than are actually needed. Later, as space becomes tighter or the space needs to be reconfigured, the corridors are modified in order to recoup this "wasted" space, and dead end corridors are created. These dead end corridors cannot be longer than 20 feet under most circumstances. If the corridor is of sufficient width, some of the dead end corridors can be converted into offices or other uses, as long as they conform to code requirements for the class of occupancy.

Where there would be an abrupt change in level across a corridor (or across an exit or exit discharge) of less than 12 inches, so that a stair would not be appropriate, a ramp is required to prevent persons from stumbling or tripping at the discontinuity. The ADA requirements would also mandate a ramp for at least a sufficient part of the width to accommodate disabled persons in wheel chairs or using crutches. Clearly, in such a case it would be desirable to have the ramp extend the full width of the passageway.

e. Stairs

Stairwells are also exitway components that are frequently abused. Stairs are a means of egress providing a protected way of exiting a building. In order to provide additional ventilation or to avoid having to continually open doors, a very common practice is to use wedges of various types to permanently prop doors open. The result is to not only void the protection afforded by the required fire-rated enclosure, but to create a chimney through which fire and smoke on lower floors may rise, changing the stairway from a safety device to a potential deathtrap and providing a means for problems on lower floors to spread to higher levels.

In order to provide a protected means of egress, a required interior stair must be enclosed within a fire separation meeting the fire resistance ratings given in Table 3.3. The stair enclosure cannot be used for any other purpose, such as storage underneath the stairs, or within any enclosed space under a required stair. Any doors leading into the stair enclosure must be exit doors. This precludes creating closets underneath stairs for storage. The width of the stairs and of landings at the head, foot, and intermediate levels must meet the minimum dimensions established by the calculated required exit capacity. All doors leading onto a landing must swing in the direction of egress travel. The restrictions on reduction of width of the landings due to doors opening are the same as for corridors.

Stairways which continue beyond the floor level leading to an exit discharge onto a basement level are common. In an emergency situation, unless the stairs are interrupted at this floor, it would be possible that persons evacuating the building would continue downward, even though there is an additional requirement that each floor level be provided with a sign indicating the number of the floor above the discharge floor, for stairways more than three floors high. The persons continuing downward might be sufficiently confused as to reenter the building on the lower level before their mistake was recognized, or in an even more serious situation cause congestion at the lower end, making access to the exit difficult or impossible for those from the lower floor. The floor level sign should be about 5 feet above the floor and readily visible whether the door is open or closed.

People are accustomed to standard stair treads and risers, and this is especially important for an exitway to be used as an evacuation route. The treads and risers in laboratory buildings shall be a minimum of 11 inches for a tread and a maximum of 7 inches and a minimum of 4 inches for a riser. The maximum variation in the actual widths for a tread or riser are to be no more than ±3/16 of an inch for adjacent steps and 3/8 inch for the maximum variation. This seems a trivial point at first glance, but the importance of it should be clear to anyone who has ever stumbled over uneven ground in the dark. As one goes up or down a flight of stairs, one quickly grows accustomed to the step configuration, and a substantial unexpected change could easily lead a person in a hurry to stumble.

A similar rationale exists for the continuance of a handrail beyond the ends of a stairway as a provision for ensuring that the person traveling the stair has something to grasp to help avoid a fall, if they cannot see that the stairs have ended. Both at the top and bottom, the handrail should turn to be parallel to the floor for at least 1 foot (plus a tread width at the bottom). There must be a handrail on both sides of a stair, and intermediate handrails must be provided so that no point over the required width is more than 30 inches from a handrail.

f. Doors

Doors are perhaps the most abused exitway component. The fire separation they are intended to provide is often defeated when they are wedged open (note the same comment for entrances to stairwells above) in order to improve the ventilation or to eliminate the inconvenience of having to open them every time the passageway is used, especially if it is frequently used for moving supplies and equipment. In some instances, doors required to be shut since they represent openings into a fire resistance-rated corridor are left open simply because individuals wish to leave their office doors open to be easily accessible to persons wishing to see them, as openness and accessibility are viewed as desirable behavioral traits. The hinge assemblies of doors are often damaged when they are prevented from closing by the use of the now-ubiquitous soft drink can forced into the hinge opening. Even maintenance departments may not be aware that a required fire resistance rating is achieved by the entire door assembly, including the frame and hardware, not just by the door itself, so that a repaired door may no longer meet code specifications.

Conversely, doors required to be operable may be blocked or rendered inoperable for a variety of reasons, one of the most common being to increase security. Compact, easily portable, and salable instrumentation, especially computers and computer accessories, represents a tempting target for theft. As a result, doors which should be readily operable are fitted with unacceptable hardware to provide additional security, in many cases by the occupants themselves. In other cases, doors are blocked simply because individuals are careless and do not consider the consequences of their actions, such as locating a piece of equipment in such a way that a rarely used door cannot swing open properly.

Doors which are required exits must be prominently indicated as such, while doors which do not form part of a legal exitway should not have signs designating them as exits, although they may have signs indicating that they provide an additional means of egress, although the quality of the passages beyond may not be sufficient to meet the requirements for a protected exitway.

For laboratory buildings of the type being discussed here, the minimum width of a door used as an exit must be at least 32 inches (most common doors are 36 inches), and the maximum width of a single leaf of a side-hinged, swinging door must be no more than 48 inches (except for certain storage spaces). If a door is divided into sections by a vertical divider, the minimum and maximum widths apply to each section. A normally unoccupied storage space of up to 800 square feet can have a door of up to 10 feet in width. The minimum height of a door is 80 inches. If two doors are to be placed in series, as might be the case where a separation of a facility from a corridor is required to be maintained, such as the airlock discussed earlier in this chapter for a high risk facility, the doors must be separated by a minimum of 7 feet.

In general, it is recommended that all doors for laboratory structures should be of the side-hinged, swinging type, opening in the direction of exit travel. For doors opening onto stairways and for an occupant load of 50 or more, or for a high hazard occupancy, doors with these characteristics are required.

It must be possible to open a door coming from the normal direction of egress without using a key. Draw bolts, hooks, bars, or similar devices cannot be used. An essential element of a door is that it cannot be too difficult to open. The opening force for most interior doors must not exceed a force of 5 pounds. To open a door that is normally power assisted must not require more than 50 pounds with the power off. Panic hardware must require no more than 15 pounds force to release, and a door not normally provided with power assistance cannot require more than 30 pounds force to initiate motion and swing to a full-open position with application of a 15-pound force. These restrictions on the force required to operate a door can easily be exceeded should the ventilation system be modified without taking this concern into account. A very moderate atmospheric pressure differential of just over 0.3 inches (water gauge) would result in a force of more than 30 pounds force on a door of the minimum acceptable size. Addition of hoods to a laboratory, without provision of additional makeup air, could easily cause this limit to be exceeded on a more representative 3 foot by 7 foot door.

Doors opening from rooms onto corridors and into stairways and forming part of a required fire resistance-rated assembly must be rated. Most doors, such as those from offices opening on a corridor, are required to have at least a 20-minute fire rating, while doors leading from rooms of 2-hour fire resistant construction, as determined from Table 3.3, must be at least 1.5-hour fire doors, as should those entering stairways. Wired glass, one quarter inch thick, specifically labeled for such use, may be used in vision panels in 1.5-hour fire doors, provided that the dimensions do not exceed 33 inches high and 10 inches wide, with a total area of no more than 100 square inches. If the potential injuries and damage resulting from dropping chemicals as a result of being struck by a swinging door are considered, there is clearly merit in taking advantage of the provision for vision panels in laboratory, corridor, and stair doors.

Doors opening onto fire resistant-rated corridors and stairways must be self-closing or close automatically in the event of a fire. The first of these requirements usually includes offices opening directly off corridors, and is the case alluded to earlier as representing one of the most commonly violated fire regulations. For whatever reason, most individuals usually prefer to work with their office door open. Unfortunately, in an emergency evacuation, many do not remember to close their doors, and so the integrity of the fire separation is breached at these points.

g. Exit Signs, Lights, Emergency Power

The need for emergency lighting within laboratories has already been discussed independently of code issues. However, the need for lighting of exitways and identification of exits in emergencies is as critical outside the laboratory proper as it is inside. It is essential in a laboratory building that evacuation not be hindered by lack of lighting, especially in multistory buildings where stairways and corridors typically do not provide for natural lighting.

Internally illuminated exit signs are a key component of an evacuation system. In every room or space served by more than one exit, as is recommended for most laboratory rooms,* all the required means of egress must be marked with a sign with red letters on a contrasting background at least 6 inches high, with a minimum width of 3/4 of an inch for each segment making up the letters. The light intensity at the surface of other than self-luminous signs must be at least 5 foot candles.

There are self-illuminated signs, containing radioactive tritium (an isotope of hydrogen with a half-life of 12.33 years) that are acceptable, both under usual fire codes and to the NRC. The radiation from tritium is exceedingly weak (18.6 KeV beta) and, since these signs are completely sealed, no radiation can be detected from them. The transparent enclosure completely absorbs the radiation. As long as they remain sealed, they represent no hazard. However, in order to provide the required level of illumination, the amount of radioactivity in each sign is substantial. If they were to be broken, in an accident or in a fire, an individual handling them could, if the unit were broken, inhale a quantity of activity substantially in excess of the permissible amount. Therefore, as a precautionary policy, radiation safety committees in a number of organizations have taken the position that these signs are not permissible at their facilities. It might be well to consider the risk versus benefit whenever the use of such units is contemplated.

Note that normal glow-in-the-dark signs do not contain radioactive material. They depend upon phosphorescence, a completely different physical phenomenon, and usually remain sufficiently visible for 1 to 4 hours after activation by exposure to light.

In addition to signs at the exit, it may be necessary to put up supplemental signs to assist in guiding persons to an exit, where the distance is substantial or the corridor curves or bends. If a sign incorporates an arrow, it should be difficult to modify the direction of the arrow. However, when a sign is damaged, maintenance personnel have been known to

---

* While two exits are recommended for most laboratories in order to provide the maximum degree of safety, neither building code provisions nor OSHA regulations require two exits unless the laboratory represents a high hazard area or is occupied by more than 50 people.

inadvertently install a replacement sign with the arrow pointing in an incorrect direction. It is well not to take anything for granted. Users of the building should verify that all exit signs indicate the correct direction to the intended final exit point. A program of continuing inspection of all fire related safety devices, such as these signs, should be in place in every building.

Exit signs and means of egress must be lighted whenever a building is occupied, even if the normal source of power fails. The level of illumination at the floor level must be at least 1 foot candle. There are a number of methods in which power can be provided to emergency lighting circuits. They all must provide sufficient power to the lights and paths of egress to meet the required lighting standards for at least 1 hour, so that the building occupants will have ample time to evacuate.

For relatively small facilities, battery-powered lights, continuously connected to a charging source and which automatically come on when the power fails, are often used. Units are available which have extended useful lifetimes of 10 years or more. As with any other standby device, it is necessary to test them on a definite schedule. These battery-powered units are relatively inexpensive, currently ranging in price (in quantity) from about $40 to $150. These are especially useful in individual rooms such as a laboratory and are an inexpensive way to retrofit older facilities with emergency lighting.

There are battery-powered units designated as "uninterruptible power supplies" which switch over within milliseconds. These are often used to maintain power to computers or electronic equipment where a loss of power can cause data to be lost. Such units can be sized to also support emergency electrical lighting.

Standby generators are another alternative to provide energy to the required emergency lights and other equipment which may need to be supported during a power failure. A generator would be preferred over a battery system for other than small buildings to provide the fairly substantial amount of power needed. However, these generators should be checked once a month under load to ensure that they will come on within the required 10 seconds for emergency lighting and within 60 seconds for other loads. Architects often fail to consider all circumstances in designing such systems. In one case, the architect designed an excellent system, but for economic reasons, the exhaust of the generator was located immediately adjacent to the building air intake, on the premise that in an emergency the ventilation system would shut down. However, because diesel exhaust fumes were drawn into the working ventilation system during tests, this effectively prevented the scheduled operational tests from being performed until the problem was corrected. Failure to provide proper maintenance and tests can lead to embarrassing and costly incidents when outside power fails and stairways and corridors are not lighted. Academic institutions are more vulnerable than corporate facilities since, in a given building, there are more likely to be a higher percentage of individuals that are relatively unfamiliar with the evacuation routes.

A last option, but one which must be used with considerable caution, is to provide outside power from two completely separate utility power feeds. Such an arrangement can be approved by code authorities if it can be shown that it is highly unlikely that a single failure can disrupt both sources of power. For example, if the local distribution system is fed by several alternative power lines and has alternate local lines to provide power to a building, it is conceivable that local building officials would approve the system, but one must remember that entire states, and even larger regions, have suffered total power losses in recent years. Among other occurrences which could lead to such a failure are natural disasters such as hurricanes, tornadoes, blizzards, ice storms, fires, and floods. Few, if any, localities are immune to all of these. In the author's area wide-spread, lengthy power outages have occurred frequently in recent years due to ice and wind storms.

h. Other Exitway Issues

A number of other topics, related to exitways, have not been touched upon here for the same reason that has been given before. This handbook is not intended to be comprehensive. The intention is to cover those topics most meaningful to a person working in a laboratory building with enough detail so that reasonable persons can evaluate their facilities to ensure that their safety is not reduced by renovations, or the actions of individuals during normal usage. The reader should also be able to follow the reasoning for many of the architectural decisions made during the planning of a facility and should be able to actively participate in the planning process. However, provisions and specifications for components such as exterior stairs, fire escapes, access to roofs, connecting floors, vestibules, and lobbies, which are all relevant topics under the general subject "means of egress," would be important to architects, but probably less so to laboratory personnel.

E. Construction and Interior Finish

The discussion of laboratory facilities has been limited to the building occupancy class in order to avoid having to go into all the parameters which would be needed if this discussion were to be extended to high hazard, or educational classes, the next most likely possibilities. However, laboratory facilities do represent a degree of risk greater than many other uses which would also be considered appropriate to the same classification so extra care is needed to ensure that construction practices and materials used in the interior finish do not add to the risks or defeat the intended level of protection.

In addition to fire protection there are other potential hazards which may also be reduced by construction details and choices of materials. In Section C of this chapter, under the topic "Laboratory Facilities" for each class of laboratories, many of the features stipulated characteristics of finish materials. As a general principle, laboratory floor coverings, wall finishes, and table and bench tops should be durable, easy to clean, and resistant to the common reagents.

A normal vinyl tile floor meets many of these requirements, but the seams around each tile form cracks in which materials such as mercury and other materials can lodge. As an example, mercury can remain *in situ* in these cracks for extended periods and create a substantial mercury vapor pressure when, ostensibly, all spilled mercury may have been "cleaned up." Radioactive materials and biological agents can similarly be trapped and pose a continuing, persistent problem unless very thorough cleaning is performed on a regular basis. One argument frequently used in favor of tile floors has been that damaged or contaminated tiles can be easily and cheaply replaced. However, common vinyl asbestos tile is now included within the category of asbestos materials by the EPA and must be removed according to the procedures for removing asbestos. Replacing a tile floor, in which either the tile or the mastic affixing it to the sub-floor contains asbestos fibers is expensive. Note that it is more likely that 9-inch tiles will contain asbestos than the larger, 12-inch tiles. Newly manufactured tiles do not contain asbestos, but unfortunately, the mastic often still does.

Any material used must meet required standards for fire spread and smoke generation, in addition to having the other properties for which it is selected. When a material has been selected that has both the desired properties and has the requisite ratings, the construction contract should contain peremptory language stipulating that NO substitutes for the specified materials will be acceptable without specific approval. In too many instances, where vague language such as, "equivalent materials may be substituted," is incorporated into a contract, substitutes have been introduced which do not meet the original specifications, possibly inno-cently, because the supervisor on site did not realize the difference. It usually is not possible to simply look at an item and determine what its properties will be from appearances alone.

Although not particularly attractive, a plain, sealed concrete floor or one painted with a durable paint probably is the best for most laboratories, while wood and carpeting would be the worst. Many different kinds of floor finishes are available, designed to prevent slipping, generation of sparks, and resistance to corrosives or solvents. Simple concrete block walls are often used for interior partitions in buildings or, alternatively drywall on steel studs. Both of these are relatively cheap for original construction and can be modified easily as well. The surface of concrete blocks is relatively porous, which can pose decontamination problems, but it can be painted to eliminate this problem. Paints for interior surfaces are available which will provide waterproofing, resistance to corrosives and solvents, and enhance fire resistance. Where biological cleanliness is an important criteria, there are paints approved by the EPA which will inhibit the growth of biological organisms. Incidentally, the sand used in concrete blocks in many parts of the country is a source of radon, a concern to many persons.

1. Construction Practices

The intent of the fire code as it applies to the interior finish and to acceptable construction practices is to prevent spread of a fire from one fire area to another, i.e., to make sure that the fire walls and other fire separation assemblies are constructed in such a way and of such materials as to maintain the fire resistance rating of the structure. When architects prepare the plans and specifications, they must include documentation for all required fire resistance ratings.

It is not possible to provide total separation of fire areas and still provide for air intake, exhaust, or return air plenums, unless these are themselves separated from the surrounding spaces by required fire-resistant shaft and wall enclosures, plus properly engineered, and labeled fire dampers meeting UL 555 specifications must be installed where ducts pass through fire separations. Unless the architect, a representative of the building owner, or a contractor's inspector provides careful supervision of the workers during construction, dampers may be installed improperly, perhaps at the end of a convenient piece of duct work, even if this happens to be in the middle of the room, far away from the fire separation wall. As noted earlier, it is possible to use the space between the ceiling of a corridor and the floor above if the space is properly separated. However, ceiling spaces used for this purpose cannot have fuel, fixed equipment, or combustible material in them. The requirement for a fire damper does not apply to ducts used for exhausting toxic fumes, as from a fume hood, since in a fire it is often desirable for purposes of protecting both the normal building occupants and firefighting personnel for toxic materials to continue to be exhausted. This requires that ducts carrying toxic fumes be continuously enclosed within a shaft of the proper rating (normally 2-hour) to the point of exhaust from the building.

It is essential that the integrity of the fire separation walls not be significantly diminished by penetrations or modifications. For example, walls less than 8 inches thick are not to be cut into after they are constructed in order to set in cabinets or chases. Among the most common violations of the integrity of fire separations are penetrations in order to run utilities and, today, cable chases for electronic services such as video signals, data cables, and computer lines. Often, these penetrations are roughly done, leaving substantial, unfilled gaps surrounding the cables, conduits, or piping. Even in new construction, especially if the penetrations are in difficult to inspect or otherwise awkward locations, the gaps around the ducts, pipes, and conduit are frequently left incompletely or poorly filled. Where retrofitting of spaces to accommodate such devices is done by maintenance and construction personnel, this deficiency is even more likely to occur if the work is inadequately supervised, since the average worker may not be aware of this requirement. Whenever unfilled spaces are found, the gaps must be filled by materials meeting fire resistance standards. If a renovation or new construction involves setting in a structural member into a hollow wall, the space around the member must be filled in for the complete thickness of the wall with approved fire stopping material.

Openings can exist in a fire wall, or else how could doors and windows exist? However, there are limits on the size of the openings —120 square feet (but no more than 25% of the length of the fire wall), except in buildings with an approved automatic sprinkler system. Larger openings (240 square feet) can exist on the first floor of a building, again with an approved automatic sprinkler system. The openings must be protected with an appropriately rated assembly, which may be a fire door. If the wall is of a 3-hour rating, the rating of the door must be 3 hours as well. For walls of 1.5- or 2-hour ratings, the doors must be 1.5 hours. Around shaft and exit enclosure walls with a fire rating of 1 hour, the door assembly must be 1 hour, also. For other fire separations with a required separation of 1 hour, the fire door need only be a 3/4-hour rated door. Unless the interior space is rated, doors to rooms such as offices opening onto a one hour corridor need only be rated at 20 minutes.

Fire walls shall extend completely from one rated assembly to another, such as the floor to the ceiling, extending beyond any false or dropped ceiling which may have been added. The joint must be tight.

This section has been primarily concerned with the interiors of a building, but measures are required to prevent a fire from spreading due to the exterior design of a building as well. The exterior walls must be rated to withstand the effects of fires within the building. Windows arranged vertically above each other in buildings of three or more stories for business, hazardous, or storage uses shall be separated by appropriate assemblies of at least 30 inches in height from the top of a lower window to the bottom of the one above. Although there are a number of exceptions, if the exterior wall is required to have a fire rating of one hour or more, then a parapet of 30 inches or greater in height above the roof is required for nonexempt structures.

2. Interior Finish

Materials used for interior trim or finishing must meet standards for flame spread and smoke or toxic fume generation. Materials are rated in accordance to how well they perform on tests made according to the ASTM E84 procedure, with lower numbers corresponding to the better materials. Class I materials have a rating between 0 and 25; class II, 26 to 75; and class III, 76 to 200. As far as smoke generation is concerned, materials used for interior finish must not exceed a rating of 450 as tested according to the provisions of ASTM E84. Based on these ratings, the interior finish requirements for the categories of interest are class I for vertical exit and passageways, exit access corridors and class II for rooms and enclosed spaces.

As usual, there are numerous exceptions based on special circumstances, for the current discussion, the most notable being: if there is an automatic fire suppression system, the minimum requirement for interior finish is class II. The propensity for materials to burn may be different depending upon the physical configuration. For example, a match placed on a piece of carpeting lying on the floor may smolder and go out, while a match applied to the bottom of the same piece of carpeting, mounted vertically may result in a vigorous fire. Most common floor coverings employed in laboratories, such as wood, vinyl, or terrazzo, are exempt from being rated.

Where interior finish materials are regulated, they must be applied in such a way that they are not likely to come loose when exposed to temperatures of 200^0F for up to 30 minutes. The materials must be applied directly to the surfaces of rated structural elements or to furring strips. If either the height or breadth of the resulting assembly is greater than 10 feet, the spaces between the furring strips must be fire stopped. Class II and III finish materials, less than 1/4 inch thick, must be applied directly against a noncombustible backing, treated with suitable fire-retardant material, or have been tested with the material suspended from the noncombustible backing. This seems to be a fairly minor restriction, but most of the inexpensive paneling available today from builder supply houses is either 3/16 inch or 4 mm thick. Many organizations have departments which have their own technicians that often do the departmental remodeling as an economy move and build improper partitions of this noncomplying material. Rated paneling is available, 1/4 thick or more, which looks exactly the same on the surface. The only realistic options available to prevent violations of the code requirements is to totally prohibit purchases of building material, strictly enforce policies of no "home-built" structures, to the extent of tearing down such constructions, or to provide a source of rated material which must be used.

Roofing materials are not interior finishing materials, but also must meet standards in order to maintain adherence to classes of construction. Class I roofing materials are effective against a severe fire exposure and can be used on any type of construction. Class 2 materials are effective against moderate fire exposures, and Class 3 materials are effective only against light fire exposures. Typical materials meeting Class I requirements would be cement, slate, or similar materials, while metal sheeting or shingles would meet Class 2. Class 3 materials would be those that had been classified as such after testing by an approved testing agency.

REFERENCES

1.   Occupational Exposure to Hazardous Chemicals in Laboratories, 29 CFR 1910.1450.
2.   Means of Egress, 29 CFR 1910, Subpart E.
3.   Hazardous Materials, 29 CFR, 1910 Subpart H, § 106.
4.   Personal Protective Devices, 29 CFR, 1910 Subpart I, § 132-139.
5.   Medical and First Aid, 29 CFR, 1910 Subpart K, § 151
6.   Bloodborne Pathogens, 29 CFR, 1910 .1030.
7.   Standard on Fire Protection for Laboratories Using Chemicals, NFPA 45, 1989.
8.   **Ashbrook, P.C. and Renfrew, M.M.,** *Safe Laboratories, Principles and Practices for Design and Remodeling,* Lewis Publishers, Chelsea, MI, 1991.
9.   The Southern Building Code, Southern Building Code International, Birmingham , AL, 1997.
10.   The BOCA Basic National Building Code/1993, 12th ed., Building Officials and Code Administrators International, Country Club Hills, IL, 1993.
11.   Standard Test for Surface Burning Characteristics of Building Materials, ASTM E84-98e1, American Society for Testing of Materials, West Conshohocken, PA, 1998.

F. Ventilation

Few research buildings at either corporate or academic institutions are constructed today without central air handling systems providing heating, cooling, and fresh air. Experience seems to indicate that relatively few of these are designed completely properly to provide suitably tempered air where it is needed and in the proper amounts, at all times. High energy costs mandate that the energy expended in heating or cooling the air supplied to a facility be optimally minimized. Laboratory buildings, however, have highly erratic needs for tempered air. In academic buildings, for example, when both faculty and students cease working in the laboratory to meet classes or attend to other responsibilities, fume hoods, which typically exhaust around 1000 cfm per minute, may or may not be individually off. In a medium-sized research building containing 50 hoods, the required capacity for makeup air could theoretically vary as much as 50,000 cfm. The occupants rarely conform to a sensible daytime work regimen. In academic institutions especially, individuals are almost as likely to be working at 4:00 a.m. as at 4:00 p.m., or while the majority may be taking a Christmas vacation, there are always a few continuing work on a project that cannot be interrupted. Under such circumstances, it is very difficult to continuously provide the right amount of air all the time to every laboratory economically. Economy is the easiest parameter to forego since engineering technology is capable, at least technically, of maintaining proper ventilation under almost any circumstance, even though it may be expensive to do so. Further, the health and safety of individuals should never be compromised for economic reasons.

Most written material on laboratory ventilation concentrates almost exclusively upon fume hoods. Ventilation does play an important part in the proper performance of hoods, and they, in turn, usually have the most significant impact of any piece of laboratory equipment on the design and performance of laboratory building air handling systems. However, there are many other aspects to laboratory ventilation. Hoods will be treated as a separate topic in Section 3.2.2, and some aspects of ventilation will be deferred to that section. Those portions of hood performance which involve the general topic of space ventilation will be covered in the following material.

Active laboratory areas should be provided with 100% fresh air. No air should be re-circulated. There are laboratories for which this would not necessarily be essential, but as noted earlier, the character of research conducted in a given space may change. Ventilation, which depends upon supply and exhaust plenums to the space being built into the building structure, is one of the more expensive services to provide as a retrofit. It is better to design for the most demanding requirements and use controls to modify the supply if the actual needs are less. If the active laboratory space can be adequately isolated from administrative, classroom, and service areas, the requirements for these other spaces may be met with a recirculating system, where the portion of fresh air introduced into the total air supply could be as little as 10%. For reasons associated with building air quality in non-laboratory buildings, it is often desirable to recirculate a large fraction of a building's air, as long as sufficient fresh air is provided to accommodate the basic needs of the occupants.

The amount of fresh air to be provided to a laboratory space should depend upon the activities within the facility, but there are little data to support a given amount. Epidemiological data, gathered by OSHA, indicates that there are health risks associated with working in a laboratory. In five studies cited by OSHA involving chemical workers, although the overall mortality rate appeared to be lower among chemists than in the general population, there was some evidence that indicated additional dangers from lymphomas and leukemia, development of tumors, malignancies of the colon, cerebrovascular disease, and prostate cancer, although virtually every study indicated a lesser rate of lung cancer. The general good health might be attributable to the generally high economic and educational status of the groups being studied, which probably translates into more interest in their health and being in a position to afford to maintain it. The general consensus that it is not a good idea to smoke in laboratories could impact on the number of observed cases of lung cancer. A survey among the members of the California Association of Cytotechnologists, investigated the use of xylene in the laboratories in which they worked; of the 70 who responded, 59% felt their ventilation was inadequate, 22.6% worked where there was no exhaust system, and 43% stated that their ventilation systems had never been inspected. In several recent health hazard evaluations conducted by the National Institutes for Occupational Safety (NIOSH), it was found that ineffective exhaust ventilation was a major contributor to the hazardous conditions. If proper procedures are employed and all operations calling for the use of a fume hood are actually performed within a hood, the general room ventilation would be expected to have relatively little bearing on the health of laboratory workers. However, sufficient hood space is not always available and, even where it is available, is not always used. Consequently, general laboratory ventilation should be sufficient to provide good quality air to the occupants.

In the absence of specific requirements, there are guidelines. *Prudent Practices for Handling Hazardous Chemicals in Laboratories* recommends between 4 to 12 air changes per hour. Guidelines for animal care facilities recommend between 10 to 20. Storage facilities used for flammables are required by OSHA to have at least six air changes per hour and would appear to be a baseline minimum level. If the air in the room is thoroughly mixed, six air changes per hour would result in more than 98.4% of the original air being exchanged. Increasing this to seven would result in a less than 0.8% further gain at the expense of a further increase of 14.3% in the loss of tempered air. A critical consideration is whether, in fact, the air does become thoroughly mixed. This depends upon a great many factors,

Figure 3.4   The air entering the room, while meeting the quantitative requirements for the amount of fresh air, does not in fact provide sufficient fresh air to the occupants.

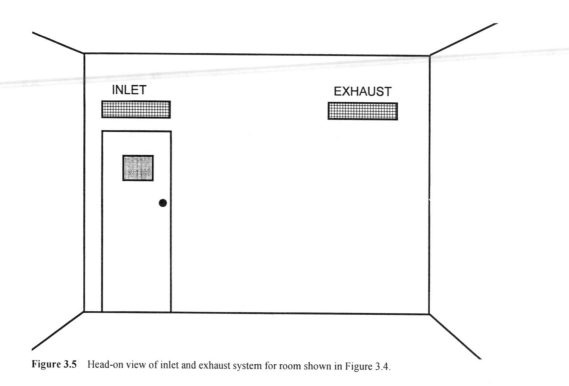

Figure 3.5   Head-on view of inlet and exhaust system for room shown in Figure 3.4.

including the location of the room air intakes and exhaust outlets, the distribution of equipment and furniture, and the number, distribution, and mobility of persons in the room. At any given time, any gases or vapors present in the air will eventually diffuse and attain a fairly uniform mix, even in an unoccupied space. Substantial amounts of movement in the room will tend to redistribute air within a room more quickly, but there will still be spaces and pockets in almost any room in which, because of the configuration of the furniture and the air circulation, mixing of the air will be slow. Because thorough mixing cannot be assured, recent trends have been to specify higher exchange rates, typically 10 to 12 air changes per hour. The American Society of Heating, Refrigerating and Air Conditioning Engineers (ASHRAE) in standard 62-189 recommends 20 cfm per person of fresh air in laboratories, or 1200 cubic feet per hour. If four persons worked in the standard module, this would require approximately 5000 ft³. A single 5-foot fume hood would exhaust this amount of air in about 5 minutes, so the make-up air for the hood would supply an ample amount to meet this criterion.

Poor design of the air intake and exhaust system can have a significant negative effect on the needed air exchange. In Figure 3.4, the results of tests of a particularly bad system are depicted. In this facility, fresh air is delivered from a unit ventilator, mounted on a roof above a corridor and then ducted through the laboratory wall at a height of about 9 feet. Along the same wall, about 12 feet away, is an exhaust duct leading back to the roof (see Figure 3.5). Air is blown into the room horizontally toward the opposite wall. It then, supposedly, traverses the room twice and leaves through the exhaust duct or a hood. Smoke bomb tests of this system, however, showed almost no vertical mixing of air in the room. Half an hour after the smoke was released, a clear line of demarcation about 8 feet above the floor between clear air and smoky air could still be discerned, the latter being partially replenished by exhaust air that had been recaptured and reentered the building. The occupants in this facility benefitted virtually not at all from the air being introduced through the standard air intake, nor did it serve the fume hoods in the room. In order for sufficient air to be provided to the hoods, additional air had to be drawn in from either the doors or, when weather permitted, through open windows. Using the corridors as a source of air sufficient to supply even one hood violates code restrictions. Further, using open windows often results in an erratic air supply due to wind gusts, and in cold weather is clearly impractical.

Ideally, air entering the laboratory should enter gently and in such a way that the air in the breathing zones of the individuals working in the laboratory is maintained free of toxic materials and that the air flow into hoods in the room will not be interrupted or disturbed by the intake air flow. Studies indicate that air directed toward the face of a hood or horizontally across its face will cause the most serious problems in meeting this latter condition, while air introduced through a diffuse area in the ceiling or from louvered inlets along the same wall on which the hood is situated will be affected the least by movement of personnel. However, recent studies conducted by the National Institutes for Health, Office of Research Services, Division of Engineering Services, in cooperation with a firm, Flomerics Limited, using Computational Fluid Dynamics software showed that the location of the hoods within the facility and with respect to the air diffuser, strongly affects the success of the hood in retaining fumes. The study also showed the effects of the supply ventilation on the air patterns within the facility. At the time of this writing, the report has just been made available on the Internet (available at Internet address http://www.des.od.nih.gov/farhad2/pdf/vol2 4of4.pdf). The study confirmed that the best location for a hood was in the back corner of a laboratory rather than along one of the walls. If more than one hood is used, it would be best if they were on perpendicular walls, at least two or more feet apart. The diffuser air flow should be small. A bulkhead for the hood would be desirable, reaching all the way to the ceiling. Surprisingly enough, it was helpful if the diffuser were in line with the center of the hood and close to the bulkhead, unless the facility allowed placing the diffuser a substantial distance from the hood.

The majority of laboratory fumes and vapors are heavier than air and will preferentially drift toward the floor, although some will diffuse throughout the room air and some will be carried upward by warm air currents. Room exhausts should be located so as to efficiently pick up the fumes. Placing exhausts in the ceiling, or high up on walls is not efficient and, as in the case described previously in this section, can serve to "short-circuit" the supply of fresh air to the room. Even if high air exhaust outlets were effective, they would tend to pull

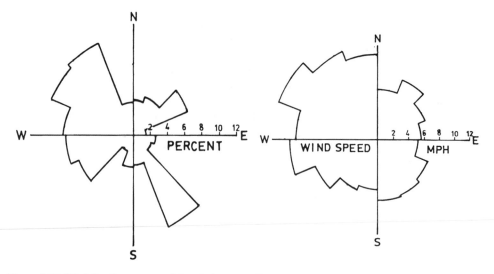

Figure 3.6A Wind direction, percent of time during year. **Figure 3.6B** Average wind velocity vs direction.

noxious fumes through the occupants' breathing zones. Exhausts placed near the floor or at the rear of workbenches would prove more effective, as long as they remain unobstructed, and the direction of air movement from a source would be away from the occupant's face. Localized exhausts, using local pickups exhausting through flexible hoses, can be used to remove fumes from well-defined sources of fumes, but they must be placed close to the source. The air movement toward the nozzle is reduced to less than 10% of the original value within a distance equal to the nozzle diameter. Outside this distance, it is unlikely that a localized exhaust would be very effective in removing fumes. If all work with hazardous materials were to be done in hoods, and the hoods ran continuously, it would be possible to rely on hoods to provide the exhaust ventilation to a room. However, this is normally not the case. Sometimes hood sashes are closed and the hoods used to store chemicals. On other occasions, hoods are turned off while apparatus is installed, or they are off while being serviced. Therefore, the design of the air exhaust system from a laboratory must be done carefully to provide continuing replacement of fresh air in the room. The fume hood system and the supplementary exhaust system should be interlocked to ensure a stable room air balance at all times. This balance can be at a lower level of fresh air delivery if the room is unoccupied. There are advanced computer control systems which do a very effective job of maintaining appropriate ventilation in laboratories automatically.

If there are administrative, classroom, or service areas within the same building as laboratories, the entire laboratory area should be at a modest negative pressure with respect to these spaces so that any air flow that exists will be from the non-research areas into the space occupied by laboratories.

It is important that the source of air for a building be as clean as possible, and that the chances for exhaust air to reenter the building be minimized. In most locations, there are preferred wind directions. In Figure 3.6, directional and velocity wind data are shown, averaged over a year for a typical building site. Such data can be obtained for a region from airports and weather bureaus. However, wind data are strongly affected by local terrain, other nearby buildings, trees, and other local variables (note the anomalously high percentage of time the wind comes from a sharply defined southeastern direction here). Where reliable data are available or can be obtained, the air intakes should be located upwind as much as possible with respect to the building. At this site, locating the air intakes at the northwest corner of the

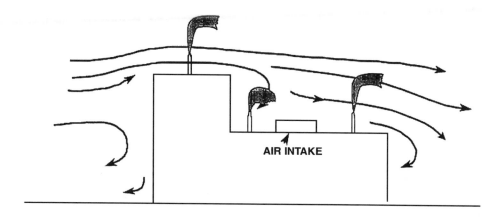

Figure 3.7 Effect of building shape on wind flow in its vicinity.

building would clearly be desirable, both because this is the predominant wind direction and the higher wind velocities from this direction would assist minimizing recapture of fumes. Building exhausts, again from the data, should be to the south of the air intakes. Obviously, the prevailing wind for the greater portion of the year is from slightly North of West. The anomalous spike from the SE is due to channeling from a nearby building complex. As shown of the left, the wind speed, averaged over the year is greatest when the wind is from the NW quadrant. These data show that the intake for a chemistry building should be on the Northwestern side of the building. There will be periods when this configuration will lead to the exhaust fumes being blown toward the air intakes, but other measures can be taken to also aid in the reduction of recirculation.

One situation which must be avoided is the situation shown in Figure 3.7. Here, the air intake is located in a penthouse on the lee side of a raised portion of a building, and in the midst of several fume hood exhaust stacks. As shown in the figure, the air moving over the top of the obstruction tends to be trapped and circulate in an eddy on the downwind side of the obstruction. If fumes are swept into this volume, either from the roof above or in the space contiguous to the obstruction slightly further downwind, they will tend to remain there. For an air intake located in this space, the entrapped fumes are likely be drawn back into the building. Ideally, the air intake should be located on the prevailing upwind side of the building near the roof.

The exhausts from the building should discharge fumes outside the building "envelope," i.e., the air volume surrounding the building where air may be more readily recaptured. Physically, this can be done with tall individual hood exhaust ducts, or the exhausts from individual hoods can be brought to a common plenum and discharged through a common tall stack. The needed height of individual stacks often make the "tall stack" alternative a physically unattractive concept. Two "rules of thumb" are employed to estimate needed heights. For one- or two-story buildings, the stack height above the roof should be about 1.5 times the building height. For taller buildings, this "rule" would lead to very high stacks, so that a height equal to 0.5 times the buildings width is often used in such cases. We will return to the "common plenum" concept again later because there are a number of design details which are required to ensure that bringing individual exhausts to a common duct can be safely used. In the former case, if it is desired to have the air escape from the vicinity of a building, inverted weather caps above duct outlets clearly should not be used since they would direct the air back toward the building. Updraft exhaust ducts with no weather caps are preferable, in which the outlets narrow to form a nozzle, thereby increasing the exit velocity. Since the exhaust air has a substantial vertical velocity, it will initially continue to move upward, so that the effective height of the duct will be higher than the physical height. The gain in the effective height will depend upon a number of factors; *viz.,* the duct outlet diameter, d; the exit velocity of the gas, v; the mean wind speed, μ; the temperature difference between the exhaust gas and outside air temperatures, δT; and the absolute temperature of the gas, T. The effective height gain is given by:

$$\text{Height gain} = d[v/\mu]^{1/4}[1+ \delta T/T].$$

The following is an example of the results of using an updraft exhaust. For simplicity's sake, assume that the indoor and outdoor temperatures are the same. For a duct diameter of 8 inches, a nozzle velocity of 4900 fpm, and a mean wind speed of 700 fpm (equal to the annual average of approximately 8 mph of the site for which data were given in Figure 3.7), the height gain would be about 10 feet. Under some weather conditions, the plume would continue to rise and in others it would fall. In gusty winds, it could be blown back down upon the building roof. In any case, the effective height above the roof of about 13 feet (duct height plus height gain) would be helpful in reducing the amount recaptured by the building and obviously is far better than the alternative of using weather caps, in which the fumes are always directed down toward the roof.

An examination of the equation used to determine the height gain shows that if the exit velocity could be maintained, it would be advantageous to have a larger duct diameter. For example, if several hoods could be brought to a common final exhaust duct 2 feet in diameter so that the exit velocity remained the same, the net gain would be 30 feet instead of 10 feet, and it would be more acceptable to have a single tall chimney rather than a forest of exhaust stacks. With this arrangement, it would be possible to be reasonably certain that the fumes would not return to the level of the air intake until the plume left the vicinity of the building for a considerably larger portion of the time.

Although some concern is usually expressed about chemical reactions due to mixing the fumes from different hoods, generally the fumes from each hood are sufficiently diluted by the air through the faces of the hoods so that the reactions in the plenum will not be a significant problem on a short-term basis. There, perhaps, could be long-term cumulative effects. The most serious operational problem is maintaining the balance of the system as the number of hoods exhausting into the common plenum varies. If all the hoods ran continuously, this would not be a problem, but for energy conservation, as well as other reasons, this mode of operation is not the most desirable. There are certain conditions that must be met. Each contributing hood exhaust must be kept at a negative pressure with respect to the building as a whole, so that fumes would not leak into the building through a faulty exhaust duct. In order to ensure that no fumes from the common plenum are forced back into the laboratory, the plenum must always be at a negative pressure with respect to the indi-vidual ducts, so that the plenum must be serviced by a separate blower system. It would be difficult to meet both the balance and energy conservation requirements simultaneously with a single plenum exhaust motor. Multiple motors, which go on and off line automatically can compensate when the number of hoods which are actually on varies. Thus, a reasonably constant negative pressure differential, as determined by pressure sensors, between the plenum and the individual hoods is maintained. The negative pressure in the plenum would increase the effectiveness of the individual hood exhaust fans. However, some common plenum designs do not include individual fume hood fans. In such a case, the individual hoods would always be feeding into a lower pressure plenum, but the face velocity of the individual hoods would change as sashes were opened and closed throughout the system. The system would have to accommodate this variation.

A risk in a common plenum system with individual hood motors is that the motors serving the plenum might fail while the individual motors serving the individual hoods do not. In this event, the fumes in the common plenum would mix and the chances would be good that some fumes would be returned to the laboratories in which the hoods had been turned off. Since the hoods would be exhausting into a volume at a higher than normal pressure, the effectiveness of individual hood systems also would be diminished so that the probability of fumes spilling from the hoods would increase, even for those hoods which continued to operate. If multiple motors serve the common plenum, the problems would not be as serious if an individual motor failed, since the system should be designed to compensate until the motor was returned to service. However, if electrical power were to fail so that the entire plenum system were to go down, the potential would exist for serious problems within the laboratories. It is essential that such a system be provided with sufficient standby electrical power, as well as an alarm system, to permit the system to continue to serve all operations that cannot be temporarily terminated or reduced to a maintenance level. A standard close-down procedure for all the individual hoods should be developed to be implemented in such a situation.

1. Quality of Supplied Air

Quantity of air is important, but so is the quality. Humans and equipment work best within a fairly narrow range of temperature and humidity. The term "fresh air" implies that it is at least reasonably free of noxious fumes, but it says nothing of the temperature and humidity. In 1979, emergency building temperature regulations were imposed which required that the temperature set points be set at a minimum of 78^0F in the summer and a maximum of 68^0F in the winter in order to conserve energy. Although these temperatures were eminently satisfactory to some individuals, a large number were vociferously unhappy. Similarly, very low humidity in the laboratory is frequently encountered during the winter as outside air at low temperatures, containing very little moisture, is brought inside and heated, resulting in desert-like humidity characteristics. As a result, many people develop respiratory problems. During the summer, unless sufficient moisture is wrung from the hot outside air while it is being reduced to more comfortable temperatures, the interior humidity may rise to very high levels. Persons will feel clammy and uncomfortable, since they will be unable to perspire as readily. Moist, warm air is also conducive to microbial growth, to which many individuals are allergic in varying degrees. Under very humid conditions, workers are less likely to wear personal protective equipment, such as lab coats, chemical splash goggles, and protective gloves. Unless the temperature and humidity do stay within a relatively narrow range, people become less productive and make more errors. Laboratories are not work environments where error-producing conditions should be acceptable. Although individuals differ, a comfortable temperature to a large number of people seems to lie between $68°F$ and $75°F$, and a comfortable humidity, between 40% and 60%.

Because of all the potential negative results of having poor quality air, it is clearly desirable to have a properly designed and maintained system to make the air as conducive to comfort as possible. In order to assure this, the initial building contract should contain clauses defining explicitly the specifications for the temperature, humidity, and volume of air for each space within the building, and the contractor should be required to demonstrate that the building meets these specifications before the owner accepts the building. This is as important, from both a usage and health and safety standpoint, as any other part of the design.

A laboratory building needing 100% fresh air for space ventilation and large quantities of air discharged by each fume hood is an energy inefficient building almost by definition. Laboratory equipment in the building also is likely to be a very substantial heat load. Buildings housing laboratories, therefore, are logical candidates for energy recovery and energy management systems. In implementing an energy recovery system, care must be taken to ensure that the system is a true *energy* exchange system, where the incoming air characteristics are moderated by the air being exhausted from the building but no air is recirculated. It is difficult to manage an energy system within a building if the occupants have the capability of modifying it locally. Thus, in a managed facility, it is likely that any windows will be permanently sealed so that they cannot be opened and thus disturb the local air balance. The use of ceiling spaces as return plenums is less desirable than the use of fixed ducts, since the former permits an individual to modify the air circulation in his space by simply making an opening into the ceiling. Modifications and connections even to hard ducts need to be done by a qualified HVAC engineer, not by a local shop crew, so that the air provided remains within needed design limits for the spaces served by the system.

A building designed to meet all the requirements of a well-designed and managed building in terms of air quality is usually a "tight building" with few chances for air to leak into and out of the building. Experience has shown that such a building may lead to the "tight building syndrome,"[*] where a significant fraction of the populace of an entire building appears prone to developing environmentally related illnesses, sometimes suddenly and acutely. Such problems may, on occasion, be triggered in a building by an unfamiliar odor, by the overall air quality moving out of the comfort zone, by an individual suddenly and unexplainably becoming ill, or for no apparent reason. Laboratory buildings, with their common and frequently unpleasant, pervasive chemical odors, could be vulnerable to this problem, although occupants of such buildings may be accustomed to "strange" odors and be more willing to accept odors which are not acceptable to other personnel. Where the building occupants have little or no means of modifying their environment, i.e., they cannot open their windows, the frustration of having no control over the problem seems to exacerbate the likelihood of the perceived problem developing and worsening the impact when it does. Of course, a bad odor does not always trigger a tight building syndrome response from a building's occupants, but the inability to personally do something about it seems, at the very least, to increase a person's irascibility. Often, even after prolonged investigation, no real cause of a tight building syndrome incident is ever discovered, and it is attributed to stress or other psychological causes, especially when the problem seems to disappear when no corrective measures were taken.

Some tight building syndrome incidents represent real health problems, with individuals persistently complaining of discomfort and showing evidence of physical distress. Most commonly the symptoms are respiratory distress, headaches, fatigue, dizziness, nausea, skin and eye irritation complaints regarding odors or a chemical "taste" in their mouths. These problems are similar to allergenic reactions, and if some individuals exhibit the problems while others do not, this is typical of the widely varying sensitivity of individuals to allergens. If no other source of the problems can be located, an evaluation of the HVAC system is in order. There is a tendency by many to ascribe the blame initially to low-level volatile compounds, but microbial contaminants may actually be the real cause. Relatively few systems are as well maintained as they should be, and even fewer are cleaned routinely. Chemistry and biological laboratory buildings are notorious for their odors in any event which also would cause one to look elsewhere than low level volatile compounds for causes of health complaints. This is not to say that the prevailing odors may not cause problems to individuals. They may do so. Some individuals may be sensitive when exposed to several different compounds simultaneously, although the levels of each individual component may be far below any acceptable exposure levels. This type of problem has been given the name of multiple exposure sensitivity. Some have refused to accept this as a problem but others feel that these synergistic reactions do occur.

Another area which should be evaluated in the case of health complaints are external to the facility factors. One source of complaints in a facility were sporadic complaints which did not

---

[*] The term "tight building syndrome" is not synonymous with the term "sick building syndrome." NIOSH estimates that about 20% of a buildings occupants are adversely affected when a building is "sick." The term "tight building syndrome" perhaps should be applied more aptly to buildings where conditions give rise to a perception of a problem. Both of these terms are used too loosely, in the author's opinion. Conditions may exist where, in a space as small as a single room, an individual may become ill due to environmental conditions. To this individual, the area is "sick."

appear to be due to any activities in the building. Eventually the problem was traced to an air intake placed just above a loading dock. Trucks would back up to the building and leave their motors running while making deliveries. The exhaust fumes would enter the building and would be distributed throughout the interior.

All HVAC systems include filtration systems, and some provide humidification and dehumidification functions. The filters may become dirty and the amount of fresh air may decrease to an insufficient level, or the filters may begin harboring dust mites or fungal growth. If the filters in different parts of the system become less able to pass air preferentially, the building air balance may shift so that areas which at one time had sufficient air may no longer do so. This alone can cause problems. The moisture used for humidification or the drain pans into which moisture from the dehumidifying process goes may become contaminated with biological organisms. Many individuals are allergic to dust, dust mites, fungi, and other microorganisms which would be distributed by a contaminated air supply. In severe cases, the air handling system could harbor *Legionella* organisms, which would require massive and disruptive decontamination efforts, if these were possible at all.

In order to avoid contamination of HVAC systems and to maintain a proper level of performance, a comprehensive and thorough maintenance program is required for all centralized HVAC systems. Filters should be changed or cleaned on a regular and frequent schedule. The efficiency and quality of the existing filters should be evaluated. Inefficient roll filters should be changed to a better quality, higher efficiency types. Many systems are initially equipped with filters which are not appreciably better than furnace filters used in the home, with efficiencies of as little as 25%. If the capacity of the fans can cope with higher filter efficiencies, these should be upgraded to 65% to 85%. Water employed for humidification and chillers, or cooling towers, should be checked frequently for biological growth. Condensate pans should be cleaned regularly. Decontamination of an afflicted ventilation system is difficult, time consuming, and consequently very costly. Preventive maintenance is very cost effective for ventilation systems. Fortunately laboratory areas are usually provided with 100% fresh air and do not experience the problems associated with recirculating systems, where reuse of up to 80% of the building air not only allows contaminants to build up, but allows contaminants in one area to be distributed to the entire building.

REFERENCES

1. **Li, E. et al.,** Cancer Mortality Among Chemists, in *J. Nat. Cancer Institute,* 43, 1159 -1164,1969.

2. **Olin, R.,** Leukemia and Hodgkin's disease among Swedish chemistry graduates, in *Lancet,* (ii) p. 916, 1976.

3. **Olin, R.,** The hazards of a chemical laboratory environment: a study of the mortality in two cohorts of Swedish chemists, in: *Amer md Hygiene Assn.* 1, 39, 557 -562, 1978.

4. **Olin, R. and Ahibom, A.,** The cancer mortality rate among Swedish chemists graduated during three decades, in: *Envir Res.,* 22, 154 -161, 1980.

5. **Hoar, S.K. and Pell, S.,** A retrospective cohort study of mortality and cancer incidence among chemists, *J. Occup. Med.,* 23. 485 - 495, 1981.

6. Comment by the California Association of Cytotechnologists re: Xylene exposure re: the OSHA Proposed Performance Standard for Laboratories Using Toxic Substances, 51 FR 26660, July 24, 1986.

7. U.S. Department of Health and Human Services, Centers for Disease Control, National Institute for Occupational Safety and Health, Health Evaluation Reports HETA 83-048-1347, HETA 830076-1414, HETA 81-422-1387, Cincinnati, OH.

8. *Prudent Practices for Handling Hazardous Chemicals in Laboratories.* National Academy Press, Washington, D.C., pp. 193 - 212, 1981.

9. *Prudent Practices* in the Laboratory, Handling and Disposal of Chemicals, National Academy Press, Washington, D.C., 1995.

10. The Guide for the Care and Use of Laboratory Animals, National Institute of Health Publication No. 85-23, 1985.

11. OSHA General Industry Standards, 29 CFR Part 1910, §106.(d)(4)(iv).

12. **Caplan, K.J. and Knutson, G.W.,** The effect of room air challenge on the efficiency of laboratory fume hoods, in *ASHRAE Trans.,* 83, Part 1, 1977.

13. **Fuller, EH. and Etchells, A.W.,** Safe operation with the 0.3m/s (60 fpm) laboratory hood, in *ASHRAE .J,* 49, October, 1979.

14. Laboratory Ventilation for Hazard Control, National Cancer Institute Cancer Research Safety Symposium, Fort Detrick, MD, 1976.

15. **Wilson, D.J.,** Effect of stack height and exit velocity on exhaust gas dilution, *ASHRAE Handbook,* American Society of Heating, Refrigerating and Air Conditioning Engineers, Atlanta, GA, 1978.

16. **Cember, H.,** *Introduction to Health Physics,* Pergamon Press, New York, 334 - 339, 1969.

17. Tight building syndrome: the risks and remedies, in *Amer Indust. Hyg. J.,* 47, 207 - 213.

18. ASHRAE 90-1980, *Energy Conservation in New Building Design,* 345 E. 47th St., New York, NY 10017.

19. ANSI Z9.2-1979, Fundamentals Governing the Design and Operation of Local Exhaust Systems, American National Standards Institute, New York, NY 10018.

20. *Industrial Ventilation: A Manual of Recommended Practice,* 20th ed., American Conference of   Governmental Industrial Hygienists, Cincinnati, OH, 1988.

21. **DiBerardinis, L.J., Baum, J., First, M.W., Gatwood, G.T., Groden, E., and Seth, A.K.,** *Guidelines for Laboratory Design: Health and Safety Considerations,* John Wiley & Sons, New York, 1987.

22. Biosafety in Microbiological and Biological Laboratories, U.S. Department of Health and Human Services, Centers for Disease Control, National Institute for Occupational Safety and Health, Publication No. (CDC)   93-8395, U.S. Government Printing Office, Washington, D.C., 1993.

23. ASHRAE 62-1989, *Ventilation for Acceptable Indoor Air Quality,* American Society of Heating, Refrigerating and Air Conditioning Engineers, Atlanta, GA, 1989.

24. **Burton, D. Jeff,** Laboratory Ventilation Workbook, 2nd Ed., IVE, Bountiful, UT, 1994.

25. **Memarzadeh, Farhad,** Methodolgy for optimization of laboratory fume hood containment- Volumes I & II.Ventilation design handbook on animal research facilities using static microisolators, Volumes I & II, National Institutes of Health. Office of Research Services, Division of Engineering Services, 1998.

26. Examining fume hood performance, Engineered Systems, Vol .13, No. 3, March 1997.

27. **Rock, J. C. and Anderson, S. A**., *Benefits of designing for ventilation diversity in a large industrial research laboratory – A case study.,* Appl. Occup. Environ. Hyg., 11 (10), Oct.1996.

G.   Electrical Systems

Electrical requirements for laboratories are relatively straightforward. The entire system must meet the NFPA Standard 70 (the National Electrical Code) and must be properly inspected before being put into service. As far as the laboratory worker is concerned, the details of the service to the building are relatively unimportant, but it is important to them that there are circuits of sufficient capacity to provide enough outlets for all of the equipment in the laboratory. In most new facilities, the designers usually provide enough, but in many older facilities, where less use of electrically operated equipment was anticipated, the number of outlets is often inadequate. Many of the older electrical systems were originally designed based on two wire circuits instead of the three wire circuits that are currently required by code. Any circuits of the older type should be replaced as soon as possible, and laboratory activities should not be assigned to spaces provided with such electrical service until this is done. All circuits, whether original equipment or added later, should consist of three wires, a hot or black wire, a neutral or white wire, and a ground or green wire. The ground connection should be a high-quality, low-impedance ground (on the order of a few ohms), and all grounds on all outlets should be of comparable quality. A poor-quality ground, perhaps due to a worker failing to tighten a screw firmly, can result in a substantial difference of potential between the grounds of two outlet receptacles. This can cause significant problems in modern solid state electronic equipment which typically operates at voltages of less than 24 volts (often at 3, 5, or 6 volts). A further problem which can result from a poor ground is that leakage current through the high-impedance ground connection can develop a significant amount of localized resistance heating. This is often the source of an electrical fire, rather than a short circuit or an overloaded circuit.

Electrical circuits should be checked using a suitable instrument capable of providing a quantitative measure of the ground impedance. Commonly available, inexpensive plug-in "circuit-checkers" which indicate the condition of a circuit by a combination of lights can give a valid indication of a faulty circuit, but a "good" reading can be erroneous. If the distance to another connection is substantial, capacitive coupling of the ground wire can result in a false indication of a low impedance reading.

The female outlets, as noted, have three connections with openings located at the points of an equilateral triangle. For the common 110 volt circuit for which most common consumer and lower wattage laboratory equipment is designed, the ground connection is round, while the other two openings are rectangular. If the two rectangular openings are of different sizes, the neutral connection will be longer and the hot connection shorter. A male plug has matching prongs, and should only fit in the outlet in a single orientation. A "cheater" or adaptor can be used to allow a three-prong male plug to be inserted into a two-wire circuit, but this is not desirable. Some types of portable equipment, such as a drill, are available with only two connectors, but these are required now to be "double-insulated" so the external case cannot provide a circuit connection to the user. All connectors, switches, and wiring in a circuit must be rated for the maximum voltage and current they may be expected to carry. It is recommended that the female sockets be protected with ground-fault interrupters as an additional measure of protection. These devices compare the current through the two-current carrying leads, and if they are different, as would be the case where current was being diverted through an electrical short or a person, they will break the circuit in a very few milliseconds, usually before any harm occurs.

There should be enough outlets appropriately distributed in a laboratory so that it should not be necessary to use multiple outlet adapters, plugged into a single socket, or to require the use of extension cords. Where it is necessary for additional circuits to be temporarily added, the circuits should either be run in conduits or in metal cable trays, both of which should be grounded and installed by qualified technicians or electricians. Even though their use should be discouraged, extension cords will continue to be used. However, as a minimum, they must be maintained in good condition, include three wires of sufficient size to avoid overheating (preferably of 14-gauge wire or better for most common uses. Better means a smaller gauge number.), and must be protected against damage. They should not be placed under stress, and should be protected against pinching, cutting, or being walked upon. Where abuse may occur, they must be protected with a physical shield sufficient to protect them from reasonably anticipated sources of damage.

Circuits must be protected by circuit breakers rated for the maximum current to be carried by the circuit. Normally, many breakers for a room or group of rooms are located together on a common breaker panel. All circuits should be identified, both within the facility and at the breaker panel, so that when required, the power supply to a given circuit may quickly and easily be disconnected. This is especially important when it is necessary to disconnect power due to an emergency. There should be no ambiguity about the breaker that needs to be thrown to kill the power to a given receptacle. Laboratories with high voltage and/or high current sources should consider a readily accessible master disconnect button, which anyone can use to kill all of the circuits in the facility if someone becomes connected to an active circuit.

Where breaker panels and electrical switches are placed in separate electrical closets or rooms, there is some question about the propriety of individual laboratory workers having access to the space. If there are electrically live parts with which individuals might accidentally come into contact, access to the spaces must only be by qualified, authorized individuals, according to code regulations. Normally, live components on breaker panels are completely covered (if not, then prompt action to replace the cover should be initiated). If, in addition, the breaker panels are segregated by a locked or otherwise secure barrier from areas containing electrically active components, then laboratory workers should be allowed to have access to the panels in order to control the circuits. However, access to these spaces must not be abused by considering them as extra ''storage space.'' Access to the electrical panels, switches, and other electrical equipment in the space must not be blocked by extraneous objects and materials.

The location of electrical circuits and electrically operated equipment in a room should be such that they are unlikely to become wet and they should not be in an area susceptible to condensation or where a user might be in contact with moisture. As unlikely as it may appear, instances have been observed where equipment has been located and electrical circuits have been installed where water from deluge showers would inundate them. For some equipment, such as refrigerators, freezers, dehumidifiers, and air conditioning units, moisture is likely to be present due to condensation, and these equipment items must be well grounded.

1. Hazardous Locations

Most laboratories do not represent hazardous locations in the context of requiring special electrical wiring and fixtures, although there may be individual equipment items which may need to be treated as such. The classification of a facility as hazardous in a regulatory context, depends upon the type of materials employed in the facility and whether flammable fumes or gases, electrically conducting materials, or explosive dusts are present in the air within these facilities in the normal course of routine activities or only sporadically due to some special circumstance. Explosion proof wiring and fixtures are substantially more expensive than ordinary equivalents and should be used only if there are no acceptable alternatives. Where the need does exist, however, they definitely should be used.

The National Electrical Code defines three different categories of hazardous locations: classes I, II, and III (note that these are not the same as the classification of flammable liquids into classes I, II, and III). Class I represents locations where flammable vapors or gases may be in the air. Class II locations involve facilities where electrically conducting or combustible dusts may be found, and class III locations contain ignitable fibers. Each class is split into two divisions, 1 and 2. In division 1 for each class, the hazardous conditions are present as a normal course of activities, or are sufficiently common due to frequent maintenance, or may be generated due to equipment failure, such as emission of dangerous vapors by the breakdown of electrical equipment. Division 2 includes locations which involve hazardous materials or processes similar to those in division 1, but under conditions where the hazardous gases, fumes, vapors, dusts, or fibers are normally contained, or the concentrations maintained at acceptably low levels by ventilation so that they are likely to be present only under abnormal conditions. Division 2 locations are also defined to include spaces adjacent to but normally isolated from division 1 locations from which problem materials might leak under unusual conditions.

Within class 1, there is a further division by groups into A, B, C, and D, depending upon the materials employed, with the distinction between the groups being based essentially on the flammable limits in air by volume. A long list of chemicals, with the groups identified to which they may belong, is given in NEPA 497M. Information on other chemicals should be available to any researcher in the Material Safety Data Sheets that laboratory employees should have readily available. However, in this listing, acetylene with a flammability range of

2.5% to 81% is the only chemical listed in group A. Group B chemicals with flammability ranges between about 4% to 75%, and with flash points less than 37.8°C or 100°F include acrolein, ethylene oxide, propylene oxide, hydrogen, and manufactured gas (>30% hydrogen by volume). However, if equipment in the facility is isolated by sealing all conduits 1/2 inch or larger in diameter, according to specifications in the National Electrical Code, the first three of these may be placed in groups of lesser risk. Among those of lesser risk are flammable liquids with flash points above 37.8°C or 100°F, but less than 60°C or 140°F. Allyl glycidal ether and n-butyl glycidal ether would also be in group B, but with the same exception as to conduit sealing.

Groups C and D have flammability ranges between about 2% and 30% and 1% and 17%, respectively. A few common chemicals falling into group C are acetaldehyde, carbon monoxide, diethyl ether, ethylene, methyl ether, nitromethane, tetrahydrofuran, and triethylamine. A partial list of common group D chemicals includes acetone, benzene, cyclohexane, ethanol, gasoline, methanol, methyl ethyl ketone, propylene, pyridine, styrene, toluene, and xylene.

Generally if the location uses chemicals in any of these four groups which have flash points less than 37.8°C or 100°F and otherwise meets the specifications of a class I, division 1 location, special electrical equipment would normally be required. Special electrical equipment would normally be required for class II flammables (flash points equal to or above 37.8°C or 100°F, but less than 60°C or 140°F) only if the materials are stored or handled above the flash points, while for class IIIA flammables (flash points equal to or above 60°C or 140°F, but less than 93.3°C or 200°F), special electrical equipment is needed only if there are spaces in which the temperature of the vapors may be above the flash points.

Group II includes conductive and combustible dusts, with some dusts falling into both categories. Dusts with resistivities above 10^5 ohm/cm are not considered conductive, while those with lesser resistivities are. If operations were such as to generate significant levels of dust in the ambient air in the work location, the decision as to whether special electrical equipment would be required would be based on whether a cloud of the dust in question would have an ignition sensitivity equal to or greater than 0.2 and an explosion severity equal to or greater than 0.5. Both of these are dimensionless parameters based on a comparison to a standard material, Pittsburgh seam coal.

The definitions of these two parameters are:

$$\text{Ignition sensitivity} = \frac{(P_{max} \times P)_2}{(P_{max} \times P)_1}$$

$$\text{Explosion severity} = \frac{(T_c \times E \times M_c)_1}{(T_c \times E \times M_c)_2}$$

where

P_{max}	= maximum explosive pressure
P	= maximum rate of pressure rise
T_c	= minimum ignition temperature
E	= minimum ignition energy
M_c	= minimum explosive concentration
Subscript 1	= standard dust
Subscript 2	= specimen dust

There are a number of metals and their commercial grades and alloys, which could give rise to the need for special electrical equipment, some of which are listed in NFPA 497. Metals, such as zirconium, thorium, and uranium, which would be found in some special laboratories, have both low ignition temperatures, around 20°C (68°F), and low ignition energies so that work with these materials would require special precautions and safeguards.

A large number of nonconductive dusts have an ignition sensitivity of 0.2 or higher and an explosion severity of 0.5 or greater. Many agricultural products, such as grains, can form dusts in this category. Other nonconducting materials with similar characteristics would be many carbonaceous materials, chemicals, dyes, pesticides, resins, and molding compounds. A long but incomplete list is given in NFPA 497M. OSHA lists many industrial operations which could be classified as class II, and in the context of laboratory safety there are pilot or bench scale research operations which would emulate these industrial locations.

Class III locations represent hazardous locations because of easily ignitable fibers. However, it is unlikely that they would be present in sufficient concentrations to produce an ignitable mixture. Thus, no special electrical wiring requirements would normally exist. However, research facilities which generate substantial airborne quantities of fibers of cotton, synthetic materials such as rayon, wood, or similar materials should take care to avoid the potential for fire.

The special electrical equipment or wiring procedures needed to satisfy the requirements of a hazardous location are specified in the National Electrical Code, NFPA Standard 70. In general, fixtures suitable for use in hazardous locations will be rated and certified as safe by a nationally recognized testing laboratory such as Underwriters Laboratory or Factory Mutual Engineering Corporation. In some laboratory installations, equipment items may be found which are nonstandard and are not listed as acceptable by any appropriate organization. In such cases, the equipment may be certified as safe if an appropriate agency charged with enforcing the provisions of the National Electrical Code finds the equipment in compliance with the Code so as to ensure the occupational safety of users. Where certification is obtained by this means, records need to be kept showing how this determination was made, from manufacturer's data or actual tests and evaluations. Special electrical equipment for use in class I locations is not necessarily gas-tight, but if an explosion does occur within it, it should be contained and quenched so that it does not propagate further.

REFERENCES

1. NFPA 70-96, National Electrical Code (latest version), National Fire Protection Association, Quincy, MA.
2. NFPA 497-1997, *Recommended Practices for Classification of Class I Hazardous Locations for Electrical Installations in Chemical Plants,* National Fire Protection Association, Quincy, MA.
3. *Prudent Practices for Handling Hazardous Chemicals in Laboratories,* National Academy Press, Washington, D.C., 1981, pp. 179-192.
4. *Prudent Practices in the Laboratory, Handling and Disposal of Chemicals,* National Academy Press, Washington, D.C., 1995.
5. The following ANSI Standards cover electrical apparatus for use in hazardous locations (by number only): 781-1992; 844-1991; 1002-1987. These are issued jointly with Underwriters Laboratories. Since these are updated on an irregular basis, the latest edition should be obtained when used as reference.

H. Plumbing

There are two aspects of laboratory plumbing system design, as opposed to operations, which are relevant to safety and health. The first is the capacity of the system to withstand the waste stream which the system may be called upon to handle. The second is the need to prevent the operations within the laboratory facility from feeding back into and contaminating the potable water system that supplies the building.

1. Sanitary System Materials

Under the standards governing the disposal of toxic and hazardous chemicals into the sanitary sewage system and into the public waters, much smaller quantities of chemicals should be going into sink drains than in the past. A very large number of the chemicals used in laboratories are now classified as hazardous waste and should be collected and disposed of according to the provisions of the Resource Conservation and Recovery Act. However, a substantial amount of acids, bases, solvents, and other chemicals still go into the sanitary system, even if from no other source than cleaning of laboratory glassware. Although the National Clean Water Act is not intended to cover small individual laboratories, or even groups of laboratories, a large research facility may add more chemicals into the sewer system than many small industries. However, there probably will be very little comparatively of any single one. Laboratory chemicals should be highly diluted in large quantities of water but, given time, can corrode ordinary plumbing systems.

Plumbing materials used to service laboratory facilities must be resistant to a large range of corrosives, physically durable, relatively easy to install and repair, and comparatively inexpensive. Metal plumbing materials are generally unacceptable because they are vulnerable to inorganic acids. Plastics, such as are used in residential systems will not withstand many common organic solvents or will absorb other solvents and not remain dimensionally stable. Glass would serve very well but is brittle and not inexpensive in large systems. Its desirable properties make it suitable and acceptably cost effective for the components most vulnerable to the actions of chemicals - the sink trap and the fittings which connect the sink to the sanitary system. A plastic resistant to a large range of inorganic and organic waste streams dimensionally stable, durable, and relatively inexpensive is poly-propylene. Other materials with comparable physical properties may be either more expensive or more difficult to install and maintain.

One characteristic not mentioned in the preceding paragraph is the necessity for the material to have established fire resistance ratings as specific structural members, i.e., tested by appropriate testing laboratories in the configurations in which they might be used. Polybutylene plumbing components are available which meet this requirement.

2. Back Flow Prevention

Many jurisdictions have specific legal requirements for devices preventing contamination of the potable water supply to be installed on each service line to a building's water system wherever the possibility exists that a health or pollution hazard to the waterworks system could exist. This will include any facility, such as a laboratory, where substances are handled in such a manner as to create a real or potential risk of contaminating the water supply external to the building. Although the regulations may only address the problems of cross-contamination between buildings, the risk may exist as easily within a building. For example, if by some means, contaminated water is drawn from a laboratory sink back into the potable water system, the contamination could easily express itself in the water available from the water fountain immediately outside in the hall, if the latter were to be fed from the same line supplying the faucets in the sink. The contamination would not necessarily only be close to the contaminating source, but could be anywhere in the system downstream from the origin of the pollution.

Illustrated in Figure 3.8 is a direct connection between a contaminated system and a clean system, connected by a valve which could be opened. Obviously, this is not a desirable situation, but it can occur. The second possibility is that a pressure differential between the

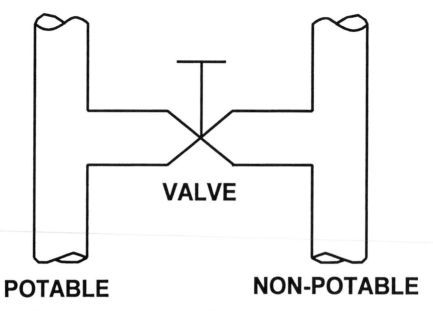

Figure 3.8 Simple cross-connection between liquid systems.

clean and contaminating systems may be established such that contamination may be forced or drawn into the clean system through some linkage. This linkage may not necessarily be a valve left open or a leaky valve, although these are two of the most likely sources. An example of an inadvertent cross-connection would be a heat exchanger, which has developed small leaks between the primary and secondary loops over a period of time. As long as the primary, or clean side, is at a higher pressure than the secondary, or contaminated side, any water flow would be from the clean to the contaminated side; as long as the leaks were small, they would probably go undiscovered. If, however, the water supply on the primary side were to become reduced and pressure were to fall below the secondary, the flow would be reversed and the clean water supply would be polluted. This would be an example of a forced back flow problem.

A very common situation existing in a laboratory is to find a section of plastic tubing draining the effluent from a piece of apparatus, lying in the bottom of a far-from-clean sink. Perhaps the sink is being used to wash dirty glassware and the plastic tubing is under several inches of water. If the water pressure should suddenly fail, perhaps due to a reduced supply because of maintenance, coupled with the simultaneous flushing of several toilets, it would be possible for the water in the sink to be siphoned back through the system and reach the potable water supply. Any system in which a connection to the potable water supply can be flooded with contaminated water would be subject to the same type of problem. Inexpensive vacuum breakers, as shown in Figure 3.9, to install on the sink faucets are available from most major laboratory supply firms.

The best correction to these problems is not always obvious. In the first example of the heat exchanger, a pressure sensor with redundancy features could be installed between the primary and secondary loops such that when the positive pressure differential falls below a stipulated value (well above the point at which the pressures are reversed), operations should be shut down and the secondary loop emptied. A better solution would be for the heat exchanger to incorporate an intermediate loop so that the secondary and primary sides could not be directly coupled.

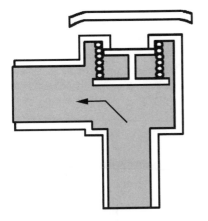 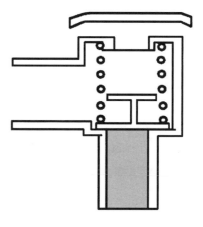

NORMAL FLOW   **ACTIVATED**

Figure 3.9 Vacuum breaker.

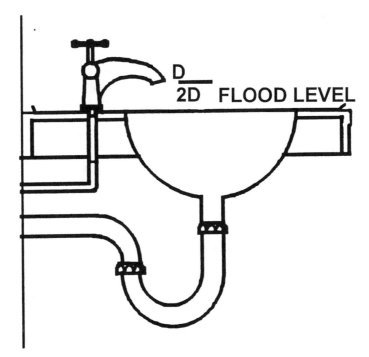

Figure 3.10 Minimum desirable air gap on sink.

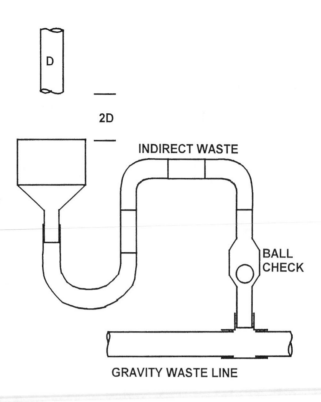

Figure 3.11 Air gap in line subject to back pressure.

In order to prevent back flow due to excess pressure, the vacuum breaker shown in Figure 3.9 cannot be used, and an air gap provides the most protection. Figure 3.10 on the preceding page and 3.11 above illustrate two versions of air gaps. The second of these is very straightforward: simply do not connect anything to the water source which could become flooded. Any formal back flow or anti-siphonage device which is installed must be an approved type which has been tested by a recognized laboratory testing agency and be of satisfactory materials.

REFERENCES

1. **Woodruff, P.H.,** Prevention of contamination of drinking water supplies, in CRC *Handbook of Laboratory Safety* 2nd ed., Steere, N.V., (Ed.), CRC Press, Cleveland, OH, 1971.
2. ANSI A-40-1993, *Safety Requirements for Plumbing*, American National Standards Institute, New York, NY.

I. Other Laboratory Utilities

There are a number of other possible utilities which may be provided to a laboratory. Among these are natural gas, compressed air, distilled water, vacuum, steam, refrigerated brine, and other gases. In some instances, there are safety issues, such as limitations on the pressure available from a compressed air line and the need to incorporate provisions for pressure relief, to ensure that personnel will not be injured by explosions due to excess pressure. Often the quality of the air is more of a problem. The compressor supplying the system should be capable of supplying air which is clear of oil and moisture. Some facilities which have a large, pressurized liquid nitrogen tank at hand, use the vapors from the tank as a source of ultra clean compressed "air" to clean work surfaces.

The dangers of natural gas are well understood. This does not prevent numerous accidents each year due to gas explosions. Natural gas used to heat experimental devices should not be left unattended, unless a heat-sensitive automatic cutoff device is connected to the system. Otherwise, if gas service were to be interrupted, gas escaping from burners or heaters left on could easily cause a major explosion. If the gas device is set up in a properly working fume hood, the gas would be exhausted through the fume hood duct system, but as noted earlier, fumes exhausted outside a laboratory building can be recaptured under some conditions. Whenever the odor of gas is perceived, under no circumstances should electric light switches, other spark-producing electric devices, or any other possible source of ignition be operated. If the gas concentration is low enough to make it safe for personnel to enter the room, all devices connected to the gas mains in the laboratory should be immediately turned off. Any windows in the room which are not fixed shut and are capable of being opened should be opened. Consideration should be given to having all nonessential personnel evacuate the entire building while volunteers check the remainder of the building or facility for other systems which might also be leaking. If the concentration is already high, everyone should evacuate the building at once, the gas service turned off at the service entrance to the building, and the utility service company notified. Even small gas leaks, barely perceptible by odor, should be sought out and repaired. Gas can seep slowly through cracks and seams and accumulate in confined spaces in which the air interchange is very slow and can explode if an ignition source presents itself. Numerous explosions of this type have occurred.

OSHA has specific regulations on four gases, acetylene, nitrous oxide, hydrogen, and oxygen, in the general industry standards. Generally the requirements in the OSHA regulations are those provided for by the Compressed Gas Association (CSA), and for the first two of the gases mentioned above, the OSHA standards simply refer to the appropriate pamphlet issued by the CSA (see G-1 (1996) for acetylene and G-8-2 (1994) for nitrous oxide). For hydrogen and oxygen, the regulatory requirements are spelled out in detail in Sections 29 CFR 1910.103 and 29 CFR 1910.104, respectively. The requirements for oxygen generally pertain to bulk systems of 13,000 cubic feet or larger. Very few research laboratory facilities would involve a system of this size.

For hydrogen, the regulations do not apply to systems of less than 400 cubic feet, so that laboratory systems involving a single typical gas cylinder would not be covered. However, in the context of the present chapter which is intended to involve the design of a building, it is entirely possible that a hydrogen gas system could be larger than 400 cubic feet and hence would be covered by the provisions of the OSHA standard. All of the requirements of the standard will not be repeated here, but among these are (1) provisions that cast iron pipes and fittings shall not be used (note that for acetylene, brass, and copper pipes, containers, valves or fittings must not be used); (2) that the system shall be above ground; (3) restrictions on the electrical system to be a class I system; (4) provisions that outlet openings shall be at the high points of wall and roof; and (5) requirements for explosion venting, among many others.

Every utility that is provided should be properly identified with a clear, unambiguous label. Color-coded discs with engraved name labels which screw into each service fixture are available from at least one company for most common service utilities. All fittings and connectors should be provided according to appropriate standards. For example, all gas fittings should comply with the provisions of the Compressed Gas Association Standard V-1(1994). Every utility, if used improperly, can cause safety problems, the only difference being that some require more of an effort on the part of a user to do so than do others.

REFERENCES

1.   OSHA General Industry Standards, 29 CFR Part 1910, Sections 101-104.

2.   *Handbook of Compressed Gases (1990),* Compressed Gas Association,   1725   Jefferson Davis Highway, Arlington, VA.

3.   *Prudent Practices for Handling Hazardous Chemicals in Laboratories,* National Academy Press, Washington, D.C., 1981, 75 - 89.

4.   *Prudent Practices in the Laboratory, Handling and Disposal of Chemicals,* National Academy Press, Washington, D.C., 1995.

J.   Maintenance Factors

Every building and every piece of equipment installed in a building should be designed with the realization that each item of equipment and most building components will eventually need service. As noted at the very beginning of this chapter, the current *modus operandi* of most research facilities is to change the nature of the research in a given space fairly frequently. Thus the space itself, treated as a piece of equipment, will need relatively frequent service. It should be possible to bring any needed services to virtually any position within the laboratory relatively easily. Good access to services should be provided in the initial design, making it possible for maintenance personnel to work on the equipment conveniently and safely.

All ducts, electrical circuits, and utilities need to be clearly identified with labels or other suitable means. Fume hood exhausts on the roof of a building should be identified with at least the room number where the associated hood is located and a hood identification if there is more than one hood in the room. It would also be desirable for the general character of the effluents from the exhaust duct to be indicated by means of a written label or a color code. Similarly, every electrical circuit should be clearly labeled so that there would be no confusion as to which breaker controls the power to the circuit. A significant amount of space (15 to 25%) is set aside in every major building for mechanical service rooms, electrical closets, and other building services. They are rarely seen by most of the building occupants, but they are critical to the good operation of the building. Most of the major equipment used to provide building services is either located within these spaces or the control panels are located within them. These areas need to be well maintained and not be used as storage spaces. Unless a major component, such as a compressor for the air conditioning system, is out or it is necessary to turn off the water to an entire area, the operations within these rooms are usually outside the experience of the usual occupants of the building, and access to these spaces should not normally be available to them. Maintenance to most occupants of the buildings involves work either in their immediate area or directly affecting their area. An exception is the electrical panel which controls the breakers to electrical circuits within a laboratory. As noted earlier, there are reasons why employees other than maintenance personnel will need (sometimes urgently) to deactivate electrical circuits, and if proper isolation of the panels is done to prevent access to active electrical wiring, the breaker panel closets can be made safely accessible.

The building plans should show all utilities, ducts, and electrical circuits correctly as they physically exist throughout the facility. Frequently during construction projects, it is necessary to make some adjustments to the original plans. Sometimes these are documented with change orders, if they are sufficiently major, but at other times they are considered minor and the building plans are not modified to show them. In principle, all changes should be reflected on the final "as built" drawings and specifications provided to the building owner. This should include not only physical changes in locations, but also any changes and substitutions made in materials as well. For example, a complicating factor in dealing with asbestos in existing buildings stems from the uncertainty as to whether asbestos was used or not, under provisions in many contracts which allowed substitution of "equivalent" materials. The resulting uncertainty has resulted in significant delays and major additional costs in renovation projects while insulating materials are tested for asbestos. Asbestos was used in construction when it was perfectly legal and desirable to do so, and it is necessary now to know where it is located in order to remove it or treat it so as to render it harmless. If such information were reliably available from building plans, it would be much easier to estimate removal costs and design asbestos abatement projects.

Unfortunately, after a period of occupancy, it is common for many changes to be initiated by the facility's users themselves. These are intended to serve a specific purpose and are often built without consideration for anything except this purpose in mind. As a result, maintenance may be significantly impaired. Perhaps the most common occupant-initiated changes involve electrical circuits which may or may not be installed or labeled properly. Similarly, the interior configuration of a laboratory space may be changed, which can significantly affect the distribution of air from the ventilation system serving the area. It should be institutional or corporate policy to prohibit such modifications and to require that they be removed once found. As a minimum, any such remodeling on the part of the occupants should be required to be reviewed by appropriate personnel and changed where necessary to comply with building codes and maintenance programs. They should not be allowed to be left in place where they are disruptive to good design and use practice. An excellent option, available to larger organizations, is to establish a dedicated component of the engineering, design and maintenance staff to deal specifically with remodeling projects. However, due to the often perceived as unreasonable costs of these engineered remodeling projects, do-it-yourself projects will continue to be done.

Where there are unusually dangerous maintenance operations, it is necessary that appropriate safety provisions be made for the purpose. A good example is the need to change contaminated high efficiency particulate and aerosol (HEPA) filters, where the contaminant might be a carcinogen or another possibly injurious biologically active agent. Ample access room to the exhaust duct must be provided so that bag-out procedures, where the contaminated filters are withdrawn into a sealed bag, may be done without difficulty and without risk of the agent of concern affecting the maintenance personnel or spreading to adjacent areas. Another example, representing an acute danger rather than a delayed one, would be repair to a perchloric acid ejector duct after a period of operation when the wash-down cycle failed to work. In this case, there would need to be provision for washing the interior of the duct for several hours prior to beginning maintenance operations, as well as developing contingency plans to protect the workers in the event flooding the duct was not totally effective in removing dried perchloric acid. Fortunately, perchloric acid is no longer used as frequently as it once was so this particular danger is diminishing.

Maintenance operations often considered as routine, such as replacement of a fume hood motor or repair to a laboratory sink, must be considered hazardous operations if the potential exists for exposure of the workers to toxic chemicals. An example is exposure to chemicals during repairs to a hood exhaust system because of fumes from nearby exhaust ducts of operating hoods or chemical residues on the motor, fan, or other components of the unit under repair. Even maintenance activities totally unrelated to laboratory operations, such as patching a roof, can permit the workers to be exposed to possibly toxic effluents from the air handling system or fume exhausts. The converse is also true. Fumes from roofing materials, paints, welding, etc. can be drawn into a building via the air intake to the accompanying discomfort of the building's occupants. Under the 1986 OSHA Hazard Communications Standard, it is required that any individual, potentially exposed to dangerous chemicals, must be informed of the risks and of the measures needed and available to protect them from the dangerous effects. Thus, when anyone is asked to perform maintenance on a piece of laboratory apparatus which could be contaminated or is asked to work under such conditions that chemical exposures could result, it is now mandatory that the situation be evaluated for potential dangers. It may be necessary to plan for a period of time when operations can be postponed or modified to eliminate the release of toxic materials, or alternatively the workers may be provided with appropriate protective equipment which they must wear during the maintenance operations. Since workers typically have a minimal chemical educational background, they tend to be more concerned about the dangers of exposures than an ordinary laboratory employee. On the other hand, such workers are often resistant to wearing "uncomfortable" protective safety equipment or may feel that it makes them appear "sissy" to their fellow workers. It is important that an effective training program be established for them to ensure that they are informed of the actual risks to which they might be exposed, and the need to wear protective gear. However, care needs to be taken so the training should neither exaggerate or minimize these risks.

II. FIXED EQUIPMENT AND FURNITURE

In Section 3.I.C, requirements for furniture in the various classes of laboratories were briefly given, *viz.,* "Furniture should be designed to be sturdy and designed for convenient utilization," and "Bench tops should be resistant to the effects of acids, bases, solvents, and moderate heat, and should not absorb water." Flammable material storage cabinets were among some of the special items of equipment needed if solvents were to be used. This section will discuss these items and others in detail. The quality of the laboratory furniture should be as good as can be afforded and should be as versatile as possible, unless it is certain that usage will always involve only a limited range of reagents in a few applications. Similarly, the specifications on equipment should be written to assure that it will perform well and *safely.* A refrigerator or freezer may be bought on sale for perhaps as little as one quarter of the price of a flammable material storage model. However, the possible consequences of storing solvents in the former make it a very poor bargain in most laboratories. Even if the immediate need is not there, few of us are so certain of the future that we can be confident that in the typical 15- to 20-year life span of a refrigerator that no one will store flammable solvents in one.

A. Laboratory Furniture

Just as ordinary furniture can be bought in many grades of quality not always discernible to a casual examination, laboratory furniture also varies in quality of construction. The least expensive may look attractive and appear to offer the same features as better units, but it will not be as durable and will have to be replaced frequently. It is also likely to not be as safe. If shelves are not firmly attached, the weight of chemicals on them can cause them to collapse. If cabinet doors and drawers do not have positive catches, they can fail to close and possibly rebound, leaving them partially open. The contents in a partially open drawer may be damaged through the opening, or the protruding drawer can cause an accident to someone who does not observe the obstruction. Poorly protected work surfaces can corrode or can become contaminated and be difficult to decontaminate.

Good quality, modular laboratory furniture is available today in a variety of materials and can be installed in configurations to fit almost any need. Units can be obtained precut to accommodate connections to utilities. In most cases, the utilities can be brought to the correct locations prior to installation of the furniture, which makes it simple to perform maintenance. Units are available which allow after-installation changes to be made easily.

1. Base Units and Work Tops

Base units can be obtained in steel, wood, or plastic laminates. The steel in the steel units should be heavy gauge, e.g., 18 gauge, with a pretreatment to reduce the corrosive effects of chemicals. Painting all surfaces with a durable, baked on, chemically resistant paint finish will also help minimize chemical effects. In better units, this is an epoxy coating. Some individuals continue to prefer wooden laboratory furniture. Because of cost, solid wooden furniture is not an economic choice, but durable wooden (or plastic) veneer furniture is available which can meet most safety requirements. Although wood may be more absorbent to liquids than steel, it is less reactive and more resistant to a very wide range of chemicals than many materials. There may be some surface degradation, but the furniture will remain usable. There are exceptions, such as facilities using perchloric acid, where wood would not be a good choice because of the vigorous reaction of perchloric acid with organic materials.

Bench tops can be made of several different materials. Some are satisfactory for light duty while others can be used for almost any purpose. Again, unless it is certain that both immediate and long-range usage will permit the use of a lesser material, it is suggested that the bench tops for a new facility be selected from among the more rugged and versatile materials. The types of materials offered for heavy usage applications by most vendors today are bench tops of either stainless steel or epoxy resin. The first of these is more often used as the interior work surfaces of specialty fume hoods or cabinets, while the latter is the most common for general, heavy usage applications. Wood or plastic laminates are used where the level of usage will permit. Wood is actually a good material for many reasons, but high quality hardwood furniture is very expensive.

When using radioactive materials or perhaps some unusually toxic materials, where it would be necessary to decontaminate the work surface after a spill, the choice clearly should be limited to those materials such as stainless steel that are least likely to absorb materials. However, if the use of such materials is minimal and limited to very low levels of activity or concentrations (as it should be when used on an open bench), the work surface can be protected with an absorbent paper with a chemically resistant backing. The higher cost of the premium materials could be avoided if the need does not otherwise exist.

Wherever possible, in order to avoid seams in which toxic materials could become trapped, the back splash panel should be an integral part of the work top. Service shelves above the back splash panel can be of lesser duty materials in most cases since they will be primarily used to store limited quantities of reagents in closed containers, intended for short-term usage. The shelving must be solidly built and well supported to ensure that it will bear the weight of materials stored upon it.

Laboratory sinks, incorporated in the work tops, are usually made of stainless steel or epoxy resin to provide the needed resistance to corrosives, solvents, and other organic and inorganic materials. Older sinks are often stoneware. However, current restrictions on disposing of chemicals into the sanitary system should reduce the burden on the laboratory sink and the remaining plumbing components.

In most instances, the tendency is to utilize all of the space underneath a work top for cabinetry and other forms of storage space. It is a good idea to leave at least one portion open and available for storage of movable carts and other items of equipment which must be left on the floor, in order to avoid reducing the aisle space below acceptable limits.

The following references to the two firms listed is not to be construed as a recommendation or endorsement of them but as simply to identify two widely accessible sources of relevant information. There are a number of other manufacturers and distributers of laboratory furniture.

REFERENCES

1. Fisher Scientific Company, 585 Alpha Drive, Pittsburgh, PA.
2. Kewaunee Scientific Corporation, 2700 West Front Street, Statesville, NC.

2. Storage Cabinets

Facilities for storage of research materials in the laboratory should be selected with as much care as any other item of equipment. Many chemicals may be stored on ordinary shelves or cabinets, with only common sense safety provisions being necessary. Obviously, the shelves or cabinets must be sturdy enough to bear the weight of the chemicals. Storage should be such as to make it unlikely that the materials will be knocked off during the normal course of activities in the room. Shelves should not be overcrowded. It should not be necessary to strain to reach materials or to return them to their places. Incompatible chemicals should be stored well apart. Finally, the amount of storage should not be excessive, in order to restrict the amount of chemicals not in current use that would otherwise tend to accumulate within the facility. Periodically, one should "weed" the shelves of chemicals which have not been used for some time and for which no immediate use is foreseen. Older chemicals are rarely felt to be suitable for critical research needs, but can be used in less critical situations such as instructional laboratories. Although these recommendations are, as noted, common sense procedures, it is surprising how frequently many laboratories violate one or more of these points and how many accidents occur as a result.

A number of classes of materials should be stored in units designed especially for them because of their dangerous properties, or because there are restrictions on the use of materials which require that provisions be made for keeping them locked up. Among these are flammables, drugs, explosives, and radioactive materials. In addition, some materials, such as acids, are best stored in cabinets designed to resist the corrosive action of the materials. Most of the storage units sold for a specific purpose are intended to be free standing, but they also may be purchased to be compatible with other modular components, to be equipped with work tops, or to be used as bases for fume hoods so that work space will not be lost.

a. Flammable Material Storage

Flammable materials are among the most commonly employed chemicals in laboratories and represent one of the most significant hazards because of their ignitability characteristics. Even a small quantity when spilled can cover a surprisingly large area, and with the increased surface area, provide copious amounts of vapor which can spread even further. Since the vapors of most flammable solvents are heavier than air and tend to remain as a relatively coherent mass unless substantial air movement is occuring, they can flow for significant distances. Should the vapors encounter an ignition source and the concentration be within the flammable limits, they can ignite and possibly flash back to the original source. Everyone who was ever a boy or girl scout surely remembers the caution about using gasoline to start a fire due to that problem with gasoline. If substantial amounts of flammable solvents are openly stored on shelves or workbenches, it is possible for a small spill to quickly escalate into a large fire, possibly involving the entire facility due to the availability of the additional fuel. Flammable material storage cabinets are primarily designed to prevent this from happening. Except for the quantities needed for the work *immediately at hand*, all of the reserves should be stored in flammable material storage cabinets, and, once the needed amounts are removed from the container, the containers should be returned to the cabinets.

Flammable liquids are divided into various classes, as given in Table 3.4. The definitions depend upon the flashpoints and in some cases the boiling points of the liquids. The flashpoint of a liquid is legally defined in terms of specific test procedures used to determine it, but conceptually it is the minimum temperature at which a liquid forms a vapor above its surface in sufficient concentration that it may be ignited. In Table 3.4, the first temperature is in degrees Celsius and the temperature in parentheses ( ) is the equivalent Fahrenheit temperature.

Neither Combustible II or Combustible IIIA materials may include mixtures in which more than 99% of the volume is made up of components with flashpoints of 93.3° (200) or higher. The OSHA Laboratory Safety Standard supersedes the OSHA General Industry Standard except for a few specific instances. However, Section 1910.106(d)(3) of the industrial standards provides guidelines on the maximum amounts of flammable liquids allowed to be stored, dependent upon class, in flammable material storage cabinets within a room and

Table 3.4 Definitions and Classes of Flammable and Combustible Liquids

Class	Boiling Points	Flashpoints
Flammable		
A	<37.8(100)	<22.8(73)
lB	≥37.8(100)	<22.8 (73)
IC		22.8(73)≤ and <37.8(100)
Combustible		
II		37.8 (100) ≤ and <60(140)
IIIA		60 (140) ≤ and < 93.3 (200)
IIIB		~93.3 (200)

defines in Section 1910.106(d)(2) the maximum size of individual containers for the various classes of flammables. NFPA Standard 45 provides guidelines as to the maximum amounts of flammable liquids that should be allowed in the three classes of facilities A, B, and C defined in that standard. This standard has been mentioned, but the three laboratory classes have not been stressed in this volume, in favor of concepts involving wider varieties of hazards than that due to the amount of flammable materials in the laboratory alone. For the purposes of this section, the low- and moderate-risk facilities described in Sections 3.I.C.2.a and 3.I.C.2.b, respectively, may be taken to be approximately equivalent to an NFPA Standard 45, class C facility, a substantial risk facility (Section 3.I.C.2.c) to be roughly equivalent to a class B facility, and a high risk facility (Section 3.I.C.2.d) to include a class A facility but with more restrictions than the latter would require. Note that OSHA has not adopted the restrictions of NFPA Standard 45 and does not address the issue of the total amount of flammables permitted in a laboratory area, although the amount permitted in an interior storage room is defined. The OSHA regulations regarding container sizes are based on sections of the 1969 version of the NFPA Standard 30. Before returning to the topic of flammable material storage cabinets, Table 3.5 defines the various classes of flammable and combustible liquids and the maximum container sizes permitted by OSHA for each class. Table 3.5 is equivalent to Table H-12 from the OSHA General Industry Standards. Table 3.5 A provides similar data from NFPA 30-1996. Recall, however, that unless adopted by a local jurisdiction, the NFPA standards are only recommendations, not regulations.

There are several exceptions to Table 3.5 which would permit glass or plastic containers of no more than 1 gallon capacity to be used for class IA and class lB liquids: (a) if a metal container would be corroded by the liquid; (b) if contact with the metal would render the liquid unfit for the intended purpose; (c) if the application required the use of more than one pint of a class IA liquid or more than one quart of a class lB liquid; (d) an amount of an analytical standard of a quality not available in standard sizes needed to be maintained for a single control process in excess of 1/16 the capacity of the container sizes allowed by the table; and (e) if the containers are intended for export outside the United States.

In section 1910.106(d)(3) of the General Industry Standards, OSHA limits the amounts of class I and class II liquids in a single flammable material storage cabinet to 60 gallons and the amount of class III liquids to 120 gallons. Thus, even if the integrity of a single storage cabinet were breached in a fire or if an accident occurred while the cabinet was open, no more than 60 gallons of class I and II liquids or 120 gallons of a class III liquid could become involved in the incident. This is not to imply that these are insignificant amounts. They are

Table 3.5 Maximum Allowable Size of Containers and Portable Tanks

Container Type	Class IA	Class IB	Class IC	Class II	Class III
Glass, or approved plastic	1 pt	1 qt	1 gal	1 gal	1 gal
Metal (other than DOT drums)	1 gal	5 gal	5 gal	5 gal	5 gal
Safety cans	2 gal	5 gal	5 gal	5 gal	5 gal
Metal drums (DOT specs)	60 gal	60 gal	60 gal	60 gal	60 gal
Approved portable tanks	660 gal	660 gal	660 gal	660 gal	660 gal
Polyethylene spec 34 or as authorized by DOT exemption	2 gal	5 gal	5 gal	60 gal	60 gal

Table 3.5a Limits on Container Sizes Specified by NFPA 30, 1996

Containers	Flammable Liquids						Combustible Liquids			
	Class IA		Class IB		Class IC		Class II		Class III	
	Liters	Gallons	Liters	Gallons	Liters	Gallons	Liters	Gallons	Liters	Gallons
Glass	0.5	0.12	1	0.25	4	1	4	1	4	1
Metal or approved plastics	4	1	20	5	20	5	20	5	20	5
Safety Cans	7.5	2	20	5	20	5	20	20	20	5

quite large and would be ample, if ignited, to create an extremely serious fire beyond the capacity of portable fire extinguishers to extinguish. Sounding an alarm and immediate evacuation would be the proper course of action if such an incident should occur.

The purpose of a flammable material storage cabinet is to postpone the involvement of the materials within the cabinet in a fire long enough to allow persons in the immediate area to evacuate the area, or in some cases to permit the fire to be extinguished. Technically, for the cabinet to pass the 10-minute fire test according to NFPA Standard 251, the interior temperature of the cabinet should not exceed 164.1°C (325°F), all joints and seams would remain intact during the fire, and the cabinet doors would remain closed.

Most commercially available flammable material storage cabinets are made of metal and would have to meet at least the following specifications:

1. Bottom, top, and sides of at least 18-gauge sheet iron and double-walled with a 1.5-inch air space.
2. Joints shall be riveted, welded, or made tight by equally effective means.
3. The cabinet door would have to be provided with a three-point lock.
4. The door sill would have to be raised at least 2 inches above the bottom of the cabinet.
5. The cabinets must be labeled in conspicuous lettering "Flammable—Keep Fire Away."

Note that there are no requirements for automatic door closers to be provided with a fusible link. Such a feature is clearly desirable since it ensures that the cabinet will close in the event of a fire, even if it were inadvertently left open. There are also no requirements for vent connections, but most flammable material storage cabinets have provisions for installing ventilation ducts. The value of these is not universally accepted. If the containers within the cabinets are always tightly sealed and there is really no excuse for putting containers away that are not tightly sealed, then there is no reason to make provisions for exhausting fumes. However, should the containers not be tightly closed, it would be possible for volatile fumes to accumulate within the cabinet that should be exhausted outside the building. Although it is possible that on some occasions containers will be returned to the cabinet improperly closed, the author's own preference is to not utilize the vents, since they are often not properly vented into an acceptable exhaust duct and volatile vapors could escape into the facility. Users may also install the venting ducts using ducts made of plastic or flimsy metals which would be destroyed quickly by fire, thus reducing the protection offered by the cabinet.

Wooden flammable material storage cabinets, if properly constructed according to the provisions of Section 1910.106(d)(3)(ii)(b) of the OSHA General Industry Standards, are also acceptable. These provisions are:

"The bottom, sides, and top shall be constructed of an approved grade of plywood at least 1 inch in thickness, which shall not break down or delaminate under fire conditions. All joints shall be rabbited and shall be fastened in two directions with flathead wood screws. When more than one door is used, there shall be a rabbited overlap of not less than 1 inch. Hinges shall be mounted in such a manner as not to lose their holding capacity due to loosening or burning out of the screws when subjected to the fire test."

Although wood will eventually burn, thick sheets of plywood such as required by the standard can withstand a substantial amount of heat and is a much better insulator than metal. In 1959, the Los Angeles Fire Department performed a number of comparative tests using various combinations of metal and wood to simulate the walls of a storage cabinet. The experimental walls were fastened to the opening of a furnace operating in the range of 704°C to 788°C (1300°F to 1450°F). A thermocouple was attached to the opposite face of the simulated wall cross-section. The experimental mockups were the following:

Metal
1. A double walled, metal structure of 18-gauge CR steel, approximately 7 x 10 inches with 1.5-inches air space.
2. A similar cross-section to number 1 made with a core of 5/8-inch sheet rock suspended midway in the 1.5-inch air space.
3. A similar cross-section to number 1 with a core of untreated 1/2-inch Douglas Fir plywood suspended in the air space.
4. A metal-walled structure insulated with 1 inch of 1-pound density fiberglass blanket in the 1.5-inch air space.
5. A metal-walled structure insulated with 1.5-inch of mineral rock wool (density un-available).

Wood

1. Two layers of 1-inch Douglas Fir plywood.
2. One layer of 1-inch Douglas Fir plywood.
3. A laminated formed of 1/2-inch plywood on each side of 1/2-inch sheet rock.

Although the experimental arrangement does not simulate a storage cabinet perfectly, the increase in temperature data as a function of time, presented in Table 3.6, does show the rate at

Table 3.6   Simulated Storage Cabinet Wall Configurations.
(Time [in minutes] vs. Temperature Rise Data)

Sample	5		10		15		20	
	•F	•C	•F	•C	•F	•C	•F	•C
Metal								
1	430	221	500	260	510	266	550	343
2	150	66	180	82	210	99	240	116
3	130	54	140	60	160	71	170	77
4	310	154	430	221	470	243	–	–
5	270	132	433	233	466	341	–	--
Wood								
1	100	38	100	38	100	38	100	38
2	120	49	133	56	166	74	200	93
3	90	32	100	38	110	43	130	54

which heat is transmitted through the various combinations. They show that the poorest of the wooden panels transmitted heat at a slower rate than the best of the metal combinations. Thus, a wooden flammable material storage cabinet, although it probably would need to be custom constructed in a laboratory facility, offers superior fire protection for short term exposures. Eventually, however the wood itself, even if protected by fire-retardant paints, would represent a source of fuel for a protracted fire. For prolonged exposure to a fire, neither would offer protection.

b. Cabinets for Drug Storage

Security is the primary concern for the storage of controlled substances or drugs because of the potential for theft and misuse. The Drug Enforcement Agency in CFR Title 21, Parts 1301.72 through 1301.76 delineates the security requirements for Schedule I through V controlled substances, including provisions for both cabinets and storage vaults. For individuals who are practitioners (there are several different categories of practitioners, including physicians, dentists, pharmacists, and some institutional personnel, all of whom are legally allowed to dispense drugs), the requirements are simple: the storage cabinet needs to be a securely locked, substantially constructed cabinet. Most institutional programs involving drugs can be easily designed to meet this level of required security. Where the programs would fall under the requirements for non-practitioners, Parts 1301.72 to 1301.74 describe the needed security provisions. Although these are given in considerable detail, Part 1301.71 provides that actual security requirements which offer structurally equivalent protection would be acceptable. The simplest manner in which protection can be provided for small quantities is a steel safe or cabinet. Among other requirements, these should be sufficiently durable to prevent forced entry for at least 10 minutes, should be either sufficiently heavy (750 pounds or more) or rigidly bolted to a floor or wall so that it cannot readily be carried away and should be equipped with an alarm which will sound in a central control station manned at all times. All of these are not required for a practitioner's storage cabinet, but the implication of these provisions for tight security should serve as a guide in defining what is meant by a "securely locked, substantially constructed cabinet." It is extremely important that accurate records are kept for the amounts received, dispensed, and the current inventory.

c. Storage of Radioactive Materials

Storage cabinets for radioactive materials need to meet multiple needs. As with drugs, there is the problem of security, although for a totally different reason. There is a widespread fear of radioactive materials among the general population. As a result, the public is extremely sensitive to the possibility that radioactive material could be lost or taken from the areas in which it is normally used, which might cause members of the public to be exposed to radiation. The quantities of materials normally used in most research laboratories are often extremely small and hence would rarely cause any exposure to the public significantly above that due to natural radiation, but to address the public's concern, the rules and regulations governing the use of radioactive materials are extremely strict, including those on security. The use of radioactive materials and radiation will be dealt with at length in Chapter 5, as well as the other major concerns involving storage of radioactive materials, shielding against radiation from the original research materials and of radioactively contaminated waste. Some of the concepts which will be developed more fully there will be used here without immediate explanation. In the following material, if there are terms which are unfamiliar, definitions or explanations will be found in Chapter 5.

Title 10, Part 20.1801 and 20.1902 of the Code of Federal Regulations covers the basic security requirements very briefly, and this brevity may appear to reflect that security is not a major issue. However, directives have been issued by the NRC to emphasize the importance which the NRC places on security. All radioactive materials must be kept in a secure restricted area to which access is controlled or in a securely locked cabinet or other type of storage unit unless a qualified, authorized user is in the immediate area. This means that when an experiment is in progress involving radioactive material and the researcher leaves the laboratory, the researcher must either lock the facility behind him or return the radioactive material to a locked storage unit, should there be no other authorized user in the laboratory unit. Users may think that the requirement for security would allow an individual to leave a facility with accessible radioactive material unattended for the few minutes it would take to go to the bathroom, obtain a soft drink or snack, but even for these short intervals, the rules apply. The concern for security extends to the need to challenge an unfamiliar person in areas where radioactive materials are in place. The stranger may be a salesman, visitor, an unannounced inspector from the NRC, or, although very unlikely, someone who might take advantage of the access to the radioactive material and remove it.

If an NRC inspector were to conduct an unannounced inspection of the facility and find radioactive material in use and unattended, it would be considered a violation of the radioactive material license. Depending upon the circumstances, this could result in a fine or even a loss of the license to use radioactive materials. In an extreme case, if the institution or corporation were to be operating under a broad license and if it could be shown that the institution was not enforcing the rules, the license for the entire corporation or institution could be lost. This is a very drastic punishment which is unlikely to occur, but many well known and reputable research facilities have received very substantial fines. Clearly, then, it is essential that every laboratory using radioactive materials have a sturdy, lockable, storage cabinet (or refrigerator/freezer) in which to keep the material. An accurate log of the material on hand is also required. A significant discrepancy, revealed upon a comparison of the materials present and the amounts that should be present could also subject the holder of the license to comparable penalties. Frequently, a radioactive material is incorporated in a research material having physical or chemical properties which would require specialized storage, such as flammable materials, because of these properties as well.

In addition to security, the radiation emitted by some of the isotopes used in research may require that shielding be incorporated into the storage cabinet. For the radiation from some isotopes, the choice of material used for shielding is critical because the wrong type, depending upon whether the shielding is of a low atomic number or not, can exacerbate the radiation problem. Although almost any radioactive material can be used in some applications, in the majority of most chemical and biological laboratories, a relatively small number of isotopes represent the major usage. Several of these are low energy, pure beta emitters and can be stored safely in any cabinet, since the betas will be completely stopped by virtually any shielding. Among these isotopes, tritium (^{3}H), ^{14}C, ^{35}S, ^{32}P, ^{45}Ca, and ^{63}Ni (used primarily as a source in gas chromatographs) are used most frequently and are among the safest to use. On the other hand, while ^{32}P is a beta emitter, the betas are very energetic and shielding is required. As the energetic betas from this isotope are slowed down and brought to a stop in the shielding, the deceleration creates a penetrating type of electromagnetic radiation called "bremstrahlung." This effect is much more pronounced for shielding made of higher atomic number materials, such as lead, than in materials such as plastics, composed primarily of carbon and hydrogen. Therefore, for this particular beta emitter, which is one of the more frequently used isotopes because of its chemical and biological properties, the shielding in the storage cabinet should be of plastic or some similar material.

Several isotopes that are commonly used also emit a penetrating electromagnetic radiation called gamma radiation. Among the more commonly used isotopes in this category are ^{125}I, ^{22}Na, ^{51}Cr, ^{65}Zn, ^{60}Co, and ^{137}Cs. The appropriate shielding material to be used in or around the storage unit in such a case would be lead.

One last type of radioactive material for which special storage criteria would be needed is in the relatively rare instances where neutron radiation sources would be used. Neutron radiation would be found primarily in reactor facilities, but neutron sources are also used in moisture density probes in a number of research areas. If the source were to be taken from the instrument, the shielding would need to consist of a layer of plastic or paraffin, 5 to 20 cm thick, either impregnated with boron or surrounded by cadmium. Some gamma shielding also might be required in an unrestricted area, which could be provided by lead or in some cases by concrete blocks.

d. Corrosive Materials

Storage cabinets can be obtained to serve either as flammable material storage cabinets or as cabinets for the storage of acids. In the latter case, the shelves must be provided with protection against the effects of corrosion. One of the more economical ways of achieving the required protection is to use polyethylene shelf liners. If they are not needed or need to be replaced, they can readily be taken out or can be replaced with stock polyethylene. The interior walls can be protected from the effects of acid vapors and fumes by being painted with an acid-resistant paint.

e. Records Protection

Approximately thirty years ago, when the first edition of the CRC Handbook of Laboratory Safety was published, the only practical way to store laboratory records, was on paper, or in log books, which then had to be stored in file cabinets or other bulk storage areas. The protection of research records was absolutely critical to a professional scientist and a chapter was devoted to record storage safety. A comment made to the author by a senior scientist at approximately that time was to the effect that, in the event of a fire, the scientist would have to be physically restrained from reentering the facility to retrieve the records because they represented the scientist's entire professional life. If they were lost, he stated "his career would be over." This is still true today but the feasibility of maintaining records in two or more separate locations has been so dramatically improved by current technology that there is no reason not to keep complete records in, at least, more than one location. Today, data normally are currently either acquired directly by a computer or promptly entered into one. Long lasting storage of these records can be transferring them to small (3.5 inch) removable magnetic disks holding anywhere from 1.44 to 250 megabytes of data (and still increasing in size). Transfer of data to optical disks is already feasible and the capacity of these disks is currently up to nearly 10 gigabytes of data. This large capacity makes it entirely feasible to scan in images of the pages of entire laboratory notebooks so that even handwritten and annotated data records can be stored as conveniently as the magnetic disks. Certainly, there are large numbers of paper records still being maintained but this shall surely decrease as time passes. Even letters, and memoranda which still represent possibly the largest volume of paper records are rapidly decreasing with the exponentially increasing use of electronic mail via the Internet. Of course, one does have to take the trouble of transferring these documents to more permanent media as discussed above, for those worthy of being kept. Actually, it is just as easy to electronically transfer duplicate copies of any set of records from a computer to a second location anywhere in the world and retrieve them at will as it is to create any kind of hard copy. In short, protection of records is still important but the means of protecting an archival set is so easy today that there is very little reason not to do so, and to use the space recovered by discarding most file cabinets, storage boxes, etc., for other productive purposes, such as laboratory equipment. The only caveat is that one must take the time to take advantage of the innovative means to store records safely. Today, virtually everyone reading this has had the experience of having the power fail at some point or their computer crash unexpectedly with the subsequent loss of the information yet unprotected. Frequent backups and transfer of data to a portable disk or an external location is essential to minimize data losses to only a brief period of time. As has been said, if there is no record of a datum, it doesn't exist.

REFERENCES

1. OSHA General Industry Standards, 29 CER 1910, § 106(d).
2. NFPA 30, *Flammable and Combustible Liquid Code,* National Fire Protection Association, Quincy, MA, 1996.
3. NFPA 45, *Fire Protection for Laboratories Using Chemicals,* National Fire Protection Association, Quincy, MA, 1996.
4. NFPA 251, *Standard Methods of Fire Tests of Building Construction and Materials,* National Fire Protection Association, Quincy, MA, 1995.
5. Steere, N.V., *Fire-protected storage for records and chemicals,* in *CRC Handbook of Laboratory Safety,* 2nd ed., Steere, N. V. (Ed.), CRC Press, Cleveland, OH, 1971, 179 - 186.
6. NFPA 232, *Standard for the Protection of Records,* National Fire Protection Association, Quincy, MA, 1995.
7. *Prudent Practices for Handling Hazardous Chemicals in Laboratories,* National Academy Press, Washington, D.C., 1981, pp. 226 - 227.
8. Food and Drug Administration, Chapter II, Drug Enforcement Agency, Title 21 CFR Part 1301, Section 72-76.
9. Nuclear Regulatory Commission, 10 CFR Part 20 Section 1801,1802.
10. *Prudent Practices in the Laboratory,* National Academy Press, Washington, D.C., 1995

B. Hoods

A key item of the fixed equipment in most research laboratories other than those employing only the least hazardous materials is a work enclosure, usually denoted, at least for chemicals, by the common name "fume hood." The OSHA Laboratory Safety Standard, while stopping short of requiring a fume hood in each laboratory, does point out that work with almost any hazardous substances can be done safely if done in a suitable, properly functioning, hood. Some laboratory facilities have been constructed recently with no open bench space, with all work within the facility being done within hoods. Others are being built or planned around the same concept. OSHA recommends that in designing laboratories, one hood should be provided for every two and one half research personnel. However, since the activities in a laboratory vary so widely, there can be no absolute standard for a suitable number for a given laboratory other than that there should be sufficient hoods available so that no work which should be performed in a hood need be done on an open bench. In the standard laboratory module described in Section 3.A.1, only one hood is shown. However, additional hoods could be added as needed. Studies made at the National Institutes of Health have shown that it would be desirable for the second hood to be placed on a facing wall instead of in the corner adjacent to the first hood. In such a case, the doors on the side of the room, should be moved to locations opposite each other as well rather than located diagonally across the room but closer to the front of the laboratory than to the far end. The air supply intakes to the room should be designed to permit location of these additional hoods so that the flow of air into the hoods will be minimally disturbed. In general, the air speed at the face of the hood must be somewhat less than the face velocity of the hood, in the range of less than 20 to 30 fpm.

The purpose of a hood is to capture, retain, and ultimately discharge any noxious or hazardous vapors, fumes, dusts, and microorganisms generated within it. It is not intended to capture contaminants generated elsewhere in the room, although since a hood exhausts so much of the laboratory air when operating, it does serve as a major component of the ventilation system for the room. A few specially designed hoods are intended to confine moderate explosions, but most are not. Unless it is designed, built, used, and maintained properly a hood will not perform its intended function.

Some individuals have been observed to be so hypnotized by the concept of a hood that they continue to use hoods which are not functioning, still counting on them to provide a normal level of protection. It actually has been necessary on occasion to padlock the sashes of hoods closed to prevent this. Unless a hood is fully functional, it should not be used. The OSHA Laboratory Safety Standard recommends that hoods be equipped with a monitor to determine if the air is moving through the face of the hood properly or not. New hoods should be equipped with velocity detectors, and older hoods should be retrofitted. The cost of such sensors is not exorbitant.

There are many different hood designs, such as chemical or biological, even within a single category. Some configurations perform better than others, while the desirability of some features depend upon individual preferences. A good quality general-purpose hood should be able to withstand corrosion, be easily decontaminated (especially for some uses), be suitable for the use of flammable materials, and be capable of withstanding the effects of a fire for a reasonable period of time, sufficient to either allow an attempt to put out the fire or to initiate an orderly evacuation. The design should be such as to minimize the possibility of initiating a fire or explosion. Since research programs change frequently, it would be highly desirable to install a hood used for heavy risk applications rather than attempt to match the specifications solely to current usage. As indicated above, a hood not performing its function adequately represents a risk to the user, rather than acting as a safety device.

1. Factors Affecting Performance

This has been covered briefly before in discussing the location of the hood within a laboratory and the effects of doors, windows, air supply inlets, and traffic. In recent years, there has been substantial research on the factors that affect the performance of fume hoods, what are suitable designs, sizes, means of controlling them, what external factors influence their efficacy, the implications on building energy efficiency by their use, the dispersal of their exhausts, the materials of which they are constructed among others. The following sections will address some of these issues.

a. Face Velocity

One of the major factors affecting the performance of a hood is the face velocity. There have been any number of figures given for the required safe face velocity for the air moving

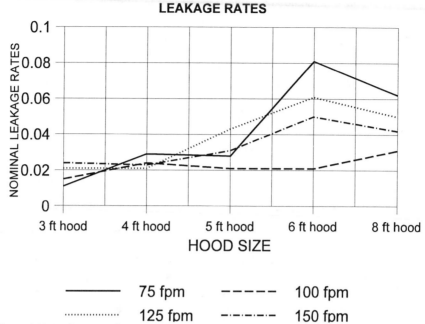

Figure 3.12 Leakage rates for various face velocities for different size bench hoods. The rates are not absolute but are values in ppm for the experimental challenge rate for the contaminant gas.

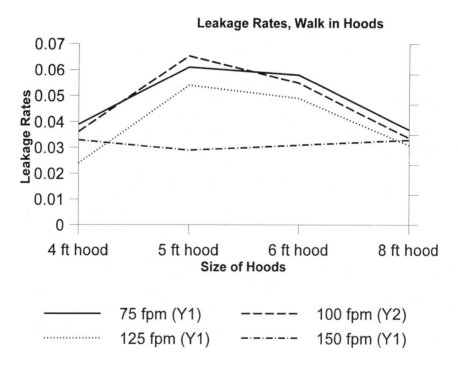

Figure 3.13 Leakage rate data for various size walk in hoods for different face velocities.

through the entrance to the hood ranging from 60 fpm to 125 fpm or greater. In actual fact, virtually any of these would serve for specific configuration of the hoods and placement of the equipment in the hood. Figure 3.12 shows in graphic form the results of a careful study performed on behalf of the Fisher Hamilton, a major manufacturer of fume hoods. The data illustrated in this chart are for bench-type fume hoods, with the sash fully open, under stable state conditions, with no items in the hood and with no one standing in front of the hood. The tracer gas was introduced in three different face locations. The data were acquired for each square foot of the hood face. All the data were acquired and processed by a computer. The data shown thus represent an average over a large number of individual measurements. Based on these data, the following conclusions appear to be possible:

1. There is little difference in the retention rates for the three smallest hoods although on average, for all face velocities. The retention rate worsens sightly as the hood size increases from 3 to 5 feet.
2. The retention rate worsens significantly for all face velocities for the 6 foot hood size.
3. Surprisingly, the retention rate improves for all face velocities for an 8 foot hood compared to the 6 foot hood, but still remained worse than for the smaller hoods.
4. Although not shown, the leakage rates for 10 and 12 foot hoods worsen again for flow rates of 75 and 125 feet face velocity.
5. For the hood sizes shown, 100 fpm appears to be close to the optimum face velocity.

Figure 3.13 shows data similar to that of Figure 3.12 but for walk-in hoods instead of bench hoods. The results are parallel in some respects. In general, a 75 fpm face velocity is less satisfactory on the average than higher speeds and the smallest hood (in this case, a four foot hood) and the largest one perform approximately equally, while the six foot hood does not perform well here either. However, there is one significant difference, on average, the apparent best choice for a face velocity for a walk-in hood is 150 fpm. Based on these two figures, if a conventional bench-type hood is to be used, the use of either a four or five foot hood and operation with a face velocity of 100 fpm would be indicated. For a walk in hood, a four foot hood with a face velocity of 100 fpm would be optimal, although for any larger size hood, the face velocity should be increased to 150 ppm. Clearly, as one goes to 75 fpm or lower, the tolerance of the system to spillage decreases. It would be better to seek other solutions to improving energy efficiency than to try to operate routinely at low face velocities.

The data shown are for a modern conventional type hood with vertical sashes and with aerodynamically designed edges to the openings to the hood interior to facilitate the even flow of air into the hood, but are otherwise operated with no other changes to improve their performance. There are two simple operational changes that would significantly decrease the amount of leakage, sash location, and location of experimental apparatus. Lowering the sash to half height or a bit more has been shown to significantly affect the aerodynamics of the air flow through the hood in a positive manner. In one major, recently constructed facility, the hoods were fitted with stops at a sash height of 18 inches as a normal operating position. A second operational procedure easily done is to require that any experimental apparatus be placed at least 20 cm (8 inches) inside the sash opening. Experimentalists can be easily reminded of this last by painting a bright contrasting line on the interior hood base at this point. Note that, although most hoods today are sized to fit standard casework, it is possible to buy hoods of different depths. Where the choice is an option, a deeper hood provides better containment and obviously allows the experimental apparatus to be based further from the sash opening.

There are other design factors which affect the containment properties of fume hoods. The position of baffles within the hood is important, and should be adjustable to accommodate either lighter than air or heavier than air fumes. Side and corner slots within the hoods help exhaust fumes, and reduce dead air locations within the hoods. A very important factor is the location of the air supply to the room. Recent use of computer modeling has shown that this is a critical factor, and has modified some ideas of where the best locations of supply intakes are and the locations of hoods with respect to each other. The best location for the supplied air typically would be in the ceiling, but possibly not as far as possible from the hoods, but this will need to be evaluated depending upon the configuration of the case work and other equipment normally in the laboratory. It is critical that the air entering the hood face be disturbed as little as possible, and extraneous air currents near the face should not exceed more than one-third of the face velocity, with even less being preferable. Placing two hoods at right angles to each other in one corner is not desirable due to potential interaction of each hood on the intake air flow patterns of the two hoods. A recent National Institutes of Health Study showed the most desirable locations of two hoods within a single laboratory would be on facing walls.

Operational factors have a major impact on hood performance. Two, sash height and internal experimental apparatus location, have already been mentioned. The presence of a heat source can influence the containment, as in effect, the heated air or fumes are likely to take on the property of being lighter than air. Certainly, operable windows with their impact on room air balance and air flow patterns within a room would make management of air distribution and balance difficult. Similarly, leaving doors open or having doors opening and shut near a hood would also affect the room air balance and severely affect hood containment. Traffic patterns near hoods are very important. An individual walking by at a speed of only one mile per hour could create localized disturbances in the face velocity comparable to the steady state face velocity flow rates. Even rapid arm motion on the part of a worker at the hood could do the same. The location of the hood at the back corner of the laboratory module should minimize the effects of traffic.

A major factor is the actual presence of a worker standing in front of the hood. A person standing in front of a hood represents a significant barrier to the free flow of air into the hood. This can not only create turbulence in the air flow on either side of the worker but will tend to create a low pressure zone directly in front of them. As the face velocity is increased, this property becomes more pronounced and may result in fumes from within the hood being drawn back toward the worker, perhaps into the workers breathing zone if the sash is operated fully open. Lowering the sash below the workers face, to the 18 inch level or using a horizontal sash hood where the worker routinely stands behind a section of the sash would obviate this problem.

For a basic hood with an individual, dedicated exhaust system, with no variable control on the hood exhaust, over a period of time, the performance of the exhaust system will normally decrease. The drive belts to the fan will gradually relax a bit, and the interior surface of the exhaust ducts will deteriorate so as to offer slightly more resistance to the air flow through them. Periodically, maintenance will be required to regain the proper performance of the system such as adjusting the dampers, tightening belts or, if necessary, by changing pulley sizes. It may be desirable to design in additional capacity to the exhaust system, to say 125 fpm, which then would allow 100 fpm (if that value is chosen as the most desirable) to be maintained over a longer period. When adjustments of this type are made, the room air balance can be significantly affected so maintenance staff should take care to check the balance as well to ensure proper function of the entire ventilation system.

A common error which frequently leads to low face velocities and poor performance is for the scientist to personally select the exhaust motor from the vendors catalog. Each hood installation needs to be configured by a ventilation engineer. The length and diameter of the duct, the number of bends and turns, the type of fan, and the termination of the duct all will influence the size of the blower motor required. Few research personnel are qualified to correctly select the size fan required. A short review of the factors involved in the calculations in any recent edition of *Industrial Ventilation, A Manual of Recommended*

Practice of the American Council of Governmental Industrial Hygienists should be sufficient to convince most scientists to defer to ventilation engineers to design the exhaust system and specify the components.

As noted earlier, individuals will continue to use hoods that are not working because they appear that they should be working. The air speed of about 1 mph is so slight that unless there are visible fumes, it is not readily apparent whether the hood is working or not. A light found on the outside of many hoods usually only indicates that power is being supplied to the motor. It does not provide a positive indication of air flow. In at least one instance, a perchloric acid hood was checked and found to have zero face velocity. When the fan on the roof was examined, the motor was running, but the blades had corroded so badly that they had fallen off and were strewn across the roof. A positive velocity sensor of some type should be incorporated into every hood, as recommended in the Laboratory Safety Standard. This should be done as part of the original purchase or as a retrofit. It should provide a visual and audible alarm when the air velocity falls below a predetermined safe level and give an indication should the sensor itself fail. Note that this conforms also with the intent of the ADA requirements for disabled persons. The loudness of the alarm and the brightness of the visual indicator may have to be especially selected for persons with corresponding disabilities.

Hoods should be checked upon installation, when any maintenance is done upon them, when there is any modification to the room which could affect airflow patterns in the vicinity of the hood, when any significant maintenance or modifications are made to the building HVAC system, and on a regularly scheduled basis. This last should be done at least annually and preferably quarterly or semiannually. The anemometers or velometers employed for testing the performance of the hoods should be calibrated before use. Measurements should be made at a number of places over the sash opening to assure that the velocity does not vary by more than ±25% (preferably less) at any point. For a conventional hood, it may be desirable to mark the sash heights on the side of the hood opening at the positions where the face velocity drops below the selected acceptable level and above a level of 150 fpm, where increased turbulence and decreased pressure due to a person standing in front of the hood could lead to spillage. Smoke generators are recommended to supplement the air speed measurements to check for spillage under various configurations of face velocity, baffle positions, and experimental apparatus positioning.

b. Construction Materials

The previous section dealt with factors that were concerned with the ability to capture and retain noxious and dangerous fumes and vapors within a stream of air passing through a hood. This section is concerned with the physical ability of a hood to contain and to withstand the corrosive actions of the materials used in it. In principle, the selection of the materials used in the fabrication of the hood should depend upon the types of chemicals intended to be used in the research program. However, hoods are major fixed pieces of equipment that would be difficult and expensive to change as the nature of a research program changes, so, as has been the general tenor of the recommendations in this chapter, the hood should be selected to be as versatile as funds will permit. If a hood is intended for a dedicated use which is not expected to change for an extended period, then it is only common sense not to spend more than necessary. The materials used for the walls and lining, base, sash, and some of the interior fittings will be discussed separately. In the following list, the materials discussed are the ones (with the exception of the first) normally available from commercial vendors. It is rare for hoods to be fabricated by the user. The list is approximately in order of versatility.

Transite: Transite is a material in which asbestos fibers are bonded with a resin and until recently was probably the most popular lining material for general-purpose, heavy-duty hoods. Although it may become discolored with use, it is highly resistant to a large number of chemicals. In recent years, because of the concern for the health effects of airborne asbestos fibers, its use has decreased rapidly and is generally no longer available for new equipment. Although the asbestos fibers are tightly bonded in the transite, some older hoods have been observed in which the fibers had become friable and likely to become detached. When transite is broken or cut, asbestos fibers may again become airborne and pose a potential health hazard to anyone in the vicinity at the time. As older hoods using transite are taken from service, they should not be replaced with the same material, even if hoods using this material were to remain available. The linings of hoods using transite constitute an asbestos hazard when they are removed and disposed of. Their disposition must be done in conformance with EPA and OSHA regulations involving asbestos handling. It is legal when doing asbestos abatements to encapsulate the asbestos containing materials. If a transite lined hood is still usable, its service life may be extended economically by painting it with a chemically resistant epoxy paint.

Stainless steel: Stainless steel may be attacked by some chemicals but type 316 stainless or equivalent is commonly used for the lining of perchloric acid hoods. Type 304 stainless may be used for radioisotope hoods which need to be easily decontaminated. Because of its vulnerability to some chemicals and its relatively high cost, it is not recommended for general purpose fume hood use. Among the problem chemicals for stainless steel are acids and compounds containing halides.

Fiberglass-reinforced polyester: This material is popular for lining general-purpose hoods and is highly resistant to a large number of materials. However, as with stainless steel, there are some materials which may cause some problems with heavy usage. Care should be used in selecting this material to avoid applications involving chemicals for which it is not suited. Among those chemicals for which it is suitable for limited service are acetone, ammonium hydroxide, benzene, and hydrofluoric acid.

Smooth surfaced, glass reinforced cement: At least one major firm has substituted this material for transite as a general-purpose hood lining material. It has now been in service for several years and appears to be giving satisfactory service.

Epoxy resin: This material is comparable to fiberglass in versatility and is affected by some chemicals. If a wide range of chemicals are to be used, advice should be sought from the vendor to ascertain if there are any significant problems for its intended primary use. Among the chemicals for which it may be unsatisfactory are benzene, fatty acids, concentrated hydrochloric acid, and nitric acid.

Polyvinyl chloride (PVC): PVC offers good protection for a wide range of chemicals but is affected by some, including liquid ammonia, amyl acetate, aniline, benzene, benzaldehyde, bromine, carbon disulfide, carbon tetrachloride, chloroform, ether, fluorine, nitric acid, and fuming sulfuric acid. Some perchloric acid hoods use an unplasticized version of this material.

Epoxy painted steel: Light duty hoods, such as may be used sporadically in a classroom situation, are sold made of this material for a liner.

Cold rolled steel: This material is intended for light duty liner applications only. However, carefully prepared and treated to resist corrosion, by a coating of epoxy, this is the most common material for the exterior of hoods.

c. Fume Hood Bases

Hoods which need to be decontaminated frequently, such as radioisotope hoods, often come with a stainless steel base surface which forms an integral part of the hood interior. Other models are available with integrated bases of different materials. However, many hoods are sold without bases and they can be selected separately. The most popular material used for fume hood bases is a molded, modified epoxy resin. Since the material is molded, it is easy to incorporate a shallow depression in the work surface to function as a water tight pan to catch and retain spills. Stainless steel bases are also used.

d. Sashes

The most popular sash configuration is the vertically sliding type. The sash must be counter weighted, especially if the sash window is made of heavy glass. In order to avoid a "guillotine" effect should the counterweight cable break, a safety device must be incorporated in the sash. As for transparent materials used for the sash window, only a few are used in good quality, currently available commercial fume hood models.

Laminated safety glass: This is probably the best material for a sash because of the resistive properties of glass to most chemicals and the safety features provided by laminating tempered glass. It is not as effective as is tempered glass for higher temperature applications (good only up to about 70°C). In the event of a moderate explosion, it is possible that the sash will remain within the frame. If the sash remains essentially intact, the employees in the laboratory will continue to be afforded some protection against fires within the hood and escape hazardous fumes, although if the sash were to be expelled more or less intact it could represent a significant danger.

Tempered glass: Although tempered glass breaks into fragments that are not sharp in the event of an explosion, the result will be a loss of glass from the sash and the loss of protection afforded by the enclosure. Tempered glass will withstand more heat (up to about 200°C) than laminated glass.

Clear high impact PVC: PVC will provide impact protection comparable to laminated safety glass, but will not be as resistant to the effects of chemicals in the hood.

Plexiglass: Plexiglass is not as durable as the other materials and is primarily provided in economy models.

Polycarbonate: Polycarbonate sashes are recommended when heavy use of hydrofluoric acid is involved. Polycarbonate plastics offer good physical strength.

e. Internal Fixtures

The lights in the hood should be shielded from the hood body in, as a minimum, a vapor-proof enclosure. These can also be selected as explosion-proof units should the usage include substantial amounts of very flammable materials. In fact, if a large portion of the work in the hood is expected to involve very volatile flammable materials, the hood should be designed completely as an explosion-proof unit. In any case, all electrical outlets and switches should be located outside of the hood on the vertical fascia panels. Similarly, utilities should be provided by remotely controlled valves in the side panels operated by handles outside the hood. The connections to the utilities inside the hood must be chemically resistant and should be clearly identifiable as to the utility provided.

2. Types of Chemical Fume Hoods

There are several different types of fume hoods: (1) conventional hood, vertical sash, (2) conventional hood, horizontal sash, (3) bypass hood, (4) auxiliary air hood, (5) walk-in hood, and (6) self-contained hood. The differences in types 1, 4, and 6 are especially important in terms of the amount of tempered air lost during operations, while 1, 2, and 3 differ primarily in the airflow patterns through the sash openings. Figures 3.14 to 3.19 illustrate each of these types and the air currents through them during typical operations. In addition, there are specialty fume hoods for perchloric acid and radioisotopes, which will be treated separately. All of the hoods discussed in this section will be updraft units, where the exhaust portal is at the top of the hood, with of course, the exception of the self-contained type.

One class of hood which will not be discussed here is the canopy hood. These have their uses, where it is desired to capture and exhaust hot fumes carried upward by convection currents until they come close enough to the canopy so that the fumes become entrained within the hood. The speed of the air movement in the vicinity of the hood face, due to the air flowing through the canopy, falls off very rapidly to about 7.5% at a distance equal to the effective size of the canopy opening. If the canopy is at a reasonable distance away from the bench top, the airflow at the work surface due to the hood will be on the order of the average air movement speed within the room, or less. Because of this, canopy hoods are very wasteful of the tempered air within the room for the amount of toxic fumes that they discharge. A further disadvantage would be that the fumes, if drawn upward, would pass through a worker's breathing zone. For these reasons, canopy hoods are not recommended as general usage laboratory fume hoods. They can be used to capture hot gases vented from some types of equipment.

In addition to hoods, there are a number of types of localized exhaust systems which can be very effective for specialized applications. A major reason for using these systems is that they do not waste as much tempered air as would a hood. These will also be discussed briefly in the following sections. Also some equipment has its own dedicated exhaust system. For efficiencies sake, these systems should be exhausted through a fume hood.

a. Conventional Fume Hood

A conventional fume hood with the sash open and closed is illustrated in Figure 3.14. Note that a section of the internal baffle essentially remains in contact with the sash at all opening positions. It is equipped with a vertically opening sash and an interior baffle so arranged so that some of the air sweeps the base of the hood and is directed up behind the baffle to the exhaust opening. There can be additional slots in the baffle, perhaps one in the middle and one near the top, through which air can pass to provide more uniform airflow. The remaining air passes through the hood interior and is directed into the exhaust portal over the top of the interior baffle. The volume of air through the hood is relatively constant, although some losses occur as the sash opening becomes smaller. A more important consequence of the decreasing sash opening is the increasingly high velocity of the air through the narrowing opening. The increased air speed could disturb operation of the experiment within the hood, and the effects of turbulence around apparatus sitting on the bottom of the hood would be increased. A person representing an external obstruction could also give rise to increased turbulence so that spillage from the hood would occur, although with the bottom of the sash well below face height, the possibility of toxic fumes entering directly into the breathing zone would not be high.

A variation on the conventional hood is one with a horizontal sash, as illustrated in Figures 3.15 and 3.16. These are made with either two or three sliding sections. One with two sections does not allow more than half of the hood face to be open at any one time, while one with three would permit up to two-thirds to be open. With three sections as shown in Figure 3.16, one could have the fume generating equipment placed behind the center section and yet be able to reach the equipment from either side. Either arrangement would provide a larger working area with lesser hood exhaust volume requirements than a conventional, vertical sash hood of the same width. Such an arrangement would allow an individual to stand behind the center section to work and be completely protected from fumes being drawn back outside the hood. Standing in this location would also afford a good degree of protection against moderate explosions or chemicals thrown from runaway reactions within the hood.

A significant commercially available variation of the horizontal sash hood is the "HOPEC" hood which combines both horizontal and vertical sash movement, using a two section horizontal sash. In this hood, the vertical movement of the sash is limited to a

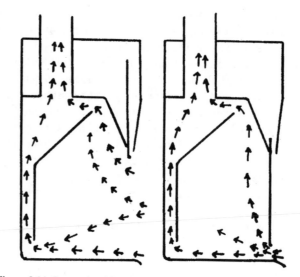

Figure 3.14 Conventional hood, sash open and closed.

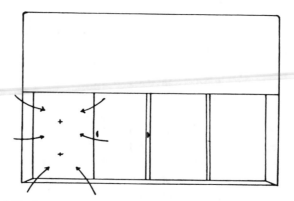

Figure 3.15 Conventional hood, horizontal sash.

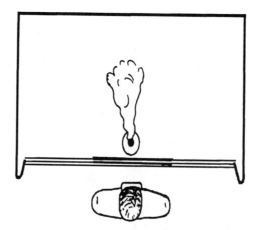

Figure 3.16 Protection offered by overlapping sashes.

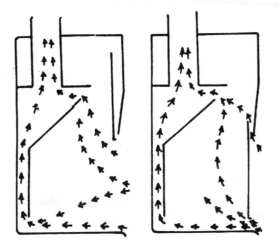

Figure 3.17 Bypass hood. Sash open and closed.

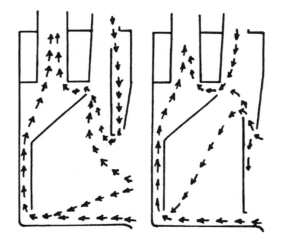

Figure 3.18 Auxiliary or add-air hood, sash open and closed.

maximum of halfway. Thus, the maximum open face is limited to no more than half, resulting in a maximum hood exhaust capacity of half that of a conventional hood with the same working area. Many newer facilities have selected these hoods because of the combination of energy efficiency coupled with the safety factors mentioned in the previous paragraph.

b. Bypass Hood

The bypass hood (Figure 3.17) is designed so that a portion of the air entering the face of the hood may pass over the top of the sash opening as well as below it. This has two consequences. The first is that the air velocity near the work surface remains reasonably constant, so that excessive air speeds which could be detrimental to delicate apparatus or experiments will not occur. The second is that there is less static pressure and hence less frictional resistance to the flow of air than with the conventional hood, so that the volume of air through the hood remains more nearly the same at different sash heights, permitting better control of the laboratory air balance. This is probably the best basic choice for a laboratory fume hood that is not equipped with a Variable Air Volume (VAV) device.

c. Auxiliary Air Hoods*

This type of hood is somewhat controversial. The original comments of Horowitz et al. which appeared in the 1974 edition of this handbook are still echoed in today's literature.

"The auxiliary air hood attempts to reduce air-conditioning requirements by providing a separate supply of air that has not been cooled and dehumidified in the summer or fully heated in the winter. The supply of air for such a hood may be drawn from outside or from the service chases within the building, which are, in turn supplied by air from attic or mechanical equipment rooms. Such hoods can substantially reduce the air-conditioning equipment capacity required to make up losses through fume hoods; operating costs can likewise be reduced. However, there are a number of disadvantages to such a hood. One type of auxiliary air-supply hood discharges untreated air just in front of the hood, usually at the head. A scientist working at the hood must work in unconditioned air. The disadvantages are obvious, and the annoyance of scientists has been evidenced by their very human attempts to invent means of foiling the intended mode of operation. One such effort consists of securing cardboard over the outlets with adhesive tape, thus closing or reducing the auxiliary supply. Attempts to rectify this problem by partially cooling or heating the air supply, depending upon the season, substantially reduce any economic advantage of this type of hood. Another type of hood introduces the auxiliary air within the hood enclosure and is inherently unsafe because the face velocity is reduced below the rate necessary to capture fumes."

There have been changes in hood design since the material quoted above was published which permits the supplementary air to be brought down outside the sash, but not in such a way that the operator will be standing in the airflow. Figure 3.18 illustrates such a design modification. Specifications on this type of hood usually permit up to 70% of the air to be provided by the auxiliary air supply with at least 85% of the supplied air passing through the hood face. Untempered air, especially during the winter, when the outside temperature may be very cold, would mandate that this supplied air be heated to at least 10 to 15°C, eliminate much of the savings for this type of hood. The installation and configuration of this type of hood is critical and proper operation is difficult to maintain. Although still manufactured and sold, their popularity has decreased and most laboratory designers of the author's acquaintance would not specify them in new construction.

d. Walk-In Hood

Walk-in hoods (Figure 3.19) are hoods which usually rest directly upon the floor or on a pad resting directly upon the floor. They are designed to accommodate tall apparatus which will not fit in a standard hood sitting upon a base unit or work bench. Because their height is usually somewhat out of proportion to their width, the airflow characteristics may not be as favorable for avoiding spillage as with a standard hood, for the same face velocity as noted in the earlier section on the effect of dimensions on hood performance. Because their height would require an abnormally long sash travel, these hoods are sometimes provided with dual

---

*   It should be noted that the example given by Horowitz, *et al.,* where the users would continue to use a hood which they themselves have caused to function improperly represents a serious attitude problem among some, although by no means a majority of scientific workers, who put expediency and comfort above safety.

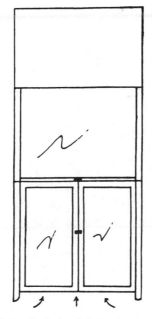

Figure 3.19. Typical walk-in hood.

sashes, each of which would cover half the opening, or with a single sash which would come down only about half way, with swinging doors being used to provide access to the lower portion. These doors may terminate a few inches above the floor to ensure that there is always some airflow through the entire length of the vertical space. As with other types of hoods, they may be obtained in different configurations, such as bypass or auxiliary air units.

e. Self-Contained Hoods

These are listed last because they are the least desirable alternative as chemical fume hoods. A self-contained chemical unit should not be confused with a biological safety cabinet, which also is often self-contained. Some of the newer designs give a reasonable approximation to a conventional hood in performance, although they are not recommended for general usage. However, there are circumstances where they can be used relatively safely. Good quality units can cost as much as a conventional hood, so their use is not mandated upon the basis of purchase price. Their use is usually contemplated when a hood is needed, but no exhaust duct is accessible and it is impractical to install one. The commercial units available are generally intended for use in histology and cytology procedures involving materials such as xylene, formalin, toluene, alcohol, etc. The newer units provide for a choice of filters which makes them suitable also for applications other than organic solvents. A typical self-contained unit pulls room air through the face of the unit over the work surface and through a filter selected for the material intended to be used in the hood. The fan unit does not become contaminated because the air is filtered before it reaches it. A typical filter will absorb several pounds of the solvent before it becomes saturated and must be replaced. For average use, the filters may last 1-to-2 years, at which time they must be replaced. A major problem with most of the units available is the inability to tell when the filter has become saturated. At least one vendor has solved this problem by placing a material at the back of the filter which reacts with the solvent when it is close to no longer being absorbed by the filter and then emits a pungent odor. The same manufacturer also provides an electronic detector for solvent vapors. The unit provided by this vendor probably comes about as close to the performance of a regular fume hood as any self-contained unit on the

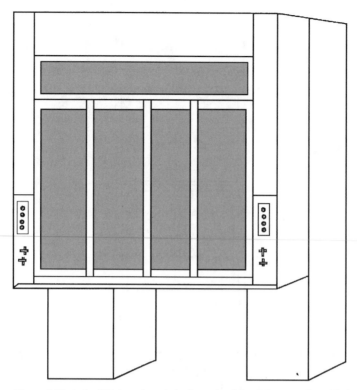

Figure 3.20   A simplified version of a horizontal sash hood that complies with the provisions of the Americans with Disabilities Act. The space between the two pedestals is wide enough and high enough to accommodate an individual in a wheel chair. When this feature is not needed, a storage cabinet may be put in this space. The pedestals are also usable for storage.

market, but it is also comparably expensive. The face velocity is set at 80 fpm, which is comparable to that of a standard hood, although it is in the range in which spillage of fumes from the hood can occur unless the hood is used with care (fume generators, 20 cm from the sash opening or better) and located where external air movement problems will be minimal. However, even this relatively well performing unit is no real substitute for an installed, ducted, fume-hood which should be the choice whenever possible. The physical design is similar to a standard conventional hood. Purchase and use of these units should be contingent upon the prior review and permission of the Environmental Health and Safety Department.

f.   Hoods for Compliance with the Americans with Disabilities Act (ADA)

The ADA is a federal regulation that requires employers to make reasonable accommodations for individuals with disabilities which would permit them to secure employment in the same activities as those without disabilities. Although there are many types of disabilities that are covered by this act, one that often comes to mind is the one requiring a wheelchair for mobility. Fortunately, this disability is one of the most straightforward to accommodate. A wheelchair will require wider aisles and a place in which to turn around, work benches with variable height adjustments, and controls within easy reach of someone sitting down. The workbench should allow space for the individual's legs under the bench, in order for the individual to reach a reasonable area on the work surface. One of the most expensive items to adapt would be a fume hood and special hoods are now commercially available for individuals in wheelchairs. An example of a horizontal sash ADA compatible hood is shown in Figure 3.20. Note the base level has been lowered to a height appropriate to accommodate a person working sitting down. The controls are placed

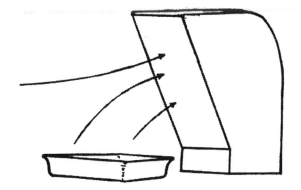

Figure 3.21 Local fixed exhaust.

on the bottom of the side wings of the hood to allow a person to use them comfortably while sitting. A fixed horizontal transparent panel is placed above the movable sash so a standing person could view the work area within the hood.

g. Other Modes of Exhaust

The trend is to do more laboratory work in hoods, but the use of hoods is a substantial burden on the energy budget of a building and alternatives should be carefully considered. An alternative is to use spot ventilation. Many laboratory supply houses provide small exhausters in which an adjustable inlet can be placed very close to well-defined spot sources of noxious fumes. The fumes are then typically discharged into a fume hood. If the work is repetitive enough to warrant setting up a permanent spot exhaust, then the designs shown in Figures 3.21 and 3.22 might be usefully employed for limited risk work. In the first design, a repetitive operation involving pipetting into test tubes or sample vials could be done virtually in the face of the cowl shaped spot exhauster. With this physical relationship, the air intake should be very effective in capturing any aerosols or fumes which might be generated. Note that the airflow would be, as with a normal fume hood, away from the research worker.

The second design would be more effective for a tray-type operation. Here, one or more narrow slots, with air being drawn through them at a relatively high inlet velocity, are placed at the rear or, occasionally to the side of the workbench close to the level of the bench top. This design takes advantage of the fact that most solvents are heavier than air. Air flowing across the surface would entrain the vapors from the tray and exhaust them through the slots at the rear of the workbench. The fumes would not have an opportunity to rise into the worker's breathing area, being pulled back and away from the worker. A modified version of this system is used for silk screening. In this version, the entire circumference has either an aerodynamic slot around the edge or the last several inches close to the edge of the table top are perforated with hundreds of holes, through which in both cases air is pulled down prior to being exhausted. There are consumer range tops for cooking built on this principle that work very well. Smoke from the food being prepared rarely rises more than an inch or two above the cook-top surface before being captured and exhausted. Chemical fumes should act in a similar manner but the fume source should be placed as close as possible to the slots.

There are several other types of local exhausters. A flexible hose connected to an exhaust fan could be used but the work must be placed within a few inches of the end of the flexible hose for it to be effective and the use of a 4 or 5 inch hose could be very wasteful of energy. There are a variety of slot type exhausts, configured much like the common vacuum cleaner crevice tool. Again to be effective, the fume source must be quite close to the exhaust device In any event, the use of a localized exhaust, if used properly, will be much more energy efficient than a fume hood.

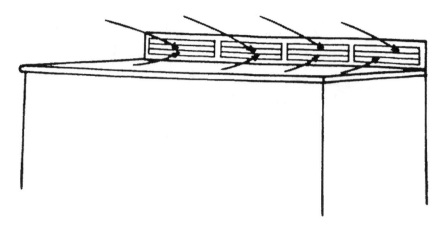

Figure 3.22 Rear plenum exhaust system.

h. Perchloric Acid Hood

Individuals working with perchloric acid and perchlorates must be trained in procedures necessary to conduct their research with maximum safety. These are extremely dangerous materials.

There is a section later on the problems of working with perchlorates and perchloric acid; this section will be restricted to a discussion of the critical factors applicable to the proper performance of perchloric acid hoods. It is sufficient to say at this point that perchloric hoods are designed to avoid accumulation of precipitates from perchloric acid or to avoid perchloric acid from coming into contact with materials with which it may react vigorously and explosively. Hoods designed for hot perchloric acid use should not be used for research using other types of materials. The hood should be prominently labeled with a sign stipulating that it is for perchloric acid work only. Exhaust systems serving hoods used with heated perchloric acid should not be manifolded into a common exhaust plenum.

For conventional hoods and their variants which have been covered earlier, the discussion of appropriate ducts and exhaust fans has been deferred to separate sections. However, perchloric fume hood systems are uniquely dangerous and will be treated as an integrated topic.

Perchloric hoods are usually constructed with an integral liner of a single piece of stainless steel, such as 316 stainless, which will resist the effects of the acid, although PVC can also be used as a liner. The liner should have coved corners and as few seams as possible to allow ease of decontamination. In order to avoid buildup of perchloric precipitates in the hood and duct system, a hood intended to be used for perchloric acid work must he equipped with a rinse system which will make it possible to thoroughly flush the interior of the hood and duct work with water. This may be done with a manual control system or by an automatic system that will come on and rinse the system for 20 - 30 minutes at the end of a work session. A combination of an automatic system which can be bypassed for additional rinses is preferable so the researcher may choose to clean the system if necessary.

The duct work should also be stainless steel or PVC. Caution must be exercised to ensure that workers on installation do not use standard organic caulks to seal the joints. Organic materials, when contaminated with perchloric acid, are highly flammable and dangerous, and such joints will also tend to leak, allowing perchloric acid to escape outside the duct work. Under circumstances which allow this errant material to be exposed to heat or to receive a sudden shock, the result could be a fire or an explosion in the space outside the duct. The most desirable procedure for stainless steel ducts is to weld the sections of stainless steel ducts together. This will require heliarc welding, which is a relatively expensive procedure compared to welding ordinary steel duct work. Some fluorinated hydrocarbon materials can be used as sealants if welding is not feasible.

It is recommended that the interior fittings of a perchloric acid hood should be non-sparking and the lights should be explosion-proof. This concept should be extended to any apparatus placed in the hood. With the dangers already represented by perchloric acid, there should be no contributory factors that could initiate an explosion. PVC ductwork can be employed instead of stainless steel, but it would be much less likely to remain intact in the event of a significant fire exposure. However, if the duct work is enclosed within a 2-hour fire-rated chase, as it usually should be, this would not be a serious drawback.

The ductwork for a perchloric acid hood should have as few bends as possible and be taken to the roof in the shortest, most direct *vertical* path. No horizontal runs should be permitted, and even slopes of less than 70° to 80° should be avoided wherever possible. For aesthetic reasons, architects prefer to place exhaust ducts away from the edge of a building so that they cannot be seen easily. As a result, horizontal runs of 100 feet or more of perchloric fume hood exhaust ducts have been observed in some older designs. Even if a wash-down mechanism were incorporated in the design, it would be unlikely to come into contact and clean the upper portion of the duct in the horizontal section. In one instance in which a perchloric hood was installed below grade in the basement of a building, the exhaust duct first was run horizontally for approximately 75 feet under the floor of an adjacent section of the building and an exhaust fan was installed in this horizontal run. The duct then ran vertically for three floors. When this was discovered, the horizontal section of the ductwork beyond the fan had corroded through and perchloric acid crystals were observed on the external surface of the duct and on the ground below it. This became a major and costly removal project.

As a minimum, the blades and any other portion of the exhaust fan coming into contact with the perchloric fumes should be coated with PVC, Teflon, or another approved material that will resist the effects of the perchloric acid. An induction exhaust fan, where none of the fumes actually pass through any part of the motor or fan is recommended. Under no circumstances should the exhaust fumes be directed down upon the roof to be absorbed in the roofing material. The contaminated roofing material could itself constitute a danger. The exhaust point should be well above the roof (at least 10 to 15 feet) to avoid the fumes readily reaching any portion of the roof prior to dilution by the outside air. The wash-down mechanism should be capable of cleaning the entire duct from the point of exhaust all the way back to the hood. The wash down system plumbing should automatically drain when shut off to avoid rupturing the supply lines due to freezing in the winter. The rinse water may be permitted to drain directly into the sanitary system where it will be quickly diluted.

Normally, in this handbook, specific manufacturers and brand names are avoided but Labconco makes an excellent perchloric acid ejector duct with all of these desirable features that exhausts the perchloric fumes at a point about 10 feet above the roof level. It is not inexpensive but performs exceptionally well.

Maintenance personnel, as well as the laboratory employees, should be trained in the dangers inherent in the use of perchloric acid and the potential for injury represented by any residual material in crevices or other places where perchloric acid or byproducts might accumulate. This training should be done in a positive way to instruct individuals how to work with such material properly, and not in such a way as to unduly frighten anyone.

i. Radioisotope Fume Hood

Radioisotopes are frequently used in the life sciences and in nuclear medicine in diag-nostic applications. Only rarely are the amounts employed large enough to be of immediate danger to the laboratory worker, if used properly. In addition, many of the more commonly used radioisotopes emit low energy beta radiation only, which will not penetrate the skin and some have short half-lives as well. However, not all radioactive materials have this last favorable property and many even relatively safe materials may cause delayed injury, perhaps 20 years later, if ingested and inhaled. The word "may" is not to be construed here in any definite sense of "they will definitely cause injury." It is intended to indicate only that the possibility of a health effect may be increased. At very low levels, there is no direct evidence of either immediate or delayed injury although there has been an enormous number of studies trying to resolve the issue. Most studies, seeking to prove the point either way, have typically been controversial and not accepted universally. The possibility of health effects mentioned above is based on a very conservative linear extrapolation from known detrimental effects at high levels to possible detrimental effects at the much lower levels usually encountered in research. However, because of the possibility of a finite risk and because of public concern, it is public policy that laboratory use of radioactive materials be stringently controlled to minimize exposures. A carefully monitored license is required. There is a separate major section in Chapter 5 devoted solely to radiation safety.

A radioisotope fume hood is designed to minimize risks of exposure to the laboratory worker by making it easier to maintain the hood in an uncontaminated condition. The liner is usually made of a single piece of stainless steel, as with perchloric acid units and for the same reason, for ease of decontamination. There should be a minimal number of seams or hard to clean areas. The major classes of research employing radioisotopes are often intended to retain the compounds containing radioisotopes in the end product, since a frequent purpose in employing radioactive materials is as a tracer. Therefore, although there may be some radioactive fumes generated, there may be less than with some other dangerous materials. One of the largest sources of concern is that many of the procedures used in the life sciences tend to generate aerosols, i.e., very fine droplets of material which can escape the work area unless care is taken to contain them. To retain any releases in whatever form, it is usually recommended that the duct work for a radioactive fume hood be of stainless steel since this is easily decontaminated.

Where relatively high levels of radioactive materials are used or where the levels of fumes (or aerosols) generated could be substantial, it may be necessary to install an absorbent filter or for particulate fumes, a two-stage high efficiency particulate air (HEPA) filter unit, which will filter out 99.97% of all particles 0.3 microns in size or larger, in line in the exhaust duct to ensure that the legal minimum concentrations of activity can be maintained at the point where the fumes are discharged to the outside. The most appropriate location for this filter is at the exit portal of the hood, since this will prevent any of the ductwork from being contaminated and the location will make it convenient to service the filter. Where a HEPA filter is used in a fume hood involving chemicals, it is essential that a device to monitor the air velocity through the face of the hood be installed. The air passages in a HEPA filter used in a chemical system where particulates are generated will soon become clogged, increasing the difficulty of drawing air through the filter so that the efficacy of the fan unit will fall off rapidly. A less expensive prefilter will significantly extend the life of the HEPA filter. A device to measure the pressure drop across the filter unit may be used to monitor the condition of the filter, but this will not necessarily measure the velocity of the airflow through the hood and duct. Gaseous radioactive materials will not be stopped by a HEPA filter. An activated charcoal or alumina absorbent filter should be used for these materials.

Although the stainless steel liner is relatively easy to clean, a number of measures make it easier to keep it in an uncontaminated condition. Among these are the use of trays to contain any spills which might occur and the use of plastic-backed absorbent paper (which may be discarded as waste) on the work surfaces.

As noted above, many of the procedures used in the life sciences, where the preponderance of radioactive materials used in research laboratories are employed, generate aerosols, so in cleaning the interior of the hood, attention needs to be paid to the interior walls and sash surfaces.

The property of emitting radiation which makes radioisotopes useful in research and potentially dangerous is also the property which makes their use relatively easy to control by research personnel who conscientiously follow good laboratory practice. The work surface, the hands and clothing of the persons performing the work, and the tools and equipment employed in the work can be easily checked for contamination by the use of appropriate instrumentation. It is unfortunate that the dispersion of many other dangerous chemical and biological agents in the laboratory cannot be monitored and controlled so readily.

The only other unique feature in a radioactive fume hood may be the need of the base to support shielding materials. The base may need to be stronger than usual to support the concentrated weight of the lead shielding which may be needed to protect the workers from radiation, such as when synthesizing a compound, where relatively large amounts may be employed and hence substantial amounts of shielding may be required. Most of the time, only small quantities of radioactive materials are in use at a given time, so personnel shielding needs normally would be small. However, substantial amounts of lead may be needed to provide adequate shielding against background radiation for the sensitive detectors used to detect minute traces of the experimental radioisotopes in the material being studied.

Fume hoods used for radioactive materials should be marked **"RADIOISOTOPE HOOD"** and in addition should be labeled with a **"CAUTION—RADIOACTIVE MATERIALS"** sign bearing the standard radiation symbol. The isotopes being used should be identified on the label. Under some circumstances, specialized additional signs may be needed.

j. Carcinogen Fume Hood

Clearly, work with a carcinogen mandates a high-quality fume hood. The features discussed in the previous two sections, which minimize spaces for materials to be trapped and facilitate decontamination, are strongly recommended. In the OSHA General Industry Standards, Subpart Z-Occupational Health and Environmental Control, many of the specific standards for the carcinogens regulated in this section contain the following standard paragraph relating to fume hoods:

> "'Laboratory type hood' is a device enclosed on three sides and the top and bottom, designed and maintained so as to draw air inward at an average linear face velocity of 150 feet per minute with a minimum of 125 feet per minute; designed, constructed and maintained in such a way that an operation involving (name of regulated carcinogen) within the hood does not require the insertion of any portion of any employees' body other than his hands and arms."

Note that the face velocity recommendations just cited may be too great if the data cited earlier are correct. Generally, 100 fpm seems to have developed to be a consensus standard and is supported by the data in Figure 3.12.

3. Exhaust Ducts

Exhaust ducts are necessary to take the fumes from the hood to the point at which the fumes are to be exhausted. For the purpose of this section, it will be assumed that the duct will exhaust directly to the outdoors, rather than to a plenum.

It is recommended that if it is desired to manifold more than one hood into a common duct, prior to the entry into a common plenum at negative pressure to the individual ducts, that this practice be limited to hoods within the same room. Otherwise, it is less likely that individuals using different hoods would be aware of each other's activities, and one might make changes which would affect the performance of hoods in the other room.. For example, a pressure differential might be established between one laboratory and another, so that fumes could be exchanged between the two areas. In addition, maintenance of the air balance in the

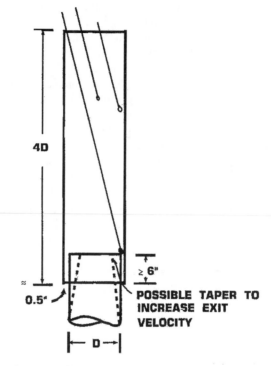

4D

≈

0.5"

≥ 6"

POSSIBLE TAPER TO INCREASE EXIT VELOCITY

D

Figure 3.23 Exhaust duct rain shield.

individual rooms may be made more difficult, and there may be problems in complying with fire codes, if fire walls are penetrated.

a.   Materials

Many of the comments in Section 3.II.B.1.b regarding materials are relevant here as well. At one time, transite was a very popular duct material, but it is no longer recommended due to the concerns regarding asbestos and is no longer used in new installations. The joints between sections required cutting by the maintenance personnel preparing or installing the ductwork, with a consequent release of asbestos airborne fibers which could be inhaled. Stainless steel is used for special types of applications, such as for perchloric acid systems, but is not universally suitable for all chemicals. PVC can be used for many applications. It is easy to install and custom fit, and is comparatively inexpensive. It is possibly the most commonly used material currently. Steel ductwork coated with a chemically resistant material, such as an epoxy coating, is popular because it is relatively inexpensive and is especially adaptable for custom installations. For those applications where it is needed, stainless steel is also used.

b.   Dimensions[*]

There are numerous sources of frictional losses for the airflow through the hood and ductwork to be overcome by the exhaust fan. The air entering the hood must be accelerated from a minimal velocity in the room to the velocity within the duct. Due to physical factors, there is always some turbulence created during this process, so that the pressure difference created by the fan must be sufficient to provide the desired airspeed in the duct and overcome

[*]   For a more thorough discussion of the material covered briefly in this section, the reader is referred to ANSI Z9.2 or in the ACGIH, *Industrial Ventilation, A Manual of Recommended Practice,* for an even more complete treatment.

the losses due to turbulence. In the specifications for a hood, the hood static pressure data provided are a direct measure of the total of the energy needed for acceleration and to overcome turbulence losses. The static pressure will be proportional to the square of the velocity of the air entering the hood face.

There are significant losses along even a straight, relatively smooth section of duct due to air friction. This is due to the energy required to maintain a velocity gradient ranging from the essentially stagnant air in contact with the walls to the rapidly moving air near the center of the duct. For example, for a nominal 10-inch internal diameter PVC duct, through which 1500 cfm of air is passing, each 10-foot section contributes about 0.1 inch, water gauge, pressure loss. This also varies rapidly with the volume of air movement; by increasing the speed by one third, the pressure loss is increased by about two thirds. The diameter of the duct is also important. For the same volume of air, 1500 cfm, the losses in a nominal 8-inch I.D. duct will be more than three times greater than in the 10-inch duct, while the losses in a 12 inch duct will only be about 40% of the amount in a 10-inch duct. In general, the total amount of duct friction in a round duct varies directly proportionally to the length, inversely to the diameter, and is proportional to the square of the velocity of the air moving through it. The equations below give an approximate value for the skin resistance of a duct.

$$wg = \frac{2\,LV^2}{CDT} \qquad \text{for circular ducts}$$

where wg = pressure loss in inches water gauge
L = length of duct in feet
V = velocity of air in feet per second
D = diameter of the duct in inches
T = absolute temperature on Fahrenheit scale (460 + °F)
C = constant = 55 for new steel ducts and 45 for older steel ducts

Similarly for a rectangular duct, sides A and B:

$$wg = \frac{LV^2}{CABT}(A + B) \qquad \text{for rectangular ducts}$$

A shock loss occurs whenever there is a sudden change in the air velocity caused by a change in the direction of the air or a change in the diameter of the duct. Every bend in the duct dramatically decreases the efficacy of the fan motor. The sharper the bend, generally the more severe the loss becomes. This loss may be estimated from the following equation:

$$wg = \frac{0.1188V^2k}{T}$$

where k varies as follows:

Mean Radius of Bend/Duct Diameter or Width (width is dimension of side measured along radius of bend)	Circular Ducts k_c	Rectangular Ducts k
Right angle elbow	--	1.25
0.50	0.75	0.95
0.75	0.38	0.33
1.00	0.25	0.17
1.50	0.17	0.09
2.00	1.50	0.08
3.00	0.13	0.07
6.00	0.10	0.05

The last equation also can be used to estimate the loss in changing duct sizes by substitution of k_c for k in the equation. The ratio given below is the ratio of the smaller flow area to the larger:

Ratio	0.1	0.2	0.3	0.4	0.5	0.6	0.7	0.8	0.9	1.0
	1.25	1.20	1.15	1.05	0.95	0.80	0.65	0.45	0.25	0

Similarly, if a secondary branch joins another duct, the angle at which it joins critically affects the fractional velocity pressure loss of the air through the secondary branch, ranging from a very small percentage at shallow angles to over 40% at 60°. Branches should not enter at right angles to the primary duct or opposite each other.

It has already been pointed out that a deflecting weather cap is inappropriate for a fume hood since it is highly undesirable for the exhaust fumes to be deflected back toward the roof. It would also cause significant pressure losses. The design shown in Figure 3.23 on page 156 would cause minimal pressure losses and yet would protect almost as well against rain, unless the rain were falling virtually straight down. Unless the fan were off, the normal speed of the discharged air would keep any rain from entering the duct as well.

c. Fan Selection

Centrifugal fans are the most commonly used type of fume hood exhaust fans. Within this category are several different variants, the choice depending upon the requirements of the individual installation (Figure 3.24).

Forward curved or squirrel cage fans are primarily suited for relatively low pressure applications, from 0 to 2 or 3 inches water gauge. They typically have a relatively large number of small blades, set close together around the periphery of the wheel, each blade being curved forward in the direction of wheel rotation. Because the spaces between the blades in this type of fan are small, they are prone to collecting dust and becoming clogged.

Paddlewheel or radial blade fans are used where high pressures of 15 inches water gauge or more are required. The radial blades, typically six, are relatively heavy and resist corrosion and abrasion well. Since they have large spaces between the small number of blades, they are the least likely to become clogged. Fans with backward curved blades are best for medium pressures, from approximately 1 to 8 inches water gauge. This type of fan resembles the forward curved type in that the blades are placed around the periphery of the wheel, but they usually have fewer blades, typically less than 16, but more than radial blade fans. This type of fan operates more efficiently than the other two types.

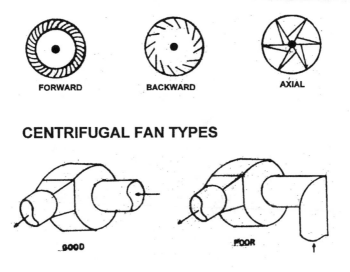

CENTRIFUGAL FAN TYPES

FAN CONNECTIONS

Figure 3.24 (Top) Types of centrifugal fn designs. (Bottom) Exhaust duct fan connections.

The inlets to the fans should be designed to take maximum advantage of the fan's performance capabilities. Either the duct should feed directly into the fan intake or be brought into it with a smooth bend to the duct. Connections which require the air to make a right angle turn will significantly and adversely affect the performance of the fan.

For general purpose fume hoods, as a minimum, the fan blades should be nonsparking. Usually this is achieved by using coated aluminum or stainless steel fan impellers. If the fume hood is certain to be used heavily for highly flammable solvents, the motor should be selected to be explosion-proof as well. If the exposure to corrosion is expected to be severe, materials with good corrosion resistance, such as PVC or fiberglass reinforced polyester materials, should be selected for the fans. If, for reasons of economy, ordinary steel fans are used, they should be coated with Teflon, or if the corrosion problem is somewhat less severe, PVC or polypropylene will serve.

For flexibility, the fan should be driven by a belt from the motor, since within limits, this will allow the speed of the fan to be changed to make up losses in efficiencies in the exhaust system over a period of time, or to make planned design changes. If a deliberate change is to be made in the amount of air discharged through a given duct, the ramifications of the change on the room air balance and the performance of other hoods in the facility should be considered in advance.

4. Energy Management

With all the variation allowed by all of these configurations, the room air supply system must be so designed as to maintain the room air balance, no matter what sash arrangement is used, or for that matter, whether the hoods are in operation or not. The hoods and HVAC systems need to be interlocked to ensure this. In addition, if the HVAC system should fail, the hoods need to be equipped with an alarm to alert users of the failure. The hoods in such a case should be shut down if this not done automatically and the sashes closed to prevent fumes from escaping into the laboratory. In newly built facilities, especially larger ones, the

HVAC systems are managed by a computer system to ensure the room air balance is maintained at all times, and so designed as to provide an alarm should the air velocity fall below the specified value in the exhaust ducts.

The primary consideration is maintaining the proper functioning of the facility so as to maintain a safe working environment. Personal comfort is a secondary, albeit important consideration, but an increasingly important consideration is energy efficiency. This has been touched upon in several of the preceding topics but it is important enough to devote some space to the topic.

The hoods are not the only source of energy consumption, possibly not even the major one. The electrical load for lighting and equipment, the heat load represented by the occupants themselves, the comfort factors of temperature, humidity, all represent energy utilization but with all that, the discharge of tempered air from all of the systems represents a very substantial energy use factor. For this reason, reductions of air discharge of the hoods and other discharge sources should be very carefully considered. The use of localized exhausts instead of hoods, the use of horizontal sash hoods, or requiring stops on hoods at significantly less than full sash openings all would be favorable steps to take. Use of an automatic, computer controlled system to automatically lower sashes when the laboratory lights are turned off would be a straightforward step to take, and with automation of HVAC systems, this type of modification would be relatively easy to do. The use of full scale Variable Air Volume (VAV) systems would extend this concept but would typically cost somewhat more. In theory, a procedural process of asking everyone to close their sashes, or turn off the hoods entirely when it was safe to do so would be equally acceptable, but the results likely would be very erratic due to the variance in compliance by individuals. One suggestion, applicable primarily to academic institutions, would be to take advantage of the periods during the year when colleges and universities virtually shut down, such as during Thanksgiving and Christmas breaks. Unfortunately, the key word is "virtually." In academic institutions, there are always a few individuals who do not leave but continue working, as they do also during odd hours when others are asleep or are otherwise not present. This pattern usually mandates continuing operation of the HVAC system and results in little energy savings available from this source, but is worth investigating.

There is one significant method of reducing the energy costs of hoods, which is applicable to those systems where hood exhausts are brought to a common discharge point and that is to incorporate an energy recovery system at that point, such as an "energy wheel."

In summary, the entire laboratory operation, facility design and equipment choices should be evaluated, preferably during the initial design phase, to provide the maximum energy conservation consistent with safety.

REFERENCES

1. **Horowitz, H., Heider S.A. and Dugan, C.N.,** Hoods for science laboratories, in *CRC Handbook of Laboratory Safety,* 2nd ed., Steere, N.V (ed.), CRC Press. Boca Raton, FL, 1971, 154-165.

2. American Society of Heating, Refrigerating, and Air Conditioning Engineers, *Applications Handbook,* 15,1978.

3. Committee on Industrial Ventilation, *Industrial Ventilation, Section 4, Hood Design Data,* 15th ed., American Conference of Governmental Industrial Hygienists, Cincinnati, OH, 1978.

4. Industrial Ventilation, A Manual of Recommended Practices, 20th ed., American Conference of Governmental Industrial Hygienists, Cincinnati, OH, 1988.

5. **Fuller, EH. and Etchells, A.W.,** Safe operation with the 0.3 m/s (60 fpm) laboratory hood, *Am. Soc. of Heating,Refrigerating, and Air Conditioning Eng. J.,* October 1979, p.49.

6. **Caplan, K.J. and Knutson, G.W.,** The effect of room air challenge on the efficiency of laboratory fume hoods, ASHRAE Trans. 83, Part 1, 1977.

7. Steere, N.V, Ventilation of laboratory operations, in *CRC Handbook of Laboratory Safety* 2nd ed., Steere, N.V. (Ed.), CRC Press, LLC, Boca Raton, FL, 1971, pp. 141—149.

8. *Useful Tables for Engineers and Steam Users,* 8th ed., Babcock & Wilcox Co., 1963.

9. *Laboratory Chemical Fume Hoods Standards, Manual 232.1,* United States Department of Agriculture, Administrative Services Division, December 1981.

10. ANSI Z9.2- 1991, *Fundamentals Governing the Design and Operation of Local Exhaust Systems,* American National Standards Institute, New York, NY.

11. ASHRAE 110, *Method of Checking Performance of Laboratory Fume Hoods,* American Society of Heating, Refrigerating, and Air Conditioning Engineers, Fairfax, VA, 1995.

12. *Corrosion Resistance Chart,* Sheldons Manufacturing Corp., 1400 Sheldon Drive, Elgin, IL, 60120.

13. **Koenigsberg, J.,** The HOPEC IV Laboratory Fume Hood, Heating, Piping and Air Conditioning, November 1992.

14. **Koenigsberg, J.,** Laboratory ventilation and VAV technology, Chemical Health and Safety, American Chemical Society, March/April 1996.

15. **Sharp, G. P.,** How airflow control affects laboratory safety, Chemical Health and Safety, American Chemical Society, March/April 1996.

16. **Koenigsberg, J.,** Local Laboratory Ventilation Devices, Heating/Piping/Air conditioning, Magazine Systems Engineering, October, 1995.

17. **Koenigsberg, J.,** The Laboratory Fume Hood, Debunking Ventilation Myths, Energy Saving Options, Upgrading Existing Units, presented at Laboratory Safety and Environmental Management Conference, Prizim, Inc. July 1998.

18. **Zboralski, J.,** The Effects of Face Velocity on Fume Hood Containment Levels, Technical Paper No. 90.01, Hamilton Industries Inc. Infobank, Two Rivers, WI, May 1995.

19. **Schill, R.J.,** The Americans with Disabilities Act Requires New Awareness in Laboratory Environments, Technical Paper No. 90.07, Hamilton Industries Inc. Infobank, Two Rivers, WI, April 1992.

20. **Farho, J.H., Goryl, W.M., and Anderson, S.A.,** Laboratory fume hood control, Heating/Piping/Air Conditioning, February, 1984.

21. Americans with Disabilities Act, 29 CFR 1630, 56 FR 35726, July 26, 1991.

5. Biological Safety Cabinets

A major purpose of most biological safety cabinets is to provide a work area free of contaminants, especially biological contaminants, for the work in progress. The primary safety role of a biological safety cabinet is intended to protect the laboratory worker from particulates and aerosols generated by microbiological manipulations. To accomplish both of these aims, the designers of these cabinets depend in part upon the ability of HEPA filters to remove the contaminants from air passing through them. Since HEPA filters are ineffective against gaseous chemicals, biological safety cabinets are generally not intended to be used for protection against gaseous chemical hazards. In general, they are also not really intended for other chemicals either since chemicals other than gases can quickly clog the passages of the HEPA filter. One type, in which the air is totally exhausted after a single pass through the work area, may be used to a limited degree for chemical applications and has been used for volatile materials. However, the electrical components are not designed to be safe for such materials as are the electrical components in chemical fume hoods.

For a more detailed discussion of the selection, installation, use and maintenance of Biological Safety Cabinets, the reader is referred to the following web site managed by the Centers for Disease Control and the National Institutes of Health:

http://www.niehs.nih.gov/odhsb/biosafe/bsc/bsc.htm.

There are basically three Classes of biological safety cabinets, Classes I, II, and III. Class II units are divided into two main sub-Classes, IIA and IIB. There are several versions of the latter of these two. All biological safety cabinets are intended to run continuously.

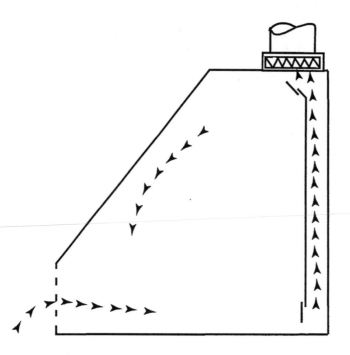

Figure 3.25 Class I Biosafety Cabinet.

a. Class I Cabinet

The Class I cabinet, illustrated schematically in Figure 3.25, is essentially a variation on the chemical fume hood, resembling closely the self-contained type. As with a chemical fume hood, worker protection is provided by air flowing inward through the work opening to prevent escape of biologically active airborne agents. Unlike a chemical hood, the view screen, equivalent to the sash in a chemical fume hood, is usually designed to provide a fixed opening, typically 8 to 10 inches high (20 to 25 cm). The view screen is generally made so that it can swing upward to permit putting equipment into the unit, although some units have an air-lock type door on the end of the unit to serve the same purpose.

The airflow through the face may range between 75 and 100 fpm. An aerodynamic shape at the entrance or suction slots near the edge to ensure the flow of air is inward aids in the performance of the cabinet, while a baffle at the rear provides that some of the inlet air will flow directly across the work surface as the remainder is drawn upward through the cabinet to the exhaust portal. It is primarily the directional flow of air which ensures that the agents of concern stay within the cabinet.

As with the chemical fume hood, the performance of the cabinet can be adversely affected by a number of factors. The velocity of air movement within the room can result in a degradation of the performance of the hood, as can too-rapid movements by the worker, location of the work too close to the entrance, or perhaps the effect of a thermal source, e.g., a Bunsen burner, within the unit. It is recommended that flame type heat sources not be used within this type of cabinet. The effectiveness of the unit can be enhanced by placing a panel over the opening in which are cut arm holes. The airflow through these arm holes will be much higher than the designed air speeds through the front, especially with the worker using them, but the relatively small gaps normally should keep the turbulence caused by the higher air speeds through the portals from causing materials to escape from the cabinet and causing problems for the users.

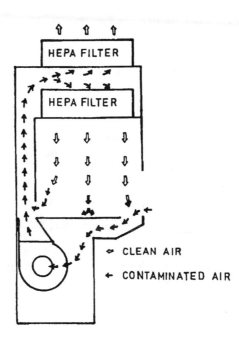

Figure 3.26 Class II biosafety cabinet.

A Class I cabinet, used properly, can provide excellent protection for the research worker, but it does not provide any protection for the active work area within the cabinet, since the air flowing into the cabinet is "dirty" air, i.e., ambient air from the room that has not been specifically cleaned. For chemical work, this was not discussed since it is rarely a concern of the chemists, but for biological research, elimination of contamination of research materials is likely to be very important.

The air from the cabinet is exhausted through a HEPA filter placed above the exit portal and before the exhaust fan. If the discharged air does not contain any chemical agent or other non-particulate material which could be expected to pass through a HEPA filter, it is not absolutely essential that the exhaust be to the outdoors. Absorbent filters, such as that used in the self-contained chemical hood, can be used to supplement the HEPA filter if there are possible chemical effluents. However, a HEPA filter can begin to leak or chemicals may build up in the filter and cause the airflow to be diminished resulting in increased spillage through the sash opening. Hence, it may be desirable to take the exhaust to the outside if the contaminants can cause problems to the workers. It is normally not as critical that the entire duct work be maintained at a negative pressure as with a chemical hood, so the exhaust fan can be integrated into the cabinet if desired. The exhaust can be into an air discharge system designed for chemicals as well.

A unit sometimes confused with a Class I biological safety cabinet is a horizontal laminar flow cabinet or work table. This type of unit serves precisely the opposite function of a class I cabinet. Clean air which has been HEPA filtered is blown across the work surface *toward* the worker so that the research or product materials are protected against contamination, but the worker is not protected at all. Such a unit is unsuited for microbiological work, except for applications which would cause no harm to the users, such as a work involving noninfectious or non-allergenic materials.

b. Class II Cabinets

Class II cabinets provide protection for both the researcher and for the research materials within the cabinet. Figure 3.26 shows an idealized version of a Class II cabinet. In this design, room air is drawn in the front opening at a minimum of 75 fpm, but instead of passing over the work surface, air is pulled down into a plenum by a fan unit under the work surface. Then the output of the fan is passed up through a channel at the rear of the cabinet into a space between two HEPA filters. A portion of the air is exhausted at this point through one of the filters and the other portion passes through the second filter, where it is then directed down as clean air to the work surface. This stream of air is intended to be a laminar flow stream that resists encroaching air and remains clean, so that the work surface is in a clean environment. Part of this air enters the intake air grill at the front of the cabinet and the remainder goes through an exhaust grill at the rear of the unit. Both of these airstreams continue going through the cycle. The result is that "dirty" room air or air that has passed over the work surface, and hence is also "dirty," is restricted to the air on its way to the filters. The work surface is in a clean air environment, and pathogens are blocked from escaping into the room by the inward flow of air at the work opening and by being removed by the HEPA exhaust filter. As noted in the beginning of this section, there are several variants on this concept, but all provide protection for both the employee and the work.

Class II cabinets are intended to be used for work with microorganisms that would be permitted in laboratories designated as biological safety levels, 1, 2, and 3. Normally they are not used for work with volatile, toxic chemicals. If these are used in the cabinet, the exhaust must be to the outdoors.

i. Class IIA Biosafety Cabinets

Class IIA units are very similar to the basic Class II biosafety cabinet just described. In order to keep the work area free of room air and air that has been contaminated by the work, the plenum containing the blower and the channel through which air is delivered to the HEPA filters under a positive pressure must be carefully sealed to be leak tight. The Class IIA unit is designed so that about 30% of the air is exhausted from the cabinet and about 70% recirculated each cycle. Thus, the make-up air through the front must provide an amount of air equal to that exhausted, or 30%. The air being supplied through the front opening must be carefully balanced with the amount of air being exhausted. If too small an amount of air is exhausted to the outside, the air in the working volume could be at a positive pressure and pathogens could be forced out into the operator's area. If too small an amount of make-up air is provided at the face, the pressure in the working volume could become negative, allowing the work area to be contaminated by room air. This type of unit is especially sensitive to anything in the work area that could perturb the laminar flow of air. Examples would be equipment blocking portions of the duct, rapid arm motions, and gas flames.

Because of the sensitivity of the unit to maintenance of the air balance, Class IIA cabinets are not ducted directly to the outside but, if it is desired to exhaust the effluent from the cabinet to the outside, a small canopy hood is used, surrounding the exhaust from the cabinet but separated from it by an air gap of about an inch. The air flow through this canopy hood is sufficient to capture the cabinet exhaust but because it is not connected directly to the cabinet exhaust, will not affect the sensitive air balance within the cabinet.

In consequence of both a smaller front opening, 8 to 10 inches, and the fact that only 30% of the air is being discharged at any one time, the amount of tempered air needed to be provided for a laboratory using this type of cabinet is much less than for a chemical fume hood, typically in the range of 250 cfm for a 4-foot unit as compared to close to 1000 cfm for the same size chemical hood. It could even be more favorable if the biosafety cabinet exhaust were to be discharged into the room, in which case no additional tempered air would need to be supplied to the room other than that needed to provide the recommended amount for personnel comfort and well-being of the number of personnel normally within the room and to compensate for other heat loads.

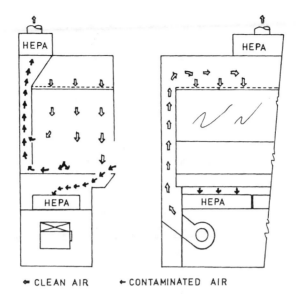

Figure 3.27 Side and partial front view of Class II B biosafety cabinet.

ii. Class IIB Biosafety Cabinets

Class IIB biological safety cabinets differ in several ways from Class IIA cabinets. Two major differences are the amounts of air recirculated (a much smaller proportion for a type B unit than for a type A) and that the air in the plenums surrounding the work area is filtered or clean air rather than contaminated air.

Shown in Figure 3.27 is a generalized drawing of a representative Class IIB biosafety cabinet designed originally by the National Cancer Institute. As with other Class II units, air entering the cabinet is immediately drawn into an air intake. Also entering this intake is filtered air from above, aiding in blocking the entrance of room air further into the working volume. In this unit, 100 fpm of room air is provided as supply air. The air drawn through the front grill is drawn through HEPA filters below the work surfaces by blowers situated below the filters. The resulting clean air is forced up plenums on each end of the unit and into a diffuser area above the work area. The clean air is circulated into the work zone through a diffuser panel at a reduced speed of 50 fpm. Most of this air, 70%, is exhausted as contaminated air through two rows of slots at the rear of the cabinet, through a HEPA filter, while 30% is recirculated through the front grill. Blowers, external to the cabinet, which are typically a part of the laboratory system, provide the exhaust pressure.

Another version of the Class IIB biosafety cabinet is the total exhaust unit shown in Figure 3.28. All of the air entering the cabinet makes only one pass through the cabinet before being discharged through a HEPA filter. One hundred (100) fpm of room air enters through the front opening, providing protection for the operator against pathogens escaping from the cabinet, and is drawn down into a plenum below the work surface by a blower. All of this air is exhausted. None is recirculated. In order to provide clean air for the work zone and to block the entrance of room air into the cabinet interior, air is drawn in from above the cabinet by another blower through a HEPA filter into the work zone. Part of this air passes through the inlet opening at the front of the cabinet and part through a slot at the rear as contaminated air. All of the air in the exhaust plenum is discharged through another HEPA filter.

Since none of the air that has passed through the work zone is recirculated, at least in principle, this type of cabinet could be used for moderate chemical applications. However care

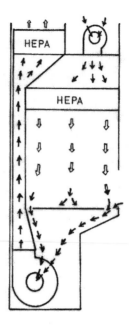

Figure 3.28 Total exhaust biosafety cabinet

would have to be taken to ensure that the exhaust filter would not suffer from loading of the filter by chemicals. The different rates of loading of the inlet and exhaust filters are a problem for all types of Class II cabinets, but the more complicated airflow systems of Class IIB biosafety cabinets, where two or more blowers are involved, exacerbate the problems.

Table 3.7 provides a summary of the characteristics of the three types of Class II biosafety cabinets. Class II biosafety cabinets are intended for low- to-moderate-risk hazards. As a minimum, they should be required to meet the National Safety Foundation (NSF) Standard 49 for Class II (laminar flow) a biohazard cabinetry. The working enclosures and plenums through which air moves should be constructed of materials that are easy to decontaminate, such as stainless steel or a durable plastic.

c. Class III Cabinets (Glove Boxes)

Class III units are totally sealed cabinets in which the user performs the manipulations with the research materials inside the chamber, using attached arm length, impermeable gloves. Materials are usually placed in the cabinet prior to work beginning, or cycled into the interior through an air lock. Materials taken from the chamber are usually taken out through a double-door autoclave, through a second air lock where decontamination procedures may be carried out, or through a chemical dunk tank. These units are intended to be used for high-risk materials. Protection for the work is that provided by preliminary cleaning of the interior. Because most are totally sealed, glove boxes typically have minimal ventilation requirements. At least one vendor, however, Baker Co. of Sanford, Maine, makes a Class III unit in which air enters through a HEPA filter and is exhausted through ultra-high-efficiency HEPA filters (99.999% efficient for particle removal) in tandem. These filters, which exceed the requirements for normal HEPA filters by better than a factor of 10, are especially tested to ensure their performance. The cabinets should be kept under slight negative pressure, so that if there are any leaks at all, the leakage would be from the outside to the inside.

d. HEPA Filters

An accepted definition of a HEPA filter is:

	Type		
Table 3.7 **Basic Characteristics of Class II Biosafety Cabinets**	**2A**	**2B**	**2B** (total exhaust)
Inlet air speed	75 fpm	100 fpm	100 fpm (minimum)
Fraction of air recirculated	70%	30%	0%
Positive pressure plenums	Contaminated	Filtered	Contaminated, but isolated
Exhaust air for a typical 4-foot unit	225 cfm	500 cfm	700 cfm

"A throwaway extended-pleated-medium dry-type filter with (1) a rigid casing enclosing the full depth of the pleats, (2) a minimum particle removal efficiency of 99.97% for thermally generated monodispersed DOP smoke particles with a diameter of 0.3 microns, and (3) a maximum pressure drop of 1 inch water gage when clean and operated at its rated airflow capacity."

A properly functioning HEPA filter is an essential component in both Class I and Class II units and may be important in some Class III biosafety cabinets.

HEPA filters used in Class II biosafety cabinets meet the requirements of Underwriters Laboratory's UL-586 and mil-spec MIL-F-5 1068. As stated in the definition, they are made by folding a continuous sheet of filter paper back and forth over corrugated separators. The separators provide strength to the assembly and form air passages between the pleats. The filter paper is made up of submicron fibers in a matrix of larger, 1 to 5 micron fibers with a small admixture of organic binder. It is a fragile material and subject to puncture or cracking if abused. The filter is usually glued to the edge of the frame by a glue that hardens, thus making this connection a highly vulnerable stress point. The frame is seated on a gasket and this seat is also vulnerable to leaks. Clearly, a HEPA filter must be treated with a great deal of care to avoid damage. If there are any questions about the integrity of the HEPA filter, operations should cease until it has been tested and certified.

A standard filter assembly of 24 x 24 inches and 5 and 7/8 inches deep has a rated capacity of 500 cfm. This filter will contain about 110 sq. ft. of filter paper. Operating at rated capacity, the air will be moving through the filter at about 5 fpm and the speed of the air leaving the face will be about 125 fpm. The filter will offer, in clean condition, a pressure drop across the filter of no more than 1 inch water gauge. When the pressure drop reaches 2 inches water gauge, it is usually time to replace the filter.

Because the desired laminar air velocity inside a Class II unit is somewhat less than 125 fpm (in the unit designed by the National Cancer Institute, described in the discussion of Class II cabinets, the downward design velocity is 50 fpm), most HEPA filters operate at well below their rated capacity. This results in a significantly increased lifetime. The resistance decreases inversely proportionally to the velocity of the air through the filter, and the amount of loading on the filter will decrease directly as the volume of air passing through the filter. The net effect is to greatly extend the life of the filter unit.

e. Installation, Maintenance, and Certification

The guidelines governing the most favorable location of laminar flow cabinets are essentially the same as for a chemical fume hood. Place them in the far end of a laboratory, in a low traffic area, and where there are no drafts. The environment of the hood should be checked to ensure that air speeds in the neighborhood of the cabinet opening are small compared to the face velocities of the cabinet.

The installation of laminar flow cabinets is a specialized skill. An examination of a cabinet shows that the components are carefully sealed together, with a large number of bolts being used to maintain sufficient pressure on the seals between the individual sections of the cabinet to ensure that they fit tightly together. Shipping or even moving the cabinet across the room may be sufficient to break some of the critical seals. Demonstration that the unit's seals are intact and that it meets all specifications at the factory simply shows that it was in good condition at that time. It does not demonstrate that it is still in a similar condition after installation. The equipment needed to test the integrity of the cabinets is expensive and requires skills and training which relatively few laboratories, institutions, or corporations have in-house. However, to be sure that a cabinet is providing the protection for the operator, product, or both for which it was designed, the cabinet needs to be tested and certified at installation, after any relocation, and periodically such as either annually or after 1000 hours of use.

A purchase order for a cabinet should include provision and funds for testing and certification of the unit after delivery and setup but before final payment is made. Few, if any, vendors routinely make provision for this service or have their own personnel to do it. In order to avoid a potential conflict of interest, this should be arranged by the purchaser with an independent contractor. The cabinet vendor should be willing to delay payment of their invoice for a reasonable period to allow this to be done. This provision needs to be made in the purchase contract so that no misunderstanding will occur. There are training programs that teach how to perform the tests properly and there are firms or individual consultants available that will perform the tests at a reasonable fee.

It has been noted that it is not *essential* that the cabinets be exhausted outdoors, but there are several reasons why it is often desirable. The most obvious reason is that no system is absolutely foolproof. A cabinet may be checked and certified in the morning, and an accident may occur during the afternoon which could cause a seal to fail or cause the HEPA filter to begin to leak. If the cabinet is exhausted outside the facility, no pathogens should be released into the facility. Decontamination may be required using formaldehyde, or a comparably undesirable chemical, that should not be released into the laboratory.   Some research may involve chemical carcinogens, radioisotopes, or other materials not eliminated by the HEPA filters and should not be discharged indoors. In fact, it may be desirable to incinerate some organisms by passing them through a flame before evacuating them even to the outside. Finally Class IIB cabinets typically do not provide an integral exhaust blower to discharge air from the unit, but depend upon the facility to provide the exhaust fan. If the cabinet is connected to the system to provide suction on the exhaust portal, there is no reason for the system to exhaust back into the laboratory or building. Where cabinets are exhausted to the roof, the height of the exhaust stack should be sufficient to ensure that the effluents are discharged above head height of any maintenance personnel who may be present on the roof. No weather cap should be provided for these exhaust ducts.

The removal and replacement of contaminated HEPA filters should be performed by trained professional personnel who take precautions against any exposure to themselves and to avoid contaminating the facility. This will require planning ahead of time to provide sufficient access to the filters. For filters anticipated to be contaminated with human pathogens (or animal, where animal exposures are a matter of concern), provision for isolating bag-out procedures should be made in advance. In general, HEPA filters should be disposed of as contaminated biological waste, preferably by incineration.

Cabinets should be decontaminated periodically. Some units include an ultra-violet tube placed inside the cabinet which will aid in disinfecting the surface, but will not significantly decontaminate the air passing through it. Cleaning after each day's use or at the end of a sequence of operations with a weak solution of household bleach is recommended, but other materials such as quaternary ammonium compounds may serve as well, be less irritating to the user, and cause less corrosion. There are procedures recommended by the National Cancer Institute and available on a slide cassette package for decontaminating cabinets. Basically, this consists of sealing the cabinet and vaporizing an amount of dry paraformaldehyde in it sufficient to provide a concentration throughout the cabinet of seven to 8.5 mg/in³. The cabinet should remain sealed for about four hours. The temperature should be between 20°C and 25°C and the humidity above 70%. The paraformaldehyde should then be exhausted to the outdoors and the cabinet ventilated for at least eight hours. It should be noted that since this procedure was recommended, questions about the carcinogenicity of formaldehyde have arisen. Because of this carcinogenic property, the OSHA 8-hour time weighted average (TWA) has been lowered to 0.75 ppm, with an action level set at 0.5 ppm. A short-term 15 minute limit of two ppm has also been set. The ACGIH recommended limits are even lower, 0.3 ppm as a ceiling limit. Persons performing the decontamination must not exceed the occupational limits. There is no question that some individuals are irritated by levels somewhat lower than the levels now permitted. An alternative decontaminant coming into use is hydrogen peroxide vapor. This material appears to be effective and represents less of a hazard to personnel and to the environment.

REFERENCES

1. National Sanitation Foundation, *Standard No. 49, Class II (Laminar Flow) Biohazard Cabinetry*, Ann Arbor, MI, 1992.
2. **Burchsted, C.A., Kahn, J.E. and Fuller, A.B., Eds.,** *Nuclear Air Cleaning Handbook*, Oak Ridge National Laboratory ERDA 76-21.
3. **McGarrity G.J. and Coriell, L.L.,** Modified laminar flow biological safety cabinet, Applied Microbiology, 28(4), 647-650, October 1974.
4. **Coriell, L.L. and McGarrity,G.J.,** Biohazard hood to prevent infection during microbiological procedures, Applied Microbiology, 16(12), 1895-1900, December 1968.
5. Selecting a Biological Safety Cabinet, (slide cassette) NAC no. 00709 and no. 01006, National Audio Visual Center (GSA), Washington, D.C., 1976.
6. Effective Use of the Laminar Flow Biological Safety Cabinet, (slide/cassette) NAC no. 00971 and no. 003087, National Audio Visual Center (GSA), Washington, D.C., 1976.
7. *Certification of Class II (Laminar Flow) Biological Safety Cabinets*, (slide/cassette) NAC no. 003134 and no. 009771, National Audio Visual Center (GSA), Washington, D.C., 1976.
8. *Formaldehyde Decontamination of the Laminar Flow Biological Safety Cabinet*, (slide/cassette) NAC no. 005137 and no. 003148, National Audio Visual Center (GSA), Washington, D.C., 1976.
9. **Jones, R., Drake, J., and Eagleson, D.,** *Using Hydrogen Peroxide Vapor to Decontaminate Biological Safety Cabinets. Acumen*, Baker Co, 1.1, Sanford, ME, 1993.
10. **Rake, B.W.,** *Influence of Cross Drafts on the Performance of a Bioloical Safety Cabinet*, Appl. Env. Microbiol. 36:, pp 278-83.
11. **Jones, R.L., Jr., Stuart, D.G., Eagleson, D., Greenier, T.J., and Eagleson, J.M., Jr.,** The Effects of Changing Intake and Supply Air Flow on Biological Safety Cabinet Performance, Appl. Occup. Environ. Hyg. 5, 370. 1990.

3. Built-In Safety Equipment

In a very real sense, the hoods and safety cabinets that have been the subjects of the last several sections can be considered built-in safety equipment. However, the items to be discussed in the following sections are equipment which comes to mind for most persons in the context of fixed safety equipment.

a. Eyewash Stations

One of the most devastating injuries a person can suffer is loss of eyesight. There are a number of protective measures which should be taken in the laboratory to prevent eye injury.

However, should all of these measures fail and chemicals enter the eye, an effective eyewash station is an essential item of fixed equipment that should be immediately available. Although superseded by the Laboratory Safety Standard, OSHA does require in Section 1910.151(c) of the General Industry Standards that:

> "Where the eyes or body may be exposed to injurious corrosive materials, suitable facilities for quick drenching or flushing of the eyes and body shall be provided within the work area for immediate emergency use."

There are no fixed standards on the maximum acceptable distance of travel to reach an eyewash station. The American National Standard Institute (ANSI) standard Z358.1-1998 however, stipulates that it require no more than 10 seconds to reach the eyewash station from the hazard location. Realistically this criterion means no one should have to open a door to reach an eyewash station or go in a tortuous path to reach the unit. No one in pain and possibly blinded should have to overcome any additional impediments or obstacles in seeking relief. An eyewash station ideally should be centrally placed in a laboratory along a normal path of egress or in an otherwise equally logical location in a given facility. Where strong acids or bases are in normal use. The ANSI standard recommends that the eyewash station be immediately adjacent to the point of use of hazardous materials.

Small squeeze bottles containing a pint, or perhaps at most a quart, of water can supplement a plumbed eyewash station but are not acceptable as the sole eyewash devices. The basic problem is lack of volume. As a minimum, eyes suffering even a light chemical burn need to be flooded with potable water for 15 to 20 minutes. The second problem is that the water in the bottle may become contaminated. Where plumbed water lines are not available, eyewash units connected to pressurized portable containers of water are acceptable substitutes if they contain sufficient amounts of water to meet the requirements of the plumbed units for at least 15 minutes.

According to the ANSI standard, eyewash stations should provide an ample amount of water, at least 0.4 gallons (1.5 liters) per minute, at a relatively low pressure, at least 30 psi, in such a manner as to flood both eyes with aerated potable water. If the unit is intended to cover the entire face, then to meet ANSI 358.1 requirements, 3 gallons or 11.4 liters per minute would be needed. The most common type, with two nozzles facing upward and aimed slightly inward toward each other is probably the best overall design. An alternative is a drenching hose consisting of a spray nozzle, connected to a flexible hose, is not a bad supplement, but it should not be the only eye-washing device available. For one thing, they afford only a single stream of water which would make it difficult to treat both eyes simultaneously. An individual alone may be in too much pain to do much more than hold his face in flowing water and certainly could not simultaneously manipulate a hose and use his hands to hold his eyelids open.

Turning on the eyewash should require minimal manual dexterity. Any number of mechanisms to turn them on are possible but perhaps the most popular is a simple paddle that the injured person can push aside. The eyewash should remain on continuously with no additional effort after the initial activation, but if an automatic cutoff is provided, it should not activate for at least 15 minutes or until 6 or more gallons of water have been delivered. Many eyewash stations are mounted as part of the plumbing over a sink. This is convenient but not essential. In the following section on deluge showers, it will be pointed out that in the case of eye injuries, safety is more important than spilled water, which can be mopped up, so that floor drains are not strictly required for safety shower installations. Preferably, eyewash stations and deluge showers should be installed as a package since it is likely that if the eyes and face have been exposed to chemicals, other portions of the body quite possibly may have been contaminated as well.

The eyewash nozzles must be located at least six (6) inches from the wall and be between 33 and 45 inches from the floor. The lower level would permit a disabled person in a wheel chair to comfortably use the unit. No obstructions must be allowed to exist which would make it difficult for access to an eyewash station. The pattern of flow should cover both eyes simultaneously. The separation of individuals' eyes varies somewhat but typically ranges between 3 and 4 inches. Purchase of an eyewash unit meeting the ANSI standard would flood both eyes within this range.

A major problem with most eyewash stations and deluge showers is that they are usually connected to the cold water line. The ANSI standard requires that the water be tepid but does not define what tepid is. Typically, tap water temperatures are in the 60°F to 70°F (15.5° to 21 °C) temperature range, but in colder climates can be much less during the winter. Water at temperatures in the 50's or lower can be painful itself, and in extreme cases can cause the injured person to go into shock. Although relatively few eyewash installations are capable of conveniently providing it, lukewarm water with temperatures close to body temperatures between 90° and 95°F (32 to 35°C) would be ideal.

A permanently installed eyewash station is an essential component in or very near every laboratory, but, if one is not available, any source of water, provided it is not too hot or extremely cold, should be used in an emergency. A sink faucet, a shower, even a large basin of water in which the injured person could immerse their eyes may be used.

All eyewash stations should be checked under full flow conditions on a definite schedule such as weekly. Any deficiencies should be corrected immediately.

Brief instructions on how to activate and use an eyewash fountain should be placed immediately adjacent to the unit in addition to simplified instructions on how to help the patient keep their eyes fully open so that the water will be able to reach the injured tissues but employees should routinely receive training in their use.

b. Safety (Deluge) Showers

According to ANSI 358.1-1998, safety showers should also be placed so that it should not be necessary to travel for more than 10 seconds to reach one in the event of an accident, nor should there be any obstacles in the way. However, showers often are placed in hallways, usually where they can service more than one laboratory. This also serves to avoid creating a massive water flood in a crowded laboratory in favor of a much more easily cleaned hallway. This latter point is not an insignificant advantage since it may be difficult to locate a shower in a small facility so that water will not splash into sophisticated and easily damaged equipment. Nevertheless, the potential of a serious life-threatening or maiming injury in a chemical spill is sufficiently likely so that the concern for personal safety should override any other factors. The design and equipping of the laboratory should take into account the prevention of damage of equipment by the shower. In addition, units located in public corridors are more subject to vandalism than those inside rooms; therefore, it is recommended that each laboratory be individually equipped with a combination safety shower and eyewash station. As noted in the previous section, accidents involving facial splashes are also likely to involve other parts of the body. Clearly, if the units are separated, it is not practical to travel from one to the other when both are needed. Both should be in the same location.

The demands of a safety shower on the water supply are much more severe than for an eyewash fountain. Their alternate name, deluge shower, is not idly applied. The water supply should be able to provide a minimum of 20 gallons per minute for at least 15 minutes. The water pressure should not be so high that the sprays are painfully vigorous. The widths of the spray at approximately shoulder height for an average person should be at least approximately two feet wide at that height. For normally available water pressures, these parameters would usually indicate that the shower water supplies are at least a one inch line. The problem of temperature is, again, a serious one, but even more so because of the greater area of the body involved. Usually the showers are connected to a cold water supply line. In colder climates, the stress of inundating the entire body with cold water at temperatures perhaps in the 50° to 60°F (10° to 15.5°C) range may be sufficiently severe to cause a person to go into shock. Ideally water temperatures should be near normal body temperatures, around 90° to 95°F (32° to 35°C) but not over 100°F (38°C).

Instructions for activating and deactivating a shower should be prominently posted near the shower. The mechanism for turning a shower on can be a paddle, which is simply pushed out of the way, or a chain which can be pulled (note, if a chain, the chain must be comfortably within the reach of a disabled person). Both are simple and require minimal physical control or the manual dexterity that may be important in this type of an emergency. The shower should continue to run until it is deliberately turned off, although it may be equipped with an automatic cutoff, especially if it is installed where there is no drain, but it should deliver at least 100 gallons before any automatic cutoff activates. Showers should be checked at least once a year, which is conveniently done by catching the flow in a large funnel connected to a fire hose which is discharged into two 55-gallon drums. This provides both a rate and volume check simultaneously, while avoiding creating a mess to be cleaned up.

It should not be necessary to point out that safety showers should not be located near any source of electricity with which the flowing water from the shower could come into contact, but numerous instances have been observed where this has occurred. Usually it has been due to the laboratory workers themselves moving portable equipment too near the shower, but occasionally one is found improperly installed by maintenance personnel or because of changes made by renovation crews.

As noted, drains are not strictly essential, although if it is feasible to make them available, they are desirable. Without a drain, a sizable mess will be created which will have to be mopped up when a shower is used, but this should be sufficiently rare that it may not justify the cost of installing additional drains to accommodate the safety showers, especially if a retrofit is necessary.

REFERENCES

1.   ANSI Z358.1, *American National Standard for Emergency Eyewash and Shower Equipment,* American National Standards Institute, New York, NY, 1998.

2.   **Stearns, J.G.,** Safety showers, in *CRC Handbook of Laboratory Safety* 2nd ed., Steere, N.V, CRC Press, Boca Raton, FL, 1971, pp. 12 1-123.

3.   **Srachta, B.J.,** Safety showers and eyewashes, in *Safety and Health,* National Safety Council, Chicago, IL, August 1987, pp. 50-51.

c.   Fire Suppression Systems

Many laboratory buildings have automatic fire suppression systems, but there are still many that do not, even though data show that the most common type, a water sprinkler system, is effective well over 90% of the time in either extinguishing a fire or controlling it until firefighting personnel arrive. In a very high percentage of the cases in which the sprinkler was not effective, the reason the system did not succeed was due to either poor installation or human error. Some scientists do not want a fire suppression system in their laboratories because they feel that it will lead to problems. The possibility that a fire in one area will cause the entire system throughout the building to activate is one basis for concern, although in most systems only those sprinkler heads in the vicinity of a fire activate. In over one third of the fires in which the sprinkler system was successful, only one sprinkler head was activated in controlling the fire. There is some concern that water will react with chemicals in the laboratory, will spread the fire due to burning solvents being carried away by excess water, or will damage sensitive equipment. If there are problem chemicals, there are alternative fire suppression systems that do not use water and will not damage even delicate equipment.

The potential for fires is higher in most research facilities than in a typical building because of the variety and character of the chemicals employed. Often, much of the equipment is home-built or temporarily rigged and has not been checked for safety by an accredited testing laboratory. Many of the operations involve heat. Some type of fire suppression system is recommended for laboratory buildings, if for no other reason than that the presence of one in a building can lead to substantial savings for insurance costs. We are in a time of insurance premiums rapidly escalating at a rate that does not appear to show signs of abating. Unless the building is insured for its full replacement value (a coverage that is becoming more difficult to obtain), including the cost of its contents, the increased cost of construction, and the high cost of sophisticated research equipment, it is likely that insurance will pay only a fraction of the actual losses. Further, insurance cannot replace in most cases the intellectual properties lost, nor can it eliminate the time to create a new facility, a probable period of some years.

The time to stop a fire is when it is very small. This is the purpose of portable fire extinguishers, which laboratory personnel can use before a fire becomes too large to control. Unfortunately, most laboratories are not staffed at all hours, while some heat-generating equipment such as stills or heat baths may be left functioning continuously. A sprinkler system serves essentially the same purpose as a person with an extinguisher 24 hours a day, 365 days a year. It can put water, carbon dioxide, or other fire suppression media directly onto a fire still small enough to be easily extinguished. There are fires which expand so rapidly that a fire suppression system will be overwhelmed but, as noted above, statistically the number of fires in this category are very small.

OSHA does not mandate automatic fire extinguishing systems, but the General Industry Standards, Subpart L, Sections 1910.155-165, do provide regulations covering the essential requirements which installed systems must meet.

The principles of an automatic fire suppression system are straightforward, but designing an actual system requires substantial engineering skills to be sure that the system will adequately and efficiently serve its purpose. The following brief sections on built-in fire suppression systems are intended only to supply basic information on the essential features of the various types of fire suppression systems to provide some insight in this critical area to concerned laboratory facility managers.

i. Water Sprinkler Systems

There are several essential components of an automatic water sprinkler system. The supply system to provide water to the sprinkler system must have sufficient pressure to deliver water to the units highest up in the building. There must be a control valve to serve the suppression system. A distribution system must carry the water to the spaces to be protected by the system, when needed, to respond to the presence of a fire. In each area where the sprinkler system is to provide protection, a carefully engineered pattern of sprinkler heads is required to distribute the water to ensure that complete coverage is obtained. The sprinkler heads themselves must be selected to meet the needs of the location. Generally, the heads themselves incorporate the heat sensing devices, in the form of fusible links, which will cause them to activate. The design of the system must take into account the maximum normal temperature that will be reached in the vicinity of the sprinkler heads and specify an operating temperature for the fusible links a reasonable amount above this temperature. In some systems, the sensing devices are separate from the sprinkler heads. Finally, when a sprinkler system is activated, a sensing device is needed which will transmit an alarm to the occupants of the building, preferably to a manned location which can immediately summon firefighters, or to a fire station directly. Any automatic water sprinkler system should be installed according to the NFPA Standard 13, as most recently amended.

The water supply system may be from a water works system, a gravity tank, or a pressure tank. NFPA Standard 13 provides specifications for each of these to ensure that they provide

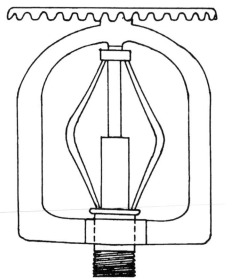

Figure 3.29 Upright sprinkler head.

for sufficient water volume and pressure to supply a sprinkler system. This standard provides seven different categories of occupancy classifications ranging from light hazard to extra hazard. These levels are only for the purpose of designing an appropriate sprinkler system and do not correspond directly to the occupancy classes under usual building codes. The extra hazard, group 2, represents a facility where, in part, there are moderate to substantial amounts of flammable and combustible liquids. As a minimum, for a light hazard occupancy the water supply must provide for a residual pressure of 15 psi at the level of the highest sprinkler and a duration of 30 minutes for the volume of water needed for the system. The requirements on duration and the amount of water needed are higher for other classifications. A gravity tank supply should be at least 35 feet above the level of the highest sprinkler head to provide an adequate pressure, while a pressure tank of sufficient size, filled two thirds, must be pressurized to at least 75 psi.

There must be at least one fire department connection through which a fire department can pump water into the sprinkler system. When a large number of sprinkler heads have become involved in a fire, or the fire department has connected to the same supply to service their own hoses, the flow of water through the sprinkler pipes may be reduced so that the required volume and pressures cannot be met. In such cases, the fire department can boost the pressure and volume through this connection, preferably from separate water mains not being used in their own fire suppression efforts.

There are two different types of water sprinkler systems, wet-pipe and dry-pipe. The lines leading to the sprinkler heads are full of water in a wet-pipe system so that water will be discharged immediately from an open sprinkler head, while in a dry-pipe system, the lines are full of air under pressure instead of water. This latter type of unit should be used where the temperature is not maintained above freezing at all times. Both types of systems include a main control valve which is designed not only to supply water to the sprinkler heads but also to provide a mechanism to cause an alarm to sound. The valves also usually provide a visual indication of whether they are open or closed. Except during maintenance, they should be in the open position. There are a number of design features for these valves which are beyond the scope of this document, except to note that some are intended to avoid false alarms due to surges or variations in the water supply pressure to the system.

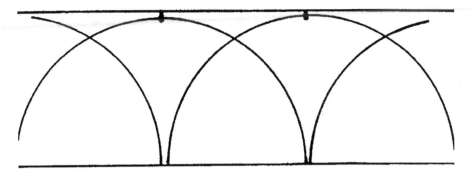

Figure 3.30 Coverage for a typical sprinkler system.

A variation of the dry-pipe system is the deluge system. The sprinkler heads are continuously open and water is prevented from entering the system by a deluge valve. When a fire is detected the valve to the water supply opens and water flows into the system and out of all the sprinkler heads. This not only wets the immediate area of a small fire, but the entire area to which the fire may spread. This type of system is usually chosen when the contents of the space are unusually hazardous. A variation on this system is the pre-action system in which the sprinkler heads are not open, but when a fire is detected the deluge valve opens and water is supplied to the sprinkler heads. The water entering the system causes an alarm to be sounded. When the heat causes the sprinkler heads to fuse, or open, water is discharged onto the fire.

After the control valve is activated, water is distributed through one or more vertical risers to portions of the system. Smaller cross-mains are connected to the risers which then service several still smaller branch lines. The sprinkler heads are connected to these branch lines. For laboratories, both the piping and sprinkler heads should be especially selected to prevent corrosion.

Sprinkler heads are very simple and rugged devices in which a valve is kept tightly closed by lever arms held in place by fusible links or other devices which fuse or open when they are heated to a predetermined level. When this occurs, the valve opens and water is discharged. There are many different versions of this simple device. Figure 3.29 shows a generalized drawing of an upright design.

In most cases, the water flow pattern is designed to be similar to that shown in Figure 3.30 to obtain a uniform overlapping distribution. The water is intended to be in the form of a fine spray. In some instances, where there are expected to be strong vertically rising convective airflow currents, the design may be modified to provide larger droplets which would be more likely to overcome the upward moving air. The orifice within the sprinkler head is usually 1/2 or 17/32 of an inch in diameter. Depending upon the design and the level of risk of the occupancy. A sprinkler head with an orifice of these sizes is intended to protect between 90 and 130 square feet for the higher levels of risk.

The temperature at which the fusible link or other device fuses depends upon the normal maximum temperature to which it may be exposed. For maximum ceiling temperatures in the vicinity of 100°F or 38°C, which would be comparable to those found in most laboratories, the fusible links should be selected to open with a temperature rating in the range of 135° to 170°F (57° to 77°C).

Water normally is not dangerous to humans, and a water sprinkler system is comparatively inexpensive and easy to maintain. Major advantages of water sprinkler systems are due to the physical properties of water which has a high heat capacity and a large heat of vaporization. It provides cooling and wetting of the fuel, which aids in puffing out the fire, and once the fire is subdued, the presence of water discourages re-ignition of the fire. However, water may cause considerable damage to equipment and can react with many materials.

ii. Halon Systems

Halon is a generic name for halogenated agents used in fire suppression systems. Until recently, two varieties, Halon 1301 and Halon 1211 were the agents of choice as alternatives to water systems where damage to equipment was a problem. Halon 1301 especially was a desirable alternative because of its low toxicity and effectiveness in low concentrations so that it could be used safely in occupied areas. The different numbers in the nomenclature correspond to the number of carbon atoms, fluorine atoms, chlorine atoms, and bromine atoms in that order in the compound. If there were a fifth number in the designation, it would refer to the number of iodine atoms. Halon 1301 is the one most commonly used in occupied spaces while Halon 1211 applications are typically in storage areas or other areas which are rarely or sporadically occupied.

Unfortunately, there is clear evidence that the class of chemicals to which the Halons belong are causing a depletion of the ozone layer in the upper atmosphere. As a result, there was an international agreement to phase out the production of these materials by 1996. Systems already in place can continue to be used but replacement of the materials, in the event the system is activated will become increasingly difficult. Alternatives, are being sought, which do not contain the critical elements, bromine and chlorine, that cause the ozone depletion, but which have similar fire suppression capabilities. There are a number of alternative systems commercially available which have been approved by the EPA. The following information is still included in this edition since it is still legal as of this date to refill existing systems and a large number are still in place. A discussion of the alternatives will follow afterwards. Halon 1301 and Halon 1211 are no longer being manufactured.

Halon 1301 Systems

Halon 1301 is a colorless, odorless gas that is noncorrosive and is electrically nonconductive. It is relatively nontoxic, and at concentrations up to at least 7% by volume in air, it is believed personnel can perform a normal evacuation of an area without significant risk. It is especially useful for fires in normally occupied spaces; where flammable and combustible liquids are present; and where electrical devices such as computers, data terminals, and electronic instrumentation and control equipment are present. These latter characteristics make it very useful for laboratory applications.

The design of Halon 1301, bromotrifluoromethane ($CBrF_3$) systems is covered in NFPA Standard 12A. This standard also contains much useful background material on the characteristics of the material, and is the major reference for this section.

There are two basic types of systems, one providing total flooding of an area and one designed for local application of the extinguishing material directly to the fire. Fixed systems will consist of a container of liquified compressed gas, pressurized with nitrogen, detectors, distribution piping, nozzles, manual releases, and a control panel incorporating an alarm. The control panel may also provide circuitry to cut off ventilation systems, and close windows, louvers and other openings to the space.

It is generally accepted that Halon agents interrupt the chemistry of a fire to extinguish it. Since fires involving the vapor phase of a material are most susceptible to contact with the extinguishing agent, Halon systems work best on this type of fire. In the case of a solvent fire for example, if the fire can be detected quickly enough and Halon 1301 applied immediately to the fire, substantial fires can be put out in seconds. Gaseous systems such as Halon work well on liquid fires, in part also, because the temperature of the liquid generally will not exceed the boiling point of the liquid, so that if the availability of oxygen is eliminated, the liquid itself will not be hot enough to reignite the fire. Other types of fires, such as fires on the surfaces of solid materials will take longer. This agent deep seated fires may not be appropriate, if a sufficient concentration can be maintained long enough, even deep seated fires can be extinguished.

In a total flooding system, the required concentration of the extinguishing agent will depend upon what fuel is present and if it is sufficient to only extinguish the fire or if it is desirable to "inert" the space, i.e., provide a sufficient concentration of Halon 1301 in the atmosphere so that the flammability range for the fuel-air combination will be non-existent at the ambient temperature. The latter concentrations are substantially greater than required for extinguishing the fire. The concentrations needed at a temperature of 77°F .or 25°C* to extinguish fires for a number of common flammable liquids are given, in Table 3.8, as well as design concentrations to provide a reasonable safety factor, and required inerting concentrations. The design concentration was chosen to be at least 5%.

Note that if ethylene were the primary fuel present, a higher concentration of the extinguishing agent than the recommended 7% would be needed. This would not preclude the use of Halon-1301, but it would mandate an effective and rapid evacuation plan The fact that only flammable liquids or gases are given in Table 3.8 is not to infer that Halon 1301 is only suitable for such materials. If a fire is confined to the surface of a solid material, chances are that Halon will be effective, i.e., a 5% concentration could possibly put it out if the fuel is allowed to "soak" in the design concentration for 10 or more minutes and if the fuel temperature falls below the level required for spontaneous ignition after the Halon is dissipated. Deep-seated fires, which in this context are defined as ones which will not be put out at a concentration of 5% with a soaking time of 10 minutes, may still be extinguished at higher concentrations, longer soaking times, or a combination of both extended time and increased concentrations. Some of the advantages of Halon 1301 diminish if higher concentrations and longer times are required and some disadvantages arise. Most of these negative factors are related to health factors.

Halon 1301 as a pure material appears to have very low toxicity. At concentrations of Halon 1301 below 7%, little if any effect on humans has been noted for test periods up to 30 minutes. In order to provide a reasonable safety margin, exposures of up to 15 minutes are considered permissible below 7% concentrations by volume. Some individuals have experienced mild effects on the central nervous system, such as dizziness and a tingling sensation in the extremities at concentrations between 7 and 10%, so to be conservative at these concentrations, exposures should be limited to 1 minute. Above 10%, the effects on the central nervous system are more pronounced; between 10 and 15%, exposures should be limited to 15 seconds. There should be no exposures to personnel above 15%. Although a number of other physical effects have been investigated, the only other significant effect noticed in tests on animals has been to the cardiovascular system, where the heart has been made abnormally sensitive to elevated levels of adrenaline, as might be present during the stress of a fire, leading to possible cardiac arrhythmia. All of the observed effects have been shown to be transitory, disappearing after the exposure ceased.

The decomposition products of Halon 1301 are dangerous in sufficient concentrations. The chief ones are the acids of the halide components, HF, HBr, and Br_2. Fluorine is too reactive to be present in substantial amounts alone. Small amounts of carbonyl fluoride and carbonyl bromide (COF_2 and $COBr_2$), respectively, have been found as well. Because of the potential for these toxic materials being present after the intervention of a Halon 1301 system in a fire, fire personnel and others entering the area of a fire should exercise caution. Where the fire involves substantial amounts of solid fuel, the use of a positive pressure, air-supplied breathing apparatus is recommended to search the area for injured personnel.

The disadvantages alluded to above, when discussing the use of Halon 1301 for deep-seated fires in solid materials, are the risks associated with the required higher concentrations of Halon 1301 due to the increased generation of decomposition byproducts with more extinguishing agent present, especially if an individual is injured or unconscious and is unable to evacuate quickly. Cost is a further and significant disadvantage. Halon 1301 is expensive and most

---

* These data are derived from information in NFPA 12A.

Table 3.8 Halon 1301 Characteristics (values in volume percent in air at 77°F [25°C] at 1 atmosphere)

Fuel	Extinguishing Concentrations	Design Concentrations	Inerting Design Concentrations
Acetone	3.3	5.0	7.6
Benzene	3.3	5.0	5.0
Ethanol	3.8	5.0	11.1
Ethylene	6.8	8.2	13.2
Methane	3.1	5.0	7.7
n-Heptane	4.1	5.0	6.9
Propane	4.3	5.2	6.7

systems are designed to completely release the supply of the agent. As a result, fixed Halon systems are usually installed where they will be cost effective, such as in large computer installations, where other expensive configurations of sensitive electronic devices are in use, or where substantial amounts of solvents are stored or in use.

To take advantage of the excellent extinguishing properties of Halon 1301, especially for very rapidly spreading fires from burning flammable liquids, a quick-acting fire detector should be selected, such as one that detects either ultraviolet or infrared light from the flames. Although precautions have to be taken for both these types of detectors to avoid false activation of the system, the rapid response capability of these units means that the extinguishing agent can be applied to the fire almost instantaneously.

Local application systems work precisely as do the fixed systems, with the exception that the agent is applied directly to the immediate area of the fire. The primary need is for the concentration to remain sufficiently high in the area of the fire long enough to ensure extinction of the fire and to permit sufficient time to elapse to allow cooling of the fuel to avoid spontaneous reignition once the concentration decreases below the level required to extinguish a fire. While in a fixed system, it is possible to calculate the amount of Halon 1301 needed with some precision, each local application is a unique situation and must be evaluated individually.

As with other common types of fire suppression systems, Halon 1301 is ineffective for fires involving a number of materials including reactive metals, metal hydrides, and materials not requiring the presence of air to burn, such as gunpowder, some organic peroxides and hydrazine.

Although Halon 1301, or it is hoped, its replacement, is not ideal for every laboratory facility, it does offer advantages over many other systems. Its toxicity is low. It is very effective on many common types of laboratory fires. It acts rapidly, which is especially critical in minimizing the spread of a fire in a laboratory situation. It will not damage the increasingly sensitive equipment found in modern laboratories. Its cost and the propensity to be set off by minor flames, if ultraviolet or infrared sensors are used, are its major disadvantages.

Halon 1211 Systems

The material in the preceding section is generally relevant to Halon 1211 bromochloro-difluoromethane ($CBrClF_2$) systems as far as effectiveness is concerned. There are some minor differences. Instead of being odorless, it has a faintly sweet smell, but otherwise many of the physical properties are similar. Given in Table 3.9 are the extinguishing characteristics for Halon 1211. Again, the two materials are very similar, as can be seen by a comparison of the data in Table 3.9 to the corresponding information in Table 3.8 for Halon 1301.

The fact that the design characteristics are not very different could be taken to imply that the two materials could be used virtually interchangeably. However, the onset of toxic effects on personnel for the most part do not become significant for Halon 1301 until concentrations of 7% or higher are reached, a level above most of the required design concentrations. For Halon 1211, the onset of problems, such as dizziness, become definite within a few minutes

Table 3.9 Halon 1211 Characteristics (values in volume percent In air at 77°F [25°C] at 1 atmosphere).

Fuel	Extinguishing Concentrations	Design Concentrations	Inerting Design Concentrations
Acetone	3.6	5.0	NA
Benzene	2.9	5.0	5.0
Ethanol	4.2	5.0	NA
Ethylene	7.2	8.6	13.2
Methane	3.5	5.0	10.9
n-Heptane	4.1	5.0	NA
Propane	4.8	5.8	7.7

at exposures above 4%, below the required extinguishing concentrations. For this reason, the use of Halon 1211 is not approved for occupied spaces or for normally unoccupied spaces where an evacuation time of more than 30 seconds would be required.

Even though there are restrictions on its use, because of its excellent fire extinguishing properties Halon 1211 is frequently used in flammable storerooms, especially small stock rooms where self-contained units can be simply hung from the ceiling. Most portable Halon extinguishers are filled with Halon 1211.

iii. Halon 1301 Substitutes

There are three fluorinated hydrocarbon materials that are suitable for flooding occupied areas for fire suppression. These are CHF_3 (FE-13™), $CF_3CH_2CF_3$ (FE-36™), and CF_3CHFCF_3 (FM-200™). The last of these appears the most suitable as a replacement. In some cases the containers used for Halon 1301 can be used for the replacement FM-200™. The design concentration for FM-200™ is 7 to 8.5 % while, as will be noted in Table 3.8, the design concentration for most substances for Halon-1301 is only 5%. However, FM-200™ has a number of favorable characteristics, it has only about half the global warming characteristics and an anticipated atmospheric lifetime of 37 years as compared to 65 years for Halon 1301. It does require about two/thirds more mass to achieve the same extinguishing characteristics as Halon-1301. FM-200™ has a zero ozone depletion potential, is safe for human exposure and leaves no residue. One significant negative characteristic is that at temperatures of 1292°F and above, it creates dangerous HF as a decomposition product.

iv. Inergen

Inergen is an attractive product in several ways. It is a blend of three inert gases, nitrogen, argon, and carbon dioxide and is discharged as are the halons. The concentration required to be effective is low enough so that the oxygen level remains high enough to support human life but not high enough to support combustion. It does not leave a residue, has zero ozone depletion, global warming, and atmospheric lifetime properties. It requires special discharge nozzles in order to reduce turbulence in the protected areas. Because of the inherent safety of the inert gases, this is perhaps the safest of the extinguishing materials.

v. Carbon Dioxide Systems

Carbon dioxide is a colorless, odorless, nonconductive, chemically inert gas. It is commonly used in small portable fire extinguishers for putting out class B fires, i.e., fires that involve solvents, petroleum products, grease, and gases. It can also be used for class C fires where the nonconductivity of the extinguishing agent is important. It is not as effective against fires involving ordinary combustible class A materials, such as wood and paper. Carbon dioxide is also used in total flooding systems. The characteristics of carbon dioxide as an extinguishing

Table 3.10   Carbon Dioxide Characteristics (values in volume percent in air)

Fuel	Extinguishing Concentrations	Design Concentrations
Acetone	27	34
Benzene	31	37
Ethanol	36	43
Ethylene	41	49
Methane	25	34
Propane	30	36

agent are included in NFPA Standard 12, which also provides information on the requirements of an approved system.

A major distinction between the use of carbon dioxide as an extinguishing agent and the Halon systems described in earlier sections is that the extinguishing mode for carbon dioxide is primarily simple smothering of the burning fuel, with no chemical action involved. There is little cooling action, with an effectiveness for carbon dioxide of about one tenth that of an equivalent amount of water. As a result, once the carbon dioxide has dissipated, the possibility of reignition exists if there are any sufficiently hot areas still present.

Table 3.10 gives the minimum extinguishing concentrations by volume and the design concentrations to provide a margin of safety. These should be compared to the equivalent data in Tables 3.8 and 3.9.

It is clear that far higher concentrations are required than for Halon systems, so high in fact that an individual trapped in a space flooded by the minimum recommended amount of 34% in the table would quickly become unconscious due to lack of oxygen, since in such a case the oxygen concentration would fall to 13.8%.

When a fixed system is activated, a major portion of the rapidly expanding gas will become carbon dioxide vapor, while the remainder will become very fine particles of dry ice. There will also be condensed water vapor due to the cooling action of the expanding gas. As a result, visibility may be limited and individuals trapped in an area may have difficulty finding their way out. An area equipped with an automatic carbon dioxide extinguishing system should be posted with warning signs, such as:

WARNING
AREA EQUIPPED WITH CARBON DIOXIDE FIRE SUPPRESSION SYSTEM
EVACUATE IMMEDIATELY WHEN ALARM SOUNDS

Pre-discharge alarms are essential as is personnel training where a carbon dioxide system is installed; individuals in adjacent spaces where the gas could flow as well as those working in the immediate area protected by the system should be included in the training. Any aisles providing a path of egress should be amply wide and kept clear at all times. Doors should swing in the direction of exit travel. Any automatic door closing systems, used to minimize leakage of the suppressant from the fire area, should be equipped with delay circuits. Self-contained, positive pressure breathing apparatus should be maintained nearby to make it safe for personnel to conduct rescue efforts for individuals who may be trapped in the area. Firefighting and rescue forces entering an area where a total flooding system has been triggered should exercise caution and, unless they are confident that the gas has been dissipated, should wear breathing apparatus.

Because of the risk to personnel, carbon dioxide systems should not be used in most laboratory situations, although they could be used in storage areas or where electrical and electronic devices are employed. It does not leave a residue and will not damage equipment.

vi. Dry Chemical Systems

Dry chemical fire extinguishing systems use a variety of dry powders as the firefighting agent. They can be stored in pressurized containers and discharged, when needed, very much like the water and gaseous materials previously covered. Most of the agents are primarily effective against fires involving solvents, greases, and gases and can be used around active electrical circuits and electrical equipment, since the chemicals are usually nonconductive. They do leave a residue, which can be a problem for delicate electronic equipment. In general, however, the residue can be readily brushed off surfaces and vacuumed or swept up. There are multipurpose formulations which can be used on ordinary combustibles such as wood and paper. Most dry chemical formulations use monoammonium phosphate, sodium bicarbonate, potassium chloride, or potassium carbonate as their fire extinguishing agent.

Dry chemical agents are nontoxic, but personnel in the area when they are being discharged may experience respiratory difficulties and vision problems due to the copious amounts of powder in the air.

The primary extinguishing mechanism is similar to that of the halogenated agents. They disrupt the chemistry of a fire so that it will not propagate. Other mechanisms include reducing the oxygen concentration in the flame zone, heat absorption by the chemical agent, and for liquid fires, reducing the amount of vapor entering the air from the liquid by reducing the amount of energy radiated by the flame reaching the surface and causing evaporation. The action of the dry chemicals is very rapid, as with the halogenated agents, which make them desirable where it is essential to prevent a fire from spreading. If the problem with the residue is acceptable, a dry chemical system would appear to be a good choice for laboratories, for chemical stock rooms, and for storerooms.

Monoammonium phosphate ($NH_4H_2PO_4$) is described as a multipurpose agent. As with other common dry chemical agents, it is effective on class B and C fires, but unlike most others, which are not recommended for class A fires, this compound does provide a mechanism for putting out fires in ordinary combustible materials. When heated, it decomposes and forms a solid residue that adheres to the heated fuel surface and excludes oxygen, thus eliminating an essential component for a fire, except in the cases of those materials which will sustain combustion without additional oxygen. Because of the formation of the solid residue adhering to equipment, it is less desirable for laboratory applications, but it would appear to be a good choice for a stockroom where materials are often stored in their original paper and wood shipping containers.

vii. Foam Systems

Foam systems are usually intended to be used to extinguish fires involving flammable liquids rather than as general purpose fire extinguishing agents, although some versions are useful on class A fires. The foam is intended to float on the surface of the liquid and extinguish the fire by excluding air and by cooling the fuel and hot objects that may have been heated by the action of the flames. It also prevents reignition by suppressing the production of flammable vapors which could come into contact with heated solid objects. Most foam systems are not intended to be used on three-dimensional fires, i.e., fires involving solid materials of a significant height. There are high expansion foam systems which are designed to smother a fire by flooding the area with a foam layer about 2 feet thick, that can be used in storage areas.

To be effective, the foams must remain intact and not mix with the burning liquids. Some foams work well on ordinary liquid hydrocarbons, but readily mix with and lose their effectiveness on polar liquids such as alcohols, acetone, and ketones. However, there are different formulations effective for each type of liquid, and at least one, a synthetic alcohol foam, may be used on any of these solvents.

Unless the laboratory is a very specialized one utilizing large quantities of flammable liquids, it is unlikely that a foam fire suppression system would be selected as a first choice. Foam-generating equipment could be useful in large facilities storing significant amounts of flammable liquids.

d. Fire Detection and Alarm Systems

Every laboratory facility should be equipped with at least a manually activated alarm system, although an automatic system is preferable since it will continue to function when the facility is unoccupied. Automatic alarms are especially useful in academic institutions since there are break periods when the population of the campus is very low. In many cases, depending upon local code requirements and the occupancy classification, an automatic fire alarm system may be required rather than optional. Every component of a system should be approved by Underwriters Laboratories, the Factory Mutual System, or other nationally recognized accrediting and testing organizations.

i. Detectors

The first essential component in any system is a device used to detect and initiate an alarm. In a manual fire alarm system, this essential device is an individual who will recognize the probability of a fire. This is not a trivial point. If a fire is behind closed doors or in a concealed chase or plenum, it may not be readily apparent. Persons in buildings where the likelihood of a fire exists should be familiar with the normal state of affairs and be able to recognize discrepancies. For example, there are fairly common situations in chemistry laboratories and operations such as welding where visible smoke and fumes may be generated. Certainly no fire alarm should be sounded in such situations. However, where smoke is observed under unusual circumstances, it should be investigated. When starting to enter a room and finding the door or doorknob is warm, one can reasonably assume that a fire is burning on the other side of the door (the door should not be opened in these circumstances, since this could provide a fresh supply of air to the fire and result in a possibly violent increase in the vigor of the fire). Under similar suspicious circumstances which cannot readily be clarified, it is usually desirable to activate an alarm.

In a manual fire alarm system, the device used to activate the alarm in most cases is a pull box or pull station. The mechanism is very simple—pulling the switch either makes or breaks an electrical circuit which in turn causes an alarm to sound. As will be discussed more fully in the next section, the alarm may sound only in the individual building, or the alarm could initiate a signal at a remote location as well.

In most research buildings, vandalism is rarely a problem. If local circumstances are such that this is not the case, there are protective measures which can be taken. The use of a glass cover, to be broken by a small hammer, is to be avoided since the person pulling the alarm can be cut by the broken glass. An alternative that has proven very effective is a cover for the pull station which sounds a local alarm when removed. Usually this frightens away a person intending to initiate a false alarm.

In an automatic system, devices are used that depend upon physical phenomena uniquely characteristic of fires. These can be classified under four categories: heat, visible products of combustion, invisible products of combustion, and electromagnetic energy output.

Heat is the most obvious choice of a characteristic by which a fire can be automatically recognized. In the section on fire suppression systems, the fusible links in the sprinkler heads represented one type of heat detector. Alloys have been developed that will have reproducible melting points. When the temperature at the detector site exceeds the melting point of the alloy, contacts are allowed to move so that the device can either make or break a circuit, just as with a manual alarm system. There are plastics which can perform in the same manner. Fixed temperature systems are very stable and not prone to false alarms, but are relatively slow to respond. There are several other versions of these fixed temperature detectors, including bimetallic strips, where the differential rate of expansion of two different metals causes the strip to flex or bend to either make or break the contact. Others depend upon the thermal expansion of liquids.

One version of a heat-sensing detector that has recently had wide application and depends upon the properties of materials changing with temperature are cables which can be run in cable trays or conduit along with data and video cables to detect excessive heat or fires within the chases. These serve to isolate within the length of each section of the heat-sensitive cable where a problem has occurred or perhaps even provide warning of an impending problem. Two wires in the cable are normally insulated from each other, but heat causes the insulator to change so that a short develops between the wires. Transmittal of data and linkage of computers is such an important part of modern technology that the use of these cables as part of a heat-sensing system is highly recommended.

Rate-of-rise heat sensors react more quickly to rapid changes in temperature than the fixed temperature units. Most of these units use the thermal expansion of an enclosed volume of air to activate a pair of electrical contacts. They include a small vent to allow some air to leak slowly in and out to avoid false alarms due to slow changes in the ambient temperature or due to changes in barometric temperatures. The vent is designed to be too small to accommodate rapid changes in air volume due to a sudden rise in temperature. Some units that use thermocouples to detect temperature changes take advantage of the property of two dissimilar metals in contact with each other generating a small electric current when a temperature difference is established between one junction and another. A rapid change in the electric current can be detected and used to initiate an alarm, while slow changes in the ambient temperature can be discriminated against electronically.

Smoke is another obvious visible product of combustion that usually accompanies a fire, although the smoke-generating properties of fires vary greatly. In recent years, the use of local smoke detectors has become very widespread. Smoke detectors include two components, a light source and a light sensor. Because of the latter component, an alternate name is "photoelectric" detector. In most units of this type, the detector is usually shielded from the light source. The detector is activated by the light being scattered into the sensor by the smoke which enters into the unit. The location of the unit should be such that there is some air movement in the area which will cause smoke to be carried into the detector. They should not be placed in areas of stagnant air.

A major failing of smoke detectors is that they can be triggered by extraneous light scattering materials. These can be aerosols, such as generated by spray cans or vapor from showers, dust from sweeping, maintenance operations such as welding or construction activities, or insects. They can be set off accidentally by cooking, persons smoking near them, or, unfortunately by individuals deliberately introducing smoke into them. Some of these deficiencies can be avoided by careful location of the detectors or by training in which maintenance personnel are instructed to cover the units when working nearby. The design should include a very fine grid over the passages by which smoke enters the unit so that only the smallest insects can enter them, or they can be constructed with dual chambers which will require both chambers to provide a positive indication. Pest strips on the outside can be used to kill any insect crawling into them. Finally, a frequent cleaning (at least twice per year) program will assist in reducing false activations.

The actions recommended above can keep the number of false alarms to a reasonably acceptable level, but deliberately initiated false alarms are a more difficult problem. Fortunately, at corporate facilities and in the academic and service buildings at academic institutions, deliberate alarms are not a major problem, although they are a significant problem in dormitories at academic institutions. One of the few recourses available is to use verification circuits to test the detector a brief time later (30 seconds, for example) after the initial triggering event. If the device has cleared, then the alarm will not be activated, while if it has not, an alarm will be sounded.

Ionization detectors are alternatives to smoke detectors. In this type of detector, a very weak radiation source emitting beta particles (electrons) is placed near two plates, one of which is charged positively and the other negatively. As the beta particles pass through the air between the plates, they create ion pairs which are collected by the plates, creating a weak electric current. When ionized combustion products enter the air space between the plates, they partially neutralize the ionization current and cause it to decrease or cease. The lack of current then triggers an alarm circuit. Some of the same problems, such as cooking or sources of ionized particles from some laboratory operations, can also trigger these units and cause false alarms. Both this type of detector and the smoke detector have good sensitivity

Sensors can be built to detect the light from a fire. However, in occupied spaces, it would not be feasible to use the visible light region between about 4000 and 7000 Angstroms. Sensors can be designed to detect energy generated by the fire in the ultraviolet region below 4000 Angstroms and in the infrared region above 7700 Angstroms.

Other sources of ultraviolet and infrared light must be prevented from entering these two types of light sensors. Welding generates ultraviolet light. Lightning reflected from a polished floor has been known to trigger an ultraviolet sensor. For infrared detectors, fairly sophisticated filters are needed to eliminate the background infrared radiation. Electronic filtering circuits can be used to help systems employing both types of these detectors avoid triggering false alarms.

ii. Automatic Alarm Systems

These systems can range from being very simple to very sophisticated. In the simplest possible system, an alarm triggered by any of the automatic sensors in the previous section will sound in the building and nowhere else. It will continue alarming until shut off. At the other extreme, the system will not only sound a local alarm, initiate a number of local measures to promote life safety and to confine and extinguish the fire, but will also send a large amount of useful information to a central location where firefighting resources are available. No matter what level system is present, the control panel should be located at the most convenient point of entrance for firefighting personnel so that they can quickly gain as much information as possible from the indicators on the panel. It should not be hidden away in a location difficult to reach, such as in a locked electrical closet.

A significant improvement in a basic system is for the building to be divided into zones with a separate module in the control panel for each zone. A zone will normally include a number of detectors within a fairly compact, contiguous area of a building. Firefighters reporting to the scene can tell from the panel in which zone the alarm had been initiated. A map of the building, indicating the boundaries of the individual zone, should be located in the immediate area of the control panel. Although the individual zone where the alarm was initiated will be identified on the panel, there still would have been only a single alarm which would have sounded throughout the building. If provision had been made to transmit the signal to a remote location, this type of system can directly transmit the zone information to the emergency personnel, or it can be limited to a single indication of a problem in the building. In some systems, this signal may go instead to a locally manned station, perhaps a security area manned 24 hours per day, which will in turn notify a fire safety crew.

The alarm signal may be any number of different devices, such as a bell, horn or a recorded voice message. It should be different from any other similar signal. For example, in an academic institution, there should be no confusion between the fire alarm signal and class bells. In addition, the requirements for the disabled need to be met. In the case of an alarm signal, it is necessary to make provision for individuals with a hearing handicap by means of a visual signal, usually a strobe light. The Americans with Disabilities Act now requires that installed alarms provide sound levels that are at least 15 decibels above ambient sound levels, or 5 decibels above the maximum sound level for a 60 second duration with a maximum level of 120 decibels. Visual alarms must be of the xenon strobe type, clear or nominal white in color. The minimum acceptable intensity is 75 candela, with a pulse duration of a maximum of 0.2 seconds and a maximum duty cycle of 40%. The flash rate must be between 1 and 3 Hertz.

In addition to initiating an alarm in the building that may or may not also be sent to a central station, frequently the control system will be capable of initiating a number of other actions. The system can activate mechanisms to close fire doors, turn off ventilation systems, close dampers, send elevators to the ground floor, lock the elevators out of service, and activate emergency lights for the evacuation routes.

The advent of small but increasingly powerful computers shows promise of revolutionizing automatic fire alarm systems. If enough contact points are available and with the proper interface to the computer, a multiplexing system would make it possible to identify the individual detector in alarm at a remote computer console. With enough memory, it would be possible to display a map of the building on a screen locating the alarm source and, if the information has been put into a database, list the physical condition or the types and amounts of hazardous materials expected to be present. Under current OSHA and EPA regulations, firefighting groups must be notified of hazardous materials (note, there are some important exemptions for laboratory operations) they may encounter at facilities in their jurisdiction. In addition, they must be provided with Material Safety Data Sheets for these materials, so that they may be informed of the risks to which they may be exposed and the appropriate emergency measures they may need to take.

The required hazardous material information is primarily being provided to the fire departments as printed material presently but this is rapidly changing. Generic databases containing thousands of Material Safety Data Sheets are now available on optical disks (CD-ROMs) which provide encyclopedic amounts of data. Small computers have also developed the capability of inexpensively placing information on these disks so that locally specific information can be added. The rapid growth of the Internet and the expansion of high speed data access also makes it feasible to obtain MSDS information by this means as well. It should not be too long before companies selling computerized alarm systems will incorporate these most recent advances into their products and be able to provide even small organizations with extremely sophisticated information resources to aid firefighters in coping with fires in research buildings and other complex facilities.

Often fire alarm system needs can be accommodated on computer or communication systems installed for other purposes. For example, several companies sell small systems, primarily designed for energy management which are designed to accept inputs from other types of systems such as security and fire safety and they provide software packages to support them.

The cost of laboratory facilities is so great and the progress in science is so rapid that no competitive corporate or academic research organization can afford the loss of a major facility. Even if insurance is available to cover the physical losses, how can the value of lost intellectual properties, perhaps irreplaceable, be determined? It takes a minimum of 2 to 3 years to construct a major facility. What would the experimental research personnel, formerly housed in the lost facility, do in the meantime? What would be the position of granting agencies that might be supporting the research? The cost of a good fire safety system, including appropriate detectors, an alarm system, and a fire suppression system, is inexpensive compared to the potential losses including, in addition to those cited, injuries to personnel.

REFERENCES

1. NFPA 11-1998, *Low Expansion Foam and Combined Agent Systems,* National Fire Protection Association, Quincy, MA.
2. NFPA I 1A-1999, *Medium and High Expansion Foam Systems,* National Fire Protection Association, Quincy, MA.

3.   NFPA 12-1998, *Carbon Dioxide Extinguishing Systems*, National Fire Protection Association, Quincy, MA.

4.   NFPA 12A-1998, *Halon 1301 Fire Extinguishing Systems,* National Fire Protection Association, Quincy, MA.

5.   NFPA 2001-1996, *Alternatives to Halon-1301*, National Fire Protection Association, Quincy, MA.

6.   NFPA 15-1996, *Water Spray Fixed Systems for Fire Protection,* National Fire Protection Association, Quincy, MA.

7.   NFPA 13-1996, *Sprinkler Systems, Installation of,* National Fire Protection Association, Quincy, MA.

8.   NFPA 17-1998, *Dry Chemical Extinguishing Systems,* National Fire Protection Association, Quincy, MA.

9.   IFSTA, *Private Fire Protection and Detection,* Laughlin, J.W., Ed., International Fire Training Association, Fire Protection Publications, Oklahoma State University, Stillwater, OK, 1979.

10. OSFIA, *General Industry Standards,* 29 CFR 1910, Subpart L, Section 155-165.

11. **Senecal, J.A.,** Halon replacement chemicals: perspective on the alternatives, *Fire Technology* 28(4), 332, November, 1992.

12. **Hall, J.,** The U.S. experience with sprinklers: who has them? how well do they work?, *NFPA Journal* 87(6), 44, November/December ,1993.

13. **Zurer, P.S.,** Looming ban on production of CFC's, halons spurs switch to substitutes, *Chemical & Engineering News,* 71, 46, November 15, 1993.

14. **Cummings, R.B. and Jaeger, T.W.,** ADA sets a new standard for accessibility *NFPA Journal,* 87(3), 43 May/June, 1993.

15. Americans With Disabilities Act, FR July 26, 1991.

16. EPA Title 40 CFR Parts 9 and 32. Significant New Alternatives Policy (SNAP)   Program, April, 1994. Also, see: http://www.epa.gov/doc/spdpublic/title6/snap.ht,l#frm.

7. Other Fixed Equipment

There are no regulations demanding it but laboratory buildings should be equipped with two separate elevator systems—one for personnel and one for freight. In order to make spaces available to the disabled, many buildings without elevators are being retrofitted with them, and others are having former freight elevators converted to be used for passengers. Because it is often too expensive or there may be no reasonable way to install two separate elevators, many of these elevators are being used for both purposes. This is not desirable, especially in research facilities where hazardous materials are in use. Passenger elevators should not be used to transport hazardous materials, nor should passengers, other than those essential to managing the materials, use a freight elevator at the same time it is being used to move hazardous items between floors. If only one elevator exists, the transport of hazardous materials should be deferred until no ordinary passengers wish to use it.

Both types of elevators should be equipped with a means to signal to a manned location in the event of an emergency, preferably by telephone (unfortunately, the experience with misuse of elevator telephones in some locations has not been positive). In the event of a fire in a building, elevators should automatically be sent to the ground floor (or the floor representing the normal entrance level to the building), and should be constrained to remain there unless fire or other authorized emergency personnel override the interlock. The Americans with Disabilities Act also affects elevators. If elevators are employed to make areas of a building accessible to the disabled, other provisions must be made to evacuate them since elevators are to be made unusable under normal procedures. In the event of an emergency requiring evacuation, designated places of refuge are to made available to which disabled persons can go while waiting assistance. The emergency plan for a building must include emergency procedures to assist the disabled . Emergency groups should be informed if there are disabled persons routinely in a facility who could require assistance.

III. CHEMICAL STORAGE ROOMS

The OSHA requirements for inside storage rooms for chemicals are given in 29 CFR Part 1910.1 06(d)(4) and are based on NFPA Standard 30-1969. This latter standard has been amended since 1969, but at this time, the OSHA requirements have not been changed to

Table 3.11 Storage in Inside Rooms

Fire Protection Provided	Fire Resistance (hours)	Maximum Size (ft²)	Total Allowable Quantities (gal/ft² of floor area)
Yes	2	500	10
No	2	500	4
Yes	1	150	5
No	1	150	2

Note: 1 square foot = 0.0929 square meter. 1 gallon = 3.785 liter.

reflect the later changes. Building codes are normally based on the changed standards. The major difference in the more recent NFPA standard involves changes in the ventilation requirements, although there are minor differences elsewhere. In a few instances, additional safety precautions which are generally accepted practices have been added as recommendations in the following material.

A. Capacity

The amount of flammable and combustible liquids permitted in inside storage rooms depends upon the type of construction and whether an automatic fire suppression system is installed or not. The permissible amounts under both the OSHA and more recent NFPA standards are the same and are given in Table 3.11.

It should be noted that these limits are generous for most operations, permitting between 300 and 5000 gallons of flammable and combustible liquids overall in an inside storage room. However, NFPA Standard 30 at the current time does not permit more than 660 gallons of lA liquids, 1375 gallons of lB liquids, 2750 gallons of lC liquids, and 4125 gallons of class II liquids, so these limits should not be exceeded individually. Excess storage should be avoided, regardless of the legal maximums permitted. For other materials which normally will be kept in the same storage room, the OSHA standard permits other materials to be stored in the same space, provided that they create no fire hazard to the flammable and combustible liquids. Materials which react with water must not be stored in the same room with flammable and combustible liquids.

B. Construction Features

In the earlier material on building codes, the factors that govern the classification of the storage space as to the degree of hazard and the implications of the classification were covered in detail. Although the local building codes will govern the actual construction, they must be at least as stringent as the OSHA regulations. The minimum construction standards which are specified by OSHA are:

1. Inside storage rooms shall be constructed to meet the governing fire-resistive ratings for their use. Doors are to be approved self-closing fire doors. Windows opening on the room, exposing other parts of the building or other properties, are to be protected according to NFPA Standard 80, *Standard for Fire Doors and Windows.*
2. The room must be liquid-tight where the floor meets the walls. Spilled liquids are to be prevented from running into adjacent rooms by one of three methods: (a) at least a 4-inch (10.16 cm) high sill or ramp at the opening; (b) the floors can be recessed at least 4 inches; or (c) an open-grated trench can be cut in the floor within the room which drains to a safe location.

3. If the room is to be used for class I liquids, the wiring must be adequate for class I, division 2, hazardous locations. If only class II and class III liquids are to be stored in the room, the wiring need only meet standards for general use.
4. Shelves, racks, scuff boards, and floor overlay may be made of wood if it is at least 1 inch (nominal) thick.

C. Ventilation

1. Every inside storage room must be provided with ventilation, either gravity or mechanical. Either type system must be capable of six complete air changes per hour according to the OSHA standard. Instead of six air changes per hour, NFPA 30 specifies an exhaust ventilation rate of 1 cfm per square foot of floor space (0.093 m^2), but not less than 150 cfm (45.7 m^3/min). For a ceiling height of 10 feet (3.05 meters), the two requirements are the same.
2. If a mechanical system is used to provide the ventilation, OSHA requires that it be controlled by a switch on the outside of the door to the room. The lights in the room are to be operated from the same switch, and if class I liquids are dispensed within the room, a pilot light must be installed adjacent to the operating switch. In order to accommodate hearing handicapped users, a strobe alarm light is recommended as well.
3. If the ventilation is provided by a gravity system, both the intake air inlets and the exhaust air outlet must be on the exterior of the building.
4. Exhaust air should be taken from a point no more than 1 foot (0.3048 meters) from the floor and exhausted directly to the roof of the building. The air intake in the room should be on the opposite side of the room from the exhaust. Since, in general, flammable and combustible liquid vapors are heavier than air, this design is intended to sweep the floor clean of vapors before they accumulate and pose a hazard. The aisles of the room should be such as not to block this sweeping action. If ducts are used, they must meet requirements of NFPA 91, Standard for the Installation of Blower and Exhaust Systems for Dust, Stock, and Vapor Removal or Conveying, and should not be used for any other purpose. It would be preferable for the air intakes to be on the upper portion of the building and upwind from the most prevalent wind direction.

D. Fire Safety

1. There shall be no smoking or open flames in a flammable or combustible material storage room. A prominent sign should be posted on the outside of the door to the facility stating:

FLAMMABLE MATERIAL STORAGE
NO SMOKING

2. At least one 12 B or larger portable fire extinguisher must be located outside the door to a flammable material storage area, no more than 10 feet from the door.
3. Any fire suppression system installed in the storage room must meet the standards of 29 CFR Parts 1910.155-165.
4. There shall be at least one clear aisle at least 3 feet (0.9144 meters) wide in every flammable and combustible material storage room. No container should be more than 12 feet (3.66 meters) from an aisle. Containers of 30 gallons (113.5 liters) capacity or larger must not be stacked more than one layer high.

REFERENCES

1. NFPA 30, *Flammable and Combustible Liquids Code,* National Fire Protection Association, Quincy, MA.
2. *Prudent Practices for Handling Hazardous Chemicals in Laboratories,* National Academy Press, Washington, D.C., 1981, 218-222.
3. ANSI Z9.2, *Fundamentals Governing the Design and Operation of Local Exhaust Systems,* American National Standards Institute, New York, 1979.
4. NFPA 91, *Standard for the Installation of Blower and Exhaust Systems for Dust, Stock, and Vapor Removal or Conveying,* American National Standards Institute, New York.

IV. MOVABLE EQUIPMENT

Many items of movable equipment represent special safety problems in the laboratory. In some instances, it is the probability of initiating an explosion or fire that is the major concern. In other cases, the major problem may be generation of toxic fumes or aerosols, and in still others, the equipment is inherently dangerous due to the physical injuries which improper maintenance or use may cause. Laboratory equipment often includes electric motors, switches, relays, or other spark-producing devices. In the presence of vapors from flammable materials, a spark can initiate a fire if the concentration of the vapor is between the upper and lower flammable limits and the temperature is above the flash point at which the given vapor can be ignited. Motors used in laboratory equipment should be induction motors which are non-sparking. Series-wound motors with graphite brushes, such as those used in many home appliances, should not be used in laboratory equipment, and appliances such as hot plates, vacuum cleaners, blenders, and power tools designed for the home should not be brought into laboratories where flammable liquids are actively used. Switches and contacts for electrical controls should be located in flammable vapor-free areas wherever possible. Equipment should be purchased which is designed to minimize the possibility of flammable vapors entering internal spaces where sparks may occur or where the vapors may come into direct contact with heating elements.

Electric shock is another hazard common to many pieces of laboratory equipment. Any electrically powered item of laboratory equipment which is subject to spillage of chemicals or water or exhibits signs of excessive wear should be used carefully. All equipment should be provided with three-wire power cords (some tools may be double-insulated as an acceptable alternative), which should be replaced if the insulation is cracked or frayed. Metallic parts of the equipment should be grounded separately, if necessary. Care must be taken to ensure that any ground is, in fact, a good one. An alligator clip on a water pipe is not sufficient. A poor ground connection can generate a high temperature if sufficient current passes through the high resistance contact, and instead of being a safety feature, actually represents a major fire hazard. The potential difference between two poorly grounded pieces of electronic can be enough to damage sensitive electronic components. Portable equipment can be connected through ground-fault interrupters, devices which detect a diversion through an alternate path (such as a person) to the standard connecting wiring and which, upon detection, shut down the circuit within a very few milliseconds.

A large variety of devices are left on continuously or for long periods, operating while unattended. Any device which could overheat to a degree that it could result in a fire within the facility should be equipped with redundant controls, heat sensors, or overload protection which will cause the equipment to shut off if excess heat is generated or to fail in such a way as to minimize heat generation. Although much concern about fires stems from the presence of flammable vapors, many fires start in overheated ordinary combustible materials, with flammable and combustible liquids becoming involved at a later stage.

Many items of equipment result in significant amounts of fumes being generated in laboratory operations. Many of these are typically being used outside of fume hoods. In general, these items of equipment, if used properly, do not cause fumes to be generated at levels exceeding or even approaching the allowable limits of exposure established by OSHA. However, these levels are subject to revision as more information becomes available, usually being lowered, and there are individuals of more than average sensitivity for whom even the original limits may be too high. Under these circumstances, it would appear prudent for laboratory managers to adopt as an informal policy within their facility, a chemical policy similar to that used to minimize exposure to radiation. This policy, which is intended to achieve exposures As Low As Reasonably Achievable (ALARA), would appear to be a good working policy as well as a good basis for equipment design. The key word is "reasonable" in deciding what precautionary measures are indicated. In the following sections, attention is directed to a number of specific items of equipment for which explicit problems may arise.

A. Refrigeration Equipment

Refrigeration units represent a hazard as an item of laboratory equipment for a number of reasons. For example, improper use of laboratory refrigerators for food to be consumed by the laboratory workers is a continuing problem but is readily solvable by firm enforcement of policies defining acceptable practices by managers. Under no circumstances should laboratory refrigerators and freezers, used for toxic chemicals and pathogenic biological agents, ever be permitted to be used for food storage. However, the major problem with refrigeration units is the tightly sealed space within them.

The confined space within refrigeration units permits vapors from improperly sealed containers to accumulate. In some instances, the vapors may be toxic, and an individual peering in to find the container desired has the potential to breathe in fumes which may substantially exceed acceptable safe levels. Unless the material has a distinctive or offensive odor, it may not even occur to the person using the unit that a problem may exist. It may be desirable to have more than one refrigerator in a laboratory, one of which would be designated and prominently labeled for the storage of dangerous materials only, which would encourage users to be especially careful in using it and would encourage them to make sure that everything placed in it was properly sealed. Unfortunately, both space and funds are often limited in laboratories, and unless the laboratory manager is careful to maintain a strict policy on segregation of storage, the possibility exists, as storage needs increase, that the unrated refrigerator could be improperly used. Careful training in how to seal containers placed in refrigerators and freezers and insistence that these procedures be followed should be a part of every laboratory's management program.

Beakers, flasks, and bottles covered with aluminum foil or plastic wrap are unacceptable for storage of volatile materials in a refrigeration unit. Corks and glass stoppers also may not form a good seal. Screw-cap tops with a seal inside are much better, when screwed on firmly. However, no type of top is foolproof when used in haste.

The problem of confined flammable vapors is fortunately, one for which an engineering solution has been developed and is commercially available, since the consequences of an accumulation of flammable vapors in a normal refrigeration unit are potentially life threatening. The vapors of flammable liquids may be ignited by sparks (or other heat sources), but some liquids are more easily ignited than others over a wider range of concentrations. Data for some flammable liquids are given in Table 3.12.

The materials in Table 3.12 were selected as examples because they have a wide range of concentrations in which their vapors could be ignited and flash points[*] well under temperatures found in most household refrigerators and freezers. The reason this information is important is

---

[*] The flash point is the lowest temperature at which a liquid gives off vapor in a high enough concentration to form an ignitable air-vapor mixture above the surface of the liquid under standardized conditions.

Table 3.12 Flammability Characteristics of Some Common Solvent Vapors

Chemical	Flash Point °F (°C)	Ignition Temperature °F (°C)	Flammable Limit (% by volume In air)	
			Lower	Upper
Carbon disulfide	- 22(-30)	176 (80)	1.3	50
Diethylether	-49 (-45)	320 (160)	1.9	36
Ethylene oxide	<-18 (<0)	804 (429)	3.6	100
Ethyl nitrite	-31(-35)	194 (90)	4.0	50
Propylene oxide	-35(-37)	840 (449)	2.8	37
Vinyl ethyl ether	-50 (-46)	395 (202)	1.7	28

that the interiors of household refrigerators contain a number of electrical contacts which could generate sparks to ignite the contained vapors. Among these are the light switch, temperature control, defrost heater (in "frost-free" models), and fan. Many frost-free models also have a drain which could allow the vapors to reach the space occupied by the compressor. Models are available in which all of these sparking devices have been eliminated, modified to be explosion-proof (such as the compressor), or moved to a safer location outside the refrigeration unit.

The confined vapors in a refrigerator or freezer, if ignited, can create a major explosion and fire within a facility. Anyone standing in front of an exploding unit would be in very real danger of being seriously injured or even losing their life. It is highly likely that any containers of flammable liquids nearby not directly involved in the explosion would be broken and contribute fuel to the fire which would probably follow the explosion. The freely flowing flammable liquids could spread to other stored materials within the laboratory. In an unprotected (by sprinklers or other fire suppression system) facility or in an older structure, the final result could be the destruction of an entire building. There could also be other toxic or hazardous materials within the refrigerator or laboratory which could be spread by the incident even if the fire did not spread. In at least one instance, the refrigerator which blew up was in a very active radiochemical laboratory. An entire floor of the building had to be decontaminated at a considerable expense in time and money.

Refrigerators which have been commercially modified to be safe for the storage of flammable materials are designated as "Flammable Material Storage Units" and meet NFPA Standard 56C and Underwriters Laboratories, Inc. standards. These units are not explosion-proof. They have only had components removed which could cause sparks *within* the interior of the unit. Only refrigerators or freezers intended to be used in hazardous locations where a spark inside *or outside* the units could cause a fire or an explosion are designated as explosion proof. The electrical power wiring to this latter class must be installed in conformance with "Commercial Refrigerator/Freezers for Hazardous Locations" class 1, groups C and D code requirements. Although in theory, it is possible to modify an ordinary refrigerator to be acceptable as an explosion-safe unit, in practice it is difficult to be sure that it has been done properly. It is strongly recommended that all laboratory refrigerator and freezer units be purchased already modified to be safe for storage of flammables, with the exception of ultra-low temperature units which operate at temperatures lower than the flash points of any commonly used flammable liquids.

The costs of "Flammable-Material Storage" refrigerators and freezers are usually two to four times higher than comparable products to be used in the home. As a result, many individuals object to a blanket policy requiring the purchase of these safer units, especially those who are not using the refrigerators in their laboratories to store solvents or other flammable liquids. There are several valid reasons to override these objections. The most important are based on the exceptionally long useful life of refrigerators. Only rarely do they last less than 10 years, and many continue to work well for more than 20 years. Few research programs endure for comparable periods and few individuals remain in the same position as long. Thus, although assurances can be given and signs can be placed on doors of the units forbidding the use of the refrigerator in question for the storage of flammables, there are no feasible means of guaranteeing that they will not be used at some time during the useful life of the refrigerator or freezer for the storage of flammables. Although the initial cost is high, over the total life of a unit it is an extremely inexpensive price to pay to totally eliminate a major source of fire and explosions.

Ordinary sized refrigerators and freezers, as well as combination units, are available from a number of sources. As the size of the units become larger, however, only a few suppliers offer flammable material safe units and then usually as ordinary refrigerators modified at the factory to be explosion safe for confined vapors at an additional cost.

There are legitimate reasons to make a few exceptions. If the use is for a basic departmental function which would never entail the use of flammable liquids and the department is a stable, established discipline or research field, then there is reason to accept ordinary units to avoid paying the additional costs involved. Refrigeration units to be placed permanently in isolated, normally unoccupied locations also might be candidates for quality consumer units. Refrigerators to be used only for the storage of food and beverages for the convenience of the employees should also be permitted, but they should *not* be allowed to be placed directly within the working area of a laboratory.

Large walk-in refrigerators and freezers, or cold rooms, pose an additional problem with condensation of water vapor on the equipment when electrically operated equipment is placed inside them, due to the very high humidity usually present. Care should be taken to avoid shorts and electrical shocks to personnel. All of the equipment should be well grounded and any electrical cords should be insulated with waterproof insulation. A recommended precaution would be to have all of the electrical sockets in the interior wired with ground-fault interrupters or to require that any equipment used inside must be connected through one.

REFERENCES

1. NFPA 325-1994, *Guide to Fire Hazard Properties of Flammable Liquids, Gases, and Volatile Solids*, National Fire Protection Association, Quincy, MA.
2. NFPA 49-1994, *Hazardous Material Data*, National Fire Protection Association, Quincy, MA.

B. Ovens

Electrically heated ovens are other devices found in many laboratories for which the problem of ignition of flammable fumes may exist. The ovens are used for baking or curing materials, out-gassing, removing water from samples, drying glassware, or in some cases providing a controlled, elevated temperature for an experiment. Very few are provided with any provision for preventing any of the materials evaporated from the samples from entering the laboratory. Ovens should be designed so that any fumes generated in the interior do not have an opportunity to come into contact with the heating elements or any spark-producing control components. A single pass-through design in which air is drawn in, heated, and then exhausted is a relatively safe design as long as nothing in the oven impedes the flow of air. Most ovens intended for home kitchens are not constructed in this manner and should not be used in the laboratory environment.

Every oven should be equipped with a backup thermostat or temperature controller to either control the unit should the primary one fail or shut the oven down. If the secondary controller permits the oven to continue to operate after the primary device fails, it should provide a warning that the failure has occurred so that the researcher can make the decision as to whether or not to continue the operation. In any event, as soon as practicable and before beginning a new run, the oven should be repaired. No unit with only a single thermostat should be used for long, unattended programs.

Because most laboratory ovens do exhaust directly to the laboratory, they should not be used to heat any material from which a toxic vapor or gas would be expected to evolve, unless provisions are made to exhaust the fumes outdoors, as would be done with a fume hood. Since the exhausted gases would be warmer than the ambient air, they should rise. This is one instance in which a canopy hood placed over an oven, or at the least the oven exhaust, could be a satisfactory choice for a hood.

Ovens can be purchased which are suitable for heating materials which contain flammable liquids. One commercial model, designed to be used for small amounts of solvents, purges the interior with several complete air changes prior to turning the heat on in order to remove any residual gas which may be present. It also automatically turns off the heat if either the exhaust fan fails or the temperature rises above the maximum temperature for which the unit is designed. The door to the unit has an explosion venting latch, which allows it to blow open in an explosion. However, a recommended feature of ovens used for solvents are explosion vents on the rear of the unit, so that any explosion would be vented away from the laboratory and its occupants.

Where asbestos has been used as insulation in laboratory ovens, some concern has been voiced about the potential exposures of service personnel performing maintenance on the units. Although under normal circumstances such operations will involve minimal contact with the insulation, it is desirable to purchase units which use other insulating materials. Eventually the ovens containing asbestos will have to be removed from service, and the asbestos insulation would represent a possibly expensive disposal problem. The oven legally could not simply be taken to the typical local municipal landfill.

C. Heating Baths

Heating baths are used to heat containers partially immersed in them and to maintain them at a stable temperature, on some occasions for extended periods. Heating baths should be equipped, as in the case of ovens, with redundant heat controls or automatic cutouts should the temperature regulating circuits fail. The material used in the bath may be flammable, and excessive temperatures could result in a fire.

A number of materials are used in heating baths. Water can be used up to about $180°$ ($82°$C). Mineral oil and glycerine are used up to about 300°F (about 150°C). Paraffin is employed in the range up to about 400°F (about 200°C). These last three materials are flammable, although the NFPA rating of each is 1, on a scale of 0 to 4. Silicone oils are recommended at temperatures up to about 570°F (about 300°C). These are also moderately flammable and are more expensive than organic oils. In a 1957 tabulation of materials in use for heating baths, R. Egly listed tetracresyl silicate as an expensive material, but one which had very good characteristics. It was listed as nontoxic, noncorrosive, fire resistant, and suitable for use from near room temperatures to approximately 750°F (about 400°C).

Heating baths should be in durable, nonbreakable containers and set up on a firm support so that they will not be likely to tip over. They should not be placed near flammable and combustible material, including wood and paper which, if exposed to continuing heat over a sufficient period of time, could reach kindling temperatures, or near sources of water (particularly deluge showers) which could cause the bath liquid to splatter violently from the container. In most cases, the bath temperatures are high enough to cause severe burns. If it is necessary to move the full container, it should be done while the liquid is cool, again to avoid the risk of burns.

If the container itself does not include a heating element, any immersion heater should be insulated to avoid the potential of electrical shock and should include a cut-out device if the temperature exceeds the set point. Alternatively, a second temperature sensor should be placed in the heat bath to act as a circuit breaker to cut off power to the heater if its thermostat fails. The thermostat clearly should always be set well below the flash point of the heating liquid in use. A thermometer placed in the bath at all times it is in use is recommended to provide a visual indication of the actual temperature of the bath. Digital controllers normally provide this additional information.

REFERENCE

1.   *Techniques in Modern Chemistry,* Vol. 3, Part 2, Weissberger, (Ed.) Interscience, New York, NY, 1957, p. 152.

D. Stills

Individual stills are frequently set up in laboratories to provide distilled water to the facility and are usually left running for extended periods unattended. The concern here, as with many of the other devices discussed in this section, is the possibility of over-heating with the subsequent initiation of a fire. Some units use two water sources, pre-treated water for the boiler, from which various impurities are removed, and ordinary tap water for cooling, while some use tap water for both. The still should be equipped with an automatic cutoff should it overheat, in case of either water pressure failure or the boiler becomes dry. Both the water supply and heat should cut off if the collector bottle becomes full; if the power fails, a valve should shut the water supplies off.

E. Kjeldahl Systems

Kjeldahl units and other digester and distillation units used for nitrogen determinations and trace element analyses can be sources of potentially unacceptable fume levels if not vented properly. Larger units are constructed so several digestions can take place simultaneously, and thus substantial amounts of corrosive fumes can be generated. The fumes in such units are usually drawn through a manifold to a discharge point at one end of the unit. At this point, the fumes are either exhausted by an integral blower or drawn to an aspirator where they are diluted and condensed by the water spray and disposed of into the sanitary system. Because the blower is integral to the unit, the exhaust duct downstream from the unit would be at a positive pressure. If the duct were to corrode, fumes could leak into the surrounding spaces. The total volume of fumes disposed of into the sanitary system would normally be small enough to be well diluted in the sanitary waste stream, so the aspirator method has some advantages. An unexpected hazard associated with Kjeldahl units, reported in the August 10, 1992 issue of *Chemical & Engineering News,* was the discovery of significant amounts of mercury (20 milliliters) in the condensation tubes while the unit was being cleaned. As the individual reporting this finding, David Lewis, Chemist for the City of Lompoc, CA, noted, the possibility exists for workers using a mercury sulfate catalyst to have a long-term exposure to mercury vapor.

F. Autoclaves

Pressurized sterilizing chambers or autoclaves are used primarily in the life sciences. Glassware, instruments, gloves, liquids in bottles, biological waste, dressings, and other materials are sterilized in them by steam under pressure, typically at a pressure of a little under 2 atm, at temperatures of up to 275°F (135°C). Since they are heated pressure vessels, they should be checked periodically to ensure that the seals to the closures are in good condition, and they should he equipped with safety devices to prevent excessive temperatures and pressures. There are a number of potential problems associated with their use. The requirements for treating infectious waste, under the bloodborne pathogen standard and other federal and state regulations, will increase the amount of materials processed through steam sterilizer units, with more accidents occurring as a consequence. All users should be thoroughly trained in safe techniques and acceptable practices.

Fortunately, most autoclaves are designed so that they cannot be opened while the chamber is under pressure. However, the materials inside will still be very warm, and removing them too hastily or forgetting to wear insulating gloves could very likely cause the item being handled to be dropped. In some cases, this would only cause a loss of sterility in the dropped material, but in other cases, a bottle containing a liquid might be broken.

Liquids placed inside in sealed bottles may explode, and liquids in ordinary glass bottles instead of Pyrex containers designed for the temperatures and pressures may rupture. If the unit is set to exhaust rapidly, as might be done for instrument sterilization, boiling may take place in bottles of liquids, with a subsequent loss of liquid into the autoclave. Flammable liquids or chemicals which could become unstable at the temperatures reached in the autoclave should not be run through the sterilizing cycle.

Operating instructions and a list of good safety practices should be posted near any autoclave for ready reference.

REFERENCE

1. **B.M. Miller, et al.** (Eds.), *Laboratory Safety Practices,* American Society for Microbiology, Washington, D.C., 1986.

G. Aerosol Generating Devices

There are a substantial number of devices, primarily used in biological laboratories, which generate very small particulate droplets (aerosols) which can remain airborne for long periods of times. These may or may not be dangerous but the possibility exists that they may be. Devices such as sonicators, ultracentrifuges, blenders, pipettes, and even the lowly hot wire scraping across a petri dish can generate significant quantities of aerosols. Where feasible, these devices should be used in a hood and employed in such a manner as to minimize the generation of aerosols. Ultracentrifuges are likely to be too large to operate in a hood. If a tube in the centrifuge were to be broken while being processed, the possibility of aerosol creation which could escape when the centrifuge cover were opened could be quite high. Since one never knows in advance when a tube might be broken, the vapors should be allowed to settle for several minutes after operating a centrifuge prior to opening it.

There will be an extensive discussion of microbiological laboratories in a later section.

V. ANIMAL LABORATORIES—SPECIAL REQUIREMENTS*,**

Much of the requirements on such physical facilities are identical to those found in the

---

* Much of this material was prepared by David M. Moore, D.V.M., for the third edition of the handbook but this section now includes some material deriving from the bloodborne pathogen standard and infectious waste regulations.

** Additional information will be given as an Appendix to the end of the material dealing with animal facilities from the Centers for Disease Control Guidelines for Vertebrate Animal Biosafety.

material in Chapter 5 for microbiological laboratories. The floors, walls, casework, and equipment should be designed to be easily decontaminated. Storage facilities should be available to store medically regulated waste (current term for infectious waste), pending collection and disposal. Ventilation recommendations can be met in more than one way, either as air changes per hour or as volume per animal housed. Generally animal facilities are provided with 10 to 20 air changes per hour. Current recommendations for human-occupied spaces have increased in recent years to 10 to 12 air changes per hour, so laboratory spaces can be designed toward a common goal. Animal holding facilities are usually designed with both temperature and humidity controls, while human spaces often do not include the latter in the design. The comfort levels for both temperature and humidity for animals are essentially the same as for humans, although certain species have more stringent requirements. Typically, 30 to 70% relative humidity is desirable while temperatures of 18 to 26°C (~64° to 79°F) will be satisfactory.

Workers who work with items possibly contaminated with diseases communicable to humans, including tissue, fluids, fecal materials, and equipment which has come into contact with any of these, should be offered appropriate immunizations, if safe effective vaccines are available. Tetanus shots are recommended for all who work with animals, while those who work with wild animals should be offered rabies vaccinations. A preemployment medical examination is mandatory and should include medical and work histories. Periodic examinations may be desirable and should be considered. Any worker who may come into contact with human or primate tissue, blood, and fluids must receive training to meet the standards of the OSHA bloodborne pathogen standard and be offered shots for Hepatitis-B.

Individuals may work with animals which have diseases that are communicable to humans or to other animals. In such cases, these animals should be kept in isolation areas and provision made for decontamination of personnel and equipment leaving the area. Of course, those who work with such animals must be especially careful to avoid exposures.

Nuclear medicine facilities also represent a special hazard. Normally, nuclear injections use a short-lived isotope ^{99}Tc with a halflife of a little over 6 hours. If the injection is in a large animal, such as a horse, the personnel exposure level can be substantial and workers should be provided with personal radiation monitors. Collection of the feces and urine may be required. In addition, the animal may not be released to a member of the public until the exposure levels fall below the legal limits for unmonitored persons (more on this later). X-rays of animals also represent a special hazard since one cannot simply tell an animal to hold still. Often taking of x-rays will require a person to hold the animal still. Holders will often receive substantial levels of scattered x-ray radiation so this duty should be spread among a number of individuals.

A. Fixed Equipment in Animal Holding Facilities

In addition to items such as fume hoods and casework, for which the requirements will be similar to those described previously in this chapter, there are some special sanitation equipment items.

1. Cage Washers

Most facilities have equipment for sanitization of cages and cage racks. There are three major types of mechanical cage washers:

1. Rack washer - This is a unit that can hold one or more cage racks or racks containing cage boxes. It can have a single entrance or two entrances, allowing movement of cages/racks from a "dirty" processing area through to a "clean" area, with the areas separated by a wall. Spray arms in the unit direct water at high pressure on all sides, and the unit reaches the recommended temperature of 180°F for a minimum of 3 minutes, which is sufficient to destroy pathogenic (disease causing) microorganisms. It can handle a larger number of cages/racks than a cabinet washer and is thus less labor intensive.
2. Cabinet washer - The cabinet washer has smaller internal dimensions than the rack washer

and can accommodate cages, but not racks. For larger, heavier cages (i.e., rabbit, nonhuman primate, dogs), this unit proves to be more labor intensive and more time consuming. Racks would require some other method for sanitization: steam generators, high-pressure spray units, or chemical disinfection.

3. Tunnel washer - "Shoebox" rodent cages, water bottles, cage pans, and other small equipment can be placed on this unit's conveyor belt. It is more efficient than the cabinet washer for sanitizing small items.

These units generate quite a bit of heat, and the ventilation for the cage wash area should be adjusted accordingly.

2. Autoclaves

Autoclaves were described earlier, but are especially important in animal facilities. Autoclaves provide support for animal surgical facilities and may be used in barrier facilities to sterilize food, bedding, water and water bottles, cages, and other equipment prior to entry into the barrier. Steam autoclaves can potentially dull sharp surgical instruments, and the heat can reduce vitamin levels in feed. Special autoclavable diets are manufactured with higher levels of vitamins to assure that appropriate levels remain after autoclaving. Ethylene oxide sterilizers are used for materials damaged by the temperatures in steam autoclaves (i.e., surgical instruments, plastic tubing and catheters, electronic devices). However, ethylene oxide gas poses a health risk for humans, and materials should be allowed to "off gas" for 24 to 48 hours before coming in contact with animals. Care should be taken in designing a gas scavenging system, and standard operating safety procedures should be established for use of the ethylene oxide sterilizer. Steam sterilization is one of the universally accepted methods of treating medically regulated waste. The use of autoclaves for tissue and related materials is not as desirable as it might be, since the end product can still decompose and be objectionable when placed in landfills. Autoclaving is more practical for labware, syringes, and other "sharps." However, the current trend is to render these types of items unrecognizable when disposed of in public landfills due to adverse publicity engendered by hospital waste washing up on beaches in recent years.

3. Incinerator

Many facilities dispose of solid wastes and animal carcasses by incineration. This is another universally accepted method of treating medically regulated wastes. The federal government and subsequently, states are adopting very stringent operating conditions on incinerators employed for this purpose. Mixed waste, which contains animal material, chemicals meeting EPA regulatory criteria, or radioactive materials may or may not be incinerated according to the characteristics of the waste. The reader is referred to the section in Chapter 4, on the current standards on incineration and the various alternatives which currently exist for processing medically regulated waste.

REFERENCE

1. NCI, *Chemical Carcinogen Hazards in Animal Research Facilities,* Office of Biohazard Safety, National Cancer Institute, Bethesda, MD, 1979, 15 - 17.

B. Equipment for Animal Laboratories and Holding Areas

Caging for small and large laboratory animal species may be either fixed or movable. Sanitization of fixed caging and the room environment is less easily accomplished than is sanitization of movable racks and cages. Fixed caging might also provide safe haven for vermin and reduces the flexibility of use of that holding room.

Animal caging is designed for the convenience of the investigator and the husbandry staff, but more importantly, for the comfort, safety, and well-being of the animal. *The Guide for the Care and Use of Laboratory Animals* provides recommendations for caging materials and cage sizes for a variety of species. Adherence to these recommendations will also assure compliance with the Federal Animal Welfare Act requirements regarding cage sizes. A description of various caging systems is given by Hessler and Moreland (1984).

REFERENCES

1. **Hessler, J.R., and Moreland, A.E.,** Design and management of animal facilities, in *Laboratory Animal Medicine,* Fox, J.G. et al. (Eds.), Academic Press, Orlando, 1984, pg. 517-521.
2. ILAR. *The Guide for the Care and Use of Laboratory Animals.* NIH Publication 85-23, NIH Bethesda, MD, 1983.

Chapter 4

LABORATORY OPERATIONS

I. GENERAL CONSIDERATIONS

The attitude of laboratory personnel toward safety is the most important factor affecting the safe conduct of research. It is more important than the quality of the equipment, regulations, managerial policies, the inherent risks associated with the materials being employed, and the operations being conducted. If the safety attitude of everyone in the laboratory is positive, and this attitude is clearly supported by either the corporation or the academic institution, then it is highly probable that a strong effort will be made for the research program to be conducted safely. Conscientious individuals will try to follow the standards of behavior established by their organization to ensure for themselves that their operations are as safe as possible, and will attempt to comply with regulations and policies which have been established for their protection. On the other hand, no matter how strict management policies are and how many regulations have been established, individuals with an attitude that safety concerns are not important and that nothing will ever happen to them will manage, somehow, to circumvent any inconvenient restrictions. Occupational Safety and Health Administration (OSHA), in its performance oriented laboratory safety standard, recognized the importance of the local laboratory manager by placing the responsibility for developing and implementing a sound safety plan for the laboratory squarely on this individual.

Rarely do you have as black and white a situation as implied by the two extremes in the preceding paragraph. No one is so careful that they avoid taking any risks, nor is any one totally unconcerned about their own safety. The goal should be to avoid taking unreasonable risks, and it is the responsibility of laboratory managers to establish, by policy and example, reasonable standards of conduct to ensure that this goal is met.

Generally, a safe laboratory operation is usually a well-run operation. For example, labeling of secondary containers of reagents is not only a good safety practice to avoid accidental reactions leading to injuries, but it serves to prevent errors which could negatively affect the research program as well.

The failure of a laboratory manager to establish the right atmosphere of safety and to enforce established safety and health policies can render the manager vulnerable to litigation on the part of an injured employee, especially if it can be shown that the failure was due to willful negligence. The OSHA Laboratory Standard does require a written hygiene plan for the laboratory facility but if written policies were not available, if a reasonable individual can be shown to have been likely to have anticipated a problem, and if due care to protect an employee under the individual's supervision was not exercised, a civil court suit against the manager by the injured party could very well be successful. On the other hand, employees (at least in an academic institution), who deliberately does not comply with safety precautions of which they have been informed and which are normally expected to be followed, may weaken their case due to contributory negligence to the extent that the suit would not succeed or the award be substantially diminished. In the corporate world, there are workman's compensation laws that govern situations in which an employee is injured and usually provide for compensation to the employee regardless of who is at fault (although accepting workman's compensation usually means waiving recourse to legal claims in court), although there are differences in coverage depending upon many factors in the different states. The whole concept of liability is constantly being modified by court actions. However, for financial as well as ethical reasons, the prudent manager or employer should be sure that the research programs for which the manager is responsible are conducted according to good safety practices, as defined by laws and regulations, corporate and institutional policies, and reasonableness.

It is symbolic of our society that this chapter, intended to provide guidelines to assist in making laboratory operations safer, should start with such a strong legal tone. Formal safety standards have been established because of concerns for the rights of individuals and society, due to abuses by the very small minority that may place results or profits ahead of the well - being of the persons involved. Individuals are no longer willing to accept what they believe to be excessive risks on behalf of their employer and are willing to go to court to protect themselves, to the extent that this prerogative is at risk of being abused. However, even without the need for laws and regulations, such a chapter in a book on laboratory safety would still be needed to provide guidelines to research personnel on how to avoid or minimize the risks associated with the conduct of research.

Much has been made of the professional expertise, experience, and judgment of scientists which should allow them to be the best judge of the safety program needed in their research. In chemistry laboratories in the academic world, however, where competent, enlightened scientists should be found, it has been estimated that the accident rate is 10 to 50 times higher than that in industrial laboratories. The broad range in the estimate is attributed to the reluctance of academic personnel, particularly students, to report accidents. The disparity between the two situations may be explained by the greater likelihood in industry that scientists might be required to do a careful hazard analysis and follow strict safety precautions. The touted expertise of scientists is often confined to the scientific object of the research program. Very few scientists have taken formal courses in safety, health, and toxicology. Most of the relevant safety articles are published in journals devoted to topics outside of their major field of interest. They are likely to have no better judgment or common sense, on average, than any comparably well-educated and intelligent group. They may, in fact, because of the intensity of their interest in a very narrow field, have only a limited awareness of information extraneous to those interests which would assist them in making research decisions. In the academic area, many profess to be concerned that academic freedom could be abridged by rules imposed from the outside. Academic freedom, however, should not be confused with issues governing the health and safety of individuals and the environment transcend this desirable concept.

There are legitimate concerns that research laboratories may become over regulated by too-specific a set of rules, since they do not fit the standard mold for which the original OSHA and other regulatory standards were designed. Instead of working with a few chemicals, a single laboratory may work with hundreds over the course of time, often for limited periods. Safety and health information may be extremely limited or nonexistent for newly synthesized substances or for many of the materials with which a scientific investigator may work. In general, research laboratory safety and health policies should not be regulated on a chemical by chemical basis except for specific, known serious risks, but this does not mean that otherwise there should be no safety rules. Health and safety programs should be based on well-defined general policies, sufficiently broad in scope, conservatively designed to encompass any *reasonable* hazard to laboratory personnel. They should be administered uniformly as institutional or corporate policies, tempered by local circumstances, to assure that all laboratory workers, including students, are equitably treated.

II. OSHA LABORATORY SAFETY STANDARD

The OSHA Laboratory Safety Standard, 29 CFR Part 1910.1450, addresses the issue of local responsibility by requiring that each laboratory develop an individual chemical hygiene plan as part of an overall organizational plan. Thus, it is the responsibility of individuals responsible for the laboratories to take time to consider the safety factors applicable to their work. The plan must be written to ensure that it is available to all the employees, and so documentation will exist that the effort has been made. The new standard is a performance plan, superseding the General Industry Standards for working with chemicals with a few exceptions, which reduces the number of explicit requirements to a very few. It also replaces, for laboratory operations, the Hazard Communication Standard, 29 CFR Part 1910.1200. This second standard addresses many of the same issues as does the Laboratory Safety Plan. The details of many of the topics found in the following sections, such as a discussion of the contents of Material Safety Data Sheets, definitions of toxic, acutely toxic, etc., are given in later sections of this chapter in order that the general provisions of the Laboratory Safety Standard not be obscured at this point by a profusion of details.

The entire Laboratory Safety Standard, as published in the *Federal Register* is only about nine pages long (not including the non-mandatory sections). Although it is a performance standard, with few explicit requirements, it does not relieve the laboratory manager of any safety responsibility. It simply leaves up to that individual, supported by the organization, the best method for creating a safety program at least as effective for the laboratory's employees as would have the General Industry Standard. The next several sections will deal with the requirements of the OSHA *Occupational Exposure to Hazardous Chemicals in Laboratories* standard, to use its official title. Information which facilitates compliance with these requirements represents the bulk of the first through fourth chapters of this book.

A. The Chemical Laboratory

The standard applies only to laboratory use of chemicals and their hazards. The definition of hazard is very broad - "a hazardous chemical means one for which there is statistically significant evidence based on at least one study conducted in accordance with established scientific principles that acute or chronic heath effects may occur in exposed employees. The term 'health hazard' includes chemicals which are carcinogens, toxic or highly toxic agents, reproductive toxins, irritants, corrosives, sensitizers, hepatoxins, nephrotoxins, agents which act on the hematopoietic systems, and agents which damage the lungs, skin, eyes or mucous membranes." The standard also mentions physical hazards for materials that are flammable, combustible, compressed gases, explosives, oxidizers, organic peroxides, pyrophoric, reactive or unstable, or water reactive. Not all uses of chemicals with these properties are covered by the standard but only those uses which occur in a "laboratory" on a "laboratory scale." Note that the list of hazardous properties does not include radioactive, ionizing and nonionizing radiation, or contagious diseases. Operations involving these types of hazards are covered under other standards or regulated by other agencies. The definitions are somewhat circular but the standard is clearly intended to exclude workplaces where the intent is to produce commercial quantities of a substance or where procedures are part of a production process or which simulate a production process. A laboratory is where small quantities of hazardous chemicals are used on a nonproduction basis. Laboratory-scale operations are those in which containers used in the work are designed to be safely and easily manipulated by one person. Also, a laboratory uses a variety of chemicals and procedures. The scale is such that standard laboratory practices and equipment can be used to minimize the exposure to the chemical hazards. The utilization of chemicals with similar hazardous properties in a nonlaboratory environment falls under the OSHA hazard communication standard.

B. Chemical Hygiene Plan

A key component of the OSHA standard is the Chemical Hygiene Plan (CHP). This is an explicit requirement for laboratory activities that conform to the definitions given in the preceding section. The facility must develop and carry out a written CHP which satisfies several criteria. The first three are generalizations but are nevertheless essential. It is not required in these three sections to define how one is to accomplish them.

1. Capable of protecting employees from health hazards associated with hazardous chemicals in the laboratory.
2. Capable of keeping exposure levels below the Permissible Exposure Levels (PELs) as listed in the General Industry Standards, 29 CFR 1910, Subpart Z.
3. The CHP shall be readily available to employees, employee representatives, and on request to OSHA.

The remaining elements of the plan are much more explicit in their requirements. The standard states "The Chemical Hygiene Plan shall include each of the following elements and shall indicate specific measures that the employer will take to ensure laboratory employee protection."

4. Standard operating procedures to be followed when working with hazardous chemicals.
5. Criteria the employer will use to select and implement measures to reduce employee exposures. This covers engineering controls, personal protective equipment, and hygiene practices. Control measures to reduce exposures to extremely hazardous chemicals are considered especially important.
6. Fume hoods and other protective equipment must be functioning properly and a program must exist to ensure that this is so.
7. Employee safety information and training must be provided.
8. Defining a program to determine the need for and procedures for a pre-initiation approval process for some operations.
9. Provisions for medical consultation and medical surveillance for employees when conditions exist in which exposures in excess of the PELs or action levels may have occurred or may routinely occur.
10. Designation of personnel responsible for implementation of the CHP, to include designation of a chemical hygiene officer (CHO) and, if appropriate, a chemical hygiene committee. Most organizations with a variety of laboratories would normally choose to form such a committee.
11. Special provisions for additional protection for work with particularly hazardous materials such as carcinogens, reproductive toxins, and acutely toxic substances.

If the scientific worker, for whom this handbook is intended, follows the recommendations in this handbook, the requirements to meet the desired outcome of the standard should be met, but a written plan is required. The next section will define what must be covered by the plan to meet the 11 requirements listed above. The topics will not be covered in the order in the list.

1. Goals

The introduction to the plan should succinctly state that the organization, for the specific laboratory plan, is committed to providing a program that reduces exposure of employee to hazardous chemicals to below acceptable limits by (1) providing them with adequate facilities for their work; (2) provision of appropriate en203gineering controls or, if that is not feasible for valid reasons, with personal protective equipment; (3) providing them, in a timely manner, with appropriate training in procedures which they are to follow, access to information about the chemicals with which they are working, the risks associated with the chemicals, how to recognize hazards which may arise, and emergency responses; (4) providing medical consultation and surveillance as needed; (5) providing ready access to the plan; and (6) monitoring the continuing efficacy of the plan.

2. Organization

The organization responsible for implementation of the plan, including key individuals, by title, should be identified, along with a brief description of the responsibility assigned to each. An organizational chart should be provided with the following positions (or groups) identified:

A. The senior person in the organization who is charged with the overall responsibility for safety and health programs in the organization. This position should be at a sufficiently high level to ensure that the program receives adequate support.

B. The organization under the executive authority charged with actual implementation of the plan. Normally this would consist of the Environmental Health and Safety Department and the chemical hygiene committee.

C. The CHO for the organization. This person could be the head of the Health and Safety department or the chairperson of the chemical hygiene committee. However, neither of these persons would normally be able to devote full time to this work and it is a critical, full-time position. The responsibility may be delegated to another person, most probably in the health and safety organization. The chemical hygiene committee should function to define policies and provide oversight of the program, while the health and safety staff should provide the daily operational support. The duties of the CHO should include:

1. Assist the individual laboratory managers to develop their own chemical hygiene programs. The CHO should not be, and indeed is not likely to be, sufficiently familiar with the operations of individual laboratories to be expected to write the plans for specific laboratories. They should provide a template or format for the persons locally responsible for a specific facility.

2. They should develop a "train-the-trainer"program to assist the local managers in providing the appropriate training for their personnel.

3. The CHO should develop a CHP covering the entire organization, containing basic policies for chemical procurement, storage, handling, disposal, facility standards, basic training, availability of Material Safety Data Sheets and other chemical information, personal protective equipment guidelines, emergency planning for the organization, and auditing and inspection protocols.

4. The CHO should conduct, or have done under their supervision, laboratory inspections of equipment, specifically including fume hoods and other fixed safety equipment, maintenance and housekeeping, chemical storage, and compliance with the organization and laboratory-specific safety plans.

5. The CHO should see that a medical consultation and surveillance program is available to the employees in the event of overexposure conditions and conduct environmental monitoring as required to support this program.

D. The local laboratory management line of authority. This could be one or more persons, dependent upon the size of the operation, with one individual designated as the senior person to whom responsibility ultimately devolves. The latter individual is responsible for seeing that the facility develop a CHP for that facility. A recommended approach would be to make the laboratory plan a second part of a document of which the organization's CHP would be the first. This would serve two purposes: every employee would have access to the policies of the organization, and would eliminate repetitious and possibly conflicting interpretations of these broad, basic policy areas. The laboratory management might choose to designate a laboratory hygiene officer, if the number of employees is large enough, to perform some of the following responsibilities and to liaise with the organizational CHO. Regardless of how it is done, the local laboratory management has the responsibility to:

1. See that the physical facilities are adequate and in good working order.
2. See that maintenance and housekeeping are satisfactory.
3. Develop and implement safe standard operating procedures for the activities conducted within the facility. These should be written and maintained in a suitable form to which the employees would have ready access.
4. Conduct training programs or see that training programs are provided to the employees to inform them of the contents and location of the CHP for the facility, the location and means of accessing chemical information, such as Material Safety Data Sheets, the standard operating procedures for the facility, the risks associated with the chemicals in active use, warning characteristics of the chemicals in use, including possible symptoms indicating over exposures or possible adverse reactions, emergency response or evacuation plans, and availability of the medical program.
5. Ensure that chemicals are stored, handled, and disposed of properly.
6. Conduct in-house inspections of the facility, conduct, or have conducted, inventories of the chemical holdings of the laboratory, and make sure that suitable personal protective equipment is available and employed as needed.

E. As discussed in Chapter 1, the employee is the one ultimately responsible for complying with safety policies, in this instance, as contained in the CHP and standard operating procedures. They have the responsibility for developing good personal safety habits.

3. Training and Information Program

The CHP must contain a description of the organization's information and training program. The training and educational programs are to be made available at the time of the employee's initial assignment to potential exposure situations. Refresher training is to be provided at a frequency determined by the employer. The information to be provided to the employees must include:

1. The contents of the laboratory safety standard. Since the standard, including its appendices, is quite short, this may be accomplished by including a copy as an appendix to the CHP.
2. The location and availability of the organization's and laboratory's CHP. This is most easily accomplished by maintaining a master copy of the basic CHP for the organization at a central location, such as the Environmental Health and Safety department, with copies of the laboratory CHP in the individual laboratories. However, the latter should include the basic plan. Access to a computerized information system is becoming widely available in many commercial laboratory organizations and larger academic institutions. The basic unit can be part of this information system and be available to anyone with access to the system at any time. A computer version has a distinct advantage in that it can be updated at any time without distribution of many

hard paper copies.

3. The OSHA PELs or action levels for the chemicals in use in the employee's work area. The entire list of PELs can be made an appendix to the CHP to satisfy this requirement rather than having to modify this information whenever a new material is brought into the facility. Not every chemical has an established PEL or action level, but the American Conference of Governmental Industrial Hygienists (ACGIH) publishes a more comprehensive list, updated annually, and the National Institutes of Occupational Safety and Health (NIOSH) also publishes lists of recommended exposure limits, and these must be made available in the absence of OSHA PELs. The three sets of levels do not always agree. Where they differ, the OSHA PELs and action levels are the legally applicable limits. Copies of the ACGIH and NIOSH limits are available as published documents and can be provided as reference material, available in the workplace. A cautionary statement should accompany the list of PELs or alternatives, stating that the limits are not absolute in the sense that a fraction below them is safe while a fraction above is not. Exposure limits should be kept well below the PELs. There are individuals with greater sensitivity for whom the legal PEL would be excessive.

4. The location and availability of reference material on the hazards, safe handling, storage, and disposal of the chemicals found in the laboratory. Note that OSHA uses the word "found," not the phrase "in use." For laboratories that have accumulated a large inventory of rarely used materials, this alone is an excellent reason to dispose of excess and obsolete materials. The minimum means of complying with this requirement is to maintain a file of the Material Safety Data Sheets (MSDSs) provided by the manufacturer of the chemicals. The MSDSs will satisfy the previous requirement for PELs or other recommended exposure levels since they include this information. As will be discussed later, maintaining an up-to-date copy of MSDSs in every laboratory is very difficult, but computer versions of these data are available which can serve as an alternative. MSDSs should be supplemented by other compilations of data. One weakness in the MSDS system is that in order to avoid liability due to recommending a less than necessary level of care, many manufacturers have gone to the other extreme and recommend very conservative measures. Manuals such as *The Merck Manual and Properties of Industrial Chemicals* by Sax would be good supplements to the MSDS data. Chemical vendors and distributors also usually maintain this information on their Internet pages. Labels on commercial chemicals provide much information. The standard requires that these labels not be defaced or removed. All of this material need not be in each laboratory, but the employee must be told where it is and how to obtain access to it. This access should be readily convenient.

5. Indicators and symptoms associated with exposure to chemicals used in the laboratory.

All of the above is basic information which can be provided as part of the basic plan for the organization, if the employees know where the material is and have reasonable means to obtain access to it. Some organizations accomplish this by computers, and as the use of computers approaches universality, this is likely to become the favored approach.

The required training program must include the following elements:

1. The employees must be informed of the methods used to detect releases or the presence of hazardous chemicals in the workplace. Some of these are available to the employee directly, such as information concerning warning properties of the chemicals (odor, visual indicators) or symptoms which might be experienced (irritation, nausea, or dizziness). Other means of detecting materials which may be used would include

fixed alarms, such as gas monitors, or environmental monitoring by safety and health support staff. Among equipment which might be available would be detector tubes, ambient gas meters, passive dosimeters, and sophisticated devices such as portable infrared, atomic absorption, or gas chromatograph instruments. Detection methods which are available and might be employed should be listed in the CHP. Where access to these methods is through nonlaboratory personnel, the training should include how to obtain the required aid and the telephone numbers of support personnel. Some of this material, such as the environmental monitoring services, should be in the organization's basic plan, but the indicators such as odor or the presence of local fixed gas monitors should be part of the laboratory's own plan.

2. The chemical and physical hazards of the chemicals in the workplace. This is almost the same as the basic information on PELs and MSDSs listed in the previous section. Those requirements basically defined limits of exposure and the sources of data. This requirement provides that the employees be given chemically specific hazard information on the chemicals in their work area. It is most important that the chemicals in actual use are the principal ones for which this information is provided. However, generic hazard information by class for chemicals present but not in use should be provided as well. There is always the potential for an accident involving chemicals not in current use. The employees must be informed that they are not to deface or remove the labels on commercial containers of chemicals, since they represent a primary source of information. It is not required by the standard, but following the requirement from the Hazard Communication Standard 29 CFR 1910.1200, that secondary containers intended for use beyond a single work shift should be labeled, it is highly recommended that this be required.

3. The employees must receive training on the measures they can take to protect themselves. The content of this training should be made part of the CHP for each individual laboratory. Among these measures are:

   a. Work practices specific to the laboratory. These include the standard operating and administrative procedures developed so that the work can be carried out safely and efficiently.

   b. Emergency procedures. This can include a wide variety of measures, including how to put out a small fire, how to evacuate an area (including identification of primary and secondary escape routes), steps to take to bring a reaction under control if time permits, how to relieve pressure on pressurized equipment, how to clean up minor spills, how to report larger spills and secure help in responding to them, how to use personal protective equipment available to them, first aid, and close-down procedures in the event of a fume hood failure or failure of any other item of protective equipment. Means of initiating a general evacuation from a facility, or the building in which the laboratory facility is located must exist and should be identified in this section.

4. The details of the CHP applicable to their area, including the basic organizational plan.

The items listed above are for normal laboratory work. If there are some operations which require prior approval by a more senior individual or external group, then these must be included in the training program as well. This need not be a special and possibly more hazardous laboratory evolution, although that is the primary intent of this requirement, but it could represent the purchase of selected items of equipment which must meet certain standards of performance, such as refrigeration units, fume hoods, heating devices, storage cabinetry for flammables, certain classes of chemicals such as carcinogens, etc.

Additional training is also needed for working with extremely hazardous materials. The training must include:

1. Where the work must be done. An area must be designated. This can be an isolated suite of laboratories with controlled access or an area as small as a fume hood, explicitly defined as the area where the work is to be done.
2. The use of special containment devices such as hoods or fully contained glove boxes.
3. Standard operating procedures for the work with the material, including use of appropriate personal protective equipment.
4. Means of safe removal and disposal of contaminated material.
5. Procedures to decontaminate the work area.

4. Medical Program

The CHP must define the means by which the facility will comply with the medical requirements of the standard. In most cases, this procedure should be the same for all laboratories within an organization, so the means should be spelled out in the basic plan. There are four specific requirements:

1. Employees working with hazardous chemicals must be provided an opportunity to have a medical examination, and follow-up examinations if necessary, under any of the following circumstances:

    a. The employee develops any signs or symptoms associated with the chemicals to which they may have been exposed in the laboratory.
    b. For specific substances regulated by OSHA, e.g., formaldehyde, for which exposure monitoring and medical surveillance requirements exist in the standard for that substance, the employee must be offered the prescribed medical surveillance program if environmental monitoring shows a routine exposure level above the action level (or PEL, if an action level is not specified).
    c. An incident occurs such as a spill, leak, or explosion and there is a likelihood that the employee might have received an exposure to a hazardous substance; the employee must be offered an opportunity for a medical consultation. The consultation is for the purpose of determining if a medical examination is needed.

2. "All medical examinations and consultations shall be performed by or under the direct supervision of a licensed physician and shall be provided without cost to the employee, without loss of pay, and at a reasonable time and place."
3. The employer must provide the following information to the referral physician, if available:

    a. The identity of the hazardous chemical(s) to which the employee may have been exposed.
    b. A description of the conditions under which the exposure occurred, including quantitative exposure data.
    c. A description of the signs and symptoms of exposure the employee is experiencing, if any.
4. The examining physician must provide a written opinion to the *employer* in a timely manner which shall include or conform to the following requirements:

    a. Any recommendation for further medical follow-up.
    b. The results of the examination and any associated tests.
    c. Any medical condition (not limited to the ones that may have resulted from the

exposure) revealed in the course of the examination which may place the employee at increased risk as a result of exposure to a hazardous chemical found in the workplace.

    d. A statement that the employee has been informed of the results of the consultation or medical examination and any medical condition that may require further examination or treatment by the physician.*

    e. The written opinion *shall not* reveal to the employer specific findings or diagnoses unrelated to occupational exposure. This obviously is to protect the employee's privacy rights.

5. OSHA does not include the use of respirators under the medical program, but it is closely related since under the General Industry Standard 29 CFR 1910.134 the ability to use a respirator depends upon the employee's health. A basic requirement is the ability of the employee to pass a pulmonary function test, but the employee must not have any other health problems which would preclude the use of respirators if they are needed or required to protect the employee. A statement must be included in the CHP that the organization has a respirator protection program which meets the requirements of the general industry standards. This program should be a written one and included in the employee's training.

5. Laboratory Produced Chemicals

A characteristic of many research laboratories is that chemicals may be produced or synthesized in the course of the research. If the composition of the chemical is known and it is a hazardous material, all of the training requirements and other provisions of the standard apply. If the composition is not known, it shall be assumed to be hazardous and, with the exception of the requirements for MSDSs and similar information sources, the provisions of the CHP apply. If the chemical is produced for a user outside the laboratory, the provisions of the Hazard Communication Standard (29 CFR 1910.1200) apply, including the requirement for providing an MSDS and proper labeling of the material. Compliance with these requirements will be the responsibility of the individual laboratory and a commitment to this compliance should be in the laboratory CHP.

6. Record Keeping

The employer must commit to establishing and maintaining for each employee an accurate record of any measurements taken to monitor employee exposure and any medical consultations and examinations, including tests or written opinions required by the standard. Further, the employer shall assure that such records will be kept, transferred, and made available in accordance with 29 CFR 1910.20.

7. Summary

The sections immediately preceding this one detail the requirements of the OSHA Laboratory Standard and suggest general means by which an organization and/or laboratory can comply with it. Appendix A of the standard provides many recommendations of how compliance can be achieved. These recommendations are not mandatory and are in several instances out of date. The standard is a performance standard which allows a great deal of flexibility on the part of the employer and employee. As noted earlier, the bulk of this

---

*   At the author's institution, when the medical surveillance program began several years ago, 22% of first-time participants had significant untreated health problems of which, they stated, they were not aware. Very few of these were related to occupational exposures, but some did require adjustments in their duties.

handbook (with the exception of Chapter 5, which covers laboratories generally working with materials not covered by the laboratory standard) is designed to provide specific information on how to achieve the appropriate level of performance in all facets of laboratory safety, including designing and equipping of facilities, covered primarily in Chapter 3, as well as operations. The remainder of this chapter starts from the point of an assumption that an adequate facility is available and proceeds from that point to the very beginning of planning a program to be done in the facility.

III. OPERATIONAL PLANNING

A typical research proposal submitted to a funding source goes into great detail on the significance of the proposed research, the approach to be taken, and the results sought. Typically, the proposal always provides a thorough justification for the technical manpower and equipment resources needed to carry out the planned program. The hazards which will be encountered and the means by which they will be controlled are likely to receive much less attention, and then only if these are sufficiently dangerous or unusual. Unless the research involves very stringently regulated materials, the reviewer of the proposal often must take on faith, if the question arises at all, that a basic infrastructure has been established to ensure that the research can be carried out safely and in compliance with contemporary regulatory standards. This situation does show signs of changing in some areas, such as when human or animal subjects are involved. The needed infrastructure does not just happen, it requires careful planning. It is the intent of this chapter to provide essential information to guide planning for safe operations in the laboratory.

The first order of priority, after authority to proceed on a specific program is obtained, is to order all of the essential items of equipment which will be needed. Orders for major items of equipment frequently take extended periods to be processed and delivered, 9 to 10 weeks being as short an interval as might reasonably be expected, especially in a facility supported by public funds, encumbered by an abundance of bureaucratic requirements. If installation is required, such as when an additional hood is needed, this period could be extended for months since the installation will have to be carefully planned to ensure, among other things, that the air handling system has sufficient capacity and that fire code requirements can be met, especially if the duct work must penetrate multiple floors. Scheduling and pricing of the actual work cannot be done in such instances without working plans. This delay may be critical when the work is scheduled to be completed within a fixed contract period with annual renewals depending upon progress, as are many academic research contracts.

If new employees need to be hired, a number of factors must be considered in addition to technical skills. As noted earlier, attitude is extremely important. A research laboratory is not the place for a casual attitude toward safety. Skills and experience are, of course, important, but a vital consideration should be a compatible personality. It is critical in any group effort for personnel to be able to work together. It is not necessary to be "popular," but it is important for individuals to be receptive to the ideas of others and tolerant of differences in points of view. A group of persons working under the stress of strained relationships is likely to be an unproductive and unsafe group. Obviously, it is important that an individual to be hired is safety conscious and willing to comply with the employer's safety policies. A principal investigator needs to establish a clear line of authority for the laboratory personnel, both for day to day operations and for emergencies. These may not be the same. The individual trained to manage the scientific aspects of the research may not have as appropriate a background to handle an emergency situation as would a senior technician who might have received special training in safety areas, such as chemical spill control or emergency first aid. Where there is the possibility of ambiguity, responsibility for various duties needs to be clearly assigned, especially those duties associated with safety. It would be well, for example, to designate a relatively senior person as the laboratory chemical hygiene officer (LCHO) and if necessary, provide access to additional safety training to that individual. This individual could be responsible, under the laboratory CHP, for such items as safety orientation of new employees and safety training of all employees when new materials or procedures are incorporated into the laboratory operations. They might be asked to perform or review a hazard analysis of any new laboratory operations, and to secure any authorizations or clearances which might be needed. It could be this individual's duty to assign other persons the responsibility of being sure that chemicals are shelved according to compatibility and to maintain safety items such as first aid kit supplies, personal protective equipment, spill kit materials, Material Safety Data Sheets, or maintenance of equipment in safe condition. The LCHO and/or the laboratory supervisor needs to act as liaison with the safety department to provide access to any new information which might affect the laboratory's operations. A knowledgeable employee, who could be the LCHO, needs to be designated as the person responsible for ensuring the safe disposal of hazardous materials. This individual needs to be responsible for seeing that all surplus and waste materials are properly identified and segregated if waste materials are combined into common containers. Individuals handling hazardous waste must receive training in the risks associated with that operation.

The emergency planning required under the OSHA Laboratory Safety Standard requires an effective emergency plan to be developed for each individual laboratory, which is consistent with and integrated into the plan for the entire building and that of the corporation or institution. It needs to take into account procedures for temporarily interrupting the research operations or for automating uninterruptible operations if possible to allow employee evacuation during an emergency. An operation can and should be allowed to fail where necessary to protect personnel from serious injury. This plan should be reviewed periodically to ensure that it is still appropriate. As has been noted many times earlier in this handbook, research programs, especially in academic institutions, change rapidly, not only in the materials in use and the operations being conducted, but also in the participating personnel (due to student involvement). Evacuation plans need to be tested periodically to ensure that they are effective. It was remarked upon earlier that the transient nature of a building's population in the academic environment creates difficulties in ensuring participation of all of the occupants. Drills held at least once a year should include enough "permanent" occupants to help those who are less familiar with the evacuation procedure.

Every aspect of the laboratory operations should be evaluated to see if it could be made more efficient and safer. Purchasing of reagents, for example, should be reviewed to see how much is actually needed on hand at a given time. If all chemicals are ordered early in the program and the program needs shift, a substantial and wasted investment in surplus chemicals could result. Today, where disposal of waste chemicals has become such a major legal issue, the cost of disposing of surplus chemicals often exceeds the original costs. The quality of partial containers of chemicals may have become dubious, and the initial investment in the excess will represent a drain on the currently available funds. Anticipation of needs is critical, especially where equipment is involved. As noted earlier, delivery of essential items of equipment may be delayed for extended periods. The temptation is to "make-do" with equipment not specifically designed to meet the actual needs, with serious safety implications being involved on occasion.

The regulations, and the information on which they are based, change sufficiently frequently that it is unreasonable to expect every purchaser to be able to keep up with the current regulations. Further, the entire body of relevant information regarding laboratory safety has become so extensive and so complex that again it is unlikely that a single individual can be sufficiently knowledgeable to adequately consider every factor. For example, the review of the purchase of a fume hood is usually not based so much on the characteristics of the hood, but on the installation. Has the location been reviewed for availability of sufficient make-up air? Has the path of the exhaust duct been selected and has the exhaust blower been sized appropriately? Will fire separations have to be penetrated? The flagging of the purchase order so that the Purchasing Department will look for sign-offs to see that these questions have been answered, and if they had not been considered, ensure that they are before the order is processed. It is highly likely that the order will have to be modified if these factors have not been addressed, and it is highly desirable that specifications be changed prior to ordering unsuitable equipment.

An evaluation of the potential exposures of individuals to hazardous materials should be made as early as possible. It may be necessary to consider selectively placing individuals in work assignments, although one has to be very careful in such cases to avoid triggering charges of discrimination based on factors such as gender or disability. Still, if there are known risks, for example, of teratogenic effects from a chemical, it would certainly be surprising if an expectant mother did not have some concerns about working in an area where it was in use, even if the levels were well below the acceptable OSHA limits for the average worker. Any work regimen would need to be fully discussed between the individual and the supervisor in such a case and be based on knowledge, not speculation. Often, once the exposure potential or lack of one is clearly understood, concerns may disappear. Failure to consider the employee's rights to a working environment free of recognized hazards could lead to a complaint to OSHA or another regulatory agency which could, in extreme cases, cause the program to be interrupted pending resolution of the safety issues.

Prior planning is needed, especially in facilities in which students are expected to be working. Legal safety standards usually have been designed for permanent employees, and although many graduate students and students on work-study programs receive stipends for their efforts, they may not be considered or treated as "real" employees by others in the work area. They typically have less experience and a different purpose in being in the laboratory than do permanent personnel. The pressures associated with completing the various hurdles of a degree program, especially those accompanying completing a research program for a thesis or dissertation within a tight schedule, often lead to students working long hours, going without enough sleep, and eating odd diets. The result may be working without adequate supervision and being affected by factors that could cause impaired judgment. The laboratory safety program should take these factors into account and make a special effort to see that these younger persons understand the goals of the safety program, as it bears upon the operations of the laboratory and the need to comply with the safety polices of the organization and the laboratory.

A. Quantities

The recommendation that volumes of reagents kept on hand be kept to the minimum needed for a reasonably short working period is found in virtually every laboratory safety manual. However, a visit to almost any laboratory will reveal many bottles and other types of containers accumulating substantial layers of dust. Many of the more recently acquired reagents very likely will be duplicates of these older materials. There must be good reasons for this apparently needless duplication.

It would appear to make a great deal of sense to order what you need and replace it when it appears that more will be needed. There are at least three reasons why this common sense approach is so rarely followed, two of which are attributable to factors in the purchasing process:

1. It takes time to process an order. Unless a central stores facility maintains a stock of chemicals at the research facility, the processing of a requisition, receipt of an order by the vendor, and delivery are unlikely to take less than 1 month, unless an alternative buying process has been established, such as a blanket order system or previously cleared requisitions for low-value purchases. Under these circumstances, a purchaser tends to buy more than is currently needed in order to avoid having to order frequently and to avoid delays in receipt of the needed material.

Container Size	Cost/Liter
One liter, each	1.000
6 x 1 liter, case	0.558
4 liter, each	0.526
4 x 4 liter, case	0.359
10 liter	0.303
20 liter	0.225

2. Unit chemical costs decrease rapidly with the increasing size of the container. For example, for one grade of sulfuric acid, the following pricing schedule has been established by one major vendor (note that these have been normalized to set the price per liter of the smallest size to equal 1). Obviously, if the volume of usage justifies the purchase, the largest size is the most economical to buy. However, there are several reasons why such a purchase is probably unwise for more reasonable levels of usage. It increases the potential risk, as in this example, to have more material than is actually needed, and storage space will have to be found for the excess material. If it is not used relatively quickly, the quality may become suspect, and users will be reluctant to use it in their research programs. The cost of disposal of any eventual surplus material is likely to eliminate any initial economic gain from buying in volume, unless the surplus can be used by someone with less critical applications.

3. In addition to the two reasons given above, sometimes a researcher wants to be sure of the consistency of the reagent, so he buys enough for his needs from one lot. However, some chemical firms will, upon request, set aside an amount of a given lot and maintain it at their regional warehouse to accommodate a larger user.

An examination of the purchases of the various kinds of research reagents by most university or corporate research facilities will probably reveal that a relatively small proportion of them are bought in substantial quantities. At the author's institution, fewer than 75 of the more than 1200 different chemicals purchased during a typical year exceeded an amount of 50 kg. Where this is true, it would appear feasible to set up a central stores for at least a limited list of chemicals. Stocking of these stores areas should probably emphasize the middle ranges of sizes. If, in the example given above, multiple-case lots of 4-liter containers were the primary sizes purchased from the vendor for stocking, most of the cost savings of volume purchases could be passed on to the local purchaser. Smaller sizes would have the advantage of being likely to be completely emptied, thus eliminating the cost of waste disposal completely for these containers, but forcing the users to buy small sizes could lead to buyer resistance because of a perceived inconvenience. It might be desirable to restrict the purchases of larger sizes to those who can establish a need or for those items for which it is feasible, disburse chemicals from drums into smaller containers by stores workers.

Except for the high-volume materials, most remaining chemicals are bought in relatively small quantities to meet specific needs of individual programs. Some chemicals pose unusual hazards, such as ethers that degrade over a short period of time. It is desirable to keep track of which group is ordering them and where they are to be found. A central stores area would make a convenient distribution center for these special materials and would facilitate maintenance of records of their use.

Bar code technology has now made it possible to conveniently mark every container received and distributed from a stores area with a unique identification code which includes the name of the chemical, the date received, the quantity, and the recipient. The last of these can be tied to a specific facility, and a specific laboratory within the facility. The availability of powerful desktop computers now makes it possible, with appropriate software that is commercially available, to establish a tracking program for every container from the point of purchase to its final disposition. Networking software even makes it possible to have more than one point of receipt and still accomplish the same task. There are obvious implications with such a program to enable volume purchasing, control of total amounts on hand, and disposal of chemicals approaching dates at which point they may no longer be safe to retain. Bar code technology, using reading devices to scan the container codes into "notebook" size computers that can be as powerful as the desktop units, makes it possible to quickly inventory all of the containers in a laboratory and to keep track of chemical containers if they are transferred from one laboratory to another.

On April 22, 1987, the EPA Community Right-to-Know standard (40 CFR, Part 370) became law (also known as SARA Title III), requiring users of hazardous materials to inform nearby communities when they had significant holdings of any of several hundred hazardous chemicals. The definition of significant holdings varies from 1 pound (0.454 kg) to 10,000 pounds (4539 kg), depending upon the chemical. Where the amount exceeds another, usually larger, threshold, the law requires that emergency planning programs be established. It is also required to report within 60 days any time these two levels are exceeded. Clearly, it is desirable to maintain amounts in storage less than the trip-point levels.

There are several exemptions to the Community Right-to-Know standard, one of which provides important relief to laboratories from the inventory and reporting provisions of the standard. The EPA, in its final rule, provided an exemption for "any substance to the extent that it is used in a research laboratory or a hospital or other medical facility under the direct supervision of a technically qualified individual." The research laboratory exemption applies only to the chemicals being used in the laboratory, not the laboratory itself, under the direction of a person meeting the specified criteria. Basically, the same limitation which qualified a laboratory-scale operation under the laboratory standard applies here. The exemption does not apply to pilot plant-scale operations or production-like programs. The difficulty of preparing the reports required under SARA Title III make this exemption extremely useful. It is surprising how many of the chemicals can be found within individual laboratories in excess of the reportable or emergency planning thresholds, and if the total number of laboratories in a larger research organization is considered, it would be very difficult to comply. Implementation of a chemical tracking program, using the bar coding concept and suitable software, will make it possible to comply with the law should the exemption be removed.

Although the exemption is very useful as a practical matter, it is philosophically some-what troubling to have to depend upon since the risks that evoked passage of the Right-to-Know act are real. Many universities and industrial research facilities are located in smaller towns and may represent a significant chemical release risk to the community, perhaps the largest risk. The individual containers are small, but if a fire involved a large chemical using building, the total amount of chemicals released into the air and perhaps running off in the water being used to fight the fire could be very large. Not only could there be a large quantity of chemicals involved, the release would be very complex because of the very large variety present, with the toxicity of the release being impossible to predict. In a large release from a burning chemical, one could find oneself facing the problem of evacuating thousands of students and employees from an academic research building and adjacent facilities within a very short time. The available emergency resources could be easily overwhelmed. This scenario for laboratory organizations may be the most pressing factor in developing a chemical management program. It is recommended that, where research-oriented firms and institutions do represent a significant environmental hazard, a representative participate on the local emergency planning committees established under the EPA standard, even if technically exempt from the regulations.

In summary, it is desirable to order and maintain in stock as small amounts of chemicals as practicable in order (a) to minimize the risks in the event of an incident, (b) to reduce the overall expense by reducing the amount requiring disposal as hazardous waste, and (c) to minimize the problem of complying, at least in spirit, with the Community Right-to-Know standard. However, in order to encourage a laboratory manager to buy and stock smaller containers, purchasing procedures need to be established to conveniently provide smaller sizes at a reasonable cost.

B. Sources

One of the more difficult tasks associated with the purchase of equipment and materials meeting acceptable safety standards is to do so in a system which requires acceptance of the low bid. Many of the safety standards or guidelines are minimal standards. It is often more desirable to exceed these minimal specifications. Usually chemicals from any major company or distributor will be acceptable, but the same is not necessarily true of equipment. In order to obtain the quality desired, purchase specifications must be carefully written to include significant differences which will eliminate marginally acceptable items. In some cases, it is virtually impossible to write such a specification, and it is necessary to include a performance criteria. This often requires considerable effort on the part of the purchaser. As an example, chemical splash goggles are sold by many companies, at prices that differ by an order of magnitude or more. All of these will usually meet ANSI Standard Z-87 for protective eye wear but many are sufficiently uncomfortable or fog up so rapidly that they will not be worn. Thorough comparative testing under actual laboratory conditions will identify a handful of the available models that offer superior performance. With documented data, it is usually possible to obtain permission of the Purchasing Department to limit purchases to sources meeting acceptably high safety and performance criteria, rather than minimal standards. This applies not only to smaller items but also to major ones, such as fume hoods. Where there is a significant difference in quality which will enhance the performance and/or safety of any unit at a reasonable price, a cooperative effort should be made by the purchaser, the Purchasing Department, and the Safety Department to obtain needed items from these sources.

C. Material Safety Data Sheets

The federal government enacted a hazard communication standard in 1984. Chemical manufacturers, importers, and distributors were required to comply with the standard by November 25, 1985, and affected employers by May 25, 1986. Originally, the standard applied only to Standard Industrial Code Classifications 20 through 39. After September 23, 1987, it has been required that Material Safety Data Sheets (MSDS) be provided to nonmanufacturing employees and distributors with the next shipment of chemicals to these groups. As of May 23, 1988, all employers in the nonmanufacturing sector must have been in compliance with all provisions of the standard. The laboratory safety standard specifically mentions that at least MSDSs need to be available to laboratory employees. Many states have enacted similar standards which extended the coverage within their own jurisdiction. Some specifically extended coverage to public employees, which included individuals at public universities and colleges.

Under the hazard communication standard, chemical manufacturers and importers must obtain or develop a MSDS for each hazardous chemical they produce or import. These

MSDSs must reflect the latest scientific data. New information must be added to the MSDS within 3 months after it has become available. The manufacturer or importer must provide an MSDS to a purchaser the first time a given item is purchased and an updated version after the information becomes available. A distributor of chemicals must provide MSDSs to their customers.

The MSDSs can be in different formats as long as the essential information is included, although a standard format may be adopted. The minimal information to be provided, which must be in English, is:

1. The identity of the chemical as used on the label of the container.
   a. For a single substance, the chemical name and other common names.
   b. Mixtures tested as a whole: The chemical and common names of all ingredients which contribute to known hazards, and common names of the mixture itself.
   c. Mixtures untested as a whole: Chemical and common names of all ingredients which are health hazards and which are in concentrations of 1% or more, or carcinogens in concentrations of 0.1% or more. Carcinogens are defined to be those established as such in the latest editions of (a) National Toxicology Program (NTP) *Annual Report on Carcinogens,* (b) International Agency for Research on Cancer (IARC) Monographs, or (c) 29 CER Part 1910, Subpart Z "Toxic and Hazardous Substances," OSHA.

If any of the ingredients which do not exceed the concentration limits in the previous paragraph could be released from the mixture such that they could exceed an established OSHA PEL, or an ACGIH threshold level value, or could represent an occupational health hazard, their chemical and common names must be given as well. The same information is also required for any ingredient in the mixture which poses a physical hazard (as opposed to a health hazard).

2. Physical and chemical characteristics of the hazardous chemicals.
3. Physical hazards of the hazardous chemical, specifically including the potential for fire, explosion, and reactivity.
4. Known acute and chronic health effects and related health information. This information is to include signs and symptoms of exposure and any medical conditions which are generally recognized as being aggravated by exposure to the chemical.
5. Primary routes of entry into the body (exposure control).
6. Exposure limits data.
7. If the hazardous material is considered a carcinogen by OSHA, LARC, or the NTP (see 1.c above).
8. Precautions for safe handling, including protective measures during repair and maintenance of apparatus employed in using the equipment and procedures for cleanup of spills and leaks.
9. Relevant engineering controls, work practices, or personal protective equipment.
10. Emergency and first aid procedures.
11. Ecological information (environmental impact) if known.
12. Transport restrictions or guidelines.
13. Date of MSDS preparation or latest revision.
14. Name, address, and telephone number of the entity responsible for preparing and distributing the MSDS.
15. Any other useful information.

Although this list appears straightforward, the MSDSs provided by different companies vary significantly in quality. Many are incomplete, perhaps not always due to lack of information. There are generic sources of MSDSs which are prepared by firms independently of the original manufacturers and are possibly more free of bias. On June 3, 1993, the American National Standards Institute approved a voluntary consensus standard for MSDSs developed by the Chemical Manufacturers Association in an effort to provide more uniformity in the documents. Several industries claimed that there were problems with the new form which were not fully considered. As a result, at the time of this writing, no consensus standard has been adopted. A suggested ANSI list is available at an Internet location included in the references.

Provision of a MSDS at the time of the initial purchase of a chemical is a responsibility of the chemical vendor, and if the vendor fails to provide it, it is the responsibility of the purchaser to take the necessary steps to require the vendor to do so. A typical MSDS can be up to several pages long, and a comprehensive file of hundreds of these, which might be required in a typical laboratory, or thousands if the file is maintained at a central location in an organization, will be bulky and difficult to maintain.

Both the distributor of a chemical and the purchaser have a major problem in complying with the requirement that a MSDS be provided to the user, where purchasing authority is widely distributed, as it often is on a university campus. Many institutions permit direct delivery to the actual location ordering a given material, while in others there is a central receiving point. In the former situation, a chemical vendor may supply an MSDS to the first purchaser of a chemical at the institution, but subsequent purchasers may not receive one, because they did not receive a copy of the first one sent to the initial purchaser. Where all the separate purchasers of a chemical are part of the same institution and located within contiguous confines of a single site, it is probable that the vendor technically can meet the legal requirement of furnishing an MSDS to the institution as an entity by providing a single MSDS to the individual laboratory first ordering a substance. In a large research institution, this would result in a very incomplete distribution of MSDSs. Designation of a single department, such as the safety department, to receive all MSDSs from the chemical vendors and to establish a master file of them, with perhaps some partial or complete duplicate files at other locations, will partially alleviate the problem. These files would need to be in places that are easily accessible to the users for a large portion of the day in order to approximate compliance with the requirement of being readily available to the employees. The laboratory standard does not contain the language "readily accessible," but only requires that the employees know where they are being held by their employer. However, the organization should still make arrangements to facilitate access. Unless the information as to which unit actually ordered the material accompanies the MSDS, it would be impossible to distribute them further internally, unless an individual department requests a specific MSDS which they wished to maintain in their local file. However, unless each department received a notice of the receipt of any revised MSDS and took the initiative to upgrade their own files, the local files would soon become obsolete. This could lead to possible liability problems if an employee assumed that the local files were current.

Some chemical manufacturers or distributors have avoided the entire problem, as far as they are concerned, by sending an entire set of MSDSs for all of their products to corporations or institutions with whom they do a substantial business. It is then up to the university or corporation to decide how to distribute them properly to comply with the regulatory requirement that any needed MSDS be readily available to employees.

If all chemicals are delivered to a central receiving location, a fairly straightforward, but labor intensive, solution to the problem exists. A master file of all MSDSs can be maintained at the central receiving location, as well as a list for each chemical of all departments or other definable administrative units which have previously ordered the material. If a department is not on the latter list for a given chemical, then a copy of the MSDS can be made and sent along with the material when it is delivered. A revised MSDS would be sent to every department listed as having the specific chemical in their possession. It would require that a copy of every purchase order and/or invoice were sent to the department maintaining the file in order to maintain the departmental lists. Although this sounds relatively easy, the amount of record maintenance required and the time spent in checking the files would be substantial. For a major research institution, the amount and variety of materials ordered, coupled with the large number of independent administrative units, would probably mandate at least a full time equivalent clerical employee for the program.

Computer technology has provided solutions to all or part of the management problem for distribution of MSDSs within complex organizations in which the variety of chemicals is numbered in the hundreds or thousands, instead of a few.

Although hard copy compilations of MSDSs are available either in print or on microfiche, the most flexible approach is to obtain access to an on-line source of MSDSs or subscribe to a vendor that will send an updated CD-ROM disk on a quarterly basis (this meets the 3-month up-date requirement). There are several firms which provide one or the other of these services. Some of the same firms also fulfill a requirement for the users of hazardous materials that they have access to a 24-hour emergency services on a per-call basis, although some provide a limited amount of free time each month. Access to a server computer housing the CD-ROM data base through modems or a network is a useful service, but the user must be sure that license requirements are met. If license agreements are available, then access can in principle be made available to every user of chemicals in a facility with access (rapidly becoming the norm) to a computer or terminal 24 hours per day through a network or modem. Both of these means can provide access to a very large MSDS data base that is current and reliable. Providers of generic data bases do assume the liability of ensuring that their information is correct, and this factor contributes in part to the relatively high cost of computer MSDS data bases. The other major reason is the substantial amount of research needed to keep up with all of the current published material available.

The references which follow are unlike the normal journal citations in that they are Internet addresses. These simply represent sites which provide, free of charge, access to a very large number of MSDS. To access almost any manufacturers MSDSs and commercial providers of MSDSs, one can enter use any Internet browser, access a search engine, and Type "Material + Safety + Data + Sheet, or MSDS" and one will receive many pages of Internet links to which to go. The following two references are simply two of the most comprehensive.

REFERENCES

1. http://hazard.com/msds/
2. http ://www.msdssearch.com/

D. Purchase of Regulated Items

There are a number of classes of items for which purchases must be carefully monitored for compliance with safety and security regulations. Several of these can be purchased only if a license is held by the individual or by the corporation or institution. There are many restrictions, in addition, on the transportation of hazardous materials. Usually, the purchaser will expect the vendor to be responsible for meeting these shipping requirements. However, there will be occasions when the institution or corporation will initiate a shipment. It is recommended that a subscription to a hazardous materials transportation regulatory advisory service be taken out by anyone who ships any hazardous material frequently, due to the relatively rapid changes in shipping regulations. Such information is also rapidly becoming available from on-line or CD-ROM computer services. Updated data is often being provided by the regulatory agencies themselves.

1. Radioisotopes

The purchase of radioactive materials, with certain exceptions, is generally restricted to those persons who are licensed to own and use the materials under one of the sections of Title 10, Code of Federal Regulations, usually Part 30. In this context, the word "person" is used quite broadly. In Part 30, which provides the rules for domestic licensing of byproduct material, "person" is defined as: "Any individual, corporation, partnership, firm, association, trust, public or private institution, group, Government agency other than the Commission or Department..., any State, any foreign government or nation or any political subdivision of any such government or nation, or other entity; and any legal successor, representative, agent or agency of the foregoing." Clearly, virtually any assemblage of persons can qualify to be licensed to own and use radioactive byproduct materials, if they can fulfill the licensing conditions provided by Part 30 and have an approved radiation management program meeting the standards of Part 20. In approximately half of the states, the oversight function to ensure compliance with the standard is done by the state rather than the Nuclear Regulatory Commission (NCR). These are "agreement states."

There are a few more definitions which will be useful. The federal regulations in Part 30 usually apply only to "byproduct material." This refers to "...radioactive materials, other than special nuclear material, yielded in or made radioactive by exposure to the radiation incident to the process of producing or utilizing special nuclear material." The NRC definition of special nuclear material is lengthy, but essentially it means plutonium, or uranium enriched in the fissionable isotopes U-233 or U-235. There are naturally occurring radioactive materials which are mostly unregulated and there are radioactive materials made radioactive by using accelerators. These latter materials are regulated by the states independently, not by the NRC. Exposure to some natural radioactive materials, such as radon, are federally regulated under some circumstances.

There are a number of classes of radioactive materials which do not require a license. If the amount is less than the exempt quantity for a given material, as listed in Paragraph 30.71, Schedule B of the regulations, a license is not required. The amount meeting this criteria is given in Table 4.1 for a few of the radioisotopes most commonly used in research. The units are in microcuries where 1 microcurie is equal to 37,000 nuclear disintegrations per second, since this is the way they appear in the regulations. A set of units different from these has been recommended by the International Commission on Radiological Protection, and is the one commonly used in professional journals. In the International System of Units (SI units), the unit of activity is the Becquerel (Bq) and is equal to 1 disintegration per second. A microcurie, therefore, equals 37,000 Bq.

There are a number of other classes described in paragraphs 30.15-20 of 10 CFR, in which the persons purchasing certain items containing radioactive materials are exempt from having a license, although the original manufacturer must have had a specific license to allow production of the unit. Among these are self-luminous devices and gas and aerosol detectors.

The amounts in Table 4.1 are very small and are usually exceeded in most research applications. For practical research using radioactive materials, it is necessary to obtain a license; a discussion of this will be deferred to Chapter 5. However, assuming that a license has been obtained and a radiation safety program has been established satisfying the NRC (or its equivalent in an agreement state; henceforth, when the NRC is mentioned, it will be understood to include this addendum), there are still formal steps to go through in purchasing and receiving radioactive materials.

In a research facility, it is common practice to establish a license to cover all users of radiation at the organization. This is called a broad license and provides limits on the total amount of each isotope that can be in possession of the licensee at any specific time. These limits are normally chosen by the institution and approved by the NRC. If there are several separate users, as is usually the case, the sum of all their holdings for each isotope,

Table 4.1 Exempt Quantities of Some of the Most Often Used Radioisotopes

Isotope	Quantity (pCi)	Isotope	Quantity (pCi)
Calcium 45	10	Iodine 131	1
Carbonl4	100	Iron59	10
Cesium 137	10	Mercury 203	10
Cobalt 60	1	Molybdenum 99	100
Chromium 51	1000	Nickel 63	10
Hydrogen 3	1000	Phosphorus 32	10
Iodine 125	1	Sulfur 35	100

including unused material, material In use, and material as waste, must not exceed these limits. Since each individual user cannot keep track of the holdings of other independent users, it is essential that all purchase orders, as well as all waste materials, be passed by or through a radiation safety specialist, whose responsibility (among many others) is to ensure that the license limits are not violated. Adherence to this and all other radiation safety regulations is essential. At one time, the primary threat in the event of a violation was the possibility of suspension of a license. This was such a severe penalty that it was invoked very infrequently. In recent years, substantial fines have been levied against universities and other users who violate the regulations and the terms of their licenses. On March 12, 1987, a city attorney filed 179 *criminal* charges against a major university within the city's jurisdiction and 10 individual members of its faculty for violations of the state standards. This established a major precedent. More recently, another university reached a settlement with the surrounding community to conduct a $1,300,000 study of the possible dispersion of radioactive materials into the community in addition to a substantial fine, because of their management of the use of radioactive materials. As the previous edition of this book was being written, a major study on the use of radioactive materials in "research" on possibly unsuspecting or involuntary participants shortly after World War II was underway after release of hitherto secret papers. Even at this late date, such information is still being discovered with significant political repercussions about the propriety of such studies.

Unless a vendor has a valid copy of the license for a person ordering radioisotopes, they are not allowed to fill an order. Since the radiation safety specialist is such a key person in the process in any event, it should also be this person's responsibility to maintain current copies of licenses, including any amendments, in the hands of prospective suppliers of radioactive materials. At many facilities, the radiation safety specialist has been assigned virtually all responsibility for ordering and receipt of radioactive materials. Title 10 CFR, Part 20.1906, requires each licensee to establish safe procedures for receipt and opening of radioactive packages. Although mistakes are rare in filling and shipping radioactive material orders, they do happen, so it is highly desirable that the radiation safety specialist directly receive each package of radioactive materials, check that its paperwork is correct, check the external radiation levels, and check the containers for damage. It has happened that all of the paperwork conformed to the expected material, but the wrong material or the wrong amounts of the ordered material were shipped. Where it is impossible for the radiation specialist to always receive all packages, provision needs to be made for temporary secure storage of packages until they can be checked.

Many radioactive materials are used in the form of labeled compounds, often prepared specifically to order. In some of these, the half-life of the isotope used in the compound is short so that procedures need to be established to ensure prompt handling and delivery to the user. In other cases, the compound itself will deteriorate at ordinary temperatures. These packages are usually shipped packed in dry ice and must be delivered immediately upon receipt or stored temporarily in a freezer until delivery. If it is necessary to ship radioactive material, the material must be packaged according to Title 49, CFR 173. Again, the radiation safety specialist is the individual who normally would be the expected to be familiar with all current standards affecting shipment and be able to arrange for transportation according to the regulations.

2. Controlled Substances (Drugs)

The purchase, storage, and use of many narcotic, hallucinogenic, stimulant, or depressive drugs are regulated under Title 21, Code of Federal Regulations, Part 1300 to the end. In addition, these substances are usually regulated by state law, which in many cases is much more stringent than federal law. The controlled substances covered by the Controlled Substances Act are divided into five schedules. Schedule I substances have no accepted medical use in the United States, have a high potential for abuse, and are the most tightly controlled, while Schedule V substances contain limited quantities of some narcotics with limited risk. For these materials, the Drug Enforcement Agency (DEA), which is the federal agency regulating the use of these substances, does not permit a broad agency license, but requires a single responsible individual in each functionally independent facility to obtain a separate license, which spells out which schedules of controlled substances are permissible for the facility to possess. This individual can permit others to use the controlled substance under his direction or to issue it to specific persons for whom he will take the responsibility, but there is no required equivalent to the radiation safety officer to monitor programs internally. Thus, the individual license holder is responsible for ordering, receiving, and maintaining an accurate current inventory for the drugs used in his laboratory.

One institutional responsibility that should be assigned to an individual or department is monitoring the expiration dates of licenses. Although the DEA has a program which should remind each licensee in ample time that their license is on the verge of expiring, experience has shown that the program has not been entirely successful. An individual within the organization should maintain a file of all licenses held by employee's of the organization and take appropriate steps to see that applications for renewals are filed in a timely manner to avoid purchasing of controlled materials on expired licenses. In organizations that have a pharmacy or pharmacists on their staff, the senior pharmacist would be the logical person to perform these limited regulatory roles.

Packages containing controlled substances must be marked and sealed in accordance with the provisions of the Controlled Substances Act when being shipped. Every parcel containing these sensitive materials, must be placed within a plain outer container or securely wrapped in plain paper through which no markings indicating the nature of the contents can be seen. No markings of any kind are permitted on the parcel which would reveal the nature of the contents. The purpose, of course, is to avoid temptation for those who would steal the contained drugs for illegal purposes.

3. Etiologic Agents

Hazardous biological agents are classified as "etiologic agents." An etiologic agent is more specifically defined as (1) a viable microorganism, or its toxin, which is listed in Title 42 CFR 72.3 or (2) which causes or may cause severe, disabling, or fatal human disease. The importation or subsequent receipt of etiologic agents and vectors of human diseases is subject to the regulations of the Public Health Service, given in Title 42, Section 71.156. The Centers for Disease Control (CDC) issues the necessary permits authorizing the importation or receipt of regulated materials and specifies the conditions under which the agent or vector is shipped, handled, and used. The interstate shipment of indigenous etiologic agents, diagnostic specimens, and biological products is subject to applicable packaging, labeling, and shipping requirements of the Interstate Shipment of Etiologic Agents (42 CER Part 72). Packaging and labeling requirements are illustrated in Figure 4.1.

In addition to the regulations of the Public Health Service, the Department of Transpor-

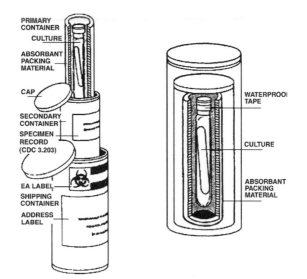

Figure 4.1 Packaging or etiologic substances showing required details and labeling.

tation has additional regulations in Title 49, CER Section 173.386-388. Shipments are limited to 50 milliliters or 50 grams in a passenger carrying airplane or rail car, and 4 liters or 4 kilograms in cargo aircraft. The U.S. Postal Service provides regulations covering the mailability of biological materials in the Domestic Mail Manual, Section 124.38. All of these agencies provide explicit instructions on how etiologic agents can be shipped. There are additional restrictions for international shipments, covered by the International Mail Manual. The ability to make foreign shipments is restricted to laboratories, by approval of the General Manager, International Mail Classification Division, USPS Headquarters, Washington, D.C. 20260-5365.

Whether a person or laboratory purchases a given etiologic agent should depend upon a review of the facilities available for the research program, the training and experience of the laboratory employees, and the type and scale of the operations to be conducted. If, as reviewed in that material, the etiologic agent is one that would require the planned operations to be conducted in a laboratory meeting Biological Safety Standard level 3 or 4, the purchase should require the prior approval of the institutional biosafety committee. Operations and classification of Microbiological and Biomedical laboratories will be covered in some detail in Chapter 5.

There are comparable restrictions for the importation, possession, use, or interstate shipment of certain pathogens of domestic livestock and poultry, administered by the U.S. Department of Agriculture.

For additional information regarding etiologic agents of human diseases and related materials, write to:

Centers for Disease Control
Attention:
Office of Biosafety
1600 Clifton Road, N.E.
Atlanta, GA 30329

For additional information regarding animal pathogens, write:

Table 4.2 OSHA Regulated Carcinogens

Asbestos	4-Aminodiphenyl	Benzene
Coal tar pitch volatiles	Ethyleneimine	Coke oven emissions
4-Nitrobiphenyl	β-Propiolactone	Cotton dust
α-Naphthylamine	2-Acetylaminofluorene	1,2-dibromo-3-chloropropane
Methyl chloromethyl ether	4-Dimethylaminoazobenzene	Acrylonitrile
3,'-Dichlorobenzidine	N-Nitrosodimethylamine	Ethylene oxide
(And its salts)	Vinyl Chloride	Formaldehyde
bis-Chloromethyl ether	Inorganic Arsenic	Methylenedianiline
β-Napthylamine	Lead	1,3 Butadiene
Benzidine	Cadmium	Methylene Chloride

Chief Staff Veterinarian
Organisms and Vectors
Veterinary Services
Animal and Plant Health Inspection Service
U.S. Department of Agriculture
Federal Building Room 810
Hyattsville, MD 20782
or call (301) 436-8017

4. Carcinogens

There are no restrictions on ordering known carcinogens. However, for the carcinogens covered by the regulations in Title 29 CFR Part 1910, Subpart Z their purchase for research should be limited to individuals who formally commit themselves to complying with the terms and conditions of the standards. As noted earlier, although the laboratory standard does preempt most of the usual OSHA standards, where there exist specific regulations for individual materials, these regulations still apply. To ensure that this is done, every requisition for purchase of one of the regulated carcinogens should be referred to the institutional safety department for review. This will normally involve a review of the research protocols to ascertain if the use is liable to meet any criteria exempting the proposed program from some of the more stringent and often expensive requirements. If the program does not appear to qualify for exemptions, then the investigator and the safety reviewer should go through each of the requirements under the standard to confirm that they can be met. Although this will seem excessive to some users, it serves not only to protect the employees, but also to minimize the potential for litigation for the research director and the academic institution or corporation.

There are a number of known carcinogenic materials, and the list is growing as the necessary studies of suspected carcinogens are completed. It is recommended that purchases of these be limited and exposures minimized as much as possible to promote the safety of everyone exposed to the materials and in consideration of potential future regulatory restrictions. As discussed in Section 4.III.C, for the purpose of the MSDSs, a listing as a carcinogen by either the NTP, or the IARC is sufficient to be considered as one for the

STATE OF CALIFORNIA
ENVIRONMENTAL PROTECTION AGENCY
OFFICE OF ENVIRONMENTAL HEALTH HAZARD ASSESSMENT
SAFE DRINKING WATER AND TOXIC ENFORCEMENT ACT OF 1986
CHEMICALS KNOWN TO THE STATE TO CAUSE CANCER OR REPRODUCTIVE
TOXICITY
November 6, 1998

The identification number indicated in the following list is the Chemical Abstracts Service (CAS) Registry Number. No CAS number is given when several substances are presented as a single listing.

Table 4.3A CHEMICALS KNOWN TO THE STATE TO CAUSE CANCER

Chemical	CAS No.
A-alpha-C (2-Am ino-9H-pyrido[2,3-b]indole)	26148685
Acetaldehyde	75070
Acetamide	60355
Acetochlor	34256821
2-Acetylaminofluorene	53963
Acifluorfe n	62476599
Acrylamide	79061
Acrylon itrile	107131
Actinomycin D	50760
Adriamycin (Doxorubicin hydrochloride)	23214928
AF-2 ;[2-(2-furyl)-3-(5-n itro-2-furyl)]acrylam ide	3688537
Aflatoxins	---------
Alachlor	15972608
Alcoholic beverages, when associated with alcohol abuse	---------
Aldrin	309002
Allyl chloride	107051
2-Aminoanthraquinone	117793
p-Aminoazobenzene	60093
ortho-Am inoazotoluene	97563
4-Aminobiphenyl (4-aminodiphenyl)	92671
1-Amino-2,4-dibromoanthraquinone	81492
3-Amino-9-ethylcarbazole hydrochloride	6109973
1-Amino-2-methylanthraquinone	82280
2-Amino-5-(5-nitro-2-furyl)-1,3,4-thiadiazole	712685
Amitrole	61825
Analgesic mixtures containing phenacetin	--------
Aniline	62533
Aniline hydrochloride	142041
ortho-Anisidine	90040
ortho-Anisidine hydrochloride	134292
Antimony oxide (Antimony trioxide)	1309644
Aramite	140578
Arsenic (inorganic arsenic compounds)	---------
Asbestos	1332214
Aurammine	492808

Azacitidine	320672
Azaserine	115026
Azathioprine	446866
Azobenzene	103333
Benz[a]anthracene	56553
Benzene	71432
Benzidine [and its salts]	92875
Benzidine based dyes	--------
Benzo[b]fluoranthene	205992
Benzo[j]fluoranthene	205823
Benzo[k]fluoranthene	207089
Benzofuran	271896
Benzo[a]pyrene	50328
Benzotrlchloride	96077
Benzyl chloride	100447
Benzyl violet 4B	1694093
Beryllium and beryllium compounds	---------
Betel quid with tobacco	---------
2,2.Bis(bromomethyl)-1,3-propanediol	3296900
Bis(2-chloroethyl)ether	111444
N,N-Bis(2-chloroethyl)-2-naphthylamine (Chlomapazine)	494031
Bischloroethyl nitrosourea (BCNU) (Carmustine)	154938
Bis(chloromethyl)ether	542881
Bitumens, extracts of steam reflned and air reflned	---------
Bracken fern	---------
Bromodichloromethane	75274
Bromoform	75252
I,3-Butadiene	106990
I,4-Butanediol dimethanesulfonate (Busulfan)	55981
Butylated hydroxyanisole	25013165
beta-Butyrolactone	3068880
Cacodylic acid	75605
Cadmium and cadmium compounds	--------
Caffeic acid	331395
Captafol	2425061
Captan	133062
Carbazole	86748
Carbon tetrachloride	56235
Carbon-black extracts	--------
Ceramic fibers (airborne particles of respirable size)	--------
Certain combined chemotherapy for lymphomas	--------
Chlorambucil	305033
Chloramphenicol	56757
Chlordane	57749
Chlordecone (Kepone)	143500
Chlordimeform	614983
Chlorendic acid	115286
Chlorinated paraffins (Average chain length, C12; approximately 60 percent chlorine by weight)	108171262 --------

p-Chloroaniline	106478
p-Chloroaniline hydrochloride	20265967
Chlorodibromomethane	124481
Chloroethane (Ethyl chloride)	75003
1-(2-Chloroethyl)-3-cyclohexyl-1-nitrosourea (CCNU) (Lomustine)	13010474
1-(2-Chloroethyl)-3-(4-methylcyclohexyl)-1-nitrosourea (Methyl-CCNU)	13909096
Chloroform	67663
Chloromethyl methyl ether (technical grade)	107302
3-Chloro-2-methylpropene	563473
4-Chloro-ortho-phenylenediamine	95830
p-Chloro-o-toluidine	95692
p-Chloro-o-toluidine, strong acid salts of	--------
5-Chloro-o-toluidine and its strong acid salts	--------
Chlorothalonil	1897456
Chlorotrianisene	569573
Chlorozotocin	54749905
Chromium (hexavalent compounds)	--------
Chrysene	218019
C. I. Acid Red 114	6459945
C. I. Basic Red 9 monohydrochloride	569619
Ciclosporin (Cyclosporin A; Cyclosporine)	59865133
	79217600
C.I. Direct Blue 15	2429745
C.I. Direct Blue 218	28407376
C.I. Solvent Yellow 14	842079
Cinnamyl anthranilate	87296
Cisplatin	15663271
Citrus Red No.2	6356536
Clofibrate	637070
Cobalt metal powder	7440484
Cobalt [II] oxide	1307966
Coke oven emissions	--------
Conjugated estrogens	--------
Creosotes-	-------
para-Cresidine	120718
Cupferron	135206
Cycasin	14901087
Cyclophosphamide (anhydrous)	50180
Cyclophosphamide (hydrated)	6055192
Cytembena	21739913
D&C Orange No.17	3468631
D&C Red No.8	2092560
D&C Red No.9	5160021
D&C Red No.19	81889
Dacarbazine	4342034
Daminozide	1596845
Dantron (Chrysazin; 1,8-Dihydroxyanthraquinone)	117102
Daunomycin	20830813
DDD (Dichlorodiphenyldlchloroethane)	72548

DDE (Dichlorodiphenyldichloroethylene)	72559
DDT (Dichlorodiphenyltrichloroethane)	50293
DDVP (Dichlorvos)	62737
N,N'-Diacetylbenzidine	613354
2,4-Diaminoanisole	615054
2,4-Diaminoanisole sulfate	39156417
4,4'Diaminodiphenyl ether (4,4'.Oxydianiline)	101804
2,4-Diaminotoluene	95807
Diaminotoluene (mixed)	--------
Dlbenz[a,h]acridine	226368
Dibenz[a,j]acridine	224420
Dibenz[a,h]anthracene	53703
7H-Dibenzo[c,g]carbazole	194592
Dibenzo[a,e]pyrene	192654
Dibenzo[a,h]pyrene	189640
Dibenzo[a,l]pyrene	189559
Dibenzo[a,j]pyrene	191300
1,2-Dibromo-3-chloropropane (DBCP)	96128
2,3-Dibromo-1-propanol	96139
Dichloroacetic acid	79436
p-Dichlorobenzene	106467
3,3'-Dichlorobenzidine	91941
3,3'-Dlchlorobenzidine dihydrochloride	612839
1,4-Dichloro-2-butene	764410
3,3'-Dichloro-4,4'-diaminodiphenyl ether	28434868
1,1-Dichloroethane	75343
Dichloromethane (Methylene chloride)	75092
1,2-Dichloropropane	78875
1,3-Dichloropropene	542756
Dieldrin	60571
Dienestrol	84173
Diepoxybutane	1464535
Diesel engine exhaust	--------
Di(2-ethylhexyl)phthalate	117617
1,2-Diethylhydrazine	1615801
Diethyl sulfate	64675
Diethylstillbestrol	56531
Diglycidyl resorcinol ether (DGRE)	101906
Dihydrosaf role	94586
Diisopropyl sulfate	2973106
3,3'-Dimethoxybenzidine (ortho-Dianisidine)	119904
3,3'-Dimethoxybenzidine dihydrochloride	20325400
(ortho-Dianisidine dihydrochloride)	--------
Dimethyl sulfate	77781
4-Dimethylaminoazobenzene	60117
trans-2-[(Dimethylamino)methylimino]-5-[2-(5-nitro-2-furyl)vinyl]-1,3,4-oxadiazole	55738540
7,12-Dimethylbenz(a)anthracene	57976
3,3'-Dimethylbenzidine (ortho-Tolidine)	119937

3,3'-Dimethylbenzidine dihydrochloride	612828
Dimethylcarbamoyl chloride	79447
1,1-Dimethylhydrazine (UDMH)	57147
1,2-Dimethylhydrazine	540738
Dimethylvinylchloride	513371
3,7-Dinitrofluoranthene	105735715
3,9-Dinitrofluoranthene	22506532
1,6-Dinitropyrene	42397648
1,8-Dinitropyrene	42397659
Dinitrotoluene mixture, 2,4-/2,6-	--------
2,4-Dinitrotoluene	121142
2,6-Dinitrotoluene	606202
Di-n-propyl isocinchomeronate (MGK Repellent 326)	136458
1,4-Dioxane	123911
Diphenylhydantoin (Phenytoin)	57410
Diphenylhydantoin (Phenytoin), sodium salt	630933
Direct Black 38 (technical grade)	1937377
Direct Blue 6 (technical grade)	2602462
Direct Brown 95 (technical grade)	16071866
Disperse Blue 1	2475458
Epichlorohydrin	106898
Erionite	12510428
Estradiol 17B	50282
Estrone	53167
Estropipate	7280377
Ethinylestradiol	57636
Ethyl acrylate	140885
Ethyl methanesulfonate	62500
Ethyl 4,4'-dichlorobenzilate	510156
Ethylene dibromide	106934
Ethylene dichloride (1,2-Dichloroethane)	107062
Ethylene oxide	75218
Ethylene thiourea	96457
Ethyleneimine	151564
Folpet	133073
Formaldehyde (gas)	50000
2-(2-Formylhydrazino)-4-(5-nitro-2-furyl)thiazole	3570750
Furan	110009
Furazolidone	67458
Furmecyclox	60568050
Fusarin C	79746815
Ganciclovir sodium	82410320
Gasoline engine exhaust (condensates/extracts)	--------
Glasswool fibers (airborne particles of respirable size)	--------
Glu-P-1(2-Amino-6-methyldipyrido[1,2-a:3',2'-d]imidazole)	67730114
Glu-P-2 (2-Aminodipyrido[l,2-a:3',2'-d]imidazole)	67730103
Glycidaldahyde	765344
Glycidol	556525
Griseofulvin	126078

Gyromitrin (Acetaldehyde methylformylhydrazone)	16568028
HC Blue 1	2784943
Heptachlor	76448
Heptachlor epoxide	1024573
Hexachlorobenzene	118741
Hexachlorocyclohexane (technical grade)	--------
Hexachlorodibenzodioxin	34465468
Hexachloroethane	67721
Hexamethylphosphoramide	680319
Hydrazine	302012
Hydrazine sulfate	10034932
Hydrazobenzene (1,2-Diphenylhydrazine)	122667
Indeno [1,2,3-cd]pyrene	193395
1Q(2-Amino-3-methyflimidazo[4,5-f]quinoline)	76180966
Iprodione	36734197
Iron dextran complex	9004664
Isobutyl nitrite	542563
lsoprene	78795
Isosafrole	120581
Lactofen	77501634
Lasiocarpine	303344
Lead acetate	301042
Lead and lead compounds	---------
Lead phosphate	7446277
Lead subacetate	1335326
Lindane and other hexachlorocyclohexane isomers	----------
Mancozeb	8018017
Maneb	12427382
Me-A-alpha-C (2-Amino-3-methyl-9H-pyrido[2,3-b]indole)	68006837
Medroxyprogesterone acetate	71589
MeQ(2-Amino-3,4-dimethylimidazo[4,5-f]quinoline)	77094112
MelQx(2-Amino-3,8-dimethylimidazo[4,5-f]quinoxaline)	7500040
Melphalan	148823
Merphalan	531760
Mestranol	72333
Metham sodium	137428
8-Methoxypsoralen with ultraviolet A therapy	298817
5-Methoxypsoralen with ultraviolet A therapy	484208
2-Methylaziridine (Propyleneimine)	75558
Methylazoxymethanol	590965
Methylazoxymethanol acetate	592621
Methyl carbamate	598550
3-Methylcholanthrene	56495
5-Methylchrysene	3697243
4,4'-Methylene bis(2-chloroaniline)	101144
4,4'-Methylene bis(N,N-dimethyl)benzenamine	101611
4,4'-Methylene bis(2-methylaniline)	838880
4,4'-Methylenedianiline	101779
4,4'-Methylenedianiline dihydrochloride	13552448

Methylhydrazine and its salts	--------
Methyl iodide	74884
Methylmercury compounds	--------
Methyl methanesulfonate	66273
2-Methyl-1-nitroanthraquinone (of uncertain purity)	129157
N-Methyl-N'-nitro-N-nitrosoguanidine	70257
N-Methylolacrylamide	924425
Methylthiouracil	56042
Metiram	9006422
Metronidazole	443481
Michler's ketone	90948
Mirex	2385855
Mitomycin C	50077
Monocrotaline	315220
5-(Morpholinomethyl)-3-[(5-nitro-furfurylidene)-amino]-2-oxazolidinone	139913
Mustard Gas	505602
Nafenopin	3771195
Nalidixic acid	389082
1-Naphthylamine	134327
2-Naphthylamine	91896
Nickel and certain nickel compounds	--------
Nickel carbonyl	13463393
Nickel refinery dust from the pyrometalurgical process	--------
Nickel subsulflde	12035722
Niridazole	61574
Nitrilotriacetic acid	139139
Nitrilotriacetic acid, trisodlum sait monohydrate	18662538
5-Nitroacenaphthene	602879
5-Nitro-o-anisidine	99592
o-Nitroanisole	91236
Nitrobenzene	98953
4-Nitrobiphenyl	92933
6-Nitrochrysene	7496028
Nitrofen (technical grade)	1836755
2-Nitrofluorene	607578
Nltrofurazone	59870
1-[(5-Nitrofurfurylidene)-amino]-2-imidazolidinone	555840
N-[4-(5-Nitro-2-furyl)-2-thiazolyl]acetamide	531828
Nitrogen mustard (Mechlorethamine)	51752
Nitrogen mustard hydrochloride (Mechlorethamine hydrochloride)	55867
Nitrogen mustard N-oxide	126852
Nitrogen mustard N-oxide hydrochloride	302705
Nitromethane	75525
2-Nitropropane	79469
1-Nitropyrene	5522430
4-Nitropyrene	57835924
N-Nitrosodi-n-butylamine	924163
N-Nitrosodiethanolamine	1116547
N-Nitrosodiethylamine	55185

N-Nitrosodimethylamine	62759
p-Nitrosodiphenylamine	156105
N-Nitrosodiphenylamine	86306
N-Nitrosodi-n-propylamine	621647
N-Nitroso-N-ethylurea	759739
3-(N-Nitrosomethylamino)propionitrile	60153493
4-(N-Nitrosomethylamino)-1-(3-pyridyl)1-butanone	64091914
N-Nitrosomethylethylamine	10595956
N-Nitroso-N-methylurea	684935
N-Nitroso-N-methylurethane	6155323
N-Nitrosomethylvinylamine	4549400
N-Nitrosomorpholine	59892
N-Nitrosonornicotine	16543558
N-Nitrosopiperidine	100754
N-Nitrosopyrrolidine	930552
N-Nitrososarcosine	13256229
o-Nitrotoluene	88722
Norethisterone (Norethindrone)	68224
Ochratoxin A	303479
Oil Orange SS	2646175
Oral contraceptives, combined	--------
Oral contraceptives, sequential	--------
Oxadiazon	19666309
Oxazepam	604151
Oxymetholone	434071
Panfuran S	794934
Pentachlorophenol	87865
Phenacetin	62442
Phenazopyridine	94760
Phenazopyridine hydrochloride	136403
Phenesterin	3546109
Phenobarbital	50066
Phenolphthalein	77098
Phenoxybenzamine	59961
Phenoxybenzamine hydrochloride	63923
o-Phenylenediamine and its salts	95545
Phenyl glycidyl ether	122601
Phenylhydrazine and its salts	-------
o-Phenylphenate, sodium	132274
PhiP(2-Amino-1-methyl-6-phenylimidazo[4,5-b]pyridine)	105650235
Polybrominated biphenyls	-------
Polychlorinated biphenyls	-------
Polychlorinated biphenyls (containing ≥60% chlorine by molecular weight	-------
Polychlorinated dibenzo-p-dioxins	-------
Polychlorinated dibenzofurans	-------
Polygeenan	53973981
Ponceau MX	3761533
Ponceau 3R	3564098
Potassium bromate	7758012

Procarbazine	671169
Procarbazine hydrochloride	366701
Procymidone	32809168
Progesterone	57830
1,3-Propane sultone	1120714
Propargite	2312358
Pronamide	23950585
beta-Propiolactone	57578
Propylene oxide	75569
Propylthiouracil	51525
Quinoline and its strong acid salts	--------
Radionuclides	--------
Reserpine	50555
Residual (heavy) fuel oils	--------
Saccharin	81072
Saccharin, sodium	128449
Safrole	94597
Salicylazosulfapyridine	599791
Selenium sulfide	7446346
Shale oils	68308349
Silica, crystalline (airborne particles of respirable size)	--------
Soots, tars, and mineral oils (untreated and mildly treated oils and used engine oils)	--------
Spironolactone	52017
Stanozolol	10418038
Sterigmatocystin	10048132
Streptozotocin	18883664
Styrene oxide	96093
Sulfalate	95067
Talc containing asbestiform fibers	--------
Tamoxifen and its salts	10540291
Terrazole	2593159
Testosterone and its esters	58220
2,3,7,8-Tetrachlorodibenzo-paradioxin (TCDD)	1746016
1,1,2,2-Tetrachloroethane	79345
Tetrachloroethylene (Perchloroethylene)	127184
p-a,a,a-Tetrachlorotoluene	5216251
Tetrafluoroethylene	116143
Tetranitromethane	509148
Thioacetamide	62555
4,4'-Thiodianiline	139651
Thiourea	62566
Thorium dioxide	1314201
Tobacco, oral use of smokeless products	--------
Tobacco smoke	--------
Toluene diisocyanate	26471625
ortho-Toluidine	95534
ortho-Toluidine hydrochloride	636215
para-Toluidine	106490

Toxaphene (Polychorinated camphenes)	8001352
Treosulfan	299752
Trichlormethine (Trimustine hydrochloride)	817094
Trichloroethylene	79016
2,4,6-Trichlorophenol	88062
1,2,3-Trichloropropane	96184
Trimethyl phosphate	512561
2,4,5-Trlmethylaniline and its strong acid salts	-------
Triphenyltin hydroxide	76879
Tris(aziridinyl)-para-benzoquinone (Triaziquone)	68768
Tris(1-aziridinyl)phosphine sulfide (Thiotepa)	52244
Tris(2-chloroethyl) phosphate	115968
Tris(2,3-dibromopropyl)phosphate	126727
Trp-P-1 (Tryptophan-P-1)	62450060
Trp-P-2 (Tryptophan-P-2)	62450071
Trypan blue (commercial grade)	72571
Unleaded gasoline (wholly vaporized)	--------
Uracil mustard	66751
Urethane (Ethyl carbamate)	51796
Vinyl bromide	593602
Vinyl chloride	75014
4-Vinylcyclohexene	100403
4-Vinyll-cyclohexene diepoxide (Vinyl cyclohexenedioxide)	106876
Vinyl fluoride	75025
Vinyl trichloride (1,1,2-Trichloroethane)	79005
2,6-Xylidine (2,6- Dimethylaniline)	87627
Zineb	12122677

TABLE 4.3 B CHEMICALS KNOWN TO CAUSE REPRODUCTIVE TOXICITY

Developmental toxicity

Acetohydroxamic acid	546883
Actinomycin D	50760
All-trans retinoic acid	302794
Alprazolam	28981977
Amikacin sulfate	39831555
Aminoglutethimide	125848
Aminoglycosides	--------
Aminopterin	54626
Amiodarone hydrochloride	19774824
Amoxapine	14028445
Angiotensin converting enzyme (ACE) inhibitors	--------
Anisindione	117373
Arsenic (inorganic oxides)	--------
Aspirin (NOTE: It is especially important not to use aspirin during the last three months of pregnancy, unless specifically directed to do so by a physician because it may cause problems in the unborn child or complications during delivery.)	50782
Atenolol	29122687

Azathioprine	446866
Barbiturates	------
Beclomethasone dipropionate	5534098
Benomyl	17804352
Benzene	71432
Benzphetamine hydrochloride	5411223
Benzodiazepines	--------
Bischloroethyl nitrosourea (BCNU) (Carmustine)	154938
Bromoxynil	1689845
Butabarbital sodium	143817
1,4-Butanediol dimethylsulfonate (Busulfan)	55981
Cadmium	--------
Carbon disulfide	75150
Carbon monoxide	630080
Carboplatin	41575944
Chenodiol	474259
Chinomethionat (Oxythioguinox)	2439012
Chlorambucil	305033
Chlorcyclizine hydrochloride	1620219
Chlordecone (Kepone)	143500
Chlordiazepoxide	58253
Chlordiazepoxide hydrochloride	438415
1-(2-Chloroethyl)-3-cyclohexyl-l-nitrosourea (CCNU) (Lomustine)	13010474
Cladribine	4291638
Clarithromycin	81103119
Clobetasol propionate	25122467
Clomiphene citrate	50419
Clorazepate dipotassium	57109907
Cocaine	50362
Codeine phosphate	52288
Colchicine	64868
Conjugated estrogens	--------
Cyanazine	21725462
Cycloheximide	66819
Cyclophosphamide (anhydrous)	50180
Cyclophosphamide (hydrated)	6055192
Cyhexatin	13121705
Cytarabine	147944
Danazol	17230885
Daunorubicin hydrochloride	23541506
o,p'-DDT	789026
p,p'-DDT	50293
Demeclocycline hydrochioride (internal use)	64733
Diazepam	439145
Dicumarol	66762
Diethylstillbestrol (DES)	56531
Dihydroergotamine mesylate	6190392
Dinocap	39300453
Dinoseb	88857

Diphenylhydantoin (Phenytoin)	57410
Doxycycline (Internal use)	564250
Doxycycline calcium (intemal use)	94088854
Doxycycline hyclate (Internal use)	24390145
Doxycycline monohydrate (internal use)	17086281
Endrin	72208
Ergotamine tartrate	379793
Estropipate	7280377
Ethionamide	536334
Ethyl alcohol in alcoholic beverages	--------
Ethylene dibromide	106934
Ethylene glycol monoethyl ether	110805
Ethylene glycol monomethyl ether	109864
Ethylene glycol monoethyl ether acetate	111159
Ethylene glycol monomethyl ether acetate	110496
Ethylene thiourea	96457
Etoposide	33419420
Etretinate	54350480
Fluazifop butyl	69806504
Flunisolide	3385033
Fluorouracil	51218
Fluoxymesterone	76437
Flurazepam hydrochloride	1172185
Flutamide	13311847
Fluticasone propionate	80474142
Fluvalinate	69409945
Ganciclovir sodium	82410320
Goserelin acetate	65807025
Halazepam	23092173
Halothane	151677
Hexachlorobenzene	118741
Histrelin acetate	--------
Hydroxyurea	127071
Ifosfamide	3778732
Iodine-131	10043660
Isotretinoin	4759482
Lead	--------
Leuprolide acetate	74381536
Lithium carbonate	554132
Lithium citrate	919164
Lorazepam	846491
Lovastatin	75330755
Medroxyprogesterone acetate	71589
Megestrol acetate	595335
Melphalan	148823
Menotropins	9002680
Meprobamate	57534
Mercaptopurine	6112761
Mercury and mercury compounds	--------

Methacycline hydrochloride	3963959
Metham sodium	137428
Methim azole	60560
Methotrexate	59052
Methotrexate sodium	15475566
Methyl bromide as a structural fumigant	74839
Methyl mercury	--------
Methyltestosterone	58184
Midazolam hydrochloride	59467968
Minocycline hydrochloride (internal use)	13614987
Misoprostol	59122462
Mitoxantrone hydrochloride	70476823
Nafarelin acetate	86220420
Neomycin sulfate (internal use)	1405103
Netilmicin sulfate	56391572
Nickel carbonyl	13463393
Nicotine	54115
Nitrogen mustard (Mechlorethamine)	51752
Nitrogen mustard hydrochloride (Mechlorethamine hydrochloride)	55867
Norethisterone (Norethindrone)	68224
Norethisterone acetate (Norethindrone acetate)	51989
Norethisterone (Norethindrone)/Ethinyl estradiol	68224/57636
Norethisterone (Norethindrone)/Mestranol	68224/72333
Noroestrel	6533002
Oxadiazon	19666309
Oxazepam	604751
Oxymetholone	434071
Oxytetracycline (internal use)	79572
Oxytetracycline hydrochloride (internal use)	2058460
Paclitaxel	33069624
Paramethadione	115673
Penicillamine	52675
Pentobarbital sodium	57330
Pentostatin	53910251
Phenacemide	63989
Phenprocoumon	435972
Pipobroman	54911
Plicamycin	18378897
Polybrominated biphenyls	--------
Polychlorinated biphenyls	--------
Procarbazine hydrochloride	366701
Propylthiouracil	51525
Quazepam	36735225
Resmethrin	10453868
Retinol/retinyl esters, when in daily dosages in excess of 10,000 IU or 3,000 retinol equivalents. (NOTE: Retinol/retinyl esters are required and essential for maintenance of normal reproductive function. The recommended daily level during pregnancy is 8,000 IU.)	
Ribavirin	36791045

Secobarbital sodium	309433
Streptomycin sulfate	3810740
Tamoxifen citrate	54965241
Temazepam	846504
Teniposide	29767202
Testosterone cypionate	58208
Testosterone enanthate	315377
2,3,7,8-Tetrachlorodibenzo-para-dioxin (TCDD)	1746016
Tetracycline (internal use)	60548
Tetracyclines (internal use)	--------
Tetracycline hydrochloride (internal use)	64755
Thalidomide	50351
Thioguanine	154427
Tobacco smoke (primary)	--------
Tobramycin sulfate	49842071
Toluene	108883
Triazolam	28911015
Trilostane	13647353
Trimethadione	127480
Trimetrexate glucuronate	82952645
Uracil mustard	66751
Urethane	51796
Urofollitropin	26995915
Valproate (Valproic acid)	99661
Vinblastine sulfate	143679
Vinclozolin	50471448
Vincristine sulfate	2068782
Warfarin	81812

Table 4.3C Female reproductive toxicity

Aminopterin	54626
Amiodarone hydrochloride	19774824
Anabolic steroids	--------
Aspirin (NOTE: It is especially important not to use aspirin duning the last three months of pregnancy, unless specifically directed to do so by by a physician because it may cause problems in the unborn child or complications during delivery.)	50782
Carbon disulfide	75150
Clobetasol propionate	25122467
Cocaine	50362
Cyclophosphamide (anhydrous)	50180
Cyclophosphamide (hydrated)	6055192
o,p'-DDT	789026
p,p'-DDT	50293
Ethylene oxide	75218
Flunisolide	3385033
Goserelin acetate	65807025
Lead	--------
Leuprolide acetate	74381536
Levonorgestrel implants	797637

Oxydemeton methyl	301122
Paclitaxel	33069624
Tobacco smoke (primary)	--------
Uracil mustard	66751

Table 4.3D Male Reprodroductive toxicity

Amiodarone hydrochloride	19774824
Anabolic steroids	-------
Benomyl	17804352
Benzene	71432
Cadmium	--------
Carbon disulfide	76150
Colchicine	64866
Cyclohexanol	108930
Cyclophosphamide (anhydrous)	50180
Cyclophosphamide (hydrated)	6055192
o,p'-DDT	769026
p,p'-DDT	50293
1,2-Dibromo-3-chloropropane (DBCP)	96128
m-Dinitrobenzene	99650
o-Dinitrobenzene	528290
p-Dinitrobenzene	100254
Dinoseb	88857
Epichlorohydrin	106898
Ethylene dibromide	106934
Ethylene glycol monoethyl ether	110805
Ethylene glycol monomethyl ether	109864
Ethylene glycol monoethyl ether acetate	111159
Ethylene glycol monomethyl ether acetate	110496
Ganciclovir sodium	82410320
Goserelin acetate	65807025
Hexamethylphosphoramide	680319
Lead	--------
Leuprolide acetate	74381536
Nitrofurantoin	67209
Oxydemeton methyl	301122
Paclitaxel	33069624
Sodium fluoroacetate	62748
Tobacco smoke (primary)	--------
Uracil mustard	66751

purpose of the Laboratory Safety Standard or the Hazard Communication Standard. However, the only carcinogens specifically regulated as such in 29 CFR Part 1910, Subpart Z are those for which individual regulatory standards have been issued. Materials currently regulated as carcinogenic by OSHA are given in Table 4.2.

In addition to the list in Table 4.2, a so-called "California" list of chemicals has been developed as a result of passage of Proposition 65 in California in 1986, which requires that the governor of the state publish each year a list of chemicals known to cause cancer or reproductive toxicity. This is the broadest list of suspect carcinogens and chemicals suspected of affecting reproduction established by any governmental entity. The California list as of

November 6, 1998, is reproduced in Table 4.3 which was given on the preceding pages. This list, represents only information, not a regulation but it can serve as a source of information for individuals concerned about their health or possible reproductive problems.

In the table, there are large variations in the degree to which they exhibit the properties which cause them to be included in the tables. In some cases, there are positive benefits which would have to be considered in individual uses or exposure situations. Inclusion in the Table should be caused as a flag to cause an individual to carefully evaluate situations where an exposure might occur and weigh all the pros and cons. One should also keep in mind the emotional content which the word carcinogen evokes, or the sensitivity of a couple expecting a child or trying to become parents.

5. Explosives

A substantial amount of laboratory research involves materials considered, in the legal sense of the term, as explosives rather than simply chemicals which can explode under appropriate conditions. The term "explosive" in this relatively narrow sense is defined as any material determined to be within the scope of Title 18, United States Code, Chapter 40, "Importation; Manufacture, Distribution and Storage of Explosive Materials," and any material classified as an explosive by the Department of Transportation in the Hazardous Material regulations (Title 49 CFR, Parts 100-199). A list of the materials that are within the scope of Title 18, United States Code, Chapter 40 is published periodically by the Bureau of Alcohol, Tobacco and Firearms, U.S. Department of the Treasury.

Classification of explosives by the U.S. Department of Transportation, Title 49 CFR Chapter I is as follows (the wording is that used by OSHA in Title 29, CFR Part 1910.109 (a)(3)):

1. Class A - "Possessing detonating or otherwise maximum hazard, such as dynamite, nitroglycerin, picric acid, lead azide, fulminate of mercury, black powder, blasting powder, blasting caps, and detonating primers."
2. Class B - "Possessing flammable hazard, such as propellant explosives (including some smokeless propellants), photographic flash powders, and some special fire-works."
3. Class C - "Includes certain types of manufactured articles which contain Class A or Class B explosives, or both, as components but in restricted quantities."
4. Forbidden or Not Acceptable Explosives - "Explosives which are forbidden or not acceptable for transportation by common carriers by rail freight, rail express, highway, or water in accordance with the regulations of the U.S. Department of Transportation, 49 CFR, Chapter 1."

Some activities involving explosives require a federal license or permit under Title XI, 18 United States Code, Chapter 40. Those activities not covered in these regulations are covered by NFPA 495, in jurisdictions where the latter standard has been adopted as a legal requirement. Under NFPA 495, no explosive materials shall be sold or transferred in any way to a person without a valid permit to have them, and no one is to conduct any operations involving explosives without an appropriate permit. Laboratories that are engaged in research with an explosive would require a "Permit to Use" under NFPA 495, unless test blasts are involved, in which case an additional "Permit to Blast" would be required.

The OSHA standards are based on the 1970 version of NFPA 495 and differ in some respects from the current version. However, they still require stringent safety precautions which must be followed by any research facility employing or investigating explosive materials. In consequence, any acquisition of explosive materials should be internally reviewed to ensure the ability on the part of the intended recipient to provide adequate facilities and safeguards to comply with the standards.

6. Equipment

In Chapter 3, a large number of items of equipment were discussed in terms of the requirements which should be met in order to ensure that the equipment meets appropriate *safety* standards. These standards change as technological improvements occur, as new information becomes available, and as new or revised regulatory standards become effective. Acquisition of many items of equipment affecting safe operations of a facility should be internally reviewed, in part to ensure that the equipment being acquired meets current safety specifications, and in part to alert units within the organization that need to know of the purchase. In the latter case, it may be necessary to confirm that sufficient attention has been given to the installation. Among those items of equipment which need to be routinely reviewed, usually by the safety department, sometimes by Facilities and Maintenance, and if separate, Planning and Engineering, are

- Biological safety cabinets
- Chemical fume hoods
- Electron microscopes
- X-ray equipment
- Equipment containing radiation sources
- Lasers
- Safety equipment
- Water stills
- Refrigeration equipment

E. Free Materials

It is tempting to accept free materials. However, with the advent of regulations providing for the safe and environmentally responsible disposal of hazardous chemicals under the Resource Conservation and Recovery Act (RCRA), the acceptance of free materials poses the risk that the "free" material may eventually generate unexpected and substantial disposal costs. In many cases, however, where the study of an experimental material is involved, while free materials are the basis for the intended research program, the amount proffered often is substantially more than is actually needed. The excess amount is typically unusable for other purposes and will require disposal by expensive legal means. Unfortunately, a few organizations have shifted disposal costs to the recipient organization by giving inexperienced persons unwanted surplus or out-of-date materials. Even if this is not the case, when normal purchasing procedures are bypassed, it could increase the possibility of unrecorded dangerous materials arriving on site. All of these negative possibilities can be avoided by the adoption of a local policy similar to the following, governing the receipt of free materials:

1. The amount of free material which may be accepted by an individual, laboratory, or other administrative unit to be used in a program of research must be limited to the amount which is likely to be actually needed in the proposed program.
2. The donor must agree in writing to accept the return of any unused amounts or pay for the legal and safe disposal of the material. The recipient may agree to waive this requirement if they are prepared to pay for the disposal from their own funds.
3. If the utilization or storage of the free material is likely to pose any substantive risk to personnel or property, the safety department and the risk management (insurance) office must be informed prior to completion of the agreement to accept the material, to allow time for a determination of whether the risks are acceptable or not, if adequate facilities are available, and to ensure that the proposed research has been reviewed for health and safety implications.

In some cases, undesirable materials may be a part of or within an item of equipment. In one example, a University engineering department accepted a large "free" transformer, which eventually turned out to be unusable due to a leak. The transformer contained nearly 500 gallons of high-purity polychlorinated biphenyl (PCB) insulating oil. The total disposal cost of the transformer and its contents was well over $10,000 because of the PCBs. Some states have passed laws or published policies requiring any state agency to audit any property the agency might acquire, either as a purchase or as a gift, for evidence of any prior hazardous waste dumping.

In another case, a commercial chemical firm gave over 5,000 containers of obsolete chemicals to a small research facility, virtually all of which turned out to be unusable. The eventual cost of disposal was prohibitively high and far outweighed the value of the usable materials.

An agricultural department at a university routinely received free chemicals for use in experimental test programs. More often, they only needed a relatively small fraction of the amount provided, and the rest was put into storage. When the accumulated surplus was finally discovered, the disposal cost was over $11,000.

Occasionally, the material is not wholly free but is provided at a very nominal cost. In one such case, several hundred drums of a "fertilizer" material, was bought at a cost of approximately $1,200. After it had become the property of the new owner, they were cited by a regulatory agency and were forced to pay a disposal cost of over $100,000.

These policies may appear unnecessarily formal to many individuals. However, as the examples given above clearly show, the cost of disposing of hazardous materials and the rate of increase of these costs have reached the point that hazardous waste disposal has become a major problem for many corporations and institutions. Most cannot afford to incur further costs by accepting free materials without consideration of the future obligations which the gift may engender. In the following references, where referring to a regulation or standard, always refer to the latest version.

REFERENCES

1. **Bilsom, R.E.**, Torts Among the Ivy: Some Aspects of the Civil Liability of Universities, University of Saskatchewan, Saskatoon, Saskatchewan, Canada.

2. **Gerlovich, J.A. and Downs, G.E.**, Eds., Legal liability, in *Better Science Through Safety,* Iowa State University Press, Ames, IA, 1981, 17.

3. **Kaufman, J.,** Laboratory Safety Workshop, Curry College, Milton, MA.

4. Occupational Safety and Health Administration, Hazard Communication; Final Rule 29 CFR Parts 1910, 1915, 1917, 1918, 1926, and 1928, *Fed Reg., 52(163),* 31852, 1987.

5. Occupational Health Services, New York, NY.

6. Nuclear Regulatory Commission, Title 10, Code of Federal Regulations; Parts 20, 30 and 33, Washington, D.C., 1988.

7. Food and Drug Administration, Title 21, Part 1300, Washington, D.C.

8. U.S. Public Health Service, Title 42, Parts 71.156 and 72, Washington, D.C.

9. Department of Transportation, Title 49, Code of Federal Regulations, Sections 173.386 to 173.388, Washington, D.C.

10. Occupational Safety and Health Administration, Title 29, Code of Federal Regulations, Subpart Z, Washington, D.C.

11. Title 18, United States Code, Chapter 40, Importation, Manufacture, Distribution, and Storage of Explosive Material, 1988.

12. Department of Transportation, Title 49, Code of Federal Regulations, Parts 100 to 199, Washington, D.C., 1988.

13. Occupational Safety and Health Administration, Title 29, Code of Federal Regulations, Part 1910, §109, Washington, D.C., 1988.

14. Code for the Manufacture, Transportation, Storage and Use of Explosive Materials, NFPA 495, National Fire Protection Association, Quincy, MA, 1996.

15. California Code of Regulations, Title 22, Section 12000, Safe Drinking Water and Toxic Enforcement of 1986, Chemicals Known to the State to Cause Cancer or Reproductive Toxicity. Also, on the Internet at http://members.aol.com/calprop65/prop65.html

IV. PURCHASING OF ANIMALS[*]

A. Introduction

Nothing can sabotage an animal-related experiment more quickly than the use of animals with latent (hidden) or overt signs of disease. The use of healthy unstressed animals is critical to obtaining quality experimental results. The following information will aid in the selection of appropriate vendors and provide criteria for selecting "clean" animals.

1. Selection Criteria for Rodents and Rabbits

Animals purchased from commercial vendors could have latent bacterial, viral, or parasitic diseases which can radically affect experimental results.[1-3] To avoid the use of infected ("dirty") animals as well as the contamination and subsequent infection of "clean" animals in a facility, selection criteria for the purchase of animals should be established.

Commercial vendors should be required to submit copies of their monthly or semiannual quality assurance health testing reports for review prior to purchase. Testing includes serological evaluation for the presence/absence of microbial pathogens (disease-causing bacteria), and evaluation for internal and external parasites. Small[4] lists the organisms which should *not* be present in mice, rats, hamsters, guinea pigs, and rabbits prior to purchase.

Since the vendor animal health report represents a snapshot in time and does not guarantee avoidance of subsequent contamination in the vendor's facility or during transport to the research facility, it would be wise to test a representative sample of the animals after receipt while they are housed in the quarantine area. Should serological testing reveal a latent viral infection, the entire group of animals should be destroyed and the quarantine area thoroughly disinfected. Studies using inbred strains of rats and mice can be decimated by genetic impurity. Vendors also test inbred strains for genetic purity, and a copy of their most recent test report can be requested prior to animal purchase.

2. Laws Affecting Animal Purchasing

The Federal Animal Welfare Act (PL 89-544) governs the purchase of dogs and cats for use in research and teaching. Under this law, these species may be purchased only from (1) a dealer licensed with the U.S. Department of Agriculture Animal/Plant Health inspection Service (USDA/APHIS), (2) a commercial breeder licensed with USPA/APHIS, (3) local or county animal shelters (recognizing that, at this time, a number of states prohibit the sale of pound animals for research or teaching), or (4) another research institution which has obtained the animals from any of these sources.

Research facilities must keep the following records on dogs and cats, usually on the

---

[*] This section was written by Dr. David M. Moore, D.V.M., University Veterinarian for Virginia Polytechnic Institute and State University.

USDA Individual Health Certificate and Identification Form (VS Form 18-1):

1. The name, address, and license number of the dealer from whom the animal was purchased.
2. The date of acquisition of each live dog and cat.
3. The official USDA tag number or tattoo assigned to each animal.
4. A description that includes species, sex, date of birth or approximate age, color and distinctive markings, and breed or type.
5. Any identification number assigned to that animal by the research facility.

3. Transportation of Animals

Transport of animals from the vendor to the research facility is also regulated by the Animal Welfare Act. Regulations on ambient temperature limits might prohibit shipment of animals on days that are too cold or too hot. Since most shipments are by common carrier (airplane, truck), one cannot be sure that animals were not subject to environmental stressors (heat or cold) while in loading areas or during transport. Heat stress can cause debilitation or death in rodents and other species. Additionally, animal transport boxes from several vendors might be shipped together, with the potential for cross-contamination of "clean" animals by "dirty" animals during shipment. Some vendors ship animals in their own environmentally controlled vehicles to preclude this problem, but this service is not available to all parts of the country.

4. Additional Laws Affecting Animal Purchase

Some states, such as California, prohibit entry of certain species (i.e., gerbils, ferrets) into the state without appropriate approval forms. Contact the state veterinarian in your state to ascertain whether similar regulations/restrictions exist.

REFERENCES

1. **Bhatt, P.,** Virus infections of laboratory rodents, *Lab. Anim.,* 9(3), 43, 1980.
2. **Orcutt, R.P.,** Bacterial diseases: agents, pathology, diagnosis, and effects on research, *Lab. Anim.,* 9(3), 28, 1980.
3. **Hsu, C.K.,** Parasitic diseases: how to monitor them and their effects on research, *Lab. Anim.,* 9(3), 48, 1980.
4. **Small, J.D.,** Rodent and lagomorph health surveillance-quality assurance, in *Laboratory Animal Medicine,* Fox, J.G. et al., Eds., Academic Press, Orlando, FL, 1984, p. 709.
5. **Smith, K.P., Hoffman, H.A., and Crowell, J.S.,** Genetic quality control in inbred strains of laboratory rodents, *Lab. Anim.,* 11(7), 16, 1982.
6. **Weisbroth, S.H.,** The impact of infectious disease on rodent genetic stocks, *Lab. Anim.,*13(1), 25, 1984.

V. STORAGE

Laboratory storage practices may enhance or diminish overall laboratory safety. There are many factors to be considered in addition to those concerned with flammable materials, briefly touched upon in Chapter 3 in the design and selection of facilities and equipment for those specific substances. Among these are the amount, location, and organization of the stored chemicals. The types of vessels in which they are contained and the information on the container labels are important. Some types of materials represent special hazards for which specific protective measures may be indicated or perhaps are mandatory due to regulations. The following sections will address these topics.

A. Compatible Chemical Storage

Many laboratories, if not the majority, find it convenient to store a large portion of their chemicals alphabetically although there are often partial exceptions, even in these facilities. The most frequently used flammable liquids, for example, are frequently placed in a common area. Other less frequently used flammable liquids may still be stored on shelves with other chemicals. Comparable usage may occur for acids, bases or other heavily used reagents. For the most part, however, general chemical storage in a laboratory is often done without regard to compatibility, although virtually everyone agrees this is not a good practice. Determination of compatibility does require some effort, as reflected by Table 4.4, which represents a composite of a number of lists of common incompatible chemicals, compiled from Material Safety Data sheets. Obviously, this is not a comprehensive list but does include a substantial number of combinations. A user, when in doubt should check the MSDS sheets for questionable combinations.

There are longer and more detailed lists of compatible chemicals which could be used to determine appropriate storage, but even the system represented by Table 4.4 may be too elaborate to encourage individuals to use it. At least one facility is in the process of developing an incompatibility list with varying degrees of incompatibility. Experience in many areas of safety has demonstrated that to be effective, systems must be kept as simple to implement as possible. In recognition of this, several of the major chemical firms have developed less complicated systems using a color code to define the groups which should be stored together. Unfortunately, although there are some similarities, the schemes of the different companies are not wholly compatible. The color code systems of two major chemical companies are shown in Figure 4.2, for comparison with each other and with Table 4.4. The color codes used to define the major groups for chemical vendors are prominently incorporated into the label. This makes it easy to decide where to place a new container and return it after use. Although providing less selectivity in segregating different materials than a system listing individual chemicals that should not be stored together, the ease of use should make color coding acceptable to most laboratory managers and employees.

Both of these systems provide an alternative for the colorblind. In the Fisher system, the first letter of the color is displayed prominently in the color bar, and is spelled out for clarity. In addition, companies may also place internationally recognizable pictograms identifying various hazards on their labels. With all this information, compatible storage should be feasible.

Other firms have similar systems, which also vary slightly in detail. Usually they agree for chemicals coded red, blue, yellow, and white, but differ in the way they denote exceptions or indicate a material for which there are few or no storage problems. For example, the J.T. Baker Company, one of the pioneers in developing a color-coded storage system, uses striped colors to denote exceptions and orange to denote chemicals which may be stored in the general storage area. Some companies depend entirely on the recognizable pictograms for quick warnings, but these can be effective if used by the laboratory workers.

Until an accident occurs in which the contents of the different containers come into contact with each other, failure to store materials according to compatibility may have few, if any, repercussions. In some instances, mixing of spilled chemicals will result in no reaction or relatively nonviolent reactions which can be controlled easily but in other cases, the result could be the insidious release of deadly toxic fumes or perhaps a violent explosion or fire. In any event, failure to store chemicals according to their properties is too much of a risk to personnel, to property, and possibly even to the intellectual value of accumulated research data files that may represent the product of years of effort. It is inexpensive insurance to make the relatively modest effort to segregate chemicals according to a color code system, or even to follow a more complicated program using groups such as those defined in Table 4.4.

Table 4.4 Incompatible Chemical Combinations

Chemical	Incompatible for Chemical Storage
Acetic Acid	aldehyde, bases, carbonates, chromic acid, ethylene glycol, hydroxides, metals, oxidizers, perchloric acid, peroxides, permanganates, phosphates, xylene
Acetone	acids, e.g. concentrated nitric and sulfuric, amines, oxidizers, plastics
Acetylene	copper metal, halogens, mercury, potassium, silver, including their compounds, oxidizers
Alkalis	acids, carbon dioxides, chlorinated hydrocarbons, chromium, mercury, oxidizers, salt, sulfur, water
Anhydrous Ammonia	acids, aldehydes, amides, calcium hypochlorite, hydrogen fluoride, mercury, oxidizers, sulfur
Ammonium nitrate	acids, alkalis, chlorates, fine organic powders, metals, nitrates, oxidizers, sulfur
Aniline	acids, e.g. nitric, aluminum, dibenzoyl peroxide, hydrogen peroxide, oxidizers
Azides	acids, heavy metals, oxidizers
Bromine	acetaldehyde, acetylene, alcohols, alkalis, amines, butadiene, butane, ethylene, fluorine, hydrogen, ketones, metals (finely divided), sodium carbide, sulfur, turpentine
Calcium oxide	acids, ethanol, fluorine
Carbon (activated)	alkalis, all oxidizing agents, calcium hypochlorite, halogens
Carbon tetrachloride	benzoyl peroxides, ethylene, fluorine, oxygen, silanes
Chlorates	acids, ammonium salts, carbon, metal powders, sulfur, finely divided combustibles and organics
Chromic acid	acetic acid, acetone, alcohols, alkalis, ammonia, bases, camphor, flammable liquids, glycerine, turpentine
Chromium trioxide	benzene, phosphorus, hydrocarbons, metals, other organics,
Chlorine	acetylene, ammonia, benzene, butadiene, ethylene, hydrazine, hydrogen, hydrogen peroxide, iodine, sodium hydroxide, turpentine, other petroleum components, finely powdered metals
Chlorine dioxide	ammonia, hydrogen, hydrogen sulfide, mercury, methane, phosphine, phosphorus, potassium hydroxide
Copper	acetylene, calcium, hydrogen peroxide, oxidizers
Cyanides	acids, alkalis, strong bases
Flammable liquids	ammonium nitrate, chromic acid, hydrogen peroxide, nitric acid, sodium peroxide, halogens
Fluorine	ammonia, halocarbons, halogens, ketones, metals, organic acids, hydrocarbons, other combustible material
Hydrocarbons	acids, bases, oxidizers
Hydrofluoric acid	glass, organics, sodium
Hydrogen peroxide	acetylaldehyde, acetic acid, acetone, alcohols, aniline, carboxylic acid, flammable liquids, metals (or their salts), nitric acid, nitromethane, organics, phosphorus, sodium, sulfuric acid
Hydrogen sulfide	acetylaldehyde, oxidizers, e.g. fuming nitric acid, oxidizing gases, sodium
Hypochlorites	acids, activated carbon
Iodine	acetaldehyde, acetylene, ammonia, hydrogen, sodium
Mercury	acetylene, aluminum, amines, ammonia, calcium, fulminic acid, lithium, oxidizers, sodium
Nitrates	sulfuric acid, other acids, nitrites
Nitric acid (concentrated)	acetic acid, acetonitrile, amines, ammonia, aniline, bases, benzene, brass, chromic acid, copper, cumene, flammable liquids and gases, formic acid, heavy metals, hydrogen sulfide, ketones, organic substances, sodium, toluene
Nitrites	acids
Nitroparaffins	amines, inorganic bases

Oxalic acid	mercury, oxidizers, silver, sodium chlorite
Oxygen	acetylaldehyde, alkalis, alkalines, ammonia, ammonia, carbon monoxide, ethers, flammable gases, liquids, solids, hydrocarbons, phophorus
Perchloric acid	acetic acid, acetic anhydride, alcohols, aniline, bismuth and bismuth alloys, combustible materials, dehydrating agents, ethyl benzene, hydriotic acid, hydrochloric acid, grease, iodides, ketones, other organic materials, oxidizers, pyridine
Peroxides, organic	acids (inorganic, organic)
Phosphorus	air, alkalis, oxygen, reducing agents
Potassium	acetylene, acids, alcohols, carbon dioxide, carbon tetrachloride, halogens, hydrazine, mercury, oxidizers, selenium, sulfur
Potassium chlorate	acids, e.g. sulfuric, ammonia, combustible materials, fluorine, hydrocarbons, metals, organic substances, sugars
Potassium perchlorate	acids, e.g. sulfuric, alcohols, combustible materials, fluorine, hydrazine, metals, organic materials, reducing agents
Potassium permanganate	benzaldehyde, ethylene glycol, glycerol, sulfuric acid
Selenides	reducing agents
Silver	acetylene, ammonia, ammonium compounds, fulminic acid, oxalic acid, oxidizers, ozonides, peroxyformic acid
Sodium	acids, carbon tetrachloride, carbon monoxide, hydrazines, metals, oxidizers, water
Sodium nitrate	acetic anhydride, acids, metals, organic matter, peroxyformic acid, reducing agents
Sodium nitrite	ammonium nitrate and ammonium salts
Sodium peroxide	acetic acid (glacial), acetic anhydride, benzene, benzaldehyde, carbon disulfide, ethyl acetate, furfural, glycerin, hydrogen sulfide metals, methyl acetate. peroxyformic acid, phosphorus
Sulfides	acids
Sulfuric acid	potassium chlorate, potassium perchlorate, potassium permanganate like compounds of sodium and lithium
Tellurides	reducing agents

B. Labeling

There are two important types of labels in laboratories. The labels on commercial containers are usually extremely comprehensive, providing not only information on the nature, amount, and quality of the product but also a very large amount of safety-related data. Typically a commercial label will readily meet the requirements of the hazard communication standard. On the other hand, labels placed on secondary containers in the laboratory by employees may be something such as "soln. A" or even less. This may be sufficient if all of the material is to be promptly used by the individual placing the label on the container, but otherwise it is not. In most instances, secondary containers of hazardous chemicals should be marked with labels identifying the chemical in the container and providing basic hazard warnings. The secondary label should be affixed before the container is put into use.

Although, as noted earlier, the laboratory standard preempts the hazard communication standard for laboratory employees, in order to provide equivalent protection it is difficult to imagine an alternative labeling procedure which would be equivalent without meeting the requirements of the hazard communication standard. Under the latter standard, the container label containing the hazard information must be in English as the primary language. As long as an English version is on the label, the same information may be provided in other languages as well to meet the needs of the personnel in the area. The intent of the labeling requirements under the hazard communication standard is primarily to protect the immediate users of the material by ensuring that they have access to the identity of the material with

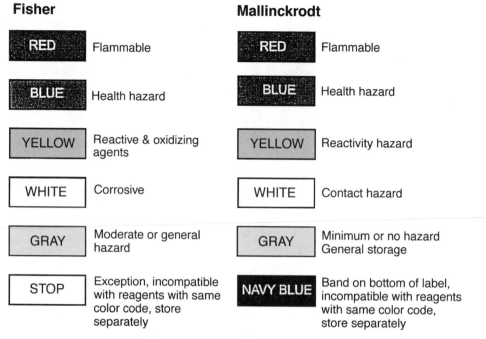

Figure 4.2   Sample hazard color codes for chemical containers.

which they are working. However, superficial and uninformative labels cause a major problem for the legal disposal of containers possibly holding hazardous chemicals. It is expensive to dispose of unknown materials under RCRA rules. This is a serious problem in in academia, due to the rapid turnover in research personnel, especially graduate students as they finish their degree programs. Every graduate student should be required to participate in a safety orientation course of at least three hours (preferably more) where this and other topics are carefully explained. Full compliance with the hazard communication standard should greatly alleviate this problem.

Laboratory employees should familiarize themselves with the commercial labels on the chemical containers that they are using. The labels will reflect the information available to the manufacturer at the time the material was packaged. The information will not be as current as that provided by an up-to-date MSDS. These are revised as often as significant new information becomes available. However, in most cases the information for a specific chemical will change sufficiently slowly so that the information on the label may be used with considerable confidence.

An examination of typical chemical labels will reveal that the following safety-related information is included on the labels of virtually all manufacturers:

- Name of compound
- GAS number
- Impurities, other components
- Flash point (if applicable)
- NFPA hazard diamond
- Risk descriptor (Danger, Warning, Caution)

- Risk descriptive statement
- Storage color code
- Handling advice
- Recommended fire extinguisher class
- First aid emergency medical advice
- UN number

Many labels contain additional useful information. The following items also are found frequently on chemical container labels:

- DOT symbols
- Hazard ratings (may be different from NFPA ratings for laboratory applications)
- Recommended protective equipment (some do this in words, some by means of stylized pictographs)
- Target organs
- Symptoms of exposure
- Spill response procedures
- Location for dating receipt of material
- Bar code
- Space for local information

C. Regulated Materials

Many regulated materials require special storage facilities, primarily for security reasons, although the security requirements frequently are based upon safety or environmental concerns. The following materials were discussed in Chapter 4, Section III.D concerning restrictions on their purchase. This list is approximately in order of the security required, although there is considerable overlap.

- Explosives
- Controlled substances (drugs)
- Radioisotopes
- Etiologic agents
- Carcinogens

1. Explosives

Explosives pose the most immediate danger to individuals of any of the items on the list in the preceding section, and those planning to use them definitely should receive appropriate training in how to use them safely. OSHA covers the storage requirements for all Class A, B, and C explosives, any special industrial explosives, and any newly developed and hence un-classified explosives in 29 CFR Part 1910.109(c). All of these materials must be kept in magazines which meet the specified requirements in the same section. As noted earlier, the OSHA standard is based on a 1970 version of NFPA Standard 495, although it does include some amendments added in 1978. Where the OSHA requirements do not cover a specific point, the most recent version of NFPA 495 should be consulted. Some state and local jurisdictions may have adopted later versions as part of their codes or may have added special regulations of their own.

The OSHA storage requirements differentiate between storage of less than 50 pounds ($\approx$22.7 kg) and more than 50 pounds. Some large research projects may require more than 50 pounds to be available at a given time, but most programs typically involve much smaller amounts, often on the order of a few grams. Class 1 magazines which are required for the larger quantities have structural requirements appropriate to room size spaces, while Class 2 magazines, those appropriate for smaller operations, may be mounted on wheels and are mobile. This section will be limited to a discussion of Class 2 magazines, which are the most suitable for typical laboratory-scale use of explosives.

Normally, explosive magazines are expected to be located outdoors. However, a Class 2 magazine may be permitted within a building (the OSHA standard specifically references warehouses and wholesale and retail establishments) when it is located on a floor which has an entrance at outside grade level and the magazine is not located more than 10 feet from the entrance. Also, it is not normally expected to have two magazines within the same building, but if one is used solely to store no more than 5000 blasting caps and it is at least 10 feet from the other, it is permissible to do so.

Class 2 magazines may be constructed primarily of wood or metal and will normally be a combination of the two. The wood of the bottom, sides, and cover of a primarily wooden magazine must be of 2-inch hardwood. The corners must be well braced and the magazine must be covered with sheet metal of not less than 20-gauge thickness. In order to avoid contact of the stored explosive with metal, any exposed nails in the interior of the magazine must be well countersunk.

Primarily metal magazines must have their bottom, top, and cover constructed of sheet metal, and lined on the interior with 3/8 inch plywood or the equivalent. The edges of the metal covers must overlap the sides by at least 1 inch. Both metal and wood magazines must be lockable. The covers must be attached securely with substantial strap hinges.

Class 2 magazines must be painted red and labeled on each of the four sides and the top with white letters at least 3 inches high:

EXPLOSIVES—KEEP FIRE AWAY

When Class 2 magazines are kept inside a building, they must be equipped with substantial wheels or casters to make it possible to remove them from the building in the event of a fire. When necessary due to climate, Class 2 magazines must be ventilated.

The general OSHA standards for the method of storage of explosives within magazines are generally intended to apply to larger units, but a number are appropriate for smaller Class 2 units. Among the more relevant are:

1. Packages must be stored lying flat with the top side up. Piles of packages or containers must be stable.
2. The oldest material of a given type of explosive must be used first to minimize the risks associated with instability upon aging.
3. Packing and unpacking of explosives must not be done within 50 feet of a magazine.
4. Except for metal tools to cut open fiberboard boxes, all other tools must be non-sparking.
5. The magazine must not be used for the storage of metal tools or for other general storage.
6. Smoking, matches, open flames, spark-producing devices, firearms (other than those in the possession of guards), and combustible materials must not be permitted within 50 feet of the magazine.
7. A competent individual must be charged with the care of the magazine at all times and is responsible for enforcement of all safety precautions. It is most important that access to the explosives in the magazine be limited to those who have demonstrated or documented experience in the safe use and handling of the explosive materials stored in the magazine. It is also important that an accurate record of the contents be maintained. This record should include an identification of each separate package in the magazine, when it was placed inside, the contents of the package, and the date and amounts of any materials that have been removed. The log should also identify individuals by name, not initials, who have been permitted to remove explosives. The log should be audited periodically by an independent person.

In addition to the materials specifically classified as explosives, there are many chemicals which, under appropriate conditions, can act as explosives, i.e., react or decompose very rapidly, accompanied by a large release of energy. The violence of an explosively rapid reaction is largely dependent upon the gas pressures produced in the reaction, enhanced by any thermal energies produced.

The best precaution in working with potential explosives is to minimize the amounts

Table 4.5 Highly Reactive Shock & Heat Sensitive Materials

Ammonium perchlorate	Dibenzoyl peroxide
Ammonium permanganate	Dilsopropyl peroxydicarbonate
Anhydrous perchloric acid	Dinitrobenzene (ortho)
Butyl hydroperoxide	Ethyl methyl ketone peroxide
Butyl perbenzoate	Ethyl nitrate
t-Butyl peroxyacetate	Hydroxylamine
t-Butyl peroxypivalate	Peroxyacetic acid
I -Chloro-2,4-dinitrobenzene	Picric acid
Cumene hydroperoxide	Trinitrobenzene
Diacetyl peroxide	Trinitrotoluene

actually present. In part, this may be achieved by maintaining proper inventory control in order to dispose of chemicals which tend to form unstable materials with age such as isopropyl ether, in which peroxides form, or perchloric acid or picric acid, which become dangerous if they are allowed to dry.

Other safety measures which should be taken in storing these laboratory materials are:

1. Keep the minimum quantities needed in a cool, dry area protected from heat and shock.
2. Potentially explosive materials should be segregated during storage from materials with which they could react, as well as flammables, corrosives, and other chemicals which are likely to interact with each other.
3. Potentially explosive: materials should be stored and used in an area posted with a sign stating in prominent letters:

CAUTION
POTENTIAL EXPLOSIVE HAZARD

4. If the material is being kept because of its potentially explosive properties, it should be treated as an explosive of the appropriate class and kept in a magazine or the equivalent.
5. Make sure that all occupants of the laboratory are aware of the potential risks and are trained in emergency procedures, including evacuation procedures, fire containment, and emergency first aid for physical injuries from an explosion.

As noted, many laboratory chemicals may cause an explosion under appropriate conditions. Table 4.5 represents a brief list of chemicals which have a reactivity rating of 3 or 4 (mostly 4) according to the NFPA No. 704M system for the identification of hazardous chemicals, and are sensitive to shock, heat, and/or friction. Materials on this list and those with comparable properties should always be treated with extreme care.

2. Controlled Substances (Drugs)

It was pointed out in Chapter 3, Section II.B.1.b.ii that the major concern where controlled substances are involved in research is security or control of the materials. This is particularly important for those materials which are usable by individuals for personal use or for sale as narcotics. The storage unit must be sufficiently strong to prevent forced entry for at least 10 minutes or more and must be either sufficiently heavy (750 pounds or more) or be rigidly bolted to the floor or wall to prevent the entire storage unit from being carried away. This is relatively easy to provide, but, as with explosives, it is equally important that a complete current record be maintained of all materials in the storage cabinet. The log should contain the following information: the date and amounts of each substance placed in the storage cabinet, and the dates and amounts disbursed from it. The amounts must be accurately quantified. The name, not, initials, of the person to whom the materials are issued should be recorded. Distribution of the material should be under either the direct control of the principal investigator identified as the holder of the license allowing possession and use of controlled substances, or an alternate designated by that person in writing.

Unfortunately, because human nature is involved, narcotics intended for research purposes are occasionally abused by those persons who may legally possess then. Provision should be made for a periodic audit of the utilization of controlled substances by each licensee to ensure that any such behavior be quickly detected, for the licensee's own protection and for that of the organization.

3. Radioisotopes

Radioactive byproduct materials are probably the most closely regulated research materials in wide use. It is rare for any organization involved to a significant degree in research, especially in the life sciences, not to have a license to use radioactive materials. The possession of the license and use of radioactive material under the provisions of the license commits the organization to establishing a radiation safety committee and to designate a radiation safety officer (RSO). Among the duties of this officer, under the terms of the license, are to make sure that each person authorized to have or use radioisotopes or sealed sources of radiation have no more than they are authorized to possess at one time, and that the total amount of each radioisotope does not exceed the overall limits provided for under the organization's license. While it was once the practice for individuals to have their own license, the task of managing perhaps as many as one hundred licenses was cumbersome for both the organization and for the NRC. It is now much more common for an organization to have a broad license for the entire organization.* It is necessary for each authorized user to make sure that all radioactive materials in their possession are securely stored. Normally, unlike controlled substances, most radioisotopes have no practical use outside the research laboratory so that the strength of the storage facilities is not comparably demanding, but whenever a legitimate user is not physically present, even for short intervals, the radioactive material must be under lock and key. Failure to provide this security would probably result in a serious citation if discovered by a NRC inspector, with the organization being fined. It is necessary to keep track of the materials used, although the type of use sometimes makes it difficult to do so with high accuracy. For example, if the use of ^{14}C results in generation of carbon dioxide, a somewhat uncertain amount may escape into the exhaust system in a fume hood, while some of the remainder may be retained within the experimental materials or another part in the apparatus.

Many radioactive materials used in the life sciences are incorporated in materials that require refrigeration to prolong their usefulness. Not only must a refrigerator or freezer used for radioactive storage be capable of being locked, but care must be taken to avoid contaminating the unit and its contents by spillage or leakage of material from a container due to freezing of the contents. In such a case, the lost material usually will be trapped in the ice or frost within a freezer. The trapped material could represent a personnel problem or the possibility of an uncontrolled release of radioactivity to the sanitary system when the freezer is defrosted. If this possibility exists, the defrost water should be collected and checked prior to disposal.

Not only must the experimental radioactive materials be stored properly, but all of the waste products which could contain any residual active materials must be kept and stored within the laboratory until they can be disposed of safely, usually by a radiation safety specialist. The

---

* Most larger organizations and many smaller ones have a "broad" license to use radioactive materials covering the entire organization, or at least that part of it at a single location. Although it is possible to have individual users directly licensed by the NRC, it is still required, in the latter case, to establish a radiation safety committee and safety officer to monitor the use of radioisotopes.

temporary storage of radioactive waste gives rise to the possibility of accidental removal of the radioactive waste as ordinary trash. Any trash containers containing radioactive waste should be distinctively marked, and custodial personnel should receive special training to recognize the containers and instructions not to combine the contents with ordinary trash. Laboratory person-nel must be sure to place radioactive waste and potentially contaminated waste in these special receptacles. Written labels or signs are not sufficient to prevent accidental losses of radioactive waste in this way. An estimate of the amount of activity going into the waste containers needs to be maintained. If, by chance, a waste can is disposed of inappropriately it will be necessary to make a reasonable estimate of the amount lost in order to judge the potential risk to the public. Of course, an effort would be expected of the organization to recover the lost radioactive material.

The inability to account for the radioactive material in the possession of a licensee is likely to be taken as a serious event by the NRC or its local surrogate in an agreement state. Each loss of radioactive material, if in such quantity and under such circumstances as to potentially pose a hazard to persons in unrestricted areas, must be reported immediately by telephone and telegraph to the Director of the NRC for the region in which the facility is located. More details on the reporting process will be provided in Chapter 5. Even if the amounts lost do not appear to pose a significant risk, operational procedures for the laboratory in question should be reviewed by the internal radiation safety committee, and the NRC notified as an item of information even if the amounts involved do not trigger required reporting levels. Operations of facilities where continuing problems of accountability occur, even for minor problems, should be carefully monitored because they can lead to a loss of credibility for the oversight program of the entire institution or commercial research laboratory. A pattern which appears to reflect poor governance of the facilities using radiation, with a subsequent failure of the radiation safety program to take prompt, effective, corrective actions has resulted in the NRC imposing substantial fines in recent years.

Another security issue that has been discussed during recent NRC inspections has been the question of ensuring that only authorized persons have reason to be in the facilities where radioactive materials are in use. Most radioisotope labs do not use radioactive materials in such quantities that they meet the requirements for restricted areas under 10 CFR Part 20. Usually when personnel are at work, the laboratory doors are not locked. It is not incorporated in the standards, but it has been made clear that it would be desirable for holders of NRC licenses to challenge the presence of any person not explicitly known to the employees as being in the area appropriately. In industry, most employees in a facility wear badges identifying them, but in academic institutions where most scientists also teach, such a policy would make it difficult for them to maintain free access to them by their students. NRC inspections are almost always unannounced, so all security and compliance measures must be current and ready to withstand an inspection at any time.

Many laboratories conduct operations which do not require the use of radioactive materials, in the same space as those in which radioactive materials are employed. In many of these instances, different personnel are employed in the two programs. Nonusers of radioisotopes should be made sufficiently aware of the procedures required for the safe, legal use of radiation so that they will neither inadvertently violate any safety requirements for the use of byproduct radioactive materials, nor misunderstand any actions of the employees involved with radiation. While the licensed users of radioactive materials are present, they can and must take precautions to avoid exposing the other persons in the laboratory unnecessarily to radiation, but when they are not present, the nonusers need to be aware of the areas where radioactive sources and waste are stored and areas they should avoid if there is any possibility of contamination. If proper security procedures are followed, the latter should theoretically not be possible. Any area containing radioactive materials should be clearly marked with signs bearing the radiation symbol and the label:

CAUTION
RADIOACTIVE MATERIALS

4. Etiologic Agents

Laboratories which employ etiologic agents in their research in addition to chemicals are different in a number of ways from ordinary chemical laboratories, primarily because the agents which are involved may be infectious to humans. There are data to support that there does exist a higher rate of incidence to the laboratory personnel of the diseases associated with the organisms which are involved with the research, but data also exist which show that there are virtually no secondary infections from the primary laboratory infections. There is a wide variation in the level of professionalism among the employees in laboratories employing biological agents, so it is extremely important that the laboratory manager and all employees set and maintain high standards of safety, as will be discussed in detail in Chapter 5.

There is less agreement in this area than most as to the actual risks posed by a number of operations and specific etiologic agents. In many cases, the scale of the operation determines the level of safety required, and in other cases there are other factors such as the availability of a safe, effective vaccine. The Centers for Disease Control (CDC) in Atlanta has published a safety guide to assist the laboratory director in making appropriate decisions as to the level of biosafety precautions required and has defined sets of standard practices, special practices, containment equipment, and laboratory equipment appropriate for four different levels, with increasingly stringent controls as the level goes from 1 to 4. A very slightly condensed version of the salient features of these guidelines will be found in Chapter 5, Section III. Note that the word "guidelines" was used. The CDC manual is a set of recommended practices that do not have the force of law; however, they do form a body of information which has been carefully reviewed by a large number of highly trained, experienced, professional individuals. It is strongly recommended that they be incorporated into a formal biosafety program at any academic or corporate research institution which has a substantial level of research in the life sciences.

Although a biosafety committee is normally not required by law, unlike for radiation safety, the creation of one to monitor the use of etiologic agents within an organization is highly desirable. For specific areas such as human subjects, recombinant DNA, and research involving the use of animals, regulatory committees are required. If a committee is provided with an appropriate charge, it can be of considerable help in expenditures to provide safety resources and services for the biological laboratories as well as providing a uniform set of safety and performance standards for the organization's laboratories.

In the context of this section, any biologically active materials which could result in human infections should be stored and used in such a way as to preclude or minimize the probability of infections for all laboratory employees, as well as any other employees, such as custodial workers, maintenance personnel, and visitors to the laboratory. This can best be done by limiting access to the facility to those who have a specific need to be present, by keeping materials in active use within the appropriate enclosure, by keeping materials not currently required in storage, and by decontaminating work surfaces frequently. Strict controls are recommended in the guidelines for the two most restrictive biological safety levels. Specific regulations are dictated by the OSHA bloodborne pathogen standard for exposures to human blood, tissue, mucus, and other bodily fluids. This standard is intended to control exposure to the HIV (AIDS) and hepatitis B viruses, but the same regulations could be used for any human pathogen. Due to an increase in tuberculosis as a secondary infection related to the spread of AIDS, it is likely that there will be additional regulations in which tuberculosis will be targeted. Infectious waste regulations also arise from the concern about AIDS but apply to any human pathogen. There will be separate sections later in this chapter on both the bloodborne pathogen and infectious waste issues.

5. Carcinogens

Cancer in its many forms is the second largest cause of death in the United States and most other developed countries. Anything which increases the likelihood of initiating the eventual onset of cancer therefore, is a matter of significant concern to those working with such materials. Although there are a few specific materials which have been known or suspected for some time to be carcinogenic agents, the long latency period for many materials, ranging from as few as 5 years to as many as 30, has delayed identification of causal relationships in many cases. OSHA has in the past regulated possible carcinogens on a case by case basis and continues to do so. Chemically specific regulations listed in Subpart Z of the OSHA regulations pertaining to exposure levels and medical surveillance requirements are not preempted by the laboratory safety standard. This is the basis of the list in Chapter 4, Section III.

Before a research program is begun using any of the currently regulated carcinogens, or any which may become regulated in the future by either the individual standard-setting process or the generic standard, there are a number of factors which should be considered:

1. Are there suitable alternative, noncarcinogenic materials which could be used?
2. Are there acceptable permissible exposure limits and ceiling limits below which it would be acceptable to use the material, with resultant exemptions from some restrictions upon the use of the material?
3. If the answer to question 2 is no, then can the provisions of the standard be met? An examination of the standards for the currently regulated carcinogens needs to be done prior to acquiring the material.

It would be inappropriate in this section, which is concerned primarily with storage (and security) of materials, to pursue the subject of research with carcinogens further at this time. However, unless there is a firm commitment to compliance with all regulations pertaining to carcinogenic research, it makes very little sense to acquire a carcinogenic chemical, especially if it is regulated or appears likely to become regulated, so that it cannot be used without complying with a perhaps onerous set of restrictions. The responsibility of assuring that it is not used and the potential liability if someone exposed to the material were to eventually develop cancer, would appear to preclude acquiring the material unless a definite research program exists of sufficient merit which would justify the risk and the compliance effort.

If a program mandating the use of a carcinogen does exist, the following recommendations should be considered as a basis for minimizing the risks to the employee. The materials should be kept in either a secure cabinet in sealed, unbreakable containers or in a sealed, protected system. A current inventory of the quantities held should be maintained. The containers should be marked:

DANGER, CONTAINS_____
CANCER HAZARD

Containers for a potential carcinogen in which the evidence of carcinogenicity is based on limited but suggestive evidence should be labeled:

POTENTIAL CANCER HAZARD

The laboratory standard has modified the area control requirements to simply stating that an area must be designated as the area in which the work is to be done, and suggests that this area could be as small as a fume hood. By also stipulating that the laboratory and the organization describe the additional protective measures needed to work with extremely dangerous materials, with carcinogens being used as an example of such a material, it is implied that special care needs to be taken by anyone in the area where this work is to be done. Therefore, it would not be inappropriate for these areas to be identified by signs, e.g.,

DANGER,_____
CANCER HAZARD
AUTHORIZED USERS ONLY

Under the laboratory safety standard, additional safety measures are expected to be provided to any employee working with carcinogens. As with other dangerous materials, they must be informed of the potential risk and procedures to be followed in an emergency. An emergency plan specific to the area and the research program must be developed in order to prevent exposures of the normal occupants of the area and to prevent the accidental exposures of others outside the immediate research facility. The training program should include everyone who may work within the designated area and handle the materials, including persons who may only stock the shelves and those individuals performing custodial and maintenance services, unless all such services are performed only with operations suspended and in the presence of a qualified laboratory employee who is fully trained.

REFERENCES

1.   *J. Hazardous Materials,* 1, 334, 1975.
2.   Allied Fisher Scientific, Fairlawn, NJ, 1986.
3.   LabGuard Safety Label System, Mallinckrodt, Paris, KY, 1986.
4.   Baker Saf-T-Data Guide, J.T. Baker Chemical Company, Phillipsburg, NJ, 1987.
5.   Chemical Labeling Standard, ANSI Z129. 1 (pending), American National Standards Institute, New York, 1987.
6.   Occupational Safety and Health Administration, General Industry Standards, Title 29, Code of Federal Regulations, Part 1910.109(c).
7.   Occupational Safety and Health Administration, General Industry Standards, Title 29, Code of Federal Regulations, Part 1910.1200.
8.   Occupational Safety and Health Administration, General Industry Standards, Title 29, Code of Federal Regulations, Part 1910.1450.
9.   Hazardous Chemical Data, NFPA 49, National Fire Protection Association, Quincy, MA.
10. Nuclear Regulatory Commission, Title 10, Code of Federal Regulations, Part 20, §207, Washington, D.C.
11. Biosafety in Microbiological and Biomedical Laboratories, HHS Publication No. 88-8395, Dept. of Health and Human Services, Washington, D.C., 1988.
12. **Favero, M.S.,** Biological hazards in the laboratory, in *Proc. Institute on Critical Issues in Health Laboratory Practices,* Richardson, J.H., Schoenfeld, E.S., Tulis, J.J., and Wagener, W. W., Eds., DuPont, Wilmington, DE, 1985.
13. **Pike, R.M.,** Laboratory-associated infections: incidence, fatalities, cases, and prevention, *Annu. Rev. Microbiol.,* 33, 41, 1979.
14. Prudent Practices in the Laboratory, Handling and Disposal of Chemicals, National Academy Press, p. 54, Washington, D.C., 1993.

INTERNET REFERENCES

1.   http:/www.c-f-c.com/charts/chemchart.htm
2.   http://research.bwh.harvard.edu/incompatchems.html
3.   http://response.restoration.noaa.gov/chemaids/react.html

D. Ethers

Ethers represent a class of materials which can become more dangerous with prolonged storage because they tend to form explosive peroxides with age. Exposure to light and air enhance the formation of the peroxides. A partially empty container increases the amount of air available, and hence the rate at which peroxides will form in the container. It is preferable, therefore, to use small containers which can be completely emptied, rather than take the amounts needed for immediate use from a larger container over a period of time, unless the rate of use is sufficiently high so that peroxides will have a minimal time in which to form.

Some of the following material is taken from the second edition of this handbook from an article by Norman V. Steere, "Control of Peroxides in Ethers." It has been edited to conform with the format of the current edition and has been added to from other sources. The sections on detection and estimation of peroxides and removal of peroxides have been substantially shortened, in line with the philosophy espoused elsewhere in this section to keep on hand only amounts that will be quickly used, and in order to reduce the risks in handling possibly contaminated materials.

Ethyl ether, isopropyl ether, tetrahydrofuran, and many other ethers tend to absorb and react with oxygen from the air to form unstable peroxides which may detonate with extreme violence when they become concentrated by evaporation or distillation, when combined with other compounds that give a detonatable mixture, or when disturbed by unusual heat, shock, or friction. Peroxides formed in compounds by autoxidation have caused many laboratory accidents, including unexpected explosions of the residue of solvents after distillation, and have caused a number of hazardous disposal operations. Some of the incidents of discovery and disposal of peroxides in ethers have been reported in the literature, some in personal communications, and some in the newspapers. An "empty" 250-cc bottle which had held ethyl ether exploded (without injury) when the ground glass stopper was replaced. Another explosion cost a graduate student the total sight of one eye and most of the sight of the other, and a third explosion killed a research chemist when he attempted to unscrew the cap from an old bottle of isopropyl ether.

Appropriate action to prevent injuries from peroxides in ethers depends on knowledge about formation, detection, and removal of peroxides, adequate labeling and inventory procedures, personal protective equipment, suitable disposal methods, and knowledge about formation, detection, and removal of peroxides.

1. Formation of Peroxides

Peroxides may form in freshly distilled and undistilled and unstablized ethers within less than 2 weeks, and it has been reported that peroxide formation began in tetrahydrofuran after 3 days and in ethyl ether after 8 days. Exposure to air, as in opened and partially emptied containers, accelerates the formation of peroxides in ethers, and while the effect of exposure to light does not seem to be fully understood, it is generally recommended that ethers which will form peroxides should be stored in full, air-tight, amber glass bottles, preferably in the dark.

Although ethyl ether is frequently stored under refrigeration, there is no evidence that refrigerated storage will prevent formation of peroxides, and leaks can result in explosive mixtures in refrigerators since the flash point of ethyl ether is -45°C (-49°F). The literature contains extensive information on autoxidation of ethyl ether.

The storage time required for peroxides as H_2O_2 to increase from 0.5 to 5 ppm has been reported to be less than 2 months for a tin-plate container, 6 months for an aluminum container, and over 17 months for a glass container. The same report stated that peroxide content was not appreciably accelerated at temperatures about 11°C (20°F) above room temperature. Davis has reported the formation of peroxides in olefins, aromatic and saturated hydrocarbons, and ethers particularly with initial formation of an alkyl hydro peroxide which can condense on standing or in the presence of a drying agent to yield further peroxidic products. Davis refers to reports that the hydroperoxides initially formed (e.g., from isopropyl ether and tetrahydrofuran) may condense further, particularly in the presence of drying agents, to give polymeric peroxides and that cyclic peroxides have been isolated from isopropyl ether.

Isopropyl ether seems unusually susceptible to peroxidation and there are reports that a half-filled 500-ml bottle of isopropyl ether peroxidized despite being kept over a wad of iron wool. Although it may be possible to stabilize isopropyl ether in other ways, the absence of a stabilizer may not always be obvious from the appearance of a sample, so that even opening a container of isopropyl of uncertain vintage to test for peroxides can be hazardous. Noller comments that "neither hydrogen peroxide, hydroperoxide nor the hydroxyalkyl peroxide are as violently explosive as the peroxidic residues from oxidized ether."

2. Detection and Estimation of Peroxides

Appreciable quantities of crystalline solids have been reported as gross evidence of formation of peroxides, and a case is known in which peroxides were evidenced by a quantity of viscous liquid in the bottom of the glass bottle of ether. If similar viscous liquids or crystalline solids are observed in ethers, no further tests are recommended, since in four disposals of such material, there were explosions when the bottles were broken.

Potassium Iodide method:

Add 1ml of a freshly-prepared 10% solution of potassium iodide to 10 ml of ethyl ether in a 25 ml glass-stoppered cylinder of colorless glass protected from light; when viewed transversely against a white background, no color is seen in either liquid. A resulting yellow color indicates the presence of 0.005% peroxides.

Ferrous Thiocyanate Detection Method

Prepare a solution of 5 ml of 15 ferrous ammonium sulfate, 0.5 ml of 1N sulfuric acid and 0.5 ml of 0.1 N ammonium thiocyanate (if necessary decolorize with a trace of zinc dust). Shake with an equal quantity of the solution to be tested. If peroxides are present, a red color will develop.

Test strips:

These test strips are available from EM Scientific, cat. No. 10011-1 or from Lab Safety Supply, cat. No. 1162. These strips quantify peroxides up to a concentration of 25 ppm. Aldrich Chemical has a peroxide test strip, cat. No. Z10,168-0, that measures up to 100 ppm peroxide. The actual concentration at which peroxides become hazardous is not specifically stated in the literature. A number of publications use 100 ppm as a control valve for managing the material safely.

3. Inhibition of Peroxides

No single method seems to be suitable for inhibiting formation in all types of ethers, although storage and handling under an inert atmosphere would be a generally useful precaution.

Some of the materials which have been used to stabilize ethers and inhibit formation of peroxides include the addition of 0.001% of hydroquinone or diphenylamine, polyhydroxyl-phenols, aminophenols, and arylamines. Addition of 0.0001 g of pyrogallol in 100 cc ether was reported to prevent peroxide formation over a period of 2 years. Water will not prevent the formation of peroxides in ethers, and iron, lead, and aluminum will not inhibit the peroxidation of isopropyl ether, although iron does act as an inhibitor in ethyl ether. Dowex- 1© has been reported effective for inhibiting peroxide formation in ethyl ether, 100 parts per million (ppm) of 1-naphthol for isopropyl ether, hydroquinone for tetrahydrofuran, and stannous chloride or ferrous sulfate for dioxane. Substituted stilbene-quinones have been patented as a stabilizer against oxidative deterioration of ethers and other compounds.

4. Removal of Peroxides

If a bottle of ether appears to contain dried crystals on its interior surfaces or the liquid appears to contain a slurry of crystals, no attempt should be made to remove the peroxides which have in all probability formed but instead the container should be <u>carefully</u> disposed of as a dangerous, shock sensitive material. Immediately notify the organization's hazardous waste disposal staff. If the material is reasonably fresh and does not show any sins of peroxides, then you may wish to follow the following procedures to remove any peroxides present. However, if one is talking about a small container, one should consider whether the financial savings would be worth the risk.

Reagents which have been used for removing hydroperoxides from solvents are reported to include sodium sulfite, sodium bisulfite, stannous chloride, lithium tetrahydroalanate (caution: use of this material has caused fires), zinc and acid, sodium and alcohol, copper-zinc couple, potassium permanganate, silver hydroxide, and lead dioxide.

Decomposition of ether peroxides with ferrous sulfate is a commonly used method; 40 g of 30% ferrous sulfate solution in water is added to each liter of solvent. Caution is indicated since the reaction may be vigorous if the solvent contains a high concentration of peroxide.

Reduction of alkylidene or dialkylperoxides is more difficult, but reduction by zinc dissolving in acetic or hydrochloric acid, sodium dissolving in alcohol, or the copper-zinc couple might be used for purifying solvents containing these peroxides.

Addition of 1 part of 23% sodium hydroxide to 10 parts of ethyl ether or tetrahydrofuran will remove peroxides completely after agitation for 30 minutes; sodium hydroxide pellets reduced but did not remove the peroxide contents of tetrahydrofuran after 2 days. Addition of 30% of chloroform to tetrahydrofuran inhibited peroxide formation until the eighth day with only slight change during 15 succeeding days of tests; although sodium hydroxide could not be added because it reacts violently with chloroform, the peroxides were removed by agitation with 1% aqueous sodium borohydride for 15 minutes (with no attempt made to measure temperature rise or evolution of hydrogen).

A simple method for removing peroxides from high-quality ether samples without need for distillation apparatus or appreciable loss of ether consists of percolating the solvent through a column of Dowex-1[©] ion exchange resin. A column of alumina was used to remove peroxides and traces of water from ethyl ether, butyl ether, dioxane, and petroleum fractions and for removing peroxides from tetrahydrofuran, decahydronapthalene(decalin), 1,2,3,4-tetrahydro-naphalene (tetralin), cumene and isopropyl ether.

Because they have a limited shelf life, as noted earlier, ethers should be bought in the smallest practicable containers appropriate to the rate of usage within the facility, preferably in 500 ml containers. In Chapter 4, Section III.A, it was acknowledged that buying in small sizes does invoke a significant financial penalty, but some of this price disadvantage can be eliminated by buying in case lots or having a central stock room buy in multiple case lots. It also eliminates much of the need for frequent checking of the conditions of the contents of a large container. No matter what size container is purchased, each container should be dated when it is received and placed in stock. The schedule given at the end of Table 4.6 for storage and disposal should be followed. Opened containers should be tested after 1 month and continue to be tested until emptied, or at frequent intervals. If only modest amounts of peroxides are found, the ethers can be decontaminated. If an alumina column is used, the contaminated alumina can be treated with an aqueous solution of ferrous sulfate and discarded as chemical waste. However, it may be difficult to get a hazardous waste disposal firm to accept it. They are very reluctant to accept any waste material if there is any possibility of an explosion. The concern about possible explosions, the cost of the manpower involved in the needed frequent checking, and the possibility of having to pay a premium price for disposal of excess materials as a potential explosive should obviate the argument of a lower unit cost for the larger sizes. The implementation of a computerized chemical tracking program, previously mentioned, allows this responsibility to be placed at the point of receipt rather than in the individual laboratory. The information can be used by those responsible for disposing of waste and surplus chemicals to retrieve these materials before a target date so that the commercial disposal firms will accept the materials for transportation.

An alternative management strategy if an organization wide computerized chemical

Table 4.6
A. Chemicals that form explosive levels of peroxides without concentration

Butadiene[a]	Divinylacetylene	Tetrafluoroethylene[a]
Chloroprene[a]	Isopropyl ether	Vinylidene chloride

B. Chemicals that form explosive levels of peroxides on concentration

Acetal	Diacetylene	2-Hexanol	2-Phenylethanol
Acetaldehyde	Dicyclopentadiene	Methylacetylene	2-Propanol
Benzyl alcohol	Diethyl ether	3-Methyl-1-butanol	Tetrahydroforan
2-Butanol	Diethylene glycoldimethyl ether	Methlcyclopentane	Tetrahyronaphthalene
Cumene	(diglyme)	Methyl isobutyl ketone	Vinyl ethers
Cyclohexanol	Dioxanes	4-Methyl-2-pentanol	Other secondary alcohols
2-Cycloexen-1-ol	Ethylene glycoldimethyl ether	2-Penten-1-0l	
Cyclohexene	glyme	4-Penten-1-ol	
Decahydro naphthalene	4-Heptanol	1-Phenylethanol	

C. Chemicals that may autopolymerize as a result of peroxide accumulation

Acrylic acid[b]	Methyl methacrylate[b]	Vinyl chloride
Acrylonitrile[b]	Styrene	Vinyl pyridine
Butadiene[c]	Tetrafluoroethylene[c]	Vinyladiene chloride
Chloroprene[c]	Vinyl acetate	
Chlorotrifluoroethylene	Vinyl acetylene	

D. Chemicals that may form peroxides but cannot clearly be placed in sections A-C

Acrolein	tert-Butyl methyl ether	Di(1-propynyll) ether[f]
Allyl ether[d]	n-Butyl phenyl ether	Di(2-propynyl) ether
Allyl ethyl ether	n-Butyl vinyl ether	Di-n-propoxymethane[d]
Allyl phenyl ether	Chloroacetaldehydediethyl acetal	1,2-Epoxy-3-isopropoxy propane[d]
p-(n-Amyloxy)benzoyl chloride	2-Chlorobutadiene	1,2-Epoxy-3-phenoxypropane
n-Amyl ether	1-(2-Chloroethoxy)-2-phen- oxyethane	Ethoxyacetophenone
Benzyl n-butyl ether[d]	Chloromethylene	1-(2-Ethoxyethyl)ethyl acetate
Benzyl ether[d]	Chloromethyl methyl ether[e]	2-Ethoxyethyl acetate
Benzyl ethyl ether[d]	§-chlorophenole	(2-Ethoxyethyl)-o-benzoyl benzoate
Benzyl methyl ether	o-Chlorophenetole	1-Ethoxynaphthalene
Benzyl 1-naphthyl ether[d]	p-Chlorophenetole	o,p-Ethoxyphenyl isocyanate
1,2-Bis(2-chloroethoxy) ethane	Cyclooctane[d]	1-Ethoxy-2-propyne
Bis(2-ethoxyethyl) ether	Cyclopropyl methyl ether	3-Ethoxyopropionitrile

Bis[2-(methoxyethoxy)ethyl] ether	Diallyl ether[d]	2-Ethylacrylaldehyde oxime
Bis(2-chloroethyl) ether	p-Di-n-butoxybenzene	2-Ethylbutanol
Bis(2-ethoxyethyl) adipate	1,2-Dibenzyloxyethane[d]	Ethyl §-ethoxypropionate
Bis(2-ethoxyethyl) phthalate	p-Dibenzyloxybenzene[d]	2-Ethylhexanal
Bis(2-methoxyethyl) carbonate	1,2 -Dichloroethyl ethyl ether	Ethyl vinyl ether
Bis(2-methoxyethyl) ether	2,4-Dichlorophenetole	Furan p-Phenylphenetone
Bis(2-methoxyethyl) phthalate	Diethoxymethane[d]	2,5-Hexadiyn-l-ol
Bis(2-methoxymethyl) adipate	2,2-Diethoxypropane	4,5- Hexadien-2-yn-1-ol
Bis(2-n-butoxyethyl) phthalate	Diethyl ethoxymethylene malonate	n-Hexyl ether
Bis(2-phenoxyethyl) ether	Diethyl fumarated	o,p-Iodophenetole
Bis(4-chlorobutyl) ether	Diethyl acetal isoamyl benzyl ether[d]	Sodium ethoxyacetylide[f]
Bis(chloromethyl) ether[e]	Diethylketene[f]	Isoamyl ether[d]
2-Bromethyl ethyl ether	m.o.p-Diethoxybenzene	Isobutyl vinyl ether
§-Bromophenetole	1,2-Diethoxyethane	Isophorone[d]
o-Bromophenetole	Dimethoxymethane[d]	p-Isopropoxypropionitrile[d]
p-Bromophenetole	1,1-Dimethoxymethane[d]	Isopropy-l 2,4,5-trichlorophen oxy acetate
3-Bromopropyl phenyl ether	Dimethylketene[f]	Limonene
1,3 Butadiyne	3,3-dimethoxypropene	1,5-p-Methadiene
Buten-3-yne	2,4-Dinitrophenetole	Methyl p-(n-amyloxy) benzoate
tert-Butyl ethyl ether	1,3-Dioxepne	

a. When stored as a liquid monomer.
b. Although these chemicals form peroxides, no explosions involving these monomers.
c. When stored in liquid form, these chemicals form explosive levels of peroxides without concentration. They may also be stored as a gas in gas cylinders. When stored as a gas, these chemicals may autopolymerize as a result of peroxide accumulation.
d. These chemicals easily form peroxides and should probably be considered under part B.
e. OSHA-regulated carcinogen.
f. Extremely reactive and unstable compound.

Safe Storage Period for Peroxide Forming Chemicals

Description	Period
Unopened chemicals from manufacture	18 months
Opened containers	
Chemicals in Part A	3 months
Chemicals in Parts B and D	12 months
Unihibited chemicals in Part C	24 hours
Inhibited chemicals in Part C	12 months[a]

a. Do not store under inert atmosphere, oxygen required for inhibitor to function.

Sources: Kelly, Richard J., Chemical Health & Safety, American Chemical Society, 1996, Sept, 28-36 Revised 12/97.

management program is in place is to identify areas where ethers are used and try to establish a sharing program so that the ethers purchased will be fully used by the global set of laboratories, prior to the expiration date for individual containers.

REFERENCES

1. **Douglas, I.B.,** *J. Chem. Educ.,* 40, 469, 1963.
2. **Steere, N.V.,** Control of hazards from peroxides in ethers, *J. Chem. Ed,* 41, A575,1964.
3. Accident Case History 603, Manufacturing Chemists Association (reported in part in Reference 2).
4. **Fleck, E.,** Merck, Sharp & Dobme Company memo, Rahway, NJ, May 11, 1960.
5. **Noller, C.R.,** *Chemistry of Organic Compounds,* W.B. Saunders, Philadelphia, 1951.
6. **Rosin, J.,** *Reagent Chemicals and Standards,* 4th ed., D Van Nostrand, Princeton, NJ, 1961.
7. Prudent Practices in the Handling and Disposal of Chemicals, National Academy Press, pp 54-55, Washington D.C., 1995
8. **Davies, A.G.,** Explosion hazards of autoxidized solvents, *J. R. Inst. Chem.,* 386, 1956.
9. **Lindgren, G.,** Autoxidation of diethyl ether and its inhibition by diphenylamine, *Acta Chir. Scand.,* 94, 110, 1946.
10. **Kelly, R.J.,** Review of safety guidelines for peroxidizable organic chemicals, *Chemical Health & Safety,* American Chemical Society, Sept/Oct 1996.
11. **Dugan, P.R.,** *Anal. Chem.,*33, 1630, 1961.
12. **Dugan, P.R.,** *Ind Eng. Chem.,* 56, 37, 1964.
13. **Feinstein, R.N.,** Simple method for removal of peroxides from diethyl ether, *J. Org. Chem.,* 24, 1172, 1969.
14. **Kirk, R.E. and Othmer, D.F.,** Eds., *Encyclopedia of Chemical Technology* Vol. 5, New York, 1950, 142, 871.
15. **Kirk, R.E. and Othmer D.F.,** Eds., *Encyclopedia of Chemical Technology,* Vol. 6, New York, 1950, 1006.
16. **Jones, D.G.,** British Patent 699,179, *Chem. Abstr.,* 49, 32, 1953.
17. **Moffett, R.B. and Aspergren, B.D.,** Tetrohydrofuran can cause fire when used as solvent for LiAlH, *Chem. Eng. News,* 32, 4328, 1954.
18. Chemical Safety Data Sheet - SD 29, Ethyl Ether, Manufacturing Chemists Association, Washington, D.C., 1956.
19. **Dasler, W. and Baner, C.D.,** Removal of peroxides from ether, *Md Eng. Chem. Anal. Ed,* 18, 52, 1946.
20. Manuals of Techniques, Beverly, MA, 1964.
21. **Ramsey, J.B. and Aldridge, F.T.,** Removal of peroxides from ethers with cerous hydroxide, *J. Am. Chem. Soc.,* 77, 2561, 1955.
22. **Steere, N.V.,** Control of peroxides in ethers, in *CRC Handbook of Laboratory Safety,* 2nd ed., Steere, N.V, Ed., CRC Press, Cleveland, OH, 1971, 250.
23. **Mateles, R.I.,** Cumenehydroperoxide explosion, Letter to the editor, *Chem. Eng. News.,*71(22), 4, 1993.
24. **Dussanlt, P.,** Potential hazard with 2-methoxyprop-2-yl hydroperoxide, Letter to the editor, *Chem. Eng. News.,* 71(32), 2, 1993.

INTERNET REFERENCES

1. http://www.orcbs.msu.edu/chemical/chp/appendix.html
2. http://offices.colgate.edu/chemmgt/CHP/n-peroxide.html
3. http://www.dec.utexas.edu/safety/l...ontents/appendices/appendix12.html
4. http://sbms.pnl.gov/amanuals/ma43/a043d880.htm

E. Perchloric Acid

Perchloric acid is used primarily in laboratories in the life sciences for digestions of organic materials. It is used less often than was once the case but, if used improperly, can represent substantial risks. Within the past two years, a small laboratory facility undergoing renovation of a perchloric acid using area was destroyed by a fire originating in a hood exhaust. Over several years, there have been many instances where explosions and fires have been attributed to perchloric acid reactions with organic materials and with shock or friction initiated incidents involving perchloric acid crystals.

In a typical perchloric acid MSDS, the following hazards or precautions to be taken associated with this material are usually listed:

1. It is highly corrosive to tissues. It can cause severe burns when in contact with the skin, eyes, respiratory tract and other parts of the body.
2. Cold perchloric acid at concentrations of 70% or less is not a strong oxidizing agent but as its temperature and concentration increase, its oxidizing power increases and it becomes a strong oxidizing agent. Because of this, perchloric acids are not sold commercially at concentrations above 72% by weight.
3. Anhydrous perchloric acid is unstable even at room temperatures and ultimately decomposes spontaneously with a violent explosion. Contact with oxidizable material such as many organics can cause an immediate explosion. Among these are alcohols, ketones, aldehydes, ethers, and dialkyl sulfoxides. Heavy metal perchlorates and organic perchlorates are very sensitive explosives.
4. Vapors from the evaporation of hot perchloric acid form crystals which are very shock sensitive. Fortunately, they are water soluble and perchloric acid used in a properly installed and operated perchloric hood can be used safely.

The following are listed among the causes of fires and explosions involving perchloric acid:

1. The instability of aqueous or of pure anhydrous perchloric acid under various conditions.
2. The dehydration of aqueous acid by contact with dehydrating agents such as concentrated sulfuric acid, phosphorous pentoxide, or acetic anhydride.
3. The reaction of perchloric acid with other substances, to form unstable materials.

Combustible materials, such as sawdust, excelsior, wood, paper, burlap bags, cotton waste, rags, grease, oil, and most organic compounds, contaminated with perchloric acid solution are highly flammable and dangerous. Such materials may explode on heating, in contact with flame or by impact or friction, or they may ignite spontaneously. Care must always be exercised in working in areas where perchloric acid has been used, even seemingly innocuous tasks may create just enough of an interaction with perchloric residue to create a problem.

1. Perchloric Acid Storage

Within the laboratory: The maximum advisable amount of acid stored in the main laboratory should be no more than two 8 pound (3.6 kg) bottles. A 450-gram (1 pound) bottle should be sufficient for individual use. Storage of perchloric acid should be in a fume hood set aside solely for perchloric acid use and stored on a ceramic or glass dish. The acid should be inspected monthly for discoloration; if any is noted, the acid should be discarded.

Outside of the laboratory, a perchloric acid container should be stored on a glass or ceramic dish on an epoxy-coated metal shelf, preferably in a metal cabinet away from organic materials and flammable compounds. Discolored acid should be discarded.

Storage of anhydrous perchloric acid is strongly discouraged. If stored for any significant length of time, on the order of 10 days or even less, it can degrade and spontaneously explode.

None of the furniture used with perchloric acid should be wood. The laboratory case work should have as few seams as possible in which perchloric acid could enter and then dry, forming sensitive crystals. Similarly, the floor should be a seamless epoxy coated floor and the casework should not be bolted to the floor, again to avoid cracks where perchloric acid vapors can collect.

Because of the risk of explosions, even small ones, due to small accumulations of crystals, solvents and other dangerous materials which could become involved in fires or secondary reactions, should not be stored with perchloric acid or in close proximity to areas where perchloric acid is used or stored.

REFERENCES[*]

1. **Everett K., Graf, F.A., Jr.,** *in CRC Handbook of Laboratory Safety, 2ⁿᵈ Ed.,* Steere, N.V., Ed., CRC Press, Cleveland, OH, 1971.

2. Prudent Practices in the Laboratory, Handling and Disposal of Chemicals, National Academy Press, pp 374-375, Washington D.C., 1995.

INTERNET REFERENCES

1. http://www.chem.utah.edu/MSDS/P/PERCHLORIC_ACID%2C_69-72%25
2. http://www.orcbs.msu.edu/chemical/safepercloricuse.html

F. Flammable Liquids

Should a fire or an explosion occur in a laboratory, a major concern is to reduce the amount of fuel available to support a fire. Many solvents commonly used in laboratories are highly flammable, and even a small quantity involved in the fire could have the capacity to significantly increase the probability of the fire spreading.

The OSHA General Industry Standard, in Section 1910.106(d)(3), provides restrictions on the maximum amounts of flammable liquids allowed to be stored, dependent upon class, in flammable material storage cabinets within a room, and defines in Section 1910.106(d)(2) the maximum size of individual containers for the various classes of flammables. NFPA Standard 45 provides guidelines as to the maximum amounts of flammable liquids that should be allowed in the three classes of facilities (A, B, and C) defined in that standard. This standard has been mentioned but the three laboratory classes have not been stressed in this volume, in favor of a four-level standard, paralleling legally defined classes of risk for biological facilities, and involving wider classes of hazards than that due to the amount of flammable materials in the laboratory alone. For the purposes of this section, the low- and moderate-risk facilities described in Chapter 3, Sections I.C.2.a and I.C.2.b, respectively, may be taken to be approximately equivalent to an NEPA Standard 45 Class C facility, a substantial-risk facility, Section I.C.2.c to be roughly equivalent to a Class B facility, and a high-risk facility, Section I.C.2.d to include a Class A facility, but with more restrictions than the latter would require. Note that OSHA has not adopted the restrictions of NFPA Standard 45 and does not address the issue of the total amount of flammables permitted in a laboratory area, although the amount permitted in an interior storage room is defined. The OSHA regulations regarding container sizes are based on sections of the 1969 version of the NFPA Standard 30. Tables 4.7 and 4.8 define the various classes of flammable and combustible liquids and the maximum container sizes permitted by OSHA for each class. Table 4.9 lists a number of Class IA solvents in common use in the laboratory.

The definitions depend upon the flashpoints, and in some cases, the boiling points of the liquids. The flashpoint of a liquid is legally defined in terms of specific test procedures used to determine it, but conceptually is the minimum temperature at which a liquid forms a vapor above its surface in sufficient concentration that it may be ignited.

Table 4.8 is equivalent to Table H-12 from the OSHA General Industry Standards. In Table 4.7, the first temperature is in degrees Celsius and the temperature in parentheses ( ) is the Fahrenheit temperature. Note that neither Combustible II or Combustible IIIA materials include mixtures in which more than 99% of the volume is made up of components with flashpoints of $93.3°C$ ($200°F$) or higher.

---

[*] The user should use the most recent MSDS accompanying the material, or one from a commercial service that is updated frequently.

Table 4.7 Definitions, Classes of Flammable and Combustible Liquids
°C (°F)

Class	Boiling Points	Flash Points
Flammable IA	<37.8 (100)	<22.8 (73)
Flammable lB	≥37.8 (100)	<22.8 (73)
Flammable IC		22.8 (73) ≤ and < 37.8 (100)
Combustible II		37.8 (100) ≤ and <60 (140)
Combustible lllA		60 (140) ~ and < 93.3 (200)
Combustible IIIB		> 93.3 (200)

Table 4.8 Maximum Allowable Size of Containers and Portable Tanks

Container Type	IA	IB	IC	II	III
Glass or approved plastic	1 pt	1 qt	1 gal	1 gal	1 gal
Metal (other than DOT drums)	1 gal	5 gal	5 gal	5 gal	5 gal
Safety cans	2 gal	5 gal	5 gal	5 gal	5 gal
Metal drums (DOT specs.)	60 gal	60 gal	60 gal	60 gal	60 gal
Approved portable tanks	660 gal	660 gal	660 gal	660 gal	660 gal
Polyethylene spec. 34 or as authorized by DOT	1 gal	5 gal	5 gal	60 gal	60 gal

Polyethylene containers have become widely available and are included in later versions of NFPA 30, so information on this type of container has been appended to the end of the table.

Several exceptions to Table 4.8 permit glass or plastic containers of no more than 1-gallon capacity to be used for Class IA and Class IB liquids: (1) if a metal container would be corroded by the liquid, (2) if contact with the metal would render the liquid unfit for the intended purpose, (3) if the application required the use of more than one pint of a Class IA liquid or more than one quart of a Class IB liquid, (4) an amount of an analytical standard of a quality not available in standard sizes needed to be maintained for a single control process in excess of one sixteenth the capacity of the container sizes allowed by the table, and (5) if the containers are intended for export outside the U.S.

If NFPA standard 45 is applicable in the reader's area, the reader is referred to the original standard for the total amounts that would be permitted in laboratories.

In the author's opinion, the total amounts of flammables in a research facility or equivalent such as described in Chapter 3 should be kept to a minimum. Except in unusual cases where a very active laboratory can justify the use of large quantities in a sufficiently short period to justify their presence, flammable materials in large amounts, other than those contained in research apparatus, should be kept in protected storage. In cases where continuing large quantities are truly needed, arrangements should be made with the organization's purchasing department to maintain an adequate reserve so that delivery to the facility can be made within a single working day.

In Section 1910. 106(d)(3) of the General Industry Standards, OSHA limits the amounts of Class I and Class II liquids in a single flammable material storage cabinet to 60 gallons and the amount of Class III liquids to 120 gallons. Thus, even if the integrity of a single storage cabinet were broached in a fire, or if an accident occurred while the cabinet was open, no more than 60

Table 4.9 A Brief List of Some Common Class IA Solvents

Acetaldehyde	Furan	Methyl sulfide
2-Chloropropane	Methyl acetate	N-Pentane
Collodian	Isoprene	Pentene
Ethyl ether	Ligroine	I-Propylamine
Ethanthiol	Methylamine	Propylene oxide
Ethylamine	2-Methylbutane	Petroleum ether
Ethyl vinyl ether	Methyl formate	Trimethylamine

gallons of Class I and II liquids or 120 gallons of a Class III liquid could become involved in the incident. Even these amounts would add an enormous amount of fuel to a fire and could quite possibly result in the loss of the building in which the laboratory was located unless an effective fire suppression system existed..

G. Refrigeration Storage

Two of the most dangerous storage units in any laboratory are the ordinary refrigerator, and to a somewhat lesser extent, the freezer. This is primarily due to the storage of flammable materials within them, although there are also problems due to individuals using them as a place to store food which they have brought to the laboratory for their own consumption, as well as for their intended function. Refrigerators intended for the storage of laboratory chemicals and biological materials should not be used for personal items, particularly food and beverages.

Table 4.10 lists the flash points for a number of common solvents with flash points below or close to the normal operating temperature (about 38°F or 3.3°C) of a common refrigerator. Also given are the flammable limits in percent by volume in air for these same solvents. Most of these evaporate rapidly so that they quickly reach equilibrium concentrations in a small confined space. Some of the most dangerous are acetaldehyde, carbon disulfide, diethyl ether, and ethylene amine, which have broad explosion limits.

Storage of flammable materials in refrigerators or in other confined spaces in which the vapors can be trapped and which also contain sources of ignition represents a potential explosion hazard. Carelessly closed containers, e.g., screw caps that are not firmly tightened or beakers containing solvents covered only with aluminum foil or plastic wrap, will allow vapors to escape from the container and, given sufficient time, build up in the confined space until they may reach a concentration in excess of the lower flammable limit. A spark may then cause ignition, and because the reaction is temporarily constrained, very high pressures can build up until the refrigerator door latch fails and a powerful explosion ensues. Many such cases have been documented, and in most cases, workers in the vicinity in front of the refrigeration unit likely would have sustained serious if not fatal injuries. Fortunately, it appears in many cases that the propensity of laboratory workers to place improperly sealed containers in refrigerators is greatest at the end of the work day, when they may be in a hurry to leave. This pattern, coupled with the materials remaining undisturbed for extended periods of time after normal working hours, tends to make night hours the most likely time for a refrigerator explosion to occur.

A normal refrigerator has many sources of ignition within it— the thermostat, interior light, the light switch on the door, the defrost heater, the defrost control switch, the compressor unit, and the air circulation fan. Most of these are located within the space being maintained cool, but self-defrosting units contain an internal drain that can permit the internal vapors to flow into the compressor space below the usable space.

It is possible to modify a normal home refrigeration unit to remove the internal sources of ignition, but unless it is done by a person who knows precisely what to do and does it very carefully the result may not be as safe as one initially designed to be used for flammable material storage. The liability which could result from the failure to prevent an explosion by an improperly, locally modified unit makes this an imprudent economy measure. In addition, the

Table 4.10 Flammability Characteristics of Some Common Solvents

Chemical	Flashpoint (^{0}C)	Flammable Limits (%)	
		Lower (%)	Upper (%)
Acetaldehyde	-37.8	4	- 60
Acetone	-17.8	2.6	12.8
Benzene	-11.1	1.3	7.1
Carbon disulfide	-30.0	1.3	50
Cyclohexane	-20.0	1.3	8
Diethyl ether	-45.0	1.9	36
Ethyl acetate	-4.0	2.0	11.5
Ethyleneimine	-11.0	3.6	46
Gasoline (approximate)	-38.0	1.4	7.4
n-Heptane	-3.9	1.05	6.7
n-Hexane	-21.7	1.1	7.5
Methyl acetate	-10.0	3.1	16
Methyl ethyl ketone	-6.1	1.8	10
Pentane	-40.0	1.5	7.8
Toluene	4.4	1.2	7.1

labor costs of making the modifications may eliminate a substantial part of the savings. In most cases, refrigeration units need only be rendered safe for prevention of ignition by components of the refrigerator themselves, i.e., be designated as safe for the storage of flammable materials, instead of meeting standards for total explosion safety, which would permit them to be operated in locations where flammable vapors and gases exist outside the refrigeration units. The additional cost of the latter units plus the cost of making the proper electrical explosion proof connections are unnecessary expenses for most laboratories.In virtually all cases, refrigeration units which operate as "ultra-lo" units in which the internal temperatures are of the order of -60°C to -120°C need not be flammable material storage units or explosion safe. Note that none of the materials in Table 4.10 have flash points in this range.

As noted earlier, refrigerators last for as many as 20 to even 30 years. It is not feasible to accept assurances by laboratory managerial personnel that no flammable materials will ever be placed in an ordinary refrigerator, because neither the individual making the promise nor the program for which the refrigerator is purchased is likely to occupy the same laboratory space for such an extended period. It is also not reasonable to depend upon marking laboratory refrigerators, no matter how prominently, as not to be used for flammable material storage and count on total compliance with the restriction. If there is an ordinary consumer-quality refrigerator in the laboratory, it is virtually certain that someone will eventually use it improperly. Therefore, it is recommended that all refrigerators to be used in laboratory areas be required to be initially constructed for flammable material storage, and bear an appropriate label on the front that it meets such standards. Exceptions should be very few, and restricted to those programs that are, in fact, essential to the basic operations of a stable department, and not depend upon the program of an individual or-a limited number of persons constituting a temporary research group. Older units not meeting the standards for flammable material storage should be phased out as rapidly as possible, or moved from laboratories in which usage of flammables is a normal activity to noncritical areas where current usage is not likely to involve storage of flammable liquids, and replaced with a suitable refrigerator or freezer. This procedure will enable all laboratories to be equipped with safe units relatively economically over a period of time and improve the general level of safety at the same time.

There are circumstances in which individuals will object strenuously to imposition of an explosion safe unit being required for their laboratory. This is most likely to occur in an academic institution where equipment is usually owned by the principal investigators as part of a grant. They may be unwilling or simply do not have the funds to purchase an appropriate unit. In such instances, the institution might consider establishing a program to subsidize the purchase of a unit as an insurance policy to forestall loss of a building.

REFERENCES

1. Occupational Safety and Health Agency, Title 29 Code of Federal Regulations, Section 106(d)(3), General Industry Standards, Washington D.C., 1988.
2. Flammable and Combustible Liquids Code, NFPA-30, National Fire Protection Association, Quincy, MA, 1981.
3. Standard on Protection for Laboratories Using Chemicals, NFPA-45, National Fire Protection Association, Quincy, MA, 1982.
4. Flammability Ratings of Flammable Liquids, Fisher Chemical Company, Chicago, IL.
5. **De Roo, J.L.,** The Safe Use of Refrigerator and Freezer Appliances for Storage of Flammable Materials, Union Carbide, South Charleston, WV.
6. **Langan, J.P.,** Questions and Answers on Explosion-Proof Refrigerators, Kelmore, NJ.
7. Properties of Common Flammable and Toxic Solvents, Division of Industrial Hygiene, New York State Department of Labor, Albany, NY.

H. Gas Cylinders

Gas cylinders are used for many purposes in the research laboratory. Most individuals probably think of gas cylinders in the context of the standard industrial gas cylinder, which is 23 centimeters (9 inches) in diameter and 140 centimeters (55 inches) high. However, there are many other sizes available. All of these cylinders can represent a significant hazard. A standard cylinder weighing about 64 kilograms (140 pounds) often contains gas at pressures of about 21 mega-Pascals or 3000 pounds per square inch or even more. Should the valve connection on top of the cylinder be broken off, such a loose cylinder would correspond to a rocket capable of punching a hole through most laboratory walls and would represent a major danger to all occupants in any area where such an incident occurred. The contents of cylinders also frequently represent inherent hazards. These pressure-independent hazards associated with the contents include flammability, toxicity, corrosiveness, excessive reactivity, and potential asphyxiation if the volume of air displaced by the contents of the cylinder is sufficient. Obviously, measures need to be taken to ensure that the integrity of the cylinder is totally maintained. Compressed gas cylinders can be used safely, if due care is taken with them and with the accessories and systems with which they may be combined.

A compressed gas is defined by the federal Department of Transportation (DOT) as "any material or mixture having in the container either an absolute pressure greater than 276 kPa (40 lbf/in.²) at 21°C (69.8°F), or an absolute pressure greater than 717 kPa (104 lbf/in.²) at 54°C (129.2°F) or both, or any liquid flammable material having a Reid vapor pressure greater than 276 kPa (40 lbf/in.²) at 38°C (100.4°F)."

The actual pressure in a cylinder will depend upon the type of gas in the cylinder and the physical state of the contents. Cylinders containing gases which are gaseous at all pressures practicable for the cylinder, such as nitrogen or helium, will have a pressure which reflects the amount of material in the cylinder, while those that are in equilibrium with a liquid phase, such as ammonia, carbon dioxide, or propane, will be at the pressure of the vapor as long as any of the material remains in the liquid phase, provided that the critical temperature is not exceeded. The weight of the cylinder in excess of its empty weight is used to measure the amount of gas in the cylinder for the latter type of gas.

As shown in Figure 4.3, compressed gas cylinders usually will be stamped near the top of the cylinder with the DOT code appropriate to the specification under which the cylinder was manufactured and the pressure rating at 21°C, usually in lbf/in^2. The last date on which the cylinder was tested usually will be stamped near the upper end of the cylinder. In most cases, the test interval for a steel cylinder is 10 years. It is the responsibility of the company distributing the cylinders to be sure that they are within the appropriate test span. Unfortunately, it appears that this is not always done, A spot-check of a large group of cylinders at one facility revealed that more than 10% were significantly beyond the required test date.

Very few laboratory facilities have the capability to refill cylinders, nor in most cases is the use heavy enough to make it economical to acquire the capacity to so safely. Most users of compressed-gas cylinders have an arrangement with a vendor to periodically replace empty cylinders with full ones of the same type on a regular basis, paying a demurrage charge on the number which they maintain in use. As a result, the ones on hand are continually changing and care must be taken to ensure that the identity of the gas in the cylinder is known. Color codes are unreliable, especially for the caps, since these are always taken off in use. Tags attached to the cap are not appropriate for the same reason. A name stenciled on the side of the cylinder or an adhesive label placed on the side of the cylinder would form the basis of a satisfactory system, but the contents of the replacement cylinders should be confirmed upon each delivery. A cylinder for which the contents are not certain should not be accepted. Any units for which the identification label or stencil has become defaced should also be returned to the supplier. Unidentified cylinders are very difficult and expensive to dispose of through most commercial waste disposal firms. Special problems are posed by those that are still pressurized but have physically damaged or corroded valves.

OSHA at one time had specific standards for the inspection of compressed gas cylinders and for the safety relief devices for the cylinders. However, these standards were revoked as of February 10, 1984. OSHA still does have general standards on gas cylinders incorporated into Section 1910.101(a) requiring that visual and other inspections shall be conducted as prescribed in the hazardous materials regulations of the Department of Transportation (40 CFR Parts 171-179 and 14 CFR Part 103). Where those regulations are not applicable, visual and other inspections should be conducted in accordance with Compressed Gas Association (CGA) pamphlets C-6-1968 and C-8-1968. Section 1910.252 of the OSHA standards, which concerns welding, cutting, and brazing, also contains regulations for compressed gas cylinders used in these operations. Users should also seek further guidance from the most current edition of these publications and from the most current version of the CGA publication "Safety Relief Device Standards."

1. Bulk Storage

Most building codes have restrictions on the location and arrangement of bulk storage facilities of compressed gas cylinders. Storage areas for gas cylinders should not intrude on a required path of egress, such as stairs and hallways, or be in an outside area where occupants evacuating a building would be required to pass them or congregate. Cylinders in storage containing oxidizing agents should be separated by at least 25 feet from those containing reducing agents, or these two different types should be separated by a fire wall with a minimum of a 30-minute fire rating at least 5 feet high. If possible, flammable gas cylinders should be stored separately from other cylinders, even those containing inert gases; so that in the event of a fire, their contribution of additional fuel would not increase the possibility of other nearby cylinders rupturing. Nominally, empty cylinders should not be stored in the same location as full ones and all "empty" cylinders should be clearly marked as empty. If separate storage is provided, a simple marking of a cylinder with the letters "MT" is usually sufficient. Empty oxidizing- and reducing-agent cylinders should be separated as if they were full, since they should never be emptied to

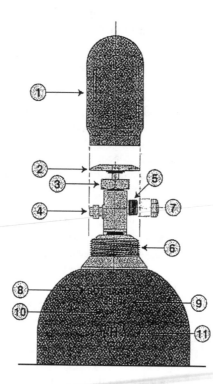

Figure 4.3 Cylinder parts and markings, (1) cylinder cap; (2) cylinder valve wheel to open cylinder; (3) valve packing nut, contains a packing gland and packing nut around the valve stem; (4) pressure relief device; (5) valve outlet connection; (6) cylinder collar to which cap is to be attached; (7) valve outlet cap, may not be present; (8) Specification number shows cylinder conforms to DOT-3AA specifications; and design service pressure is 2265 psig @ 70°F; (9) indicates date, month and year first tested. Test is to repeated every 5 years and test date stamped on cylinder. This figure and Figures 4.4 and 4.5 courtesy of Matheson Gas Products.

Figure 4.4   Hand truck for safe transportation of gas cylinders

Figure 4.5   Safe storage for multiple cylinders.

less than about 172 kPa (25 lbf/in²), in order to prevent contamination of the interior of the cylinders.

In addition to any code restrictions, there are a number of common sense guidelines which should be followed in providing a storage location. They should be stored in well-ventilated areas, and air should be able to circulate freely around them, so that any leakage of gases will be quickly dissipated. The area in which they are stored should not be damp, and the surface on which they sit should be dry to minimize corrosion of the steel. Outdoor storage areas should be protected from the weather, and cylinders should not be stored in areas where heavy objects might fall on them, nor in areas with heavy vehicular traffic where they might be struck by a moving vehicle. Indoor storage areas should be of fire-resistant construction. Since many compressed gases are heavier than air, the storage location should preferably be above grade, and should not readily connect to spaces where it would be likely for escaping heavier-than-air gases to flow and collect. Cylinder caps should always be on the cylinders while they are in storage and at any time they are moved. Signs listing all gases stored at the location should be prominently displayed at the storage area. All cylinders containing a specific gas variety should be grouped together. Compressed gas cylinders should be handled carefully when moving them to and from the bulk storage area and during normal use. Although they appear sturdy, the cylinders are designed as shipping containers for the gases within them and hence are designed to be as light as possible, consistent with reasonable margins of safety for pressure and physical handling. Cylinders should always be moved using a transporting device, e.g., strapped to a common hand dolly (see Figure 4.4) or using any number of commercial units especially designed to transport cylinders. They should never be moved by supporting the valve cap with one hand while rotating and rolling the base of the cylinder along the ground with the other. There are also manual hand trucks specially designed for moving heavy objects up and down stairs. If a freight elevator is not available when moving a cylinder from one floor to the next, one of these devices should be employed. A cylinder should never be hoisted by attaching a cable to the cylinder cap.

Except those cylinders designed to hold toxic gases, most cylinders incorporate a rupture disk as an over pressure safety device, which will melt at the relatively low temperatures of between 70°C and 95°C (158°F and 203°F). Because of these safety devices, the temperature in an area where cylinders are stored should not exceed 52°C (125°F), nor should the cylinders be exposed to localized heating.

Cylinders should be stored in an upright position, with the valve end up, never on their side (see Figure 4.5). The storage area should contain facilities which would make it possible to firmly secure the cylinders in an upright position and prevent them from falling or being knocked over. Parallel bars with space between them sufficient to accommodate a cylinder, with a chain holding them in place make a secure storage area, although stored compactly in a mass with a chain drawn tightly around the entire group also would be satisfactory. In the latter case, however, the length of the chain holding them upright should be adjustable so that if the number stored decreases substantially, the chain will not become so slack as to permit cylinders to fall over.

2. Laboratory Storage

Storage of cylinders in a laboratory at a given time should be restricted to those in actual use or attached to a system ready for use. If this is not feasible, the actual numbers of cylinders present should be maintained at an absolute minimum. No cylinder should be in a laboratory which is not securely fastened to a support so that it cannot fall over. No freestanding cylinder should be allowed to be present in a laboratory, even a nominally "empty" cylinder. There is always a possibility that the empty cylinder has been mislabeled. A good rule of thumb is to treat a compressed gas cylinder as you would a gun; unless confirmed otherwise, always assume that it is loaded and treat it as such.

As long as cylinders in a laboratory are not connected to a system and potentially in use, the guidelines in the preceding section (1. Bulk Storage) should apply. Such restrictions as the physical separation of oxidizing and reducing agents should not be abrogated unless circumstances are appropriate.

Additional information on operations involving gas cylinders will be provided in Chapter 4, Section VI.B.8.

REFERENCES

1. *Prudent Practices in the Laboratory, Handling and Disposal of Chemicals,* National Academy Press, Washington, D.C., 1995, pp 74-75.
2. Specialty Gas Data Sheets, Air Products and Chemicals, Emmaus, PA.
3. *Safety In Academic Laboratories,* 5th ed., American Chemical Society, Washington, D.C., 1991.
4. **Pinney, G.,** Compressed gas cylinders and cylinder regulators, in *CRC Handbook of Laboratory Safety* 2nd ed., Steere, N.V, Ed., CRC Press, Cleveland, OH, 1971, 565.
5. *Safety Relief Device Standards—Cylinders for Compressed Gases,* Compressed Gas Association, New York, 1965.
6. *Standard Compressed Gas Cylinder Valve Outlet and Inlet Connections,* Vol. 1, Compressed Gas Association, New York, 1965.

INTERNET REFERENCE

For the most complete up-to-date material on the subject of Compressed Gas, use the following Internet reference.
1. http://www.cganet.com/Default.htm

3. Animal Food and Supply Storage[*]
a. Animal Food

The Federal Animal Welfare Act (PL 89-544) requires that animal food be stored in facilities which protect it from infestation or contamination by vermin (wild rodents, birds, and insects). Food can be stored in individual animal rooms in vermin-proof containers with lids, such as plastic garbage containers. Ideally, bulk-food shipments should be stored in a room or warehouse where the temperature can be maintained at less than 70°F and the relative humidity at 50% or less. The room should have doors that prevent the entry of rodents or birds. Vermin control is important since wild rodents, birds, and insects can contaminate stored feed with bacteria, viruses, or parasites which could adversely affect laboratory animal health. Pesticides should not be used to control vermin in this area while food supplies are present; contamination of food with pesticides can seriously affect experimental results in animals. Boric acid powder can be placed along the walls to control cockroaches, without the negative experimental impact of organophosphate insecticides.

Most commercial laboratory diets contain preservatives and stabilizers which maintain nutrient quality in the diet for up to 6 months. However, diets containing vitamin C (e.g., guinea pig chow and nonhuman primate chow) have a limited shelf life of 90 days because of the instability of vitamin C in the diet. Feed sacks are coded at the manufacturer with the date of milling, and this date should be recorded upon receipt of the shipment, recognizing that the food should be used within 90 days after the milling date to avoid vitamin C deficiency problems in guinea pigs and monkeys. Most facilities use the "first-in, first-out" method of warehousing feed, stacking the pallets/bags such that the oldest feed is most accessible for transport to the individual animal rooms.

---

[*] This section was prepared by Dr. David M. Moore, D.V.M, University Veterinarian, Virginia Polytechnic Institute and State University.

The ingredients in purified or chemically defined diets are not as stable as those in most commercial diets, and the NIH Guide for the Care and Use of Laboratory Animals recommends that these diets be stored at 39°F or colder.

Diets which contain potential or known hazardous compounds (carcinogens, mutagens) should not be stored in the same area as control diets.

b. Supply Storage

Potentially hazardous compounds such as detergents, chemical disinfectants, and insecticides should not be stored in the same area with bulk-feed stores to prevent contaminating the latter. The storage area should be clean and orderly, with appropriate precautions taken to keep it free of vermin.

c. Animal Carcass Storage

Animal carcasses that are not immediately incinerated should be kept refrigerated at 44°F or lower. Those to be kept for an extended period should be frozen. Refrigerator units should not be used to store food if used for carcass retention.

REFERENCES

1. Guide for the Care and Use of Laboratory Animals, NIH Publ. 85-23, U.S. Department of Health and Human Services, Washington, D.C., 1985, 22.
2. Animal Welfare Act (PL *89-544)*, Title 9, Subchapter A, Subpart B, Section 3.25—Facilities, General, Paragraph C - Storage.
3. **Hessler J.R. and Moreland, A.F.,** Design and management of animal facilities, in *Laboratory Animal Medicine,* Fox, J.G. et al., Eds., Academic Press, Orlando, FL, 1984, 509.

VI. HANDLING AND USE OF CHEMICALS: LABORATORY OPERATIONS

Laboratory personnel work in a *potentially* extremely hazardous and unforgiving environment. The substances with which they work may be toxic, flammable, explosive, carcinogenic, pathogenic, or radioactive, to mention only a few unpleasant possibilities. The hazards may cause an immediate or acute reaction, or the effects may be delayed for several years. A worker may be lulled into a false sense of security because of the seeming safety of a material; according to current knowledge, but eventually evidence may develop that continued exposure may cause unexpected or cumulative and irreversible effects.

The equipment in the facility, if used improperly or if it becomes defective, could represent physical hazards that could result in serious injuries or death. Electric shock, cuts, explosions due to rupture of high pressure systems (or implosion of large vacuum systems), exposure to cryogenic materials, excessive levels of exposure to ionizing and nonionizing radiation, heat, mechanical injuries due to moving systems, equipment or supplies simply falling on a person, among many other possibilities, may occur in a laboratory environment.

Of course, the laboratory environment is not the only place an injury can occur. Scoffers who do not put a high probability on the possibility of an accident happening in the laboratory often point out that they could have been injured while driving to work. Certainly this is possible, as is being struck by lightning or any number of other possibilities. If we were to brood about all the things that could happen each day, we might choose to not get up each morning. However, it is necessary that we do so and that some risks be taken. It is prudent, however, to follow practices which will minimize the risks. Most of us would not choose to drive with faulty brakes, deliberately drive on the wrong side of the road, or through red lights and stop signs, nor would most of us deliberately violate similar common sense rules governing practices which would lead to ill health. It is impossible to achieve absolute safety, but in the presence of hazards, it is only reasonable to take those steps which will efficiently and effectively reduce the risks to acceptable levels. Laboratory workers should follow ALARA principles (using the parlance of radiation safety) and reduce the risks to a level *as low as is reasonably achievable.*

Laboratory operations are so varied that it would be totally impractical to attempt to exhaustively cover the topic. There are, however, some basic considerations which should be used to enhance the safety of laboratory operations. Some of these are common sense and some have been made mandatory by regulatory requirements, because some safety-related practices are too important to be left to choice. Prior to addressing specific topics, the following list of simple rules, if followed religiously, should dramatically reduce the number of laboratory accidents, or would diminish the consequences of those that do occur.

1. Plan the work carefully. At the beginning of an extended project, formally analyze the proposed program for possible hazards and consider the consequences of possible failures or errors. Ask a colleague to review the hazard analysis with you. Being too close to a subject often leads to overlooking potential problems. Unfortunately, even with the best plans, eventualities will exist which no one thinks of, and these are just the ones which may result in accidents.

2. Make sure the right equipment is available and in good condition. All too often, makeshift equipment or deteriorated equipment is the cause of an accident. Rarely is it worth the risk to take chances. Most persons with more than a few years of experience can think of a number of examples where this has proven true, sometimes tragically.

3. Make sure all systems are assembled in a stable and solid manner, making sure that accommodations for the specific limitations or failure modes of the individual components are factored into the operation of the total system.

4. If the release of a toxic or hazardous substance may occur, the work should be done in a fume hood appropriately designed for the operation. Use the fume hood so as to maximize its effectiveness.

5. Use an explosion shield, other protective enclosures, and/or personal protection equipment such as goggles. and a face mask if there is a possibility of a violent reaction. Do not overlook the possibility that scaling up a process will change the safe operating parameters.

6. Chemicals should be handled carefully at all times, using appropriate containers and carrying devices. Open containers should be closed after use, and unneeded reagents should be returned to secure storage.

7. Do not hurry unnecessarily or compromise on safety. Take the time to do things properly, e.g., label temporary containers as they are employed. Many accidents are due to unnecessary haste or the use of "shortcuts."

8. Follow good safety practices with electrical circuits and equipment. Avoid use of extension cords, multiple plugs, and devices to defeat the need to use three-wire connectors.

9. Avoid working alone if possible. As a minimum, a second person should be aware of an individual working alone in a laboratory and definite arrangements should be made for periodic checks. Excessively long working hours increase the likelihood of mistakes due to fatigue.

10. Follow good housekeeping practices. Maintain the work area in an orderly fashion.

11. Do not set up equipment so as to block means of egress from the work area. Consider the activities of others sharing the facility with you in establishing your own work space.

12. Conscientiously use any required protective equipment and wear appropriate clothing.

13. Make sure that you are familiar with and conscientiously follow all safety and emergency procedures.

14. The work area in a laboratory is not a restaurant or a place to socialize. Coffee and meal breaks should be taken at a desk outside the active work area or in a lounge set aside for the purpose. Especially, do not use laboratory beakers for beverages.

15. Anyone indulging in horseplay or practical jokes within a laboratory should be excluded from the facility.
16. Never work while under the influence of drugs or alcohol nor allow others to do so.

REFERENCES

1. *Safety in Academic Laboratories,* 5th ed., American Chemical Society, Washington, D.C. 1991.
2. *Prudent Practices in the Laboratory, Handling and Disposal of Chemicals,* National Academy Press, Washington, D.C., 1995.

A. Physical Laboratory Conditions

Many of the points to be made in this section were alluded to in Chapter 3, where the factors that should be considered in laboratory design were discussed. If the layout of the laboratory is similar to that of the standard laboratory module shown in Figure 3.1, and repeated on the next page as Figure 4.6, many safety practices which depend upon the physical configuration of the facility will almost automatically follow. However, in many cases, laboratories are often placed in structures originally designed for other purposes and ill adapted for the intended use. Even in this latter case, safety can be significantly enhanced by following a few straightforward guidelines as closely as the available space permits.

1. Organization of the Laboratory

The basic premise in laying out the interior design of a laboratory facility or allocating space for the various activities within an existing facility is to separate areas of high risk from those of low risk as much as possible, and to place high-risk operations where there will be the least traffic and the least probability of blocking escape from the laboratory in case of an accident. Escape routes should, wherever possible, lead from high- to low-risk areas. A high-risk component may not always be obvious. For example, storage of chemicals in appropriate cabinets does not represent a high risk under most circumstances, but if left open, a flammable material storage cabinet along a path of egress can become a major danger if the liquids stored inside become involved in a fire. If the configuration of the laboratory permits, the laboratory furniture should be selected to permit two alternative evacuation paths from any point in the room. One of these, constituting a secondary escape path, may not necessarily lead directly from a high- to a low-hazard area, but even a poor alternative is better than none at all.

Fume hoods are intended to be used to house activities that should not be done on an open bench because of the potential hazard which the activities represent, usually the generation of noxious fumes. The ability of fume hoods to capture and retain fumes generated within them is especially vulnerable to air movement, either due to traffic or other factors such as the location of air system ducts, windows, doors, or fans. Clearly, they should be located, as in the standard laboratory module, in a remote portion of the laboratory selected for low traffic and minimal air movement. Other fume generating apparatus, such as Kjeldahl units, should also be placed in out of the way places where errant air motion will not result in dispersion of the fumes generated into more heavily occupied areas of the room. A point that needs to be considered is the work habits of laboratory employees. Data on the possible health effects of long-term exposures to the vapors from most laboratory chemicals is relatively scant, although there are beginning to be more epidemiological data for various modes of exposures, indicating some general problems. There are very little data on the synergistic effects of combinations of general laboratory chemicals. It is known, however, that the sensitivity of individuals to materials varies widely. Again, in the spirit of ALARA, as applied to chemical usage, when evidence to the contrary is missing, a conservative approach is recommended to limit exposures. Research personnel should spend no more time than is essential to the work in progress in areas where

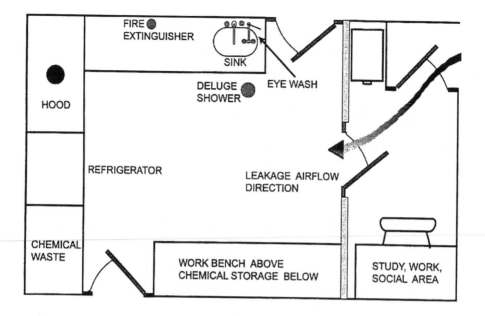

Figure 4.6   Standard Laboratory Module.

the generation and concentration of chemical vapors is likely to be higher (although certainly not above the PEL or action levels established by OSHA), as compared to other spaces in the facility. The practice of allowing or requiring laboratory technicians to have desks in the work area either for convenience or so that they can keep an eye on the work in progress should be discontinued for both health and safety reasons. The standard module provides for this oversight safely by making the barrier between the desk area and the laboratory proper transparent.

The location of the various items of equipment in a laboratory should depend upon a number of factors, such as frequency of use, distances to be traveled, and the need to transport chemicals to and from the primary work location and the storage areas. The distances traveled to and from the most heavily used apparatus should be minimized, as should the frequency and distances involved in the use of chemical reagents. Specialized work, such as the use of radioactive materials, should be isolated from the other activities in the laboratory, especially if only some of the laboratory's employees are involved in the activity while others are not. Any equipment which generates fumes or vapors, but not of the character or concentration that would mandate use of a fume hood, should take into account the air distribution patterns within the room so that the dispersion into heavily occupied areas would be minimized.

Dangerous apparatus should be placed in areas in which protection can be afforded to the maximum number of laboratory employees. For example, a temporary glass system containing a highly reactive material under pressure might be located to one side of the laboratory with explosive barriers placed on either side of the system so that if the system did rupture, the flying particles of glass would be directed toward a normally unoccupied area of the laboratory. Of course, if the probability of an explosion is significant, instead of being only a comparatively remote possibility, the work should take place in a laboratory built with the proper explosion venting and explosion-resistant barriers.

Safety showers and deluge showers should be conveniently located within the facility so that the approach to them is uncomplicated and unlikely to be blocked. They should also be located close to the primary entrance to the laboratory so that persons rendering assistance to an injured individual should not have to enter the laboratory any farther than necessary, to reduce the possibility of having to enter a contaminated area. Note that in the standard laboratory module, both shower and eyewash station are located at the end of the workbench closest to the entrance and immediately adjacent to a secondary entrance. Also note that the portable fire extinguisher is on the wall behind the workbench, near the fume hood and close to the combination eyewash station and deluge shower, nearest the areas where it might be needed and useful. A first aid kit should be maintained in the desk area.

2. Eating, Studying, and Other Social Activities

Unless provision is made for acceptable alternatives or extremely tight discipline is maintained in the laboratory, it should be recognized that the work area of the laboratory will be used for eating, studying, and social activities. Such activities, however, should not take place, and as a minimum the laboratory should be clearly marked where these activities will not be allowed under any circumstances. The NRC, for example, considers failure to prevent eating, drinking, and smoking in the active work area sufficient cause for a citation. In recognition of the need to control these activities, the standard laboratory module does provide a convenient and acceptable location within the laboratory for eating, record and lab book maintenance, studying, and social activities by placing a desk space immediately inside the laboratory facility, separated from the rest of the laboratory by a partition. The location of this office class space also provides direct access to the exit way corridor and to the reminder of the building without having to reenter the laboratory in the event of an emergency. The upper half of the partition separating the laboratory and desk spaces in the laboratory module is intended to be transparent, made either of tempered safety glass or a plastic material such as Lexan™. This makes it possible to keep an eye on laboratory operations while taking a coffee or lunch break in safety, or to perform any other desired activity without interfering with laboratory operations or being disturbed by others still in the laboratory. Since the laboratory, in most instances, should be at a modest negative pressure with respect to the corridor, the passage of vapors from the laboratory into the desk area should be inhibited, reducing the routine exposure to the laboratory atmo-sphere to those at their desks significantly. The low speed of the air through the door, on the order of 10 to 20 fpm, will still make it possible for traffic through the door to allow a much reduced amount of laboratory odors into the office compartment. The two doors in sequence also serve an additional safety function, representing a simplified air lock, separating the corridor from the laboratory, thus adding some stability to the HVAC demands within the room. If they are closed following an evacuation, they would provide an additional barrier to any fire or noxious gases spreading from the laboratory to the remainder of the building.

3. Maintenance

Topics generally overlooked in laboratory safety are safety factors involved in providing needed maintenance and custodial services. Access to equipment needing service must be pro-vided to service personnel under conditions which make it possible for them to perform their work safely. Generally, equipment maintenance in the laboratory by support personnel should be coordinated by an individual who is familiar with current and recent research programs, and can advise the workers who arrive on the scene of possible risks in handling the various components. An example is maintenance on fume hood exhaust fans. Instances are known of workers servicing fume hood exhausts who suffered severe reactions to contaminants on the equipment and the roof in the vicinity of the exhaust duct, even though the hood was not in use at the time. It is not enough to warn the maintenance department upon the initial request for services. Direct information needs to be provided to the service persons on the scene. Some materials can remain a problem for extended periods of time. In such cases, the workers need to be protected while doing the work, using appropriate items of protective clothing such as gloves, respirators, coveralls, and goggles or full-face respirators, depending upon the level of risk. The levels will depend upon the contaminants that might exist due to work in the laboratory, and these risks should be evaluated conservatively to enhance safety of the workers.

Where fume hood exhausts are brought to the roof through individual ducts, the area in which maintenance is needed may be surrounded by exhaust ducts still in use, since in most cases it is impractical to shut off operations for an entire building or even a significant portion of one, because it is too disruptive to the research programs. Therefore, it is probably desirable to have a standard personnel protective equipment package for the maintenance workers to use, consisting of half- or full-face respirators providing protection against solvents, particulates, and inorganic acids, chemically resistant coveralls, and gloves selected to provide a broad spectrum of protection against chemicals. Requiring personnel to wear these may appear to be excessively cautious but, as noted earlier, there have been instances where unanticipated severe and long-lasting health effects have occurred. Not all employees who work on the roof of a chemistry building may be employees of the organization. Outside contractors also are used to do a variety of maintenance duties and, under the hazard communication standard, they must be apprized of the risks to which they might be exposed. The mix of materials exhausted through ducts is typically so complex that meeting this requirement is difficult, if not impossible. Recommending to them to wear equivalent protection should fulfill the spirit of the standard. Unfortunately, maintenance personnel may scoff at the need to wear protective equipment, or alternatively, be so fearful of exposure that they may refuse to perform the needed task. It is the responsibility of the organization to provide sufficient indoctrination and enforcement of their personnel protection policies that both of these eventualities can be avoided.

Fume hood maintenance is one of the more active areas in which maintenance personnel have concerns and where both support and laboratory personnel need to assume responsibility for seeing that the work is properly coordinated. Some simple suggestions that have been found useful are to ensure that each exhaust duct on the roof is properly labeled with the room location of the hood itself. Workers have been known to turn off power to motors on hoods in active use. Where hoods are dedicated to special uses which represent unusual hazards, such as radioactive materials, perchloric acid, exceptionally toxic gases, or any other especially unusual risk, the duct should also be labeled with the application involved or a color code employed to identify these unusual risks. The latter program would alert maintenance personnel to definitely contact the laboratory from which the duct came before working in the vicinity of the duct. Power to the motors on the roof should also be provided in such a way as to ensure that the workers on the roof can completely control the circuits while working to avoid accidental activation of the circuits from the laboratory. However, should the exhaust motor be turned off by maintenance workers without prior notification of laboratory personnel, an alarm should sound in the laboratory warning that the hood is not functional. A tagging and lock-out procedure should also be employed during the maintenance operation.

Once hoods are removed from service to perform maintenance, they should not be returned to use until it is verified that they are performing according to required standards. It is easy to erroneously wire a three-phase motor so that the fan rotates opposite to the desired direction. Belts may need to be tightened or a pulley size changed to achieve the proper face velocity.

Fume hoods have been used to illustrate some of the problems that can arise from lack of coordination of maintenance and laboratory personnel, but there are many other possible problems. Explosions can occur if gas service is turned off without everyone being aware of it and they leave gas jets open, flooding a facility with gas when service is restored. Stills can overheat if condenser water supplies are interrupted. Electrical service to an area should be discontinued and restored only with full prior notification to all persons that might be affected.

Today, with the large amount of computer automation being used, interrupting the power to the facility may disrupt the entire operation and bring down the wrath of the scientist on the worker who caused the problem. Any computer equipment with such a critical function should be equipped with an uninterruptible power supply with sufficient capacity to allow a safe managed shutdown procedure. These have become relatively inexpensive. Individual laboratory technicians or students often modify their facility without informing the groups responsible for maintenance, thereby raising the possibility of an injury to an unsuspecting service-person, or make a repair which will not be based on an accurate assessment of the conditions which could affect their work. Also, the as-built drawings for many buildings do not reflect reality due to change orders during construction which are not added to the drawings. If discrepancies are known, maintenance personnel should be notified prior to beginning work if the differences could impact the job.

4. Housekeeping

Another maintenance issue is what reasonably should be expected of custodians. Experience has shown that there is a tremendous variation in the level of expectations and wishes among laboratory supervisors. There are those who do not wish custodians to enter their laboratory at all, while there are those who have no qualms in asking custodians to clean up a hazardous chemical spill. Most safety and laboratory personnel would agree that the latter is asking too much, while most would also agree that, if they wish, facility personnel should be allowed to take care of their own housekeeping, as long as reasonable standards of cleanliness are maintained. Most laboratory groups, however, fall somewhere between these two extremes.

The salary levels of most custodial positions are usually among the lowest in most organizations and limit the skill levels one can expect from the persons filling the positions. Unfortunately, literacy rates are often less than average and, in many cases, it certainly would be unrealistic to expect a custodial worker to have a significant level of technical training which would permit an understanding of the problems that they might encounter in a laboratory. As a result, custodial workers are often quite afraid of the laboratory environment. However, alternative positions are also usually hard to find for these employees, so they frequently are very concerned about losing their jobs. Most cannot afford to do so. As a result of these conflicting pressures, they may attempt to do things they really do not understand and are afraid to ask about, and may make mistakes in consequence. It is the responsibility of the laboratory supervisor, working with custodial management, to carefully establish safe constraints on the areas of responsibility for the custodians in the laboratory.

Among things a custodian can reasonably be expected to do in most laboratories are:

1. Clean and maintain the floor area.
2. Dispose of ordinary trash. However, if other than ordinary solid waste is generated in the laboratory, it should be placed in distinctively shaped and/or colored containers. If the custodians are still expected to handle it, then the circumstances and procedures should be carefully delineated and training given. This latter responsibility is not recommended.
3. Wash windows. If they are expected to wash bench tops or other laboratory furniture, it should be only when additional supervision is provided by laboratory personnel.

Among items which they should not be expected to do are:

1. Clean up chemical spills. They are not trained to do it according to established regulatory guides nor to do it in such a way as to ensure that they do not expose themselves to the potential injury.
2. Dispose of broken glass, syringes, or "empty reagent containers." These items can be disposed of by them if they are carefully prepared by the laboratory workers in advance. For example, broken glass should be disposed of by custodians only if it is placed in a sturdy kraft board box (or equivalent), sealed, and labeled as "broken glass." Other items, such as syringes, are now considered to be regulated medical wastes in most instances, as, for example, if they fall under the provisions of the bloodborne pathogen standard. Syringes are to be placed in a leak- and puncture-proof container specifically intended for them. The needles are not to be sheared, broken, or removed in any way. The containers are to be rendered biologically safe and disposed of by techniques which will be discussed in detail in the section on infectious waste disposal. Custodians are explicitly not to handle these wastes. Empty reagent bottles should be triple rinsed and then placed

in a box labeled "triple rinsed reagent containers." All of the restrictions on glassware should be thoroughly explained to the custodial employees. There are specific requirements under the bloodborne pathogen standard that this be done. If custodians believe that they are being asked to handle unsafe waste, they should ask their super-visors to intercede for them.

3. Handle special wastes in any way including radioactive materials, chemical wastes, or contaminated biological materials. All of these require special handling by specialists and precautions must be taken to ensure that these materials are not accidentally collected by custodians. The custodians should be given awareness training to ensure that they have sufficient knowledge to allow them to recognize these special wastes.

4. Clean the work surfaces and equipment in the laboratory, except in special circumstances and under the direct supervision of a responsible laboratory employee. Even in this case, a preparatory program should have been carried out in advance by laboratory personnel to remove or secure items which could be dangerous in the area being cleaned.

Housekeeping also means maintaining the laboratory in a reasonably organized fashion on a day-to-day basis. This is the responsibility of all laboratory personnel, but individuals will follow the laboratory manager's own performance as a guide. Reagents not in use should be returned to proper storage. Secondary containers should be labeled according to the requirements of the hazard communications standard. Glassware should be cleaned and put away. Trash should not be allowed to accumulate. Equipment should not be allowed to encroach upon aisles. Cables and temporary electrical extensions should not become a tripping hazard. Periodically, refrigerators and other storage units should be gone through and cleaned out. An audit of materials should be made periodically to dispose of old, degraded, and obsolete materials before they become a hazard. Chemicals stored inappropriately outside of their hazard class should be returned to their proper locations. Bottles heavily covered with dust, indicating a lack of use for an extended period, are likely to remain unused and should be eliminated. No one should expect a busy laboratory to be spotless, but neither should it be a disaster area. Unless a concerted effort is made, eventually housekeeping problems tend to slowly accumulate. An effective mechanism used by the author to combat this erosion of order was to schedule a quarterly "field day" during which all personnel, including faculty, staff, and students, ceased research and returned everything to reasonable order. This rarely took more than a few hours and furthered a sense of cooperation between the various groups of people.

5. Signs and Symbols

Many situations exist in which a person entering an area needs to be made aware that a hazard exists in the area or needs to know of restrictions placed on persons entering the area. In addition, there are signs which are intended to provide information to individuals in an emergency. There are literally hundreds of specialized safety signs and symbols which can be purchased for the laboratory. Given below is a partial list of some of the more important ones, along with a brief description of the types of applications for which they would be needed. In many cases, the signs in this list are mandated by regulatory requirements, while in other cases they represent common sense safety practices. In most cases, the hazard signs will be prefaced by a risk descriptor, defining the level of risk represented in the specific instance. The three cautionary words in normal use, in decreasing order of risk are **DANGER, WARNING,** and **CAUTION.**

1. **AREA UNSAFE FOR OCCUPANCY** - This is used to indicate a contaminated area or an area otherwise rendered unsafe, temporarily or otherwise, for normal use.

2. **AIRBORNE RADIOACTIVITY AREA** - Some applications involving radioactive materials result in the generation of airborne radioactive materials in excess of those permitted by the standards of the NRC, or of the equivalent state agency in an agreement

Figure 4.7 Radiation safety symbol.

Figure 4.8 Biological hazard symbol.

state. Should such an operation exist, the boundaries of the room, enclosure, or operating area where the airborne material may exist must be posted with this sign. The legend will be accompanied by the standard radiation symbol shown in Figure 4.7.

3. **ASBESTOS** - Asbestos is still used in a number of products employed in laboratories. If there is the potential while using these products that asbestos fibers may become airborne, the area needs to be marked with a sign:

CAUTION
ASBESTOS-CONTAINING MATERIAL PRESENT

Note that a sign such as this does not say that there are asbestos fibers in the air. The sign is intended to alert people that their actions could result in the generation of airborne asbestos fibers. If there is a risk that airborne asbestos fibers may be present, the appropriate department needs to be notified (usually either physical plant or health and safety) to correct the problem.

4. **AUTHORIZED ADMISSION ONLY** - This sign may accompany many other signs or it may stand alone in restricting access to an area to those who have legitimate reasons to be there, or who are aware of the risks within the area to which they may be exposed.

5. **BIOLOGICAL HAZARD** - The sign will be accompanied by the standard biological hazard symbol shown in Figure 4.8, indicating that an agent which may prove infections to human beings is present within the area.

6. **CARCINOGENIC AGENT** - The laboratory safety standard requires that areas in which carcinogenic agents are in use be designated as such. This can be done with a sign such as:

CANCER-SUSPECT AGENT
AUTHORIZED PERSONNEL ONLY

Where the agent might be unusually dangerous, the agent would be specified and any special protective measures needed would be appended.

7. **CHEMICAL SPLASH GOGGLES REQUIRED WHILE WORK IN PROGRESS** It is recommended that this sign (Figure 4.9) be used at the entrances to all active laboratories where chemicals are employed and are actively being used. In order to enforce the requirement, care must be taken to select goggles which resist fogging, do not become oppressively warm while being worn at comfortable room conditions, and

Figure 4.9   Eye Protection required Pictograph.

do not exert uncomfortable pressure on the face. They should also accommodate wearing normal size prescription glasses at the same time. Many goggles which meet the minimum regulatory standard based on ANSI Z87.1 for impact protection do not meet all of these practical considerations, but there are several brands that do. When work is not in progress, or when a person is in an area well separated from the active work area, it may be permissible, for reasons of comfort, to allow goggles to be removed.[1]

8. **CRYOGENIC LIQUIDS** - All containers which contain cryogenic liquids, most commonly liquid nitrogen (as in the example below) but also other gases maintained at very low temperatures, should be prominently labeled:

CAUTION
LIQUID NITROGEN

The container of the cryogenic fluid, usually a large flask with walls separated by a vacuum called a dewar, will also usually be labeled with the cautionary information:

FRAGILE CONTAINER UNDER VACUUM
MAY IMPLODE.

9. **EMERGENCY INFORMATION SIGNS** - Prominent signs, such as those shown in Figure 4.10, should be posted near the safety device mentioned to aid in locating them in an emergency. Symbols can be used in place of or in addition to some of these.

10. **EXPLOSIVES** - If explosives are stored in Class 1 magazines, or in outdoor Class 2 magazines, the property must be posted with signs stating,

EXPLOSIVES—KEEP OFF

Class 2 magazines must have labels on all sides except the bottom in letters at least 3 inches high,

EXPLOSIVES—KEEP FIRE AWAY

11. **FLAMMABLE MATERIALS** - Cabinets containing flammable materials and areas or rooms where flammable materials are stored or used must be posted with this sign, which may also be indicated by the symbol shown in Figure 4.10. This sign should always be accompanied by the **NO SMOKING** sign, which may be augmented by a standard no smoking symbol.

---

* A point needs to be made here, which will not be repeated for reasons of brevity, that signs and rules must take into account human factors. Otherwise, they are likely to be ignored and weaken compliance with other rules overall. Unfortunately, the strictures on wearing goggles while performing work in the laboratory seems to be one less often followed and enforcement is often lax.

12. **HIGH VOLTAGE DANGER** - Spaces which contain accessible high voltage panels, such as switch rooms and electrical closets, should be locked and provided with these signs to warn persons lacking training and experience in working with high voltage circuits not to enter. Equipment containing high voltage circuits should also bear the same warning label.

13. **HYDROGEN -FLAMMABLE GAS, NO SMOKING OR OPEN FLAMES—** This sign must be posted in all areas where hydrogen is used or stored.

14. **INTERLOCKS ON** - Equipment with internal hazards, such as X-ray diffraction cameras, or areas in which the space is rendered unsafe to enter by the presence of a hazard, are often provided with a fail safe circuit, or interlock, which will turn off the equipment representing the problem if the circuit is broken. The sign provides a warning that the interlock is on to prevent access to the hazard.

15. **LASERS** - Labeling of lasers should follow 21 CFR 1040, the Federal Laser Product Performance Standard. The spaces in which lasers are located should also have a similar warning at the entrance. The label will depend upon the class of laser involved. All of the labels will include a stylized sunburst symbol, with a tail extending to the left (see Chapter 5). The signal word **CAUTION** is to be used with Class II and IIIA laser systems while the signal word **DANGER** is to be used for all Class IIIB and Class IV systems.

16. **MACHINE GUARDS IN PLACE** - OSHA requires that many machines, such as vacuum pumps or shop equipment, be provided with guards over the moving parts. Signs should be posted near these machines to remind employees not to use the equipment if the guards are not in place.

17. **MICROWAVES** - This sign must be posted in any area where it is possible to exceed the current occupationally legal limit of exposure to microwave electromagnetic radiation.

18. **NO EATING, DRINKING, SMOKING, OR APPLYING COSMETICS** -This sign should be posted wherever toxic materials are used, in the working areas of wet chemistry laboratories, or in biological laboratories using pathogenic substances.

19. **NO SMOKING** - A **NO SMOKING** sign, as shown in Figure 4.10 must be posted wherever flammables are in use; where there is a risk of explosion due to the presence of explosives or from gases, vapors, or dusts; and where toxic materials are in use.

20. **RADIATION AREA** - Areas where the radiation exceeds a level established by the NRC must be posted with this sign (Figure 4.10). If the level exceeds a higher level set by the NRC, the area must be posted with a **HIGH RADIATION AREA** sign. Most of these areas will be within an area posted with a **RESTRICTED AREA— AUTHORIZED ADMISSION ONLY** sign. Specifics on these requirements will be given in Chapter 5. Signs defining radiation areas should not be used to post areas where radioactive materials are stored unless radiation levels equal or exceed the stipulated limits. Areas where radioactive materials are stored should be posted with a **CAUTION — RADIOACTIVE MATERIALS** sign.

21. **RADIOACTIVE WASTE** - This is not a sign specifically required by the NRC but is recommended to denote areas within a laboratory where radioactive waste is temporarily stored prior to being removed for permanent disposal, in order to help avoid accidental removal of radioactive waste as part of normal laboratory waste. Much radioactive waste resembles ordinary trash, such as paper.

22. **REFRIGERATOR (FREEZER) NOT TO BE USED FOR STORAGE OF FLAMMABLES** - All refrigerators or freezers not meeting the standards permitting the storage of flammable materials (see Chapter 4, Section V.G.) should be marked with this sign.

23. **REFRIGERATORS NOT TO BE USED FOR STORAGE OF FOOD TO BE USED FOR HUMAN CONSUMPTION** - Laboratory refrigeration units used for the storage of chemicals and biological materials must be posted with this sign to prevent the use of units to store lunches and other food.

24. **RESPIRATORY PROTECTIVE EQUIPMENT REQUIRED** - Wherever airborne pollutants are present which exceed the PELs established by OSHA, respiratory protection is required (see Figure 4.10). In many cases, AGGIH threshold limit values (TLVs) are lower than the OSHA PELs and respiratory protection is recommended when the levels approach these lower limits. It is recommended that in most cases an action level of half or less of the TLV values be set to accommodate in part the different sensitivity of individuals to materials.

25. **SAFETY GLASSES REQUIRED** - This sign is to be posted wherever there is a risk of eye injury due primarily to impact.

26. **TOXIC GAS** - Areas where toxic gases are used or stored must be posted with this warning sign.

27. **ULTRAVIOLET LIGHT EYE PROTECTION REQUIRED** - This sign should be posted wherever there is a risk of eye injury due to ultraviolet light emission.

There are many other signs and symbols identifying hazards or denoting specific requirements to aid in reducing a specific risk. The following generic signs are representative of many of these.

28. **(SPECIFIC ITEM) PERSONAL PROTECTIVE EQUIPMENT REQUIRED** - Many other risks exist which would require specific items of protective equipment. Where these items are needed, the area should be appropriately posted.

29. **(SPECIFIC) TOXIC OR HAZARDOUS MATERIAL** - There are a number of materials that pose known risks, and the areas in which these materials are used should be posted with an appropriate sign.

30. **(SPECIFIC) WASTE CHEMICALS ONLY** - Disposal of waste chemicals according to RCRA standards requires that wastes be identifiable, in some cases by class only, but in most cases it is desirable that wastes not be mixed. Posting of areas or containers with this sign where several waste streams of different character exist will aid in legal disposal of waste.

B. Working Procedures

This section can only touch upon the broad topic of safe laboratory working procedures because of the immense scope of the subject. The procedure to be followed here is to provide generic approaches to most of the hazards covered rather than discuss specific instances in which a given hazard could occur. Some of the more common areas which offer the potential for mishaps will be covered in some detail, but undoubtedly there will be areas that are considered comparably important by many that will be touched upon lightly or not at all. Sections will be devoted to a small number of the more hazardous chemicals to illustrate the precautions that need to be taken when working with such materials. In addition to physical hazards, such as fire, electrical hazards, and explosions, health risks will be discussed in some detail, since in many cases these are more insidious and less often recognized by many laboratory workers. The next several sections will be concerned primarily with physical hazards and the latter part of the chapter will be devoted to short- and long-term aspects of laboratory operations on workers health.

1. Protection Against Explosions

Unusually careful planning must take place whenever there is any reason to suspect that work to be undertaken may involve the risk of an explosion. However, not all potentially explosive

Figure 4.10 Representative group of commercially available emergency, warning, and cautionary signs.

situations are recognized in advance. Letters from experimenters describing work in which unexpected explosions occurred can be found in a substantial proportion of issues of *Chemical and Engineering News.* Because these incidents were unanticipated, sufficient protective measures often were not employed; consequently injuries which could have been avoided are reported in these letters. Explosions may occur under a variety of conditions, the most obvious being a runaway or exceedingly violent chemical reaction. Other situations could include the ignition of escaping gases or vapors, ignition of confined vapors with the subsequent rupture of the containment vessel, rupture of a system due to over-pressure caused by other mechanisms, or a violent implosion of a large vessel operating below atmospheric pressure. Partial confinement within a hood can actually enhance the dangerous effects of an explosion; areas in front of the open face may be damaged more severely than if the explosion were not confined.

Injuries can occur due to the shock wave from a detonation (if the release of energy occurs at supersonic speeds) or deflagration (if the energy release occurs at subsonic speeds). Most laboratory reactions belong to the latter class. Hearing loss may result if the shock wave causes a substantial over-pressure on the eardrums. According to Table C-3.1(a) of NFPA 45, Appendix C, the equivalent of as little as 1 gram of TNT can rupture the eardrum of a person within 0.75 m (~2.4 ft), while 10 g is likely to rupture the ears of 50% of persons within 67 cm (2.2 ft) of the explosion. The shock wave, as a wave, can "go around" barriers or be reflected and reach areas that would be shielded from direct line-of-sight interactions. Injuries can occur due to the heat or flames from the explosion. Fume hood materials should be selected to contain fires occurring within them. However, if the sash is severely damaged, flames or burning material can escape through the front opening and the flames may spread to other fuels in the vicinity. Due to this possibility, flammable materials should not be stored in the open in close proximity to fume hoods. Respiratory injuries can occur due to inhalation of fumes and reaction products. However, the most serious hazard is usually flying debris, including fragments of the contain-ment vessel, other parts of the experimental apparatus, or nearby materials or unreacted chemicals which can inflict physical injuries. The risk of the latter type of injuries can be reduced by eliminating the possibility of line-of-sight or single-ricochet paths for missiles from likely sources of an explosion to workers or to equipment which could be damaged and result in secondary harmful events. The possibility of extraneous material becoming involved in an explosion is a powerful argument in favor of not using a hood as a storage area, especially in experimental activities. The reflected shock wave can act in much the same way as a piece of physical debris in causing damage external to a fume hood. Overreaction of a worker or involuntary reflexive actions to even minor explosions can also lead to quite serious secondary injurious incidents.

In addition to immediate injuries, an insufficiently contained explosion can lead to fires or cause damage sufficient to wipe out expensive apparatus, destroy months or years of research effort, or even destroy an entire facility. Conservative precautionary measures to reduce the likelihood of these repercussions are worthwhile from this aspect alone. Ordinary fume hoods offer a fair amount of protection to the sides and rear of the hood, if they are of good quality with substantial walls. However, most fume hoods are not intended to provide really significant explosion protection against a major explosion for a user standing immediately in front of the hood, although sash materials are usually designed not to contribute to the hazard. This is accomplished by having the sash material made of either laminated or tempered glass so that the broken sash will not cut persons standing in front of the hood. The laminated glass may remain intact but may result in the entire sash being expelled which could represent a hazard. The tempered glass is more likely to be shattered and contain the explosion less but the small glass fragments are normally relatively less dangerous than larger ordinary glass fragments. A hood with a three-section horizontal sash, where the user stands behind the central section, provides superior protection to the more common vertical sash hood. If the work to be done involves a known explosive risk, certainly a hood specifically designed to contain any anticipated explosion, or to provide safe explosion venting should be employed.

For the majority of laboratories equipped only with ordinary fume hoods, supplementary measures should be taken to minimize the type of risks described above if a careful analysis of the planned operation reveals any significant potential for an explosion.

A simple way to reduce the potential risks is to minimize the amount of material involved in the experiment. The smallest amount sufficient to achieve the desired result should be used. The trend toward microscale experimentation supports this option. Care should be taken in scaling up from a preliminary trial run in which minimal quantities were employed. Increasing the amount of material in use could significantly change the physical parameters so that insufficient energy removal, inadequate capacity for the reaction products, or excessive pressures could develop in the scaled-up version of the work and lead to a dangerously unsafe condition. One of the more violent explosions in the author's experience was of this last type.

A number of other measures can be taken to enhance the protection of workers against explosions. Provision of barriers is a straightforward measure. The selection of an appropriate barrier will depend upon the circumstances. A variety of factors should be considered.

The strength of the barrier material is clearly an important factor. Tests have been made of many materials commonly used in laboratory protective barriers and available either in commercial units or readily amenable for fabrication of custom shielding. Table 4.11 is adapted from a study by Smith in which each material tested was 0.25 inches or 6.4 millimeters thick. The relative susceptibility to fracture was measured by either the ASTM D 256 test method or by dropping balls from various heights. It required 12 to 16 foot pounds of energy to fracture the polycarbonate material in the ASTM D 256 test. Additional protection can be obtained by increasing the thickness of the materials used in fabricating the shield, approximately proportional to the thickness added. An equal thickness of steel would have a relative effectiveness on this scale of about 40. Resistance to fracture is not the only consideration. Wired glass, for example, may represent an additional hazard due to the presence of the wire, if shattered. Ordinary glass should not be used due to the danger of cuts from the flying debris. Methyl methacrylate is not suitable where high temperatures may occur. However, sheets of methyl methacrylate are commonly available at moderate cost and can readily be fabricated into custom shields. Polycarbonate obviously offers considerable strength, but can be damaged by organic solvents. Steel is resistant to both heat and solvents, but does not offer the desired transparency. However, there are alternatives to this deficiency such as mirrors, optical devices, or closed-circuit television. Remotely controlled manipulating devices can be used to control apparatus behind any shield material.

The simplest types of supplementary protection suitable for moderate risks are commercial shields which are available from most laboratory supply firms. Shields usually found in catalogs are of transparent material, most commonly polycarbonate, weighted at the bottom to increase their stability. Since these are free standing, they often will not remain upright in explosive incidents and, if the explosion is severe enough, may actually be hurled through the air and cause injury themselves. Since the scale of an explosion cannot always be accurately estimated, it would be desirable to secure these shields firmly to the work surface. For small-scale reactions, they offer a worthwhile degree of added protection. The shields should be located so as to provide the maximum protection against flying debris, chemicals, or, as noted earlier, external shock wave interactions. Individuals in the laboratory should be trained to use the shields correctly and not to move or modify them to improve their convenience in performing tasks, if these changes could reduce the level of protection.

For larger-scale experiments, especially if the operation is to continue for a significant length of time, the small commercial shields should be replaced with custom-fabricated shields. Because of its availability in a wide range of thicknesses, relatively high impact resistance, moderate cost, and ease of fabrication, methyl methacrylate is a convenient material to use in constructing custom transparent shields. The same essential design considerations that apply to commercial shields

Table 4.11 Shock Tests on Transparent Shields

Material	Thickness		Drop Ball		ASTM D 256
	mm	in.	kg/m	in./lb	
Double-strength glass..	3.2	0.125	446	25	
Laminated glass	6.4	0..25	1,964	110	
Plate glass	6.4	0.25	1,964	110	
Wired glass	6.4	0.25	2,000	112	
Tempered glass	6.4	0.25	10,393	582	
Methyl methacrylate	6.4	0.25	19,400	1086	0.4 to 0.5
Polycarbonate	6.4	0.25			12 to 16

also apply to custom shields: (1) select the material that has the appropriate mechanical and physical properties, and (2) place the shield to provide the maximum protection against flying debris and other external effects and to minimize direct injuries to personnel and secondary events. As the size or hazard of the experiment increases, consideration should be given to relocating the work to a more appropriate containment facility, preferably one specifically designed and engineered to limit damage should an explosion occur. If it is essential that the work be done in a specific location not explicitly designed for work with potentially explosive situations, it is strongly recommended that the design of the proposed renovation or modification be done by a qualified engineer and fabricated by professionals, not graduate students.

There are differences in the cost of different types of shielding materials and in the approach that is taken to provide adequate protection. A cost-benefit analysis is always appropriate in selecting or designing any experimental system. On the other hand, selecting too inexpensive an approach can be a false economy. The person making the decision may be the one injured or killed if the protection is insufficient. There is always the question of liability if others are injured and, finally, there is always the ethical question of what should have been done if one only did what was minimally required to be done and a person was injured as a result.

a. Personal Protective Equipment

Personal protective equipment will be covered in more detail in Chapter 6, but an important aspect of enhancing the safety of workers where the potential for explosions exists is to provide and require the use of protective equipment. Every laboratory worker in a facility where any potentially injurious chemical is in active use should always wear protective eyewear. Common spectacles with side shields are not nearly as effective as properly selected chemical safety goggles in protecting workers' eyes. The latter not only fit all around the eyes and thus protect against direct impact, but also protect the eyes against flying liquids. Since a properly fitting pair of goggles offers a snug fit to the face, it generally provides superior protection from lateral impacts which could knock off an ordinary pair of safety spectacles. Additional protection to the eyes, face, and throat should be provided, as circumstances warrant, by the use of face masks in addition to goggles. A mask which protects the throat is preferable to one which does not, due to the vulnerability of the carotid arteries on each side of the neck. There are commercial face masks which provide excellent overall protection to the head area.

In addition to eye protection, sturdy laboratory smocks, preferably made of chemical resistant and flame-retardant or flame-proof material, should be worn. Short-sleeved shirts, T-shirts, shorts, or sandals should not be allowed in any laboratory where the potential for exposure of the skin to chemicals exists, or where even minor explosions can occur. Sturdy gloves or gauntlets, selected for the immediate requirement of manual dexterity, should be worn if manual manipulation of apparatus or materials with the potential for explosion is needed. Although the work should be done in a specially equipped facility if the potential for a major explosion exists, in principle, even bullet-proof vests could be used in certain situations. The potential for hearing damage can be greatly reduced if research personnel use good quality earmuffs designed to provide hearing protection. If these are to be issued, the workers should be provided an audiometric hearing evaluation beforehand to determine if a hearing loss has actually occurred subsequent to an accident.

b. Summary

Laboratory explosions occur frequently in situations in which they are least expected. Simple caution should dictate a conservative estimate of the probabilities of such an event and encourage the use of appropriate preventive measures in all laboratory work. It is far better to prevent an explosion than to attempt to confine one or reduce the severity of the damage resulting from one. However, appropriate barriers and personal protective devices can aid in reducing the seriousness of consequences when explosions do occur, if these ameliorative measures are used properly. Training all personnel to understand the risks associated with their work, coupled with encouragement by management and the cooperation of all laboratory workers to follow safe procedures, can significantly reduce the risks of explosions.

REFERENCES

1. Fire Protection for Laboratories Using Chemicals, Appendix C., NFPA Standard 45, National Fire Protection Association, Quincy, MA.
2. **Smith, D.T.,** Shields and barricades for chemical laboratory operations, in *Handbook of Laboratory Safety* 2nd ed, Steere, N.V., Ed., CRC Press, Cleveland,OH, 1971, 113.

2. Corrosive Chemicals

The definition of corrosive chemicals is very broad. However, in the sense that the action of the chemical will result in an immediate, acute erosive effect on tissue as well as other materials, strong acids and bases, dehydrating agents, and oxidizing agents are commonly considered to be corrosive materials. These terms may not be mutually exclusive.

Accidents with corrosive materials in which the material may splash on the body are very common in the laboratory. The eyes are particularly vulnerable to injury, and injuries to the respiratory system may range from moderate irritation to severe injury. Skin injuries may be very slow to heal. Ingestion can cause immediate injury to the mouth, throat, and stomach, and in severe cases can lead to death. Some common household chemicals that are equivalent to laboratory corrosives, such as drain cleaners, are common causes of child fatalities. Work with strongly corrosive materials should be done in a fume hood. Personal protective equipment, such as gloves, laboratory aprons, and chemical splash goggles should be used and, if the possibility of inhalation is significant, appropriate respirators fitted with specific cartridges for the type of materials being used. Every laboratory, especially those using these materials, should be individually equipped with deluge showers and eyewash fountain combinations.

Corrosive chemicals should always be handled with the greatest care. Where available, they should be purchased in containers coated with a protective plastic film. If coated containers are dropped the probable result should be at most a leak through the film instead of a potentially dangerous splashing of the chemical onto the skin of the person transporting the material and possibly others. Although there is a surcharge in most cases for ordering containers with protective films, the additional protection afforded is substantial. Even with the film, the material leaking out of the container can still represent a nasty mess which must be cleaned up promptly. Spill kits for various corrosives are available from most laboratory chemical and equipment supply firms.

Safety carriers are available to use in transporting containers of dangerous chemicals. These should be used whenever the chemical containers are unprotected, breakable bottles. Unfortunately, it is unlikely that they will be used for the movement of chemicals a few feet within a laboratory, which is much more common than transporting materials from one laboratory to another, or from the stockroom to the laboratory. Just as most car accidents occur within a few miles of home, most chemical accidents occur where the workers spend most of their time. The use of protected containers can largely ameliorate the seriousness of these accidents, since it does not require an extra effort to use them.

When it is necessary to move chemical containers a significant distance, the use of safety carriers, even for protected containers, is strongly recommended. Although the result is likely to be much less serious when using coated bottles, accidents can still occur and it is always desirable not to have to clean up after one. If several bottles are to be moved at once, the bottles should be moved on a low cart, with a substantial rim around the edge, so that if the cart is struck the chemicals will be likely to stay on it, or if not, have only a short distance to fall.

Wherever possible chemicals should be moved from one floor to another on a freight elevator rather than carrying them manually up and down a flight of stairs. If only a passenger elevator is available, the use of the elevator should be delayed until no other passengers are using it or passengers desiring to use it should be courteously asked to wait until the chemicals have been moved. Although elevator accidents are rare, should a dangerous material be released in one while the passengers were trapped inside, the results could be catastrophic. The fumes could easily cause serious injury before it would be possible to leave the elevator.

Laboratory personnel should be trained to quickly limit the area affected by a spill, and the necessary supplies to enable them to do so should be immediately available within the laboratory. Kits containing absorbent pillows, neutral absorbent materials, or neutralizing materials are commercially available and will enable knowledgeable persons to safely contain most small accidents, such as spills of the contents of a single bottle of reagent, until a final cleanup in which the materials in the container as well as the materials used to contain the spill can be disposed of, usually as hazardous waste.

If the average person who has taken chemistry class retains one safety rule, it is probably the one about always adding acid to water, never the reverse (often brought up in the context of discussing the use of sulfuric acid, which is a strong dehydrating agent). This precaution is taken to avoid splashing of acid on the body due to the localized generation of heat as the two substances mix.

There are a number of other basic safety procedures involving corrosives. Keep the container sizes and quantities on hand as small as possible, consistent with the rate of use. Store each class by themselves. Keep containers, not in use, in storage, and store the containers either in separate cabinets or on low shelves rather than high ones. Remember that reactions involving these substances will usually generate substantial heat, so that closed containers in the area of a spill involving these highly reactive materials could become hot and rupture due to excessive pressures.

Some brief comments about some of the classes of corrosives are given below. In the sections which follow, some of these as well as others will be covered in more detail.

1. Strong Acids - Concentrated strong acids can cause severe and painful burns. The pain is due in part to the formation of a protein layer which resists further penetration of the acid. In general, inorganic acids are more dangerous than organic acids, although the latter can cause deep-seated burns on extended contact with the skin. Leakage from

containers and material remaining on the outside of the containers following a sloppy transfer can cause corrosion of the shelving and, if the acids are stored with materials with which they may react, accidents can result if the chemicals come into contact.

2. Strong Alkalis - Alkali metal hydroxides are very dangerous when allowed to come into contact with tissue. The contact with the skin is likely to be less painful than a comparable exposure to acid because a protective protein barrier is not formed. Damage may extend to greater depths as a result of the lesser pain because the injured person may not be as aware of the seriousness of the incident. Any area exposed to a strong alkaline material should be flooded with water for at least 15 minutes or longer. This is especially important in eyes since the result of an exposure can be a rupture of the global structure of the eye.

3. Nonmetal Chlorides - Compounds such as phosphorous trichloride and corresponding bromides react violently with water and are a common cause of laboratory accidents.

4. Dehydrating Agents - Strong dehydrating agents such as sulfuric acid, sodium hydroxide, phosphorous pentoxide, calcium oxide, and glacial acetic acid can cause severe burns to the eyes because of their strong affinity to water. When they are added to water too rapidly, violent reactions accompanied by spattering can occur.

5. Halogens - Halogens are corrosive on contact with the skin, eyes, and the linings of the respiratory system as well as being toxic. Because they are gases, they pose a greater danger, especially by inhalation, of coming into contact with sensitive tissue.

The following Internet reference provides a concise list of common laboratory corrosives.

INTERNET REFERENCE

1. http:// www.orcbs.msu.edu/Chemical/chp/appendixf.html

3. High-Energy Oxidizers

Oxidizing agents such as chlorates, perchlorates, peroxides, nitric acid, nitrates, nitrites, and permanganates represent a significant hazard in the laboratory because of their propensity under appropriate conditions to undergo vigorous reactions when they come into contact with easily oxidized materials, such as metal powders and organic materials such as wood, paper, and other organic compounds. Elements from group 7A of the Periodic Table, fluorine, chlorine, bromine, and iodine, react similarly to oxygen and are also classified as oxidizing agents.

Most oxidizing materials increase the rate at which they decompose and release oxygen with temperature. The rate of decomposition of hydrogen peroxide goes up by a factor of about 1.5 with each $10°F$ ($5.6°C$). Because of this ability to furnish increasing amounts of oxygen with temperature, the reaction rate of most oxidizing agents is significantly enhanced with increasing temperature and concentrations. The hazard associated with these agents therefore increases as well. For example, cold perchloric acid at a concentration of 70% or less has little oxidizing power, while at concentrations above 73% has significant oxidizing power at room temperatures, and which increases still further at higher concentrations. Hot, concentrated perchloric acid is a very strong oxidizing agent. Containers of oxidizing agents may explode if they are involved in a fire within a laboratory.

The quantities of strong oxidizing agents stored within the laboratory should be minimized, and their containers should be rigorously segregated from materials with which they could react. The containers, preferably the original shipping containers, should be protected glass, with inert stoppers instead of rubber or cork.

Quantities of potentially vigorously reacting materials, such as strong oxidizing agents, used in a given research operation or evolution should be kept to the minimal quantities needed in a cool, storage area, isolated from other materials. The work should always be performed in a hood, with appropriate safety features (such as the wash-down system, recommended for research involving hot perchloric acid digestions). Oxidizing agents should be heated with glass-heating mantles or sand baths. The use of personal protection devices, including sturdy gloves and eye protection which provides both chemical splash and impact protection, should be mandatory. If the potential for explosions is determined to be significant, the operator as well as others within the facility should be protected with explosion barriers. If the risk is sufficiently high, the research should be performed in an isolated facility especially designed for the program, which would include explosion venting and explosion-resistant construction.

The following are brief comments regarding the hazardous properties of a number of representative common, powerful oxidizing reagents. Most form explosive mixtures with combustible materials, organic substances, or easily oxidizable materials and most yield toxic products of combustion. Current MSDSs will provide additional data on each of these materials.

Ammonium Perchlorate (NH_4ClO_4) - Similar in explosion sensitivity to picric acid. Explosive when mixed with organic powders or dusts. Highly sensitive to shock and friction when mixed with powdered metals, carbonaceous materials, and sulfur.

Ammonium Permanganate (NH_4MnO_4) - May become shock sensitive at $60°C$ ($140°F$) and may explode at higher temperatures. Avoid contact with readily oxidizable, organic, or flammable materials.

Barium Peroxide (BaO_2) - Combinations of this compound and organic materials are sensitive to friction and contact with small quantities of water.

Bromine (Br) - Highly reactive material. Causes serious burns to tissue; toxic; when inhaled can cause serious damage to respiratory system.

Calcium Chlorate ($Ca(ClO_3)_2$) - Explosive mixtures are ignitable by heat and friction.

Calcium Hypochlorite ($Ca(ClO)_2$) - Ignites easily when in contact with organic and combustible material. Chlorine evolved at room temperatures when mixed with acids.

Chlorine Trifluoride (ClF_3) - Vapor at room temperature and is dangerously reactive. Most combustible materials ignite spontaneously on contact. This is an exception to the use of glass containers. The material reacts strongly with silica, glass, and asbestos. Extremely toxic. Causes severe burns to tissue.

Chromium Anhydride or Chromic Acid (CrO_3) - Ignites on contact with acetic acid and alcohol and may react sufficiently vigorously with other organic materials to ignite.

Dibenzoyl Peroxide ($(C_6H_5CO)_2O_2$) - Extremely explosion sensitive to shock, heat, and friction. Comparatively low toxicity.

Fluorine (F_2) - Extremely reactive gas, reacting vigorously with most oxidizable materials at normal room temperatures, often vigorously enough to ignite. Burns with invisible flame. Causes severe burns to tissue. Severe danger of damage to respiratory tract.

Hydrogen Peroxide (H_2O_2) - Commercial products usually sold inhibited against decomposition. At concentrations between 35 and 52%, shares properties of hazards of other oxidizing agents associated with coming into contact with easily oxidizable materials, but may violently decompose when coming into contact with many common metals and their salts, e.g., brass, bronze, chromium, copper, iron, lead, manganese, silver, etc. At higher concentrations, most combustible materials will ignite on contact. Mixing of organics with concentrated hydrogen peroxides may create very sensitive explosive combinations. Solutions at concentrations above 8% must be stored with vented caps. Otherwise, the pressure within the container from oxygen released from the decomposing hydrogen peroxide could rupture the container.

Magnesium Perchlorate ($Mg(ClO_4)_2$) - Sensitive to ignition by heat or friction.

Nitric Acid (HNO_3) - Explosively reactive with carbides, hydrogen sulfide, metallic powders, turpentine. Causes severe burns to tissue.

Nitrogen Peroxide (in equilibrium with nitrogen dioxide (N_2O_4; NO_2)) - May cause fire, on contact with clothes and other combustible materials. Reactions with other fuels and chlorinated hydrocarbons may be violent. Vapors are life threatening at very low concentrations. Severely dangerous to tissue.

Nitrogen Trioxide (N_2O_3) - May cause fire on contact with combustible materials. Very damaging to tissue, especially respiratory tract, where fatal pulmonary edema may result, although onset of symptoms may be delayed for several hours.

Perchloric Acid ($HClO_4$) - Very dangerous oxidizing agent at high concentrations and elevated temperatures (see Chapter 4, Section VI.B.4).

Potassium Bromate ($KBrO_3$) - Sensitive to ignition by heat or friction. Relatively moderate health hazard.

Potassium Chlorate ($KClO_3$) - Similar explosive properties to potassium bromate but is toxic and fumes liberated by combustion are toxic.

Potassium Perchlorate ($KClO_4$) - Similar to potassium chlorate above. Yields toxic fumes in fires and is irritant to eyes, skin, respiratory system.

Potassium Peroxide (K_2O_3) - Reacts vigorously with water. Mixtures with combustible, organic, or easily oxidizable materials are explosive. They ignite easily with heat, friction, or small quantities of water. Toxic, if ingested.

Propyl Nitrate (normal) ($CH_3(CH_2)_2NO_3$) - Very dangerous material. Forms explosive mixtures with air. Very wide flammable limits (2% to 100%), Flash point = 20°C (68°F); very low energy required for ignition comparable with acetylene and hydrogen. Vapors are heavier than air and may travel some distance to ignition source and flash back. Material itself is toxic by either inhalation or ingestion and combustion products highly toxic.

Sodium Chlorate ($NaClO_3$) - Properties similar to potassium chlorate above.

Sodium Chlorite ($NaClO_2$) - Releases explosive, extremely poisonous chlorine dioxide gas upon contact with acid.

Sodium Perchlorate ($NaClO_4$) - Properties similar to potassium perchlorate above.

Sodium Peroxide (Na_2O_2) - Properties similar to potassium hydroxide above.

REFERENCES

1. Hazardous Chemical Data, NFPA 49, National Fire Protection Association, Quincy, MA.
2. **Armour, M.A., Browne, L.M., and Weir, G.L.,** *Hazardous Chemicals, Information and Disposal Guide,* 2nd ed., University of Alberta, Edmonton, Canada, 1984.
3. **Lewis, R.J., Sr.,** *Sax. Dangerous Properties of Industrial Materials,* 8th ed., Van Nostrand Reinhold, New York, 1993.

INTERNET REFERENCE

1. How do I work safely with oxidizing liquids and solids? Canadian Centre for Occupational Health & Safety (CCOOHS). http://www.coohs.ca/oshanswers/prevention/oxidizing.html.

4. Perchloric Acid*

Perchloric acid is a clear liquid, has no odor, and boils at 203°C (397°F). Hot perchloric acid is a very strong oxidizing agent used for the complete digestion of organic materials. Room temperature perchloric acid at concentrations of 70% or less does not have significant oxidizing

---

* This section contains excerpts from the material in the 4th Edition of the Handbook of Laboratory Safety, by E. A. Graf, Jr.. Perchloric acid is used less often than it once was so the space for this topic has been considerably shortened, but essential material has been retained.

power although it is still a strong non-oxidizing acid, but at higher temperatures and higher concentrations it develops strong oxidizing properties.

a. Recommendations for the Safe Handling of Perchloric Acid[*]

Graf provided updated recommendations in a paper in an article published in 1966. Material Safety Data Sheets are provided by the manufacturer when perchloric acid is purchased. An excellent summary of safety practices is given in the Internet reference at the end of this sections. The recommendations from these and other sources are combined and summarized below.

i.   Facility and Equipment

Floors - Perchloric acid should not be handled in a building with wooden floors but instead on concrete protected with epoxy paint. The epoxy does not react significantly with room temperature perchloric acid, and is relatively simple to clean up, following spill cleanup procedures specific for perchloric acid.

No equipment of any kind should ever be bolted to a floor by using bolts that screw into the floor. Perchlorates can enter and form hazardous metallic perchlorates that can initiate a detonation when the bolt is removed. Studs, firmly and permanently set into the floor to which the equipment can be bolted, are far safer. The nuts can then be flushed with water and sawed off with a hacksaw under a constant water spray if necessary to remove that equipment.

Laboratory benches - Laboratory benches should be constructed of resistant materials, not wood, to prevent acid absorption, especially at the bottom surface which rests on the floor and would be subject to the greatest exposure from acid spills. Bench tops of resistant and nonabsorbent materials such as chemical stoneware, tile, epoxy composites, and polyethylene are recommended.

Shelves and cabinets - Shelves and cabinets of epoxy-painted steel are highly recommended over wood.

Heating source - Hot plates (electric), electrically or steam-heated sand baths, or a steam bath are recommended for heating perchloric acid. Direct flame heating or oil baths should not be used.

Vacuum - Vacuum pumps from which all traces of petroleum lubricants have been flushed and refilled with halocarbon, Kel-F, or fluorolube are recommended.

Glassware. The hazards that may ensue if an apparatus cracks or breaks due to thermal or mechanical shock are sufficient to make it desirable that quartz apparatus be considered, especially as it is necessary in many experiments to chill rapidly from the boiling point.

Glass-to-glass unions, lubricated with 72% perchloric acid, seal well and prevent joint freezing arising from the use of silicon lubricants. Rubber stoppers, tubes, or stopcocks should not be used with perchloric acid due to incompatibility. Repeated exposure of the motor windings to perchloric acid vapor could result in a fire, unless the motor is an explosion-proof type.

Sundry items - Tongs—The choice of tongs for handling hot flasks and beakers containing perchloric acid mixtures should be given due thought. Since the use of radioactive materials has become commonplace, much thought has been put into the design of indirect handling equipment. The cheap, commonly used crucible tongs are most unsuitable for picking up laboratory glassware. If possible, tongs with a modified jaw design should be used to ensure that a safe grip is obtained.

Stirrers. —Pneumatically driven stirrers are recommended rather than the electric motor type.

ii.   Operations with Perchloric Acid

1. Use chemical splash and impact-rated goggles or face shield, gloves and protective apron

---

[*]   See Table 4.12 for materials resistant to perchloric acid.

Table 4.12 Materials Compatible With and Resistant to 72% Perchloric Acid

Compatible	
Material	**Compatibility**
Elastomers	
Gum rubber	Each batch must be tested to determine compatibility
	Slight swelling only
Vitons[1]	
Metals and alloys	Excellent
Tantalum	Excellent
Titanium (chemically pure grade)	
Zirconium	Excellent
Niobium	Excellent
Hastelloy*	Slight corrosion rate
Plastics	
Polyvinyl chloride	
Teflon*	
Polyethylene	
Polypropylene	
Kel~F**	
Vinylidine fluoride	
Saran***	
Epoxies	
Others	
Glass	
Glass-lined steel	
Alumina	
Fluorolube	

Incompatible
Plastics
Polyamide (nylon)
Modacrylic ester, Dynel (35 - 85%) acrylonitrile
Polyester (dacron)
Bakelite
Lucite
Micarta
Cellulose-based lacquers, Metals
Copper
Copper alloys (brass, bronze, etc.) for very shock-sensitive perchlorate salts
Aluminum (dissolves at room temperature)
High nickel alloys (dissolves), Others
Cotton
Wood
Glycerin-lead oxide (letharge)

* duPont Trademark, **3M trademark, ***Dow Chemical Trademark

whenever the acid is handled.

2. Always transfer acid over a sink in order to catch any spills and afford a ready means of disposal.

3. Perchloric acid digestions or any procedure involving heating of perchloric acid must always be done in a fume hood designed as a perchloric acid hood. This type of hood is made of stainless steel or PVC and includes a wash down system which includes the entire exhaust system.

4. No organic materials are to be in the hood when a digestion is taking place.

5. No organic materials should be stored in a perchloric acid hood.

6. Do not allow perchloric acid to come into contact with strong dehydrating agents (concentrated sulfuric acid, anhydrous phosphorous pentoxide, etc.).

7. Perchloric acid should be used only in standard analytical procedures from recognized analytical texts involving perchloric acid.

8. Perchloric acid waste is not to be mixed with any other waste but returned to the original container, if feasible, and disposed of as hazardous waste.

9. The amount of perchloric acid used should be kept to the minimum required.

10. Perchloric acid must be stored by itself, in an appropriate cabinet, and on a ceramic or plastic pan. Organic acids, bases, flammable material or other organics should not be stored in the same location.

11. No maintenance work should be done on a perchloric hood unless it has been thoroughly decontaminated first.

iii. Perchloric Acid Spills

Perchloric acid spilled on the floor or bench top represents a significant hazard, especially if allowed to dry. It should not be mopped up, nor should dry combustibles be used to soak up the acid. The spilled acid should first be neutralized and then soaked up with wet rags or spill pillows. The contaminated wipes must be kept wet to prevent combustion upon drying. They should be placed in a plastic bag and sealed and then placed in a flammable waste disposal can. Unless the publically owned water treatment works (POTW) does not allow it, since perchloric acid dissolves easily in water, the spill can be safety rinsed into the sanitary drain, followed by substantial quantities of water. The workers should wear a chemical splash goggles or a face mask, chemical gloves, coveralls, and protective shoe covers.

iv. Disposal

Stir the acid into cold water until the concentration is less than 5%; follow by neutralization with aqueous sodium hydroxide; then dispose of the resulting mixture in the sanitary system, accompanied by abundant water. Larger quantities in the original unopened containers may be acceptable to a commercial hazardous waste vendor. If it is potentially explosive, the best option available is to hire a firm specializing in disposal of exceptionally hazardous materials. This will be expensive.

b. Decontamination/Removal of a Perchloric Acid Fume Hood System

If perchloric acid has been used in a modern perchloric acid hood system, consisting of a stainless steel hood and duct system, welded seams, an automatic wash down system, which includes the exhaust duct system, equipped with an induction blower so the fumes never pass through the exhaust blower, and discharged through a continuously purely vertical exhaust duct ending at least 10 feet or better above the roof, there should be minimal safety problems with the operation. However, in older systems, one or all of these features may have been missing. In the author's personal experience, one system was found to have all of the fan blades corroded to the extent that they were lying on the roof. The motor was still running and the hood was being used but with zero exhaust velocity. In another case, not only was there a horizontal run in the exhaust duct but the exhaust fan was at the beginning of the run. The horizontal section was thoroughly corroded and there were loose perchloric acid crystals on the surface below it. In both cases, a firm was hired to remove the systems following closely the following recommendations which were patterned after those reported in the Graf article. The work was done without incident. It is critical that if an external firm is hired to do this kind of work, that they have demonstrated experience doing it. Again in the authors experience, an inexperienced firm was hired not having been reviewed by a knowledgeable safety person with very unfortunate results.

c. Procedures For Decommissioning a Perchloric Acid Hood

1. Risks can be minimized by performing the work during periods of low occupancy, such as evenings and weekends. The immediate area should be evacuated of unessential personnel.
2. Wash the entire system for at least 12 hours, just prior to dismantling, by introducing a fine water spray within the hoods, and at the exhaust end of the duct system with the fans operating.
3. Hose the fan downs thoroughly.
4. Thoroughly wet and then carefully remove the fan mounting bolts and connectors. Nonsparking tools should be used throughout.
5. Remove the fans to the outdoors. Keep a substantial shield between the workers and the fans during transit.
6. Dismantle the fans individually behind a substantial shield. Thoroughly rewash.
7. Remove the mounting plate bolts carefully and systematically in such a way as to minimize the chance of sparking. Any tools required should be nonsparking.
8. Wash and clean all disassembled parts. Any adhering gasket material on the flange are to be scraped off with a wooden scraper.
9. Any metal work requiring cutting , e.g. duct work, bolts , etc., should be cut with a high speed saw with a water spray used to cool the cut and wash away any residue.

REFERENCES

1. **Robinson, W.R.,** Perchlorate salts of metal ion complexes: potential explosives, *J. Chem. Ed,* 62(11), 1001, 1985.
2. **Shumacher, J.C.,** *Perchlorates: Their Properties, Manufacture and Uses,* American Chemical Society Monogr. Set No. 146, Reinhold, New York, 1960.
3. **Graf, E.A.,** Safe handling of perchloric acid, *Chem. Eng. Prog.,* 62(10), 109, 1966.
4. **Muse, L.A.,** Letter to the editor, *Chem. Eng. News,* 5 1(6), 29, 1973.
5. **Elsenbannei R.L. and Miller, G.G.,** Letter to the editor, Chem. Eng. News, 63(25) 4, 1985.
6. **Varyn, M.E., Schenoff, J.B., and Dabkowski, G.M.,** Letter to the editor, *Chem. Eng. News,* 63(27), 4, 1985.
7. **Pennington, B.E.,** Letter to the editor, *Chem. Eng. News,* 62(32), 55, 1982.
8. **Osterholm, J. and Passiniemi, P.,** Letter to the editor, *Chem. Eng. News,* 64(6), 2,1986.
9. **Raymond, K.N.,** Letter to the editor, *Chem. Eng. News,* 63(27), 4, 1985.
10. **Everett, K. and Graf, F.A., Jr.,** Handling perchloric acid and perchlorates, in *CRC Handbook of Laboratory Safety* 2nd ed., Steere, N.V., Ed., CRC Press, Cleveland, OH, 1971, 265.
11. **Breysse, Peter A.,** *Occupational Health Newsletter,* 15(2,3) 1, 1966.

INTERNET REFERENCES

1. http://www.orchs.msu.edu/chemical/safeperchloricuse.html
2. http://www.chem.utah.edu/MSDS/P/PERCHLORIC_ACID%2C_69-72%25

5. Ethers

Much of the concern in working with ethers is due to the problem of peroxide formation. Chapter 4, Section V.D., already dealt with this at some length, discussing this problem in the context of storage of ethers. If ethers are bought in small sizes, so that the containers, in use on a given day are emptied, the concerns of peroxide formation in partially full containers in storage should not arise. Peroxides can form in unopened containers as well, although due to the absence of excess available air in the restricted empty space above the liquid level, the rate should be much slower. However, the inventory of unopened containers in a laboratory should be maintained at a reasonable level so that no ethers should be kept for extended periods. All containers should be stamped with the date received and a schedule of disposal or testing established. If a partially empty container is kept, a target date for testing for peroxides should be placed on the container, as well as subsequent dates for the tests to be repeated. A centralized computer tracking program for chemicals could take this responsibility from the individual laboratory managers and largely eliminate the problem of outdated ethers. The effort needed to maintain a reliable program in the laboratory to check for the presence of peroxides could almost certainly be used to greater advantage on the basic research program. Unused portions of ethers should not be returned to the original container. Small quantities can be allowed to evaporate in a fume hood.

Because of the explosion risk associated with peroxides, older containers of ethers are not normally accepted as part of an ordinary hazardous waste shipment by a commercial chemical waste disposal firm. They would have to be disposed of as unstable explosive materials which is much more expensive than if the materials were not outdated. Any savings in buying the "large economy size" would be more than compensated for by the additional disposal costs. Attempting to treat the ethers to remove the peroxides or to dispose of them by laboratory personnel carries with it the risk of an explosion and the subsequent liability for injuries.

Although a major risk usually associated with ethers is, as noted, the problem of peroxides, they also pose additional problems because of their properties as flammable solvents. Many of them have lower explosion limits, in the range of 0.7 to 3%, and flash points at room temperature or, in several cases, much lower. Ethyl ether, the material that most frequently comes to mind when "ether," otherwise unspecified, is mentioned, has a lower explosion limit of 1.85%, an upper explosion limit of 36%, and a flash point of -45°C (-49°F). The vapors of most ethers are heavier than air, and hence can flow a considerable distance to a source of ignition and flash back. Because of their flammable characteristics, many ethers, especially ethyl ether, placed in improperly sealed containers in an ordinary refrigerator or freezer release vapors which represent a potential "bomb" that can be ignited by sparks within the confined space and explode with sufficient force to seriously injure or even kill someone in the vicinity. A fire is virtually certain to result should such an incident occur which, depending upon the type of construction and the availability of fuel in the laboratory area, could destroy an entire building. Because ethers, as well as other flammable solvents, are used so frequently in laboratory research, it is strongly recommended that all refrigeration units in wet chemistry laboratory environments should be purchased without ignition sources within the confined spaces and with explosion-proof compressors, i.e., they must meet standards for "flammable material storage," even if the current program does not involve any of these materials. At some time during the effective lifetime of most refrigeration units, it is likely that flammable liquids will be used in the facility where they are located.

A spill should be promptly cleaned up using either a commercial solvent spill kit material to absorb the liquid or a preparation of equal parts of soda ash, sand, and clay cat litter, which has been recommended as an absorbent. Since the lower explosion limit concentrations are so low for so many of the commonly used ethers, all ignition sources should be promptly turned off following a spill and all except essential personnel required to leave the area. The personnel performing the clean up should wear half-face respirators equipped with organic cartridges. The resulting waste mixture from the clean up can be placed in a fume hood temporarily until removed from the laboratory for disposal as a hazardous waste.

Class B fire extinguishing agents are to be used to combat ether fires as well as other fires involving flammable liquids. Usually the most effective are dry chemical extinguishers which interrupt the chemistry of a fire, while carbon dioxide units can be used to smother small fires. Portable Halon extinguishers are also usable if the fire is such that the fuel has time to cool before the concentration of the halogenated agents falls below the critical concentration at which it is effective (the current extinguishing agents used in most of these, Halon-1301 and Halon-1211, damage the ozone in the atmosphere and will be replaced with units using safer alterative agents). It is worth pointing out once again that unless there is a reasonable chance of putting out a fire with portable extinguishers, it is preferable to initiate an evacuation as quickly as possible and to make sure that everyone can safely leave the building, rather than engage in a futile attempt to put out the fire.

In general, the toxicity of ethers is low to moderate, although this generalization should be confirmed for each different material to be used by information obtained either from the container label or the MSDS. Prolonged exposure to some ethers has been known to cause liver damage. Many have anesthetic properties and are capable of causing drowsiness and eventual unconsciousness. In extreme cases of exposure, death can result. Four ethylene glycol ethers - 2-methoxyethanol, 2-ethoxyethanol, 2-methoxyethanol acetate, and 2-ethoxyethanol acetate have been identified as causing fetal developmental problems in several animal species, including fetal malformations and resorption and testicular damage. Studies have also shown adverse hematologic effects and behavioral problems in the offspring of animals.

6. Flammable Solvents

Much of the concern in the literature is centered on the flammable characteristics of flammable liquids, much of which has already been addressed in Chapter 4, Section V.G. This characteristic will be treated in more detail in the following section. There are many other issues relating to the health effects of these solvents which also need to be considered.

a. Flammable Hazards

The fire hazard associated with flammable liquids should be more appropriately associated with the vapors from the liquid. It is the characteristics of the latter which determine the seriousness of the risk posed by a given solvent. In previous sections on storage of liquids, two of the more important properties of flammable liquids were mentioned in terms of defining the classes to which a given solvent might belong. Definitions of Class lA, lB. 1C, 2A, 3A and 3B were based on the boiling point and flash point of the solvents. The formal definition of these two terms will be repeated below, as well as three other important parameters relevant to fire safety, the ignition temperature and the upper and lower explosion limits.

Boiling point (bp) - This is the temperature at which the vapor of the liquid is in equilibrium with atmospheric pressure (defined at standard atmospheric pressure of 760 mm of mercury).

Flash point (fp) - This is the minimum temperature at which a liquid gives off vapor in sufficient concentration to form an ignitable mixture with air near the surface of a liquid. The experimental values for this quantity are defined in terms of specific test procedures which are based on certain physical properties of the liquid.

Ignition (autoignition) temperature - This is the minimum temperature which will initiate a self-sustained combustion independent of the heat source.

Lower explosion (or flammable) limit (lel) - This is the minimum concentration by volume percent in air below which a flame will not be propagated in the presence of an ignition source.

Upper explosion (or flammable) limit (uel) - This is the maximum concentration by volume percent of the vapor from a flammable liquid above which a flame will not be propagated in the presence of an ignition source.

For a fire to occur involving a flammable liquid, three conditions must be met: (1) the concentration of the vapor must be between the upper and lower flammable limits; (2) an oxidizing material must be available, usually the air in the room; and (3) a source of ignition must be present. The management strategy is usually to either maintain the concentration of the vapors below the lower flammable limit by ventilation (such as setting the experiment up in an efficient fume hood) or to eliminate sources of ignition. The latter is easier and more certain because the ventilation patterns even in a hood may be uneven or be disturbed so that the concentrations may locally fall into the flammable range. While some materials may require an open flame to ignite, it is much easier to ignite others. For example, the ignition temperature for carbon disulfide is low enough (80° C or 176°F) that contact with the surface of a light bulb may ignite it.

In order to work safely with flammable liquids, there should be no sources of ignition in the vicinity, either as part of the experimental system or simply nearby. Use nonsparking equipment. When pouring a flammable liquid from one metal container to another, both of the containers should be grounded, as the flowing liquid can itself generate a static spark. Flammable materials should be heated with safe heating mantles (such as a steam mantle), heating baths, or explosion-safe heating equipment. Many ovens used in laboratories are not safe for heating flammables because the vapors can reach the heating element, or either the controls or thermostat may cause a spark. Any spark-emitting motors should be removed from the area. Flammable materials should never be stored in an ordinary refrigerator or freezer because they have numerous ignition sources in the confined volume. Placing the entire system in a hood where the flammable vapors will be immediately exhausted aids in limiting the possibility of the vapors coming into contact with an ignition source.

The vapors of flammable liquids are heavier than air and will flow for a considerable distance away from the source. Should they encounter an ignition source while the concentration is in the flammable range, a flame may be initiated and flash back all the way to the source. In at least one instance, a fire resulted when a research worker walked by a fume hood carrying an open beaker of a low ignition point volatile solvent. There was an open flame in the hood, and when the fumes were pulled into the hood, the fumes ignited and flashed back to the container, which was immediately dropped, with the result that the feet and lower legs of the worker and the entire floor of the laboratory became engulfed in flame. Fortunately, a fire blanket and fire extinguisher were immediately available, so that the worker escaped with only minor burns and the fire was extinguished before anything else in the laboratory became involved. This incident clearly illustrates the need to consider all possibilities of fire when using solvents, and that a hood does not totally isolate a hazardous operation.

REFERENCES

1. **Armour, M.A., Browne, L.M., and Weir, G.L.,** *Hazardous Chemicals Information and Disposal Guide,* 2nd ed., University of Alberta, Edmonton, Canada, 1984.
2. *Federal Register* F.R. 10586, April 2, 1987.
3. **Langan, J.P.,** Questions and Answers on Explosion-Proof Refrigerators, Kelmore, Newark, NJ.
4. *Prudent Practices for Handling Hazardous Chemicals in Laboratories,* National Academy Press, Washington D.C., 1981, 57.
5. **Lewis, R.J.,** *Sax Dangerous Properties of Industrial Materials,* 8th ed., Van Nostrand Reinold, New York, 1993.
6. Hazardous Chemicals Data, NFPA 49, National Fire Protection Association, Quincy, MA.

7. Reactive Metals

Lithium, potassium, and sodium are three metals that react vigorously with moisture (lithium to a lesser extent than the other two, except in powdered form or in contact with hot water), as well as with many other substances. In the reaction with water, the corresponding hydroxide is formed along with hydrogen gas, which will ignite. Lithium and sodium should be stored under mineral oil or other hydrocarbon liquids that are free of oxygen and moisture. It is specifically recommended in the literature that potassium be stored under dry xylene.

The chemical hazards of the three metals are similar in many ways. All three form explosive mixtures with a number of halogenated hydrocarbons; all three react vigorously or explosively with some metal halides, although potassium is significantly worst in this respect, and the reaction of all three in forming a mercury amalgam is violent. They all react vigorously with oxidizing materials. Potassium can form the peroxide and super oxide when stored under oil at room temperature and may explode violently when cut or handled. Sodium reacts explosively with aqueous solutions of sulfuric and hydrochloric acids. The literature provides a number of other potentially dangerously violent or vigorous reactions for each of these three materials. It is not the intent here to list all of the potentially dangerous reactions that may occur, but to point out that there are many possibilities for incidents to happen. No one should plan to work with these materials without carefully evaluating the chemistry involved for potential hazards. The materials should be treated with the care which their properties demand at all times.

In one instance, a very old can of sodium was determined by visual inspection to have "completely" reacted to form sodium hydroxide, and the worker decided to flush it with water to dispose of the residue. The bottom 2 inches were still sodium metal and consequently the can exploded. The only entrance to the laboratory was blocked by an ensuing fire so that the occupants had to escape through windows. Fortunately, they were on the first floor and the windows were not blocked. The latter point is worth noting because shortly before this incident, bars over the windows to prevent break-ins had been removed at the insistence of the organization's safety department. Unsubstantiated assumptions or misplaced priorities are a major cause of injuries.

Since the three metals all react vigorously with moisture, care should be taken to avoid skin and eye contact, which could result in burns from the evolved heat and direct action of the hydroxides. The materials should always be used in a hood, and, as a minimum, gloves and chemical splash and impact-resistant goggles should be worn while working. If the risk of a violent reaction cannot be excluded, additional protection such as an explosion barrier or a face mask should be considered. If a fire should occur involving reactive metals, appropriate class D fire extinguishers should be available within the laboratory. These vary somewhat in their contents, which usually are specific for a given material. Appropriate material suitable for extinguishing fires involving these reactive materials are dry graphite, soda ash, and powdered sodium chloride. Other materials which might be used in these class D units as extinguishing agents are pulverized coke, pitch, vermiculite, talc, and sand. These materials will usually be mixed with various combinations of low melting fluxing salts, resinous materials, and alkali-metal salts which, in combination with the other material in the extinguisher, form a crust to smother the fire. Water (obviously), carbon dioxide, or halogenated units should not be used. There are a number of other reactive metals, which while not as active chemically as these three, once ignited, require class D fire extinguishing agents as well. These include magnesium, thorium, titanium, uranium, and zirconium. Other materials for which class D units should be used include metal alkyds and hydrides, red and white phosphorous, and organometallic compounds.

As with many other materials, one of the major considerations in using these reactive metals is the problem associated with disposal of unneeded surplus quantities or waste materials. There are suggestions in the literature for treating waste for each of these three materials. For example, small amounts of potassium residues from an experiment should be treated by promptly reacting them with tert-butyl alcohol because of the danger that they will explode (even if the potassium is stored properly). This is appropriate for small quantities in the laboratory but disposal of substantial quantities of unwanted material is a different matter. Recycling or transfer to another operation needing the material should be investigated, but disposal by local treatment should be avoided. There are limits to local treatment permitted under the RCRA Act, beyond which a permit is required to become a treatment facility. In addition, there are safety and liability risks associated with processing dangerously reactive materials which must be considered. Reactive metals are among those materials that require special handling by commercial waste disposal firms. Because special procedures are required, the cost of disposal is much higher than for routine chemical waste. Quantities purchased and kept in stock should be limited.

REFERENCES

1. Hazardous Chemicals Data, NFPA 49, National Fire Protection Association, Quincy, MA.
2. **Armour, M.A., Browne, L.M., and Weir, G.L.,** *Hazardous Chemicals Information and Disposal Guide,* 2nd ed., University of Alberta, Edmonton, Canada, 1984.
3. **Lewis, R.J.,** *Sax's Dangerous Properties of Industrial Materials,* 8th ed., Van Nostrand Reinhold , New York, 1993.
4. **Sax, N.I. and Lewis, R.J., Sr.,** *Rapid Guide to Hazardous Chemicals in the Workplace,* Van Nostrand Reinhold, New York, 1986.
5. Portable Fire Extinguishers, NFPA-10, National Fire Protection Association, Quincy, MA.

8. Mercury

Mercury and its compounds are widely used in the laboratory. As metallic mercury, it is often used in instruments and laboratory apparatus. In the latter application especially, it is responsible for one of the more common types of laboratory accidents, mercury spills. Thermometers containing mercury are frequently broken; mercury is often spilled in working with mercury diffusion pumps or is lost when cleaning the cold traps associated with high vacuum systems in which mercury pumps are used. Over a period of time, the small amounts of mercury lost each time can add up to a substantial amount. In one instance, the cold traps were always cleaned over a sink. After a few years of this, a large amount of mercury accumulated in the sink trap which finally eroded the metal, and spilled on the floor. Over 15 pounds of mercury were recovered. Mercury is frequently ejected from simple manometers consisting of mercury in plastic tubing, connected to a system under vacuum. As an example of the consequences of this last type of accident, an employee, working in a room previously used for years as an undergraduate biology laboratory, was diagnosed as having a somewhat severe case of mercury poisoning, although he did not use mercury. Upon investigation, more than 50 pounds of mercury were retrieved from under the wooden floorboards in the room. Although the instructors had "cleaned up the mercury" when spills had occurred, over the years a substantial amount had obviously not been recovered, but had worked its way through the cracks in the floor. No measurements of the airborne concentration were made at the time, but the normal equilibrium vapor pressure of mercury in air at normal room temperatures is between 100 and 200 times the current permissible levels of mercury in the workplace. It should also be noted that the vapor pressure rises rapidly with temperature. At the temperature of boiling water at standard atmospheric pressure, the vapor pressure is more than 225 times higher than at 20°C (68°F) (0.273 mm Hg) and reaches 1 mm Hg at 126.2°C (259.2°F). Clearly, mercury always should be heated in a functioning fume hood instead of an open bench, yet this is not always done.

Mercury poisoning has been known to affect many individuals, among them such prominent scientists as Pascal and Faraday as well as workers in various industries, such as those exposed to mercury as an occupational hazard while using mercuric nitrate in the hat industry in making felt. A frequently cited example of the effects of mercury poisoning is the "Mad Hatter" in Lewis Carroll's *Alice in Wonderland.* In the 1950s, many Japanese in a small fishing village suffered serious permanent damage to their central nervous system and, in many cases, death or permanent disability due to eating fish containing methyl mercury as a result of the industrial discharge of mercury compounds into the sea near their village. In the 1970s, fish taken from some of the common waters of Canada and the U.S. were found to be contaminated with mercury, and fishing was banned in some areas. Some commercial swordfish and tuna were found to be contaminated with mercury and had to be withdrawn from the market In at least one instance, fish in a river in the southeastern United States were found to be contaminated by mercury-containing waste from a chemical plant. Eventually the plant closed down its operations because the expense of modifying its operation to eliminate the discharge would have been too high. Mercury compounds were at one time used as fungicides and many individuals died from mistakenly eating seed corn treated with these materials. Probably the worse such incident occurred in 1972 when at least 500 persons died in Iraq from consuming treated grain mistakenly issued to them as food. Mercury, once it enters the biosphere is slow to biodegrade. As a result of the dangers inherent in materials containing mercury, the use of mercury for most agricultural purposes has been banned, and dumping wastes containing mercury compounds in such a way as to be able to contaminate the environment is no longer permitted. Currently it is difficult to dispose of compounds containing mercury. Waste liquid mercury metal, on the other hand, can be recycled. Small batteries containing mercury should not be disposed of in common trash.

Elemental mercury is probably not absorbed significantly in the gastrointestinal tract,but many of its compounds are. Poisoning due to inhalation and absorption of mercury vapors results in a number of symptoms. Among these are personality and physiological changes such as nervousness, insomnia, irritability, depression, memory loss, fatigue, and headaches. Physical effects may be manifested as tremors of the hands and general unsteadiness. Prolonged exposure may result in loosening of the teeth and excessive salivation. Kidney damage or even failure may result. In some cases the effects are reversible if the exposure ceases, but, as noted in the previous paragraph, ingestion of some organic mercury compounds may be cumulative and result in irreversible damage to the central nervous system. Alkyl mercury compounds have very high toxicity. Aryl compounds, and specifically phenyl compounds, are much less toxic (in the latter case comparable to metallic mercury) and therapeutic compounds of mercury are less toxic still. In the case of the Minemata Bay exposure in Japan in 1953 and in Nigata in 1960, it was found that the fetus was especially vulnerable to the exposure. Mercury passes readily through the placenta from mother to child. In recognition of the seriousness of the potential toxic effects, the permissible ceiling exposure level for metallic mercury is currently set by OSHA at 0.01 mg/M^3, and for alkyl organomercury compounds, an 8-hour time weighted average of 0.01 mg/M^3 has been established by OSHA with a ceiling average of four (current ACGIH recommendation is three times that level).

The first four paragraphs in the next section are taken directly from the article by Steere from the second edition of this handbook.

a. Absorption of Mercury by the Body

In occupational studies, the primary intake is by mercury vapor in the lungs, with up to 90% of the mercury taken in by this route being absorbed. A relatively minor amount is absorbed by the skin or large droplets reaching the gastrointestinal track, perhaps 15%, although when exposures by this route occur, the exposure level is likely to be high. Inorganic mercury is transformed to some extent by microorganisms in the mouth and gut to short-chain alkyl(methyl and ethyl) forms, which are readily absorbed. Further distribution of absorbed mercury is facilitated by the blood.

Inorganic and organic mercury compounds have a strong affinity for thiol chemical groups. Most proteins and all enzymes contain these groups so that mercury readily is bound to body tissues. Most mercury compounds are potent enzyme inhibitors which affects membrane permeability, which in turn affects nerve conduction and tissue respiration.

The biological half-life of mercury in the blood is approximately three days, following an exposure but the mercury bound to body tissues clears much more slowly with a half-life of about 90 days. Thus the end of an exposure will have long lasting effects. The levels in tissue will not fall below 10% of the peak level until somewhat more than four half-lives have passed.

The kidney plays a key roll in the absorption of mercury in the body. Kidney tissue contains a thiol-rich protein called metallothionein. Exposure of the kidney to mercury and other toxic metals causes production of this protein which binds the metals tightly, and retains it in the kidney in a relatively harmless form. As long as the kidney is not overwhelmed by the influx of the toxic metal, the excretion of mercury will eventually balance intake so that worsening of adverse symptoms will be limited. However, acute levels can lead to renal failure.

Chronic mercury exposure can seriously affect fertility and the outcome of pregnancy. Mercury passes readily through the placenta and the concentration in the cord blood is elevated above the maternal blood. In men, organic forms of mercury can cause hypospermia, and a reduction in libido and cause impotence in some men. For some men, there has been an increase in the rate of spontaneous abortion in their partners.

b. Excretion of Mercury

Mercury is excreted by the body through the feces and urine, with a minute amount by the respiration. The liver excretes some in bile, which is partially reabsorbed but is eventually disposed of by the kidney. Some mercury passes directly by the body in the urine instead of being bound by proteins. For a steady state exposure, the urine level reflects body burden of mercury. Another indicator of mercury intake is the concentration of mercury in the hair. As the hair grows the mercury levels can be measured along the length of the hair by such techniques as neutron activation analysis.

c. Control Measures

Mercury is dense (specific gravity of just under 13.6 at 4°C (39.2°F)) and has a high surface tension and low viscosity. As a result, it tends to break up into small droplets when it is poured or spilled. Anyone who has tried to pick up small droplets using a stiff piece of paper can attest to the appropriateness of the alternate name "quicksilver." As the droplets are disturbed as, for example, when walked upon on the laboratory floor, they tend to break up into smaller and smaller droplets, eventually becoming too small to see. In a laboratory where mercury has been in use for an extended period of time, it is instructive to run a pen knife in the cracks in a tile floor or in the seams where cabinets and bench tops fit together. Invariably, small droplets of mercury will be found.

Although a thin film of oxide will form on the skin of mercury droplets, it is very fragile and will break. Similarly, sprinkling flowers of sulfur on the location of a mercury spill has been suggested as a control measure but the surface film which forms also apparently is also very fragile and will allow the mercury underneath the film to be readily exposed.

d. Exposure Reduction

One of the simplest control measures is to reduce the amount of mercury used in instrumentation and equipment. Mercury thermometers can be replaced with alternatives. Vacuum gauges can be changed. Mercury thermometers which used to be used almost exclusively in hospitals, and which used to be a major source of spilled mercury, have been replaced by electronic digital devices which have the further advantages of being quick and less intrusive. At one time, the number of broken medical mercury thermometers was estimated at two per bed per year. One of the major sources of dumped mercury in the United States used to be from disposal of dry cell batteries, accounting for 86% of the total. One should look for statements on batteries declaring them to be mercury-free or nearly so before making a purchase.

Wherever possible, work with mercury should be done in a fume hood, preferably one that has a depressed surface, so that a lip will aid in preventing mercury spills from reaching the floor, and with a seamless interior, as recommended for radiological work and work with perchloric acid. As noted earlier, heating mercury causes it to emit fumes at concentration levels two to three orders of magnitude above the PELs. Heating of mercury should never be done on the open bench.

The general restrictions on eating, drinking, and smoking in the laboratory should be strictly enforced in laboratories where mercury compounds are commonly used. Depending upon the level of use and the availability of fume hoods, the use of personal protective equipment is recommended in addition to the use of goggles and laboratory aprons. Respiratory protection, consisting of half-face respirators fitted with a cartridge which will absorb mercury, should be considered if there is any potential for mercury exposure. A number of mercury compounds are absorbed through the skin and are strong allergens. The skin in such cases should be protected with gloves covering the forearms as well as the hands.

Vinyl tile is a commonly used material for the floors in a laboratory because it is "easy" to maintain and inexpensive to install. Ease of maintenance is not the case for a tile floor in a laboratory using mercury, because of the propensity of the extremely small (20 microns or less) mercury droplets to collect in the cracks. A seamless vinyl or poured epoxy floor should be used instead, with the joints of the floor with the wall being curved or "coved." Similarly, the bench top should be curved where it joins the back panel. Existing tile floors, especially the smaller 9 inch x 9 inch size, frequently represent an additional maintenance problem since a large proportion contain asbestos, as may the mastic holding them to the floor. When these tiles need replacing, the work must be done in conformance with EPA and OSHA asbestos standards and can be very costly. One procedure to be avoided at all costs is to grind up the old tile. This can distribute asbestos fibers so widely that the already expensive asbestos removal can be made prohibitively so.

e. Monitoring

The fumes from mercury provide no direct sensory evidence that they are present. Where use is substantial, monitoring measures should be readily available. The least expensive means to detect mercury levels is a detector tube in which a given volume of air is pulled through a glass tube containing a material that undergoes a color change when mercury comes into contact with the material. Normally the air is drawn through by means of a hand-operated pump. This method provides a reasonable accuracy but any finding within approximately 25% of the PEL should be considered to be sufficient warning of a possible overexposure of personnel working in the area. Each measurement requires a new tube. A popular instrument used to provide a direct reading of mercury concentrations in the air is a hand-held atomic absorption spectrometer in which air contaminated with mercury is drawn through the instrument and the degree of interference with ultraviolet light of the wavelength corresponding to a characteristic line of the mercury spectrum is translated into a numerical reading of the concentration of mercury in the air. The calibration of these instruments must be carefully maintained. Another instrument which provides an accurate, rapid reading of the level of mercury in the air depends upon the property of mercury to amalgamate with a thin gold film. This latter type of instrument is probably the most accurate and reliable but is also the most expensive. Unless the work with mercury compounds is quite heavy, the laboratory may be unable to afford either of the last two types of monitoring devices. In such a case, the safety department should be provided with an instrument to be used at all locations within an organization.

f. Spill Control Measures

Large globules of mercury can be cleaned up mechanically by carefully brushing them onto a dustpan or a stiff piece of paper. Another simple device is the use of a small mechanical hand-held pump to suck the globules into a small container. This is a tedious procedure limited to small spills and, of course, to droplets big enough to be seen. Bulk mercury recovered by these procedures can be recovered and purified for reuse.

Mercury spill kits available commercially usually include a small pump, sponges impregnated with a material to absorb mercury and which can be used to wipe up the area of a small spill, and a quantity of an absorbent powder that reacts with mercury to form a harmless amalgam. The latter can be spread on cracks and seams in the floor and furniture and is effective in collecting mercury from otherwise inaccessible places. After leaving the material on the floor or contaminated surfaces for several hours in order to allow the amalgam to form, the powder can be swept or brushed up and the waste material disposed of as a hazardous waste.

Ordinary vacuum cleaners **MUST NOT** be used to clean up a mercury spill. An ordinary vacuum filter bag will not stop an appreciable fraction of small particles in the region of several microns or less and, more importantly, the mercury globules pulled into the bag will be broken up into even finer droplets and spewed out of the vacuum's exhaust into the air, substantially increasing the surface area of mercury exposed to the air and greatly enhancing the rate at which mercury vapor will be generated. There are commercial vacuum cleaners which are specifically designed to pick up mercury, however. One such unit, sold by Nilfisk of America, which also makes specialized units for other toxic materials, first draws the mercury into a centrifugal separator and collects the bulk of the material into an airtight plastic bottle. The contaminated air is then passed into a collection bag which collects bulk solid waste and then through an activated charcoal filter. Additional filters (some optional) follow the charcoal filter collector. This unit can be used to clean up virtually any spill alone but can be used with other control measures to ensure a complete clean up of the spilled material. The hose in the Nilfisk unit vacuum cleaner has an especially smooth surface to prevent mercury particles from adhering to the inside of the hose. As with most specialized units, the vacuum cleaner is not inexpensive. In some instances, a special purpose mercury vacuum cleaner is virtually indispensable for use on a spill on a porous, rough material such as carpeting. The use of carpeting as a laboratory floor covering is very rare but, in at least one instance where this was done, a large area of the carpet was thoroughly contaminated by an extensive mercury spill.

g. Ventilation

The ventilation system in a laboratory using mercury or mercury compounds should conform to the general recommendation that wet chemistry laboratories involving any hazardous material be provided with 100% fresh air instead of having a portion of the air recirculated. Local ventilation systems, such as the exhausts of mechanical pumps servicing mercury diffusion pumps, should be collected with a local exhaust system and discharged into the fume hood exhaust system in the room or to a separate exhaust duct provided to service such units. The mercury vapor is much heavier than air so it is important that the room exhausts be placed near the floor or at the back of the workbench to collect as much of the vapors as possible.

h. Medical Surveillance

It is recommended, as a minimum, that permanent employees working with mercury or mercury compounds be provided with periodic physical examinations with a test protocol selected specifically for mercury poisoning. Women who may be pregnant should be especially careful and encouraged to participate in the medical surveillance program if they cannot avoid exposure entirely.

REFERENCES

1.   Vostal. J.J. and Clarkham, T.W., Mercury as an environmental hazard,. *J. Occupational Medicine,* 15, 649, 1973.

2. **Armour, M.A., Browne, L.M., and Weir, G.L.,** *Hazardous Chemicals Information and Disposal Guide,* 2nd ed., University of Alberta, Edmonton, Canada, 1984.

INTERNET REFERENCES

1. http://www.mercury.safety.co.uk/hlthinfo.htm - Mercury Toxicity and how it affects our health, Mercury Safety Products, Ltd., 1997.
2. http://www.chem.ucla.edu/Safety/newsletterl.html - Mercury cleanup and disposal.

9. Hydrofluoric Acid

Anhydrous hydrofluoric acid (HF) (CAS 7664-39-3) is a clear, colorless liquid. Because it boils at 19.5°C (67.1°F) and has a high vapor pressure, it must be kept in pressure containers. It is miscible in water, and lower concentration aqueous solutions are available commercially. It is an extremely dangerous material and all forms, including vapors and solutions, can cause severe, slow-healing burns to tissue. At concentrations of less than 50%, the burns may not be felt immediately and at 20% the effects may not be noticed for several hours. At higher concentrations, the burning sensation will become noticeable much more quickly, in a matter of minutes or less. Fluoride ions readily penetrate skin and tissue and, in extreme cases, may result in necrosis of the subcutaneous tissue which eventually may become gangrenous. If the penetration is sufficiently deep, decalcification of the bones may result. The current OSHA PEL 8-hour time weighted average to HF is set at 3 ppm (2.5 mg/M^3), which also is the ceiling TLV currently recommended by the ACGIH. Chronic exposure to even lower levels may irritate the respiratory system and cause problems to the bones. Even brief exposures to high levels of the vapors may cause severe damage to the respiratory system, although the sharp, irritating odor of the acid will usually provide a warning to assist in avoiding inhalation in normal use. Contact with the eyes could result in blindness. If eye exposure occurs, it is urgent to flush the eyes as quickly as possible. It is especially recommended that every laboratory using hydrofluoric acid have both an eyewash station and deluge shower within the laboratory. Dilute solutions and vapors may be absorbed by clothing and held in contact with the skin, which will probably not result in an immediate sensation of pain as a warning but eventually may lead to skin ulcers which, again, may take some time to heal. A generalization might be made here about absorbent clothing. In many instances, as in this case, absorbent clothing which can retain toxic materials and maintain them in close contact with the skin may be worse than no protection at all, changing the exposure from a transient phenomena to a persistent one. This is not always a problem, but it should be kept in mind as a possibility when choosing protective apparel. All work with hydrofluoric acid should be done in a fume hood.

Hydrofluoric acid attacks glass, concrete, and many metals (especially cast iron). It also attacks carbonaceous natural materials such as woody materials, animal products such as leather, and other natural materials used in the laboratory such as rubber. Reactions with carbonates, and sulfites and cyanide will produce asphyxiants or toxic gases. Lead, platinum, wax, polyethylene, polypropylene, polymethylpentane, and Teflon will resist the corrosive action of the acid. In contact with metals with which it will react, hydrogen gas is liberated and hence the danger exists of a spark or flame resulting in an explosion in areas where this may occur.

a. Treatment to exposure

Successful treatment of severe exposures is dependent on rapid reactions by those responding to the incident and by the affected person(s). In the following sections, reference is made to various medications specific to the treatment of hydrofluoric acid exposure. It is unlikely that the typical rescue squad called to the scene will have these medications so they should be part of the first aid supplies maintained in the immediate area where exposures may occur. Have someone call for emergency medical assistance as soon as possible and direct them to arrange treatment with a physician or trauma center familiar with chemical burns. In all types of exposure, the first action recommended is prolonged flushing with copious amounts of water so an eyewash station, a shower and a source of potable water should be immediately available.

For an eye exposure, the eye should be flushed for 30 minutes, with the eyelids being kept out of contact with the surface of the eye. For a skin exposure, any clothing in contact with the affected area should be removed, with care, and the area flushed with running water for at least 20 minutes. If the affected area is large, do this in a safety shower or if restricted to a small area, with a hose or a steady stream of water from another source.

A recommended first aid treatment for an eye exposure, while obtaining medical treatment, is to apply one or two drops of 0.5% Pontocaine Hydrochloride solution. Afterwards, it has been recommended that the eyes should be washed with a 1% calcium glutonate in normal saline solution for 5 to 10 minutes. Subsequently, for the next two to three days, the eyes should continue to be treated with this solution every two to three hours.

One suggested treatment for a skin exposure is to immerse the burned area, after thorough washing, in a solution of 0.2% iced aqueous Hyamine 1622[*] or 0.13% iced aqueous Zephiran Chloride. If the area cannot be immersed conveniently, then towels soaked with these solutions should be applied. The compresses should be changed every few minutes.

Another first aid treatment for surface burns from hydrofluoric acid is to rub the affected area with a 2.5 % gel of calcium glutonate after a brief one minute washing. This can be continued for 3 to 4 days and done 4 to 5 times daily. For burns of areas greater than 50 cm^2, about 10 in^2, the patient should be hospitalized. As the area of the burn increases, the likelihood of inhalation becomes greater and the victims pulmonary function should be carefully evaluated by the attending physician.

For deep burns by greater than 20% solutions of HF, treatment by subcutaneous injections of a 5% solution of calcium glutonate (prepared by diluting 10% ampules of the material) is recommended. The injection should be limited to no more than 0.5 cc per square centimeter. One authority does not recommend that this be done on the digits of the hand, or should be done very carefully for all areas of the hands, feet and face. The same authority also states that concentrations greater than 5% tends to produce severe irritation and can lead to the formation of keloids and scarring.

If the exposure is inhalation of HF vapors, the victim should be provided with 100% oxygen as soon as possible, followed quickly by inhalation of a 2.5 to 3% solution of calcium glutonate using a nebulizer. The attending physician should watch for signs of edema of the upper airway and the airway maintained clear of obstruction.

Ingestion is less likely but if it occurs, severe burns can result which may be fatal. Call for medical assistance immediately but, while waiting, the only first aid treatment recommended is having the victim drink large quantities of water.

REFERENCES

1. **Proctor, N.H. and Hughes, J.P.,** *Chemical Hazards of the Workplace,* J.B. Lippincott., Philadelphia, 1978, 290.
2. **Knight, A.L.,** *Occupational Medicine: Principles and Practical Applications,* Zenz, C. (Ed.), Year Book Medical Publishers, Chicago, 1975, 649.
3. **Wetherhold, J.M. and Shepherd, E.P.,** Treatment of hydrofluoric acid burns, *J. Occup. Med,* 7, 193, 1965.
4. **Reinhardt, C.F., Hume, W.G., and Linch, A.L. et al.,** Hydrofluoric acid burn treatment, *Am. Med. Hyg. Assoc.J.,* 27, 166, 1966.

[*] Hyamine is a trade name for tetracaine benzethonium chloride, Merck index Monograph 1078.

5. **Gosselin, R.D., Hodge, H.C., Smith, R.P. et al..**, *Clinical Toxicology of Commercial Products: Acute Poisoning,* 4th ed., Williams & Wilkins., Baltimore, 1976, 159.

6. **Browne, T.D.**, The treatment of hydrofluoric acid burns, *J. Occup. Med,* 24, 80, 1974.

7. **Tepperman, P.R.**, Fatality due to acute systemic fluoride poisoning following a hydrofluoric acid skin burn, *J. Occup. Med,* 22, 691, 1980.

8. **Abukurah, A.R., Moser A.M., Baird, C.L. et al.**, Acute sodium fluoride poisoning, JAMA, 222, 816, 1972.

9. **Thevino, M.A., Herrmann, G.H., and Sproul, W.L.**, Treatment of severe hydrofluoric acid exposures, *J. Occup. Med.,* 25(12), 861, 1983.

INTERNET REFERENCES

1. http://www.camd.lsuedu/msds/h/hydrofluoric_acid.htm
2. http://www.qrc.com/hhmi/science/labsafe/lcss/lcss51.htm
3. http://www.filemedia.com/hf/
4. http://www.cdc.gov/niosh/npg/npgd0334.html

10. Hydrogen Cyanide

Hydrogen cyanide (HCN) (CAS 74-90-8), also called hydrocyanic acid or prussic acid, is an extremely dangerous chemical that is toxic by ingestion, inhalation, or by absorption through the skin. The current OSHA 8-hour PEL to the vapors from this chemical is 10 ppm, as is the current ACGIH ceiling limit (with a cautionary note that skin absorption could be a contributory hazard). The NIOSH recommended limit is 4.7 ppm. The material has a characteristic odor of bitter almonds, but the odor is not usually considered to be sufficiently strong to be an adequate warning of the presence of the vapors at or above the PEL. A substantial number of persons, perhaps as many as 60%, cannot detect this odor. Not only is HCN toxic, it has a very low flash point, -17.8°C (0°F), a lower explosion limit of 6%, and an upper explosion limit of 41%, so that it also represents a serious fire and explosion hazards. It has a boiling point of 26°C (79°F), so that it is normally contained in cylinders in the laboratory. Heating of the liquid material in a pressure-tight vessel to temperatures above 115°C (239°F) can lead to a violent, heat-generating reaction. The material is usually stabilized with the addition of a small amount (0.1%) of acid, usually phosphoric acid, although sulfuric acid is sometimes used. Samples stored more than 90 days may become unstable.

Hydrogen cyanide can polymerize explosively when amines, hydroxides, acetaldehyde, or metal cyanides are added to the liquid material, and it also may do so above 184°C (363°F).

Although there will be variations among individuals, a concentration of 270 ppm in air is usually considered fatal to humans. A few breaths above this level may cause nearly instantaneous collapse and respiratory failure. Exposures at lesser levels may be tolerated for varying periods, e.g., 18 to 36 ppm may be tolerated for several hours before the onset of symptoms. Initial symptoms of exposure to HCN include headache, vertigo, confusion, weakness, or fatigue. Nausea and vomiting may occur. The respiratory rate usually increases initially and then decreases until eventually it becomes slow and labored, finally ceasing. The symptoms reflect the mechanism by which the toxic action occurs. The chemical acts to inhibit the transfer of oxygen from the blood to tissue cells by combining with the enzymes associated with cellular respiration. If the cyanide can be removed, the transfer of oxidation will resume. On average, absorption of 50 to 100 mg of HCN, directly by ingestion or through the skin as well as by inhalation can be fatal.

Treatment of a person poisoned by HCN is based on the introduction of methemoglobin into the bloodstream to interact with the cyanide ions to form cyanmethemoglobin. In any area where HCN is being used, a special emergency kit should be provided, containing an ample supply of ampules of amyl nitrite, a solution of 1% sodium thiosulfate solution, and an oxygen cylinder accompanied by a face piece and tubing to permit administering the oxygen. This kit should be labeled **FOR HCN EMERGENCIES ONLY.** For this kit to be useful, several individuals in the area should be trained in how to use it effectively. Sodium nitrite might also be kept in the kit if there is someone available qualified to administer drugs intravenously introduction of sodium nitrate directly into the bloodstream has been suggested as a means to increase the rate of conversion of cyanide to the thiocyanate, which is less toxic. Treatment should begin as soon as possible after an acute exposure and after recognition of the symptoms in less intense exposures. If the exposure has been due to contamination in the air in the area, the patient should be removed from the area (if the source of vapor is from a cylinder, the valve on the cylinder should be closed). Any contaminated clothes should be removed and the skin flooded with water.

If the patient is not breathing, resuscitation should be begun. As soon as the patient is breathing, an open amyl nitrite ampule should be held under the patient's nose for 15 seconds per minute, with oxygen being administered during the remaining 45 seconds. Medical aid should be called for immediately. If there is a person available qualified to administer drugs intravenously, injection of sodium nitrite while administering amyl nitrite should prove beneficial. Subsequent intravenous injection of sodium thiosulfate also has been suggested as an ameliorative action. Rescue squad teams usually have at least one member qualified to administer drugs while under the direction (by radio) of an emergency room physician. If the patient has swallowed HCN, the recommended treatment is to get the patient to swallow one pint of the sodium thiosulfate solution, followed by soapy water or mustard water to induce vomiting. Vomiting should not be induced in an unconscious patient. Application of amyl nitrite may restore consciousness.

All work with HCN must be done in a fume hood, operating with a face velocity of at least 100 fpm and with the apparatus set well back from the face of the hood to ensure that all vapors will be captured and discharged by the exhaust system. The hood should comply in every respect to recommended good practices for the location, design, and operations of hoods in Chapter 3. The hood should have its own individual duct to the roof. If the work is a continuing program, the exhaust duct should be labeled: **DANGER, DO NOT SERVICE OR WORK IN THE VICINITY WHILE UNIT IS OPERATING.**

Protective gloves and chemical splash goggles should be worn while working with HCN. No one should work alone with this material. The laboratory entrance(s) should be posted with a warning sign: **DANGER, HCN, AUTHORIZED PERSONNEL ONLY.** As noted above, care needs to be taken to be sure that persons outside the area, such as workers on the roof, are not inadvertently exposed. All work with HCN should be done in trays or other shallow containers of sufficient capacity to retain any spill from the apparatus.

Cleaning up of spills represents a serious problem with a chemical as dangerous as HCN, so extra care should be taken to avoid accidents with the material. If a spill occurs outside a hood, the laboratory should be evacuated as quickly as possilble. All ignition sources and valves to cylinders of HCN should be turned off. If any individuals are splashed with the compound in an accident, they should immediately remove their contaminated clothes and step under a nearby deluge shower, preferably located in a space outside the laboratory in which the accident occurred. The occupants of the latter space should be warned of the accident and encouraged to evacuate as well. If the spill is substantial, the evacuation of either all the contiguous spaces or the entire building might be considered. The evacuation of additional spaces is especially important in facilities in which the reentry of fumes exhausted from the building is known to be a problem. Medical observation and care for any exposed persons should be obtained as soon as possible. It would be desirable to have sufficient self-contained escape-type breathing devices on hand to equip every occupant of the laboratory. Unless laboratory personnel have received specific training in handling hazard material incidents, the nearest hazardous material emergency response center should be called for assistance for a substantial spill.

Hydrogen cyanide is categorized as a chemical which is immediately dangerous to life and health (IDLH). As such, clean up of spills should be handled very carefully. Individuals performing the clean up should wear a self-contained, positive pressure breathing apparatus, equipped with a full face piece, rubber or neoprene gloves, and chemically protective outer-wear. A type C supplied air respirator unit operated at a positive pressure can be used as well, but a self-contained unit should be available as a backup. Anyone asked to wear this equipment must have received prior training in the proper use of the equipment. The material can be cleaned up using absorbent pillows or other absorbent materials. Waste should be placed in double heavy-duty plastic bags, which are then tightly closed by twisting the top, folding the top over and wrapping it securely with duct tape. The sealed plastic bags should then be placed in heavy plastic containers or steel drums which can be tightly sealed. Waste material should not be placed in fume hoods to evaporate or be disposed of in drains. In the latter case, the possibility of fumes collecting in sections of the drain piping and reentering a building thorough a dry sink trap is too great. The waste should not go to a normal landfill. Incineration is the preferred means of disposal.

A leaking cylinder which cannot be readily repaired should be taken to a remote location where the gas in the cylinder can be released safely. Cylinders which are damaged but not leaking should be returned to the vendor for disposal wherever possible. Disposal of gas cylinders by commercial hazardous waste firms can be very expensive.

REFERENCES

1. Occupational Health Guidelines for Chemical Hazards, Mackison, F.W, Stricoff, R.S., and Partridge, Jr., L.J., (Eds.), U.S. Department of Health and Human Services and U.S. Department of Labor. DHHS (NIOSH) Pub. No. 81-123, 1981.
2. *Prudent Practices in the Laboratory, Handling and Disposal of Chemicals,* National Academy Press, Washington, D.C., 1995.
3. **Chen, K.K. and Rose, C.L.,** Nitrite and thiosulfate therapy in cyanide poisoning, JAMA, 149, 113, 1952.
4. **Hirsch, E.G.,** Cyanide poisoning, *Arch. Environ. Health,* 8, 622, 1964.

INTERNET REFERENCES

1. http://www.cdc.gov/niosh/npg/npgd0333.html
2. http://www.qrc.com/hhmi/science/labsafe/lcss/lcss50.htm
3. http://www.state.nj.us/health/eoh/rtkweb/rtkhsfs.htm

11. Fluorine Gas

Fluorine (CAS no. 7782-41-4) is an extremely reactive gas which reacts violently with a wide variety of materials, a representative sample of which are most oxidizable substances, most organic matter, silicon-containing compounds, metals, halogens, halogen acids, carbon, natural gas, water, polyethylene, acetylides, carbides, and liquid air. Many of these reactions will initiate at very low temperatures. Because it will react with so many materials, extreme care must be taken when working with fluorine. The work area should be very well ventilated and free of combustible materials which would act as fuel in the event of a fire. A written hazard analysis should be prepared for the research program prior to beginning work and an emergency contingency plan developed as a part of the laboratory industrial hygiene plan required by the OSHA laboratory standard. Written standard operating procedures are required, and employees fully trained in the nature of the risks and protective measures necessary to avoid injury.

The OSHA PEL is 0.1 ppm or 0.2 mg/M^3. However, the 1993-94 ACGIH TWA levels are

10 times higher with short-term exposure limits (STEL) another factor of 2 higher. An exposure to 25 ppm for 15 minutes has caused severe eye symptoms. The LC_{50} (50% lethal concentration) for a 1-hour exposure for rats and mice is 185 and 150 ppm, respectively. It is highly irritating to tissue.

Fluorine will react with brass, iron, aluminum, and copper to form a protective metallic fluoride film. Circulating a dilute mixture of fluorine gas and inert gases through a system of these metals will passivate the surfaces and render them safe to use, provided the film remains intact. However, it is recommended that an inert gas be circulated through any fluorine system before the fluorine is introduced.

All systems containing fluorine should be checked frequently for leaks. Filter paper moistened with potassium iodide can be used to perform the tests. The paper will change color when any escaping gas comes in contact with it.

Work with systems using fluorine should always be done within a fume hood. The research worker should be protected by an explosive shield. The worker should also wear protective goggles and a face mask. Unless the cylinder valve is operated through a remote control device, the user should wear sturdy gloves with extended cuffs to protect his hands and arms while manipulating the valve. A protective apron should be worn as well. However, all of these may give only limited protection in the event of an accident since fluorine may react with many common items of personnel protective gear.

Self-contained escape breathing apparatus should be available for all occupants of a laboratory in which fluorine is in active use. In the event of an accident, immediate evacuation of the area should take place, being sure to close doors as personnel leave to isolate the problem as much as possible. Evacuation of nearby areas should be considered or, depending upon the scale of the accident, perhaps the entire building, especially as noted elsewhere, if there are known problems with exhausted materials reentering the building. No remedial measures should be attempted under most circumstances; the incident should be allowed to proceed until the fluorine supply is exhausted. Firefighting efforts should be aimed at preventing a fire from spreading. Applying water directly to the leak could intensify the fire. Obviously, it would be desirable to use smaller cylinders, if practicable, for the research program to reduce the scale of any incident.

Cylinders with valves that cannot be dislodged without application of sufficient force to damage the valve or the connection to the cylinder should be returned to the vendor for repair rather than take a chance on a massive rupture and release of the contents of the cylinder. Ordinary maintenance personnel should never be asked to attempt to free the valve. Although the consequences here would be exacerbated by the extremely hazardous properties of the contents of the cylinders, the same recommendation would apply to virtually any cylinder containing a substantial volume of gas under high pressure. There are firms that specialize in handling dangerous situations such as peroxides, explosives, highly reactive materials, and damaged cylinders.

REFERENCES

1. Occupational Health Guidelines for Chemical Hazards, Mackison, F.W., Stricoff, R.S., and Partridge, L.J.,Jr., (Eds.), U.S. Department of Health and Human Services and U.S. Department of Labor, DHHS (NIOSH) Pub. No. 81-123, 1981.

2. Hazardous Chemical Data, NFPA-49, National Fire Protection Association, Quincy, MA.

3. Threshold Limit Values for Chemical Substances and Physical Agents and Biological Exposure Indices, American Conference of Governmental Industrial Hygienists, Cincinnati, OH, 1993-1994.

INTERNET REFERENCE

1. http://www.cdc.gov/niosh/npg/npgd0289.html

12. General Safety for Hazardous Gas Research

The previous section was devoted to fluorine as an example of an exceptionally hazardous gas. However, many others commonly used in the laboratory pose comparable risks, some because of their own toxicity or that of gases or vapors evolved as a consequence of their decomposition, others because of their flammable or explosive properties or due to reactions with other chemicals, or a combination of all of these characteristics. Even relatively innocuous gases such as nitrogen, carbon dioxide, argon, helium, and krypton can be a simple asphyxiant if they displace sufficient air, leaving the oxygen content substantially below the normal percentage. If the content of these inert gases approaches one third or higher, symptoms of oxygen deprivation begin to occur, and at concentrations of around 75%, persons will survive for only a brief period. Any gas under high pressure in a cylinder poses a problem if the cylinder is mishandled so as to rupture the containment of the gas. In such a case the cylinder can represent an uncontrolled missile with deadly consequences for anyone in the vicinity. The ability of escaping gas to move readily throughout a volume greatly enhances the likelihood that a flammable gas will encounter a source of ignition. This problem is shared by the vapors of many volatile liquids. Many gases are heavier than air and may collect in depressions or areas with little air movement, representing a danger to unsuspecting persons. Gases often do not have a distinctive odor or are not sufficiently irritating to warn of their presence, and some that do, such as hydrogen sulfide, act to desensitize the sense of smell at levels which would be dangerous.

Cylinders connected to systems in the laboratory must always be strapped firmly to a support to ensure that they do not fall over. Not only is there a risk of breaking the connection on the cylinder side of the regulator valve, with the concomitant risk of the cylinder becoming a missile, but the connection to the system also may be broken so that gas will escape from the low pressure side of the regulator. If the amount of gas to be stored in the laboratory is substantial, it may be preferable to pipe the gas in from a remote outside storage area with control valves located both outside and within the laboratory.

Where the explosion risk is substantial, the facility may need to be designed with explosion venting so that the force of any explosion and the resulting flying debris can be released in a relatively safe direction, minimizing the risk to the occupants. Systems for smaller-scale operations should be placed within a hood, perhaps one specially designed to provide partial protection against explosions. Explosion shields are available to aid in protecting the research worker. Wherever the possibility of an explosion exists, laboratory personnel should wear impact-resistant goggles, possibly supplemented by a face mask to protect the lower part of the face and throat. Gauntlets should be worn if there is risk to the hands and forearms while conducting experimental evolutions.

Not all work with high pressures involves gas cylinders. Reaction bombs are commercially available which operate at pressures up to 2000 psi (13.8 MPa) and temperatures up to 350°C (662°F). An error on the part of the research worker could permit these design parameters to be exceeded, and although these units are designed with a substantial safety factor, the failure of one of them could lead to disastrous consequences. Not only is there the immediate danger of injury due to flying components of the system, there is the risk of reactions involving reagents from broken bottles. These secondary events could escalate the consequences far beyond the original scope of the incident. Any device in which the potential for a high-pressure accident exists should be set up in a hood to provide some explosion protection, and explosion barriers should be used to provide additional protection to the occupants of the room. Personnel working in this type of research should be especially careful to not work alone. They should wear goggles, a face mask, and sturdy gloves to protect their hands and forearms.

Systems involving toxic gases should be adequately ventilated. If possible, the systems should be set up totally within a fume hood. Large walk-in hoods often are used for this purpose. All systems should be carefully leak-tested prior to introduction of toxic materials into the system, periodically thereafter, and after any maintenance or modifications to the system which could affect its integrity.

Many gases are potentially so dangerous that access to the laboratory should be limited to essential personnel that are authorized to be present. When working with such materials, no one should work alone. It may be desirable to have one of the persons somewhat removed from the immediate area of operations, but a second person should be within the working area. Entrances to a high hazard gas research facility should be marked with a **DANGER, SPECIFIC AGENT, AUTHORIZED PERSONNEL ONLY** or a comparable warning sign. Hood exhausts also should bear a comparable warning legend. In some cases, automatic alarm sensors have been developed to detect the presence of gas levels approaching dangerous levels. It is recommended that warning trip points on these devices be set at no more than 50% of either the OSHA PEL or the ACGIH TLV value, whichever is lower. If an automatic sensing device is available, circuitry can be devised to activate a valve to cut off the gas supply as well as to provide a warning. The latter is especially important if the operation is left unattended and no signal is transmitted to a manned location, rendering an alarm ineffective. The growing application of programmed personal computers, or laboratory computer workstations dedicated to experimental control has increased the amount of sophisticated experimentation that can be automated.

In any laboratory involving the use of highly hazardous materials, an emergency plan is required by OSHA to be developed in advance of initiation of any major project or any major modification to an ongoing project based on a thorough hazard analysis. Because of the special problems associated with gases, the emergency contingency plan should make provision for rapid evacuation of the immediate laboratory using short-duration, self-contained breathing apparatus and provision for initiation of evacuation from other areas of the facility. Provision of detailed information to emergency response groups is now required under the Community Right-to-Know law for many hazardous substances when the amount involved exceeds a prescribed threshold amount. It is also recommended that employees involved in any research involving hazardous materials participate in a medical surveillance program, consisting of a comprehensive prior screening exam, acquisition of a serum sample for comparison with a sample following a possible incident, and a complete medical history, so that baseline information on individuals will be available to medical personnel called upon to treat personnel that may have been exposed, as for example in this area, to toxic gases. If there are specific organs or bodily functions that might be affected, the examination may need to include special tests in these areas.

The most common problem associated with the use of gas cylinders is the problem of leaks. The cylinder can develop leaks at any of four points, assuming that no gross rupture of the cylinder wall itself has occurred: the valve threads, valve safety device threads, the valve stem, and the valve outlet. Repairs of leaks in the first two of these would require repairs done at high pressure and are not to be attempted in most laboratory facilities. It may be possible that some adjustments are incorporated into the design of the cylinder to allow stopping leaks in the latter two areas. In either case, it is best to contact the manufacturer for advice before attempting any repair. Forced freeing of a "frozen" or corroded valve should <u>not</u> be attempted.

Leaks involving cylinders containing corrosive materials may increase in size with time as the corrosive gas interacts with atmospheric moisture and erodes the opening. Removal of leaking cylinders containing corrosives from an occupied facility to a remote location should be done as quickly as possible. The vendor should be called for advice or assistance. If time permits, a solution would be to slowly exhaust the leaking cylinder into a neutralizing material. If the protective cap has corroded into position so that no bleeder hose can be attached, two heavy-duty plastic bags can be placed over the leaking end of the cylinder and the gases conducted from the bags through a hose into a drum of neutralizing solution. If the leak is too large for easy handling, a commercial, state, or local hazardous material response group should be called upon for assistance. The wearing of protective suits and self-contained breathing apparatus may be required, which requires special training.

Leaks involving toxic and flammable gases pose the risk of personal injury to individuals attempting repairs, and repairs should not be attempted by local personnel unless it is certain that they can be done safely. Evacuation is usually recommended. Where flammable gas leaks are concerned, all ignition sources must be turned off prior to evacuating a facility, if time permits. It is foolish to risk life or personal injury unless by doing so there is a reasonable likelihood to possibly save others.

In cases where the melting point or boiling point of the contained gas is sufficiently high, the leak rate may be substantially reduced or virtually stopped by putting the body of the cylinder into a cooling bath while deciding upon the appropriate corrective measures or waiting for assistance. This should not be done if the contained material will react with the coolant.

REFERENCES

1. Handling and Storage of Flammable Gases, Air Products and Chemicals, Allentown, PA.
2. *Prudent Practices in the Laboratory Handling and Disposal of Chemicals,* National Academy Press, Washington, D.C., 1995.
3. Safe Handling of Compressed Gases in the Laboratory and Plant, Matheson Gas Products. Also see: Internet site: www.mathesongas.com.

13. Some Hazardous Gases

The properties of a number of hazardous gases will be given in this section, along with a few brief comments on noteworthy problems associated with these materials. The common chemical name, formula, and CAS number will be given for each substance, followed by a definition of the primary hazard class(es) represented by the material, the boiling point as degrees Celsius (degrees Fahrenheit)[*], the explosive range (lower explosive limit [lel] - upper explosive limit [uel]) in percent by volume in air (NA where not applicable or not available), the vapor specific gravity referred to air as 1, OSHA 8 hour time weighted average PEL in parts per million[**] and finally, salient comments about the material.

Acetylene - C_2H_2, 74-86-2; explosive, flammable, asphyxiant; bp = -84.0°C [-119°F]; er= 2-82%; sp g= 0.91; PEL = 2500 ppm (10% of lel). Acetylene in cylinders is usually dissolved in acetone and is relatively safe to handle, but purified acetylene has a very low ignition energy and a relatively low minimum ignition temperature, 300°C [571°F]. It forms explosive mixtures with air over a wide range. Utilization of acetylene should be at a pressure of 15 psi gauge or less. Piping for acetylene systems should be steel, wrought iron, malleable iron, or copper alloys containing less than 65% copper. It may form explosive compounds with silver, mercury, and unalloyed copper.

Ammonia - NH_3, 7664-41-7; flammable gas, causes tissue burns, strong respiratory irritant; bp = -33.3°C [-28°F]; er = 15-28%; sp g 0.6; PEL = 50 ppm. May cause severe injury to

---

[*] In most cases, the celsius temperature is from the literature, while the Fahrenheit temperature is calculated from the former.
[**] The ceiling value is given if the level should not exceed this value at any time or for a limited period. If a level has not been set by OSHA and an ACGIH TLV value is available, the latter will be given.

respiratory system and eyes, common 35% laboratory solution can cause severe skin burns. High concentrations may cause temporary blindness. Baseline physical should stress respiratory system and eyes. Skin should be examined for existing disorders. Tests should include pulmonary function and chest X-ray. Should wear self-contained breathing apparatus and rubber shoe covers when cleaning up a spill (by dilution with ample amounts of water and mop to a drain). On EPA list of extremely hazardous substances, 40 CFR Section 302.

Arsine - AsH_3, 7784-42-1; deadly poison by inhalation, fire and explosion hazard; bp = -62.5°C [-144.5F; er = NA; sp g = 2.66; PEL = 0.05 ppm. Recognized carcinogen; causes pulmonary edema; primarily poisonous due to interaction with hemoglobin, causes anemia; early symptoms are headache, dizziness, nausea, and vomiting. Severe exposures result in kidney damage, delirium, coma, and possible death. In increased use due to applications to semiconductor research. Should be used in a sealed system, or the system should be set up in an efficient fume hood. Leaking cylinders can be handled by allowing the leaking gas to interact with a 15% aqueous solution of sodium hydroxide. The arsine forms a water soluble precipitate. After neutralization with sulfuric acid, the precipitate is filtered out to be disposed of as hazardous waste, and the neutral solution can be disposed of into the drain, diluted by large amounts of water. Individuals doing this should wear self-contained breathing apparatus and chemically resistant gloves. Arsine will ignite in contact with chlorine and undergoes violent oxidation by fuming nitric acid. On EPA list of extremely hazardous substances, 40 CFR Section 302.

Boron Trifluoride - BF_3, 7637-07-2; inhalation poison, irritant, nonflammable; bp = -100°C [-148°F]; er = NA; sp g = 2.3; ceiling PEL = 1 ppm.; irritating to eyes and respiratory system. Animals have been shown to have kidney damage after high exposures. Baseline physical should stress respiratory system, eyes, and kidneys. Tests should include a chest X-ray, pulmonary function test. Cylinders with a slow leak can be allowed to leak into an efficient fume hood for disposal, or seal the cylinder and return to the vendor. Produces a thick white smoke by interaction with moisture in humid air, but otherwise insufficient data on warning properties. On EPA list of extremely hazardous substances, 40 CFR Section 302.

1,3-Butadiene - C_4H_6, 106-99-0; flammable, irritant to eyes and mucous membranes, suspect carcinogen; bp = -4.5°C [23.9°F]; er = 2-11.5%; sp g = 1.9; PEL = 1000 ppm (ACGIH = 10 ppm, suspect human carcinogen). Narcotic at high concentrations. May form peroxides on exposure to air, at high temperatures, may self-polymerize exothermally, forms explosive mixtures with air. Participation in a medical surveillance program recommended for users. On EPA list of extremely hazardous substances, 40 CFR Section 302. On "California" list as a carcinogen.

Carbon Dioxide - CO_2, 124-38-9; nonflammable, asphyxiant; sublimes at -78.5°C [-109°F]; er = NA; sp g = 1.53; PEL = 5000 ppm. Very common laboratory gas, also used as "dry ice." Causes problems primarily by displacement of air. At 5% concentration, respiratory volume quadrupled. Heart rate and blood pressure increases reported at 7.6%. At 11%, unconsciousness typically occurs in 1 minute or less. No warning other than symptoms: dizziness, headaches, shortness of breath, and weakness because it is colorless and odorless.

Carbon Disulfide CS_2, 75-15-0; flammable liquid (vapor pressure 400 mm at 28°C [82.4°F], poisonous; bp = 46.5°C [115°F]; er = 1.3-50%, flash point = -30°C [-22°F] and ignition temperature is only 90°C [194°F]; sp g of vapor = 2.64; PEL = 20 ppm. Central nervous system poison. Extended exposure can cause permanent damage to the CNS in severe cases. Numerous other physiological problems to heart, kidneys, liver, stomach. Has strong narcotic and anesthetic properties. Poisonous if inhaled or ingested, or with prolonged contact with skin. Baseline physical should stress central and peripheral nervous systems, cardiovascular system, kidneys, liver, eyes, and skin. Tests should include urinalysis for kidney function, liver panel, electrocardiogram, and ophthalmic exam. Vapors can be ignited by contact with an incandescent light bulb. Air-carbon disulfide mixture can explode in the presence of rust. Do not pour down a sink. Do not use where electrical sparks possible. Use nonsparking tools. Explosion-proof wiring, fixtures not necessarily effective to arrest flame. Use dry chemical or carbon dioxide to fight fires. On EPA list of extremely hazardous chemicals, 40 CER Section 302. On "California" list as a chemical with reproductive toxicity.

Carbon Monoxide - CO, 630-08-0; flammable, poisonous, experimental teratogen; bp = -191.1°C [-311.9°F]; er= 12.5-74%, sp g = 0.9678; PEL = 50 ppm. Carbon monoxide combines highly preferentially with hemoglobin to the exclusion of oxygen and hence is a chemical asphyxiant. Pregnant women and smokers more susceptible to risk. On "California" list as a chemical with reproductive toxicity. Baseline physical recommended. Medical history taken to discover history of problems involving heart, cerebrovascular disease, anemia, and thyroid toxicosis. A complete blood count should be taken. Insidious poisonous gas. Does not have adequate warning properties, odorless and nonirritating. Contact with strong oxidizers may cause fires and explosions. Dangerous fire hazard.

Chlorine - Cl_2 7782-50-5; nonflammable gas but supports combustion of other materials, toxic; bp -34.5°C [-30.1°F]; er = NA; sp g = 2.49; Ceiling PEL = 1 ppm. Forms explosive mixtures with flammable gases and vapors. Reacts explosively with many chemicals and materials such as acetylene, ether, ammonia gas, natural gas, hydrogen, hydrocarbons, and powdered metals. Many incompatibles; carefully review the literature and MSDS before working with this material. Systems should be set up in an efficient fume hood. Strong odor, noticeable well below acute danger levels. Inhalation can cause severe damage to lungs. Baseline physical exam should stress eyes, respiratory tract, cardiac system, teeth, and skin. Pulmonary function test and chest X-ray recommended. Warning of the presence of gas well below PEL due to odor and irritating properties. On EPA list, of extremely hazardous chemicals, 40 CFR Section 302.

Cyanogen - C_2H_2 460-19-5; highly flammable, toxic; bp -21°C [-5.8°F], er = 6.6-32%, sp g = 1.8; ACGIH TLV = 10 ppm. Odor resembles almonds, poisonous by inhalation and skin; symptoms are headache, dizziness, nausea, vomiting, and rapid pulse; severe exposures lead to unconsciousness, convulsions, and death; strong eye irritant; fire hazard when exposed to oxidizers, flame, sparks. Reacts with fluorine (ignites) and oxygen (combination of liquid cyanogen and liquid oxygen will explode).

Cyanogen Chloride - CClN, 506-77-4; nonflammable, inhalant poison; eye irritant; bp = 13.1°C [55.6°F]; er = NA; sp g = 1.98; ACGIH ceiling level = 0.3 ppm. Toxic properties similar to hydrogen cyanide; heat causes it to decompose and emits highly toxic and corrosive fumes.

Diazomethane - CH_2NO_2 334-88-3; explosion hazard, irritant to respiratory system, eyes, suspect carcinogen; bp = -23°C [-9.4°F]; er =NA; sp g = 1.4; OSHA PEL = 0.2 ppm. One of the most dangerous chemicals used in chemical laboratories. Strong allergen, irritating to eyes and to respiratory system. May sensitize as well as irritate. Recommend initial medical history, full chest X-ray, and pulmonary function tests to individuals planning to work with this material. Explosively sensitive to shock, heat (about 100°C, 212°F), exposure to rough surfaces (e.g., ground glass joints), alkali metals, calcium sulfate. Do not store, prepare fresh when needed. Always use in an efficient fume hood, use an explosive shield, wear impact and chemical splash-resistant goggles, possibly supplemented by a face mask. Respiratory protection recommended if levels may approach PEL.

Diborane - B_2H_6 19287-45-7; flammable, poison, suspect carcinogen; bp = -92.5°C [- 134.5°F]; er = 0.9-98% (ignites at temperatures of 38-52°C [100-125°F] or less in humid air); sp g = 0.96; OSHA PEL = 0.1 ppm. Respiratory irritant, causes pulmonary edema. Strong irritant to skin, eyes, other tissues. Keep cool and away from oxidizing agents. Baseline physical should stress lungs, nervous system, liver, kidneys, eyes. Pulmonary function test and chest X-ray recommended. Reacts with aluminum and lithium to form hydrides which may explode in air. Use only in an efficient fume hood. Use an explosion shield, impact and chemical splash protecting goggles, supplemented with a mask, and hand and forearm protection. On EPA list of extremely hazardous chemicals, 40 CFR Section 302.

Dimethyl Ether - C_2H_6O, 115-10-6; flammable, inhalation and skin irritant, narcotic properties; bp = -23.7°C [-10.7°F]; er = 3.4-27%; sp g = 1.62; OSHA PEL = NA. Explosion hazard when exposed to flames, sparks, forms peroxides, sensitive to heat.

Ethylene - C_2H_4, 74-85-1; simple asphyxiant, flammable, poisonous to plants; bp = -103.9°C [-155°F]; er = 2.7-36%; sp g = 0.98; OSHA PEL = NA. Dangerous when exposed to heat and flames.

Ethylene Oxide - C_2 40, 75-21-8; flammable, strong irritant, inhalation, and oral poison, carcinogen, bp = 10.7°C [51.3°F], er = 3-100%; sp g= 1.52; OSHA PEL = I ppm. Regulated carcinogen, 29 CFR Section 1910.1047. Increases rate of miscarriages. Very commonly used as sterilizing agent. Should be used in a sealed system and exhausted outdoors. Personnel exposures should be monitored. It is a very strong irritant to eyes, skin, and to the respiratory tract. Can cause pulmonary edema of respiratory system at high levels. Forms explosive mixtures with air. Very reactive when in contact with alkali metal hydroxides, iron and aluminum oxides, and anhydrous chlorides of iron, aluminum, and tin. Also reacts with acids, bases, ammonia, copper, potassium, mercaptans, and potassium perchlorate. On EPA list of extremely hazardous chemicals, 40 CER Section 302. On "California" list as a carcinogen, and a chemical with reproductive toxicity for women.

Formaldehyde - CH_2O, 50-00-0; irritant to eyes, skin, respiratory system, oral poison, allergen, OSHA regulated carcinogen, flammable; bp = -19.4°C [-3°F]; er = 7-73%; sp g = slightly greater than 1; OSHA PEL = 0.75 ppm. Previous physical properties are for the gas. It is normally sold as an aqueous solution of 37 to 52% formaldehyde by weight. Other solvents are also used. Liquid formaldehyde when heated can evolve the gas which will burn. Ingestion of the solution will cause stomach pain, nausea, and vomiting and can result in loss of consciousness. Severe eye irritant. Extended exposures can cause skin and bronchial problems. The gas is on the EPA list of extremely hazardous materials, 40 CFR Section 302. As a gas, is on the "California" list as a carcinogen.

Hydrogen - H_2, 1333-74-0; flammable, explosive, asphyxiant; bp = -252.78°C [-422.99°F]; er = 4. 1-74.2%; sp g = 0.0695; OSHA PEL = NA. Gaseous hydrogen systems of 400 cubic feet (11.35 cubic meters) and containers of liquid hydrogen of more than 150 liters (39.63 gallons) are regulated by OSHA, 29 CFR Section 1910.103. Systems to be used for hydrogen should be purged with inert gas prior to use. Consideration should be given for incorporation of safety systems required for larger systems, depending on the research program and facilities available. Hydrogen will burn with virtually an invisible flame. Care should be used in approaching a suspected hydrogen flame; holding a piece of paper in front of you is recommended to detect the flame. Because it is so light, it tends to escape from rooms where leaks occur.

Hydrogen Chloride (Anhydrous) - HCl, 7647-01-0; nonflammable, toxic gas by all routes of exposure and intake; bp -84.8°C [-121°F]; er = NA; sp g = 1.27; OSHA ceiling PEL = 5 ppm. Gas combines with moisture to become corrosive to eyes, skin, and respiratory system. Baseline physical exam should stress respiratory system, eyes, and skin. Pulmonary test and chest X-ray recommended. Exposure to airborne concentrations above 1500 ppm can be fatal in a few minutes. Irritating properties detectable at about PEL. Should only be used in an efficient fume hood. Workers should wear gas-tight goggles, acid, resistant aprons, gloves, and outerwear.

Hydrogen Fluoride - HF, 7664-39-3; strong irritant and corrosive to eyes, respiratory system, internal tissue, skin, via contact, inhalation, ingestion, noncombustible; bp = 19.5°C [67.2°F]; er = NA;.sp g = 0.7; OSHA PEL = 3 ppm. Readily dissolves in water to form hydrofluoric acid. Prolonged exposure can cause bone changes. Baseline physical should stress eyes, respiratory tract, kidneys, central nervous system, skin, and skeletal system. Special tests should include urinalysis, pelvic X-ray (use shielding to protect genitals as much as possible), and an ophthalmic examination. Provides warning at levels near PEL due to irritant effects. Gas on EPA list of extremely hazardous chemicals, 40 CFR Section302. For other comments, see Chapter 4, Section VIB.9 Hydrofluoric Acid.

Hydrogen Selenide - H_2Se, 7783-07-5; flammable, very toxic via inhalation, also toxic by contact with eyes and skin; bp = - 41.3°C [- 42°F]; er = NA; sp g = 2.1; OSHA PEL = 0.05 ppm. Very offensive odor, but threshold for detection above dangerous levels. Recommend initial medical screening prior to work with this substance for existing respiratory problems and impaired liver function. Causes irritation of eyes, nose, throat, and lungs. Symptoms of exposure include nausea, vomiting, followed by a metallic taste, garlic odor to breath, dizziness, fatigue. Can cause eye, liver, spleen, and lung damage. Use only in an efficient fume hood. Additional respiratory and eye protection recommended if potential exposure problem. Contact with acids, halogenated hydrocarbons, oxidizers, water may result in fire or explosion. On EPA list of extremely hazardous chemicals, 40 CFR Section 302.

Hydrogen Sulfide - H_2S, 7783-06-4; flammable, irritant, asphyxiant; bp = -60°C [-76°F]; er = 4.3-46%; sp g 1.19; OSHA ceiling PEL = 20 ppm (single 10 minute peak PEL = 50 ppm). Strong odor of rotten eggs, but sense of smell desensitized by gas after short interval (minutes) at high levels. Preliminary medical exam recommended prior to work with this material, with stress on eyes and lungs, to include chest X-ray and pulmonary function test. Severe eye and respiratory irritant at moderate concentrations. Rapidly acting systemic poison which causes respiratory paralysis at high levels. Exposures at 1000 to 2000 ppm can cause immediate death. Prolonged exposure above 50 ppm can damage eyes and cause respiratory problems. Susceptibility increases with repeated exposures. Symptoms caused by low concentrations are headache, fatigue, insomnia, irritability, gastrointestinal problems. Highly reactive with strong nitric acid and strong oxidizing agents. On EPA list of extremely hazardous chemicals, 40 CFR 302.

Methane - CH_4, 74-82-8, flammable, simple asphyxiant; bp = - 161.4°C [-258.6°F]; er = 5-15%; sp g 0.52; OSHA PEL NA. Keep away from sources of ignition.

Methyl Acetylene - C_3H_4, 74-99-7; flammable, anesthetic; bp = -23°C [10°F]; er = 1.7-11.7; sp g = 1.4; OSHA PEL = 1000 ppm. Sweet odor. Overexposure causes drowsiness. Reactions with chlorine and strong oxidizing agents may result in fires or explosions. It forms very shock-sensitive compounds with copper. Equipment components containing more than 67% copper should not come into contact with the compound.

Methyl Acetylene Propadiene Mixture (MAPP) - C_3H_4 isomers, No CAS No.; flammable, anesthetic; bp = -34.5°C [-30°F]; er = 3.4-10.8%; sp g = 1.5; OSHA PEL = 1000 ppm. Foul odor. Reactivity similar to methyl acetylene (MA). Overexposure can cause drowsiness and unconsciousness as with MA. Odor detectable well below PEL. Will attack some plastics, films, and rubber.

Methyl Bromide - CH_3Br, 74-83-9; fumigant, toxic by inhalation, contact with eyes and skin, and ingestion; cumulative poison; bp = 3.6°C [3 8.4°F]; er = 13.5-14.5% (requires high energy ignition source); sp g = 3.3; OSHA ceiling PEL = 20 ppm (skin). Severe respiratory irritant, neurotoxin, narcotic at high concentrations. Symptoms of overexposure include headache, visual disturbances (blurred or double vision), nausea, vomiting, in some cases vertigo, tremors of the hand, and in more severe cases convulsions may occur. Persistent depression, anxiety, hallucinations, inability to concentrate, vertigo may follow severe exposures. Kidney damage may occur. Contact with skin can cause skin rash or blisters. Complete pre-use physical recommended, stressing nervous system, lung function, skin condition. Chest X-ray and pulmonary test recommended. Possible carcinogen. Contact with aluminum and strong oxidizers can lead to fires or explosions. On EPA list of extremely dangerous chemicals, 40 CFR Section 302. On "California" list as a chemical with reproductive toxicity.

Methyl Chloride - CH_3Cl, 74-87-3 flammable, Moderate irritant, suspected carcinogen, poison; bp = -23.7°C [-10.7°F]; er = 8.1-17%; sp g =1.78; OSHA PEL = 100 ppm (ceiling = 200 ppm), peak = 300 ppm for 5 minutes in any 3-hour period). Dangerous fire hazard from heat, flame oxidizers. Prolonged exposures can cause psychological problems due to damage to central nervous system. Also can damage liver, kidneys, bone marrow, cardiovascular system. Faint, sweetish odor does not provide adequate warning of overexposure. Use only in an efficient fume hood.

Methyl Mercaptan-CH_4S, 74-93-1; flammable, inhalant poison, possible carcinogen; bp = 5.95°C [42.7°F]; er = 3.9—21.8%; sp g = 1.66; OSHA ceiling PEL = 10 ppm. Warning odor of rotten cabbage. Reacts vigorously with oxidizing agents. Will decompose to emit toxic and flammable vapors when reacts with water, steam, and acids.

Nitric Oxide - NO; 10102-43-9; noncombustible, strong oxidizing agent, inhalant poison; bp = -152°C [-241°F1; er = NA, sp g = 1.0; OSHA PEL = 25 ppm. Nitric oxide causes narcosis in animals. Causes drowsiness. Sharp, sweet odor provides warning well below dangerous levels. Changes to nitrogen dioxide in air (see following material), although conversion is slow at low concentrations. Reacts vigorously with reducing agents. Will attack some plastics, films, rubber. On EPA list of extremely hazardous chemicals, 40 CFR Section 302.

Nitrogen Dioxide - NO_2, 10 102-44-0; poisonous by inhalation; bp = -21°C [-5.8°F]; er = NA; sp g = 2.83; OSHA ceiling PEL = 5 ppm. Brown pungent gas. Odor detection threshold approximately the same as the OSHA PEL. Strong respiratory irritant. Exposure to 100 ppm for 1 hour will normally cause pulmonary edema and possibly death; 25 ppm will cause chest pain and respiratory irritation. Onset of symptoms may be delayed. Recovery from an overexposure may be slow and may result in permanent lung damage. Recommend a baseline physical examination, with emphasis on respiratory and cardiovascular systems. Special tests recommended are chest X-ray, pulmonary function test, and electrocardiogram. Reacts vigorously with chlorinated hydrocarbons, ammonia, carbon disulfide, combustible materials, possibly resulting in fires and explosions. High-temperature glassblowing operations may generate significant levels of this material. On EPA list of extremely hazardous chemicals, 40 CFR Section 302.

Nitrogen Trifluoride - NF_3, 7783-54-2; poisonous by inhalation; bp = -129°C [-200°F]; er = NA; sp g = 2.5; OSHA PEL = 10 ppm. May affect the capacity of the blood to carry oxygen; reacts vigorously with reducing agents. Baseline physical should stress examination of blood, cardiovascular nervous systems, and liver and kidney function. A complete blood count should be taken. Persons with a history of blood disorders should take special care to avoid exposures.

Oxygen Difluoride - OF_2 ; 7783-41-7; poison via inhalation, strong respiratory irritant, noncombustible, strong oxidizing agent; bp = -145°C [-229°F]; er = NA; sp g = 1.86; OSHA PEL = 0.05 ppm. Foul odor, but detection threshold too high to provide adequate warning. The sense of smell fatigues rapidly. Inhalation at less than 1 ppm causes severe headaches. Very corrosive to tissue. Strong irritant to respiratory system, kidneys, internal genitalia. Baseline medical exam recommended, with particular emphasis on affected systems. Use only in an efficient hood system.

Ozone - O_3, 10028-15-6; strong irritant to eyes and respiratory system; bp = -1120°C [-169.6°F]; er = NA; sp g = *1.65;* OSHA PEL = 0.1 ppm. Sharp, distinctive odor; odor detection level about the same as the PEL. Affects central nervous system. May be mutagen. Powerful oxidizing agent for both organic and inorganic oxidizable materials. Some reaction products are very explosive. Baseline physical recommended with emphasis on heart and lungs. Chest X-ray and pulmonary function test recommended. On EPA list of extremely hazardous materials, 40 CFR Section 302.

Phosgene - CCl_2O, 75-44-5; inhalant poison, nonflammable; bp = 8.2°C [46.7°F]; er = NA; sp g = 3.4; OSHA PEL = 0.1 ppm. Odor of "moldy hay" Sense of smell desensitized quickly, irritating properties well above PEL, so does not provide adequate warning. Must be used carefully in an efficient fume hood. Paper soaked in a 10% mixture of equal parts of p-dimethylaminobenzaldehyde and colorless diphenylamine in alcohol or carbon tetrachloride, then dried, makes a good color indicator. Color changes from yellow to deep orange at about the maximum allowable concentration. Severe respiratory irritant, but irritation does not manifest itself at once, even at dangerous levels. Decomposes in presence of moisture in lungs to HCl and CO. Baseline medical examination recommended, with emphasis on respiratory system. Chest X-ray and pulmonary function test recommended. On EPA list of extremely hazardous chemicals, 40 CFR, Section 302.

Phosphine - PH_3, 7803-51-2; flammable, inhalation poison; bp = -87.8°C [-126°F]; er = 1%-?; sp g 1.17; OSIIA PEL = 0.3 ppm. Fishy odor detectable well below PEL. Inhalation is a severe pulmonary irritant and systemic poison. Results of severe overexposure are chest pains, weakness, lung damage, and in some cases coma and death. Persons with prior history of respiratory problems should not work with this material without extra precautions. Baseline physical should include pulmonary function test. On EPA list of extremely hazardous chemicals, 40 CFR Section 302.

Propane - C_3H_8, 74-98-6; flammable, asphyxiant; bp = -42.1°C [-43.7°F]; er = 2.3-9.5; sp g = 1.6; OSHA PEL = 1000 ppm. High exposures (100,000 ppm) caused dizziness after a few minutes. Odorless, nonirritating, so commercially sold propane usually has a foul-smelling odorant added as a warning device. Reacts vigorously with strong oxidizing agents. Explosion hazard from heat and flames.

Propylene - C_3H_6, 115-07-1; flammable, simple asphyxiant; bp = -47.7°C [-53.9°F]; er = 2-11.1%; sp g = 1.5; OSHA PEL = simple asphyxiant. Dangerous when exposed to heat and flames. Can react vigorously with oxidizing agents.

Silane - SiH_4, 7803-62-5; flammable, respiratory irritant; bp = -112°C [-169.6°F]; er = NA; sp g = NA; ACGIH TLV =5 ppm. Moderate respiratory irritant. Repulsive odor. Easily ignites in air. Reacts vigorously with chlorine, bromine, covalent chlorides.

Stibine - SbH_3, 7803-52-3; flammable, poison by inhalation; bp = -17°C [1°F]; er = NA; sp g = 4.34; OSHA PEL = 0.1 ppm. Odor similar to hydrogen sulfide but data not available as to whether the odor is an adequate warning of the PEL. Toxic hemolytic agent, which causes injury to liver and kidneys. Probable lung irritant. Symptoms of overexposure may be delayed for up to 2 days and would include nausea, headache, vomiting, weakness, and back and abdominal pain. Death would result from renal failure and pulmonary edema. Baseline medical exam should stress blood, liver, and kidneys. Special tests should include complete blood count, urine analysis, and liver panel.

Sulfur Dioxide - SO_2, 7446-09-5; nonflammable, strong respiratory, eye, skin irritant; bp = 10.05°C [13.9°F]; er = NA; sp g 2.26; OSHA PEL = 5 ppm. Sharp, irritating odor. Detectable well below OSHA PEL. Reacts rapidly with moisture to form corrosive sulfurous acid (H_2SO_3). Mostly absorbed in upper respiratory tract. High concentrations can cause pulmonary edema and respiratory paralysis. Reacts vigorously with water, with some powdered metals, and with alkali metals such as sodium and potassium. Baseline medical exam should emphasize eyes and respiratory tract. Recommend chest X-ray and pulmonary function tests. Some individuals (10 to 20% of young adults) may be hypersensitive to the material. On EPA list of extremely hazardous chemicals, 40 CFR Section 302.

Sulfur Tetrafluoride - SF_4, 7783-60-0; powerful irritant, poisonous by inhalation; bp = - 40°C (- 40.2°F); sp g = NA; ACGIH ceiling TLV = 0.1 ppm. Reacts with water, steam, and acids to produce toxic and corrosive fumes. On EPA list of extremely dangerous chemicals, 40 CFR Section 302.

Trifluoromonobromomethane (Halon 1301) - $CBrF_3$; 75-63-8; affects heart at high levels; bp = -57.8°C[-72°F]; er = NA; sp g =5; OSHA PEL = 1000 ppm. Former popular fire extinguishing agent used for solvent fires. Production now discontinued due to effects of this gas on earth's ozone layer. At design use range no observed health effects but at high concentrations can cause cardiac arrhythmia. Individuals with heart problems should use with caution. Can emit dangerous gases and vapors on decomposition by heat.

Vinyl Chloride - C_2H_3Cl, 75-01-4; flammable, dangerous irritant, carcinogen; bp = -13.9°C [7°F]; er = 4-33%; sp g= 2.15; OSHA PEL = 1 ppm, ceiling = 5 ppm. Regulated under 29 CFR Part 1910.1017. Vinyl chloride monomer has been shown to cause a rare liver cancer angiosarcoma. Latency period 20 years or more. Participation in medical surveillance program required by OSHA standard if use meets prescribed conditions. Dangerous irritant to respiratory system, skin, eyes, mucous membranes. Reacts vigorously with oxidizers. Decomposes when heated to generate phosgene. On "California" list as a carcinogen.

REFERENCES

1. Occupational Health Guidelines for Chemical Hazards, DHHS (NIOSH)Pub. No. 8 1-123, 1981. Mackison, F.W., Stricoff, R.S., and Partridge, L.J., Jr., (Eds.), U.S. Department of Health and Human Services and U.S. Department of Labor, Washington, D.C.

2. **Sax, N.L and Lewis, R.J., Sr.,** *Rapid Guide to Hazardous Chemicals in the Workplace,* Van Nostrand Reinhold, New York, 1986.

3. **Armour, M.A., Browne, L.M., and Weir, G.L.,** *Hazardous Chemicals Information and Disposal Guide,* 2nd ed., University of Alberta, Edmonton, Canada, 1984.

4. **Lewis, R.J.,** *Sax's Dangerous Properties of Industrial Materials,* 8th ed., Van Nostrand Reinhold , New York, 1993.

5. Threshold Limit Values for Chemical Substances and Physical Agents and Biological Exposure Indices, American Conference of Governmental Industrial Hygienists, Cincinnati, OH, 1993.

6. OSHA, General Industry Standards, Subpart Z, Occupational Health and Environmental Control, 1910.1000-1500.

7. Hazardous Chemical Data, NFPA-49, National Fire Prevention Association, Quincy, MA.

8. *Prudent Practices in the Laboratory Handling and Disposing of Chemicals,* National Academy Press, Washington, D.C., 1995.

9. *NIOSH Pocket Guide to Chemical Hazards,* Hazardous Materials Publishing Co., Kutztown, PA, 1991. Also see Internet; http://www.gov/niosh/npg/npg.html

14. Cryogenic Safety[*]

Cryogenics may be defined as low temperature technology or the science of very low temperatures. To distinguish between cryogenics and refrigeration, a commonly used measure is to consider any temperature lower than -73.3°C (-100°F) as cryogenic. Although there is some controversy about this distinction and some who insist that only those areas within a few degrees of absolute zero may be considered as cryogenic, the broader definition will be used here.

Low temperatures in the cryogenic area are primarily achieved by the liquefaction of gases, and there are more than 25 which are currently in use in the cryogenic area. However, the seven gases which account for the greatest volume of use and applications in research and industry are helium, hydrogen, nitrogen, fluorine, argon, oxygen, and methane (natural gas).

Cryogenics is being applied to a wide variety of research areas, a few of which are food processing and refrigeration, rocket propulsion fuels, spacecraft life support systems, space simulation, microbiology, medicine, surgery, electronics, data processing, and metalworking.

---

* This section, except for a short amount of material appended at the end, is taken, with minor editing, directly from the article, "Cryogenic Safety," by Spencer,[15] in the second edition of this Handbook.

321

Table 4.13 Properties of Cryogenic Fluids

Gas	Normal Boiling Point °C	°F	Volume Expansion Ratio	Flammable	Toxic	Odor
Helium-3	269.9	3.2	757 to 1	No	No	No
Helium-4	268.9	4.2	757 to 1	No	No	No
Hydrogen	252.7	20.4	851 to 1	Yes	No	No
Deuterium	249.5	23.6	—	Yes	Radioactive	No
Tritium	248.0	25.1	—	Yes	Radioactive	No
Neon	245.9	27.2	1438 to 1	No	No	No
Nitrogen	195.8	77.3	696 to 1	No	No	No
Carbon monoxide	192.0	81.1	—	Yes	Yes	No
Fluorine	187.0	86.0	888 to 1	No	Yes	Sharp
Argon	185.7	87.4	847 to 1	No	No	No
Oxygen	183.0	90.1	860 to 1	No	No	No
Methane	161.4	111.7	578 to 1	Yes	No	No
Krypton	151.8	121.3	700 to 1	No	No	No
Tetrafluromethane	128	145	—	No	Yes	No
Ozone	111.9	161.3	—	Yes	Yes	Yes
Xenon	109.1	164.0	573 to 1	No	No	No
Ethylene	103.8	169.3	—	Yes	No	Sweet
Boron trifluoride	100.3	172.7	—	No	Yes	Pungent
Nitrous oxide	89.5	183.6	666 to 1	No	No	Sweet
Ethane	88.3	184.8	—	Yes	No	No
Hydrogen chloride	85.0	188.0	—	No	Yes	Pungent
Acetylene	84.0	189.1		Yes	Yes	Garlic
Fluoroform	84.0	189.1		No	No	No
11-Difluoroethylene	83.0	190.0		Yes	No	Faint ether
Chlorotrifluoromethane	81.4	191.6	—	No	Yes	Mild
Carbon dioxide	78.5	194.6	553 to 1	No	Yes	Slightly pungent

(Recent advances in high temperature superconductors will further increase the use of liquid nitrogen.)

Cryogenic fluids (liquified gases) are characterized by extreme low temperatures, ranging from a boiling point of -78.5°C (-109°F) for carbon dioxide to -269.9°C (-453.8°F) for an isotope of helium, ^{3}He. Another common property is the large ratio of expansion in volume from liquid to gas, from approximately 553 to 1 for carbon dioxide to 1438 to 1 for neon. Table 4.13 contains a more complete summary of the properties of cryogenic fluids.

a. Hazards

There are four principal areas of hazard relating to the use of cryogenic fluids or in cryogenic systems. These are flammability, high pressure gas, materials, and personnel. All categories of hazard are often present in a system concurrently and must be considered when introducing a cryogenic system or process.

The flammability hazard is obvious when gases such as hydrogen, methane, and acetylene are considered. However, the fire hazard may be greatly increased when gases normally thought to be non-flammable are used. The presence of oxygen will greatly increase the flammability of ordinary combustibles and may even cause some noncombustible materials like carbon steel to burn readily under the right conditions. Liquified inert gases such as liquid nitrogen or liquid helium are capable, under the right conditions, of condensing oxygen from the atmosphere and causing oxygen enrichment or entrapment in unsuspected areas. Extremely cold metal surfaces are also capable of condensing oxygen from the atmosphere.

The high pressure gas hazard is always present when cryogenic fluids are used or stored. Since the liquified gases are usually stored at or near their boiling point, there is always some gas present in the container. The large expansion ratio from liquid to gas provides a source for the build-up of high pressures due to the evaporation of the liquid. The rate of expansion will vary, depending on the characteristics of the fluid, container design, insulating materials, and environmental conditions of the atmosphere. Container capacity must include an allowance for that portion which will be in the gaseous state. These same factors must also be considered in the design of the transfer lines and piping systems.

Materials must be carefully selected for cryogenic service because of the drastic changes in the properties of materials when they are exposed to extreme low temperatures. Materials which are normally ductile at atmospheric temperatures may become extremely brittle when subjected to temperatures in the cryogenic range, while other materials may improve their properties of ductility. The American Society of Mechanical Engineers *Boiler and Pressure Vessel Code, Section VIII ,Unfired Pressure Vessels,*may be used as a specific guide to the selection of materials to be used in cryogenic service. Some metals which are suitable for cryogenic temperatures are stainless steel (300 series and other austenitic series), copper, brass, bronze, monel, and aluminum. Non-metal materials which perform satisfactorily in low temperature service are Dacron™, Teflon™, Kel-F™, asbestos impregnated with Teflon™, Mylar™, and nylon. Once the materials are selected, the method of joining them must receive careful consideration to ensure that the desired performance is preserved by using the proper soldering, brazing, or welding techniques or materials. Finally, chemical reactivity between the fluid or gas and the storage containers and equipment must be studied. Wood or asphalt saturated with oxygen has been known to literally explode when subjected to mechanical shock. When properties of materials which are being considered for cryogenic use are unknown or not to be found in the known guides, experimental evaluation should be performed before the materials are used in the system.

Personnel hazards exist in several areas where cryogenic systems are in use. Exposure of personnel to the hazards of fire, high pressure gas, and material failures previously discussed must be avoided. Of prime concern is bodily contact with the extreme low temperatures involved. Very brief contact with fluids or materials at cryogenic temperatures is capable of causing burns similar to thermal burns from high temperature contacts. Prolonged contact with these temperatures will cause embrittlement of the exposed members because of the high water content of the human body. The eyes are especially vulnerable to this type of exposure, so that eye protection is necessary.

While a number of the gases in the cryogenic range are not toxic, they are all capable of causing asphyxiation by displacing the air necessary for the support of life. Even oxygen may have harmful physiological effects if prolonged breathing of pure oxygen takes place.

There is no fine line of distinction between the four categories of hazards, and they must be considered collectively and individually in the design and operation of cryogenic systems.

b. General Precautions

Personnel should be thoroughly instructed and trained in the nature of the hazards and the proper steps to avoid them. This should include emergency procedures, operation of equipment, safety devices, knowledge of the properties of the materials used, and personal protective equipment required.

Equipment and systems should be kept scrupulously clean and contaminating materials avoided that may create a hazardous condition upon contact with the cryogenic fluids or gases used in the system. This is particularly important when working with liquid or gaseous oxygen.

Mixing of gases or fluids should be strictly controlled to prevent the formation of flammable or explosive mixtures. As the primary defense against fire or explosion, extreme care should be taken to avoid contamination of a fuel with an oxidant, or the contamination of an oxidant by a fuel.

As a further prevention when flammable gases are being used, potential ignition sources must be carefully controlled. Work areas, rooms, chambers, or laboratories should be suitably monitored to automatically warn personnel when a dangerous condition is developing. When practical, it would be advisable to provide for the cryogenic equipment to be shut down automatically as well as to sound a warning alarm.

When there is a possibility of personal contact with a cryogenic fluid, full face protection, an impervious apron or coat, cuffless trousers, and high-topped shoes should be worn. Watches, rings, bracelets, or other jewelry should not be permitted when personnel are working with cryogenic fluids. Basically personnel should avoid wearing anything capable of trapping or holding a cryogenic fluid in close proximity to the flesh. Gloves may or may not be worn, but if they are necessary in order to handle containers or cold metal parts of the system, they should be impervious and sufficiently large to be easily tossed off the hand in case of a spill. A more desirable arrangement would be hand protection of the potholder type.

When toxic gases are being used, suitable respiratory protective equipment should be readily available to all personnel. They should thoroughly know the location and use of this equipment.

c. Storage

Storage of cryogenic fluids is usually in a well insulated container designed to minimize loss of product due to boil-off.

The most common container for cryogenic fluids is a double-walled, evacuated container known as a Dewar flask, made of either metal or glass. The glass container is similar in construction and appearance to the ordinary "Thermos" bottle. Generally the lower portion will have a metal base which serves as a stand. Exposed glass portions of the container should be taped to minimize the flying glass hazard if the container should break or implode.

Metal containers are generally used for larger quantities of cryogenic fluids and usually have a capacity of 10 to 100 liters (2.6 to 26 gallons). These containers are also of double-walled evacuated construction and usually contain some absorbent material in the evacuated space. The inner container is usually spherical because that shape has been found to be the most efficient in use. Both the metal and glass Dewars should be kept covered with a loose-fitting cap to prevent air or moisture from entering, and to allow built-up pressure to escape.

Larger capacity storage vessels are basically the same double-walled containers, but the evacuated space is generally filled with powdered or layered insulated material. For economic reasons, the containers are usually cylindrical with dished ends, approximating the shape of a sphere, which would be expensive to build. Containers must be constructed to withstand the weights and pressures that will be encountered and adequately vented to permit the escape of evaporated gas. Containers also should be equipped with rupture discs on both inner and outer vessels to release pressure if the safety relief valves should fail.

Cryogenic fluids with boiling points below that of liquid nitrogen (particularly liquid helium and hydrogen) require specially constructed and insulated containers to prevent rapid loss of product from evaporation. These are special Dewar containers which are actually two containers, one inside the other. The liquid helium or hydrogen is contained in the inner vessel, and the outer vessel contains liquid nitrogen which acts as a heat shield to prevent heat from radiating into the inner vessel. The inner neck should be kept closed with a loose fitting, non-threaded brass plug which prevents air or moisture from entering the container, yet is loose enough to vent any pressure which may have developed. The liquid nitrogen fill and vent lines should be connected by a length of gum rubber tubing with a slit approximately 2.54 cm (1 inch) long near the center of the tubing. This prevents the entry of air and moisture, while the slit will permit release of the gas pressure. Piping or transfer lines should be double-walled evacuated pipes to prevent the loss of product during transfer. Large liquid helium systems also are usually equipped with a gas recovery system to recover the helium.

Most suppliers are now using a special fitting to be used in the shipment of Dewar vessels. Also, an automatic pressure relief valve and a manual valve prevent entry of moisture and air, which will form an ice plug. The liquid helium fill (inner neck) should be reamed out before and after transfer and at least twice daily. Reaming should be performed with a hollow copper rod, with a marker or stop to prevent damaging the bottom of the inner container. Some newer style Dewar vessels are equipped with a pressure relief valve and pressure gauge for the inner vessel.

Transfer of liquids from the metal Dewar vessels should be accomplished with special transfer tubes or pumps designed for the particular application. Since the inner vessel is mainly supported by the neck, tilting to pour the liquid may damage the container, shortening its life or creating a hazard due to container failure at a later date. Piping or transfer lines should be so constructed that it is not possible for fluids to become trapped between valves or closed sections of the line. Evaporation of the liquid in a section of line may result in pressure build-up and eventual explosion. If it is not possible to empty all lines, they must be equipped with safety relief valves and rupture discs. When venting storage containers and lines, proper consideration must be given to the properties of the gas being vented. Venting should be to the outdoors to prevent an accumulation of flammable, toxic, or inert gas in the work area.

ACKNOWLEDGMENT

Reprinted with permission from the *Journal of the American Society of Safety Engineers,* Vol. 11(8), 15-19, August 1963, Chicago, IL.

d. Addendum to Section 14

There are some applications in which Dewars with wide mouths are used, such as storage of certain biological materials. These come with a loosely fitting cap to prevent absorption of air and moisture into the liquid nitrogen, the refrigerant most frequently used in these Dewars. As briefly alluded to in the preceding article, a buildup of oxygen in liquid nitrogen containers over a period of time can become a problem if care is not taken to keep the cap on or to change the entire volume occasionally. If the liquid takes on a blue tint, it is contaminated with oxygen and should be replaced. The contaminated liquid should be treated as a dangerous, potentially explosive material. Most users fill Dewars from larger ones, usually by pressurizing the larger one with nitrogen from a cylinder, thereby forcing the liquid into the smaller one. In order to not waste liquid nitrogen by evaporation in a warm container, neither of the two Dewars are usually allowed to become totally empty, again leading to possible oxygen contamination. If these practices are continued for a sufficiently long time, the oxygen content of the cryogenic liquid may become dangerously high.

There are two relatively common ways to maintain a supply of liquid nitrogen at a facility, one being to have a large reservoir of up to several thousand liters capacity from which individual users fill their smaller Dewars. The boil-off from a large reservoir can be used to provide a supply of ultra-clean "air" to laboratories to use to clean surfaces. Liquid nitrogen is also usually available, if reasonably close to a distributor, in 160-liter pressurized containers delivered directly to the laboratory. In either case, the quantities actually needed for most small laboratories can be obtained frequently enough to avoid having an elaborate piping and control system from a large central reservoir, with the associated problems of avoiding blockage of the system by ice plugs. There are, of course, applications for which automatically controlled systems are necessary that provide safety relief and warning devices.

REFERENCES

1. *Cryogenics,* Marsh & McLendon Chicago, IL, 1962.

2. *Industrial Gas Data,* Air Reduction Sales Co., Acton, MA.

3. *Matheson Gas Data Book,* 47th ed., The Matheson Co., Inc. East Rutherford, NJ, 1961.

4. Precautions and Safe Practices for Handling Liquid Hydrogen, Linde Company, New York, 1960.

5. Precautions and Safe Practices for Handling Liquified Atmospheric Gases, Linde Company, New York, 1960.

6. **Braidech, M.M.,** *Hazards/Safety Considerations in Cryogenic (Super Cold) Operations,* Conference of Special Risk Underwriters, New York, 1961.

7. **Honre, Jackson, and Kurti,** *Experimental Cryophysics,* Butterworths, London, 1963.

8. **MacDonald, D.K.C.,** *Near Zero, An Introduction to Low Temperature Physics,* Anchor Books, Doubleday & Co., New York. 1961.

9. **Nears R.M.,** *Handling Cryogenic Fluids,* Linde Company, New York, 1960.

10. **Scott, R.B.,** *Cryogenic Engineering,* D. Van Nostrand , Princeton, NJ, 1959.

11. **Timmerhaus, K.D.,** Ed., *Advances in Cryogenic Engineering,* Vol. 7, Plenum Press, New York, 1961.

12. **Vance, R.W., and Duke, W.M.,** Eds., *Applied Cryogenic Engineering,* John Wiley & Sons, New York,1962.

13. **Zenner G.H.,** Safety engineering as applied to the handling of liquified atmospheric gases; in *Advances in Cryogenics Engineering,* 6, Plenum Press, New York, 1960.

14. Cryogenic Safety. A Summary Report of the Cryogenic Safety Conference, Air Products, Allentown, PA, 1959.

15. **Spencer E.W.,** Cryogenic safety, in *CRC Handbook of Laboratory Safety* 2nd ed., Steere, N.V., Ed., CRC Press, Cleveland, OH, 1971.

16. *Prudent Practices in the Laboratory Handling and Disposing of Chemicals,* National Academy Press, Washington, D.C., 1995, pp 128-130.

15. Cold Traps[*]

Cold traps are used in instrumentation and elsewhere to prevent the introduction of vapors or liquids into a measuring instrument from a system, or from a measuring instrument (such as a Mcleod gauge) into the system. A cold trap provides a very low-temperature surface on which such molecules can condense and improves pump-down (the achievable vacuum) by one or two magnitudes.

However, cold traps improperly employed can impair accuracy, destroy instruments or systems, and be a physical hazard. For example, many of the slush mixtures used in cold traps are toxic or explosive hazards, and this is not indicated in the literature.

The authors (of this article) became aware of the deficiencies in tunnel instrumentation, where it was necessary to measure pressures in the micron to 760 torr region (a torr is equal to a pressure of 1 mm Hg). The instrumentation system used Stratham gauges for ambient pressure down to 100 to 150 torr or about 2 to 3 psia and NRC alphatron gauges for pressures to 5×10^{-2} torr. To prevent calibration shifts and contamination of the NRC transducers by oil fumes from the vacuum pump and possible wind tunnel contaminants, a cold trap was placed in the line.

The cold trap was filled with liquid nitrogen, and the valve to the tunnel line shut off. When the valve was opened, cold gas shot out, shown by the condensation; the over-pressure developed

---

[*] This section is taken from the article "Cold Traps," by Kaufman and Kaufman in the second edition of this handbook.

in the system destroyed the Stratham strain gage bridge, although it was not sufficient to rupture the transducer diaphragm. As no satisfactory explanation was forthcoming, a glass cold trap was procured and set up in a dummy system. The cause of the phenomenon soon became apparent: air in the trap and system lines was liquified in the trap. When the valve was opened, this liquid air was being blown into the warmer lines by atmospheric pressure. The resultant volatilization of liquid into gas was practically an explosion.

Nevertheless, cold traps are often the only satisfactory means of removing contaminants, although in ordinary experimental work the charcoal trap is occasionally acceptable. A charcoal trap will remove oil and condensable vapors so that pressures to 10^{-8} tor or better may be secured, but it presents a serious restriction on pumping speed.

The errors introduced by the water vapor, when measuring low pressures, depend on the vacuum gauge used. The presence of water vapor also affects the magnitude of vacuum that can be achieved. The equilibrium point of a dry-ice-acetone slush is -78°C (-108.4°F), which, although sufficient to trap mercury vapor effectively, does not remove water vapor; a temperature of at least -100°C(-148°F) is required to eliminate water vapor or, alternatively exposure to anhydrous phosphorous pentoxide (P_2O_5). This material is usually rejected for field use because of possible biological, fire, and explosive hazards: in absorbing water it produces heat and reacts vigorously with reducing materials.

Slush mixtures using liquid air and liquid oxygen were considered and dropped, either because of the explosive hazard or toxicity of the vapors or because they were not cold enough. Table 4.14 lists many common thermal transfer and coolant fluids with their hazards and limitations.

a.   Virtual Leaks

If the cold trap is chilled too soon after the evacuation of the system begins, gases trapped will later evaporate when the pressure reaches a sufficiently low value. The evaporation of the refrigerated and trapped gases is not rapid enough to be evacuated by the system, but is enough to degrade the vacuum, producing symptoms very similar to those of a leak. To avoid these virtual leaks, keep the trap warm until a vacuum of about 10^{-2} torr is obtained. The tip of the trap is then cooled until the ultimate vacuum is reached, at which time the trap may be immersed in the coolant to full depth.

b.   Safety Precautions

If liquid nitrogen is the coolant, liquid air can condense in the trap, inviting explosion. Liquid air, comprising a combination primarily of oxygen and nitrogen, is warmer than liquid nitrogen. Depending on the nitrogen content, air liquifies anywhere from -190°C (-310°F) (5°C warmer than liquid nitrogen) to -183°C (-297.4°F) (liquid oxygen). If liquid nitrogen is used, the trap should be charged only after the system is pumped down lest a considerable amount of liquid oxygen condenses, creating a major hazard. Handle any liquid gas carefully; at its extremely low temperature, it can produce an effect on the skin similar to a burn. Moreover, liquified gases spilled on a surface tend to cover it completely and intimately, and therefore cool a large area.

The evaporation products of these liquids are also extremely cold and can produce burns. Delicate tissues, such as those of the eyes, can be damaged by an exposure to these cold gases which is too brief to affect the skin of the hands or face. Eyes should be protected with a face shield or safety goggles (safety spectacles with or without side shields do not give adequate protection). Gloves should be worn when handling anything that is or may have been in contact with the liquid; asbestos gloves are recommended (this is no longer true because of the concern about airborne asbestos fibers from products containing asbestos. Gloves made of an artificial material such as Kevlar™ or Zetex™ are recommended as an alternative), but leather gloves may be used. The gloves must fit loosely so that they can be thrown off quickly if liquid should spill or splash into them. When handling liquids in open containers, high-top shoes should be worn with trousers (cuffless if possible) worn outside them.

327

Table 4.14 Thermal Transfer Fluids[a] Used With Instrumentation Type Cold Traps

Element[b]	Temperature[c] °C	°F	Hazard[d] Inhalation Toxicity	Skin Toxicity	Explosive or fire	Remarks
Glycerine, 70% by weight, water 30%	-38.9	-38	None	Slight	Slight	2.5×10^{-3} torr at 50°C (122°F)
Ethyl alcohol, dry ice	-78	-108	Moderate	Slight	Dangerous	—
Ethylene glycol, 52% by vol., water	-40	-40	None	Slight	Slight	—
Chloroform, dry ice	-63.5	-82	Extreme	Slight	Slight	Vapor pressure 100 torr at 22°C (71.6°F)
Liquid SO_2	-75.5	-103	Extreme	Extreme	None	Very dangerous
Methyl alcohol, dry ice	-78	-108	Slight	Slight	Dangerous	Vapor pressure 100 torr at 22°C (71.6°F),ingestion very dangerous
Acetone, dry ice	-78	-108	Moderate	Slight	Dangerous	Vapor pressure 400 torr at 30°C (86°F) Very dangerous
Methyl bromide	-78	-108	Extreme	Extreme	Moderate	—
Fluoretrichloromethane (Freon 11) dry	-7S	-108	Slight	Slight	None	—
Methylene chloride, dry ice	-78	-108	Moderate	Moderate	Slight	Very dangerous to eyes, Vapor pressure 380 torr at 22°C (71.6°F)
Calcium chloride	-42	-44	Slight	Slight	None	—
Ethyl methyl ketone	-78	-108	Moderate	Moderate	Dangerous	—

a. Transfer fluids will freeze solid and become colder if subject to temperatures lower than their freezing point. A slush mixture is secured by lowering the temperature such as by introduction of limited quantities of dry ice until the mixture is quasi frozen.

b. These materials are often sold under trade names. In general, any combination of elements shown was selected for the coldest mixture obtainable.

c. If the refrigerant is dry ice, the transfer fluid will not go below -78°C (-108.4°F), the temperature of solid dry ice.

d. The consensus is that many of these liquids, while dangerous at room temperature, are not hazardous when cooled, since their evaporation at low temperatures is thirly low. For utmost safety, those noted to be dangerous should not be employed unless venting or other special precautions are taken. For greater detail, see reference 2.

Stand clear of boiling and splashing liquids and its issuing gas. Boiling and splashing always occurs when charging a warm container or when inserting objects into the liquid. Always perform these operations *slowly* to minimize boiling and splashing.

Should any liquified gases used in a cold trap contact the skin or eyes, immediately flood that area of the body with large quantities of unheated water and then apply cold compresses. Whenever handling liquified gases, be sure there is a hose or a large open container of water nearby, reserved for this purpose. If the skin is blistered, or if there is any chance that the eyes have been affected, take the patient immediately to a physician for treatment (call for emergency medical aid; normally rescue squads can be in immediate contact with an emergency room physician by radio).

Oxygen is removed from the air by liquid nitrogen exposed to the atmosphere in an open Dewar. Store and use liquid nitrogen only in a well ventilated place; owing to evaporation of nitrogen gas and condensation of oxygen gas, the percentage of oxygen in a confined space can become dangerously low. When the oxygen concentration in the air becomes sufficiently low, a person loses consciousness without warning symptoms and will die if not rescued. The oxygen content of the air must never be allowed to fall below 16%.

The appearance of a blue tint in liquid nitrogen is a direct indication of its contamination by oxygen, and it should be disposed of, using all the precautions generally used with liquid oxygen. Liquid nitrogen heavily contaminated with oxygen has severe explosive capabilities. In addition, an uninsulated line used to charge Dewars will condense liquid air; liquid air dripping off the line and revaporizing causes an explosive hazard during the charging operation.

If the cold trap mixture is allowed to freeze, and the cold trap becomes rigid, slight movement in other parts of the apparatus could result in breakage of the trap or other glassware.

If a gas trap has to be lifted out of the Dewar cold bath for inspection, it will be difficult to reinsert into the slush. Therefore, it is preferable to use a liquid that will not freeze at 78.5°C.

ACKNOWLEDGMENT

Reprinted with permission of Rimbach Publications, Pittsburgh, Pennsylvania, from *Instrument and Control Systems,* vol. 36, pp. 109-111, July 1963.

REFERENCES

1. **Strong, J., Neher, H.V., Whitford, A.E., Cartwright, C.H., and Hayward, R.,** *Procedures in Experimental Physics,* Prentice-Hall, New York, 1938.

2. **Sax, N.I.,** *Dangerous Properties of Industrial Materials,* Reinhold Publishing , New York, 1961.

3. **Dushman, L.,** *Scientific Foundation of Vacuum Techniques,* John Wiley & Sons, New York, 1962.

4. **Kaufman, A.B. and Kaufman, E.N.,** Cold traps, in *CRC Handbook of Laboratory Safety,* 2nd ed., Steere, N.V., Ed., CRC Press, Cleveland, OH, 1971, 510.

16. Care and Use of Electrical Systems

Some of the problems associated with electrical systems have been covered in previous sections, such as Chapter 3, Section I.G. There may be some unavoidable repetition in this section, which will be primarily concerned with the safe use of electricity rather than the characteristics of individual items, although there will be a brief discussion of generic problems associated with the design of equipment. Most of the hazards associated with the use of electricity stem from electrical shock, resistive heating, and ignition of flammables, and most of the actual incidents occur because of a failure to anticipate all of the ways in which these hazards may be evoked in a laboratory situation. This lack of appreciation of the possible hazards may be reflected in the original choice of suitably safe electrical equipment or improper installation of the equipment. In some instances the choice of equipment may simply involve a continued use of equipment on hand under conditions for which it is no longer suitable, so that safety specifications are not really considered. Often, this is due to a familiarity with the existing resources rather than a deliberate choice.

Part of the problem is a feeling that questionable electrical practices routinely followed at home as well as in the laboratory are actually safe unless you do something "really bad," such as standing in water while in contact with an electrically active wire, or a similar feeling about wiring, "Just hook it up with the extension cord bought on sale at the department store." When asked about a number of similar practices involving multiple connections to a single outlet or the use of extension cords, most people will answer that they know that they should not do it, but see no real harm.

Two major electrical factors need to be considered in the choice of most electrical items of equipment. The equipment needs to be selected so that it will not provide a source of ignition to flammable materials, and it should be chosen so as to minimize the possibility of personnel coming into contact with electrically live components. This latter problem will be addressed first.

2. Electrical Shock

OSHA has included the relevant safety portions of the National Electrical Code in 29 CFR 1910, Subpart S. This regulatory standard, as are many other sections of the OSHA regulations, is primarily oriented toward industrial applications, but it does speak directly to the problem of preventing individuals from coming into contact with electricity. Live parts of electrical equipment operating at 50 volts or more must be guarded against accidental contact. Indoor installations that contain circuits operating at 600 volts or more, and accessible to electrically untrained persons, must have the active components within metal enclosures or be located within a space controlled by a lock. The higher voltage equipment also must be marked with appropriate warning signs. Access points to spaces in which exposed electrically live parts are present must be marked with conspicuous warning signs which forbid unqualified persons to enter.

The effects of electricity on a person depend upon the current level and, of course, on physiological factors unique to the individual. Table 4.15 gives typical effects of various current levels for 60-Hz currents for an average person in good health.

Several things affect the results of an individual incident. The duration of the current is important. In general, the degree of injury is proportional to the length of time the body is part of the electrical circuit. A suggested threshold is a product of time and energy of 0.25 watt-seconds for an objectionable level. The voltage is important because, for a given resistance R, the current I through a circuit element is directly proportional to the applied voltage V.

$$I = V/R \qquad\qquad (1)$$

If the contact resistance to the body is lowered so that the total body resistance to the flow of current is low, then even a relatively modest applied voltage can affect the body. The condition of the skin can dramatically alter the contact resistance. Damp, sweaty hands may have a contact resistance which will be some orders of magnitude lower than dry skin. The skin condition is more important for low voltage contacts than for those involving high voltages, since in the latter case the skin and contact resistance break down very rapidly. The remaining resistance is the inherent resistance of the body between the points of contact, which is on the order of 500 to 1000 ohms. As can be noted in Table 4.15, the difference in a barely noticeable shock and a potentially deadly one is only a factor of 100. For an individual with cardiac problems, the threshold for

Table 4.15 Effects of Electrical Current in the Human Body

Current (milliamperes)	Reaction
1	Perception level, a faint tingle.
5	Slight shock felt; disturbing but not painful. Average person can let go. However, vigorous involuntary reactions to shocks in this range can cause accidents.
6 to 25 (women) 9 to 30 (men)	Painful shock, muscular control is lost. Called freezing or "let-go" range*.
50 to 150	Extreme pain, respiratory arrest, severe muscular contractions, individual normally cannot let go unless knocked away by muscle action. Death is possible.
1,000 to 4,300	Ventricular fibrillation (the rhythmic pumping action of the heart ceases). Muscular contraction and nerve damage occur. Death is most likely.
10,000 —	Cardiac arrest, severe burns, and probable death.

*The person may be forcibly thrown away from the contact if the extensor muscles are excited by the shock.

a potentially life-threatening exposure may be even lower. The major danger to the heart is that it will go into ventricular fibrillation due to small currents flowing through it. In most cases, once the heart goes into ventricular fibrillation, death follows within a few minutes.

Even if an individual survives a shock episode, there may be immediate and long-term destruction of tissue, nerves, and muscle due to heat generated by the current flowing through the body. The heat generated is basically resistive heating such as would be generated in heating coils in a small space heater, with the exception that the resistive elements are the tissues and bones in the body. The power, P, or heat is given by

$$P = I^2R \qquad (2)$$

The scope of the effects of external electrical burns is usually immediately apparent, but the total effect of internal burns may become manifest later on by losses of important body functions due to the destruction of critical internal organs, including portions of the nervous system, which is especially vulnerable.

Several means are available to prevent individuals from coming into contact with electricity in addition to exclusion of unqualified personnel from space, as mentioned in the introduction to this section. These include insulation, grounding, good wiring practices, and mechanical devices. Before addressing these latter options, it might be well to briefly discuss the concept of a qualified person.

Certainly a licensed electrician would in most cases be a qualified person, and a totally inexperienced person would just as clearly not be a qualified individual. There is no clear definition of the training required to be "qualified" to perform routine laboratory electrical and electronic maintenance. As a minimum, such training should include instruction in the consequences of electrical shock, basic training in wiring color codes (so as to recognize correct leads), familiarity with the significance of ratings of switches, wiring, breakers, etc., simple good wiring practices, and recognition of problems and poor practices, such as frayed wiring, wires underfoot, wires in moisture, overloaded circuits, use of too small or wrong type of conductors, poor grounding procedures, and improper defeat of protective interlocks. This would not make an individual a licensed electrician, which requires extensive training and experience, but would reduce the number of common electrical errors.

Insulation is an obvious means of protecting an individual against shocks. In general, good wiring insulation is the most critical, particularly that of extension cords, which are often abused. Insulation must be appropriate for the environment, which may involve extremes of temperature or exposure to corrosive vapors or solvents. The insulation itself may need to be protected by a metal outer sheath, or the wires may need to be installed in conduit.

As it ages, insulation may become brittle and develop fine cracks through which moisture may seep and provide a conductive path to another component or to a person who simply touches the wire at the point of failure. Many plastic or rubber insulating materials will soften with heat and, if draped over a metal support, may eventually allow the wire to come into contact with the metal, thus rendering the metal electrically active, if it does not first cause other problems. Extension cords, as noted above, are particularly susceptible to abuse. They are often carelessly strewn across the floor or furniture. On the floor, they may be walked upon, equipment may be rolled across them, or they may become pinched between items of furniture. Extension cords should only be used as temporary expedients, but if they are used, they should be treated as any other circuit wiring, put out of harm's way, and properly supported on real insulators separated by distances not to exceed 10 feet. Defective extension cords with badly deteriorated insulation should be discarded. Insulation is not used solely to protect wiring. Insulation in the form of panels supporting printed circuits may break if excessive force is applied. If an arc temporarily flashes across an insulating surface, a carbonized conducting path may be permanently established on the surface which could render an external component such as a chassis mounting screw "hot." Care should be taken with all electrical equipment, especially older items; to ensure that the integrity of the insulation has been maintained.

Proper grounding of equipment is another requirement to ensure that components are not electrically live. Most equipment for use with 120-volt circuits comes with a three-wire power cord, which requires a mating female connector at the power source, many of which are designed so that the neutral, hot, and ground connections can be readily identified and matched. The ground wire, which is either green or perhaps green with yellow stripes, is always connected to the female socket which accommodates the round prong on the male connector. The neutral circuit wire which normally completes the circuit for the equipment is usually white or gray. The socket and corresponding male connector are often wider than the connections for the hot wire. The hot wire is usually covered with a black insulator, although red may also be used. Where there are both red and black wires, usually both will be hot wires. Some equipment is double insulated, and does not have the third ground wire in the power connector. Usually, these will have a polarized connector, so that the neutral and hot wires will be properly oriented. Older circuits unfortunately do not always provide the proper connections and should be replaced. If this is not feasible, the third wire on the power connector, if one is present, should be directly connected to a good quality ground.

Auto-transformers, which may be used to supply variable voltages to heating devices, may be connected in such a way that either outlet line may be high with respect to ground. They should be purchased with a switch which breaks the connection of both outlet sockets to the power input line, or they should be rewired with a double-pole power switch to accomplish this.

The quality of all ground connections (and of all connections) needs to be good. This is often taken for granted, but the connection may vibrate or work itself loose, or a careless worker may fail to tighten a connection. In such cases, a significant difference of potential may arise between two different items of equipment. This can be enough to give rise to a discernible shock for a person coming into contact simultaneously with both pieces of equipment, and in some cases can cause damage to the equipment if they are interconnected. A careful researcher should have the electrical circuits checked periodically for the resistance to ground for all the wiring in his facility. A ground with a resistance of 100 ohms will be at a difference of 10 volts with respect to ground if a current of 100 milliamperes were to flow through the ground connection. Good quality grounds with resistances of a few ohms are easily achievable with care.

Adapters or "cheaters" can be used to allow power cords with three wires to be connected to sockets providing only two wires, or used to avoid connecting the third wires to ground. There are only a few exceptions which would make this an acceptable practice; one is where an alternative direct connection to ground is provided for the equipment. Another would be on a very few occasions in which even the difference of a few millivolts between the separate ground connections would affect the experimental data signals. In the latter case, ground connections can be made directly between the components, with a single connection being made to the building ground. Except in clearly defined situations where their use is clearly made safe, these adapters should not be used.

Simple devices such as fuses, circuit breakers, and ground fault interrupters are available to cut off equipment when they overload or short out or an inbalance develops between the input and output current from a device or circuit. More sophisticated devices can also be used to determine a problem, such as a redundant heat detector used to deactivate a circuit serving a still, condenser, or heat bath should the temperature become too high.

Fuses and circuit breakers are the simplest devices used to shut off a circuit drawing too much current. A fuse inserted in one of the circuit legs functions by melting at a predetermined current limit, and the breaker by mechanically opening the circuit. The latter device is more flexible in that it can be reset while the fuse must be replaced. A ground fault interrupter (GFI), on the other hand, specifically can protect an individual who comes into contact with a live component. The individual's body and the wires become parallel circuits through which a fraction of the total current flows. The amount through the body makes the two normal halves of the circuit out of balance, which the GFI detects, and causes it to break the circuit. A GFI can detect a difference on the order of 5 milliamperes and can break the circuit in as little as 25 milliseconds. A review of Table 4.15 would show that the contact might be barely noticeable but would cause no direct harm because of the short duration of the current flow. Although GFI's are generally used in the construction industry they would serve a useful purpose in laboratories, such as, where moisture would be a problem. Any laboratory containing equipment operating at high voltages should have each electrical outlet protected by a GFI, supplemented by a master disconnect switch in an obvious and easily accessible space. It is critical to remove the current source in as short a time as possible.

The best defenses against electrical shock injuries are good work practices, as invoked by using good judgment and exercising care appropriate to the risk. The basic principles embodied in the OSHA lock-out, tag-out provisions of their electrical standard should be followed. These are basic common sense. Maintenance of electrical equipment or wiring should be done only with the system deenergized unless it is essential that the circuit be active for the required maintenance. In the latter case, specific care is to be taken to come into contact (if necessary) with only one side of a circuit, so that the circuit cannot be completed through the body. Procedures should be followed to confirm that power to the system has been disabled and remains so during the duration of the maintenance activity or, alternatively, if the circuit must remain powered, that a second person is available to disable the circuit and assist in the event of an incident. Formal lock-out procedures are recommended where high voltage circuits are involved. The tools used to perform maintenance should be in good condition. Barriers may be needed to isolate live circuits in the maintenance area. Good judgment should be used to determine safe distances, to not use metal ladders, or metal devices, where it would be possible to contact a hot circuit. In some cases, it might be necessary to use rubber gloves and gauntlets, insulating mats, and hard hats certified for electrical protection.

A relatively simple protective stratagem which should be followed by anyone working with or handling live electrical circuits is to remove all conducting jewelry, specifically items on or near the hands such as rings, watches, and bracelets, or to avoid wearing necklaces which may dangle from the neck and complete a circuit to the neck. If work activities involving direct contact with electrical components are infrequent, removing these items at the time may suffice. If the work is a normal activity, the practice of avoiding wearing metallic objects should be routine to avoid having to remember to remove them. Avoiding the use of conductive items in the vicinity of electricity should extend to any object which might come into contact with the circuit. Many tools must be metallic, but any tool used in electrical work should have insulating materials in those areas normally in contact with the hands.

Interlocks should never be bypassed by the average laboratory worker. If it becomes necessary, the decision to do so must be done with the knowledge of all persons who might be affected by the decision. Bypassing an interlock should not be a decision permitted for an inexperienced graduate student or new employee. Whenever an interlock is bypassed, a definite procedure, requiring positive confirmation that all personnel are no longer at risk, must be adopted and in place. This may involve actual locks, for which only the responsible person has a key, tags which cannot be removed without deliberately breaking a seal, an alarm, or a combination of the above. No preventive procedure should depend upon the continued functioning of a single device, such as a micro-switch, which may fail in such a way as to defeat the alarm or interlock.

Each circuit should be clearly identified and labeled to correspond to a circuit breaker in a service panel. Access to these service panels should be provided to most laboratory employees, but they should not be permitted to remove the protective panel covers protecting the wiring. No closet containing an electrical service panel should be allowed to be used for a storage closet by laboratory personnel. Access to the panel should not be blocked by extraneous items, and accidental contact with the wires should not be possible, especially for an untrained person entering the space.

One of the most effective safety practices, as well as one highly conducive to productivity, is a definite scheduled program of preventive maintenance. Each item of equipment should be periodically removed from service, carefully inspected and calibrated, any faults or indications of deterioration repaired, and tagged with the date of review and the name of the maintenance person, if more than one technician could have been responsible. A permanent file or maintenance log on each major item of equipment is useful for identifying trends or weak components.

Finally, in a facility in which electrical injuries are a reasonable possibility, it is strongly recommended that at least some permanent personnel be trained in CPR and the measures to be taken should a person receive a severe shock. Individuals should also be trained to effect a rescue without themselves becoming a casualty. If, for example, a live wire is lying across a person and the circuit cannot be readily broken, they should be instructed to find a meter stick or some other insulated device to lift the energized wire from the victim, or use rubber gloves or other insulator in attempting to loosen a person from a circuit.

b. Resistive Heating

This is one of the two major electrical sources of ignition of flammable materials in a laboratory; the other being sparks. Electrical heating can occur in a number of ways - poor connections, undersized wiring or electrical components (or, alternatively overloaded wiring or components), or inadequate ventilation of equipment. Equation 2 in the previous section shows that the power or heat released at a given point in a circuit is directly proportional to the resistance at that point. A current of 100 milliamperes through a connection with a resistance of 0.1 ohms would generate a localized power dissipation at that point of only 10 milliwatts, while a poor connection of 1000 ohms resistance would result in a localized power dissipation of 100 watts. The former would normally cause no problems, while the latter might raise the local temperature enough to exceed the ignition temperature of materials in the vicinity. Poor or loose connections have, in fact, caused many fires due to just such localized heating. An alligator clip used to attach a grounding wire is a good example of a potentially poor connection. Similarly, a contact which has been degraded by a chemical, a wire that has been insecurely screwed down, or the expansion and contraction of a wire such as aluminum may in time result in this kind of problem.

Table 4.16 Electrical Characteristics of Wire per 50 Feet

Wire Size (D)	Resistance Amperes	Maximum Drop (V)	Voltage Loss (W)	Power
18	0.3318	7	4.6	32
16	0.2087	10	4.2	42
14	0.1310	15	3.9	59
12	0.0825	20	3.3	61
10	0.0518	30	3.1	93
8	0.0329	40	2.6	105
6	0.0205	55	2.3	124
4	0.0129	70	1.81	127
3	0.0103	80	1.64	131
2	0.00809	95	1.53	138
1	0.00645	110	1.42	156
0	0.00510	125	1.27	159

Overheating of switches, fixtures, and other electrical components due to electrical overloads can be avoided very simply by reading the electrical specifications for the component, usually printed or embossed on the item, and complying with the limitations. If a switch is rated to carry 7 amperes at 120 volts, it will not survive indefinitely in a circuit in which it is carrying 30 amperes.

Each size or gauge wire is designed to carry a maximum amount of current. This is based on the voltage drop per unit length and the amount of power dissipated in the wire. The voltage drop should not exceed 2 to 5% due to wiring resistance. A 5% drop in a 120 volt circuit supplying an item of equipment would mean an actual voltage at the connection to the equipment of only 114 volts. Although many items of equipment will accommodate a drop of this amount, some may not. The heat developed in an overloaded circuit may heat the wiring to a point where the insulation may fail or in extreme cases actually catch on fire. Even moderate overheating, continued long enough, will probably cause an eventual breakdown in the insulation. In addition, any energy dissipated in the wiring is wasted energy.

Table 4.16 gives the maximum current for copper wire of various sizes, the resistance, voltage drop, and power loss per 50 ft (about 15.25 m) of line (the latter two values computed for a wire carrying the maximum rated current). Most inexpensive extension cords purchased at a department store are made of either 16 or 18 gauge wire. As can be seen from the table, inexpensive extension cords do not carry sufficient current to be useful for providing power to more than a few instruments at most, when properly used. Overloading them will cause a larger voltage drop and power dissipation (heating) in the wire. Although extension cords made of wire which is too small will probably not immediately fail in most applications, they are not suitable for continued use. The lower available voltages can result in damage to equipment or failure of relays in control circuits, as their magnetic fields become weaker.

Additional electrical load is a problem for extension cords and for the permanent wiring as well, if multiple outlet plugs are used in a socket. Virtually every safety professional has at least one photograph or slide of several multiple plugs plugged into each other, all drawing current from a single socket. The result will usually be an overheated fixture and wiring, as well as a lower voltage at the plug. This process can continue all the way back to service panels and to the power supply to the facility. Examples of breaker panels almost too hot to touch are, unfortunately, fairly common. Low overall supply voltages are becoming common in older

Table 4.17 Electrical Requirements of Some Common Laboratory Devices

Instrument	Current (amperes)	Power (watts)
Balance (electronic)	0.1- 0.5	12 - 60
Biological safety cabinet	15	1,800
Blender	3 - 15	400 - 1,800
Centrifuge	3 - 30	400 - 6,000
Chromatograph	15	1,800
Computer (PC)	2 - 4	400 - 600
Freeze dryer	20	4,500
Fume hood blower	5 - 15	600 - 1,800
Furnace/oven	3 - 15	500 - 3,000
Heat gun	8 - 16	1,000 - 2,000
Heat mantle	0.4 - 5	50 - 600
Hot plate	4 - 12	450 - 1,400
Kjeldahl digester	15 - 35	1 800 - 4,500
Refrigerator/freezer	2 - 10	250 - 1,200
Stills	8 - 30	1,000 - 5,000
Sterilizer	12 - 50	1,400 - 12,000
Vacuum pump (backing)	4 - 20	500 - 2,500
Vacuum pump (diffusion)	4	500

facilities as the larger electrical loads of laboratories replace the lesser loads of classrooms in many academic institutions. Many older facilities have insufficient electrical capacity. Table 4.17 lists a number of common laboratory devices with the current and power requirements for representative units. Note that some of these normally require 208 to 240 volt circuits. Some may also require connection to three-phase current, which if not done correctly can result in an accident or, at best, poor performance (a relatively common mistake is to wire a three-phase motor so that it operates backwards). A fume blower miswired in this way would result in minimal exhaust velocity.

Devices with resistive heating elements, such as furnaces, heat guns, hot plates, and ovens should be configured in such a way that personnel cannot come into contact with an electrically active element, nor should volatile solvents be used in the proximity of such devices (or in them, as in an oven) where the temperature may exceed the ignition temperature of the solvent.

c. Spark Ignition Sources

Induction motors should be used in most laboratory applications instead of series wound electric motors, which generate sparks from the contacts of the carbon brushes. Sealed explosion-proof motors can also be used but are expensive. It is especially important to use nonsparking motors in equipment which result in substantial amounts of vapor, such as blenders, evaporators, or stirrers. Equivalent ordinary household equipment or other items such as vacuum cleaners, drills, rotary saws, or other power equipment are not suitable for use in laboratories where solvents are in use. Blowers used in fume exhaust systems should at least have nonsparking fan blades, but in critical situations with easily ignitable vapors being exhausted, it may be worth the additional cost of a fully explosion-proof blower unit. Any device in which an electrically live circuit makes and breaks, as in a thermostat, an on-off switch, or other control mechanism, is a potential source of ignition for flammable gases or vapors. Special care should be taken to eliminate such ignition sources in equipment in which the vapors may become confined, as already discussed for refrigerators and freezers. It is also possible in other equipment such as blenders, mixers, and ovens and such devices should not be permitted to be used with or in the vicinity of materials which emit potentially flammable vapors.

REFERENCES

1. OSHA, General Industry Standards, Subpart S-Electrical 29 CFR 1910.30 1-339.
2. National Electrical Safety Electrical Code, ANSI C2, American National Standards Institute, New York.
3. *Prudent Practices in the Laboratory Handling and Disposal of Chemicals,* National Academy Press, Washington, D.C., 1995.
4. **Lockwood, G.T.,** Protective lockout and tagging of equipment, in *CRC Handbook of Laboratory Safety,* 2nd ed., Steere, N.V, Ed., CRC Press, Cleveland, OH, 1971, 511.
5. Electronic Industries Association, Grounding electronic equipment, in *CRC Handbook of Laboratory Safety,* 2nd ed., Steere, N.V, Ed., CRC Press, Cleveland, OH, 1971, 516.
6. **Dalziel, C.E,** Deleterious effects of electrical shock, in *CRC Handbook of Laboratory Safety,* 2nd ed., Steere, N.V., Ed., CRC Press, Cleveland, OH, 1971, 521.
7. **Ehrenkranz, T.E., and G.W. Marsischky,** Electrical equipment, wiring, and safety procedures, in *CRC Handbook of Laboratory Safety,* 2nd ed., Steere, N.V., Ed., CRC Press, Cleveland, OH, FL, 1971, 528.
8. **Ehrenkranz, T.E.,** Explosion-proof electrical equipment, in *CRC Handbook of Laboratory Safety,* 2nd ed., Steere, N.V., Ed., CRC Press, Cleveland, OH, 1971, 540.
9. Controlling Electrical Hazards, U.S. Department of Labor, OSHA, 3075, Washington, D.C., 1983
10. Electrical Hazard, U.S. Department of the Interior, National Mine Health and Safety Academy Safety Manual No. 9.

17. Glassware

It is recommended that most laboratory glassware be made of borosilicate glass. Pyrex™ is one of the most familiar brand names under which this type of glass is known. Borosilicate glass has many of the desirable characteristics of pure silica glass. It has good chemical durability and a low temperature expansion coefficient. Exceptions to this recommendation are reagent bottles, stirring, tubing, and any other glass item where the value of a low temperature expansion glass is not needed.

Breakage is the major safety concern with glassware. Bottles often break when dropped. The seriousness of this type of accident can be reduced by purchasing reagents in glass containers coated with plastic. This largely eliminates the danger of cutting oneself on the sharp edges since the broken container normally will be held more or less together by the film, and the film should reduce the amount of liquid leaking so that the problem of harm due to splashing liquids is also minimized. Highly corrosive materials, which for any reason are not in these safety coated containers, should always be transported in rubber or plastic carriers designed for the purpose.

Laboratory systems routinely are made of glass flasks, other glass components, and glass tubing. These systems may explode if the internal pressures becomes too high or implode if the system is evacuated. Many circumstances may give rise to over-pressure: a reaction which escalates beyond that anticipated, an explosion, a too-violent reaction, or excessive heat, among others. When possible, spherical containers should be used where pressurized systems (over- or under-pressure) are constructed, or if not, then the reduced strength of nonspherical containers should be compensated for by using thicker glass. If the system is to be evacuated, it is common practice to tape the glass containers to reduce the problem of flying glass should the container implode. Flying glass fragments with razor-sharp edges are extremely dangerous. Large systems, especially should be contained in a surrounding stiff wire cage for the same reason. Glass systems can be set up within a fume hood and protected with explosion barriers as additional protection.

Glass under strain is very vulnerable to sharp blows, but strains can also release spontaneously, especially with rising temperature. In some instances, the release will cause the glass under stress to crack. Systems should be put together to minimize additional mechanical strains placed on the components. The strains in glass can also release spontaneously for no apparent cause. In one instance in which the author was personally involved, a gallon jug of a hazardous chemical was delivered to the secretaries in a research-oriented department. No one was at the laboratory to receive it so it was left in the departmental mail room. About 3 hours later, the jug spontaneously split, the chemical ran out, dissolved the plastic packaging protective material, ran into the carpet and the floor underneath, and dripped through into an office area below, damaging the floor there as well. The material was toxic, a suspect carcinogen, so evacuation was necessary; the cleanup required working in protective clothing and the use of respiratory protection. The direct cost to repair the damage was on the order of $2,000 to $3,000, with an additional comparable cost in manpower. Fortunately, only a handful of persons were in the building at the time, and no one suffered an acute exposure.

Glass systems under either positive or negative pressure are very dangerous due to the inability of the glass to withstand impact. The static force on the surface of an evacuated container may be very large. For example, the net force on an evacuated spherical container 10 cm (4 inches) in diameter is equal to the force equivalent to the weight of an 81kg (179 lb) mass. There is essentially no difference in this force for a good vacuum and a poor one. The force exerted on a container with a poor vacuum equivalent to 1 mm of mercury is within 0.13% of the force on a container with a vacuum several orders of magnitude better. The forces on high pressure systems can, of course, be much higher than at atmospheric pressures. Pressures in gas cylinders can be 200 times atmospheric pressure or more, and hence never should be attached to most glass research systems except through a regulator valve. The existing forces on pressurized glass systems make them unusually vulnerable to other factors, such as a sharp blow, which they might otherwise withstand. A warning sign should be placed on systems under pressure to alert personnel of the unusual danger. A wire cage, mentioned earlier, to contain flying glass also can protect the glass from external blows. Additional protection can be provided by setting glass systems up in fume hoods and keeping the sashes closed. Walk-in hoods can be used for larger systems.

Anything that weakens the strength of the glass in a system under pressure increases the risk of either an implosion or explosion. For example, some laboratories use metal evaporating systems under a large bell jar to deposit metal films. A current is passed through a small crucible or "boat," raising its temperature to a sufficient level to evaporate a metal placed within it. In order to get a good deposition of the metal, a good vacuum needs to be established. Although there usually is not sufficient air in the bell jar to transfer heat by either conduction or convection, substantial radiative transfer does occur. In order to get good evaporation, in most cases the metal must be heated to a brilliant white-hot temperature. This corresponds to a temperature on the order of 1,500°C (2,700°F) or better, well above the point at which borosilicate glass begins to soften. If the temperature is maintained at this level for a substantial length of time, the strength of the glass bell jar can be impaired with a resulting implosion. Metal-film evaporating systems should always be enclosed with a sturdy wire cage during use.

Broken glassware offers many additional opportunities for injury due to cuts from the sharp fragments. Washing glassware offers the opportunity for cutting oneself on a broken flask or beaker while doing so. Accidentally shattering a glass container while holding it can also represent an opportunity for injury. Cleaning up broken glass must be done carefully. Note that broken glassware placed into ordinary trash represents an unnecessary and unacceptable hazard to an unsuspecting maintenance worker. Custodial workers routinely transfer trash to plastic garbage bags which offer no protection against broken glass. All broken glassware should be placed in a sturdy kraft board box, or equivalent, taped closed, with the words **BROKEN GLASS** written prominently on the outside before being placed beside, not in, the ordinary trash containers.

Working with glass tubing offers a number of opportunities for injury. A section of tubing can be cut to length by scribing the tubing with a triangular file for approximately one third of its circumference, then wrapping it with cloth and exerting sideways pressure on the opposite side of the tubing with the thumbs placed a bit to each side of the filed line. The tubing should snap cleanly. If it does not, the file cut should be deepened and the procedure repeated. Multiple file marks may increase the chance of an irregular break which may have sharp edges. Fire polishing of the tubing should be done in every case to avoid cuts from any such edges.

Inserting glass tubing into stoppers or flexible plastic or rubber hoses is the source of one of the most common laboratory accidents with glass. Usually the cause is trying to force the tubing into the hole with no hand protection. Serious cuts can occur in such cases. The worker should either wear leather gloves or protect the hand holding the tubing by wrapping the tubing with a cloth and wrapping cloth around the hand holding the flexible tubing or stopper.

For flexible tubing, the hole size is predetermined by selecting the plastic or rubber tubing to have an internal diameter just slightly smaller than the glass tubing to which it is to be mated. To cut a hole in a rubber stopper, a well-sharpened borer just a size smaller than one which will slide over the glass tubing should be used. The stopper should be held only with the fingers. Wrapping the entire hand around it to get a "better grip" offers an opportunity to cut that hand with the borer. Holding the stopper in that manner should not be necessary The borer should be lubricated with either water or glycerol. The hole should be cut by exerting a steady pressure on the borer while using a smooth rotary motion to slice through the stopper. The index finger should be placed along the barrel of the borer and close to the stopper. In the beginning, this will limit the distance the borer will travel if a slip occurs, while it will limit the distance the borer will extend beyond the stopper when penetration occurs.

Once a clean, well-shaped hole is prepared in a stopper, or if plastic or rubber tubing is to be attached to a section of glass tubing, the following steps should be taken:

1. Lubricate the length of the glass tubing which will need to go through the stopper with stopcock grease, or other suitable lubricant. Use leather gloves to protect the hands or, as mentioned earlier, protect the hand holding the glass by wrapping a cloth around the glass tubing, and protect the other hand by wrapping it with a piece of toweling or other bulky cloth.
2. Hold the glass tubing an inch or two from the end.
3. Insert the glass into the hole, using moderate inward pressure and a slight twisting motion. Neither the forward pressure nor the twisting force should be excessive to avoid breaking the glass.
4. Continue the process in step 3 until the glass has been inserted to the desired or needed distance.

If a cut occurs, despite all the precautions, bleeding can be controlled in most cases by placing pressure directly on the wound, preferably by placing a clean dressing over the wound and holding it tightly in place while seeking first aid.

a. Other Safety Problems

It is relatively easy to heat and bend pieces of glass tubing to achieve a desired shape or to draw out a piece of tubing to diminish the internal diameter of the tube. Working glass to create complicated shapes for laboratory applications, however, is best left to the professional glassblower. In order for the shapes to be free of strains which could cause the glass to fail under stress, all of the components should be properly annealed, which requires experience and the correct equipment to do properly.

There are two special health problems associated with fabricating glass, other than the danger of cutting oneself. When a glassblower is configuring a glass component, he frequently employs a glass lathe which has some asbestos components. This asbestos tends to become quite friable with time and use, and individuals working near the lathe can be exposed to airborne asbestos fibers. Alternative insulating materials, such as Zetex™, should be used or the glass worker and nearby personnel should wear appropriate respiratory protection. The use of localized exhaust to remove the asbestos fibers is usually not practical since the moving air could cause problems with the work.

The second special problem is primarily associated with the construction of relatively large components of silica glass. The temperature at which fused silica can be conveniently worked is about 1,580°C (2,876°F), while ordinary borosilicate glass can be worked at a temperature nearly 500°C (900°F) lower. The formation of nitrogen dioxide from the air is significant at the higher temperatures involved in fabricating silica components.

Both of these problems can be alleviated by ensuring that the glassblowing shop or work area is well ventilated. Because heat rises, canopy hoods placed directly a few feet above the workstations will be effective in trapping the gases, fumes, and even particulates entrained in the rising hot air. This is a major exception to the previous recommendation against the use of canopy hoods in the laboratory.

b. Glassware Cleaning

Cleaning of glassware in laboratories is an essential part of laboratory procedures. In most cases a simple cleaning with soap and water is sufficient, but in some cases chemical cleaning is necessary. Strong chemical agents such as sulfuric acid, perchloric acid, chromic acid, nitric acid, etc. should not be used unless the need specifically exists. When employing these strong reagents, protective gloves, eye protection, chemically resistant aprons, and possibly respiratory protection is recommended.

When washing glassware with soap or detergent and water in a sink, under normal circumstances household gloves (such as Playtex HandSaver® gloves) usually will be sufficient, although gloves made of different materials may be indicated depending upon the materials originally in the items being cleaned. Scouring powder and brushes might be needed for stubborn residues. Glassware may break while being washed so care must be taken to avoid cuts. A soft mat in the bottom of the sink will reduce the chance of this occurring.

In the life sciences, contact with contaminated glassware could allow an individual to contract a contagious disease. Biologically contaminated glassware should be autoclaved and/or submerged in a sterilizing bath before further cleaning. This is a requirement under the OSHA bloodborne pathogen standard for items which may be contaminated with human blood, tissue, or other bodily fluids; under this standard, custodial personnel are specifically prohibited from cleaning up broken glassware that could be contaminated. In some larger laboratories, glassware is washed in a central location, and in such facilities it is the responsibility of the individuals using the glassware to ensure that workers in the washing facility are alerted to the possibility of the glassware being contaminated and protected from materials still in or on the containers and utensils.

REFERENCES

1. **Smith, G.P.,** Glass, in *CRC Handbook of Laboratory Safety,* 2nd ed., Steere, N.V., Ed., CRC Press, Cleveland, OH, 1971, 544.
2. *Guide for Safety in Academic Chemistry Laboratories,* 5th ed., American Chemical Society, Washington, D.C., 1990, 9.

3. **Tucker, B.,** Acid cleaning of glassware, in *CRC Handbook of Laboratory Safety,* 2nd ed., Steere, N.V, Ed., CRC Press, Cleveland, OH, 1971, 557.

4. Occupational Exposure to Bloodborne Pathogens, 29 CFR Part 1910.1030, FR *58, 235,* December 6, 1991, 64175.

18. Unattended Operating Equipment

a. Laboratory Distillations

Laboratories may have operations which may run unattended overnight. Stills to produce distilled water are perhaps the most common, although distillation columns for many other substances are often used. Some of these substances are hazardous in themselves and precautions need to be taken. With appropriate designs and automatic controls, even those involving hazardous materials can be allowed to operate unattended. Other equipment to which these same concerns occur are ovens, heating baths, and any other comparable operation.

Self contained commercial systems are available today for distillation of water which virtually eliminate all of the safety concerns. Water purifiers are available which produce water pure enough in many cases to eliminate the need for a still altogether. Many of these start with a reverse osmosis unit which eliminates up to 90% of the impurities and then this water is further purified by deionizer type units. If these units are satisfactory, then replacement of older units with either the non-still type unit or one of the self contained commercial stills should be considered.

A serious fire occurred in the author's institution due to an older water still overheating and setting some nearby combustible materials on fire. This unit had been working unattended for an uncertain but lengthy period of time without problems, but the constant exposure to the heat from the still eventually caused materials close to the still to spontaneously ignite and resulted in a major fire within the laboratory. Fortunately, the facility was constructed of largely noncombustible materials and the damage was confined to the contents of the laboratory, which was serious enough for the effect it had on the work in progress. The reason for the still being used was the distilled water supplied as a building source was unsatisfactory to the scientist. The result of the fire was an upgrade to the building system which enhanced the safety of the laboratory involved and for others as well.

Temperature is a relatively easy thing to measure and to interface with a control system. As this is being written, a firm, Dallas Semiconductor has just placed on the market a computer chip which incorporates a digital thermometer, nonvolatile memory, a clock, control logic, and serial interface to a computer. Thus, one has all the ingredients for a control system at one point. This is a new product and all of these items can be readily assembled using individual components. With this equipment one can determine if the temperature changes beyond preset limits and cause appropriate decisions to take place. This may result in any of several actions. Why did the excursion take place? Did the power to the heating unit fail? Did the cooling water flow stop or slow to the point that it was insufficient? If the water pressure escalates to a point where the connections break, it would be desirable to decrease the water flow before this could happen. If a power loss occurred, should the operation resume when the power is restored? All of these decisions, and others, can be programmed for a computer to take the appropriate action. If the initial conditions return to normal, then it may be appropriate to simply resume operation.

In order to keep the power to the experimental apparatus, and to the computer controller stable, constant voltage transformers are available which prevent fluctuations. However, these will not maintain power should the electrical service fail. A relatively inexpensive alternative is now available which not only provides a steady source of power but provides a temporary source of power for a period after the electrical power to the unit fails. Usually, total losses of power are short lived so these so-called "uninterruptible" power supplies, with capacities of 5 to 30 minutes, often provide sufficient temporary power to cover these power failures, and certainly enough time for a computer managed system to activate a shut-down procedure.

Distillations not involving water stills present additional problems because these are *often ad hoc* custom setups designed for a specific purpose. Most of these are made of glass and are usually slender elongated configurations. The supports must be secure but still not place undue stress on the glass. Stress on the glass can create strains which cause the glass to fail unexpectedly. Inexperienced persons who fabricate their own systems and do their own glass blowing may introduce strains into the glass which can fail unexpectedly. Of course, the materials being used in the distillation present problems if they have hazardous properties, such as flammability, reactivity or explosive characteristics. In such cases, special care must be taken not to allow the materials to come into contact with heat or material which could trigger a violent reaction. In some cases, materials in glass systems, not necessarily distillation systems, are sufficiently sensitive that a violent reaction may be triggered spontaneously. Several years ago, during the author's graduate training, a colleague's large glass system containing silane, exploded during the night, totally demolishing most of the contents of the laboratory, and of course ruining the work in progress. The system was not enclosed in a protective wire cage which could have contained flying glass. If anyone had been in the room during the incident, they would almost certainly have been seriously injured.

Another possible problem in working with materials which are toxic is failure of the system and releasing toxic vapors into the facility. In one case, such a failure occurred in a system set up properly in a fume hood. Unfortunately, the exhaust of the fume hood was too close to a nearby open window, and a student working near the window was overcome by fumes drawn into the window. A major contributor to this incident was an overtaxed ventilation system with insufficient makeup air so that the building as a whole was at negative pressure.

In summary, systems which are susceptible to failure, especially those left unattended while in operation should include fail-safe precautions with redundant safeguards should a primary safety feature fail.

b. Other Laboratory Operations

Not all extended operations are dependent primarily on a single variable such as temperature. In a second article by Conlon which appeared in the 2nd, 3rd, and 4th editions of this handbook, a complex apparatus was used as an illustration that involved monitoring nine different vari-ables, time, volume of product, rate of generation of the product, flow of cooling water, the rate of reflux of the material, rate of stirring, viscosity of the material, pressure or vacuum in the system, and, rate of gas or liquid flow. Some of these factors also involved temperature. He then discussed how one can monitor each of these variables. Since the original article was written, an entire scientific discipline has evolved coupling new types of sensors with computer data acquisition and management. At the end of this section, a number of recent references will be given which should be used to design a modern system. Virtually all of the devices mentioned in the Conlon article are analog devices while the newer ones typically convert an analog measurement directly into a digital form for manipulation by a personal computer.

A major advantage of the use of computers is that all of the variables can be measured simultaneously and coupled to cause a coherent response based on all the variables. The response can be based on Boolean logic, AND, OR, NAND, NOR, IF, etc., so that, for example, if the coolant flow stops or deceases significantly, or if the product vessel becomes too full, then the heat can be turned off. The system can then be designed to restart automatically or only after a manually activated restart sequence, or it can send an alarm to whomever the individual in charge wishes for further actions. The status of the system variables while operating can be logged continuously so a determination can also be made as to what caused the problem and what corrective steps need to be made, if any. If an exothermic reaction is involved, the computer program can be designed to reduce the heat applied to the system or even to provide additional cooling. In other words, the computer can be designed to initiate responses automatically to a large number of contingencies and to do so very quickly.

Another advantage to using sensors that can directly interface with a computer is that one can program the operation based on the data from the sensors. If for example, one wanted to change the temperature at different points, this sequence can readily be programmed. Of course not all variables are amenable to automatic controls although the vast majority are. With computer power expanding as rapidly as it is today at increasingly lower costs, automated control of chemical processes should increase correspondingly rapidly.

REFERENCES

1. **Gruhn, P., and Cheddie, H.,** *Safety shutdown systems, design, analysis and justification,* Instrument Society of America, Research Triangle Park, NC, 1998.

2. **Georgakis, C.,** *Dynamics and control of process systems 1998,* DYCOPS-5: A proceedings volume from the 5th IFAC symposium, Corfu, Greece, 8-10, June 1998, Kidlington, Oxford, UK; New York, Published for the International Federation of Automatic Control by Pergamon Publishing, 1999.

3. **Corripio, A.B.,** *Design and application of process control systems,* International Federation of Automatic Control, Research Triangle Park, NC 1998.

4. Sensors Magazine, Helmers Publishing, 174 Concord Street, Peterborough, NH, 03458.

5. Instrument Society of America, P.O. Box 12277, Research Triangle Park, NC, 27709, also see http://www.isa.org/

VII. HAZARD AWARENESS (RIGHT-TO-KNOW)

In recent years, a movement has been growing to legally require that individuals working so that they would be exposed to chemicals in the workplace are informed about the risks which these chemicals pose to their health. A number of states passed what are commonly referred to as "Right-to-Know" laws applicable to various groups of employees in their states. Eventually, after extended hearings, OSHA published a national version of this requirement as a hazard communication standard, 29 CFR 1910.1200. Comments were received from many sources, including industrial and trade associations, labor unions, academic groups, individuals, and others as to the content of the standard and which employees should be covered by the standard. As originally issued, the standard covered only the employers and employees in the Standard Industrial Classification (SIC) Codes 20 through 39. These classifications are given below:

20 Food & Kindred Products	30 Rubber & Plastic Products
21 Tobacco Manufacturers	31 Leather & Leather Products
22 Textile Mill Products	32 Stone, Clay & Glass Products
23 Apparel & Other Textile Products	33 Primary Metal Industries
24 Lumber & Wood Products	34 Fabricated Metal Products
25 Furniture & Fixtures	35 Machinery, Except Electrical
26 Paper & Allied Products	36 Electrical Equipment & Supplies
27 Printing & Publishing	37 Transportation Equipment
28 Chemicals & Allied Products	38 Instruments & Related Products
29 Petroleum & Coal Products	39 Miscellaneous Manufacturing Products

Laboratories in these industries were partially covered by some of the more critical portions of the standard. Laboratories in other industries and in academic institutions were not covered by the standard. Comments received during the hearings revealed marked differences of opinion on how the coverage should be applied to research laboratories. Some comments reflected an opinion that research laboratories, particularly academic laboratories, were among the most critical areas in which a definitive, explicit standard was needed while others felt that as knowledgeable, responsible individuals, research scientists needed at most only a voluntary standard.

Since many employees were omitted in the original version of the standard, court action followed almost immediately to force OSHA to cover all employees using or exposed to chemicals. After some delay in the court system, the legal issue was resolved in favor of OSHA being required to extend the coverage of the standard. The final rule was published in the *Federal Register* on August 24, 1987, to go into effect on May 23, 1988.

In a few areas the standard was applied to some research laboratories outside of the original list of industries. Virginia, for example, in adopting the federal standard added all public employees in the state to the list of those covered, which included laboratory personnel in the state's public universities and colleges as well as state-supported laboratories.

On May 1, 1990, 29 CFR 1910.1450, the OSHA standard for occupational exposure to hazardous chemicals in laboratories, went into effect. This standard preempted the general industry standards, including the hazard communication standard for laboratory scale use of chemicals, with a few exceptions. If the use of chemicals does not meet the definition of laboratory scale, even if the exposure takes place in a laboratory, the employer and employee must comply with the relevant part of 29 CFR 1910, Subpart Z. If the laboratory standard does preempt the hazard communication standard, in some key respects they are much alike. The definitions of hazardous chemicals in the laboratory standard are specifically referenced to Appendices A and B of the Hazard Communication Standard for guidance. The OSHA PELs and action levels are still the same as those listed in Subpart Z. The MSDS requirements are basically the same. The training requirements differ very little. The requirements for informing the employees of the hazards associated with the materials are equivalent. The key differences under the laboratory standard are (1) the additional provision that a written chemical hygiene plan must be available for a laboratory, (2) standard operating procedures must be available, and (3) a chemical hygiene officer must be appointed. Since the use of chemicals in a laboratory is assumed to involve small quantities, the provisions in the hazard communication standard for notifying authorities of releases are not spelled out.

Although the hazard communication standard does not apply to employees involved with laboratory-scale use of chemicals, some of their activities may invoke the standard, even within the laboratory. Employees managing bulk chemical stores and individuals handling disposal of chemical wastes would not necessarily be engaging in laboratory-scale activities. Other support groups of a research facility would also be covered by the hazard communication standard in many of their activities. Therefore, the following sections will cover the basics of that standard. The standard as finally written is a performance standard, as is the laboratory safety standard. The required program is not spelled out except in very broad strokes, but the goals and objectives to be achieved are well defined. It is to be understood that the following material applies to the use of chemicals affecting non-laboratory employees and to laboratory employees engaged in tasks not covered by the laboratory safety standard.

A. Basic Requirements[*]

Portions of the first few sections of 29 CFR 1910.1200, directly quoted below, spell out the basic concept of the standard.

---

[*] The phrasing is not identical to the language of the laboratory safety standard, but these requirements are virtually identical to the corresponding ones in that standard.

(a) PURPOSE

(1)   The purpose of this section is to ensure that the hazards of all chemicals produced or imported are evaluated, and that information concerning their hazards is transmitted to employers and employees. This transmittal of information is to be accomplished by means of comprehensive hazard communication programs, which are to include container labeling and other forms of warning, material safety data sheets, and employee training.

(2)   This occupational safety and health standard is intended to address comprehensively the issue of evaluating and communicating the potential hazards of chemicals, and communicating information concerning hazards and appropriate protective measures to employees, and to preempt any legal requirements of a state, or political subdivision of a state pertaining to this subject....

(b) SCOPE AND APPLICATION

(1)   This section requires chemical manufacturers or importers to assess the hazards of chemicals which they produce or import, and all employers to provide information to their employees about the hazardous chemicals to which they are exposed, by means of a hazard communications program, labels and other forms of warning, material safety data sheets, and information and training. In addition, this section requires distributors to transmit the required information to employers.

(2)   This section applies to any chemical which is known to be present in the workplace in such a manner that employees may be exposed under normal conditions of use or in a foreseeable emergency.

(3)   This section applies to laboratories only as follows:

(i)   Employers shall ensure that labels on *incoming* containers of hazardous chemicals are not removed or defaced;

(ii)   Employers shall maintain any material safety data sheets that are received with *incoming* shipments of hazardous chemicals, and ensure that they are readily accessible to laboratory employees; and,

(iii)   Employers shall ensure that laboratory employees are apprised of the hazards of the chemicals in their workplace in accordance with paragraph (h) of this section....

(h) EMPLOYEE INFORMATION AND TRAINING

(1)   Employers shall provide employees with information and training on hazardous chemicals in their work area at the time of their initial assignment, and whenever a new hazard is introduced into their work area. Information may be designed to cover categories of hazards, e.g., flammability, carcinogenicity, or specific chemicals. Chemical-specific information must always be available through labels and material safety data sheets.

(2)   INFORMATION. Employees shall be informed of:

(i)   The requirements of this section;

(ii)   Any operations in their work area where hazardous chemicals are present; and,

(iii)   The location and availability of the new written hazard communication program, including the required list(s) of hazardous chemicals, and material safety data sheets required by this section.

(3)   TRAINING. Employee training shall include at least:

(i)   Methods and observations that may be used to detect the presence or release of a hazardous chemical in the work area (such as monitoring conducted by the employer, continuous monitoring devices, visual appearance or odor of hazardous chemicals when being released, etc.);

(ii) The physical and health hazards of the chemicals in the work area;

(iii) The measures employees can take to protect themselves from these hazards including specific procedures the employer has implemented to protect employees from exposures to hazardous chemicals, such as appropriate work practices, emergency procedures, and personal protective equipment to be used; and,

(iv) The details of the hazardous communication program developed by the employer, including an explanation of the labeling system and the materials safety data sheet, and how employees can obtain and use the appropriate hazard information.

B. Written Hazard Communication Program

Based on the requirements outlined above, employers must develop a written program. The program must include provisions to ensure proper labeling of chemicals, to maintain a chemical inventory, to maintain a current and accessible MSDS file for all incoming chemicals, and to provide training and information to the employees in a number of relevant safety and health areas.

In order to develop an effective hazard communication program, it is desirable to have input from all of the groups that need to cooperate to make the program work. A straightforward way of accomplishing this is to establish a hazards communications committee to help define the program and to monitor the performance of the program once it has become operational. This can be a duty assigned to the organization's Laboratory Safety Committee as well. There should be representatives on the committee from the administrative departments, including health and safety, personnel, purchasing, and physical plant, and from the research operations. In an academic setting, each college should be represented by a member from a major chemical-using department in the college, while in an industrial setting each major chemical-using division should be represented. One department should be assigned the leadership role in developing and implementing the program, probably the health and safety department, but it should be clear that all major constituencies share in the responsibility of formulating the program and also share the responsibility of making it work.

There are a number of things which must be in a written program:

1. Although not explicitly required, all the employees that are exposed to chemicals in the organization during the normal course of their employment must be identified so that they can participate in the program. Note that this does not necessarily mean that the identified individuals use chemicals. An electronics technician might not use chemicals himself, but could be considered to be exposed to those chemicals used by others working in the same room. An affected employee in this context would not be limited to salaried employees, but also to workers paid an hourly wage. On the other hand, a custodial worker would not be considered to have a significant exposure to a laboratory's chemicals if they only took out nonhazardous trash, but would have to be included if the cleaning materials which they use themselves contain hazardous chemicals. However, it would probably be desirable for a custodial employee to be informed of some of the general risks associated with chemicals in laboratories in their work area so they would understand the need to be careful while in the laboratory. If done, providing this information should be done carefully to avoid frightening an individual who probably has very little technical knowledge.

2. A list of hazardous chemicals in the workplace must be compiled. The provisions in the standard only require laboratories keep track of incoming chemicals subsequent to the effective date of the standard for employers, May 25, 1986. Existing inventories were in a sense "grandfathered." Eventually, as older stocks are disposed of or used, the list

will come to reflect the actual holdings in a facility. The list should be kept as current as possible. If one person is assigned the responsibility to maintain the list, and the data kept in a personal computer data base, it is only necessary to keep track of additions and deletions in order to maintain a complete, current list. For the purposes of complying with this portion of the standard, the quantities of each chemical in the laboratory are not needed, although these data would be important for a sound management program and would be helpful in planning a safety program. The list of chemicals, in combination with the list of employees, will serve to help define the training program.

3. The written program must define how the employees are to be informed of the requirements of the standard. This will include: details of (1) how the employees are to be informed about the contents of the standard; (2) the contents of the written plan; (3) how they are to meet the labeling requirements; (4) how they are to learn of the methods available to them to warn them of exposures; (5) how to obtain and interpret a MSDS for a given chemical; (6) the hazards associated with the chemicals to which they are exposed; (7) how they are to be trained in procedures which will eliminate or reduce these chemical hazards; and (8) how they are to react in an emergency.

4. Many laboratory uses of chemicals involve repetitive tasks, while others do not. Employees must be made aware of the risks associated with the latter type of activities as well as those accompanying the more routine uses of chemicals, and the same basic type of information provided as in item 3.

5. Although pipes are not considered containers for the purpose of this standard and need not be labeled, the plan must include education of employees about the hazards associated with any unlabeled pipes containing chemicals in their work area and how to deal with these hazards.

6. There must be a procedure or statement in the plan as to how transient employees, such as persons working on contract, are to be informed of the chemical hazards to which they may be exposed, and for provision of information of protective measures for these transient employees. It is not specifically spelled out in the standard but there is a need for the converse as well. Contractors are often called in to do renovations, perform an asbestos abatement project, or to conduct a pest control program, as examples, and use hazardous chemicals in the process or expose personnel to airborne hazards. Provision should be made in the contracts for these groups for them to provide information to the occupants of the spaces where their work is being done.

1. Personnel Lists

This appears to be relatively straightforward, but in fact can be rather complicated. In a large academic institution, the actual duties associated with a given job classification often become blurred over a period of time. For example, a job title of laboratory technician might appear to logically relate to chemical exposure, but the duties of the individual may have changed so that the job may never bring the individual into contact with chemicals at all. It is not possible to simply have the personnel department list all persons in specific job classifications as professional staff or faculty in research areas.

As an initial step to determine which employees need to participate in a formal hazard communication program, a questionnaire can be sent to each department or other internal division asking them (1) to define those areas in which chemicals are used in their department, (2) to list each employee in those areas with their job title, and (3) for their appraisal of the involvement of these individuals with chemicals. This should be followed up with a second questionnaire to the managers of the individual areas, asking for the same information, and then the area should be visited to confirm the data provided. This sounds unnecessarily involved, but experience has shown that all three steps are necessary. In many cases, through oversights, individuals are not identified who should have been included and, occasionally someone is listed who has no exposure, usually because it was easier to list everyone rather than consider each individual case.

It is important to identify the position (most positions now have internal identification codes) so that when the position becomes vacant, a mechanism can be established to ensure that the new person filling the position receives a proper orientation program. Often, such a position might qualify for participation in a pre-employment medical screening examination as well so the effort to correlate positions with exposure to chemical hazards might be justified for more than one purpose.

All persons being considered for positions covered by the hazard communication program should receive a brief written statement concerning the program so that they may ask appropriate questions at the time of their job interview. If they are selected, a more extensive document should be provided so that they will be aware of the explicit requirements of the standard.

2. Chemical List

A list of all hazardous chemicals in the workplace is required as part of the standard. It may be difficult to convince many managers to take the time to go through their stocks of chemicals to prepare a list for their facilities. (The need for this information is implied in the laboratory safety standard, but not required.) This is one of the more burdensome tasks associated with the standard. However, it is very desirable that an effort be made. Not only does it provide information on which to base the training program, but it also is needed to prepare an MSDS file for the facility, although, again, the MSDS file is only required for incoming chemicals. In practice, however, it is difficult to justify not having an MSDS for a hazardous chemical in use, based on a technicality. There is no real alternative for the initial survey as a basis for defining the scope of the program.

It would be desirable if the problem of maintaining the list of incoming chemicals could be centralized, perhaps as the purchase order is being processed. However, although surprisingly few chemicals are bought in quantity in most research institutions, even very large ones, there may be more than 1000 different substances bought during the course of a single year and several thousand purchased over a number of years. Many of these are bought under a number of synonyms, or as components of brand-name formulations. Commercial software is available by which a chemical can be identified by any number of synonyms, standard chemical name, trade names, or CAS numbers. Local information as to the purchaser and destination (building or facility) can be provided by the customer. Software is now available which can combine this information and more, e.g., date received and quantity, and to generate a unique bar code that can be affixed to each container. These data can be used to maintain a continuing inventory for an entire organization and to track a chemical from the time of receipt to eventual full consumption or disposal. It is practical for individual laboratories to use microcomputers, even without these specialized programs to perform this task for themselves, using commercial database or spreadsheet programs. At most it would require a week or so, per laboratory, by an employee to acquire the initial data and enter it, even if existing inventories were included. Maintenance of the data would involve only adding new containers and removing old ones.

3. Labeling

Current labels on original containers of chemicals as purchased from the distributor or manufacturer will almost certainly meet and exceed the requirements of the hazard communication standard (see Chapter 4, Section V.B). These requirements are:

1. The identity of the hazardous chemical
2. Appropriate hazard warnings
3. The name and address of the chemical manufacturer, importer, distributor, or other responsible party

Item 2 is the only ambiguous requirement. Most commercially sold chemicals provide this sort of information on the label in a number of ways, such as:

1. A risk descriptor, i.e., Danger, Warning, Caution
2. The NFPA hazard diamond
3. A descriptive statement of the hazards
4. By use of stylized symbols, such as a radiation or biohazard symbol

Other useful safety-related information is normally provided as well, such as the flash-point (if applicable), fire extinguisher type (if applicable), first aid and medical advice, a color code to aid in avoiding incompatible storage, and standard identifiers, such as a CAS number which can aid in referring to a MSDS data base, and a UN number which is needed in disposing of the chemical as a hazardous waste.

Facilities are specifically enjoined by the terms of the standard from removing or defacing the labels on incoming containers of chemicals. However, it is relatively common to transfer a portion of the contents from the original container to a secondary container. If this material remains under the control of the individual responsible for the transfer, and is to be used during a single work session, then it is not necessary to label the secondary container. If it is not to be used under these conditions, then the secondary container must be marked with the identity of the chemical(s) in the container and with "appropriate" hazard warnings for the protection of the employee. These "appropriate" warnings need not be as comprehensive as the original label, but must provide adequate safety information.

The most likely occasions when secondary containers are used without proper labeling would be when chemicals are disbursed from a larger container into a smaller one at a central stockroom, and when containers are to be taken from the initial workplace into the field. Personnel must be sure to label the secondary containers in these cases and in any other comparable situation. If secondary containers are labeled properly, it also will help remedy one of the more troublesome problems associated with hazardous waste disposal, inadequately identified containers of chemicals. Once in the laboratory, the tendency is to label the secondary containers less thoroughly, often with a cryptic label such as "soln. A" or some other non-informative label.

The warning labels must be in English, although they may be provided in other languages *in addition,* if appropriate. In many academic institutions, in particular, graduate students who routinely use a language other than English as a primary language, are becoming numerous, and, in some cases, consideration may be given to supplementing the commercial label with warnings in other languages. However, the majority of these graduate students can be expected to understand written English satisfactorily. Some areas of the U.S. have changed demographically so that the use of languages other than English may have become predominant. All employees using chemicals must be instructed in how to interpret the hazard information on the labels.

A specific part of the written plan must address how the employees are to be made aware of the labeling requirements and how they are expected to comply with this standard. It would be highly desirable to develop a uniform program across an organization, particularly as to labeling of secondary containers, to avoid unnecessary confusion.

4. Material Safety Data Sheets

Since the receipt of MSDSs is tied so strongly to the purchase and receipt of chemicals, they were discussed in some detail in Chapter 4, Section III.C. The exact form of a MSDS is not mandated by the standard as long as the proper information is provided. Firms use a variety of formats to provide the required information. ANSI and the CMA have recommended a standard form which may be adopted.

There are two basic requirements associated with MSDSs in the hazard communication standard. Employees must be trained in how to use the information in them and the MSDSs must be readily available to the employees.

As discussed in Section III.C., a major problem in a research institution, where the chemical users may operate virtually independently of each other and are likely to be housed in a number of different buildings, is to ensure that all users of a given chemical have ready access to a copy of the most recent version of the MSDS. The distributor is only required to send one copy of a MSDS to a purchaser and an updated version when a revision is necessary because of new information. Where several different components of an organization order independently, one purchaser may receive an update while the others do not, since the vendor technically has fulfilled its obligation by sending the MSDS to the first unit making the purchase. If a centralized mechanism for tracking chemical purchases has been established, then all MSDSs could be sent to a single location from which copies can be forwarded to all groups within the organization that need them. This is relatively labor intensive and still may not reach all users since chemicals may be transferred from one laboratory to another with no paper trail. Another alternative, which does not provide as ready access to all users but does not require the tracking mechanism referred to above, would be to have all MSDSs received at one location and maintained in a master file, with copies placed in several secondary master files at locations reasonably convenient to the users. A third alternative, but still less accessible to users, would be for a single master file to exist, with copies, of individual MSDSs provided upon request. This might not be considered to meet the accessibility requirement if the delay in receiving the MSDS is more than 1 or 2 working days. All of these mechanisms are at best cumbersome and manpower intensive.

Comprehensive generic MSDSs are now commercially available on optical discs which can be processed by a computer and accessed at any time by the users. These typically are updated quarterly so that they can satisfy the need for the MSDS file to remain current. They are not inexpensive, but the cost is much less than for the amount of manpower needed to maintain an equivalent hard copy file, and they provide a comparable level of access. There also are firms which maintain computerized MSDS files available to subscribers as a database service. These are accessed by the users from their terminals using modems, but line and access charges are incurred.

As noted earlier, MSDSs are widely available on the Internet, either directly from the chemical manufacturer or distributor (which meets the criteria of directly identifying the chemical supplier producing the MSDS) or many organizations and universities now maintain and pro-vide generic MSDSs at their Internet site.

No matter how a facility or organization sets up a MSDS file, a component of the training program for the employees must include an explanation of how an individual employee can obtain access to the file. It should be possible for an employee to obtain copies of a MSDS for a given chemical upon request. In addition to providing access to the MSDS file, part of the program training must also include instruction in how to interpret a MSDS to obtain appropriate hazard information.

The information presented in the various categories in a MSDS should pose no real difficulty to most technically trained persons. Some definitions of terms may need to be provided, such as LD_{50} (lethal dose, 50% of the time for the test species), if an individual is not accustomed to using such terms, but even these are straightforward. However, some persons will not be as scientifically sophisticated, and the training program for these individuals will need to be more thorough. In an instance at the editor's institution, grasping the distinction between a monomer and a polymer was a major problem for some clerical personnel who felt that they had been exposed to dangerous levels of a chemical due to some activities in the building where they worked. The health hazard data and the TLV values given in the MSDS for the chemical stated that they were for the monomer only and that the effects of exposure of the chemical would be serious at a few parts per billion in air. The employees exposure was not to the monomer, but to very small quantities of the stable polymer, for which the health hazards were minimal. Extensive (and expensive) tests had to be run before the personnel were convinced (some perhaps continued to have doubts) that they had not been unduly exposed. During training, an effort needs to be made to ensure that understanding has been achieved.

Compliance with the training requirements for utilization of a MSDS as a source of hazard information can be readily achieved for technically trained personnel. For example, a written handout informing the employees how they can obtain access to a needed MSDS and a short video tape explaining the contents might be all that is needed. An individual capable of explaining any confusing points should administer such a program and be available to answer questions. A statement affirming that the training was received should be signed (and dated) by the employee after any questions were resolved. This can have significant legal implications. An employee may maintain that they were never exposed to the information but a signed statement that they were present at a lecture is hard to refute. For less knowledgeable employees, a formal training session should be set up and an instructor-student format used. The handout and videotape mentioned above can still be part of the instruction program, but the instructor should go over each of the categories in a MSDS and encourage the employee to ask questions. The employee also should be provided with a written version of the concepts covered, for later reference.

Some organizations document that an employee has not only been exposed to the information, but also understands it, by requiring that each employee take a very simple written quiz on the covered material, in place of signing a simple statement. This is not required to comply with the standard, but it does provide stronger documentation of an effective training program. Individuals should not resent imposition of such a requirement, but some professionals feel, rightly or wrongly, that they have demonstrated sufficient proficiency in their area by fulfilling the required educational and professional certifications. If a quiz is made part of a program, it should be expected that a number of individuals will object to taking it. This problem occurs mostly among highly educated staff. Such a program, though, can be made to work with patience and support of the organization.

5. Employee Training and Information

Portions of the training program common to every employee: The basic concepts of the organization's program, how it is administered, the requirements of the standard, and how to read and understand information labels and MSDSs, have already been covered in the earlier sections. These can be given by the lead department in the organizations hazard communication program, usually the Environmental Health and Safety Department. However, a number of other areas will require a cooperative effort between the administrative department in overall charge of the program and the individual facilities and departments. In an organization which uses chemicals in a wide variety of activities, the local managers will need to be the primary parties responsible for providing much of the required training relating to operations specific to their program and facility.

The following topics need to be covered routinely in a training program to comply with the standard, in addition to those already discussed:

1. The physical and health effects of the specific hazardous chemicals which the employees may use or to which they may be exposed.
2. Means to detect the presence of toxic materials in the workplace. This should include means directly available to the employee, such as odor, presence of a respiratory irritant, and visual means or various symptoms such as dizziness, lassitude, etc. It also should

include types of monitoring that can be done by laboratory personnel, by the organization's Safety and Health Department, or by outside public and private agencies.

3. Means to reduce or eliminate the exposure of the employee to the risks associated with the hazardous chemicals in the workplace. This should include work practices that will reduce the exposures or the use of personal protective equipment.

4. Actions the organization has taken to minimize the exposure of employees to the chemical hazards. This can include the engineering controls which have been implemented such as ventilation or monitoring devices. It can also include policy positions which encourage or require that employees and their supervisors follow good safety practices at all times, and programs which provide incentives for them to do so or to penalize those who do not.

5. Emergency procedures to follow in the event of an accidental exposure to a hazardous material.

6. Procedures to warn nonorganizational personnel working in the area of potential exposures. Generally this will mean persons working under contract to the organization. It could also include maintenance and other support personnel.

7. Measures to provide information as to the hazards and the protective measures which both the employer and employee can take to reduce or eliminate the hazards associated with a nonroutine task involving chemicals.

8. Measures to inform personnel of the hazards associated with unlabeled pipes carrying chemicals in their work area, and the safety precautions which should be taken.

9. Availability of a medical evaluation should an over exposure have occurred or be suspected.

The responsibility for training covering these topics should be shared between the local administrative unit, usually the individual facility or department, and the department assigned the lead in implementing the organization's program.

There are a number of generic chemical topics which would be essentially the same, no matter what type of chemical exposure is involved. Among these are:

- flammable and combustible liquids
- corrosives, acids, and bases
- gases
- explosives
- toxic materials
- irritants
- carcinogens
- allergens
- pathogens

There also are a number of topics on protective measures to minimize exposures that could be made the topic of standardized presentations, such as:

- safe chemical working practices
- use of personal protective equipment
- fire safety
- safe working practices
- electrical safety
- emergency procedures

Standardized programs can be developed for these topics, as well as others, and videotapes made which can be used virtually anywhere. If the latter course is taken, much of the hazard communication standard training requirements could be met by requiring a new employee to view selected tapes, supplemented by an opportunity for the employee to ask questions. Standardized programs are especially useful where there is a continuing turnover, with customized training being required for single individuals on a sporadic basis.

A large number of commercial companies offer hazard communication training programs, many of which do provide videotapes as discussed above. They need to be reviewed prior to purchase. Some, in order to demonstrate compliance with one aspect of a safety program, do not follow good safety practices in other areas. Cost is not necessarily a guide; some of the better tapes are among the less costly.

Although much of the training obligation can be met with standardized programs, the primary responsibility for training must be borne by local managerial personnel or persons delegated by them. The actual exposures vary from place to place. Generic programs can provide an excellent foundation for a hazard communication training program, but they must be interpreted and adapted to local environments and workplace practices. When training has been completed for a given area, or even for a single chemical and if this is all that is necessary when a new chemical is brought into the workplace, the employee should be asked to document that the training has been provided by signing a dated statement to that effect. It is not essential that the employee agree that all of the information has been understood, although it is certainly hoped that this is the case, but it is important to have a record documenting that the information has been provided. The employer may be called upon to provide this documentation during an OSHA inspection or in the event of litigation.

REFERENCE

1. Hazard Communication; Department of Labor, Occupational and Health Administration, 29 CFR Parts 1910 (Section 1200), 1915, 1917, 1918, 1926, and 1928, Federal Register 52(163), 31852, August 24, 1987. Also see (for current version): http://frwebgate.access.gpo.gov/cgi-bin/get-cfr.cgi

VIII. HEALTH EFFECTS

In recent years, there has been increasing emphasis placed on the health effects of chemical exposures. However, as has been frequently noted, health effects are much more difficult to quantitatively characterize than most physical safety parameters. It is straightforward to define with reasonable accuracy a number of physical hazards, such as the upper and lower explosive limits of the vapors of a flammable material. However, the exposure levels (see Figure 4.16, taken from the Federal Register Vol. 53, No. 109, June 7, 1988, p. 21342) which will cause a given physiological effect in humans are not nearly as precise, especially if the effect of interest is delayed or is due to prolonged exposure to low levels of a toxic material.

Even individual reactions to low levels of materials that cause serious immediate or acute effects at high doses are strongly dependent upon the inherent susceptibilities of individuals. Some will exhibit a reaction at extremely low levels, while others show no signs of responding at all to relatively high levels. Part of this is due to the natural range of sensitivity in a population, but part is also due to contributory effects. There are synergistic effects: for example, a heavy smoker may have developed emphysema due to the effects of inhaling smoke for an extended period. Such an individual would be affected by airborne toxic materials which reduce pulmonary function before a person with healthy lungs. The sensitivity of individuals can change with time: an example is the common oleoresin allergen, poison ivy. The sensitivity of individuals is usually small to an initial exposure, but with successive exposures the sensitivity increases. Similarly, the sensitization of individuals to bee stings is well known. The effects of medication also can modify the sensitivity of individuals. The serious problems associated with simultaneously taking tranquilizers and drinking alcohol represent a well-known example of this phenomenon.

The natural differences in individuals due to genetic factors, age, sex, lifestyle, etc., make evaluation of laboratory tests quite subjective. On examination of the results of a typical blood panel, one notes that the patient's results are given for each parameter measured, as well as a *range* to be expected for a typical healthy person. An individual can have results somewhat outside the normal range due to hereditary or environmental conditions and still be perfectly healthy. Fasting and foregoing any medication prior to laboratory tests is an attempt to eliminate as many variables as possible which would affect the tests. Unless a baseline series of tests are available from a time prior to an exposure when, presumably the quantities to be measured are "normal" for the individual, then it is often difficult to determine with certainty whether a given result is due to an exposure or not. Even then, the results could be distorted by extraneous factors occurring earlier.

Dependence on observed symptoms to indicate an exposure to a toxic material is also very subjective. Again, the wide variation in tolerance of individuals to concentrations of toxic agents causes a corresponding wide variation in the responses to an exposure. Many of the common symptoms associated with occupational exposures - shortness of breath, headaches, nausea, dizziness, etc., often can also be the result of other problems, illnesses such as flu, lack of sleep, psychological problems such as stress due to personal problems or personality conflicts with a supervisor, overindulgence, etc.. An individual needs not only to be aware of the symptoms which could result from an exposure, but also needs to try to distinguish when these are likely to be due to an occupational exposure and when they are likely not to be. If, for example, an individual normally enjoys good health, has not done anything which might result in any of the symptoms which the person is experiencing, and there are no "bugs" going around, he might well suspect that he has suffered an exposure to some environmental hazard. In such a case, he should mention it to coworkers, report it to his supervisor, and leave the work area. Especially if others are experiencing similar symptoms, although not always, it is very likely that an exposure has occurred and appropriate steps should be taken to seek medical aid for the exposed personnel, limit the exposure of others, and correct the situation. The exception is when hysteria causes psychosomatic effects among others , although even here it is wisest to treat the situation as real, until proven otherwise. If only one person is having difficulties, but the immediate work environment differs for each person, then an exposure may have occurred limited to the individual, and precautionary steps, such as leaving the area, lying down, and observation by a colleague, should take place. Often, prompt recognition of a problem is critical in minimizing the consequences, especially when the possible culprit is a material which provides no other warning signs. Whenever an exposure has occurred which has resulted in physical effects, an evaluation by a physician should be obtained promptly. Even if no exposure has occurred, an individual complaining of an illness should be taken seriously. There could be a medical disorder requiring care, intervention, or at least documentation. Even if malingering is suspected, evaluation by a physician can help to confirm that the claim of illness is or is not valid.

Delayed effects due to prolonged exposures to relatively low levels of toxic materials or radiation rarely are reflected in immediate sensations of malaise sufficient to trigger concern about possible consequences of the exposures. If a material does not have any warning properties, then exposures may exist at unsafe levels indefinitely without the occupants of the area being aware of the exposure. The eventual consequences may be masked by naturally occurring illnesses of the same type. For example, lung cancer is, unfortunately, common and so the occurrence of lung cancer might not be recognized as due to an occupational exposure if this occurred. Birth defects occur in about 3 to 6% of natural births (depending upon how birth defects are defined). What percentage might be due to an occupational exposure of the mother or father? Similarly, infertility is a problem for about 15% of married couples. What is the role of occupational exposures for the unfortunate couple? Some neurotoxins cause deterioration of the central nervous system, but age and other illnesses may do the same. It is often difficult to establish a correlation between an occupational exposure and an illness, even statistically for a group, because it is difficult to isolate the effect from the influence of other variables or to define an equivalent control group. Anecdotal evidence citing an apparently unusual rate of a specific illness may be due to a statistically random occurrence. Often, unless the illness is rare, such as the angiosarcomas caused by exposure to vinyl chloride, it is impossible to definitely verify a causal relationship between an occupational exposure and a disease.

Not all delayed effects are due to low levels of exposure. The onset of cancer which may occur due to exposure to asbestos or radiation is often delayed for periods of 15 or more years, or, and this is a key point, they may not occur at all. By no means do all individuals exposed to even high levels of such hazards suffer the consequences.

There are basically three mechanisms by which health hazard data may be acquired: (1) epidemiological studies of groups of exposed individuals, (2) human experimentation, and (3) animal studies. There are problems associated with each of these three sources.

The major problem with epidemiological studies is that often one does not have a controlled experiment; the data is either generated by an ongoing work situation or extracted from past medical records. In some instances, case reports are sufficiently unusual that they call attention to themselves, e.g., a reduction in fertility in a group of workers is so large that only a simple study to determine the cause-effect relationship between a common exposure factor and the resulting fertility depression is required, the effect being known; it only remains to determine what experience the workers have in common. Rarely are situations as simple as this, although they do occur.

In most instances, epidemiological studies to determine if an exposure to a substance results in a given effect take the form of cohort studies, in which two separate groups composed of exposed and unexposed individuals are studied. It is critical that the study be unbiased either by the way the participants are selected or by the manner in which the outcome is tested. Another critical factor is whether the two groups are in fact similar in all essential respects, which could affect the outcome of the study, or that the differences are such that they can be taken into account either in the design of the experiment or in the analysis of the data. In order to judge the validity of a study, all of the relevant factors must be completely documented and available for review.

Most of the epidemiological studies concerning exposure to toxic substances are from the industrial sector since only in such an environment is it likely that exposures would be limited to a single chemical or class of chemicals, and where the exposures would be relatively stable over a prolonged period of time. The majority of the studies that are available tend to come from Scandinavia, where, for example, Finland maintains a computerized data base of the health records of all its citizens. Similar records do not exist in the U.S., although some categories of specialized health data are maintained. Many epidemiological studies of exposures in the U.S. depend upon records maintained by corporations, or equivalent public agencies such as the national laboratories, which are managed by industrial firms and have similar medical surveillance programs to them. The limited range of chemicals for which such work situations provide the basis for valid epidemiological studies limits the scope of this approach.

Human experimentation is limited by statute and by ethical considerations to studies in which there is no prospect of permanent harm to the volunteers participating in the study. This obviously limits the scope of the results obtained by this route, although it can be employed to determine the onset of early symptoms or to determine threshold levels for detection of odors or irritation as a potential warning mechanism. Any experiment of this type must be carefully reviewed by a human subject review committee of the institution or corporate research facility where the research is being contemplated. Any subject of such experimentation must be fully informed of any risks or benefits and normally must be given an opportunity to withdraw at any point. However, even with this restriction, many experiments using volunteers have been conducted and significant data have been obtained on symptoms initiated by modest levels of exposure. It should also be remembered, though, that "fully informing the volunteer of all known risks" can in and by itself skew the results. Therefore, published results on subjective symptoms using a small sample must be viewed as inconclusive or suspect.

There have been data obtained from direct human exposures due to accidental exposures to high levels of a number of hazardous chemicals. These, of course, are not controlled experiments and the dose levels must be inferred from the circumstances of the incident, but as direct evidence of the results of high exposures, they are extremely useful.

Since data from exposures directly to humans are limited, much of the available data on the toxic effects of chemicals is obtained from animal data. The easiest data to obtain are the median dose or the median concentration in air which is fatal to an animal under a standardized experimental protocol although animal rights individuals are taking an increasingly active role in opposing such tests. The most common animals used for this purpose are strains of rats and mice because they can be obtained with uniform characteristics relatively inexpensively, and the cost of housing and feeding them is small compared to most other species. Recently a few laboratories have succeeded in cloning individual mice to achieve a completely homogenous population. In addition to rats and mice, many other animals are used, such as primates (monkeys, chimpanzees), guinea pigs, rabbits, dogs, cats, and chickens, in efforts to obtain a model which would parallel the effect on humans. In the generic carcinogen standard, relevant animal studies were intended to specifically involve mammalian species.

The median lethal dose, written LD_{50}, is given in mg/kg and the species is given. The median lethal air concentration, LC_{50}, may be given in mg/m^3 or ppm for a given species, and the exposure time interval is usually specified. Lesser amounts of data are given in the literature at other survival fractions, such as 25% or 75%. These data must be obtained under rigorously controlled conditions to be useful, and the experimental protocol must be totally documented. Among other things, enough animals must be used to provide statistical accuracy. Where the effect to be measured is less well defined than lethality, the number of animals needed to obtain the data may become quite large. Even after the data using animals has been obtained, the question remains in many cases of whether the animal model is sufficiently close to that of a human response to use it to determine the equivalent human response. Much of the controversy of using animal data to establish human exposure effects revolves around this question.

Another procedure which leads to varying interpretations on the health effects of tested materials is the practice of using large, nonlethal doses to reduce the number of animals required when studying other effects such as carcinogenicity. The premise is that a large dose given to a small number of animals is experimentally equivalent to small doses given to a large number of animals. A linear extrapolation hypothesis usually is used to estimate the effects at low exposures. This practice is not uniformly accepted and is often used as an argument to discredit the results which are obtained, but is the basis of much of the data on these non-acute effects. Other possibilities to estimate the effects of low doses would be to assume that the response will approach zero more or less rapidly than the dose. A linear extrapolation generally is considered conservative. Data from experiments such as this are frequently used in arriving at health standards, by regulatory agencies, pharmaceutical manufacturing companies and the media.

The use of animals has provided the greatest amount of health hazard data, but the practice has come under increasing attack by animal rights activists. Much of the public support for this movement originated from widely publicized instances in which animals were not well treated and undoubtedly suffered more than was necessary. As a result of public pressure and a concern on the part of many scientists, many new safeguards have been instituted to minimize the amount of pain and suffering experienced by laboratory animals. Animal care committees are now required to review experimental protocols and must approve the procedures in order to qualify the research for federal support. The number of animals involved in the research is limited to the number required to achieve meaningful results and the pain experienced by the animals must be no more than absolutely necessary. These committees must include persons not affiliated in any other way, directly or indirectly, with the institution, and who might be expected to be caring about the well-being of the animals.

Although conceding that improvements have been made in the care of the experimental animals, the animal rights activists goal is to prevent the use of animals in any research which would adversely affect the animals. There have been instances in which animals have been "liberated" from facilities and instances where these "liberated" animals have been released into the environment. There are two practical problems with such actions, regardless of the ethics:

(1) the animals usually are not accustomed to surviving in the wild and most often do not, and (2) there generally have been no efforts to ensure that the animals are healthy and, hence, a disease could be introduced into the environment.

The argument that there is no alternative to using animals to gain knowledge to prevent disease or to cure human illnesses, since experimentation on humans cannot be done, is rebutted in two ways by the animal activists, the first being a purely moral stance of "why do we assume that it is morally right to cause pain to animals to help humans?" This is an issue that each individual must answer for himself, unless a legal restriction is imposed. The second argument is that animal experimentation is no longer necessary. It is claimed that computer modeling can provide equivalent information. Relatively few scientists accept this latter argument as a generalization, although it is agreed that computer modeling can be used in some cases and as an indication of productive research.

There is merit on both sides, although the extremists of both groups are undoubtedly too extreme, and some middle position will eventually become acceptable practice. However, animal health hazard data may be less available in the future.

A newer but effective modality which can be statistically relevant is to use cell cultures. Direct effects can be seen and measured as to benefits or toxicity. It is more complicated and has seen limited use as yet. There is also the question as to whether effects seen in individual cells can be extrapolated to a complex organism.

Recent mapping of the human genome have greatly accelerated an understanding of how diseases are caused and are opening many more options in treating diseases. It may be possible to forego animal experimentation entirely in the future. A current program taking place in Iceland where a remarkably homogenous population exists along with extensive genealogical data may prove especially helpful in determining what genes are involve with specific diseases, leading to better approaches to treating these diseases.

A recent report by the EPA indicates that research now appears to indicate that some toxic substances are less dangerous than formally supposed, based on animal studies, specifically in regard to carcinogenicity. This finding, according to EPA spokespersons, is based on a better knowledge of how the metabolism of chemicals differs in various species, a better knowledge of how much of a chemical that has been taken into the body actually reaches an organ where it may do harm, and a better understanding of how the chemical influences the mechanisms that cause cancer. This is a controversial position since, in general, it leads to higher acceptable exposure levels of the chemicals under discussion, such as dioxin and arsenic. There are scientists that feel that the level of scientific knowledge does not as yet justify moving away from a very conservative approach. Both sides should avoid treating the issue as one that can be settled by politically biased discussion, and the data on which the findings are based should be evaluated, as should any other hypothesis, on the basis of their scientific merit.

As stated in the section on the hazard communication standard, Section VII.B of this chapter, OSHA defines health effects, for the purposes of the standard, in Appendix A to 29 CFR 1910.1200. The definitions given below are from that appendix. The laboratory safety standard also specifically suggests using these same definitions for guidance in defining hazardous chemicals. The definitions below originated in 21 CFR Part 191.[*] A few modifications have been made to number 7.

For the purposes of this section any chemicals which meet any of the following definitions, as determined by the following criteria, are health hazards:

---

[*]   Some of the material quoted is rearranged, and some nonessential verbiage is deleted. The essential information is not changed.

Criteria:
1. *Carcinogenicity:* ... a determination by the NTP, the IARC, or OSHA that a chemical is a carcinogen or potential carcinogen will be considered conclusive evidence for purposes of this section.
2. *Human data:* Where available, epidemiological studies and case reports of adverse health effects shall be considered in the evaluation.
3. *Animal data:* Human evidence of health effects in exposed populations is generally not available for the majority of chemicals produced or used in the workplace. Therefore, the available results of toxicological testing in animal populations shall be used to predict the health effects that may be experienced by exposed workers. In particular the definitions of certain acute hazards refer to specific animal testing results.
4. *Adequacy and reporting of data:* The results of any studies which are designed and conducted according to established scientific principles, and which report statistically significant conclusions regarding the health effects of a chemical, shall be a sufficient basis for a hazard determination.

Definitions:
1. *Carcinogen.* A chemical is considered to be a carcinogen if:
   a. It has been evaluated by the IARC, and found to be a carcinogen or potential carcinogen; or
   b. It is listed as a carcinogen or potential carcinogen in the *Annual Report on Carcinogens* published by the NTP (latest edition); or
   c. It is regulated by OSHA as a carcinogen.
2. *Corrosive.* A chemical that causes visible destruction of, or irreversible alterations in, living tissue by chemical action at the site of the contact. For example, a chemical is considered to be corrosive if, when tested on the intact skin of albino rabbits by the method described in the U.S. Department of Transportation in Appendix A to 49 CFR Part 173, it destroys or changes irreversibly the structure of the tissue at the site of contact following an exposure period of 4 hours. This term shall not refer to action on inanimate surfaces.
3. *Highly toxic.* A chemical falling within any of the following categories:
   a. A chemical that has a median lethal dose (LD_{50}) of 50 milligrams or less per kilogram of body weight when administered orally to albino rats weighing between 200 and 300 grams each.
   b. A chemical that has a median lethal dose (LD_{50}) of 200 milligrams or less per kilogram of body weight when administered by continuous contact for 24 hours (or less if death occurs within 24 hours), with the bare skin of albino rabbits weighing between 2 and 3 kilograms each.
   c. A chemical that has a median lethal concentration (LC_{50}) in air of 200 ppm by volume or less of gas or vapor, or 2 milligrams per liter or less of gas or vapor, or 2 milligrams per liter or less of mist, fume, or dust, when administered by continuous inhalation for 1 hour (or less if death occurs within 1 hour) to albino rats weighing between 200 and 300 grams each.
4. *Irritant.* A chemical which is not corrosive, but which causes a reversible inflammatory effect on living tissue by chemical action at the site of contact. A chemical is a skin irritant if, when tested on the intact skin of albino rabbits by the methods of 16 CFR 1500.41 for 4 hours exposure or by other appropriate techniques, it results in an empirical score of five or more. A chemical is an eye irritant if so determined under the procedure listed in 16 CFR 1500.42 or other appropriate techniques.
5. *Sensitizer.* A chemical that causes a substantial proportion of exposed people or animals to develop an allergic reaction in normal tissue after repeated exposure to the chemical.
6. *Toxic.* A chemical falling within any of the following categories:

a. A chemical that has a median lethal dose (LD$_{50}$) of more than 50 milligrams per kilogram but not more than 500 milligrams per kilogram of body weight when administered orally to albino rats weighing between 200 and 300 grams each.

b. A chemical that has a median lethal dose (LD$_{50}$) of more than 200 milligrams per kilograms but not more than 1,000 milligrams per kilogram of body weight when administered by continuous contact for 24 hours (or less if death occurs within 24 hours), with the bare skin of albino rabbits weighing between 2 and 3 kilograms each.

c. A chemical that has a median lethal concentration (LD$_{50}$) in air of more than 200 ppm by volume of gas or vapor, but not more than 2000 ppm by volume of gas or vapor, or more than 2 milligrams per liter but not more than 20 milligrams per liter of mist, fume or dust, when administered by continuous inhalation for 1 hour (or less if death occurs within 1 hour) to albino rats weighing between 200 and 300 grams each.

7. Target organ effects. The following is a target organ categorization of effects which may occur, including examples of signs and symptoms and chemicals which have been found to cause such effects. These examples are presented to illustrate the range and diversity of effects and hazards found in the workplace, and the broad scope employers must consider in this area, but are not intended to be all-inclusive.

a.	Hepatotoxins	Chemicals which produce liver damage
	Signs and Symptoms	Jaundice, liver enlargement
	Chemicals	Solvents such as toluene, xylene, carbon tetrachloride, nitrosoamines
b.	Nephrotoxins	Chemicals which produce kidney damage
	Signs and Symptoms	Edema, proteinuria, hematuria, casts
	Chemicals	Halogenated hydrocarbons, uranium
c.	Neurotoxic	Chemicals which produce their primary effect on the nervous system
	Signs and Symptoms	Narcosis, behavioral changes, coma, decrease in motor functions
	Chemicals	Mercury, carbon disulfide, lead
d.	Agents which act on the blood or hematopoietic system	Decrease hemoglobin function, deprive the body tissue of oxygen
	Signs and Symptoms	Cyanosis, anemia, immune function, depression
	Chemicals	Carbon monoxide, cyanide
e.	Agents which damage the lung	Chemicals which damage the pulmonary function
	Signs and Symptoms	Cough, tightness in chest, shortness of breath
	Chemicals	Silica, asbestos, organic fibers such as cellulose, cotton
f.	Reproductive toxins	Chemicals which affect the reproductive capabilities, including chromosomal, damage (mutations) and effects on fetuses, (teratogenesis).
	Signs and Symptoms	Birth defects, sterility, functionality
	Chemicals	Lead, DBCP, some blood pressure medications
g.	Cutaneous hazards	Chemicals which affect the dermal layer of the of the body
	Signs and Symptoms	Defatting of the skin, rashes, irritation, discoloration
	Chemicals	Ketones chlorinated compounds soaps, solvents
h.	Eye hazards	Chemicals which affect the eye or visual capacity

| Signs and Symptoms | Conjunctivitis, corneal damage, blephaharitis |
| Chemicals | Organic solvents, acids, alkalis |

A. Exposure Limits[*]

Consideration and attention to the concentrations of chemicals within any worker's environment are mandatory, such as chemicals in respired air, in water (used at work or for drinking), food contamination, contact with skin or eyes either directly or by vapor, and possibly radiation from specific chemicals used. The garnering of meaningful data has been slow, and changes in acceptable levels occur, as new resources and studies are produced and policy formulations are agreed upon.

The ACGIH has published TLVs for decades, and recent international meetings have formulated and categorized the values into what are called occupational exposure limits (OEL). All of these limits refer to airborne values only. Where skin is involved an "s" may be appended. The latter values refer to time weighted averages (TWA) over a normal working day of 8 hours in a 40-hour week, extrapolated for the worker's lifetime. OSHA limits are defined as PELs, which are the legally enforceable limits for the materials for which they have been established. There are also ceiling limits dented by "C" and short-term exposure limits denoted by "STEL."

The exposure limits should not be used to define a boundary, on one side of which working conditions are acceptable and on the other side, unacceptable. First, they are for a typical person, and individuals may not react in a typical way. Low levels for which a prolonged exposure could give rise to a delayed effect would not necessarily provide any warning for individuals who might be abnormally sensitive. Secondly, levels should be kept as low as possible, and preferably well below the exposure limits. Maintenance of air concentrations below the TLV, or PEL does not guarantee that a worker will or will not be affected at any given concentration. There is a very small percentage of workers who are so sensitive to particular chemicals that even minute quantities will precipitate severe reactions. It is not possible or feasible to try to make exposures acceptable for these individuals in most cases. Removing them from the environment and avoiding contact to the material causing the reaction is best for both the worker and the facility. However, one must make reasonable accommodations for these individuals or risk violating anti-discrimination statutes. The essential OSHA goal that "no worker will be affected" is unattainable in cases involving large numbers of persons. The variability of innate propensities, as well as such factors as daily diet variations, stress levels, heath status, and circadian rhythms, often relegates efforts to provide a "safe" environment to statistics and chance.

One must differentiate between somatic effects and nuisance effects because:

1. Such a distinction is based on the nonexistent dichotomy between soma and psyche.
2. With adequate testing, a nuisance may be shown to actually be a somatic problem.
3. The nuisance may be the premonitory signs and symptoms of a slowly developing somatic problem.
4. The nuisance perception can be a statistically significant factor in diminished productivity.

Stokinger appropriately considers nuisance perceptions and signs and symptoms as inducers of disease. Efforts to specifically categorize these areas might well be a fruitless endeavor but they should be seriously evaluated.

It would be valuable to keep in mind the effects of synergism in the workplace. In considering the many exposure/response relationships, extraneous factors must be considered as

---

[*] The author wishes to acknowledge the significant contribution of Dr. Richard F. Desjardins, M.D. to this section.

background variables which could skew results. Such factors include smoking, hypertension, heart disease, asthma or pulmonary fibrosis, diabetes, and many more. The individual's preliminary examination made by the physician is made by the health facility. It has the purpose through a good present and past health history, a work history with attention to types of exposures, a perceptive physical examination, and appropriate testings to determine whether that individual can healthfully be employed in a given area. This must be done at that time to: (1) avoid harmful exposures which would compromise the person's health, and (2) avoid even the slightest suspicion that a disease was evolved within the facility.

The Committee for the Working Conference on Principles of Protocols for Evaluating Chemicals in the Environment in the U.S. defined adverse effects as "those changes in morphology, growth, development and life span that:

1. Result in impairment of functional capacity or in a decrement of the ability to compensate additional stress.
2. Are irreversible if such changes cause detectable decrements in the ability of the organism to maintain homeostasis.
3. Change the susceptibility of the organism to the deleterious effects of other environmental influences."

The "hygienic rule" states that exposure to chemicals at work should always be minimized as far as possible. Ultimately the goal is zero. Pragmatically at this time, this must be interpreted as at the lowest levels reasonably achievable, or ALARA. The airborne exposure limits, regardless of their source - ACGIH, NIOSH, or OSHA regulatory limits - are intended to represent TWA levels at which it should be acceptable for most workers to work normal 8-hour-day, 5-day work weeks without suffering any ill effects. Note the word "most." As noted earlier on several occasions, individuals vary widely in terms of susceptibility, and there may be contributory effects which would make these levels inappropriate for a specific person. The values are intended to be based on the best available data from the sources described earlier as evaluated by committees composed of professionals with appropriate expertise. The OSHA values are an exception, to a degree, in that they are adopted as part of a legal standard, and changing them is often a long, involved process which frequently includes legal contests. Therefore, changes in OSHA values tend to lag somewhat behind those data such as the limits suggested by the ACGIH, which are reviewed and revised annually, if the data warrant a revision. However, even the OSHA values were originally based on the same evaluation procedure and were in most cases actually taken from the other sources. OSHA revised its air contamination standards in 1989, setting them in most cases to the ACGIH-recommended TLVs in effect at that time. Many of the values were lowered. However, in July 1992, a U.S. Court of Appeals struck down these new standards, and currently the federal standards have reverted to the pre-1989 values, except for specific substances for which individual revisions had been made.

Many of the newer legal limits established by OSHA include an action level concept, typically 50% of the OSHA PEL, at which an employer is to take action to reduce the level or to ensure that the 8-hour time-weighted PEL is not exceeded.

Although the 8-hour TWA, by definition, includes periods during which the level exceeds the exposure limits, the departure should not be gross. In some instances ceiling levels are recommended or mandated above the 8-hour TWA limits, but they should never be exceeded. In other instances, there are short-term exposure limits which are also set higher than the TWA limits, usually intended to be for durations of no more than 15 minutes, for a limited number of occasions during a work span (ACGIH uses four as a limit) and spaced widely throughout a work day. These are usually set for processes which may be sporadic or have occasions when higher levels are normal. Sometimes these higher numbers are based on toxicological data, and sometimes they are indirectly related to typical maximum excursions which might occur in an industrial environment. The TLV along with the TWAs establish guidelines for exposure control, both for the employee's health as well as for possible corporate or institutional responsibility.

The responsibility for maintaining levels of toxic vapors and gases below acceptable limits is usually assigned to management, preferably by engineering controls unless these can be shown to be impractical, and otherwise by procedural controls and the use of personal protective devices. However, it is equally important for the employees to properly use the equipment in the laboratories so as not to defeat engineering controls, to report promptly any safeguards needing repairs, to not follow procedures which reduce safeguards, and to use and maintain personal protective equipment provided them. Neither a "macho" or a "scoff-law" attitude is appropriate for laboratory activities as regards compliance with acceptable exposure levels.

REFERENCES

1. **Ziclluis, R.L.,** Occupational exposure limits for chemical agents, in *Occupational Medicine,* Zenz, C., Ed., Chicago, 1988.
2. World Health Organization, Methods Used in Establishing Permissible Levels in Occupational Exposure to Harmful Agents, Tech. Report 601, WHO, Geneva, 1977.
3. World Health Organization, Principles and methods for evaluating the toxicity of chemicals, in *Environmental Health Criteria* 6, Part 1, WHO, Geneva, 1978.
4. **Zielhuis, R.L. and Notten, W.R.E.,** Permissible levels for occupational exposure: basic concepts, *mt. Arch. Occup. Environ. Health,* 42, 269, 1978 - 1979.
5. **Zielhuis, R.L.,** Standards setting for work conditions as risky behavior, in *Standard Setting,* Grandjean, P.H., Ed., Arbejosmiljofondet, Copenhagen, 1977, 15.
6. **Sherwin, R.P.,** What is an adverse health effect?, *Environ. Health Perspect., 52,* 177, 1983.
7. **Stokingei H.E.,** Modus operandi of Threshold Limits Committee of the ACGIH, *Am. Md Hyg. Assoc. J,* 25, 589, 1964.
8. **Stokinger, H.E.,** Criteria and procedures for assessing the toxic responses to industrial chemicals, in *Permissible Levels of Toxic Substances in the Working Environment,* International Labor Organization, Geneva, 1970.
9. Principles for Evaluating Chemicals in the Environment: A Report of the Committee for the Working Conference on Principles of Protocols for Evaluating Chemicals in the Environment, National Academy of Science, Washington, D.C., 1975.
10. American Conference of Governmental Industrial Hygienists (ACGIH), Cincinnati, OH.

B. Environmental Monitoring[*]

Since many toxic materials provide no sensory warnings, and in many other cases, materials which do have an odor or cause irritation (the two most common warning properties) do not result in a recognizable warning at or near the recommended limits or even at the presence of levels of immediate danger, the establishment of limits is of no practical value unless the capability of monitoring exists. Standard monitoring programs, based on 8-hour TWA measurements, are often not appropriate for the laboratory environment, where the duration of usage of a given material is rarely this long. However, average and peak levels for the various activities should be determined.

---

[*] This section was coauthored by A. Keith Furr and Richard F. Desjardins.

Monitoring programs are usually performed by an industrial hygienist instead of the laboratory employees, except where permanently installed fixed monitors are used. Ideally, air monitoring should measure the levels of chemical substances in the breathing zone of each employee. As will be discussed below, monitoring near the breathing zone is practical with small sampling systems which can be worn in a shirt pocket or attached to a lapel. Preferably, the employee should wear the monitoring device during an entire work day and, again preferably, over a period of several days to ensure that the results obtained will be truly representative, although in many research laboratories the variety of chemicals employed change so rapidly that there may not be a truly "typical" day. It would be good if this could indeed be done for every employee. The amount of time required would, in all likelihood, be prohibitive in most large research institutions to do this for every laboratory. However, for laboratories working with regulated carcinogens, as an example, or where other highly toxic materials are in use, it may be desirable or even necessary to establish such an intensive and thorough program.

Where the activities of many employees are essentially similar, the exposures of a typical, representative person may serve to reflect the exposures of all the persons doing the same tasks. In some instances, the areas of exposure to chemical hazards are relatively small and well defined. Sampling devices may be placed at the location used to determine the average exposures per unit time of persons working in those areas. If these devices do not show concentrations above the acceptable limits, then individuals can be reasonably assured that their exposures do not as well.

In many instances, the rate of release of toxic vapors and gases varies widely, and an extended sample taken for 8 hours may show a very low average level of exposure, but for brief intervals, the levels actually may be dangerously high. In such a case, a monitoring program should include "instantaneous" or "grab" samples taken during these periods of higher-than-normal levels. All of the organizations which publish levels recognize that for some materials, even short-term high levels should not be permitted. This concept is embodied in the ceiling (C) or short-term exposure levels (STEL) incorporated in the standards for some materials. The monitoring program should be able to address the need to document compliance with these short-term excursions.

All data accumulated in the monitoring program should be maintained in a permanent file, preferably for each employee but definitely for each facility. These records should be dated, signed by the individual responsible for the data, and documented as to the measurement parameters, i.e., instrument used, duration of sample collection, characteristics of the detection device, location of the person wearing the device, and, if possible, supporting data on the operations being conducted at the time of data acquisition.

There are many different types of sampling instruments used, and it is not the author's purpose to dwell on the characteristics of each. Some are direct reading while others require sophisticated follow-up procedures to analyze collected material. There is some legal merit in acquiring material for later analysis by accredited commercial laboratories to provide additional credibility in case the results are contested at some later date. The delay in receiving results following this practice, which may vary from a day or so to weeks, may be unacceptable, and in-house analysis or alternative methods may be chosen to obtain results more rapidly.

Direct reading instruments, by definition, provide "instantaneous" or "grab" samples which may be entirely appropriate, as when the need is present to determine if an IDLH situation exists. A good example is the need to use an oxygen meter when entering any confined space. A combustible gas meter is often used at the same time to determine the presence of organic gases, in the confined space. These two devices are often combined in the same instrument, using different sensors. Other instances are reentry to areas where the presence of a highly toxic gas may be suspected.   Direct reading instruments also are useful in conducting surveys within a facility to determine the relative levels of pollutants in different parts of a space. There may be areas, for example, where air exchange is minimal and toxic levels of gases can accumulate. In such instances, the data can be used to help determine what remedial actions can be taken. If previous measurements have not been taken, as for example when an operation is to be performed for the first time, a direct reading instrument, perhaps incorporating an alarm, might well be used. In some cases, direct reading instruments may be incorporated into fixed monitoring devices to provide a warning and to automatically shut down an operation. If, for example, experimentation with explosive gases is being done, a sensor may be used to determine when the concentration of the gas reaches a predetermined level relative to the lower explosion limit of the gas and to automatically close a valve in the gas supply.

Some direct reading instruments are extremely sophisticated, and others are very simple. Some instruments include a sensor for a single substance, while others can serve to detect and provide quantitative information on a number of materials. Among the more sophisticated units which provide direct readings are compact, portable infrared spectrometers, atomic absorption units, and gas chromatographs. These may be set up to provide a high sensitivity for a single element, such as the familiar Bachrach atomic absorption unit for mercury detection, while the Miran infrared spectrometer can provide accurate readings for hundreds of organic solvents. Individuals using these instruments need to be properly trained to obtain accurate data. Interferences are a major problem, where two or more possible chemicals may contribute to the same "window" or peak area. If the potential presence of interfering chemicals is not recognized, erroneously high readings may be reported. If the problem is recognized, these same high readings still may be interpreted as an upper limit for the level of the chemical of concern, and if the combined contribution is still well below acceptable limits, then exposures may be no problem to the employees.

All of these instruments must be well maintained and kept in good calibration in order for the data to be meaningful. In some instances, as with radiation measuring instruments, the calibration may be done by a commercial facility to provide traceability to a National Bureau of Standards standard in order to be certified, while in other cases calibration can be done locally. Documentation of all maintenance and calibrations should be maintained in a log. Assurance of accuracy is especially important if the levels to be measured turn out to be near the acceptable limits. As a minimum, the accuracy should be such as to assure that the results are within at least $\pm 25\%$ of the actual value with a confidence level of 95%. It would, of course, be desirable to exceed this minimal goal.

A simple direct reading device is the familiar detector tube intended to provide a measurement of the airborne concentration of a specific chemical, in which a known volume of air is drawn through a tube containing a material with which the chemical of concern will react. The known volume of air is usually provided by a manually activated pump attached to one end of a glass tube, and air is drawn in through the other. The amount of air may be provided by a single cycle of the air pump, or several cycles may be required. The reaction of the chemical in the air and the material in the tube will begin at the end away from the pump, and the length of the stain will indicate the level of the airborne contaminant. There are now hundreds of tubes available for individual chemicals that provide a great deal of flexibility for a relatively modest cost. The units do have a limited shelf life so that it is generally not feasible to keep tubes on the shelf for every possible contaminant (newer versions have improved shelf lives for a number of common substances). However, they can be the basis for a monitoring program for a chemical which is in relatively steady use. They are, as noted, relatively inexpensive per tube, but if a large number of measurements are required and if a direct reading instrument is available, it may be a more economical long-term investment to buy a device which does not consume an expendable tube for each measurement. The detector tubes, although for a specific chemical, may   react with more than a single chemical, so that interferences may be a problem. These interferences generally will be identified in the literature accompanying each tube. A major use of detector tubes, other than being an alternative for a more expensive device, is for testing of atmospheres for entry into an area in which high levels of a specific contaminant are expected to exist.

The devices previously mentioned are not suitable for taking extended measurements since they provide only an instantaneous measurement, unless the instrument has an output which can be connected to a device to integrate the data over a period of time. There are a number of ways in which this can be done, from a relatively crude system based on a chart recorder in which the area under the trace can be translated into an integrated reading to a digital sampling device acting to transfer the data to a computer with an appropriate interface and software, or a dedicated multichannel scaler in which the data occurring in successive intervals are accumulated in successive channels. Newer versions of many of the instruments in common use have an integrating mode of operation or a data-logger feature to acquire these data over extended periods.

Most long-term measurements depend upon using a constant-volume air pump in which the pump pulls in a given volume of air over several hours, and the contaminating material in the air is collected within a collecting device. The type of contaminant will determine the choice of the collector. Many organic gases and vapors are adsorbed readily on the surface of materials such as activated charcoal, activated alumina, or silica gel from which they can be eluted within a laboratory or driven off by heat and the amounts adsorbed quantified using a gas chromatograph. In other instances, the materials are bubbled through a liquid into which they go into solution. The collection efficiency of such a unit can approach 100%. Particulates also may be absorbed into a liquid, or, more frequently, collected by impact onto various types of filters, after which a number of techniques are available to quantify the amount of pollutant collected. The efficiency of the filter will depend upon the type of filter used.

In some cases, a collecting device called a cascade impactor is used, in which the rapidly moving contaminated air passes through a number of stages. There are slits between each stage of decreasing size. At each stage some of the particles, because of their inertia, tend to continue in a straight line, impact on a collecting plate, and adhere to it. In the first stage the larger particles which have the most inertia are collected, while smaller ones are drawn through the exit slit into the next stage and are collected there. This continues until the smaller particles are collected at the last stage on a permeable membrane. Thus, not only are the contaminants collected but they are sorted into approximate sizes. Since the quantity of air drawn through the collector is known and the amount on the filter or collectors can be measured, it is a simple matter to compute the concentration of the particulates in terms of milligrams per cubic meter for each range of sizes.

In instances in which there is only a single possible contaminant in the air, no further processing of the collected material is necessary but in most cases, analysis of the collected material is necessary to separate and quantify the amounts present. NIOSH has published explicit analytical techniques to be used for a large number of airborne pollutants. For meaningful results which are legally defensible, these procedures should be followed exactly. In many cases, the commercial laboratory will provide the appropriate collector, as well as explicit instructions on how they should be used, including, for example, the airflow rate and the length of time to be used.

Each time a pump is used it should be calibrated to ensure that the volume of air is accurately known. Most companies that sell sampling pumps also sell convenient calibration units, but if one is not available, the pumps should be calibrated using a graduated cylinder and determining, the time a soap film across the cylinder moves a measured distance. If sampling pumps are to be mounted in a fixed location, they generally can be plugged in to ordinary electrical power outlets, but if they are to be worn by an individual, they will have to depend on battery packs.

Battery packs are usually nickel-cadmium and these require some care to be able to run for long periods. Unless a nickel-cadmium battery is frequently taken through a major portion of its discharge cycle, it will lose capacity. It is said to "remember" the range over which it is used, and if it is typically drawn down only one third, for example, during each use, eventually this will be its capacity. Nickel-cadmium batteries are best suited for frequent, heavy use. Nickel hydride and lithium batteries may also be used.

A simple sampling device, well suited for personnel monitoring, uses a permeable membrane over an activated charcoal collector in the form of a badge which may be worn during the period of interest. Normal movement by the wearer and the inherent motion of molecules in air are sufficiently characteristic to bring a definite quantity of the air bearing the contaminant into contact with the membrane, through which the contaminant passes and is adsorbed by the charcoal. From this point on, the adsorbed material is eluted as with any other charcoal collector and analyzed. Normally, the company which sells the badge offers the analytical service as well, but the badges also may be processed locally. If care is taken, several organic solvents (if present in the air) can be identified with a single badge. Badges are available at this time for a relatively small number of materials, but the convenience of wearing a badge instead of a pump, no matter how quiet and lightweight, make them desirable. Sampling should be done as close to the breathing zone as practical. The results obtained as the worker moves around (and the air being sampled is representative of the individual's actual exposure) generally do not agree well with samples taken by fixed collectors. This raises some questions of the value of the grab samples obtained in surveys by portable instruments.

For various reasons, individuals asked to wear monitoring devices, such as a small sampling pump, are not always cooperative and do things which invalidate the data which are obtained. They may feel that they are being "checked on," perhaps for careless or sloppy work. They may simply feel that the pump is too noisy or that it gets in their way They may not want to know the levels to which they are being exposed, for fear that they may be too high and that they could be in danger of losing their job. In any event, if the individual responsible for the monitoring program feels that the data obtained are not reliable, then steps should be taken to investigate and correct the problem. A direct person-to-person explanation of the reason for the monitoring program may be all that is required, persuasion may be effective, or in an extreme case of lack of cooperation in which there are legal issues at stake, unfortunately it might be necessary to make it a disciplinary matter. It is hoped that the latter could be avoided since it would create the potential for a future adversarial relationship.

Any monitoring program only provides data (hoped to be valid) for the exposure levels present in the area where the program was carried out and under the circumstances as they existed at the time the program was conducted. If the circumstances change, then it may need to be repeated, extended to other materials, or modified because procedures have changed. Most laboratory organizations do not have sufficient manpower to monitor laboratories often enough to catch every variation. It is incumbent upon laboratory directors or managers or other employees to notify the industrial hygiene specialist of internal operational changes that could adversely affect the airborne levels of any hazardous material.

It also should be reiterated that the acceptable levels are those appropriate for an average person not to experience any adverse health effects. There are persons who are hypersensitive to a specific substance and the levels may not be suitable for them. It also is possible for persons not to be initially sensitive to a given material and to become more sensitive to it over a period of time. Sometimes this sensitization process may carry over to additional materials. A substantial amount about the acute effects of many chemicals at relatively high levels can be known or inferred, primarily because of animal experimentation, but we do not know a great deal about the delayed effects of a single exposure to a toxic material, the delayed effects of prolonged exposure to low levels of contaminants, or the synergistic effects of combinations of chemical exposures. It also is obvious that some individuals, cannot tolerate particular environments because of heat, vibration, lighting conditions, ergonomics requirements, a phobic response such as claustrophobia, or because of the presence of chemicals or odors. However, most people can develop an equilibration with the environment, tolerating ambient conditions without any untoward effects as long as they are not excessively extreme. Occasionally a new element, such as a change in the process, additional or different chemicals, heat, vibration, etc., occurs, resulting in a hyper-sensitization wherein the previously acceptable pollutants are no longer tolerable. This is not a synergism, but a new condition. Precipitating concentrations of any pollutant may be below the TLV or TWA. Good housekeeping measures may be of some value, but when hypersensitization has occurred, there might be no alternative but to remove the employee from the location. When asked for how long, only time will answer. Some individuals will gradually lose the sensitivity over 3 to 6 months, but some may retain it for life. Another condition can exist: the individual who has acquired the immune response or developed a lower threshold and is removed from the situation may become asymptomatic. The immune system, however may remember and upon a reexposure the previous allergic response can occur in full-force; just as it did at the time of primary acquisition. This is called the anamnestic response. Although it would be desirable to return an afflicted employee to the previous job, one must keep anamnesis in mind and proceed with caution. When reentry fails one must avoid the tendency to become angry, frustrated, or impatient with the employee. The reaction is not imaginary; it is more real and discomforting to the employee than it is to management. Therefore, the legal limits or the recommended levels should be considered as upper limits, not to be exceeded or approached if reasonable measures are available to reduce the actual levels in the workplace. There is no reason to excessively fear the laboratory environment, but it is foolish to scoff at or ignore reasonable measures to reduce levels of exposure.

C. Modes of Exposure

In order to better understand how exposure to hazardous materials in the laboratory enters into operational safety, a brief article from the 2nd edition of this handbook, pp. 314 to 316, by Herbert E. Stokinger, will be used to illustrate this point.

1. Means of Contact and Entry of Toxic Agents

Of the various means of body exposure to toxic agents, skin contact is first in the number of affections-occupationally related. Intake by inhalation ranks second, while oral intake is generally of minor importance except as it becomes a part of the intake by inhalation or when an exceptionally toxic agent is involved. For some materials, as might be inferred, there are multiple routes of entry.

a. Skin Contact

Upon contact of an industrial agent with the skin, four actions are possible: (1) the skin and its associated film of lipid and sweat may act as an effective barrier which the agent cannot disturb, injure, or penetrate; (2) the agent may react with the skin surface and cause primary irritation; (3) the agent may penetrate the skin, conjugate with tissue protein, and effect skin sensitization; and (4) the agent may penetrate the skin through the folliculi sebaceous route, enter the blood stream, and act as a systemic poison.

The skin however is normally an effective barrier for protection of underlying body tissues, and relatively few substances are absorbed through this barrier in dangerous amounts. Yet serious and even fatal poisonings can occur from short exposures of skin to strong concentrations of extremely toxic substances such as parathion and related organic phosphates tetraethyl lead, aniline and hydrocyanic acid. Moreover, the skin as a means of contact may also be important when an extremely toxic agent penetrates body surfaces from flying objects or through skin lacerations or open wounds

b. Inhalation

The respiratory tract is by far the most important means by which injurious substances enter the body. The great majority of occupational poisonings that affect the internal structures of the body result from breathing airborne substances. These substances lodging in the lungs or other parts of the respiratory tract may affect this system, or pass from the lungs to other organ systems by way of the blood, lymph, or phagocytic cells. The type and severity of the action of toxic substances depend on the nature of the substance, the amounts absorbed, the rate of absorption, individual susceptibility, and many other factors.

The relatively enormous lung-surface area (90 square meters total surface, 70 square meters alveolar surface), together with the capillary network surface (140 square meters) with its continuous blood flow, presents to toxic substances an extraordinary leaching action that makes for an extremely rapid rate of absorption of many substances from the lungs. Despite this action, there are several occupationally important substances that resist solubilization by the blood or phagocytic removal by combining firmly with the components of lung tissue. Such substances include beryllium, thorium, silica, and toluene-2,4-disocyanate. In instances of resistance to solubilization or removal, irritation, inflammation, fibrosis, malignant change, and allergenic sensitization may result.

Reference is made in the following material to various airborne substances, and to some of their biologic aspects.

i. Particulate Matter: Dust, Fume, Mist, and Fog

Dust is composed of solid particulates generated by grinding, crushing, impact, detonation, or other forms of energy resulting in attrition of organic or inorganic materials such as rock, metal, coal, wood, and grain. Dusts do not tend to flocculate except under electrostatic forces; if their particle diameter is greater than a few tenths of a micron, they do not diffuse in air but settle under the influence of gravity. Examples of dust are silica dust and coal dust.

Fume is composed of solid particles generated by condensation from the gaseous state, as from volatilization from molten metals, and often accompanied by oxidation. A fume tends to aggregate and coalesce into chains or clumps. The diameter of the individual particle is less than 1 μm. Examples of fumes are lead vapor on cooling in the atmosphere; and uranium hexafluoride (UF_6) which sublimes as a vapor, hydrolyses, and oxidizes to produce a fume of uranium oxyfluoride (UO_2F_2).

Mist is composed of suspended liquid droplets generated by condensation from the gaseous to the liquid state as by atomizing, foaming, or splashing. Examples of mists are oil mists, chromium trioxide mist, and sprayed paint.

Fog is composed of liquid particles of condensates whose particle size is greater than 10 μm. An example of fog is super-saturation of water vapor in air.

ii. Gas and Vapor

A gas is a formless fluid which can be changed to the liquid or solid state by the combined effect of increased pressure and decreased temperature. Examples are carbon monoxide and hydrogen sulfide. An aerosol is a dispersion of a particulate in a gaseous medium while smoke is a gaseous product of combustion, rendered visible by the presence of particulate carbonaceous matter.

A vapor is the gaseous form of a substance which is normally in the liquid or solid state and which can be transformed to these states either by increasing the pressure or decreasing the temperature. Examples can include carbon disulfide, gasoline, naphthalene, and iodine.

c. Biologic Aspects of Particulate Matter

The size and surface area of particulate matter play important roles in occupational lung disease, especially the pneumoconioses. The particle diameter associated with the most injurious response is believed to be less than 1 μm; larger particles either do not remain suspended in the air sufficiently long to be inhaled or, if inhaled, cannot negotiate the tortuous passages of the upper respiratory tract. Smaller particles, moreover, tend to be more injurious than larger particles for other reasons. Upon inhalation, a larger percentage (perhaps as much as tenfold) of the exposure concentration is deposited in the lungs from small particles. This additional dosage and residence time act to increase the injurious effect of a particle.

The density of the particle also influences the amount of deposition and retention of particulate matter in the lungs upon inhalation. Particles of high density behave as larger particles of smaller density on passage down the respiratory tract by virtue of the fact that their greater mass and consequent inertia tend to impact them on the walls of the upper respiratory tract. Thus a uranium oxide particle of a density of 11 and 1 μm in diameter will behave in the respiratory tract as a particle of several microns in diameter, and thus its pulmonary deposition will be less than that of a low density particle of the same size.

Other factors affecting the toxicity of inhaled particulates are the rate and depth of breathing and the amount of physical activity occurring during breathing. Slow, deep respirations will tend to result in larger amounts of particulates deposited in the lungs. High physical activity will act in the same direction not only because of greater number and depth of respirations, but also because of the increased circulation rate, which transports the toxic amounts of certain hormones that act adversely on substances injurious to the lung. Environmental temperature also modifies the toxic response of inhaled materials. High temperatures in general tend to worsen the effect, as do temperatures below normal, but the magnitude of the effect is less for the latter.

d. Biologic Aspects of Gases and Vapors

The absorption and retention of inhaled gases and vapors by the body are governed by certain factors different from those that apply to particulates. Solubility of the gas in the aqueous environment of the respiratory tract governs the depth to which a gas will penetrate in the respiratory tract. Thus very little if any of the inhaled, highly soluble ammonia or sulfur dioxide will reach the pulmonary alveoli, depending on concentration, whereas relatively little of insoluble ozone and carbon disulfide will be absorbed in the upper respiratory tract.

Following inhalation of a gas or vapor, the amount that is absorbed into the blood stream depends not only on the nature of the substance but more particularly on the concentration in the inhaled air, and the rate of elimination from the body. For a given gas, a limiting concentration in the blood is attained that is never exceeded no matter how long it is inhaled, providing the concentration of the inhaled gas in the air remains constant. For example, 100 ppm of carbon monoxide inhaled from the air will reach an equilibrium concentration in the blood corresponding to about 13 percent of carboxyhemoglobin in 4 to 6 hours. No additional amount of breathing the same carbon monoxide concentration will increase the blood carbon monoxide level, but upon raising the concentration of carbon monoxide level in the air, a new equilibrium level will eventually be reached.

e. Ingestion

Poisoning by ingestion in the workplace is far less common than by inhalation for the reason that the frequency and degree of contact with toxic agents from material on the hands, food, and cigarettes are far less than by inhalation. Because of this, only the most highly toxic substances are of concern by ingestion.

The ingestion route passively contributes to the intake of toxic substances by inhalation since that portion of the inhaled material that lodges in the upper respiratory tract is swept upwards within the tract by ciliary action and is subsequently swallowed, thereby contributing to the body intake.

The absorption of a toxic substance from the gastrointestinal tract into the blood is commonly far from complete, despite the fact that substances in passing through the stomach are subjected to relatively high acidity and on passing through the intestine are subjected to alkaline media.

On the other hand, favoring low absorption are observations such as the following: (1) food and liquid mixed with the toxic substance not only provide dilution but also reduce absorption because of the formation of insoluble material resulting from the combinatory action of substances commonly contained in such food and liquid; (2) there is a certain selectivity in absorption through the intestine that tends to prevent absorption of "unnatural" substances or to limit the amount absorbed; and (3) following absorption into the blood stream the toxic material goes directly to the liver, which metabolically alters, degrades, and detoxifies most substances.

ACKNOWLEDGMENT

Reprinted from *Occupational Diseases,* Publication No. 1097, U.S. Public Health Service, Washington, D.C., 1964, pp. 7-12.

D. Health Assurance Program

Concern for the effect of the work environment on an employee's health should be the primary purpose of any medical monitoring program. The term "health assurance" not only implies that the health of the participant will be scrutinized, but carries the positive connotation that steps will be taken to eliminate or minimize any negative impact on the employee's health of work-related factors.

1. Basis of a Health Assurance Program*

The laboratory environment, as well as that of many support activities in academic institutions and industrial organizations, has the potential of causing health problems for the employees. Since many laboratories are working on the forefront of knowledge, the persons working in these environs may be exposed to materials for which significant knowledge of the health effects are not yet known. In other cases, the exposures to known materials may be sufficient to cause an adverse effect on an individual. A very positive component of a health and safety program for any industry, but especially those industries for which the risks are incompletely known or where stress on an individual's health may be unusually high, is a health assurance program, i.e., a program in which the health of the individuals is monitored on a regular basis or as a result of any out-of-the ordinary work experience. Not only does such a program serve to alert the facility and the individual of possible health implications of their activities, but normally is viewed positively by the employees as an interest in their welfare.

It is incumbent on universities and/or industries to provide well thought out and imple-mented health assurance programs. The size and complexity of the programs obviously should be geared to the size, type, and activity of the facility. The recognition of the elements which contribute to poor environmental conditions or are harmful to employees is mandatory before control or corrective actions can be formulated and effected.

Progressive corporations and academic institutions are recognizing that a pleasant, safe, healthy environment and contented, healthy employees results in:

---

* This section was written by Richard F. Desjardins, M.D., former Director of the Health Assurance Program at Virginia Polytechnic Institute & State University.

- Decreased employee turnover - less training for new employees
- Increased productivity - both quantity and quality
- Longevity - employees retire later
- Decreased lost time and insurance costs

The corporate health service can encompass many areas - biomechanics, psychology, ergonomics, health promotion, nutrition, CPR, first aid training, industrial hygiene, statistics, epidemiology, and occupational medicine. Related areas which could be in the program or closely tied to a health assurance program could be data processing, cost analysis, insurance evaluation, and workers' industrial compensation and laws. These areas might be addressed to whatever degree applicable by nurses, hygienists, technicians, or physicians. If the corporation is small, or if a large facility is near a specialized clinic, all of these parameters could be handled by that clinic without establishing an in-house health facility, although the organization might have more control over costs with an in-house program.

The interrelated services of a health assurance department must be correlated and administered by a highly motivated, knowledgeable manager. This may vary from a nurse in a small facility to an experienced health services professional in a large facility, such as a university with a multiplicity of situations. In a large factory, the gamut of exposures is relatively static, but in a technically oriented organization with multiple research programs covering many disciplines, there are myriad factors to be considered such as: (1) radiation - isotopes, reactors, X-ray machines, lasers, electron microscopes, etc., (2) diverse chemicals - solvents, pesticides, herbicides, carcinogens, metals, etc., (3) animal handling-research, veterinarian school, and agriculture, (4) genetic recombination-engineering, (5) employees working in power plants, steam tunnels, electric services, plumbing, carpentry, custodial work, etc., (6) air quality in buildings such as gymnasiums, laboratories, laboratory hoods, "energy efficient" buildings, dormitories, and classrooms, (7) emergencies - crowd control, fire control facilities, injuries whether in an athletic performance, fights, drug or alcohol problems, etc.

All of these areas must be dealt with, whether they exist in a university or factory, as they arise in order to maintain a healthy, safe environment.

When environmental problems in the workplace arise, industry and university personnel are free to consult federal and state agencies for advice, to register complaints, or for help in controlling or solving certain conditions. Consultants who are specialists in nearly any given area can be called upon for help from such agencies as NIOSH and OSHA.

Qualifications of health physicists, safety analysts, technicians, and engineers can be obtained from the many texts relating to the organization, function, and running of a good health and safety department. There are many good references with details.[1-5, 13]

a. Health Assurance Medical Departments

It is desirable that the health assurance program be a division of a health and safety department within an organization. Their missions are strongly interdependent. No nurse or physician has the specific knowledge, experience, or facilities to perform the tests, surveil-lance, or control as do the other members of the health and safety department, nor do the latter typically have the medical knowledge to operate independently. However, both groups should be able to obtain the help required from the other.

The size and complexity of the facility ultimately determines the size and complexity of the medical branch of the health and safety department. It is the responsibility of the employer to know when, where, and how to obtain medical personnel to provide medical services for the employees. The AMA (American Management Association) and the American Occupational

Medical Association have plans and pamphlets containing much information to help devise and implement a medical facility.

In a small facility, the number of employees and the nature of the work dictate the extent of health monitoring. Clerical employees should be able to avail themselves of the medical facility, but may not need general physical or specific testing, unless, as is becoming more frequent, air quality problems arise stemming from the construction of energy efficient buildings in recent years. On the other hand, workers exposed to asbestos, toxic chemicals, or radiation would require more complete and frequent examinations, testing, and monitoring. In a small facility without undue usage of toxic chemicals or hazards, a trained nurse, paramedic, or physician's assistant might be qualified to do all that is necessary and refer more complicated problems to a contract physician or hospital.

According to Dr. Marcus Bond,[5,7] a work force of 300 that includes employees who require periodic monitoring and medical examinations can justify an in-plant medical department with a full-time nurse and a part-time physician. For 100 to 300 employees, a part-time nurse, paramedic, or physician assistant might suffice; for 300 to 800 employees, one full-time nurse is usually sufficient. A part-time physician could be on call or spend a specified number of hours at the plant. A group of employees numbering 800 to 1500, including a substantial number who require physical and environmental monitoring, will justify a full-time physician and at least two industrial nurses.

This is an opportunity for a family practice physician[6] to participate in a part-time capacity in industry. It must be of some interest for the physician or else motivation would be missing in the proper medical and psychological care of employees. It is an opportunity to further involve themselves with patients and the functioning community. Statistics indicate that primary care physicians provide about 80% of employee medical care. Only 20% of plant physicians are occupational physicians. In the U.S., financially secure companies continue to have a full-time physician for approximately each 3000 employees, although some industries are known to have only one for about 15,000 employees. Indeed, there are a few substantial companies which have no full-time physician regardless of the number of employees. This might be because of low exposure or toxicity in the facility or because a commercial medical clinic is contracted to care for their employees (HMO, PPO, emergency clinics).[7]

The acceptance of a health program by both management and by those actually manufacturing the product, whether it is paperwork, nuts, bolts, chemicals, or research, is by a slow accrual of satisfied patients. There is an inherent suspicion and skepticism by both areas. Management needs to know that they are complying with federal and state mandates and would like to see a positive productivity gain as a result of health expenditures. The workers find it hard to believe that management would venture a program other than for financial gain. It is gratifying to observe the progression of acceptance by both vital areas in programs that succeed.[8] It has been amply shown that it is cost effective to have a good health program with caring personnel. The employees are healthier, happier, and more productive; the turnover of workers decreases, lessening training expenditures; loss time for illness or injuries decreases both because of attitude and from instruction and training in "wellness;" workers who feel as though they are an integral, functional, and productive part of a facility are more content to stay on the job until retirement. Environmental stress is diminished and managed[14-17]

The physician who accepts the challenge of participation in a health program should obviously be able to perform the usual functions of a general physician, and, in addition, should be knowledgeable in the psychology of workers, the hazards and conditions in varying work sites, toxicology, and communicable diseases in the workplace.[11,12] Recognition of such problems as drug abuse, alcoholism, and the effects of smoking is mandatory. Although this is a very broad background, we must consider that patients/workers are male/female, and young/middle-aged/elderly.

The occupational health examination[8,9] may be divided into three areas:
1. Pre-employment or pre-placement
    a. medical history
    b. occupational history
    c. physical exam
    d. laboratory and X-ray (if needed)
    e. multiphasic screening
2. Periodic examination (with interval history)
    a. annual physical examination or at desirable intervals
    b. executive physicals[10]
    c. toxic or hazardous exposure bioassays
3. Special examinations
    a. food handlers
    b. job transfers (if markedly different)
    c. return to work after serious illness or injury
    d. retirement examinations (document final condition and advise as to future health and wellness)
    e. fitness classification

These areas of assessment (1) determine the immediate health state, any change since previous examinations, and the suitability to work in any area; (2) will suggest advice and modalities to enhance or improve health; and (3) may indicate conditions of stress or unhealthy situations in the environment needing attention and change.[12]

The results of an examination should be kept confidential. A layperson should not be expected to interpret the results and make decisions as to employment on that data. The physician does not hire or advise that a person should be hired. That is a corporate decision. However, the physician can categorize the pre-employee or employee into several levels:

1. Fit for general work - physically and mentally.
2. Fit for work only in specific categories -physically or mentally.
3. Unfit for employment at this time - presence of a medical condition requiring attention. When corrected may be eligible for employment.
4. Incapacitating condition - illness, injury (old or new), or mental illness of a chronic nature. These would prevent employment in either a general category or a specific category.[8]

The validity of the physician's assessment would depend on his knowledge of the required work conditions of the specific facility and his ability in disability assessment.[11] It is also an introduction to the applicant or employee of a caring medical resource within the company or university. The perception of the physician as a "company doc" is really an uncomplimentary epithet. Fostering the perception of a caring physician really interested in the patient, who also happens to work for the establishment, will contribute to a more accurate assessment as well as help maintain healthiness at work.

Many applicants and employees are educationally deficient, but this must not be interpreted as low intellectual capacity. Very often, the appreciative patient will indicate that they had always wondered about a condition, but no doctor had ever taken the time to discuss it in understandable terms. The results of a small amount of expended time are very gratifying.

The initial assessment of an employee should be comprehensive. The elements of a specific company's products should dictate the specific areas to be assessed beyond the general. Obviously, an individual who is to do heavy lifting of any kind should have musculoskeletal and neurological systems carefully examined. Whether X-rays are mandated is still a moot point; they really are of marginal value generally. If there is a suggestion of a problem which would make X-rays of value, either on the physical or from the medical history, X-rays must be taken. On the other hand, exposure to chemicals, gases, heat, cold, radiation, etc., would need more specific scrutiny in other systems. The initial medical history is completed by the applicant prior to seeing the physician, and carefully reviewed in the applicant's presence. Some areas can be amplified and clarified during this time. Encouraging questions by the patient at this time emphasizes caring. Following this, a careful physical examination is carried out.[8] It is probably a good idea to discuss findings on the physical examination as it proceeds. When the examination is completed, proceed to other testing. If specific findings during the physical mitigate against performing certain tests, the routine can be modified. Some tests may be done by the physician, nurse, or a technician. They might include color perception, visual acuity audiometry, spirometry, glaucoma, and electrocardiography. Note that some tests, such as the one for glaucoma, are not normally occupationally related, but initial screening is relatively easy to do and detecting these things will enable the employee to seek additional medical help. There are some tests, such as a blood panel or urinalysis, which require laboratory services. All of these are mainly screening tests. Positive findings may, and probably should, be referred back to the patient's physician or to an appropriate specialist. The industrial physician must not be in competition with outside physicians. By the same token, if the patient signs an authorization, a complete copy of the examination can be forwarded to them or to the physician of their choice. Under ordinary conditions, it is not wise or proper to refer an employee to a specific physician unless that physician is the only one able to perform a given function. A list of qualified names can be provided for referral. This also contributes to harmonious relations with the local medical community.

Subsequent physical examinations of employees should be spaced appropriately to the nature of their jobs. The questionnaire may be abbreviated if they indicate the absence of changes. The physical, however, should be as careful and complete as at the beginning. This will naturally reveal any changing status, i.e., needs glasses or hearing aids, dermatitis, tumors, glaucoma, asbestosis, etc. Finding any deviation from normal early is a real bonus to treatment. By the time some symptoms are obvious to the patient, it may be too late. As an old medical professor once said, "There's a lot of pathology out there. All you have to do is find and recognize it." It is the responsibility of the physician to adhere to three dictums or duties:

1. Prevent disease.
2. Diagnose and treat to the best of your ability.
3. Help the patient's demise to be with as much dignity as possible. Good rapport will allow this.

The industrial physician may want to compose or purchase pamphlets appropriate to the specific facility and leaflets or booklets on general health-promoting ideas: smoking cessation, cholesterol control, back care, weight control, why and how to exercise, etc. How much good these actually do is not well documented, but employees do pick them up and carry them home. Perhaps even small dollops of advice absorbed will contribute to the enhancement of healthfulness. Recently the author's institution added a wellness program available to all of the employees (including retired) and spouses instead of just the occupationally challenged. An excellent response was obtained, which may indicate that there is a substantial level of interest in employer-sponsored health programs, and positive results may improve the overall health of the participants.

The highly motivated industrial physician should have the facilities to address most environmental, factory, and facility problems. If management recognizes and adequately funds a health assurance program, tangible evidence of the health improvement (reflected in decreased loss time) would be found to be cost effective. Intangible evidence is hard to accrue, but well employees are likely to feel more content with their work and remain with the company longer, and are more likely to accept healthful ways of work and pursue similar attitudes at home.

Health assurance is an important element of a well thought out and implemented health and safety program. It is cost effective, humane, and generally good administrative policy to provide such a program.

REFERENCES

1. **LaDou, J.,** Ed., *Introduction to Occupational Health and Safety,* National Safety Council, Chicago, 1986.
2. Council of Occupational Health, AMA: Scope, objectives and functions of occupational health programs, *JAMA,* 174, 533, 1960.
3. Council on Occupational Health, AMA: A management table guide for occupational health problems, *Arch. Environ. Health,* 9, 408, 1964.
4. Guide to Developing Small Plant Occupational Programs, American Management Association, Chicago, 1983.
5. **Bond, M.B.,** Occupational health services for small businesses and other small employee groups, in *Occupational Medic me: Principles and Practical Applications,* Chicago, 89, 1988.
6. **Howe,H.F.,** Small industry: an opportunity for the family physician, *Gen. Pract.,* 26, 166, 1962.
7. **Knight, A.L. and Zenz, C.,** Organization and staffing, in *Occupational Medicine, Principles and Practical Applications,* Chicago, 1988.
8. **Collins, T.R.,** The occupational examination: a preventive medicine tool, *Continuing Med. Ed.,* 77, February 1982.
9. *Guiding Principles of Medical Examinations in Industry,* American Management Association, Chicago, 1973.
10. **Thompson, C.E.,** The value of executive health examinations, *Occup. Health and Safety,* 49, 44, 1980.
11. Disability Evaluation Under Social Security, A Handbook for Physicians, Social Security Administration, Baltimore, August 1973.
12. Occupational Diseases. A Guide to Their Recognition, U.S. Dept. of Health. Public Health Ser., Pub. no. 1097, U.S. Government Printing Office, Washington, D.C., 1966.
13. **Felton, J.S.,** Organization and operation of an occupational health program, *.1. Occup. Med,* 6, 25, 1964.
14. **French,J.R., Caplan, R.D., and Von Harrison, R.,** *The Mechanism of Job Stress and Strain,* Wiley Series of Studies in Occupational Stress, John Wiley and Sons, New York, 1982.
15. **House, J.S., Wells, J.A., Landerman, L.R., et al.,** Occupational stress and health among factory workers, *Health Soc. Behavior,* 20, 139, 1979.
16. **Kahn, R.L.,** Conflict, ambiguity, and overload: three elements in job stress, *Occup. Ment. Health,* 3, 2, 1973.
17. **Selye, H.,** *The Stress of Life,* McGraw-Hill, New York, 1956.
18. Description and Evaluation of Medical Surveillance Programs in General Industry and Construction, Final Report, Office of Regulatory Analysis, Directorate of Policy, Occupational Safety and Health Administration, U.S. Department of Labor, Washington, D.C., 1993.
19. **Stokinger, H.E.,** Means of contact and entry of toxic agents, in *CRC Handbook of Laboratory Safety,* 2nd ed., Steere, N.V., Ed., CRC Press, Cleveland, OH, 1971, 314.

2. A Health Assurance Program

A formal Health Assurance (HA) program should not be intended to replace an environmental monitoring program but to complement one. In an earlier section, some of the limitations of a monitoring program for chemical exposures were mentioned, but the main point, which should be appreciated, is that we do not necessarily know what are "safe" exposure levels. Indeed, safe levels may not exist, only levels in which the negative effects attributable to an exposure disappear into the statistical noise caused by the presence of other parameters. This is especially true for individuals, with their wide range of susceptibilities. In many instances, even this level of knowledge is not available since the data are not available, simply because the studies have not been performed. For example, the ACGIH tables incorporate several hundred chemicals and the OSHA PEL levels for a few less, but this is very small compared to the 50,000 to more than 100,000 (according to various estimates) commercial chemicals already in use, to which must be added hundreds of new ones developed yearly. Further, the number of studies establishing safe levels for synergistic interactions of combinations of materials is virtually nil, as often as not associating the effects of a material with an easily measurable personal habit, such as whether the exposed individual smokes. The prospect of conducting synergistic studies, with all the combinations involved, is obviously not bright.

Regulatory requirements for a number of materials now specifically mandate employee access to an employer-supported medical surveillance program designed to monitor problems associated with the specific chemical. Many authorities recommend participation in programs for anyone who works with toxic substances during the normal performance of their duties. However, there are few specific recommendation on what actually constitutes a significant involvement with toxic materials which should trigger participation in an HA program. This will be discussed further later in this section. The content of the examination will depend to a great extent on the duties associated with the job. However, as noted in the previous section, there are some components of a HA program which are universally agreed upon as essential:

1. A medical history
2. A prior work history
3. A pre-placement examination
4. Periodic reexaminations
5. An end of employment examination

The entire program should, in addition to a number of standard components, be tailored to the anticipated types of exposures. As these exposures change during alterations in the research program, the periodic reexaminations can be modified to reflect the changing conditions.

Even before the HA program is initiated, there are a number of key ethical issues which must be addressed. If participation in a HA program is to be required as a condition of employment, the advertisement for position should so state. Further, the examination must be clearly intended to determine if the duties would be such as to make it unsafe or very difficult for the employee to perform the work or would aggravate an existing health problem. If reasonable adjustments can be made in the duties or responsibilities, then the examination cannot be used to discriminate against an otherwise qualified applicant. This issue is basically the premise of the Americans With Disabilities Act. (Note that a U.S. Supreme Court Decision just made, in June 1999, says that if the condition causing disability is readily correctable, such as by wearing glasses, the employer can not be held liable for discrimination).The employee should have an assurance of confidentiality. There are factors which may be health related, but which have no bearing on the ability of an employee to do the work assigned, and which cannot harm those with whom the employee would come into contact. The employee has the right to expect that any such information remain confidential. Finally, the employee should have access to the results of the examination and any tests performed and should be able to authorize release of the data to others, such as the persons family physician, if they so desire.

a. Participation

If the organizational approach is to provide an examination based on need, then the necessity arises to define criteria as to who should be included and who should not. Individuals who do not work with chemical or biologic agents or where their duties are not unduly stressful or physically hazardous can be justifiably excluded for these reasons. On the other hand, in the first few years of operation of an HA program at the author's institution, approximately 20% of the participants were found to have medical problems of which they were not aware and which should be treated. In such cases, they were encouraged to seek medical assistance from their family physician. If this is typical of the typical employee population, it could well be to the employer's benefit to screen all prospective employees to secure a healthier and more productive workforce. However, since the intent of a HA program in the current context, as applied to research personnel, is to monitor the impact of chemicals or pathological organisms on the health of the individual (either directly or indirectly, e.g., wearing a respirator can place a burden on a person with impaired pulmonary function), exposure to chemicals and biologically active (to humans) agents should be a major factor to be considered in the participation of an individual in a HA program. OSHA requires access to a medical program for persons working with regulated carcinogens or who are exposed to human blood, tissues, and other fluids, and those required to wear respiratory protection, among others (a list of the OSHA sections requiring medical surveillance programs is provided as an appendix to this section). However for other substances the toxicity of the material, the mechanism of exposure, the duration of the exposure, the intensity of the exposure, the safeguards available to prevent exposure, the current state of an individual's health, and prior exposures all play a part in the decision.

If an individual is working with an agent which is significantly infectious to humans, there appears to be little question that participation in a medical program is needed. Although the probability of contracting a disease increases with higher exposure rates, once contracted, the characteristics of the disease are not dependent upon continued exposure or the initial level of exposure.

If a major portion of an individual's time is spent working with a regulated carcinogen, other regulated materials such as lead or cotton dust, or other materials which meet the criteria for being highly toxic, corrosive, a sensitizer, or an irritant, then again it is usually required or desirable for the individual to be in a medical program. Even if facilities are available, such as totally enclosed glove boxes in which the work is done, it is arguable that unplanned exposures could occur, and the conservative approach would be to include rather than exclude the person. It could be argued also that if the exposure levels are maintained sufficiently low, then participation in a program is not needed. It is on this basis that OSHA defines exempt levels for meeting some of the regulatory requirements for some of the regulated carcinogens. However, documentation of the low levels would appear to be required to deny access of an individual to a medical surveillance program on this basis. Of course, participation in a HA program is clearly indicated if a monitoring program provides information indicating that the individual uses materials of concern or is in an area where others use them and is actually exposed to airborne concentrations which are typically a significant percentage of acceptable levels.

Where the use of respiratory protection is indicated, then the OSHA standard for respiratory protection in 29 CFR 1910.134(c) includes the statements: (1) In any workplace where respirators are necessary to protect the health of the employee, the employer shall establish and implement a written respiratory protection program with work site specific procedures. ... The employer shall include in the program the following provisions...(ii) Medical evaluations of employees required to use respirators; and in 29CFR 1910.134(e) is the following:

"Medical evaluation. Using a respirator may place a physiological burden on employees that varies with the type of respirator worn, the job and workplace conditions in which the respirator is used, and the medical status of the employee. Accordingly, this paragraph specifies the minimum requirements for medical evaluation that employers must implement to determine the employee's ability to use a respirator."

(1)   General. The employer shall provide a medical evaluation to determine the

employee's ability to use a respirator, before the employee is fit tested or required to use the respirator in the workplace. The employer may discontinue an employee's medical evaluations when the employee is no longer required to use a respirator.

(2) Medical evaluation procedures.

  (a) The employer shall identify a physician or other licensed health care professional (PLHCP) to perform medical evaluations using a medical questionnaire or an initial medical examination that obtains the same information as the medical questionnaire.

  (b) The medical evaluation shall obtain the information requested by the questionnaire in Sections 1 and 2, Part A of Appendix C of this section.

(3) Follow-up medical examination.

  (a) The employer shall ensure that a follow-up medical examination is provided for an employee who gives a positive response to any question among questions J. through 8 in Section 2, Part A of Appendix C or whose initial medical examination demonstrates the need for a follow-up medical examination.

  (b) The follow-up medical examination shall include any medical tests, consultations, or diagnostic procedures that the PLHCP deems necessary to make a final determination.

(4) Administration of the medical questionnaire and examinations.

  (a) The medical questionnaire and examinations shall be administered confidentially during the employee's normal working hours or at a time and place convenient to the employee. The medical questionnaire shall be administered in a manner that ensures that the employee understands its content.

  (b) The employer shall provide the employee with an opportunity to discuss the questionnaire and examination results with the PLHCP.

(5) Supplemental information for the PLHCP.

  (a) The following information must be provided to the PLHCP before the PLHCP makes a recommendation concerning an employee's ability to use a respirator:

  (b) The type and weight of the respirator to be used by the employee;.

  (c) The duration and frequency of respirator use (including use for rescue and escape);

  (d) The expected physical work effort;

  (e) Additional protective clothing and equipment to be worn; and

  (f) Temperature and humidity extremes that may be encountered.

  (g) Any supplemental information provided previously to the PLHCP regarding an employee need not be provided for a subsequent medical evaluation if the information and the PLHCP remain the same.

(6) The employer shall provide the PLHCP with a copy of the written respiratory protection program and a copy of this section.

Note to Paragraph (e) (5) (iii): When the employer replaces a PLHCP, the employer must ensure that the new PLHCP obtains this information, either by providing the documents directly to the PLHCP or having the documents transferred from the former PLHCP to the new PLHCP. However, OSHA does not expect employers to have employees medically reevaluated solely because a new PLHCP has been selected.

(7) Medical determination. In determining the employee's ability to use a respirator, the employer shall:

  (a) Obtain a written recommendation regarding the employee's ability to use the respirator from the PLHCP. The recommendation shall provide only the following information:

  (b) Any limitations on respirator use related to the medical condition of the employee, or relating to the workplace conditions in which the respirator will be used, including whether or not the employee is medically able to use the respirator;

(c) The need, if any, for follow-up medical evaluations; and

(d) A statement that the PLHCP has provided the employee with a copy of the PLHCP's written recommendation.

(e) If the respirator is a negative pressure respirator and the PLHCP finds a medical condition that may place the employee's health at increased risk if the respirator is used, the employer shall provide a PAPR (powered air supplied respirator) if the PLHCP's medical evaluation finds that the employee can use such a respirator; if a subsequent medical evaluation finds that the employee is medically able to use a negative pressure respirator, then the employer is no longer required to provide a PAPR.

(8) Additional medical evaluations. At a minimum, the employer shall provide additional medical evaluations that comply with the requirements of this section if:

(a) An employee reports medical signs or symptoms that are related to ability to use a respirator;

(b) A PLHCP, supervisor, or the respirator program administrator informs the employer that an employee needs to be reevaluated;

(c) Information from the respiratory protection program, including observations made during fit testing and program evaluation, indicates a need for employee reevaluation; or

(d) A change occurs in workplace conditions (e.g., physical work effort, protective clothing, temperature) that may result in a substantial increase in the physiological burden placed on an employee.

Persons who do not have a continued exposure to chemicals, but periodically perform tasks requiring intense uses of chemicals for a brief period, such as in agricultural field experimentation, should probably be included in a HA program. Not only are many agricultural chemicals quite toxic, but the working conditions place severe physiological stress on the research personnel and their support staff. Respirators should be worn, as should clothing which will not be permeable to the chemical sprays. Respirators place stress on the pulmonary and cardiac system; protective clothing which is impermeable to fumes and vapors usually does not transpire either, and the body temperature will rapidly increase since the clothing prevents heat from being carried away from the body by evaporation, conduction, or convection of perspiration.

Persons who have known health problems that could be aggravated by the exposures involved in their job duties or who have had prior work histories where they could have had significant exposures to chemicals that could have sensitized them to chemicals in the workplace or that could have initiated delayed effects also would fall in a category which should be considered for participation in a HA program.

It is most difficult to determine whether persons for whom the exposures are marginal, i.e., where the portion of their duties in which they use chemicals is limited, but they do use materials part of the time with properties that could cause ill effects should be included in a HA program. It would be easy to establish a criterion that any use whatsoever should qualify a person for participation. However, many activities of normal everyday life involve use of such items as gasoline and household products containing toluene, acetone, phenol, isopropyl alcohol, ethyl alcohol, hydrogen peroxide, acid, and caustic materials such as lye, which certainly are toxic materials. The "any use" criterion is undoubtedly too liberal, unless one simply admits that there are no selection criteria and includes every employee. A compromise which appears reasonable, but which has no other scientific justification, is to arbitrarily select a percentage (such as 10% of a typical work week) for actual use or exposure to a chemical or combination of chemicals of average health risk as a threshold. An employee approximating an exposure of this level could be asked to fill out a form listing the chemicals which are in use in their vicinity and an estimate of the average time each are used. This form should be reviewed by a physician (preferably with a background in environmental and occupational medicine, if one is available) and his recommendation should govern the question of participation or not. However, if an individual wishes to be included, but who might not be recommended, probably should be permitted to do so.

Although every person who has duties which could give rise to health problems should be a participant, it is especially critical to include permanent employees. Many of the tests which are run on the individual have a sufficiently wide "normal" range (except in extreme cases of acute exposure, where the individual should definitely receive medical attention anyway) that a single examination may not be particularly informative. However, problems due to environmental work conditions, as will be discussed later, may be revealed by trends shown by comparison of successive examinations.

b. Medical and Work Histories

The medical history and prior work histories are key components of any health assurance program. There are any number of health-related factors for which a heredity predisposition exists, so that the medical history will normally include a segment concerning family members, particularly parents. Obviously, known prior medical conditions will be of importance. Emphysema, for example, would certainly be of concern if the employee were to have to wear a respirator frequently during the course of his duties. Hypertension and heart problems would clearly be of importance if the job involved significant physical stress, and of course there are chemicals which directly or indirectly affect the heart function. Medical history forms vary substantially in content but one used for a HA program should be comprehensive. It is part of a record which, along with the prior work history and the actual examination, including tests which may be run, will constitute the baseline against which changes in the employee's health will be compared to determine if occupational exposures are having a negative effect on the employee's health. Some prospective employees are inclined to conceal previous illnesses if it is likely to affect their chances of obtaining a desired position. This is unfortunate but quite understandable. It is important, if these conditions later manifest themselves, that they have been detected, if possible, by appropriate questions or during the pre-placement examination. The need to do so stems from a desire (1) to avoid responsibility for an occupational exposure causing the diseases and (2) to explain any problems which do develop on the possible basis of occupational exposures experienced by the individual.

The prior work history serves essentially the same purpose as does the medical history. For example, a prospective employee who would be working in agricultural research programs might have had a previous period of employment working with pesticides and herbicides, which could have had a depressant effect on his cholinesterase enzyme levels. It would be important to include a test of this parameter in the pre-placement examination. Even some nonchemical activities, such as previous work in a heavily dust-laden atmosphere, might have caused a decrease of pulmonary function to an extent that an individual might find it impossible to wear a respirator to provide protection against solvent fumes. Previous exposures to some chemicals or substances may result in effects delayed for many years, such as the latency periods generally associated with carcinogens. Among other substances for which any history of prior exposures might be elicited would be asbestos, dusts, welding fumes, heavy metals, pesticides; herbicides; acids, alkalis, solvents, dyes, inks, paints, paint thinners, paint strippers, gases, radiation, etc. If there are any specific areas of concern because the work regimen will involve materials known to have a possible impact on a given physiological function, or organ, then the physician should supplement the standard questionnaire for both the medical and work histories with questions designed to elicit as much relevant information as practicable.

There are any number of common diseases which are not necessarily related to an occupational exposure but which could be a risk in the work environment. Diabetes and hypertension are certainly not necessarily work related, but the individual, unless treated, could be a hazard to one's self and potentially to co-workers. Similarly, a disease of the eyes, such as glaucoma, could interfere with a person's ability to see properly, but many persons could be unaware of its onset as it is an insidious disease, primarily a problem to persons over 40. A relatively simple automatic instrument is available to detect pressure increases in the eyes, which is a sign of the disease, and which would permit the physician to refer persons to an ophthalmologist. Loss of hearing could be a problem if persons do not hear warnings and, again, many persons do not realize that this has become a problem or are reluctant to admit it, even to themselves, as a sign of increasing age. Although these problems should have been brought to the attention of the individuals by their family physician, a surprising number of persons do not have a family physician or do not see one frequently. As noted earlier, at the author's institution, approximately 15- to 20% of the persons participating in the HA program for the first time had reasonably serious problems of which they were unaware, and which could have placed them at risk, or at best, reduced their efficiency and productivity. The scope of the examination should be sufficient to detect these conditions which may not be job related.

c. Pre-Placement Examination

It would be highly desirable if a pre-placement exam could be given prior to any work exposure to provide a true baseline for the individual. However, unless a medical examination has been an integral part of an organization's employment procedure since the inception of the company or institution, then instituting a HA program will always catch a number of current employees already in the midst of research programs involving exposure to hazardous materials. A medical examination at this time will still have significance in the sense that future examinations can still be compared to the earlier one to detect changes during the interval between examinations. However, the information gained in the exam, including any test results, will not necessarily reflect the normal conditions for the employee. If, for example, an individual has been working within the organization using agricultural chemicals and has a very low level of cholinesterase enzyme at the time of the initial examination, it may be suspected that the employee's work has caused the depressed level of the enzyme to occur, but it is uncertain. The individual may be one who has a naturally low level. If a person is tested at the time of initial employment, then the effect of the working environment on the parameters measured in the examination will be much more apparent, although the effect of work exposures on an individual may be confused if similar exposures are likely to occur outside the workplace. In the example just used, if the initial examination revealed a normal enzyme level and a later one showed a depressed value, perhaps after a suspected exposure, then the initial conclusion, barring any alternate exposure mechanism, would be that an exposure had occurred and remedial steps taken to prevent further exposures and to prevent future incidents of the same kind. Where alternate exposure conditions exist outside the workplace which could have caused the same conditions, then there could be problems for the employee receiving financial compensation for the problems, such as workmans compensation.

The other major purpose of a pre-placement examination would be to avoid placing a person in a position in which an existing condition would be aggravated or the individual could be injured by the work environment. A color-blind person, for example, should not be placed in a position in which the ability to distinguish colors is essential to being able to work safely. An individual with severely reduced pulmonary function should not be placed in a position requiring wearing a respirator for protection. These restrictions may make it impossible for an applicant to be offered a position, and it should be clearly stated in the advertised job qualifications in such a case that passing a pre-placement examination is required as a condition of employment. A byproduct of such a restriction is that the organization may be protected against acquiring a future liability if, for example, a person with depressed pulmonary function is hired without an examination and placed in a situation in which exposures could cause the same result, it could be difficult to prove that the problem did not arise from a recent exposure. On the other hand, detection of the problem in the pre-placement examination might lead to a decision not to hire the individual because of the problem.

There are some pitfalls in using the pre-placement examination as an exclusionary device. This was alluded to in Section VIII.C of this chapter, in which a cautionary flag was raised against using it as a discriminatory device. This can occur with the best intentions in the world. An organization may decide to exclude women of fertile age from a position in which they may be exposed to a teratogen. Discrimination may be claimed if installation of engineering controls to reduce the exposures to well below the permissible limits are feasible, but not done. A woman may decide on her own not to work in an area where even low levels of an embryo toxin are present, but the decision should be clearly her own with no taint of coercion. In the recent revision to workplace rules involving radiation by the NRC, a provision has been added making it the employer's responsibility to limit the exposure to the unborn fetus, but before the employer can take the required steps, the woman has to herself declare her condition. If she should choose not to do so, no matter how obvious it is, the employer may not be permitted to take the necessary steps.

A fairly common practice in a pre-placement medical examination program, in addition to a thorough physical and a battery of tests, is to take a serum sample to be stored in an ultra-low temperature freezer. These samples take up very little space and are valuable should a question later arise where a comparison between a current serum specimen and a baseline sample would be useful. It is also possible and feasible to lyophilize the serum for storage. This might be cost effective and space saving if a large number of specimens are to be kept.

d. Reexamination

Periodic reexaminations should be scheduled for all participants of a HA program, whether it is a part of a program mandated by a standard, as is becoming more common in newer regulations, or as a result of an internal decision based upon the level of usage. Note that the new laboratory safety standard does not call for periodic examinations but only requires access to a medical surveillance program on an "as needed" basis should an over-exposure occur or might have occurred. The frequency of the reexamination need not be any fixed interval, but should be based on the level of exposure. Returning to the use of pesticides, which could cause a depression in the cholinesterase enzyme level, as an example, it might be desirable to test for this one component prior to a period of active use, a second time at the height of the spraying season, and again at the end of the period of activity (assuming the material is significantly dangerous to humans). For less toxic pesticides, this amount of testing might be excessive, but for intensive use of an exceptionally dangerous material such as parathion, it might be desirable to test daily. However, normally there would be no reason to perform a complete examination at an accelerated schedule such as this.

An annual testing interval between scheduled examinations is probably the one most often used in HA programs for individuals with typical exposures in a representative laboratory. However, for persons only marginally meeting requirements for participation, the interval between examinations might be extended to 2, 3, or 5 years. Some programs use a 5-year interval for a complete physical, but recommend special tests more often. The National Institute of Health (NIH), in their program for their animal handlers, recommends taking a new serum sample every 5 years, but does not require a complete physical each time. The medical advisor or occupational physician should evaluate the requirements for each participant to establish the optimum period between examinations.

After each examination, the physician should compare the results of the current examination to the findings of previous examinations. Except in isolated instances, such as the depression of the cholinesterase enzyme which we have been using as an example, or unless there has been a severe exposure in which acute effects might be anticipated, the primary means of detecting problems will be the comparison between the results of successive examinations. Changes in various parameters which have been measured, such as pulmonary function, might vary slightly between two successive tests, but a persistent trend toward poorer performance would indicate damage to the respiratory system. Similarly, should a persistent trend develop for the other parameter measured, the examining physician should discuss the work environment and other possible contributing factors, such as leisure time activities, with the employee. In at least one instance, a spraying program to control insects at a cottage where weekends were spent was a major threat to an individual's health rather than any personal problem or exposure from any other source. The patient did not mention this factor to the physician because he did not recognize it as a potential problem. As a result, the examining physician had major problems identifying the cause of the individual's illness and was unable to treat it successfully. By the time the problem was recognized, the patient had been highly sensitized to any similar material and had some long-term health problems, that affected his capacity to perform many activities.

e. Utilization of Results

The primary purpose of the examination is to protect the employee, with a secondary purpose being to help protect the organization from the liability associated with unwittingly allowing an individual to become ill due to the work environment. As noted earlier, a substantial number of persons involved in a HA program may have existing problems which are not job related or, as a normal course of events, develop health problems which are clearly not job related. These may be detected during the HA examination, as readily as in any other comparable comprehensive physical examination. Some organizations will assume direct responsibility for treating these illnesses, although most do not, leaving the burden of seeking treatment on the patient. Financially, there is only a moderate difference to the patient in many cases, due to the wide availability of group health insurance plans, although the increasing costs of health care has caused many employers to partially shift costs back to the employees. If the patient has the responsibility of seeking out medical treatment, however, the condition may remain untreated, although the examining physician should certainly encourage the individual to seek assistance. In such cases, the employees should have the right to authorize the release of the medical records to their own physicians and to have this done promptly by the organization for whom they work.

Where the physical examination reveals a medical condition which may be job related or is aggravated by the duties of the person's job or the environment in which the individual works, steps should be taken to protect the employee's health. One of the first things to consider is to confirm that the condition exists or to obtain additional data to better understand the problem by seeking additional tests, obtaining a second opinion, or referring the patient to a specialist. These options should be discussed with the patient. In some cases, the situation is sufficiently straightforward so that these follow-up steps would not be necessary.

Whether one postpones gathering supportive data from additional examinations depends somewhat upon the seriousness of the problem which has been discovered and the work situation. The physician, in consultation with the individual, his supervisor, and usually a representative of the department with overall responsibility for the organization's health and safety program should meet to see what temporary steps can be taken to reduce the risk to the employee. In some cases, the head of the department in which the employee works may have to become involved if the supervisor does not have the authority or the flexibility to make changes.

Once all the data are obtained, consideration of the options that are available to protect the employee should be carefully reviewed. A number of these should be routinely considered.

1. If the condition can be treated, a temporary change in duties may be all that is needed.
2. It may be feasible to make engineering changes to modify the work environment.
3. Personal protective equipment or safety devices can be used to reduce an individual's exposure if engineering changes are not practicable.
4. It may be possible to change the job activity.
5. Job responsibilities may be distributed differently among personnel in the facility if the person has a unique problem and if the duties causing the difficulties are not a problem to the others.
6. It may be possible to reschedule the activity causing the problem to another time or to modify the affected individual's schedule.
7. If there are no suitable options available within the individual laboratory, then relocation within the organization should be considered.

Some of these options are more easily applied in the industrial environment than in the typical academic laboratory, where each person may be supported on a grant and each laboratory is nearly autonomous. There may be very little flexibility available to the laboratory supervisor or laboratory director. This makes the task of treating the employee fairly much more difficult, since the work causing the problem usually must be done, and the laboratory supervisor does not have the funds to hire a new person and provide work to justify keeping the original employee as well. There may, in fact, be little flexibility of any kind if there are no positions available at the time for which the individual is qualified or is willing to accept. However, every avenue must be explored, because it is not permissible to maintain the individual in a situation in which their health may be endangered, even if the person wishes to do so. A waiver of responsibility for any future problems by the corporation or institution signed by the employee is not an acceptable alternative, nor is it likely to be legally defensible.

In extreme cases in which every option has been examined and none are feasible, individuals may have to cease to work in the organization either by resigning or being terminated for their own protection. An employee relations specialist in matching persons to jobs as well as an individual charged with seeing that employees are not discriminated against should be brought into the situation well before this drastic step is considered. In such a case a financial severance settlement, insurance such as workers compensation, or disability retirement options may be available to the employee.

f. Physician Training

Any physician involved in a health assurance program will have had the usual training and exposure to a variety of medical experiences. It also is highly desirable for the individual to have specific training in industrial medicine. Since the actual conditions of employee exposure to hazardous materials will differ with each organization, the physician should be sufficiently familiar with the types of exposures represented by the job descriptions of the employees to be able to apply his own expertise and experience to the potential exposures. The more complex and diversified the research programs in an organization, the more difficult this task will be, and unfortunately, there are relatively few physicians trained as occupational physicians. It probably would be desirable for the physician to set aside some time to visit the various research areas, and visit with both the supervisors and individual employees.

Since OSHA requires that a medical surveillance program be made available to employees working with regulated carcinogens and a number of other materials, the physician should be provided with all current information related to these standards as well as appropriate technical information relating to these materials and other hazardous materials used by the employees. The physician, should, for example, have access to a set of all current MSDSs for the chemicals used by the employees. Subscriptions should be provided to some of the excellent services available to keep track of the rapidly changing regulatory and technical information, as well as the usual medical journals. The physician should have an opportunity to attend relevant workshops, seminars, and professional meetings to ensure that his background is maintained at a high standard.

Finally, the way in which the employees perceive the physician is an extremely important component of an organization's health and safety program. He should be perceived as professionally capable. It also is important that the employees do not perceive him as a "company" man. They must feel that their health is important to the physician and that if they are having a problem on the job, the physician is concerned about it for their sake, not because it will cause a problem for the organization. Certainly the physician should be concerned about the welfare of the organization, but this can be done by working to make sure that the health and safety of the organization's employees is protected. This is one reason, as noted earlier, why the name "health assurance program" is recommended over "medical surveillance program." The former has a much more positive sound than does the latter. Since a pre-placement medical examination is recommended for individuals exposed to hazardous materials, the physician has a superb opportunity to establish from the beginning that the company or institution is concerned about the employee's well-being.

g. Records

Because many materials are now known to have long term effects and extended latency periods are known to exist for many carcinogens, it would be desirable to maintain all records pertaining to the medical examinations as well as those relating to exposures and monitoring for an extended period, even after the employee had left the organization. Many of the specific OSHA standards describe the records which must be maintained and the period for which the records must be kept. However, 29 CFR 1910.20 covers the topic of health and safety records in general. Some of the more, critical portions of this section are given below. Note that some of the language is omitted (indicated by ...) where it was felt to be non-essential. Some of the requirements are very detailed and demanding. The reader is referred to the OSHA Standards for General Industry for the complete version of the requirements for record maintenance in the event that changes were to be made.

Access to employee exposure and medical records

(a) Purpose: The purpose of this section is to provide employees and their designated representatives a right of access to relevant exposure and medical records, and to provide representatives of the Assistant Secretary a right of access to these records in order to fulfil responsibilities under the Occupational Safety and Health Act....

(2) This section applies to all employee exposure and medical records, and analyses thereof, of employees exposed to toxic substances or harmful physical agents, whether or not the records are related to specific occupational safety and health standards.

(4) "Employee" means a current employee, a former employee, or an employee being assigned or transferred to work where there will be exposure to toxic substances or harmful physical agents. In the case of a deceased or legally incapacitated employee, the employee's legal representative may directly exercise all the employee's rights under this section.

(5) "Employee exposure record" means a record containing any of the following kinds of information concerning employee exposure to toxic substances or harmful physical agents:

(i) environmental (workplace) monitoring or measuring, including personal, area, grab, wipe, or other form of sampling, as well as related collection and analytical methodologies, calculations, and other background data relevant to interpretation of the results obtained;

(ii) biological monitoring results which directly assess the absorption of a substance or agent by body systems (e.g., the level of a chemical in the blood, urine, breath, hair, fingernails, etc.) but not including tests which assess the biological effect of a substance or agent;

(iii) material safety data sheets; or

(iv) in the absence of the above, any other record which reveals the identity (e.g., chemical, common, or trade name) of a toxic substance or harmful physical agent.

(6)(i) "Employee medical record" means a record concerning the health status of an employee which is made or maintained by a physician, nurse, or other health care personnel, or technician, including:

(A) medical and employment questionnaires or histories (including job description and occupational exposures),

(B) the results of medical examinations (pre-employment, pre-assignment, periodic or episodic) and laboratory tests (including X-ray and all biological monitoring),

(C) medical opinions, diagnoses, progress notes, and recommendations.

(D) descriptions of treatments and prescriptions, and

(E) employee medical complaints.

(ii) "Employee medical records" does not include the following:

(A) physical specimens (e.g., blood or urine samples) which are routinely discarded as a part of normal medical practice, and are not required to be maintained by other legal requirements,

(B) records containing health insurance claims if maintained separately from the employer's medical program and its records, and not accessible to the employer by employee name or other direct personal identifier (e.g., social security number, payroll number, etc.), or

(C) records concerning voluntary employee assistance programs (alcohol, drug abuse, or personal counseling programs) if maintained separately from the employer's medical program and its records.

(7) "Employer" means a current employer, a former employer, or a successor employer.

(8) "Exposure" or "exposed" means that an employee is subjected to a toxic substance or harmful physical agent in the course of employment through any route of entry (inhalation, ingestion, skin contact or absorption, etc.) and includes past exposure and potential (e.g., accidental or possible) exposure, but does not include situations where the employer can demonstrate that the toxic substance or harmful agent is not used, handled, stored, generated, or present in the workplace in any manner different from typical non-occupational situations.

(9) "Record" means any item, collection, or grouping of information regardless of the form or process by which it is maintained (e.g., paper document, microfiche, microfilm, X-ray film, or automated data processing).

(d) Preservation of records. (1) Unless a specific occupational safety and health standard provides a different period of time, each employer shall assure the preservation and retention of records as follows:

(i) Employee medical records. Each employee medical record shall be preserved and maintained for at least the duration of employment plus thirty (30) years, except that health insurance claims records maintained separately from the employer's medical program and its records need not be retained for any specified period;

(ii) Employee exposure records. Each employee exposure record shall be preserved and maintained for at least thirty (30) years, except that:

(A) Background data to environmental (workplace) monitoring or measuring, such as laboratory reports and worksheets, need only be retained for one (1) year so long as the sampling results, the collection methodology (sampling plan), a description of the analytical and mathematical methods used, and a summary of other background data relevant to interpretation of the results obtained, are retained for at least thirty (30) years; and

(B) Material safety data sheets and paragraph (c)(5)(iv) records concerning the identity of the substance or agent need not be retained for any specified period as long as some record of the identity (chemical name if known) of the substance or agent, where it was used, and when it was used is retained for at least (30) years; and

(iii) Analyses using exposure or medical records. Each analysis using exposure or medical records shall be preserved and maintained for at least thirty (30) years.

(e) Access to records. (1) General. (i) Whenever an employee or designated representative requests access to a record, the employer shall assure that access is provided in a reasonable time, place, and manner, but in no event later than fifteen (15) days after the request for access is made.

(ii) Whenever an employee or designated representative requests a copy of a record, the employer shall, within the period of time previously specified, assure that either:

(A) a copy of the record is provided without cost to the employee or representative,

(B) the necessary mechanical copying facilities (e.g., photocopying) are made available without cost to the employee or representative for copying the record, or

(C) the record is loaned to the employee or representative for a reasonable time to enable a copy to be made.

Employers can charge reasonable direct expenses for additional copies of records, except that a certified collective bargaining agent for the employee can receive a copy without cost, and if new information is added to the record, this information is available to the employee without cost under the same conditions as the original record.

For certain medical records, there is protection for the privacy of the individual in making records available:

Section (e)(2)(ii)(E) Nothing in this section precludes a physician, nurse, or other responsible health care personnel maintaining employee medical records from deleting from the requested medical records the identity of a family member, personal friend, or fellow employee who has provided confidential information concerning an employee's health status.

and, under "Analyses using exposure or medical records:"

Section (e)(2)(iii)(B) Whenever access is requested to an analysis which reports the contents of employee medical records by either direct identifier (name, address, social security number, payroll number, etc.) or by information which could reasonably be used under the circumstances indirectly to identify specific employees (exact age, height, weight, race, sex, dates of initial employment, job title etc.), the employer shall assure that personal identifiers are removed before access is provided. If the employer can demonstrate that removal of personal identifiers is not feasible, access to the personally identifiable portions of the analysis need not be provided.

New employees have certain rights concerning records from the beginning of their employment.

(g) Employee information. (1) Upon an employee's first entering into employment, and at least annually thereafter, each employer shall inform employees exposed to toxic substances or harmful physical agents of the following:

(i) The existence, location, and availability of any records covered by this section;

(ii) the person responsible for maintaining and providing access to records; and

(iii) each employee's rights of access to these records.

Corporations are often bought out, merge with other firms, or cease to operate, and provision is made in the standard for the retention of records for the required periods by transfer of the records to the successor firm, or under requirements of specific standards to NIOSH. This is rarely a problem for academic research institutions which seldom cease to operate, although semiautonomous components which retain their own records may cease to exist. In such a case, their records should be subsumed into those of the parent institution.

h. CPR and First Aid Training

Subpart K - Medical and First Aid, 29 CHR 1910.151 of the OSHA Standards for General Industry describes the minimal medical care which must be available to employees, although there are references to first aid in several other sections of the standards. This short section is given below in its entirety:

Sec. 1910.151 Medical services and first aid.

(a) The employer shall ensure the ready availability of medical personnel for advice and consultation on matters of plant health.

(b) In the absence of an infirmary, clinic, or hospital in near proximity to the workplace which is used for the treatment of all injured employees, a person or persons shall be adequately trained to render first aid. Adequate first aid supplies shall be readily available.

(c) Where the eyes or body of any person may be exposed to injurious corrosive materials, suitable facilities for quick drenching or flushing of the eyes and body shall be provided within the work area for immediate emergency use.

Appendix A to Sec. 1910.151--First aid kits (Non-Mandatory)

First aid supplies are required to be readily available under paragraph Sec. 1910.151(b). An example of the minimal contents of a generic first aid kit is described in American National Standard (ANSI) Z308.1-1978 "'Minimum Requirements for Industrial Unit-Type First-aid Kits." The contents of the kit listed in the ANSI standard should be adequate for small worksites. When larger operations or multiple operations are being conducted at the same location, employers should determine the need for additional first aid kits at the worksite, additional types of first aid equipment and supplies and additional quantities and types of supplies and equipment in the first aid kits.

In a similar fashion, employers who have unique or changing first-aid needs in their workplace may need to enhance their first-aid kits. The employer can use the OSHA 200 log, OSHA 10P1s or other reports to identify these unique problems. Consultation from the local fire/rescue department, appropriate medical professional, or local emergency room may be helpful to employers in these circumstances. By assessing the specific needs of their workplace, employers can ensure that reasonably anticipated supplies are available. Employers should assess the specific needs of their worksite periodically and augment the first aid kit appropriately.

If it is reasonably anticipated that employees will be exposed to blood or other potentially infectious materials while using first aid supplies, employers are required to provide appropriate personal protective equipment (PPE) in compliance with the provisions of the Occupational Exposure to Bloodborne Pathogens standard, Sec. 1910.1030(d)(3)(56 FR 64175). This standard lists appropriate PPE for this type of exposure, such as gloves, gowns, face shields, masks, and eye protection.

[39 FR 23502, June 27, 1974, as amended at 63 FR 33466, June 18, 1998]

Effective Date Note: At 63 FR 33466, June 18, 1998, Sec. 1910.151 was amended by revising the last sentence of paragraph (b) and by adding appendix A to the section, effective Aug. 17, 1998.

The need for access to medical services in emergencies was discussed at some length in Chapter 1. Prompt action can frequently save an individual's life or can significantly reduce the seriousness of injuries. Although not intended as an instruction manual, Chapter 1 presented some first aid procedures for accidents involving chemicals and CPR techniques. It would be highly desirable for individuals working in facilities where hazards are present to be trained in both of these subjects. By working carefully so as not to tempt fate too much, and with a great deal of luck, an individual may go through an entire working career without personally experiencing an accident or being present when someone else does, but this cannot be counted upon. Although you cannot perform CPR on yourself, and you may be incapacitated so that even simple first aid is beyond you, if enough personnel in a laboratory do make the effort to become trained, it is likely that someone will be available to start emergency aid while waiting for more skilled personnel to arrive. The institution at which the author worked had an in-house volunteer rescue squad and the members of this squad as well as the town squad were scattered throughout the university. Although the rescue squad was normally present within 3 to 4 minutes, on several occasions these on-scene personnel were instrumental in ameliorating severe accidents prior to the squad's arrival. The training for basic first aid and single-person CPR is not difficult, and everyone should annually devote the few hours necessary to receive and maintain these skills. Many rescue squads, fire departments, hospitals, and other public service agencies offer the training at a minimal fee covering only the cost of the manuals and supplies.

i. Vaccinations

All of us as children probably received some vaccinations against a number of diseases. A number of common diseases afflicting children who were born in the first third of this century are now decreasing in frequency as a result of widespread vaccination programs. A recent controversy centered around whether the last smallpox virus in the world, being maintained in a laboratory, should be destroyed. Yet this used to be one of the world's great killers. Relatively recently vaccinations for other diseases have been developed, and diseases such as polio and measles are relatively rare now in the United States, although, unfortunately, there has been a modest resurgence of these two illnesses. Tuberculosis is also on the rise as a consequence of the spread of AIDS. It would appear that with the obvious benefit, vaccination against a disease would be a matter of course, providing that a vaccine exists. This is not necessarily the case.

Several factors need to be considered in determining whether vaccination is desirable or not. The first clearly is: Does a safe, reliable vaccine exist? At one time, rabies vaccine using duck embryos was the best available. However, it did not always provide a reliable immunization, and a significant fraction of the persons on which it was used had reactions, some of which were neurologically very severe. Now, a much more reliable human diploid rabies vaccine is available which provides protection for a very high percentage of persons, and the incidence of untoward reactions is very low. It is probably desirable to mandate vaccination for all personnel who face a high risk of exposure to rabies, i.e., persons who work directly with animals that might be rabid, individuals who do necropsies on such animals, and technicians who work with untreated tissue from potentially rabid animals. The second question is: What is the risk-benefit to the individual if the disease is contracted? The disease may not be sufficiently serious as to warrant the risk of a possible reaction to a vaccination. On the other hand, if the disease is sufficiently life threatening, then the use of a vaccine would be indicated. Third: Is there a satisfactory post-exposure treatment? This is really critical for life-threatening diseases. If there is not, and the exposure risk is significant, the use of even a less than totally satisfactory vaccine might well be considered. Other considerations would be the state of the individual's health. If the condition of the person is such that the possibility of an adverse reaction could have a strong negative impact on the individual, then one would question the desirability of using a vaccine, but one would also question placing such a person in an environment in which vaccination might be considered.

Booster injections are needed for some diseases to ensure an adequate protective level of antibodies. However, some patients may experience reactions to a booster. It is advisable to do a blood titer test prior to repeat injections. If the titer is adequate, no booster should be administered.

Although the laboratory supervisor should have considerable input in deciding whether a vaccine should be used or not, any decision to institute a mandatory vaccination program should be reviewed by a separate biosafety committee before implementation. Individuals must be fully informed of any possible risks.

APPENDIX

Materials that should be included in a medical surveillance program. Note that not all of these necessarily involve laboratory usage of the material, some materials not normally found in the laboratory are included for completeness.

1.   29 CFR 1910.1014: 2-Acetylaminofluorene
2.   29 CFR 1910.1045: Acrylonitrile
3.   29 CER 1910.1011: Aminodiphenyl
4.   29 CFR 1910.1018: Arsenic, inorganic
5.   29 CFR 1910.1111 and 1101 and 1926.1101: Asbestos - nonlaboratory exposure
6.   29 CER 1910.1010: Benzidine
7.   29 CFR 1910.1028: Benzene
8.   29 CFR 1910.1030: Bloodbome pathogens - post exposure
9.   29 CFR 1910.1051: 1,3 Butadidiene
10.   29 CER 1910.1027: Cadmium
11.   29 CFR 1910.1008 and 1926.1127 : bis-Chloromethyl ether
12.   29 CFR 1910.1029: Coke Oven emissions - exposure
13.   29 CFR 1910.1043: Cotton dust - exposure
14.   29 CFR 1910.1044: 1,2-Dibromo-3-chloropropane
15.   29 CFR 1910.1007: 3,3- Dichlorobenzidine and its salts
16.   29 CFR 1910.1015: 4-Dimethylaminoazobenzene
17.   29 CFR 1910.1012: Ethyleneimine
18.   29 CFR 1910.1047: Ethylene oxide
19.   29 CFR 1910.1048: Formaldehyde
20.   29 CFR 1910.120(f): Hazardous waste, emergency response personnel
21.   29 CFR 19 10.1450(g): Laboratory chemicals, exposures above action, PEL levels, incidents, possible symptoms
22.   29 CFR 1910.25 and 1926.62: Lead
23.   29 CFR 1910.1006: Methyl chloromethyl ether
24.   29 CFR 1910.1052: Methylene Chloride
25.   29 CFR 1910.1004: alpha Naphthylamine
26.   29 CFR 1910.1009: beta Naphthylamine
27.   29 CFR 1910.1003: 4-Nitrobiphenyl
28.   29 CFR 1910.1016: N-Nitrosodimethylamine
29.   29 CFR 1910.95: Noise, hearing conservation program required above action level
30.   29 CFR 19 10.1013: Propiolactone, beta
31.   29 CFR 1910.1001 and 134: Respirator use, pulmonary function

32. 29 CFR 1910.1017: Vinyl chloride
33. 29 CFR 1910 Subpart Q: Exposures to welding fumes
34. 29 CER 1928: Particulate respiratory exposures
35. 29 CFR 1928: Pesticide applicators

REFERENCES

1. **Hogan, J.C. and Bernaski, E.J.,** Developing job-related pre-placement medical examinations, *J. Occup. Med.,* 23(7), 469, 1981.
2. Health Monitoring for Laboratory Employees, Research and Development Fact Sheet, National Safety Council, Chicago, 1978.

INTERNET REFERENCES

1. http://www.frwebgate.access.gpo.gov/cgi-bin/get-cfr.cgi
2. http://www.osha-slc.gov/STLC/medicalsurveilance/index.html

E. The OSHA Bloodborne Pathogen Standard: Infection from Work with Human Specimens

OSHA published 29 CER 1910.1030, regulating exposures to bloodborne pathogens on December 6, 1991. The rule took effect on March 6, 1992. Under the standard, employers affected by the standard were to be in full compliance by July 6, 1992. An exposure control plan was to be in effect by May 5, 1992, and employee training made available by June 5, 1992. OSHA has taken a firm posture on implementation of the rule. OSHA can impose a fine of up to $70,000 per willful violation.

In the standard, OSHA defines *bloodborne pathogen* to mean "pathogenic microorganisms that are present in human blood and can cause disease in humans." These pathogens include, but are not limited to, hepatitis B virus (HBV) and human immunodeficiency virus (HIV). The emphasis has been on these two diseases, although other diseases may be found in human blood and tissue and technically are covered. The standard also does not limit itself to human blood despite the name. The standard includes *other potentially infectious materials,* which are defined by the standard to mean (1) the following human body fluids: semen, vaginal secretions, cerebrospinal fluid, synovial fluid, pleural fluid, pericardial fluid, peritoneal fluid, amniotic fluid, saliva in dental procedures, any body fluid that is visibly contaminated with blood, and all body fluids in situations in which it is difficult or impossible to differentiate between body fluids; (2) any unfixed tissue or organ (other than intact skin) from a human (living or dead); and (3) HIV-containing cell or tissue cultures, organ cultures, HIV- or HBV-containing culture medium or other solutions, and blood, organs, or other tissues from experimental animals infected with HIV or HBV.

The CDC published guidelines for prevention of transmission of bloodborne diseases and identified certain substances in which the potential for transmittal of HBV and HIV was extremely low or nonexistent. These substances were feces, nasal secretions, sputum, sweat, tears, urine, and vomitus, unless they contain visible blood.

When the standard was published, it was predicted that the standard would prevent approximately 9000 infections of HBV and approximately 200 deaths per year from this disease. Far fewer persons are known to have acquired HIV from occupational activities, less than 100 at the time the standard went into effect. By mid-1998, there have been 54 known cases of occupationally acquired HIV infection, and an additional 133 possible cases.

1. Basic Provisions

There are several key components to a program to achieve compliance with the standard:
1. An exposure control plan
2. Exposure determination

The standard identifies several areas in which explicit guides are provided to assure compliance:
1. General
2. Engineering and work practice controls
3. Personal protective equipment
4. Housekeeping

Other requirements include:
1. HBV vaccination and postexposure evaluation and follow-up
2. Communication of hazards to employees
3. Record keeping

Each of the above areas will be discussed in the following sections.

2. Exposure Control Plan

The exposure control plan is the organization's written statement of how it plans to eliminate or minimize employee exposure. It must cover each of the broad areas listed in the previous section.

a. Exposure Determination

The employer must compile a list of job classifications in which (a) all employees in those jobs have occupational exposures and (b) some employees have occupational exposures. The employer must identify a list of tasks and procedures or groups of closely related tasks and procedures in which occupational exposure occurs and which the employees in the first two lists perform. The lists do not take into consideration any use of personal protective equipment to exclude personnel.

The first list in which all employees in those jobs do have occupational exposure might include for any research-oriented organization:

1. Medical doctor (organization's health service)
2. Nurse (organization's health service)
3. Research scientists
4. Technicians
5. Glassware cleaners
6. Laundry staff
7. Police/security
8. Cleaning staff
9. Maintenance staff
10. Athletic trainers (academic organizations)

The first two of these clearly would fit the requirement that all members of the group would be occupationally exposed. The last eight could include some that would have occupational exposures while others do not.

The tasks and procedures associated with each of these groups is straightforward. A physician in the type of environment considered here, a research organization rather than a clinical hospital, would have to examine patients, perform pre-employment and periodic physical examinations, and treat minor injuries. The nurse would have to perform phlebotomies (draw blood), give shots, perform minor first aid, dress wounds, and collect contaminated clothing and dressings. The nurse would in the course of these actions handle syringes, blood vials, and possibly glass slides, and do pin pricks on fingers to obtain small blood samples. Technicians would handle vials containing blood and other fluids, pipettes, culture dishes, blenders, and sonicators, all of which could contain contaminated materials and result in an exposure if used, or handled improperly. Research scientists and those actively working with them can be exposed to virtually anything listed for a technician, but since they do not do as many things routinely, perhaps have a higher possibility for error.

Glassware cleaners, laundry staff, and cleaning staff are affected because of the operations of the professional staff. Glassware may have dried contaminated fluids or tissue still in or on the glassware. The laundry staff may be asked to wash gowns, sheets, and lab coats with dried blood on them. The cleaning staff may be exposed to broken glassware, lab instruments (e.g., syringes, scalpels, broken culture dishes), soiled bandages, and dressings improperly disposed of in the ordinary trash. The maintenance staff may be asked to repair equipment or change a HEPA filter in a biological safety cabinet.

Police and other emergency personnel, such as rescue squad members, are two groups with a high potential for exposure. The police and emergency rescue personnel are, other then incidental bystanders, normally the first on the scene of an accident and are often called upon to administer emergency assistance such as CPR, treating wounds, or performing other actions in which their hands and clothing often are contaminated. The same sort of exposures occur for athletic trainers who, especially in the rougher contact sports, often have to bandage an injured player, remove bloody clothing, and collect potentially contaminated clothing after a game.

All of these classes of tasks and procedures would need to be identified in the exposure plan.

b. Implementation of the Exposure Control Plan

General: The foundation of all the measures of compliance will be that universal precautions will be followed. In August 1987, the CDC published a document entitled "Recommendations for Prevention of HIV Transmission in Health Care Settings." This document introduced the concept of universal precautions which basically recommended that steps be taken to prevent exposure of health care workers to possibly contaminated blood, other body fluids, tissue from a human living or dead, HIV-containing cultures and other possibly contaminated items which might be found in the laboratory (see the first part of this section). The universal part of this concept comes from the assumption that all of these possible sources of infections are treated as if they *were* infected. This assumption extends to all personnel who may become infected by coming into contact with contaminated materials, from the physician or research scientist to the laundry employee. The plan must make a commitment to adopt this policy and enforce it.

i. Engineering and work practice controls.
1. Engineering controls available could include:
    Biological safety cabinets, class I, II, or III
    Hand-washing sink in the laboratory
    Employees shall wash their hands (or any other body part) as soon as possible after removing gloves or other personal equipment or after contact of any part of the body with potentially contaminated materials.

Autoclave readily available

Eyewash station, deluge shower available in the laboratory

Plastic-backed absorbent paper or other work surface protective materials available

Containers available to place under apparatus to catch spills

Surfaces of doors, walls, floors, and equipment water-resistant for ease of decontamination

Joints, fixtures, and penetrations of the preceding items sealed to prevent contaminants from accumulating and to facilitate decontamination

Vacuum lines protected by disinfectant traps and HEPA filters

Ventilation inward into the facility not recirculated but discharged directly to the outside

"Sharps" containers, as specified in the standard, readily available; in addition to syringes and needles, other "sharps" would include disposable pipettes, culture plates, capillary tubes, any broken or chipped glassware.

Biological waste containers, as specified in the standard, available

2. Work practices could include:

Wear gloves whenever handling tissues or body fluids. Disposable gloves shall not be reused. Gloves shall be changed between contact with individual patients where such contacts are involved.

Wear appropriate personal protective equipment. Protective equipment is appropriate only if it prevents blood or other potentially contaminated equipment to pass through or reach the employee's work clothes, street clothes, undergarments, skin, eyes, mouth, or other mucous membranes, or normal usage of the equipment.

Smoking, eating, drinking, applying cosmetics, or handling contact lenses shall not be permitted in the laboratory.

No items intended for human consumption are to be stored in refrigerators, in cabinets, or on shelves or counter tops with potentially contaminated materials.

Always use mechanical pipettes.

Procedures shall be used to avoid creation of aerosols, droplets, splashing; where these are unavoidable, containment shall be used.

Contaminated needles shall not be bent, recapped or removed, or sheared or broken except under conditions where no alternative is feasible.

All work surfaces will be decontaminated after a spill with a 1:10 solution of household bleach.

Clothing, lab coats, aprons, gowns, and other items of protective clothing worn in the facility will be removed prior to leaving the work area.

Avoid touching any item unnecessarily with contaminated gloves, including documents, pens, door knobs, telephones, etc. Any such items should be decontaminated after such contact.

The lists above of suggested environmental controls and work practices are not complete, since each facility will have some unique procedures, but clearly the principle behind all of these is to prevent employees from coming into contact with any contaminated materials and to prevent the spread of contamination to areas outside the work area or to other individuals.

ii. Personal protective equipment (PPE): In the preceding section on engineering controls and work practices, numerous items of protective equipment were mentioned. There are several principles in the standard that govern the selection and use of PPE.

1. It must be appropriate. To be appropriate, it must not allow contaminated materials, including all of the materials which come under this classification, from reaching any part

of the person wearing or using the PPE, or any of their clothing. In selecting PPE, the employer should consider the comfort and convenience of the wearer, as well as the equipment meeting the basic design standards for that item. Goggles, for example, need to be comfortable, not prone to fogging up, and permit wearing of prescription lenses, otherwise the employee may not wear them on every occasion when the need is necessary.

2. Provision of PPE is the responsibility of the employer. It must be provided at no cost to the employee. Among items of PPE that are definitely included are goggles, respirators, masks, face shields, gloves, surgical caps, or hoods. Since some facilities expect employees to provide their own uniforms, lab coats, and gowns, OSHA would consider these as general work clothes which the employer would not be required to provide, but if any of these were to be considered as PPE, the responsibility would be that of the employer.

3. It is the responsibility of the employer to maintain, clean, and launder (if necessary) all items of PPE. Employees are not to be allowed to take clothing home for laundering.

4. OSHA requires the employer to "ensure" that the employee uses PPE when it is needed.

5. There are circumstances that make it acceptable to not wear PPE under conditions when it would normally be used. OSHA, in 1910.1030 (d)(11) states ..."unless the employer shows that the employee temporarily and briefly declined to use PPE when, under rare and extraordinary circumstances, it was the employee's judgement that in the specific instance its use would have prevented the delivery of health care or public safety services or would have posed an increased hazard to the safety of the worker or co-worker. When the employee makes this judgement, the circumstances shall be investigated and documented in order to determine whether changes can be instituted to prevent such occurrences in the future."

iii. *Housekeeping:*

1. Employers shall ensure that the work site is maintained in a clean and sanitary condition. The employer must establish a written cleaning and decontamination schedule appropriate to the location of item to be cleaned within the facility, type of surface to be cleaned, type of soiling present, and tasks or procedures being performed in the area.

2. All equipment and environmental and working surfaces shall be cleaned and decontaminated after contact with blood or other potentially contaminated infectious materials. The standard covers five specific situations:

   a. Work surfaces shall be decontaminated with an appropriate disinfectant (i) after completion of procedures, (ii) immediately or as soon thereafter as feasible when surfaces are overtly contaminated, (iii) after any spill of blood or potential infectious materials, or (iv) at the end of the work shift if the surface may have been contaminated since the last cleaning.

   b. Protective coverings of work surfaces, e.g., plastic-backed absorbent paper, shall be removed and replaced when they become overtly contaminated or at the end of a work shift if it may have become contaminated during the work shift.

   c. Bins, pails, cans, or similar receptacles intended for reuse that have a reasonable likelihood of becoming contaminated shall be inspected and decontaminated on a regular schedule and immediately cleaned and decontaminated as soon as feasible upon visible contamination.

   d. Broken, potentially contaminated glassware shall not be picked up with the hands.It shall be cleaned up using mechanical means (brush and dustpan, forceps, or tongs).

   e. Reusable "sharps" that are contaminated shall not be stored or processed in a manner that would require employees to reach by hand into the containers where these "sharps" are placed.

Figure 4.11 Sturdy, secure, leakproof, red sharps container with biohazard symbol.

iv. Regulated wastes:

This section deals only with regulated medical or infectious waste. If waste materials have other hazardous characteristics, e.g., chemical or radioactive, handling of these mixed wastes will have to comply with the standards applicable to the other hazard characteristics as well.

1. Contaminated "sharps," discarding and containment: Contaminated "sharps" shall be discarded immediately or as soon as feasible into containers (Figure 4.11) that are closable, puncture-resistant, leakproof on sides and bottom, and labeled according to the requirements of the standard. The latter usually requires a biohazard symbol on the container and/or the container being red.[*]

   a. The "sharps" containers shall be easily accessible to personnel, located as close as is feasible to where the "sharps" are used or can reasonably be expected to be found.
   b. Maintained upright throughout use.
   c. Replaced frequently and not be allowed to be overfilled.
   d. Closed immediately prior to being moved when being moved from the area of use to prevent spillage or protrusion of the contents.

If any chance of leakage of the "sharps" primary containers exists, the container shall be placed in a secondary container which can be closed tightly, constructed to contain all the contents and prevent leakage during handling, storage, transport, or shipping, and labeled as was the original container.

2. Other regulated waste containment. The requirements for other regulated waste containers are essentially the same as for "sharps." If any contamination of the outside

---

[*] It is recommended that these containers not be used for ordinary waste, although, in light of the recent changes in incineration of infectious waste, it may be acceptable to mix the contents afterwards with ordinary solid waste for incineration.

of the primary regulated waste container occurs, the container shall be placed in a secondary container.

3. Disposal of regulated waste. This topic will be covered in more detail in a later section of this chapter, but basically it requires that the waste be rendered noninfectious, and the trend is to require it to be rendered unrecognizable as regulated medical wastes. Until recently the waste was either steam sterilized and subsequently discarded as ordinary trash or incinerated. Newer technologies are now being used in a number of locales, but steam sterilization and incineration remain the most common means of treatment of the wastes. The generator must be sure to follow the local regulations applicable to themselves. Otherwise, under the standard, OSHA can issue a citation.

v. Laundry

Contaminated laundry shall be handled as little as possible, with a minimum of agitation. It shall be bagged or placed in an appropriate container at the location where it was used. Sorting or rinsing shall not take place at the location of use. The contaminated laundry shall be placed and transported in bags that are labeled or color coded as the standard requires unless the facility uses a laundry which utilizes universal precautions in handling all of the laundry received. If such is the case, alternative labeling or color coding is permitted if it permits the employees to recognize the laundry which requires compliance with universal precautions. The employer must ensure that employees who come into contact with contaminated laundry wear protective gloves and other appropriate personal protective equipment. If the contaminated laundry is wet and may soak or leak through the bag, the laundry must be placed in a secondary container to prevent this.

vi. Hepatitis B vaccination:

1. The employer shall make available the HBV vaccine to all employees who have occupational exposures. The vaccinations shall be made available to the employee at no cost to the employee at a reasonable time and place. The vaccinations shall be performed by or under the supervision of a licensed physician or by or under the supervision of another licensed health care professional. The health care professional responsible for the vaccinations must be given a copy of the bloodborne pathogen standard. Any tests must be done by an accredited laboratory at no cost to the employee.

2. The vaccinations must be made available after the employee has received training concerning information on the efficacy safety, method of administration, benefits of vaccination, and the fact that the vaccine and vaccination will be free to the employee and within 10 working days of initial assignment to a position where occupational exposures are possible. If a titer reveals that the employee still has immunity from a prior vaccination or there are medical reasons why the employee should not receive the vaccination, the vaccination is not required.

3. The employer cannot make participation in a pre-screening program a prerequisite for receiving the vaccination.

4. The employee may decline to receive the vaccination, but the employer must provide the vaccination at a later time if the employee is still covered by the standard and decides to accept the vaccination.

5. If the employee declines to accept the vaccination, the employer must ensure that the employee signs the following statement:

"I understand that due to my occupational exposure to blood or other potentially infectious materials I may be at risk of acquiring hepatitis B virus (HBV) infection. I have been given the opportunity to be vaccinated with hepatitis B vaccine at no cost to myself. However, I decline hepatitis vaccination at this time. I understand that by declining this vaccination, I continue to be at risk of acquiring hepatitis B, a serious disease. If in the future I continue to have occupational exposure to blood or other potentially infectious materials and I want to be vaccinated with hepatitis B vaccine, I can receive the vaccination series at no charge to me."

This statement should be signed and dated by the employee and kept as part of the records.

6. Any recommended future booster shots shall be made available under the same conditions.

vii. Post exposure evaluation and follow-up.

If an employee has an exposure incident, the employee should immediately notify the person in charge of the facility or another responsible person. Following the report, the employer shall immediately make available to the exposed employee a confidential medical evaluation and follow-up including the following elements:

1. Documentation of the incident, including the route of exposure and the circumstances under which the incident occurred.
2. Identification and documentation of the source individual, unless the employer can establish that identification is not feasible or prohibited by state or local law.
3. The source individual's blood shall be tested as soon as possible. If the source individual does not consent, the employer shall document that the legally required consent cannot be obtained. If consent is not required by law, the source individual's blood, if available, shall be tested and the results documented.
4. If the source individual is known to be infected with either HIV or HBV, testing is not required.
5. The exposed employee shall be informed of the results of the tests of the source individual. The employee shall also be informed of any laws or regulations affecting disclosure of the identity of the source individual.
6. The exposed employee's blood shall be collected and tested as soon as possible after consent is obtained from the exposed employee.
7. An employee may consent to have a blood sample collected but not consent to an evaluation for HIV. The sample can be retained for up to 90 days and tested as soon as possible for HIV if the employee elects to have it done during that interval.
8. Post-exposure prophylaxis, when medically indicated as recommended by the U.S. Public Health Service, shall be made available, as well as counseling and evaluation of reported illnesses.
9. The employer must ensure that the health-care professional evaluating an employee after an exposure incident shall receive the following information: (a) a copy of the standard, (b) a description of the employee's duties relevant to the incident, (c) documentation on the route of exposure and the circumstances of the incident, (d) results of the source individual's blood tests, if available, (e) medical records relevant to the appropriate treatment, including vaccination status.
10. The employer shall obtain and provide a copy of the health care professional's written opinion within 15 days after completion of the evaluation.
11. The written opinion for HBV vaccination shall be limited to whether vaccination is indicated and if the employee has received the vaccination.
12. The health care professional's written opinion concerning post-exposure evaluations and follow-up shall be limited to the following: (a) the employee has been told about the results of the evaluation and (b) any medical conditions resulting from exposure to blood or other infectious materials which require further treatment.

viii. Hazard communication - Labels and signs.

1. Warning labels shall be affixed to containers of regulated waste, refrigerators and freezers containing potential contaminated materials, and other containers used to store, transport, or ship potentially infectious materials. The labels shall state BIOHAZARD and be accompanied by the biohazard symbol. The labels shall be fluorescent orange or orange-red or predominantly so, with lettering or symbols in a contrasting color. The labels shall be firmly attached by adhesive, string, wire, or other methods that would prevent the loss of the label or unintentional removal. Red bags or red containers can be substituted for labels. Labels required for contaminated equipment shall be the same as the above and must also state which portions of the equipment are contaminated.
2. Containers of blood, blood components, or blood products released for transfusion and labeled as to their contents are exempted from the other labeling requirements.
3. Individual containers placed in a properly labeled larger container need not be individually labeled.
4. Regulated wastes that have been decontaminated need not be labeled.
5. Work areas where work with potentially contaminated blood or other materials should be posted with the BIOHAZARD legend and symbol. This sign is required for HIV and HBV research laboratories and production facilities.

ix. Information and Training:

Employers must ensure that all employees with occupational exposures participate in a training program, at no cost to the employees and during normal working hours. The training shall be provided as follows:

1. At the time of initial assignment to tasks, where occupational exposure may take place.
2. All employees shall receive annual refresher training within 1 year of their previous training.
3. Additional training is required for new exposure situations caused by new tasks or changes in procedures.
4. The training material will be appropriate in terms of content, vocabulary level, literacy level, and language for the training participants.

The content of the training program shall, at a minimum, contain the following elements:

1. An accessible copy of the standard and an explanation of its contents.
2. A general explanation of the epidemiology and symptoms of bloodborne diseases.
3. An explanation of the modes of transmission of bloodborne pathogens.
4. An explanation of the exposure control plan and how a written copy can be obtained.
5. An explanation of how to recognize tasks and other activities that may involve exposure to blood and other potentially infectious materials.
6. An explanation of how to prevent or reduce exposure by means of engineering controls, work practices, and use of personal protective equipment.
7. Information on the types, proper use, location, removal, handling, decontamination, and disposal of PPE.
8. An explanation of the basis for selecting PPE.
9. Information on the HBV vaccine as to its efficacy, benefits and method of administration and that the vaccine and vaccination will be free.
10. What actions to take and whom to contact in an emergency involving blood and other infectious materials.
11. What procedure to follow if an exposure incident occurs, including the method of reporting the incident and the medical follow-up that will be available.

12. Information on the post-exposure evaluation and follow-up to be provided to the employee by the employer.
13. An explanation of the signs and labels and/or color coding required by the standard.
14. The employee will have opportunities to ask questions of the person providing the training. The person providing the training shall be knowledgeable about the subject matter in the context of the employee's workplace.

x. Record keeping - medical records:

The employer shall establish and maintain records for each employee with occupational exposure in accordance with 29 CFR 1910.20 (see Section VIII.B, this chapter). The records are to be kept confidential and not to be disclosed or reported to anyone without the employee's written consent to any person within or outside the organization, except as required by the standard or by law. The records are to be maintained for at least the duration of the employee's employment plus 30 years. The content of the records shall consist of:

1. Name and Social Security number of the employee.
2. A copy of the employee's hepatitis B vaccination status, including the dates vaccination were received any medical records relevant to the employee's ability to receive vaccinations. If the employee had been involved in an exposure incident, the following additional information would be required:

    a. A record of all results of examinations, medical testing, and follow-up procedures for the employee.
    b. The employer's copy of the health care professional's written opinion.
    c. A copy of the information provided to the health care professional.

xi. Record keeping - training records:
The training records shall include the following information:

    1. The dates of the training sessions.
    2. The contents, or a summary thereof, of the training sessions.
    3. The names and qualifications of the persons conducting the training.
    4. The names and job titles of all persons attending the training sessions.

The training records shall be kept for a period of 3 years from the date the training occurred. The access to records shall be as provided in 29 CFR 1910.20. If the employer ceases to do business and there is no successor employee, the employer shall notify the OSHA Director at least 3 months prior to their disposal and transmit them to the Director if required by the Director to do so, within that 3-month period.

There are additional requirements for HIV or HBV research laboratories or production facilities. If information on these requirements is needed, the reader is referred to the standard in such cases.

As a result of the increase in HIV infections, there has been an increase in the incidence of tuberculosis in the United States. Between 1985 and 1992, the number of cases increased by 20%. In 1992, there were 26,673 cases in the U.S. However, as will be noted from Figure 4.12, since then the number of cases has been declining significantly to 19,851 in 1997, a decrease from the 1985 level of 10.7%. Separate data on laboratory type occupations is not available. The disease may be becoming more resistant to treatment. In New York City, the cure rate for the two normally most effective drugs has dropped from 100% to 60%. As a result, in October 1993 OSHA issued guidelines applying specifically to health-care settings, correctional institutions

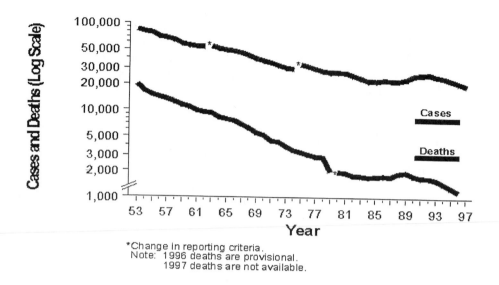

*Change in reporting criteria.
Note: 1996 deaths are provisional.
1997 deaths are not available.

Figure 4.12 Tuberculosis cases and deaths in the United States, 1953-1997.

homeless shelters, long-term care facilities for the elderly, and drug treatment centers. OSHA, as they did originally for HIV and HBV before the current standard was enacted, then stated its intention to base citations on the General Duty Clause. On Oct 17, 1997, OSHA proposed a new regulation to control exposures to tuberculosis. At this writing, the regulation has not gone into effect.

REFERENCES

1. Occupational Exposure to Bloodborne Pathogens, 29 CFR 1910.1030, FR *56, 235,* December 6, 1991, 64175.
2. Recommendations for prevention of HIV transfusion in healthcare settings, in *Morbidity and Mortality Weekly Report,* Centers for Disease Control, U.S. Department of Health and Human Services, Public Health Service, 36, 25, August 21, 1987.
3. **Domin, M.A. and Smith, C.E.,** OSHA's final rule on occupational exposure to bloodborne pathogens, *J. Health Care Material Manage.,* 10, 2, 1992.
4. Update: universal precautions for prevention of transmission of human immunodeficiency virus, hepatitis B virus; and other bloodborne pathogens in healthcare settings, in *Morbidity and Mortality Weekly Report,* Centers for Disease Control, U.S. Department of Health and Human Services, Public Health Service, 37, 24, June 24, 1988, 377-382, 387-388.
5. Guidelines for Handling Human Tissues and Body Fluids Used in Research, National Disease Research Interchange, Philadelphia, PA, 1987.

INTERNET REFERENCES

1. http://www.frwebgate.access.gpo.gov/cgi-bin/get-cfr.cgi Text of OSHA Bloodborne Patogen standard.
2. http://www.cdc.gov/nchstp/tb/surv/surv97/surv97gif/surv2.gif Tuberculosis data.
3. http://www.osha-slc.gov/FedReg_osha_data/FED19971017.html Proposed OSHA Tuberculosis regulation.

3. HIV Infection: Detection, Initial Management, and Referral*

Infection with HIV produces a spectrum of disease that progresses from a clinically latent or asymptomatic state to AIDS as a late manifestation. The pace of disease progression is variable. The time between infection with HIV and the development of AIDS ranges from a few months to as long as 17 years (median:10 years). Most adults and adolescents infected with HIV remain symptom-free for long periods, but viral replication is active during all stages of infection, increasing substantially as the immune system deteriorates. AIDS eventually develops in almost all HIV-infected persons; in one study of HIV-infected adults, AIDS developed in 87% (95% confidence interval [CI]=83%-90%) within 17 years after infection. Additional cases are expected to occur among those who have remained AIDS-free for longer periods.

Greater awareness among both patients and healthcare providers of the risk factors associated with HIV transmission has led to increased testing for HIV and earlier diagnosis of the infection, often before symptoms develop. The early diagnosis of HIV infection is important for several reasons. Treatments are available to slow the decline of immune system function. HIV-infected persons who have altered immune function are at increased risk for infections for which preventive measures are available (e.g., *Pneumocystis car/nil* pneumonia IPCPI, toxoplasmic encephalitis [TEl, disseminated *Mycobacterium aviuni* complex [MAC] disease, tuberculosis [TB], and bacterial pneumonia). Because of its effect on the immune system, HIV affects the diagnosis, evaluation, treatment, and follow-up of many other diseases and may affect the efficacy of antimicrobial therapy for some STDs. Finally, the early diagnosis of HIV enables the health-care provider to counsel such patients and to assist in preventing HIV transmission to others.

Proper management of HIV infection involves a complex array of behavioral, psychosocial, and medical services. Although some of these services may be available in the STD treatment facility, other services, particularly medical services, are usually unavailable in this setting. Therefore, referral to a healthcare provider or facility experienced in caring for HIV-infected patients is advised. Staff in STD treatment facilities should be knowledgeable about the options for referral available in their communities. While in the STD treatment facility, the HIV-infected patient should be educated about HIV infection and the various options for HIV care that are available.

Because of the complexity of services required for management of HIV infection, detailed information, particularly regarding medical care, is beyond the scope of this book and may be found elsewhere.

4. Hepatitis A and B

One of the most effective means of preventing the transmission of STDs is pre-exposure immunization. Currently licensed vaccines for the prevention of STDs include those for hepatitis A and hepatitis B. Clinical development and trials are underway for vaccines against a number of other STDs, including HIV and HSV. As more vaccines become available, immunization possibly will become one of the most widespread methods used to prevent STDs.

Five different viruses (i.e., hepatitis A-E) account for almost all cases of viral hepatitis in humans. Serologic testing is necessary to confirm the diagnosis. For example, a healthcare provider might assume that an injecting-drug user with jaundice has hepatitis B when, in fact, outbreaks of hepatitis A among injecting-drug users often occur. The correct diagnosis is essential for the delivery of appropriate preventive services. To ensure accurate reporting of viral hepatitis and appropriate prophylaxis of household contacts and sex partners, all case reports of viral hepatitis should be investigated and the etiology established through serologic testing.

---

*   This section and the following sections dealing with hepatitis A and B are exerpted from the 1998 Guidelines for the Treatment of Sexually Transmitted Diseases by the Centers for Disease Control.

a. Hepatitis A

Hepatitis A is caused by infection with the hepatitis A virus (HAV). HAV replicates in the liver and is shed in the feces. Virus in the stool is found in the highest concentrations from 2 weeks before to 1 week after the onset of clinical illness. Virus also is present in serum and saliva during this period, although in much lower concentrations than in feces. The most common mode of HAV transmission is fecal-oral, either by person-to-person transmission between household contacts or sex partners or by contaminated food or water. Because viremia occurs in acute infection, bloodborne HAV transmission can occur; however, such cases have been reported infrequently. Although HAV is present in low concentrations in the saliva of infected persons, no evidence indicates that saliva is involved in transmission.

Of patients who have acute hepatitis A, ≤20% require hospitalization; fulminant liver failure develops in 0.1% of patients. The overall mortality rate for acute hepatitis A is 0.3%, but it is higher (1.8%) for adults aged >49 years. HAV infection is not associated with chronic liver disease.

In the United States during 1995, 31,582 cases of hepatitis A were reported. The most frequently reported source of infection was household or sexual contact with a person who had hepatitis A, followed by attendance or employment at a day care center, recent international travel, homosexual activity, injecting-drug use, and a suspected food or waterborne outbreak. Many persons who have hepatitis A do not identify risk factors; their source of infection may be other infected persons who are asymptomatic. The prevalence of previous HAV infection among the U.S. population is 33% (CDC, unpublished data).

Outbreaks of hepatitis A among homosexual men have been reported in urban areas, both in the United States and in foreign countries. In one investigation, the prevalence of HAV infection among homosexual men was significantly higher (30%) than that among heterosexual men (12%). In New York City, a case-control study of homosexual men who had acute hepatitis A determined that case-patients were more likely to have had more anonymous sex partners and to have engaged in group sex than were the control subjects; oral-anal intercourse (i.e.,the oral role) and digital-rectal intercourse (i.e., the digital role) also were associated with illness.

i. Treatment

Because HAV infection is self-limited and does not result in chronic infection or chronic liver disease, treatment is usually supportive. Hospitalization may be necessary for patients who are dehydrated because of nausea and vomiting or who have fulminant hepatitis A. Medications that might cause liver damage or that are metabolized by the liver should be used with caution. No specific diet or activity restrictions are necessary.

ii. Prevention

General measures for hepatitis A prevention (e.g., maintenance of good personal hygiene) have not been successful in interrupting outbreaks of hepatitis A when the mode of transmission is from person to person, including sexual contact. To help control hepatitis A outbreaks among homosexual and bisexual men, health education messages should stress the modes of HAV transmission and the measures that can be taken to reduce the risk for transmission of any STD, including enterically transmitted agents such as HAV. However, vaccination is the most effective means of preventing HAV infection.

Two types of products are available for the prevention of hepatitis A: immune globulin (IG) and hepatitis A vaccine. IG is a solution of antibodies prepared from human plasma that is made with a serial ethanol precipitation procedure that inactivates HBV and HIV. When administered intramuscularly before exposure to HAV, or within 2 weeks after exposure, IG is >85%

effective in preventing hepatitis A. IG administration is recommended for a variety of exposure situations (e.g., for persons who have sexual or household contact with patients who have hepatitis A). The duration of protection is relatively short (i.e., 3-6 months) and dose dependent.

Inactivated hepatitis A vaccines have been available in the United States since 1995. These vaccines, administered as a two-dose series, are safe, highly immunogenic, and efficacious. Immunogenicity studies indicate that 99%-100% of persons respond to one dose of hepatitis A vaccine; the second dose provides long-term protection. Efficacy studies indicate that inactivated hepatitis A vaccines are 94%- 100% effective in preventing HAV infection.

iii. Preexposure Prophylaxis

Vaccination with hepatitis A vaccine for pre-exposure protection against HAV infection is indicated for persons who have the following risk factors and who are likely to seek treatment in settings where STDs are being treated.

- Men who have sex with men. Sexually active men who have sex with men (both adolescents and adults) should be vaccinated.
- Illegal drug users. Vaccination is recommended for users of illegal injecting and noninjecting drugs if local epidemiologic evidence indicates previous or current outbreaks among persons with such risk behaviors.

iv. Postexposure Prophylaxis

Persons who were exposed recently to HAV (i.e., household or sexual contact with a person who has hepatitis A) and who had not been vaccinated before the exposure should be administered a single IM dose of 1G (0.02 mL/kg) as soon as possible, but not >2 weeks after exposure. Persons who received at least one dose of hepatitis A vaccine ≥1 month before exposure to HAV do not need 1G.

b. Hepatitis B

Hepatitis B is a common STD. During the past 10 years, sexual transmission accounted for approximately 30% - 60% of the estimated 240,000 new HBV infections that occurred annually in the United States. Chronic HBV infection develops in 1% - 6% of persons infected as adults. These persons are capable of transmitting HBV to others, and they are at risk for chronic liver disease. In the United States, HBV infection leads to an estimated 6,000 deaths annually; these deaths result from cirrhosis of the liver and primary hepatocellular carcinoma.

The risk for perinatal HBV infection among infants born to HBV-infected mothers is 10% - 85%, depending on the mother's hepatitis B e antigen (HbeAg) status. Chronic HBV infection develops in approximately 90% of infected newborns; these children are at high risk for chronic liver disease. Even when not infected during the perinatal period, children of HBV-infected mothers are at high risk for acquiring chronic HBV infection by person-to-person transmission during the first 5 years of life.

i. Treatment

No specific treatment is available for persons who have acute HBV infection. Supportive and symptomatic care usually are the mainstays of therapy. During the past decade, numerous antiviral agents have been investigated for treatment of chronic HBV infection. Alpha-2b interferon has been 40% effective in eliminating chronic HBV infection; persons who became infected during adulthood were most likely to respond to this treatment. Antiretroviral agents (e.g., lamivudine) have been effective in eliminating HBV infection, and a number of other compounds are being evaluated. The goal of antiviral treatment is to stop HBV replication. Response to treatment can be demonstrated by normalization of liver function tests, improvement in liver histology, and seroreversion from HBeAg-positive to HBeAg-negative. Long-term follow-up of treated patients suggests that the remission of chronic hepatitis induced by alpha interferon is of long duration. Patient characteristics associated with positive response to interferon therapy include low pretherapy HBV DNA levels, high pretherapy alanine aminotransferase levels, short duration of infection, acquisition of disease in adulthood, active histology, and female sex.

ii. Prevention

Although methods used to prevent other STDs should prevent HBV infection, hepatitis B vaccination is the most effective means of preventing infection. The epidemiology of HBV infection in the United States indicates that multiple age groups must be targeted to provide widespread immunity and effectively prevent HBV transmission and HBV-related chronic liver disease. Vaccination of persons who have a history of STDs is part of a comprehensive strategy to eliminate HBV transmission in the United States. This comprehensive strategy also includes prevention of perinatal HBV infection by (a) routine screening of all pregnant women, (b) routine vaccination of all newborns, (c) vaccination of older children at high risk for HBV infection (e.g., Alaskan Natives, Pacific Islanders, and residents in households of first-generation immigrants from countries in which HBV is of high or intermediate endemicity), (d) vaccination of children aged 11 - 12 years who have not previously received hepatitis B vaccine, and (e) vaccination of adolescents and adults at high risk for infection.

iii. Pre-exposure Prophylaxis

With the implementation of routine infant hepatitis B vaccination and the wide-scale implementation of vaccination programs for adolescents, vaccination of adults at high risk for HBV has become a priority in the strategy to eliminate HBV transmission in the United States. All persons attending STD clinics and persons known to be at high risk for HBV infection (e.g., persons with multiple sex partners, sex partners of persons with chronic HBV infection, and injecting-drug users) should be offered hepatitis B vaccine and advised of their risk for HBV infection (as well as their risk for HIV infection) and the means to reduce their risk (i.e., exclusivity in sexual relationships, use of condoms, and avoidance of nonsterile drug-injection equipment).

Persons who should receive hepatitis B vaccine include the following:

- Sexually active homosexual and bisexual men;
- Sexually active heterosexual men and women, including those (a) in whom another STD was recently diagnosed, (b) who had more than one sex partner in the preceding 6 months, (c) who received treatment in an STD clinic, and (d) who are prostitutes;
- Illegal drug users, including injecting-drug users and users of illegal noninjecting drugs;
- Healthcare workers;
- Recipients of certain blood products;
- Household and sexual contacts of persons who have chronic HBV infection;
- Adopted children from countries in which HBV infection is endemic;
- Certain international travelers;
- Clients and employees of facilities for the developmentally disabled;
- Infants and children; and
- Hemodialysis patients.

iv. Screening for Antibody Versus Vaccination Without Screening

The prevalence of previous HBV infection among sexually active homosexual men and among injecting-drug users is high. Serologic screening for evidence of previous infection before vaccinating adult members of these groups may be cost effective, depending on the costs of laboratory testing and vaccine. At the current cost of vaccine, prevaccination testing on adolescents is not cost-effective. For adults attending STD clinics, the prevalence of HBV infection and the vaccine cost may justify prevaccination testing. However, because prevaccination testing may lower compliance with vaccination, the first dose of vaccine should be administered at the time of testing. The additional doses of hepatitis vaccine should be administered on the basis of the prevaccination test results. The preferred serologic test for prevaccination testing is the total antibody to hepatitis B core antigen (anti-HBc), because it will detect persons who have either resolved or chronic infection. Because anti-HBc testing will not identify persons immune to HBV infection as a result of vaccination, a history of hepatitis B vaccination should be obtained, and fully vaccinated persons should not be revaccinated.

v. **Vaccination Schedules**

Hepatitis B vaccine is highly immunogenic. Protective levels of antibody are present in approximately 50% of young adults after one dose of vaccine; in 85%, after two doses; and >90%, after three doses. The third dose is required to provide long-term immunity. The most often used schedule is vaccination at 0, 1-2, and 4-6 months. The first and second doses of vaccine must be administered at least 1 month apart, and the first and third doses at least 4 months apart. If the vaccination series is interrupted after the first or second dose of vaccine, the missing dose should be administered as soon as possible. The series should not be restarted if a dose has been missed. The vaccine should be administered IM in the deltoid, not in the buttock.

vi. **Postexposure Prophylaxis**

Exposure to Persons Who Have Acute Hepatitis B, Sexual Contacts

Patients who have acute HBV infection are potentially infectious to persons with whom they have sexual contact. Passive immunization with hepatitis B immune globulin (HBIG) prevents 75% of these infections. Hepatitis B vaccination alone is less effective in preventing infection than HBIG and vaccination. Sexual contacts of patients who have acute hepatitis B should receive HBIG and begin the hepatitis B vaccine series within 14 days after the most recent sexual contact. Testing of sex partners for susceptibility to HBV infection (anti-HBc) can be considered if it does not delay treatment >14 days.

Nonsexual Household Contacts

Nonsexual household contacts of patients who have acute hepatitis B are not at high risk for infection unless they are exposed to the patient's blood (e.g., by sharing a toothbrush or razor blade). However, vaccination of household contacts is encouraged, especially for children and adolescents. If the patient remains HBsAg-positive after 6 months (i.e., becomes chronically infected), all household contacts should be vaccinated.

Exposure to Persons Who Have Chronic HBV Infection

Hepatitis B vaccination without the use of HBIG is highly effective in preventing HBV infection in household and sexual contacts of persons who have chronic HBV infection, and all such contacts should be vaccinated. Postvaccination serologic testing is indicated for sex partners of persons who have chronic hepatitis B infections and for infants born to HBsAg-positive women.

vii. **Special Considerations**

Pregnancy

Pregnancy is not a contraindication to hepatitis B vaccine or HBIG vaccine administration.

HIV Infection

HBV infection in HIV-infected persons is more likely to lead to chronic HBV infection. HIV infection also can impair the response to hepatitis B vaccine. Therefore, HIV-infected persons who are vaccinated should be tested for hepatitis B surface antibody 1-2 months after the third vaccine dose. Revaccination with three more doses should be considered for those who do not respond initially to vaccination. Those who do not respond to additional doses should be advised that they might remain susceptible to HBV infection.

REFERENCE

1. *The 1998 Guidelines for the Treatment of Sexually Transmitted Diseases,* Morbidity and Mortality Weekly Report, Vol 47, No. RR-1, U.S. Department of Health and Human Services, Centers for Diseases Control, Jan. 23, 1997, Atlanta GA. Available online at http://www.niaid.nih.gov/publications/hivaids/4.htm

5. Zoonotic Diseases[*]

Zoonotic diseases are those transmitted from animals to humans, with a wide range of manifestations in humans from simple illness to death. Some of these are of special concern to pregnant women. Toxoplasmosis and listeriosis can cause, among other things, spontaneous abortions.

An understanding of the modes of transmission of these diseases and clinical signs observed in affected animals will help facility managers establish preventive measures to protect individuals who come into contact with the animals or their tissues.

a. Modes of Transmission

Disease agents can be transmitted either directly or indirectly. Bacterial, viral, fungal, and parasitic disease agents can be transmitted through direct contact with animal saliva, feces, urine, other body secretions, bites, scratches, aerosols, or excised body tissues. Humans can be protected through use of gloves, masks, gowns, and other protective clothing, and through the use of restraint techniques which minimize the possibility of bites and scratches.

One indirect means of disease agent transmission involves fomites, inanimate objects (boots, brooms, cages, instruments, etc.) that can transport the agent following contact with animals, secretions, or wastes. Disease agents may be short lived when outside the body, or persistent for years on fomites if the object is not cleaned or disinfected. The Orf virus, from sheep and goats, has remained viable for 15 years in dried scabs. Rooms, cages, and equipment should be adequately disinfected with virucidal disinfectant agents.

A second indirect means of transmission involves vectors, living organisms (insects) which can extract or carry the disease agent from one animal to other animals or humans. A mechanical vector extracts and carries the disease agent without any change occurring in the agent. In a biological vector, the disease agent undergoes changes in one or more stages of its life cycle before becoming an infective form.

An effective vermin and insect control program is needed to eliminate these indirect means of transmission.

b. Routes of Exposure

Barkley and Richardson[1] listed the four primary routes of exposure or entry of a disease agent:

---

[*]   This section was written by David M. Moore, D.V.M.

1. Ingestion (i.e., placing contaminated fomites in one's mouth, or contaminated hand contact with food)
2. Inhalation (i.e., aerosolized material - urine, feces, saliva, or other bodily secretions; these materials may also become aerosolized when using high pressure water hoses to clean rooms or cages)
3. Contact with mucous membranes (i.e., contact with nose, mouth, eyes through spills, contaminated hands, or aerosolized material)
4. Direct parenteral injection (i.e., bites, cuts, scratches, accidental needle sticks)

Each should be handled accordingly by prohibiting food consumption in animal holding areas, requiring the practice of good hygiene, altering sanitation procedures to lessen aerosol production, providing protective garments and safety items (safety goggles, respirators, gloves, and masks), and establishing safety awareness training programs to advise employees as to risks and preventive measures.

REFERENCES

1. **Barkley, E.W. and Richardson, J.H.,** Control of biohazards associated with the use of experimental animals, in *Laboratory Animal Medicine,* Fox, J.G. et al., Eds., Academic Press, Orlando, FL, 1984, 595.
2. **Fox, J.G., Newcomer, C.E., and Rozmiarek, H.,** Selected zoonoses and other health hazards, in *Laboratory Animal Medicine,* Fox, I.G. et al., Eds., Academic Press, Orlando, FL, 1984, 613.
3. Biological Hazards in the Nonhuman Primate Laboratory, Office of Biohazard Safety, NCI, Bethesda, MD, 1979.
4. **Richardson, J.H. and Barkley, E.W.,** Biosafety in Microbiological and Biomedical Laboratories, PHS/NIH, HITS Publication No. (CDC) 84-8395, 37, 1984.
5. **Hellman, A., Oxman, M.N., and Pollack, R.,** *Biohazards in Biological Reseatch,* Cold Spring Harbor Laboratory, New York, 1973.
6. *Am. Md. Hyg. Assoc. J., 54(3),* 113, 1993.

c. Allergies

Some investigators and animal care technicians who have prolonged contact with laboratory animals may develop allergies to animal dander, hair, urine, tissues, or secretions. Reactions to skin contact or inhalation of these materials vary from a wheal and flare phenomenon (a firm, red raised area at the site of skin contact which develops within several minutes) to life-threatening anaphylactic shock.

Olfert[2] lists the species most commonly associated with allergic reactions in a laboratory setting: rat, rabbit, guinea pig, and mouse. When transfer of personnel to a nonanimal area is not a viable option, other measures should be taken to avoid exposure to specific allergens. Lutsky et al.[1] suggests the use of gloves, masks, protective outer garments, and filtered cages as methods to reduce exposure. Additionally, eliminating recirculation of room air will decrease the levels of allergens, as will more frequent cage cleaning.

REFERENCES

1. **Lutsky, I.T., Kalbfleisch, J.H., and Fink, J.N.,** Occupational allergy to laboratory animals: employer practices, *J. Occup. Med,* 25(5), 272, 1983.

2.   **Otfert, E. D.,** Allergy to laboratory animals - an occupational disease, Lab Animal, 5(5), 24, 1986.
3.   **Krueger B.,** Lab animal allergies: a manager's perspective, *LAMA Lines,* 2(6), 16, 1987.

d. Waste Collection and Storage*

Shearing of hypodermic needles following injection of infectious or toxic agents or following routine clinical use in animals should be avoided (note that this procedure is not allowed under the bloodborne pathogen standard when humans are the subject). Aerosolization of the contents of the needle can occur during shearing, posing a hazard to humans or other animals in the room.[1]

e. Bedding

Bedding from cages housing animals treated with biohazardous microbial or chemical agents should be considered contaminated and disposed of appropriately.[2] If an incinerator is not on site or available for direct dumping of bedding, then bedding should be double bagged and tagged as hazardous material prior to transport to the incinerator to avoid contamination of personnel or of the work environment. The incineration of carcinogen-contaminated bedding requires an incinerator capable of operating at a temperature range of 1,800°F to 1,900°F with a retention time of 2 seconds.[3]

REFERENCES

1.   **Barkley, W.E., and Richardson, J.H.,** Control of biohazards associated with the use of experimental animals, in *Laboratory Animal Medicine,* Fox, J.G., et al., Eds., Academic Press, Orlando, FL, 1984, *595.*
2.   **Wedum, A.G.,** Biohazard control, in *Handbook of Laboratory Animal Science,* Melby, E.G. and Altman, N.H., Eds., CRC Press, Cleveland, OH, 1974, 196.
3.   Chemical Carcinogen Hazards in Animal Research Facilities, Office of Biohazard Safety, NCI, Bethesda, MD, 15, March 1979.
4.   **Dinimick, R.L., Vogl, W.F., and Chatigny, M.A.,** Potential for accidental microbial aerosol transmission in the biological laboratory, in *Biohazards in Biological Research,* Hellman, A., et al., Eds., Cold Spring Harbor Laboratory, Cold Spring Harbor, NY, 246, 1973.

F. Reproductive Hazards

One of the major concerns of young adults working in the laboratory is the possible effects of their environment on any children they might have. There is very little known about the subject. It is known that smoking, drug, and alcohol use by the mother while carrying a child to term can harm the child but relatively little is known about other substances, nor is there very much known about the problems affecting the male employee. Because of this concern, OSHA under its National Occupational Research Agenda (NORA) has established fertility and reproductive problems as a priority research area. The following two brief excerpts are taken directly from OSHA publications.

"Of those chemicals in the 1994 Register of Toxic Effects of Chemical Substances (RTECS) that are identified reproductive hazards, workers were found by the 1983 National

---

*   In addition to this and the following section, the reader should refer to the later material in the handbook on handling regulated medical wastes (infectious waste) for a more complete treatment of handling not only animal waste contaminated by pathogens contagious to humans, but also other medical wastes potentially contaminated with contagious organisms.

Occupational Exposure Survey (NOES) to be exposed to 1,132 of these chemicals. Some of these chemicals, as well as physical and biologic agents, are in widespread use at work, including various heavy metals (e.g., lead and cadmium), organohalide pesticides, organic solvents (e.g., glycol ethers), chemical intermediates (e.g., styrene and vinyl chloride), waste anesthetic gases, and some anti-cancer drugs. Most of the approximately 70,000 chemicals in commercial use have never been tested for reproductive effects.

Occupational exposures can produce a wide range of effects on reproduction. The effects of parental exposure before conception include reduced fertility, unsuccessful fertilization or implantation, an abnormal fetus, reduced libido, or menstrual dysfunction. Maternal exposure after conception may result in perinatal death, low birth weight, birth defects, developmental or behavioral disabilities, and cancer.

There is considerable uncertainty about the number of workers actually exposed to harmful levels of workplace reproductive hazards and the number of resulting adverse health effects. However, a substantial number of scientific studies have found these effects in specific groups of workers following both maternal and paternal exposure.

- For example, adverse effects on semen quality have been observed in forestry workers and papaya workers following exposure to fumigants containing ethylene dibromide (EDB). Furthermore, adverse effects to semen quality were observed at exposure levels near the NIOSH recommended limit for EDB and greatly below OSHA's current standard for EDB.
- In addition, adverse effects to male fertility (lowered sperm count) were also observed in workers exposed to 2-ethoxyethanol (2-EE) used as a binder slurry in a metal castings process and in shipyard painters exposed to 2EE and 2-methoxy-ethanol (2-ME)."

"Disorders of reproduction include birth defects, developmental disorders, spontaneous abortion, low birth weight, pre-term birth, and various other disorders affecting offspring; they also include reduced fertility, impotence, and menstrual disorders. Infertility is currently estimated to affect more than 2 million U.S. couples (one in 12 couples find themselves unable to conceive after 1 year of unprotected intercourse). Though not all infertile couples seek treatment, it is estimated that about one billion dollars was spent in 1987 on health care related to infertility. In 1991, physician visits for infertility services numbered 1.7 million. Although numerous occupational exposures have been demonstrated to impair fertility (e.g., lead, some pesticides, and solvents), the overall contribution of occupational exposures to male and female infertility is unknown. Moreover, observed global trends in men's decreasing sperm counts have elevated concerns about the role of chemicals encountered at work and in the environment at large.

Birth defects are the leading cause of infant mortality in the United States, accounting for 20% of infant deaths (more than 8,000) each year. Every year about 120,000 babies are born in the United States with a major birth defect—about 3 per 100 live births. The 1992 costs for 17 of the most clinically important structural birth defects and for cerebral palsy were estimated to be about $8 billion. Neural tube defects (which include spina bifida and anencephaly), affect 4,000 pregnancies each year, with each new case of spina bifida having a discounted lifetime cost of $294,000 (1992 dollars). Seventeen percent of all children in the United States have some type of developmental disability. The major developmental disabilities of mental retardation, cerebral palsy, hearing impairment, and vision impairment affect about 2% of all school-age children.

Most birth defects and developmental disabilities are of unknown cause. The overall contribution of workplace exposures to reproductive disorders and congenital abnormalities is not known. Although some specific reproductive hazards have been identified in humans

Table 4.18. Dynamics of Reproductive Toxicology

Stage of Development	Stage of Pregnancy		
	Preconception	Intrauterine	Perinatal
Vulnerable areas	Gametes (sperm, ova) organogenesis Spermatogenesis Qogenesis Fertilization	1st trimester organogenesis, 2nd, 3rd trimester fetus	Infant Lactation
Major developmental effects	Mutagenesis	Teratogenesis	ONS—late transplacental carcinogenesis
Adverse manifestations	Sterility, decreasing fertility, chromo- somal aberrations	Implantation defects, spontaneous abortions	Stillbirth; structural, behavioral, or functional alternations
Parental source of problem	Maternal and paternal	Maternal (3rd trimester)	Maternal (lactation)

(e.g., lead, solvents, and ionizing radiation), most of the more than 1,000 workplace chemicals that have shown abnormal reproductive effects in animals have not been studied in humans. In addition, most of the 4 million other chemical mixtures in commercial use remain untested. Substances and activities that upset the normal hormonal activity of the reproductive system, such as shift work or pesticides that possess estrogenic activity, also need evaluation. Similarly, the effects of physical factors, such as prolonged standing, reaching, or lifting, or the interactive effects of workplace stressors and exposures on pregnancy and fertility have not been rigorously investigated.

Although the total number of workers potentially exposed to reproductive hazards is difficult to estimate, three-quarters of employed women and an even greater proportion of employed men are of reproductive age. More than half of U.S. children are born to working mothers. The vast number of workers of reproductive age together with the substantial number of workplace chemical, physical, and biological agents suggest that a considerable number of workers are potentially at risk for adverse reproductive outcomes. Although the causes of reproductive disorders and adverse pregnancy outcomes are poorly defined, lost productivity and deep suffering by affected individuals and families are evident. The contribution that may be made by occupational factors is largely unexplored, since the reproductive health of workers has only recently emerged as a serious focus of scientific investigation. Identifying reproductive hazards in the workplace has the potential for significantly reducing the multibillion-dollar costs and alleviating the personal suffering associated with disorders of reproduction."

Perhaps the most recent list of chemicals which currently are suspected of being reproductive toxicants for females and males is found at the end of Table 4.3, the "California List "earlier in this chapter. This compilation is updated annually.

Table 4.18 above, taken from the article on Pregnancy in the Laboratory by Dr. Richard F. Desjardins, M.D. in the 4th edition of this handbook, gives a summary of the dynamics of reproductive toxicology. Links to a very large number of references on this subject may be found at the basic Internet reference given below.

REFERENCES

1. **Desjardins, R.F,** *Pregnancy in the laboratory,* in Handbook of Laboratory Safety, 4th Ed., A. Keith Furr Ed., CRC Press, Boca Raton, FL, 1995.

INTERNET REFERENCE

1. http://www.osha-slc.gov/SLTC/reproductive hazards/index.html. Reproductive Hazards.

G. Regulated and Potential Carcinogens

An individual planning to work with carcinogenic material must be prepared to ensure that the laboratory hygiene plan developed to meet the performance standards implicit in 29 CFR 1910.1450, the laboratory safety standard, meets all of the criteria set forth in the standard in terms of quality of the facility, training and information made available to the employee, operational procedures, and availability of personal protective equipment. These considerations should arise at least by the time of purchase of the research material or before. Therefore, the list of chemicals considered to be probable carcinogens was placed in the section on purchasing earlier in this chapter.

1. Carcinogens (Ethylene Oxide)

Ethylene oxide is used here as an example of a carcinogenic compound and is one for which a specific standard was adopted by OSHA relatively recently (August 24, 1984). The standard, in its appendices, provides an unusually complete guide to the safe use of this material. The laboratory safety standard may modify some of the specific requirements, but adhering to them would form the basis of a sound management strategy.

Ethylene oxide (C_2H_4O; CAS No. 75-21-8) is a gas at normal temperatures (boiling point = 10.7°C, 51.3°F). The specific gravity of the gas with respect to the density of air is 1.49. It dissolves readily in water. It has an ether-like odor when concentrations are well above the OSHA PEL, so that it cannot be considered to warn adequately of its presence by its odor. It is a significant fire hazard in addition to being a health hazard. The lower and upper explosion limits are, respectively, 3% and 100%. It will burn without the presence of air or other oxidizers, with a flash point below 0°F (-18°C) and may decompose violently at temperatures above 800°F (444°C). It will polymerize violently when contaminated with aqueous alkalis, amines, mineral acids, and metal chlorides and oxides. It would be classified as a class B fire hazard for purposes of compliance with 29 CFR 1910.155. Locations defined as hazardous due to its use would be class I locations for purposes of compliance with 29 CFR 1910.307.

Although dangerous because of its physical properties, the primary reasons for regulating the material by a separate standard were related to health effects, specifically its identification as a human carcinogen, adverse reproductive effects, and ability to cause chromosome damage. Because of the latter two problems, women who suspect or know that they are pregnant should take special care to avoid exposures above the acceptable limits. There are a number of other adverse health effects in addition to these relatively newly identified problems.

Acute effects from inhalation include respiratory irritation and lung damage, headache, nausea, vomiting, diarrhea, shortness of breath, and cyanosis. Ingestion can cause gastric irritation and liver damage. It is irritating on contact to the eye and skin and can cause injury to the cornea and skin blistering on extended contact. Contact with pressurized, expanding vapor can cause frostbite. Individuals using this material should not wear contact lenses. It has also been associated with mutagenic, neurotoxicity, and sensitization effects.

Safe work practices for ethylene oxide fall into two areas: (1) the normal practices associated with the use of a flammable gas and (2) health practices needed to reduce exposures to the vapors. For the former set of problems and the more common health problems, the procedures are relatively straightforward, the same as for other chemicals with similar properties: keep ignition sources and reactive materials away from the material; do not smoke, eat, or drink in the area; and wear personal protective equipment (goggles, gloves, respiratory protection, and protective clothing) as needed to prevent exposure.

The means of preventing exposure to the gas to protect against the carcinogenic and reproductive hazards are spelled out in considerable detail in the appendices to the OSHA standard (29 CFR 1910.1047) as well as the specific measures that would meet the needs of the standard. A major use of the material is as a sterilizing agent in medical care operations, so much of the material in the standard is concerned with means of avoiding release of the gas or to capture gas that has been released. Other portions of the appendices deal with achieving compliance with other parts of the standard. The basic features of the standard are briefly given below as representative of those of the other regulated carcinogens. Many details are omitted for which the reader is referred to the complete current standard, available from any local OSHA office or online at the Internet reference provided here.

A key provision of the laboratory safety standard is that procedures be developed to ensure that it would be unlikely that any individual on a worst-case basis will exceed a 0.5 ppm action level for an 8-hour TWA exposure. These levels are to be measured without taking into account any respiratory protection provided by personal protective equipment. When any circumstance changes in such a way that the levels of exposure may increase, it will probably be necessary to demonstrate anew that the levels are lower than the action levels. Records of these data must be kept and available for examination. If the action levels are exceeded, steps must be taken to reduce the levels, preferably by means of engineering controls. Medical surveillance provisions of the laboratory safety standard would be invoked by persistent levels above the OSHA action levels.

Probably the most critical requirement of the standard is to ensure that no employee is allowed to be exposed to an airborne concentration level of EtO in excess of 1 ppm as an 8-hour TWA. This must be demonstrated for each employee, although, where the exposure conditions are sufficiently similar, measurements need only be made for representative employees. The method used to make the measurements must be capable of providing an accuracy of ±25% in the range of the PEL of 1 ppm with 95% confidence limits and ±35% in the range of the action limit of 0.5 ppm. It should be noted that a STEL of 5 ppm has also been adopted.

If the measured levels are between 0.5 and 1 ppm, monitoring is required to be repeated at least every 6 months, while if the levels are in excess of 1 ppm, monitoring is required to be done at least every 3 months. The results of the monitoring program must be made available to the employees as well as the outcome of any corrective actions taken to reduce the levels.

A regulated area must be established wherever the airborne concentrations may exceed 1 ppm. Access to this area must be limited to authorized personnel and the number of these persons must be kept at a minimum. The entrances to this area must be clearly marked with the following sign:

DANGER—ETHYLENE OXIDE

CANCER HAZARD AND REPRODUCTIVE HAZARD
AUTHORIZED PERSONNEL ONLY
RESPIRATORS AND PROTECTIVE CLOTHING MAY BE
REQUIRED TO BE WORN IN THIS AREA

Any containers of EtO with the potential for causing an exposure at or above the action level must be labeled with the legend:

DANGER
CONTAINS ETHYLENE OXIDE
CANCER HAZARD AND REPRODUCTIVE HAZARD

The label also must warn against breathing EtO. If EtO is to be used as a pesticide, the container labeling requirements of the Federal Insecticide, Fungicide, and Rodenticide Act (FIFRA) preempt the OSHA requirements.

If the PEL of 1 ppm is exceeded, the employer must establish and implement a written program to reduce actual employee exposures to below this level. Preferably, this protective program should be based on engineering controls and work practices, but also may include the use of approved respiratory protection where alternate measures are not feasible. Approved respiratory protection means those respiratory devices specifically approved for protection for EtO exposure by either the Mine Safety and Health Administration (MSHA) or by the NIOSH, under the provisions of 30 CFR, Part 11. Employee rotation is not an acceptable means of achieving compliance. The laboratory safety plan also must include means of leak detection, and an emergency plan. The plan must be reviewed and revised as needed at least annually. The written emergency plan must provide for equipping employees with respiratory protection. It must include those elements required under 29 CFR 1910.38. Provision must be made for alerting the employees of an emergency and for evacuation of employees from the danger area.

A medical surveillance and consultation program must be available for any employee who may be exposed to EtO at or above the action level of 0.5 ppm for 30 or more days a year, or in an emergency situation. There must be a pre-employment examination, a medical and work history, an annual examination for each year the 30-day criterion is met, and a post-employment examination. In addition to these requirements, examinations may be indicated for employees exposed in an emergency and as soon as possible for any employees who believe that they are exhibiting symptoms of exposure to ethylene oxide. Employees may also request and be given medical advice about the effects of their exposures to EtO on their ability to produce a child. The physician may recommend other examinations. For example, an employee may wish to obtain fertility and pregnancy tests, and they are to be given these tests if the physician considers the tests appropriate under the circumstances. The surveillance program must include a medical and work history, with emphasis placed on the pulmonary, hematologic, neurologic, and reproductive systems, the eyes, and the skin. The physical examination must emphasize the same areas. A complete blood count is to be part of the examination as well as any other appropriate tests designated by the physician.

The results of the tests must be made available to the employer, employee, and others (such as the employee's physician) upon the employee's written authorization. The physician must provide a written opinion including the results of the examination, whether the examination revealed any conditions that the employee's occupational exposure would aggravate, and whether there should be any restrictions placed on the employee or modification to the employee's duties to reduce exposure. The physician must also state that he has discussed the results of the examinations with the employee and any follow-up actions that should ensue. If there are any extraneous medical factors not pertinent to the work-related activities of the employee discovered in the course of the examination, the employee has the right to expect complete confidentiality of this information. The medical records must be maintained, according to the provisions of 29 CER 1910.20, by the employer for the duration of the employee's employment plus 30 years. Much of the supporting data records, such as exposure information, must be kept for a similar period.

In addition to the labeling and signs that have already been discussed, a written laboratory safety plan must be established by the employer. Before an employee is assigned duties which could result in an exposure above the action level, they must be provided training and information about EtO. The information and training program must be repeated at least annually, and must be made current as needed during the course of the year. The program must include information about the requirements of the OSHA standard including: where a copy of it can be obtained, operational procedures, the medical surveillance program, methods available to detect EtO, measures taken by the employer to ensure compliance with the standard, measures the employees can take to protect themselves, the emergency plan, the hazards of EtO, where the current MSDS can be found and how to interpret the information on it, and the details of the laboratory and corporate or institutional industrial hygiene plan.

REFERENCES

1. Occupational Safety and Health Administration, General Industry Standards, Subpart Z -Occupational Health and Environmental Controls, Ethylene-Oxide, 29 CFR 1910.1047, Washington, D.C.
2. Occupational Safety and Health Administration, Occupational Exposure to Hazardous Chemicals in Laboratories, 29 CER 1910.1450, Washington, D.C.
3. Occupational Safety and Health Administration, Employee Emergency Plans and Fire Prevention Plans, 29 CFR 1910.38, Washington, D.C.

INTERNET REFERENCES

1. http://frwebgate.access.gpo.gov/cgi-bin/getcfr.cgi, Ethylene oxide standard (1998 revision)
2. http://www.osha-slc.gov/OshDoc/Fact_data/FSNO95-17.html, Ethylene Oxide Osha fact sheet

H. Neurological Hazards of Solvents

The Environmental Protection Agency (EPA) in August 1993 published a document in the *Federal Register* entitled *Draft Report: Principles of Neurotoxicity Risk Assessment,* and called for comments. As the document stated, there are data indicating that exposure to neurotoxic agents may constitute a significant health problem. Assuming that this is so, a need exists to develop guidelines to assess the hazards. The document covered in some detail current knowledge of the characteristics of neurotoxicity, methods of assessing human neurotoxicity, and the use and efficacy of animal studies in neurotoxicity investigations. The goal is to combine the three key components of a hazard assessment program - hazard identification, dose-response relationship, and exposure assessment - to characterize the risk that neurotoxins may pose to exposed individuals. In addition, under the Toxic Substances Control Act, the EPA (also in August of 1993) required companies that make, import, or produce any of ten high-volume organic solvents to perform a series of four tests on rats for the ten solvents. The tests are functional observational battery (FOB), motor activity neuropathology, and schedule-controlled operant behavior (SCOB). The ten chemicals (numbers in parentheses are the Chemical Abstract Service identifiers) are acetone (67-64-1), 1-butanol (71-36-3), diethyl ether (60-29-7), 2-ethoxyethanol (110-80-5), ethyl acetate (108-10-1), methyl isobutyl ketone (108-10-1), N-amyl acetate(628-63-7), N-butyl acetate (123-86-4), isobutyl alcohol (78-83-1), and tetrahydrofuran (109-99-9). Several major companies objected to testing the first six of these and the use of the SCOB test, and in October of 1993 filed a lawsuit to express their objections.

Many solvents have been recognized or suspected of being carcinogenic or to pose reproductive problems (see Section III.D.4 of this chapter). However, many of the recognized health effects due to solvents are neurotoxic.[*] In commonly available references, such as Sax or Merck, much of the descriptive material on health symptoms due to exposure are based on neurotoxic actions. As with most other health effects, neurotoxic problems may be divided into acute or immediate effects, or those due to chronic exposures which lead to delayed and possibly persistent health changes.

Much of the epidemiological data on health effects has been due to exposures in an industrial setting rather than the laboratory, since the exposures are liable to be at relatively stable levels and involve a relatively small number of solvents, rather than the extremely complicated and rapidly changing laboratory environment.

The primary modes of uptake of solvents by the body are inhalation and by absorption through the skin. The rate of absorption by inhalation is affected by a number of factors, some due to the properties of the material and the interaction of the solvent vapors with the lungs, and some due to other factors such as the concentration of the solvent fumes in air, the duration of the exposure, and the level of exertion at the time of exposure by the exposed individual. The rate of intake is significantly increased by elevated levels of physical activity.

The rate of intake through the skin is dependent upon the duration of the contact, skin thickness, degree of hydration of the skin, and possible breaks in the integrity of the skin, i.e., injuries or skin disorders. An example of the comparative rates of intake by the two routes is that an immersion of both hands in xylene for 15 minutes gives about the same levels in the blood as an exposure to an airborne concentration of 100 ppm for the same period. However, this cannot be assumed to be an accurate reflection of the comparative rates for other solvents.

After exposure, the material is often transformed by the liver into less toxic water-soluble compounds or in some cases into more toxic intermediate metabolites. In other cases, the solvents are lipophilic, i.e., may be taken up and accumulate in lipid-rich tissues such as the nervous system. The acute, short-term neurotoxic effects result from action of the solvent on the CNS. The effects may range from symptoms resembling intoxication to CNS depression, psychomotor impairment, narcosis, and death from respiratory failure. At intermediate levels of exposure, common effects are drowsiness, headache, dizziness, dyspepsia, and nausea. Short-term effects may cause mood changes as well, as reflected by increasing feelings of physical and mental tiredness during an exposure period corresponding to a typical workshift, for example.

The effects of extended or chronic exposures have been divided into three categories of varying severity at two relatively recent international workshops— minimal, moderate and pronounced— although the nomenclature differed slightly at the two conferences. The last category, which so far has not been observed in an occupational situation but has in persons deliberately exposing themselves to solvent fumes, would be reflected in serious deterioration of the nervous system, including mental capacity and function. The effects of such extreme

---

[*] The information in the remainder of this section is derived primarily from NIOSH Current Intelligence Bulletin 48, Organic Solvent Neurotoxicity, DHHS (NIOSH) Publication No. 87-104, National Institutes for Occupational Safety and Health, U.S. Department of Health.

exposures would be only partially reversible at best.

The least severe category would be characterized by deterioration in memory function and ability to concentrate, physical fatigue, and irritability, while the second level would involve sustained mood and personality changes, as well as further deterioration of intellectual functions, including learning capacity. At least some of the effects of chronic exposure appear to persist well beyond the termination of the exposure, and may be permanent.

Acute effects appear to be caused by the solvent itself, while the chronic effects may be associated with the intermediate metabolic reaction products. The effects of short-term exposures appear to be reversible, while the effects of prolonged exposures leading to changes in nerve tissue may be irreversible. Although the data on chronic exposure toxicity are not as abundant or definitive as desired, the data that are available definitely appear to support a conservative approach to personnel exposure. Levels to which an individual is exposed, especially over an extended period, should emulate the NRC "as low as reasonably achievable" (ALARA) philosophy. The various levels established as regulatory by OSHA (PEL), recommended exposure levels (REL) by NIOSH, or the TLVs recommended by the ACGIH should be adopted as a maximum permitted occupational level, using the lowest of the three values as a guide. An action level of 50% of the level chosen is recommended as a trip point for initiating remedial steps.

Adequate general ventilation, provision of appropriate containment equipment (fume hoods, safety cabinets, or localized exhaust systems), procedural controls, or use of personal protective equipment, respiratory protection, and protective gloves, clothing, and eye protection are all part of a possible program to reduce exposures.

It is difficult in the laboratory to monitor the exposure levels as recommended above, but monitoring should be performed wherever possible. Even partial data or data taken under nonstandard conditions are better than none at all. As a supplement to a monitoring and exposure limitation program, those individuals who actively use solvents as a routine part of their job a significant portion of their time should participate in a medical surveillance program. Individuals should receive a pre-employment examination, which should include a prior work history and medical history. The examination should emphasize the nervous, cardiovascular respiratory, and reproductive systems as well as the liver, kidneys, blood, gastrointestinal tract, eyes, and skin. A comprehensive blood panel should be run as well as a complete blood count and a urinary test. Some special tests might be suggested by the physician, such as a cholinesterase enzyme test for information on the CNS. A blood serum sample might be stored for later comparison. Periodic reexaminations should be given, annually if the exposure is heavy but no less often than every 5 years.

Table 4.19 provides, for a number of common solvents, current recommended PELs, RELs, and TLVs for 8-hour TWAs. In most cases, The OSHA PEL values were lowered, for a time to the ACGIH TLVs at the time, but this was subsequently rescinded. In the context of this table, those solvents that affect the CNS are so noted with (cns) while those that affect the peripheral nervous system are indicated with (pns).

As can be seen from Table 19, the values from the three sources frequently disagree. Usually, the ACGIH values are the most current, while the OSHA levels often go back to the original adoption of the OSHA act. The OSHA values are binding legally, however, while those of the other two are guidelines only. The NIOSH values, where available, are often the most conservative.

Table 4.19 OSHA PELs, NIOSH RELs, and ACGIH TLVs for Some Organic Solvents[a]

Compound (CAS Number)	OSHA PEL (CEILING)	NIOSH REL (Ceiling)	ACGIH TLV (STEL 15 min)
Alkanes, C5-C8			
Pentane (109-66-0)	1000	120 (610/15 min)	600 (750)
Hexane (110-54-3)	500	50 (510/15 min)	50
Heptane (142-82-5) (pns)	500	85 (440/15	400 (500)
Octane (111-65-9)	500	75 (385/15 min)	300 (375)
Ally chloride (107-5-1)	1	1 (2/15 min)	1 (2)
Benzene (71-43-2) (cns) (carcinogen)	1	0.1 (1/15 min)	10
Benzyl chloride (100-44-7)	1	(1/15 min ceiling)	1
Carbon disulfide (75-15-0) (cns,pns)	20 (30 ceiling; 100/30 min)	1 (10/15 min)	10
Carbon tetrachloride (56-23-5) (cns, carcinogen)	10 (25 ceiling; 200/5 min in 4 hours)	2 (1 hour)	5
Chloroethane (75-00-3)	1000	Handle with care	1000
Chloroform (67-66-3) (carcinogen)	50 ceiling	2 (1 hour)	10
Chloroprene (126-99-8) carcinogen	25	(1/15 min ceiling)	10
Cresol (1319-77-3) (cns)	5	2.3	5
Di-2-ethyl-hexylphthalate (117-81-7) (carcinogen)	5 mg/m^3	5 (10) mg/m^3	5 (10) mg/m^3
Dioxane (123-91-1) (carcinogen)	100	(1/30 min ceiling)	25
Epichlorohydrin (106-89-8) (carcinogen)	5	minimize occupational exposure	2
Ethyene dibromide (106-93-4) (carcinogen)	20 (50/5 min)	0.045 (0.13/15 min)	Suspect skin carcinogen

Table 4.19   OSHA PELs, NIOSH RELs, and ACGIH TLVs for Some Organic Solvents[a] (continued)

Compound (CAS Number)	OSHA PEL (CEILING)	NIOSH REL (Ceiling)	ACGIH TLV (STEL 15 min)
Ethylene dichloride (107-06-2) (cns) (carcinogen)	50 (200/5 min in 3 hours	1 (2/15/min)	10
Furfuryl alcohol (98-00-0)	50	10 (15)	10 (15)
Glycol ethers			
2-Methoxyethanol (110-86-4) (cns)	25	Reduce exposure to lowest feasible level	5 (skin)
2-Ethoxyethanol (110-80-5)	200	Reduce exposure to lowest feasible level	5 (skin)
Isopropyl Alcohol (67-63-0)	400	400 (500)	400 (500)
Ketones			
Acetone ((67-64-1)	1000	250	750 (1000)
Methyl ethyl ketone (78-93-3)	200	200	200 (300)
Methyl -n-propyl ketone (107-87-9)	200	150	200 (2500)
Methyl -n-butyl ketone (591-78-6)	100	1	5
Methyl -n-amyl ketone (110-43-0)	100	100	50
Methyl -isobutyl ketone (108-10-1)	100	50	50 (75)
Diisobutyl ketone (108-83-8)	50	25	25
Cyclohexanone (108-94-1) (cns)	50	25	25 (skin)
Mesityl oxide (141-79-7) (cns)	25	10	15 (25)
Diacetone alcohol (123-42-2)	50	50	50
Isophorone (78-59-1)	25	4	5 (ceiling)
Mercaptans			
Butanethiol (Butyl mercaptan) (109-79-5) (cns)	10	(0.5/15 min)	0.5
Etanethiol (Ethyl mercaptan) (75-08-1)	(10 ceiling)	(0.5/15 min)	0.5

Table 4.19 OSHA PELs, NIOSH RELs, and ACGIH TLVs for Some Organic Solvents[a] (continued)

Compound (CAS Number)	OSHA PEL (CEILING)	NIOSH REL (Ceiling)	ACGIH TLV (STEL 15 min)
Methyl alcohol (67-56-1)	200	200 (250)	200 (250)
Methyl bromide (74-83-9) carcinogen	(20 ceiling)	Reduce exposure to lowest feasible level	5 (skin)
Methyl chloride (74-87-3) (cns)(carcinogen)	100 (200 ceiling, 300/5 min in 3 hours)	Reduce exposure to lowest feasible level	50 (100)
Methylene chloride (75-09-2) (cns)(carcinogen)	500 (1000 ceiling, 2000/5 min in 2 hours)	Reduce exposure to lowest feasible level	50
Methyl iodide (74-88-4) (cns)(carcinogen)	5 (skin)	Reduce exposure to lowest feasible level	2 (skin)
Nitriles			
Acetonitriles (75-05-8) (cns)	40 (60)	20	40 (60)
Tetramethyl succinonitrile (3333-52-6) (cns)	0.5 (skin)	0.5 (skin)	0.5 (skin)
2-Nitropropane (79-46-9) (cns)(carcinogen)	25	25	10
Styrene monomer (127-18-4)	100 (800 ceiling, 600/5min in 3 hours)	50 (100 ceiling)	50 (100)
1,1,2,2 Tetrachloroethane (79-34-5) (cns)(carcinogen)	5	1 (skin)	1 (skin)
Tetrachloroethylene (127-18-4) (cns)(carcinogen)	200 (200 ceiling, 300/5 min in 3 hours)	Minimize workplace exposure levels, limit no. of workers	25 (100)
Toluene (108-88-3) (cns)	200 (300 ceiling, 500/10 min)	100 (150)	100 (skin)

Table 4.19   OSHA PELs, NIOSH RELs, and ACGIH TLVs for Some Organic Solvents[a] (continued)

Compound (CAS Number)	OSHA PEL (CEILING)	NIOSH REL (Ceiling)	ACGIH TLV (STEL 15 min)
1,1,1-Trichloroethane (71-55-6) (cns)	350 (450)	(350 ceiling/15 min)	350 (450)
1,1,2-Trichloroethane (79-00-5) (cns) (carcinogen)	10 (skin)	10 (skin)	10 (skin)
Trichloromethane (79-01-6) (cns)(carcinogen)	100 (200 ceiling, 300/5 min in 2 hours	25	50 (100)
Xylene (1330-20-7) (cns)	100	100 (200/10 min)	100 (150)

[a] All values are inparts per million except where noted.

REFERENCES*

1. *TLVs⁰, Threshold Limit Values and Biological Exposure Indices for 1993-94,* American Conference of Governmental Industrial Hygienists, Cincinnati, OH, 1993.

2. **Astrand, I.,** Uptake of solvents in the blood and tissues of man, *Scand. J. Work Environ. Health,* 199, 1975.

3. **Bakei E.L. and Seppalainen, A.M.,** Session 3: human aspects of solvent neurological effects, report on the workshop session on clinical and epidemiological medicine, *Arch. Md. Health,* 13, 581, 1986.

4. **Baker E.L. and Vyskocil, J.,** The neurotoxicity of industrial solvents: a review of the literature, *Am. J. Med,* 8, 207, 1956.

5. **Bird, M.,** Industrial solvents: some factors affecting their passage into and through the skin, *Ann. Occup. Hyg.,* 24, 235, 1981.

6. **Browning, E.,** *Toxicity and Metabolism of Industrial Solvents,* Elsevier Publishing, Amsterdam, The Netherlands, 1965.

7. **Cherry, N., Venables, H., and Waldron, H.,** The acute behavioral effects of solvent exposure, *J. Soc. Occup. Med,* 33, 13, 1983.

8. **Gamberale, E.,** Behavioral effects of exposure to solvents, experimental and field studies, in *Adverse Effects of Environmental Chemicals and Psychotropic Drugs,* Vol. 2, Horvath, M., Ed., Elsevier Publishing, Amsterdam, The Netherlands, 1976, 111.

9. **King, M., Day, R., Oliver, J., Lush, M., and Watson, J.,** Solvent Encephalopathy, *Br. Med. J.,* 283, 663, 1981.

10. **Knave, B., Anselm-Olson, B., Elofsson, S., Gamberale, F., Isaksson, A., Mindus, P., Persson, H.E., Struwe, G., Mennberg, A., and Westerhoim, P.,** Long-term exposure to jet fuel. II. A cross-sectional epidemiologic investigation on occupationally exposed industrial workers with special reference to the nervous system, *Scand. J. Work. Environ. Health,* 4, 19, 1978.

11. **Lindstrom, K.,** Changes in psychological performances of solvent-poisoned and solvent-exposed workers, *Am. J. Ind. Med.,* 1, 69, 1980.

12. Criteria for a Recommended Standard: Occupational Exposure to Refined Petroleum Solvents, U.S. Department of Health, Education and Welfare, Public Health Service, Centers for Disease Control, National Institute for Occupational Safety and Health, DHEW (NIOSH) Pub. No. 77-192, Cincinnati, OH,1977.

13. **Orbaek, P., Risberg, J., Rosen, I., Haeger-Aronson, B., Hagstadius, S., Hjortsberg, U., Regnell, G., Rehastrom, S., Svensson, K., and Welinder, H.,** Effects of long-term exposure to solvents in the paint industry *Scand. J. Work Environ. Health,* 11 (Suppl. 2), 1, 1985.

14. **Politis, M., Schaumburg, H., and Spencer, P.,** Neurotoxicity of selected chemicals, in *Experimental and Clinical Neurotoxicology,* Spencer, P. and Schaumburg, H., Eds., Williams and Wilkins, Baltimore, 1980, 613.

15. **Seppalainen, A.M.,** Neurophysiological aspects of the toxicity of organic solvents, *Scand J. Work. Environ. Health,* 11 (Suppl. 1), 61, 1985.

16. **Seppalainen, A.M. and Antti-Poika, M.,** Time course of electrophysiological findings for patients with solvent poisoning, *Scand. J. Work. Environ. Health.,* 9, 15, 1983.

17. **Seppalainen, A. M., Husman, K., and Martenson, C.,** Neurophysiological effects of long-term exposure to a mixture of organic solvents, *Scand. J. Work. Environ. Health,* 4, 304, 1978.

18. **Seppalainen, A.M., Lindstrom, K., and Martelin, T.,** Neurophysiological and psychological picture of solvent poisoning, *Am. J. Ind. Med,* 1, 37, 1980.

19. **Spencer, P. and Schaumburg, H.,** Organic solvent neurotoxicity, facts and research needs, *Scand. J. Work. Environ. Health,* 11 (Suppl. 1), 53, 1980.

20. **Toftgard, R. and Gustafsson, J.,** Biotransformation of organic solvents, a review, *Scand J. Work. Environ. Health,* 6, 1, 1980.

21. **Valciukas, J., Lilis, R., Singer, H., Glickman, L., and Nicholson, W.,** Neurobehavioral changes among shipyard painters exposed to solvents, *Arch. Environ. Health,* 40, 47, 1985.

22. **Waldon, H.A.,** Solvents and the brain, *Br. J. Ind. Med.,* 43, 73, 1986.

---

*   The references are the more general and comprehensive ones from the list of references in the original article with some more recent ones added.

23.   Organic Solvents and the Central Nervous System, EH5, World Health Organization and Nordic Council of Ministers, Copenhagen, 1985, 1.

24.   EPA Draft Report: Request for Comments on Principles of Neurotoxicity Risk Assessment, *58 Fed Reg., 41558,* Aug. 4, 1993.

25.   **Parkinson, D.K. et al.,** *Am. J. Indust. Med,* 17, 661, 1993.

26.   National Research Council, *Environmental Neurotoxicology,* National Academy Press, Washington, D.C., 1992.

27.   **Tilson, H. and Mitchell, C.,** Eds., *Neurotoxicology,* Raven Press, New York, 1992.

IX. SPILLS AND EMERGENCIES

A chemical spill is probably the most common type of laboratory accident and potentially one of the most serious if the material gives rise to hazardous vapors, interacts with the laboratory environment in a violent physical fashion, e.g., a fire, or is toxic or corrosive upon contact with a person's body. Most accidents involving chemical spills do not have such dramatic consequences, but they must all be handled correctly. The response to an emergency spill may invoke meeting the requirements of 29 CFR 1910.120 - Hazardous Waste Operations and Emergency Response. In most facilities using chemicals, there will need to be a plan to comply with the requirements of this standard.

A. Small- to Moderate-Scale Spills

Chemical spills generally involve only small quantities of materials, such as would be contained in a single reagent container or reaction flask. Unless the chemicals have unusually hazardous properties, procedures to correct such minor spills are relatively straightforward. If the material is a solid, it is simply swept up into a container with which it will not react for disposal. However, relatively few materials may then be disposed of as ordinary trash. Custodians should not, under normal circumstances, be expected to remove trash containing chemical materials from laboratories. Chemicals known to be harmless if placed in a municipal landfill and which are to be disposed of in this way should be placed in a separate container, labeled with the identity of the material, and marked "safe for disposal in regular trash." Some materials which are soluble in water or, if liquid, miscible in water may also be disposed of into the sanitary system, while others may be rendered harmless by reaction with other chemicals in the laboratory. However, before any of these things are done, it should be confirmed that the material does in fact not fall under the provisions of the RCRA which prohibits such disposal procedures. If restrictions exist, the materials must be treated as hazardous waste. There is a movement to allow more laboratory treatment of chemical wastes under consideration by EPA.

Individuals handling even small quantities of hazardous waste such as might result from the remediation of a small spill should be provided with basic training on what to do with the materials, how to identify and label them, suitable containers for various classes of waste materials, and, of course, basic safety measures needed to protect themselves and others. This level of training should be provided to every new employee in an area or to each new graduate student in an academic institution. The material in this section will apply to handling spills that do not involve a release of a hazardous material into the environment, and that can be handled by in-house personnel, meaning either laboratory employees or safety employees who have received appropriate training prescribed by the standard. Unless the training is provided, an employee or student should not be allowed to be involved with cleaning up a spill or handling hazardous waste. It is possible that the organization could be cited under 1910.120 otherwise. Basic training such as this may be provided using the employer's own staff.

1. Spill Response Procedures

Spilled liquids often may be diluted with water and simply mopped up or, in some cases, eliminated by spreading an absorbent material, such as vermiculite or a clay absorbent (such as calcium bentonite) on the spilled material or, a bit more neatly, but much more expensively, by placing pillows or pads containing an absorbent material on the liquid, after which the absorbent material is collected into containers for later disposal.

Spills of larger quantities of solids are usually no different from smaller spills except in scale. The spilled solids simply remain as they fell. As long as the solids do not react with materials with which they may have come into contact (such as a reactive metal like sodium might do) or the dust is not a breathing hazard, the clean up is simply mechanically larger, with more containers being set aside for evaluation as to the mode of disposal, i.e., to determine if they are to be treated as hazardous waste or not.

Spills of liquids, even of relatively innocuous materials, are messier and more likely to come into contact with other materials with which they may interact. The use of absorbent materials (such as loose absorbent, spill control pillows, and absorbent pads) which should be immediately available in the laboratory, quickly confine the area of the spill, the immediate problems are minimized, and the clean-up is considerably easier.

Relatively few laboratory spills involve totally innocuous materials. However, there are obviously degrees of hazards associated with spilled materials, from the mildly troublesome to IDLH materials. The risk also depends upon the exposure mode or the character of an ensuing physical hazard. A material which emits deadly fumes is infinitely more dangerous to most laboratory personnel than one which is highly corrosive to tissue, although the latter is bad enough. Unless an individual is directly injured by contact with the material, a corrosive hazard may be avoided by removing oneself from the immediate area, and protective clothing is often enough to protect those engaged in the ensuing clean up. However, generation of deadly vapors or gases will usually mandate an evacuation of at least the laboratory and perhaps the entire building. Individuals correcting the latter situation will require air-supplied breathing apparatus and often wear garments protecting their entire body from contact with the airborne materials. Similar and often more cumbersome protective gear will be required if a fire occurs after a spill of flammable materials. Remedying the situation involving a corrosive spill is very likely to be within the capacity of laboratory employees, but incidents involving toxic gases or fire would almost certainly mandate the participation of trained emergency personnel. Even wearing the self-contained suits will require prior formal training.

The following material will address handling spills and emergencies which are confined to the laboratory. Those emergencies not confined to at least the building in which they originate will be treated in a later section in the context of the specific requirements of the OSHA standard.

There should be some basic emergency equipment and supplies readily available to every laboratory using hazardous chemicals. It is essential that some items be available within the laboratory itself. For example, in any laboratory in which flammables are stored, OSHA standards (19 10.106(d)(7)(b)) require that at least one 12-B portable fire extinguisher be located not less than 10 feet nor more than 25 feet from a flammable material storage area within a building. If other types of chemicals are in use, other types of fire extinguishers should be available, such as class D units if reactive metals are stored or in use or pressurized water units if there are substantial amounts of ordinary combustible materials present.

Spill control materials to absorb spilled materials are available commercially. Some materials are universal, and some are intended to not only absorb materials, but also to neutralize specific materials such as acids, hydrofluoric acid, caustics, and mercury. Some of the products for coping with acids and caustics contain color indicators to determine when the spill is neutralized. If bought in convenient small packages suitable for cleaning up modest spills of about half a liter, the cost per unit of absorbent is high for these commercial products. However, they can usually be bought in small drums at a substantial saving. If bought in this way, the materials can be repackaged into convenient smaller sizes for use in individual laboratories. For many purposes, calcium bentonite to absorb liquids, bought in bulk, serves very nearly as well as most commercial absorbents and is much less costly.

Whether commercial kits of absorbent materials or cheaper substitutes are used, some capability to quickly soak up spilled materials needs to be readily available. Bulk quantities can be kept in a central location for replenishing supplies in individual laboratories or for use in atypical larger spills, but a sufficient supply should be kept within an individual laboratory to use on a spill of up to a gallon or two.

In addition to absorbent material, the following items should be maintained in an individual laboratory (Chapter 1 of this handbook also provides a more comprehensive list of items and equipment that should be available within the building, although not necessarily within the laboratory):

- Bags, large, 6-mil polyethylene
- Brooms
- Brushes, hand
- Bucket, plastic (polyethylene)
- Containers, plastic (5-gallon)
- Coveralls, chemically resistant treated, lightweight
- Dustpan
- Gloves, chemically resistant
- Goggles, chemical splash
- Mops
- Paper, plastic-backed, absorbent roll
- Paper towels, regular
- Respirators, organic, acid, dust, caustics
- Scoops, shovel
- Shoe covers, high-topped, chemically resistant
- Soap, detergent
- Tape, duct

For typical laboratory spills, many of these items will not be needed, but all of them could have a use, depending upon the type of spill. For example, an acid spill could quickly destroy a person's shoes during the clean up, and drops of acid on a person's clothes would ruin them. Since those doing this clean up work are likely to be laboratory personnel, not only should they be protected against injury, but they also should not be expected to incur any economic loss.

Large quantities of each item should not be needed in a typical spill kit. One or two persons actually working on the clean up, with one person bringing supplies and taking waste away is probably about optimum for an individual laboratory, unless the spill is unusually large. Aisle widths in the typical laboratory would preclude easy access of more than a few persons to the spill at a time, so that even if relatively complete protective clothing and equipment were needed, no more than about three sets would be needed at a time.

None of the equipment listed above requires an extraordinary level of training to use properly. It should be possible for most technically trained personnel to clean up fairly substantial spills safely. There are some straightforward guidelines to aid the user in preventing any material from getting on their persons. If any substantial amount of the equipment were needed, it would be desirable for the work to be done under the supervision of a person trained as required by 1910.120.

Chemically resistant clothing generally snaps, buttons, or zips up the front, and although the two edges overlap, material can still enter the front seam. Overlapping layers of duct tape will seal this opening. The sleeves are usually loose fitting and should be folded over around the outside of the glove, and again duct taped, with the tape in contact with both the sleeve and the glove. The trouser cuffs should be brought down over the top of the shoe covers, and folded tightly around them. Duct tape is then used to seal this opening by wrapping it around the ankle so that the adhesive is in contact with both the shoe cover and the coverall. With this done, the entire body below the neck is protected from incidental contact with materials. With the hands and feet covers attached to the body of the coverall, movement will tend to strain the garment unless there is some slack in the fit. It would be desirable if the coverall or suit were somewhat oversized to provide the needed freedom of motion. The head can be covered with a hood of the same material as the coverall and the face protected with a full-face respirator, if this is required. Otherwise, a half-face respirator and goggles will provide almost as complete protection.

There are two notes of caution about the apparel described above. Lightweight coated garments are usually not rated for chemical protection, although for the level of contact which should be experienced if one works carefully the wearer should be reasonably well protected against light exposures for the time needed to clean up a minor spill. If substantial contact with the chemical or its vapors is anticipated, a heavier coverall designed to provide protection against chemicals should be substituted. Another problem with the coated clothing is that it does not "breathe," and after a relatively short period—30 minutes to 1 hour depending upon the level of exertion— it is necessary to cease work for a period, open the coveralls, and "cool-off." Heat exhaustion can occur and even heat stroke. The latter can be fatal. However, it should be possible to clean up most small to moderate laboratory spills in less than 1 hour.

If the outside of the coverall is contaminated, there is a definite procedure to be used to avoid the outside surface from coming into contact with skin and clothes. If a hood and face mask are worn, these should be removed first. The duct tape sealing the hood to the coverall is peeled off and then the hood is grasped near the lower edge of the hood and peeled. backward so that it tends to turn inside out. The hood is then put in a plastic bag. The respirator (and goggles, if worn separately) is then removed and placed in a plastic bag, since it is likely to be worn again, whereas the coveralls possibly may not be reused. The duct tape down the front of the coverall is then removed and the front undone. The two sides are then separated and pulled back off the shoulders so that the inside is exposed, the outside doubling back over itself. The duct tape is removed from one sleeve, and the arm of the coverall on that side is pulled down over the hand, so that the sleeve is turned inside out. The operation is then repeated on the other side, to free both arms. The duct tape is removed from the cuffs of the pants and the leg of the coverall pulled down over the foot so that the leg is turned inside out. The operation is then repeated on the other leg. The coverall is now off, with the exposed surface being the clean, interior surface. The coverall is put in the plastic bag with the hood. The shoe covers can then be removed carefully (so that the outside does not come into contact with the skin and clothing), and placed in the plastic bag with the clothing. Finally, the gloves are removed in such a way that they also are turned inside out and placed into the contaminated clothing bag. If carefully done, there should be no opportunity for the contaminated outer surface of any of the protective gear from ever coming into contact with the wearer.

The clean up also has an optimum procedure. If two persons are working directly on the spill, then a third person should be available to bring fresh supplies and to take waste away. Individuals working directly on the clean up should not leave the immediate vicinity to avoid contaminating a wider area. It normally is desirable for spills of substantial sizes and involving materials of significant hazard to establish a formal entrance and exit control point, the side including the accident area which is to be considered dirty and the other maintained clean. The clean up should start at the perimeter of the contaminated area and move inward so as to steadily reduce the area involved. In principle, the workers should clean in front of themselves so that they create a clean area in which to work. A previously cleaned area should not normally become contaminated again if this procedure is followed.

Any waste from the dirty area should be placed in plastic pails and passed to the person outside the dirty area. If the pails are contaminated they should be placed in double, clean, heavy-duty (preferably 6 mil or better) plastic bags before being set on a floor area covered with plastic-backed absorbent paper.

Once the spilled chemical has been removed from the surface, it is always good practice to thoroughly scrub or mop the previously contaminated area. Occasionally, especially if the floor is permeable as a wooden or untreated concrete floor could be, the contamination may have penetrated the surface and a layer of the floor might have to be removed.

These procedures seem very formal on first reading if one is not familiar with de-contamination work. However, it takes very little additional time to do the job properly. The procedures ensure that no one is likely to be injured or that no clothes are damaged in the process, and it is unlikely that the clean up will have to be repeated.

For large spills, especially as the hazards associated with the material become more serious, it would be highly desirable for the work to be done under the guidance and assistance of an experienced safety professional. At some higher level of risk, the danger to inexperienced personnel, as most laboratory personnel usually are in the context of a serious chemical emergency, should cause one to consider not using laboratory personnel at all, other than as an information resource. These are judgment calls to be made by the organization's emergency coordinator, who should be called to the scene of any major chemical emergency, especially ones in which a building has been evacuated. Really large emergencies may be beyond even the most experienced local personnel, and hazard material response teams should be called in from outside agencies as quickly as possible. These are often available as a state resource. One of the more important factors in reducing the scope of most emergencies is prompt, effective action. If local capacity to deal with an emergency is questionable, outside aid should be called for immediately while local efforts continue to protect human life and confine the scope of the incident. Most emergency groups would rather be called for unnecessarily rather than to arrive at a scene where the situation has deteriorated to the point of being out of control. Personal considerations of assignment of blame should not be a factor in delaying a request for help.

B. Large-Scale Releases of Chemicals

In the last few years, there have been some noteworthy releases of chemicals from facilities using chemicals, which have caused substantial numbers of injuries and deaths in nearby communities. There already had been a substantial movement in many states to require operators of facilities using hazardous chemicals to provide information on these chemicals to the nearby communities, and to work with the communities on emergency planning. Many responsible chemical firms had individually begun to provide such data. The accidents which occurred in 1984 in Bhopal, India, and in 1985 in Institute, West Virginia, undoubtedly provided additional motivation for federal action in this area. On October 17, 1986, the Emergency Preparedness and Community Right-To-Know Act was signed into law. This law is more commonly known as Title III of the SUPERFUND Amendments and Reauthorization Act (SARA, Title III), and this appellation will be used henceforth in this document. This act extended and revised the authorities established under the original SUPERFUND Act (the Comprehensive Environmental Response, Compensation, and Liability Act of 1980, CERCLA). The extended act specifically establishes new authorities for emergency planning and preparedness, community right-to-know reporting, and toxic chemical release reporting.

The OSHA standard covering hazardous waste operations and emergency response, 29 CFR 1910.120, is clearly intended primarily for hazardous waste site operations, but the standard covers all operations authorized under RCRA (Resource Conservation and Recovery Act) and which are required to have a permit or interim status under EPA, pursuant to 40 CFR 270.1 or from a state agency pursuant to RCRA. If an organization or institution is a large generator of hazardous waste, then they are covered if they have personnel who collect and manage the hazardous waste prior to disposal and if they respond to chemical emergencies. A very large number of organizations fall under this definition. Training to comply with this standard is known commonly as HAZWOPER training as a convenient term for "hazardous waste operations." Many commercial firms offer HAZWOPER training.

Hazardous Waste Operations and Emergency Response and SARA Title III requirements will be covered in the following two sections. In many ways, they are two different aspects of the same effort, to ensure that hazardous materials are handled in such a way as to minimize risks to employees, the public, and the environment.

1. Hazardous Waste Operations and Emergency Response

Under 29 CFR 1910.120, Hazardous Waste Operations and Emergency Response, an employer who has employees engaged in managing hazardous wastes and responsible for responding to emergencies involving the wastes and other chemical incidents is required to have a written safety and health program. Programs already in place to satisfy other state and federal regulations are acceptable if they already cover or can be modified to cover hazardous waste operations.

In either case, they must contain the following elements:

- An organizational structure
- A comprehensive work plan
- A site-specific safety and health plan
- A safety and health training program
- A medical surveillance program
- The employer's standard operating procedures for safety and health
- Any necessary interface between the general program and site-specific activities

Relatively few commercial research organizations or academic institutions operate their own treatment, storage, disposal and recycling facilities, although a few do. These larger operations should be familiar with all of the rules and regulations affecting their personnel. However, this book is intended for laboratory facilities of all sizes. The program as described in this section will assume that the organization is a large scale generator of hazardous waste, i.e., more than 1000 kg per month but uses a commercial hazardous waste disposal firm. For a large scale generator, there will be typically at least one full-time employee, and frequently several, who will be covered by the standard.

2. Safety and Health Program
a. Organizational Structure

The "site" for most research facilities, and especially for academic facilities, can be considered as encompassing all of the buildings in which chemicals are used, since hazardous waste operations are conducted in these buildings, such as collection of the wastes, or more strictly as the area where hazardous waste is stored temporarily while awaiting being picked up. There can be a primary storage site and several satellite storage sites. Hazardous wastes are not only stored in these areas for up to 90 days, but preparation of some of the wastes for disposal, such as bulking of compatible liquids and segregation into various classes is also done. Some organizations pack their own wastes and are simply spot checked by the disposal firm. These storage areas most closely match the intended scope of the standard. Since this situation is generally much simpler than the operations intended to be covered by the standard, some aspects of the standard will be passed over. However, the hazardous waste operational group should consider carefully the entire standard in preparing their own plan.

The employees who perform hazardous waste management tasks are usually members of the environmental health and safety group or department who run the program and who also provide the required safety oversight. The organizational structure for this type of operation might identify the hazardous waste manager as the general supervisor and responsible safety and health official. If an alternate is needed, the individual to whom the safety and health manager reports can be identified as the ultimate authority for safety and health. The organizational structure should also identify all other personnel associated with the operations and their general functions and responsibilities. The organizational structure would include an organizational chart identifying the lines of authority, responsibility, and lines of communication, such as the chart shown in Chapter I, with the waste operations division delineated more fully.

b. Comprehensive Work Plan

The comprehensive Work Plan must include the following components:
1. Normal operating procedures and activities required for incident response and clean up
2. Definition of the work tasks and objectives and methods used to accomplish them
3. Personnel requirements
4. Provisions for implementation of the required training and information programs
5. Provisions for the implementation of the medical surveillance program

c. Site-Specific Safety and Health Plan

The safety and health plan shall, at a minimum, include:
1. A hazard analysis for each task and operation involved in managing the hazard waste and emergencies
2. Employee training assignments to ensure compliance
3. Personal protective equipment (PPE) to be used by the employees for the different tasks and operations
4. Medical surveillance requirement
5. Description of the air monitoring programs, including methods of calibration and maintenance of the monitoring equipment
6. Site control measures
7. Decontamination procedures
8. An emergency response plan, including necessary PPE and other equipment
9. Confined space entry procedures, if applicable
10. Spill containment program

d. Hazard Identification and Relevant Information

The areas where hazardous chemicals are stored prior to shipment and where any ancillary operations are conducted should be carefully evaluated for hazards. Access and evacuation routes should be clearly marked and checked to ensure freedom from obstruction. Any very hazardous materials, especially any IDLH materials, should be in separate, well-marked areas. Means of annunciating an emergency alarm should be readily available. Assistance agencies, e.g., fire departments, rescue squads, police, and haz-mat teams, should be identified and standard procedures for invoking their assistance prepared.

e. Personal Protective Equipment

Personal protective equipment should be available for evacuation, such as 5-minute escape air packs, and for normal operations or for reentry into an area in which a chemical emergency has occurred. Among other items needed would be positive-pressure, supplied-air units, protective clothing, goggles, gloves, head covers and shoe covers, appropriate to the conditions. These should be available so that they could be accessed quickly. However except for the escape air packs, they should be stored outside the immediate area so that they would remain accessible in the event of an incident.

f. Monitoring

Monitoring equipment appropriate to the hazards present shall be available. This may include organic vapor detectors of various types, combustible gas detectors, oxygen detectors, detector tubes and pumps, and specialized detectors such as mercury detectors.

3. Hazard Communication Program

A program embodying the requirements of the OSHA 29 CFR 1910.1200 hazard communication program shall be in effect.

a. Training

The standard has specific initial training requirements that have become known as HAZWOPER training. Hazardous waste personnel are required to receive either 40 hours of training prior to working on a site, followed by 3 days of field training under the supervision of a knowledgeable individual, or 24 hours of training followed by 1 day of supervised field experience depending upon the type of site involved and the character of their duties. Academic training and work relevant experience can be substituted for some or all of these requirements if the alternatives can be shown by the employer to be equivalent. However, it is recommended that training programs offered by professional groups or professional individuals be offered to hazardous waste employees, if possible. This would eliminate any question of conflict of interest and could reduce the liability to the employer should an accident occur.

The 40-hour HAZWOPER course is required for those working on uncontrolled hazardous waste operations mandated by governmental bodies. Employees who do not routinely work at a site but still work there occasionally must have the 24-hour course. Managers and supervisors directly responsible for clean-up operations must have an additional 8 hours of specialized training in waste management. Annual 8 hour refresher training is required for the regular site workers and managers.

For sites other than the ones just discussed, such as at a generator facility, the 24-hour course is required for most of the hazardous waste technicians and hazardous waste specialists who are covered by the standard, due to the organization being a large generator of hazardous waste. Much of the time, these employees' time is not spent in directly working with hazardous waste that is likely to cause exposures. Even during collection of materials from individual generating sites, the wastes normally are in sealed containers so that the risk of exposure is minimal, but they do need the training should an accident occur. Neither the 40-hour nor 24-hour programs qualify an individual to manage a large-scale clean up effort, although both would enable the employee to understand the need for various aspects of the clean up and be able to assist under the supervision of professionals experienced in handling emergency situations. Note that in many jurisdictions the responsibility for remediation of a hazardous material release that extends beyond the boundaries of a facility is established by law as being that of a government affiliated haz-mat team or a local fire department. Incident commanders must have at least the 24-hour training and demonstrate competence in germane areas.

Following is a syllabus for a 24-hour HAZWOPER training course that would meet the requirements of the standard:

1. A review of the requirements needed to comply with the 29 CFR 1910.120 regulation
2. A review of the employer's work and safety program and comprehensive safety plan
3. General nonchemical safety hazards, electric-powered equipment, walking surfaces, hot and cold temperature-related hazards
4. Health and safety effects of hazardous chemicals
5. Employer's medical surveillance program, recognition of signs and symptoms of overexposure
6. Means of detection of hazardous vapors and gases, monitoring technology, procedures, and strategies
7. Engineering and procedural control of environmental releases and exposures
8. Personal protection equipment, selection and use (respiratory protection, clothing, etc.)
9. Contents of an effective health and safety plan

10. Identification and recognition of hazards, site safety planning
11. Emergency response planning, emergency procedures
12. Decontamination procedures
13. Sources of information, assistance, emergency response agencies
14. Employee and employer rights and responsibilities
15. Field exercises, table-top exercises
16. Final exam

The employee who successfully completes a HAZWOPER training program must receive a written certification to that effect. Unless employees have been certified or could be shown by the employer to have had the equivalent training by other means, they must be prohibited from working in hazardous waste operations. The employees are to receive 8 hours of refresher training annually.

In academic institutions, students are often used to supplement permanent employees working with hazardous waste. Usually this participation involves little risk. However, even if the involvement is limited to transporting small containers from a laboratory to a satellite collection site within the same building, they must receive training about the potential risks and measures to minimize the risks to themselves and others.

4.   Medical Surveillance Program

The employer shall provide a medical surveillance program for employees handling hazardous materials. This program is very similar to medical programs required by other standards, but does have a few specific additional requirements. Employees are covered if (1) they may be exposed above the PELs or, if none exist, other published exposure levels, for 30 days or more per year; (2) if they wear a respirator more than 30 days per year; or (3) they are injured or develop signs or symptoms of possible overexposure to hazardous substances during an emergency response or from hazardous waste activities. The examinations are to be at no cost to the employee, without loss of pay and at convenient times and places.

The employee shall receive an examination (1) prior to assignment of work involving hazardous waste, (2) annually or at least biannually, at the discretion of the examining physician, (3) at termination of employment or reassignment unless the employee had an examination within the previous 6 months, (4) as soon as possible after an incident or overexposure, or upon developing signs and symptoms of an over exposure, and (5) more frequently if the examining physician deems it medically necessary.

The examinations are to include a medical history and work history, and are to emphasize symptoms associated with exposure to hazardous chemicals, and health hazards and the employee's fitness for duties requiring the wearing of PPE under conditions that might be expected during the employee's work assignments. The examinations are to be done by or under the supervision of a licensed physician, preferably one with a background in occupational medicine.

The employer is required to provide the physician with a copy of the standard, including appendices, and (1) the employee's duties relevant to chemical exposure, (2) the exposure levels or anticipated exposure levels, (3) PPE used or expected to be used, (4) information from prior examinations, and (5) any relevant information required by 29 CFR 1910.134.

As with other OSHA required examinations, the physician is required to provide a written opinion to the employer who shall provide a copy to the employee. The opinion shall include any medical conditions which would place the employee at a significant health hazard from work with hazardous waste operations or use of a respirator. This last requirement would imply that a pulmonary function test be given. The opinion would also include the physician's recommendations on any limitations on the employee's work. The physician's opinion must include the statement that the employee had been informed of the results of any medical findings, especially those that would require further evaluation. Any findings not related to the employee's occupational exposures are not to be included in the report, but, of course, the employee should be notified at discovery of any serious problems previously unknown to the employee so that they could seek medical help on their own The medical records are to be retained as defined by 29 CFR 1910.20.

5. Engineering Controls, Work Practices, and Personal Protective Equipment for Employee Protection

As with other OSHA standards, the preferred hierarchy of protection is (1) engineering controls, (2) work practices and procedures, and (3) use of PPE. In hazardous waste operations, all three are likely to be needed and used. The standard contains very specific recommendations based on rather rigorous exposure scenarios.

The latter half of the standard goes in depth into emergency planning and emergency response. Since the hazardous waste RCRA regulations require comparable contingency plans and emergency response, the discussion of hazardous material emergency response will be deferred until Section X of this chapter.

6. SARA Title III, Community-Right-to-Know

The SARA provisions apply to those industries covered under the OSHA Hazard Communication Act which has been extended to all chemical users in addition to the originally covered industry groups in Standard Industrial Codes 20 - 39 and which handle, use, or store certain chemicals or extremely hazardous chemicals at levels in excess of limits established by EPA under the Act. The limits are such that many facilities do not have quantities in excess of the limits.

There are currently a number of important exemptions under SARA, one of which is specifically applicable to the research facility. In Title 40 CER 370.2(5) a "hazardous chemical" means any hazardous chemical as defined under 29 CFR 191 0.1200(c), except that such a term does not include the following substances:

1. Any food, food additive, color additive, drug, or cosmetic regulated by the Food and Drug Administration
2. Any substance present as a solid in any manufactured item to the extent exposure to the substance does not occur under normal conditions of use
3. Any substance to the extent it is used for personal, family, or household purposes, or is present in the same form and concentration as a product packaged for distribution and use by the general public
4. Any substance to the extent that it is used in a research laboratory or a hospital or other medical facility under the direct supervision of a technically qualified individual
5. Any substance to the extent it is used in routine agricultural operations or is a fertilizer held for sale by a retailer to the ultimate customer

The fourth exemption in this list is obviously the one of importance in the context of the users of this book. Note that it does not exempt research facilities; it simply redefines the definition of a hazardous chemical to exclude those used in small quantities in laboratories under the direct supervision of a technically knowledgeable person. It does not exclude the same chemicals stored in a warehouse or stores area, nor does it exclude chemicals used in maintenance or support operations, except as they may fall under the other exemptions. There are a number of critical requirements under SARA for a facility which falls under the provisions of the Act. A facility is subject to the provisions of the Act if it has any, other than exempt quantities, of the extremely hazardous chemicals currently on the list which EPA originally issued (with 406 chemicals) on

April 22, 1987, in excess of the threshold planning quantity established by EPA for that substance.

Individual facilities were sent the list of 406 extremely hazardous chemicals by their state commissions in the spring of 1987 and required to inventory their holdings and make a report by May 17, 1987, of those which met the threshold planning quantity. It was also necessary to list the chemicals which the facility had in excess of higher thresholds for emergency planning. Eligible facilities were then required by September 17, 1987, to notify the local planning committee which was established under the act of their facilities emergency planning coordinator. The local committee was required to complete their local emergency plan by October 17, 1988. Large facilities may be represented directly on the planning committee since membership is required to be drawn from elected state and local officials, law enforcement, civil defense, firefighting, first aid, health, local environmental, hospital, transportation personnel, broadcast and print media, community groups, and facility owners and operators subject to SARA.

The first request for information was originally designed to identify agencies which would be likely to fall under the provisions of SARA. Organizations subject to SARA are required to prepare certain inventory forms and make them available to: (a) the appropriate local emergency planning committee, (b) the state emergency response commission, and (c) the fire department with jurisdiction over the facility.

There are basically two inventory forms containing certain levels of information, defined as Tier I and Tier II. The inventory form with Tier I information was required to be submitted by March 1, 1988, and annually thereafter. If Tier II forms are requested by the groups listed in the preceding paragraphs for a given year, then Tier I forms are not required for that year.

Tier I information required on the inventory form is:

1. An estimate (in ranges) of the maximum amount of hazardous chemicals in each category of health and physical hazards, as set forth by the Occupational Health and Safety Act, and regulations published under that act, at the facility during the preceding calendar year
2. An estimate (in ranges) of the average daily amount of hazardous chemicals in each category present at the facility during the preceding calendar year
3. The general location of hazardous chemicals in each category

Tier II information is required to be provided only upon request to the same three groups. They in turn can make the information which they have been provided available to others, such as other state and local officials or the general public under certain conditions.

Tier II information applies to each hazardous chemical at the facility The following information is required:

1. The chemical name or the common name of the chemical as provided on the MSDS for the chemical
2. An estimate (in ranges) of the maximum amount of the hazardous chemical present at the facility at any time during the preceding calendar year
3. A brief description of the manner of storage of the hazardous chemical
4. The location at the facility of the hazardous chemical
5. An indication of whether the owner elects to withhold location information of a specific hazardous chemical from disclosure to the public under Section 324 (which allows the owner to do so upon request)

Leaks, spills, and other releases of specified chemicals, into the environment require emergency notification under both SARA Title III and section 103 of CERCLA. Under CERCLA, those in charge of a facility must report any spill or release of a chemical on a list of hazardous chemicals included in that act in excess of a reportable quantity (RQ) that is substance specific. The report must be made immediately to the National Response Center (1999 telephone numbers

800-424-8802 or 202-267-2675). Under SARA Title III, the notification process now includes all the 360 extremely hazardous chemicals. In addition to the National Response Center, the state commission and the local committee must be notified. Note that if the release is confined totally within a facility and does not enter the "environment," so that it affects only the employees within the building, it does not have to be reported, unless it falls under an OSHA regulation and would have to be reported to OSHA.

Generally, the following information must be reported:

1. The name of the chemical(s) (trade name protection of the chemical is not permitted)
2. Identification of whether the chemical is on the extremely hazardous chemical list
3. The quantity released or an estimate of the quantity released
4. The location, time, and duration of the release
5. Weather conditions, wind speed, and direction
6. Medium (air, water, soil) into which the chemical(s) was released
7. Known acute or chronic risks, and any available helpful medical data
8. Precautions to take, including evacuation, if necessary
9. The names and telephone numbers of persons to be contacted for further information

Obviously, if a facility has an emergency contingency plan covering releases of materials from its facilities to the environment, these should be coordinated with emergency plans for the local region developed by the district emergency planning committee. Under the current standard, it appears that most laboratories are, in effect, exempt from the act due to the provision that chemicals in research laboratories shall not be considered hazardous if they are used under the direct supervision of a qualified individual, but it should be remembered that the facilities themselves are not exempt and there may be chemicals elsewhere on site which trigger coverage by SARA Title III. Also, these provisions may well be modified at a later date. Finally, if an emergency does occur within a laboratory such that a substantial portion of the stock becomes involved in an incident so that there is a significant release to the environment, a responsible position may be to have actively participated in the emergency response process in the local district.

REFERENCES[*]

1. Resource Conservation and Recovery Act (RCRA) of 1986, SARA, Title III, Sections 300 to 330, Washington, D.C., 1986.
2. Extremely Hazardous Substance List, Sections 302 to 304, *Fed Reg.,* 13378, 1987.
3. Emergency Planning and Hazardous Chemical Forms and Community Right to Know Reporting; Final Rule, Title III, Sections 311-312, *Fed Reg., 52,* 38344, 1986.
4. Toxic Chemical Release Reporting; Community Right-to-Know; Final Rule Title III, Section 313, *Fed Reg., 53,* 4500, 1986.

---

[*] Regulations are frequently modified, added to or deleted. Check the latest version of the cited regulations to be sure to be accurate. The latest information can be found at the two INTERNET REFERENCES following the standard references.

5. Occupational Safety and Health Administration, Hazardous Waste Operations and Employee Right-to-Know, 29 CFR 19 10.120, Washington, D.C.

6. Occupational Safety and Health Administration, Employee Emergency Plans and Fire Emergency Plans, 29 CFR 1910.38, Washington, D.C.

INTERNET REFERENCES

1. http://www.access.gpo.gov/nara/cfr/waisidx/40CFR302.htm EPA regulation
2. http://www.nrc.uscg.mil/nrxhp.htm National Response Center

X. CHEMICAL WASTES

Until a relatively few years ago, disposal of many chemical wastes from laboratories was down the nearest convenient drain. However, this practice is now illegal as a general procedure, although to a limited degree it is still permissible for certain wastes. The change has been due to the growing concern about the impact of chemicals on the environment. Virtually all the chemical wastes in this country are due to industrial sources, with less than 1% being due to laboratory wastes. However, regulations established to govern the disposition of chemical wastes include the wastes generated by laboratories, so that organizations of which the laboratories are a part must have a hazardous waste program. Locally, in a small community, a major research university or corporate research laboratory may be one of the largest, if not the largest, generators of hazardous waste so the overall 1% may be somewhat misleading.

The following sections will provide the basics of a hazardous waste program, and an overview of the regulations that apply to the program. It would be impossible to provide here a complete guideline covering every contingency. Such a document would be comparable in size to this entire handbook. Operators of hazardous waste programs should receive specialized training from any of the many commercial training programs and acquire a complete set of copies of the regulations and appropriate reference materials.

A. Resource Conservation and Recovery Act

Many individuals who are now responsible for the management of laboratory facilities received their initial training when there were no regulations governing the disposal of laboratory wastes, and they may not fully appreciate the need for the newer procedures. However, especially in the academic area, the training of students should be in the context of compliance with regulations so that when these students embark on their own careers, they will be trained to comply with legally applicable standards. Senior personnel need to adapt, if they have not already done so, and manage their operations by current standards.

Organizations involved with excess hazardous chemical materials must conform to the provisions of the Environmental Protection Agency Resource Conservation and Recovery Act (RCRA) 40 CFR Parts 260 to 265, as most recently amended. In 1984, the Hazardous and Solid Waste Amendments (HSWA) of 1984 required that the EPA ban the land disposal of over 400 waste streams unless the wastes are treated or unless it could be demonstrated that there would be no migration of the waste while the waste remains hazardous. The portion of these rules covering dioxin and solvents went into effect on November 8, 1986. Although there was a partial extension for some wastes, the extension did not include solvents. The rationale for the extension was the lack of alternate capacity for the large amounts of soil contaminated by dioxin or solvents. The second portion covering the "California" list, including several heavy metals, went into effect on July 8, 1987. The remainder of the restrictions went into effect on three different dates: August

8, 1988, June 8, 1989, and July 8, 1990. These amendments have significantly changed the options available to laboratories for disposing of their waste chemicals.

Laboratory facilities or their parent organizations have important decisions to make concerning the scope of their activities that will be subject to the provisions of the RCRA regulations. Unless they produce less than 100 kg per month, at which level they are defined as a small generator, they are subject to portions of 40 CER Part 262, which applies to generators. If they produce more than 1000 kg per month, they are "large" generators and they must comply with all of the requirements applicable to generators.

Other portions of the act cover the operations of treatment, storage, disposal, and recycling (TSDR) facilities, which require additional permitting. Many large industrial organizations, in which laboratory operations represent only a small part of their activities involving hazardous chemicals, may choose or have chosen to go through the permitting process. Some also transport their own hazardous waste and are subject to the regulations on transporters in addition to the regulations applying to generators. However, only a few academic institutions have the capacity or interest required to meet the stringent regulations covering operations other than those of the generators of the waste chemicals. The same is true of many small industrial operations and their affiliated laboratories. Those organizations which intend to treat, store, dispose, and recycle their hazardous waste normally will be sufficiently sophisticated and knowledgeable about the procedures to not need additional information, so this section will concern itself only with waste management programs for generators.

Before a program of waste management appropriate to a research organization is discussed, some additional background information will be helpful. What constitutes a hazardous waste needs to be defined and the essential portions of 40 CER Part 262 need to be addressed. It is virtually a full-time job to keep up with the changing requirements in the context of laboratory operations, even for relatively small organizations. It is desirable to transfer at least the transportation of the wastes to an outside contractor, but substantial savings in hazardous management costs can be realized by performing much of the waste management operations internally.

1. Definition of a Hazardous Waste

There are relatively frequent changes in detail as to what constitutes a hazardous waste according to the EPA, but the main features of the definitions have been consistent. Frequently, there are evaluations and rulings pertaining to the hazardous nature of various wastestreams, for example, reported in the *Federal Register* but these are based on certain guides in 40 CFR 261 Subpart D. There have already been some changes in some definitions, specifically that of toxicity. The announced requirements for treatment before land disposal are examples of the more stringent requirements. There have also been significant changes in information that the generator is required to provide the transporter. For a current list, see the first INTERNET REFERENCE to this section.

The criteria are (a) it is a waste listed in Part 261 (in 1995 there were about 600 commercial chemicals listed in 261.31 to 261.33), and (b) it meets certain criteria of reactivity, ignitability, corrosiveness, and toxicity.

Wastes which meet certain hazardous criteria may be placed on the lists under three different categories:

Hazardous wastes from nonspecific sources - Examples of laboratory wastes that would fall in this category would be spent solvents, the residue resulting from distillation recovery of used solvents, materials left over from silk screening and electroplating procedures in electronic laboratories, and other sources of used chemicals.

Hazardous wastes from specific sources - Unless the laboratory is a pilot operation simulating an industrial process, it is unlikely that most research laboratories would fall within this category. Note however, that hazardous chemicals used in a pilot plant operation normally would not be exempt from the regulatory provisions of SARA Title III, discussed in Section IX.B of this chapter.

Discarded commercial chemical products, off-specification species, containers, and spill residues thereof - There are two levels in this category (a) those materials which are acutely hazardous, and (b) those which are less so. There are more stringent limitations on the former.

The following excerpts are from 40 CFR 261.11: *Criteria for listing hazardous waste.*

(a) The Hazardous Waste Administrator shall list a solid* waste as a hazardous waste only upon determining that the solid waste meets one of the following criteria:

(1)   It exhibits any of the characteristics of hazardous waste identified in Subpart C: ignitability, corrosivity, reactivity, and toxicity

(2)   It has been found to be fatal to humans in low doses or, in the absence of data on human toxicity, it has been shown in studies to have an oral LD_{50} toxicity (rat) of less than 50 mg/kg, an inhalation LC_{50} toxicity (rat) of less than 2 mg/L, or a dermal LD_{50} toxicity (rabbit) of less than 200 mg/kg, or is otherwise capable of causing or significantly contributing to an increase in serious irreversible, or incapacitating reversible illness. (Waste listed in accordance with these criteria will be designated Acute Hazardous Waste.)

(3)   It contains any of the toxic constituents listed in Appendix VIII (in 40 CFR 261), unless after considering any of the following factors, the Administrator concludes that the waste is not capable of posing a substantial present or potential hazard to human health or the environment when improperly treated, stored, transported, or disposed of or otherwise managed:

i.   The nature of the toxicity presented by the constituent

ii.   The concentration of the constituent in the waste

iii. The potential of the constituent or any toxic degradation product of the constituent to migrate from the waste into the environment under the types of improper management considered in paragraph (a)(30)(vii) of this section of the standard

iv.   The persistence of the constituent or any toxic degradation product of the constituent

v.   The potential for the constituent or any toxic degradation product of the constituent to degrade into non-harmful constituents and the rate of degradation

vi.   The degree to which the constituent or any degradation product bioaccumulates in ecosystems

vii. The plausible types of improper management to which the waste could be subjected.

viii. The quantities of the waste generated at individual generation sites or on a regional or national basis.

ix.   The nature and severity of the human health and environmental damage that have

---

*   "Solid" waste need not be physically solid, as will be noted from some of the definitions in this section.

occurred because of the improper management of wastes containing the constituent

x. Action taken by other governmental agencies or regulatory programs based on the health or environmental hazard posed by the waste or waste constituent

xi. Such other factors as may be appropriate

Substances will be listed in Appendix VIII of the standard only if they have been shown in scientific studies to have toxic, carcinogenic, mutagenic, or teratogenic effects on humans or other life forms.

If the material is not a listed waste, it is a hazardous waste if it meets any of the criteria in Subpart C, Characteristics of Hazardous. Waste. Following are the essential sections of this part of the regulations.

261.21 Characteristic of Ignitability

(a) A solid waste exhibits the characteristics of ignitability if a representative sample has any of the following properties:

(1) It is a liquid other than an aqueous solution containing less than 24% alcohol by volume and has a flash point less than 60°C (140°F)....

(2) It is not a liquid and is capable, under standard temperature and pressure, of causing fire through friction, absorption of moisture, or spontaneous chemical changes and, when ignited, burns so vigorously and persistently that it creates a hazard.

(3) It is an ignitable compressed gas as defined in 49 CFR 173.30Q and as determined by the test methods described in that regulation or equivalent test methods approved by the Administrator under paragraphs 260.20 and 260.21....

(4) It is an oxidizer as defined in 49 CFR 173.51.

(b) A solid waste that exhibits the characteristic of ignitability has the EPA Hazardous Waste Number of D001.

261.22 Characteristic of Corrosivity[*]

(a) A solid waste exhibits the characteristics of corrosivity if a representative sample of the waste has either of the following properties:

(1) It is aqueous and has a pH less than or equal to 2 or greater than or equal to12.5....

(2) It is a liquid and corrodes steel (SAE 1020.) at a rate greater than 6.35 mm (0.250 inches) per year at a test temperature of 55°C (130°F)....

(b) A solid waste that exhibits the characteristic of corrosivity has the EPA Hazardous Waste Number of D002.

261.23 Characteristic of Reactivity

(a) A solid waste exhibits the characteristic of reactivity if a representative sample of the waste has *any* of the following properties:

(1) It is normally unstable and readily undergoes violent change without detonating

---

[*] Note that the earlier OSHA definition of a corrosive material in terms of a health hazard was explicitly limited to action of a substance on tissue.

Table 4.20 Maximum Concentration of Contaminants for the Toxicity Characteristic

EPAHW Number	Contaminant	CAS Number	Regulatory Level (mgIL)
D004	Arsenic	7440-38-2	5.0
D005	Barium	7440-39-3	100.0
D018	Benzene	71-43-2	0.5
D006	Cadmium	7440-43-9	1.0
D019	Carbon tetrachloride	56-23-5	0.5
D020	Chlordane	57-74-9	0.03
D021	Chlorobenzene	1 08-90-7	100.0
D022	Chloroform	67-66-3	6.0
D007	Chromium	7440-47-3	5.0
0023	o-Cresol	95-48-7	200.0
D024	m-Cresol	108-39-4	200.0
D025	p-Cresol	106-44-5	200.0
D026	Cresol (total)[2]		200.0
D016	2,4-D	94-75-7	10.0
D027	1,4-Dichlorobenzene	106-46-7	7.5
D028	1,2-Dichloroethane	107-06-2	0.5
D029	1,1-Dichloroethylene	75-35-4	0.7
D030	2,4-Dinitrotoluene	121-14-2	0.13[*,**]
0012	Endrin	72-20-8	0.02
D031	Heptachlor (and its epoxide)	76-44-8	0.008
0032	Hexachiorobenzene	118-74-1	0.13[*]
D033	Hexachlorobutadiene	87-68-3	0.5
D034	Hexachloroethane	67-72-1	3.0
D008	Lead	7439-92-1	5.0
D013	Lindane	58-89-9	0.4
D009	Mercury	7439-97-8	0.2
D014	Methoxychlor	72-43-5	10.0
D035	Methyl ethyl ketone	78-93-3	200.0
D036	Nitrobenzene	98-95-3	2.0
D037	Pentachlorophenol	87-86-5	100.0
D038	Pyridine	110-86-1	5.0[*]
DOIO	Selenium	7782-49-2	
Doll	Silver	7440-22-4	5.0
D039	Tetrachloroethylene	127-18-4	0.7
DOIS	Toxaphene	8001-35-2	0.5
0040	Trichloroethylene	79-01-6	0.5
D041	2,4,5-Trichlorop!,enol	95-95-4	400.0
D042	2,4,6-Trichlorophenol	88-06-2	2.0
D017	2,4,5-TP (Silvex)	93-72-1	1.0
D043	Vinyl chloride	75-01-4	0.2

(2) It reacts violently with water.
(3) It forms potentially explosive mixtures with water.

---

[*]   Quantitation limit is greater than the calculated regulatory level. The quantitation limit therefore becomes the regulatory level.

[**]   If o-, m-, and p-cresol concentrations cannot be differentiated, the total cresol (D026) concentration is used.

(4) When mixed with water, it generates toxic gases, vapors, or fumes in a quantity sufficient to present a danger to human health or the environment.

(5) It is a cyanide or sulfide bearing waste which, when exposed to pH conditions between 2 and 12.5, can generate toxic gases, vapors or fumes in a quantity sufficient to present a danger to human health or the environment.

(6) It is capable of detonation or an explosive reaction if it is subjected to a strong initiating source or if heated under confinement.

(7) It is readily capable of detonation or explosive decomposition or reaction at standard temperature and pressure.

(8) It is a forbidden explosive as defined in 49 CFR 173.51, or a Class A explosive as defined in 49 CFR 173.53, or a Class B explosive as defined in 49 CFR 173.88.

(b) A solid waste that exhibits the characteristic of reactivity has the EPA Hazardous Waste Number of D003.

261.24 Characteristic of Toxicity

(a) A solid waste exhibits the characteristic of toxicity if, using the test methods described in Appendix II [of the regulation] or equivalent methods approved by the Administrator under the procedures set forth in paragraphs 260.20 and 260.21, the extract from a representative sample of the waste contains any of the contaminants listed in Table 1 [of the regulation] at the concentration equal to or greater than the respective value given in that Table. Where the waste contains less than 0.5 percent filterable solids, the waste itself, after filtering using the methodology outlined in Appendix II is considered to be the extract for the purposes of this section.

(b) A solid waste that exhibits the characteristic of toxicity has the EPA Hazardous Waste Number specified in Table 4.20 which corresponds to the toxic contaminant causing it to be hazardous.

Since the appendices and lists referred to in the above material change as materials are added to and deleted from the list, laboratory personnel should subscribe to an information service which will provide information to permit maintenance of a current valid list or obtain the information from the Internet.

A generator's hazardous waste must be evaluated to decide if it may be exempt from being considered a hazardous waste or, if not, if it is a listed waste, as defined above, or meets any of the criteria for a hazardous waste. A large percentage of laboratory wastes are likely to fall under one of these provisions and must be treated as a hazardous waste. A major problem with many chemicals produced by laboratory activities, as opposed to chemicals that are purchased, is that they are insufficiently identified. The identity of the chemical in a bottle, for example, labeled "solution A" may have been known perfectly well by the individual who labeled it at the time the label was affixed to the bottle, but even this person may not recall the contents several months later. As often happens, the chemical is an "orphan" left behind by a departed graduate student or employee, in which case the identity of the contents may be even more uncertain. Commercial waste disposal firms will normally not accept these unknown containers until they are characterized. Other examples of unknown chemicals which also must be identified are older containers that have lost their labels or whose labels may have become damaged so that the contents cannot be determined. In all cases of unknowns, there is a certain amount of risk associated with opening an unlabeled container for any purpose, including that of identifying the contents. The container may contain, for example, degraded chemicals that have become explosive, such as ether with a high content of peroxides or dry picric acid. Procedures should be adopted that require all personnel to properly identify the contents of any container which will not be used in its entirety by the maker or the person transferring a quantity from an original container to a secondary container to avoid these problems. The latter

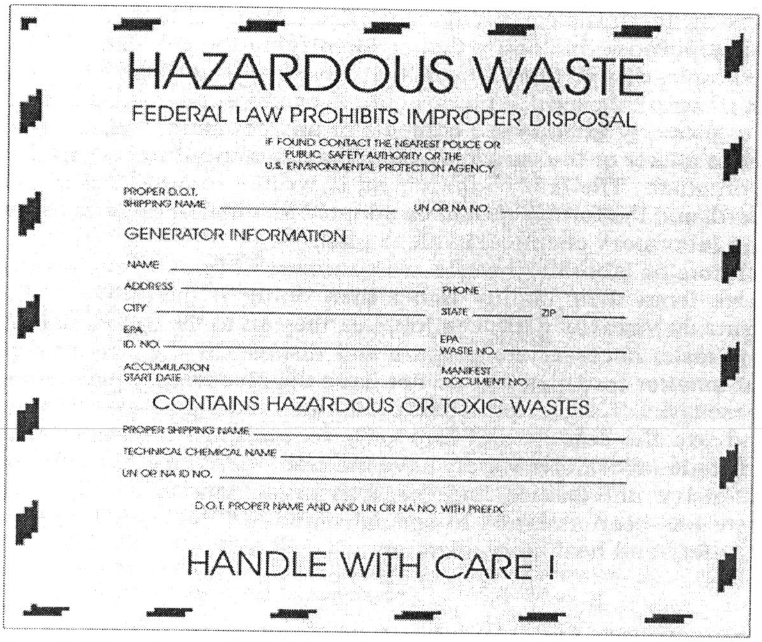

Figure 4.13 Hazardous waste label meeting EPA and DOT requirements.

requirement is written into the OSHA hazard communication standard, and the former should be adopted as a matter of good laboratory practice in developing a laboratory chemical hygiene plan. Unfortunately there are always individuals who actually seem to be compelled to not follow rules, even where it is important to do so.

Most generators of laboratory waste use commercial firms as the means of removing hazardous waste from their facility rather than doing it themselves. Sometimes, the commercial firms do virtually a turnkey job, i.e., they go to the individual laboratories and perform all the tasks necessary to prepare and dispose of the waste properly. This is appropriate for smaller facilities that do not have the resources to have their own staff to do many of these tasks. Larger operations often find it much more economical to do most of the work themselves, and use the commercial firm only to transport the material to an ultimate disposal site. Single laboratories rarely have the resources to use the latter method, and in most comprehensive universities and research organizations, the process of handling hazardous waste has been assigned to specialists within the organization, usually as an adjunct of the safety and health organization.

2.   Requirements for Generators of Hazardous Waste

Title 40 CER Part 262 defines the requirements for generators of hazardous waste. As in other parts of this section, it will be assumed that the generators referred to will be those who are not owners or operators of treatment, storage, disposal, or recycling facilities and who dispose of most of their waste by shipping it away from the generating facility, except for the modest amount of processing permitted under the law. As will be noted, the generator assumes substantial responsibilities in initiating a shipment of hazardous waste from the facility. The information in this section is taken directly from the standard. In some cases the language has been simplified or made more concise for brevity and clarity, while in other cases the exact wording of the standard has been followed where it did not lend itself to modification.

1. The generator must first decide if the waste meets the criteria for hazardous waste. The previous section describes the basic criteria which define a hazardous waste, and the

waste must be evaluated in terms of these criteria. Figure 4.13 shows this process graphically. In addition, it may be decided that a given material is a hazardous waste based on the materials or processes used. This last interpretation might apply to much of the materials synthesized or generated in a research laboratory, where the detailed characteristics of the waste might be difficult to determine, but the procedures make it likely that it is a hazardous material. The material classification can be assigned on the basis of this information and simple tests for at least the ignitability, corrosivity, and reactivity characteristics.

2. If the generator intends to initiate a shipment of hazardous waste, an EPA identification number must be assigned by the EPA Administrator.

3. The generator must be sure that anyone engaged to handle or receive his hazardous waste has a valid EPA identification number. This step is required but it is not sufficient to assure that waste generated at a facility will actually be handled legally and environmentally soundly. Although it may consume additional time and resources, visits to a waste-brokers storage facility and/or to the eventual site in which the material will be placed in a landfill, incinerated, or otherwise processed is highly recommended. All of the organizations willing to bid on laboratory waste disposal services purport to offer virtually the same degree or quality of service. Many major, reputable firms have had to close facilities, pay lines, or substantially modify their procedures because they did not comply with acceptable operating practices. In the worst possible case, where the operator of a facility does not have resources available to correct a problem which has been identified by the EPA, the waste generators can be required to share the cost of cleaning up a substandard facility. Even if a firm has proven reputable in the past, ownership and management changes may result in changes in performance quality.

On December 21, 1990, The Department of Transportation (DOT) published new "Performance-Oriented Packaging Standards." These standards essentially brought the United States into conformance with the United Nations Dangerous Goods regulations. The result is that the manner in which a generator (and other groups) package and document hazardous waste for shipment has changed in almost every detail. The modified procedures are complex, and generators who use commercial disposal firms to dispose of their hazardous waste may prefer that the disposal firm help them prepare the needed information for the documentation to accompany the waste offered for transportation, and to ensure that the waste is properly packaged. The new regulations were phased in over a period of several years. Many of them went into effect prior to October 1, 1994. All of the regulations will have been in effect by October 1, 1996. Most of the critical ones affecting generators are in effect, following several clarifications. The following section covers the basics of the new requirements under Department of Transportation, HM-181. These are subject to change so anyone affected should be careful to utilize the latest version.

a. Basics of Compliance with HM-181

The DOT hazard identification system has been changed to a numerical one. Table 4.22 shows the newer classifications and some new terms, " Packaging Groups and Hazard Zones." This information is part of that required for preparation of shipping papers.

In the Table, the definitions of packing groups are:

I — Great Danger
II — Medium Danger
III — Minor Danger

Table 4.21 DOT Hazard Identification Scheme

Name of Class/Division	Class Number	Class Division	Packing Groups	Other
Flammable gas	2	2.1	None	
Nonflammable gas	2	2.2	None	
Poisonous gas	2	2.3	None	Hazard zones, A to D
Flammable liquid	3	None	I, II, III	
Combustible liquid	None	None	III	
Flammable solid	4	4.1	II, III	
Spontaneously combustible solid	4	4.2	I, II, III	
Dangerous when wet solid	4	4.3	I, II, III	
Oxidizer	5	5.1	I, II, III	
Organic peroxide	5	5.2	II	Types A to G
Poisonous material	6	6.1	I, II, III	Hazard zones A to D, PIH vapors only, liquids/mixture
Infectious substance	6	6.2	None	
Corrosive material	8	None	I, II, III	
Miscellaneous hazardous materials	9	None	II, III	
ORM-D (consumer commodities)	None	None	None	

Packing groups, where applicable, are provided for all of the materials listed in 49 CFR 172. 101.

Hazard zones for gases that are poisonous by inhalation are:

A — $LC_{50} \leq 200$ ppm
B — $LC_{50} > 200$ ppm to ≤ 1000 ppm
C — $LC_{50} > 1000$ ppm to ≤ 3000 ppm
D — $LC_{50} > 3000$ ppm to ≤ 5000 ppm

Hazard zones for peroxides are assigned from a generic system consisting of the seven letters, A to G. The organic peroxide table in 49 CFR 173.225 contains all of the peroxides assigned to a generic type. If not listed, the material cannot be transported without written approval from the associate administrator for Hazardous Materials Safety.

The hazard zones for poisonous materials that are toxic because of inhalation of vapors from the material are (for a single toxic substance)

A — vapor concentration ≥ 500 LC$_{50}$ where LC$_{50} \leq 200$ ml/m^3

B — vapor concentration ≥ 10 LC$_{50}$ to < 500 LC$_{50}$ where LC$_{50} > 200$ ml/m^3 to ≤ 1000 ml/m^3

Both of these hazard zones would correspond to Packing group I. There are no hazard zones for Packing groups II and III. There are four hazard zones A to D for mixtures containing more than one poisonous substance.

Following are a few comments about some rationales for the various divisions in Table 4.22. Many of the divisions are self-explanatory. To adequately perform the required procedures, it is essential to have a copy of the relevant hazardous material data tables identified in HM-181.

Class 2.1 gases are gaseous at standard atmospheric temperature and pressure, or have boiling points at standard pressure less than standard temperature (20°C or 68°F) and can be ignited at air mixtures of less than 13%.

A flammable liquid is defined differently in this context from that of an OSHA flammable liquid. It is any liquid with a flash point between 141° and 200°F (OSHA definition is a liquid with a flash point of 100°F).

Infectious materials are those that can cause disease in humans or animals. Note that this is a broader definition than that described earlier in the section on bloodborne pathogens.

The definition of corrosive has been extended beyond that in 40 CFR Part 261 to include materials that cause visible destruction or irreversible alterations of the skin tissue at the site of contact when tested on the intact skin of an animal for ≤ 3 minutes besides its effect on steel.

The primary source of information needed to prepare the proper shipping descriptions and manifests is 40 CFR 172.101 Subpart B, Table of Hazardous Materials, and the appendix to the table. Table 101 is divided into ten columns.

Column 1 - Symbols. There are only five symbols and most of the materials in the table do not have anything in this column. The symbols are "+" which fixes the proper shipping name regardless of whether the material meets the criteria under the hazard class or packing group column; A or W is used to designate a material which is hazardous only when transported by air or water, unless it is a hazardous material or hazardous waste; D indicates that the proper shipping name is only correct for domestic shipments; and I indicates the case is for international shipments.

Column 2 - Proper shipping name. The only proper shipping name is the one that appears in this column. No other name is acceptable, including chemical formulas.

Column 3 - Hazard Class number, see Table 4.22. A "+" in the symbol column supersedes the hazard class/division normally associated with the material.

Column 4 - A number that specifically corresponds to the proper shipping name of the material and provides identification of the material in case of an accident. If the number is preceded by "UN" it is a United Nations number that is recognized both internationally and domestically. If preceded by NA, the numbers are only valid in the United States and Canada.

Column 5 - Packing Group, see Table 4.22.

Column 6 - Labels are required on a package filled with a material (materials if there is a secondary material). In which case the label for the secondary materials must not indicate the Class Number/Division. Only the label for the primary material may show this and it must do so. The labels would be a 4 inch x 4 inch diamond shape. There are exceptions when no label is required or, in some cases, allowed.

Column 7 - Special provisions. A letter is followed by a number; these are defined in 40 CFR 172.102.

Column 8 - Packing Authorizations. Column 8 is divided into three sub-columns; the first identifies exceptions, the second non-bulk requirements, and the third, bulk requirements. The numbers identify paragraphs in Part 173.304.

Column 9 - Quantity Limitations. The maximum amount allowed on a passenger-carrying aircraft or rail car.

Column 10 - Quantity Limitation. The maximum value allowed when shipped by cargo aircraft.

The appendix to 49 CFR 172.101 lists the reportable quantities (RQ) for many "hazardous substances." For DOT purposes, a hazardous substance must be listed in this appendix to be considered a hazardous substance and also there must be an amount equal to or greater than the RQ in one package or, if in a mixture or solution, the amount present is in concentrations by weight equal to or greater than is given in the table below from 49 CFR 172.8.

RQ (lb)	Concentration by weight %
5000	10
1000	2
100	0.2
10	0.02
1	0.002

Table 172.203 is comparable to 172.101 materials shipped as generic, not otherwise specified (n.o.s.).

Most hazardous waste generators will depend upon the waste broker to comply with all of the requirements for shipping the waste, but the generator should be familiar with the requirements as well to ensure that nothing is overlooked. There are many excellent training programs available which can easily be found on the Internet by using any of the standard search engines to search for "hazardous waste training." The following sections are intended to provide a reasonably complete idea of what is needed, but not to serve as a substitute for up-to-date-training.

49 CFR 172.200 describes the following information that must be provided on the shipping papers to describe a hazardous material offered for transportation. The required information must be in English.

1. A hazardous material identified by the letter X in a column labeled HM placed before the proper shipping name (if appropriate, X may be replaced by RQ but RQ cannot be replaced by X)
2. Proper shipping name (this may be one of the generic descriptions in 172.203)
3. Hazard class or division
4. The identification number (UN or NA) from 172.101 or 172.203
5. The Packing Group, from 172.101 or 172.203, if any
6. Quantity

The information in Table 4.22 must be given in that order with no other information interspersed, except that technical or chemical group names may be entered in parentheses between the proper shipping name and hazard class or following the basic description. Two other items must be on the shipping paper:

1. An emergency response telephone number
2. Certification - "This is to certify that the above-named materials are classified, described, packaged, marked, and labeled and are in proper condition for transportation according to the applicable Department of Transportation rules." An acceptable alternate statement is similar, but also identifies the transporter.

Other information may be on the shipping papers as long as it is placed after the basic description and is not inconsistent with it. Normally, the shipping paper will contain the name and address of the shipper and consignee. For hazardous waste shipments over highways, as most of the waste covered in this section will be, the shipping papers must be within the cab with the driver, or in the absence of the driver, on the seat or in a holder fixed to the inside of the door on the driver's side. This is to ensure that they are available during an inspection or in case of an accident, are in a standard location so that they could be found, if necessary. The requirements for labels, placards, manifests all have this underlying purpose, to provide information in case of an emergency so that emergency response personnel will have vital information to guide their response.

A manifest must be prepared according to 40 CFR Part 262.20. The original copy of the manifest must be dated by and bear the handwritten signature of the person representing (1) the generator of the waste at the time it is offered for transportation, and (2) the initial carrier accepting the waste for transportation. The manifest prepared according to 40 CFR Part 262 can also serve as the shipping paper as long as it contains all of the information required by 49 CFR 172.200-205.

Copies of the manifest must be dated and bear the handwritten signature of the person representing each subsequent carrier of the waste and the designated facility receiving the waste, upon receipt.

Copies of the manifest bearing all of the required dates and signatures must be:

1. Given to the person representing each carrier involved in the transportation of the waste. The copies must be carried during transportation.
2. Given to the person representing the designated facility receiving the waste.
3. Retained by the generator and each carrier for 3 years from the date the waste was accepted by the initial carrier. The copies must bear all of the required signatures and dates up to and including those entered by the next person receiving the waste.

b. Container Labels

The containers, normally drums, offered to the transporter or used by the commercial disposal firm must be marked with an appropriate label as defined in the tables in Parts 172.101 and 172.203. No container can be marked with a label for a material not in the container. Besides these DOT labels, EPA/RCRA regulations require each container to have a label stating prominently that the container contains a hazardous waste. The EPA/RCRA label contains much of the same information found on the shipping papers or manifests: generator name and address, accumulation start date, manifest number, proper shipping name, and UN or NA number. Additional information may be provided by the generator. An example of a hazardous waste label is shown in Figure 4.13. If the container contains a hazardous chemical regulated by OSHA in a substance-specific health standard, a label identifying the material and the hazard must be on the container. Labels are to be placed within 6 inches of each other and must be on the side of the container, not the top.

The vehicle carrying the hazardous waste must be properly placarded, as required by DOT regulations and given in the tables in 49 CFR 172.101 and 172.203, or as required under any special provisions. It is the responsibility of the generator to ensure that this is done, if necessary by providing the appropriate placards to the transporter although it is unlikely that this will be required since transporters routinely provide the correct placards.

All shipments of hazardous materials shall be transported without unnecessary delay, from and including the time of commencement of the loading of the cargo until its final discharge at its destination. The generator must receive a completed copy of the manifest signed by a representative of the TSDR facility within 35 days or must initiate action to track the wastes. If the waste is not located within 45 days, an exceptions report must be sent to the EPA or the equivalent state agency. An intermediate destination at a suitable storage facility is permitted, in which waste materials from different sources can be combined for delivery to a point of final disposition of the material. If the materials on the vehicle will not be treated identically, then the broker or processor can break up a shipment, as long as complete track of each container is maintained and records kept.

c. Local Waste Management

Unless the generator has been granted an extension by the EPA, the generator must either have a permit as a storage facility or must not collect and store hazardous waste for more than 90 days. During this storage interval, the materials must be kept in appropriate containers, the beginning of the accumulation period must be shown on each container, and they must be labeled as hazardous waste. The label shown in Figure 4.13 would be affixed when the container was put into storage.

It may be appropriate to store the material within the individual laboratories generating the waste until shortly before a waste disposal firm is scheduled to pick up an organization's waste. Under appropriate conditions, as much as 55 gallons of hazardous waste (or 1 quart of acutely hazardous waste) may be accumulated at or near the point of generation, i.e., the laboratory beyond 90 days without a permit as a storage facility. The containers used must be in good condition so that they do not leak, and they must be constructed so that the hazardous waste does not react with the container so as to impair the ability of the container to store the material. Except for periods during which materials are to be added to the container, the container must remain closed. The container must be treated to keep it from leaking. As a minimum, the containers must be labeled with the words "Hazardous Waste," but the label should provide a comprehensive description of the contents.

3. Record Keeping Required of the Generator

Copies of the manifests signed by the transporter must be maintained for at least 3 years. Retention substantially beyond this date would probably be advisable in the event subsequent problems develop due to actions by the broker or at the TSDR facility. Similarly, copies of required biennial reports (see following paragraph) and exception reports must be kept for at least 3 years beyond the date on which they were due. Any test data relevant to the shipment must also be kept for 3 years beyond the date of the shipment.

Biennial reports must be prepared by each generator initiating shipments of hazardous waste away from the facility. These must be prepared on EPA Form 8700-13A, submitted by March 1 of each even-numbered year, and must cover the previous calendar year. The report must provide: (a) the EPA identification number, name and address of each facility to which waste was shipped during the year, (b) the name and EPA identification number of each transporter used during the year, and (c) a description, EPA identification waste number and amount of each hazardous waste shipped off-site. This information must be identified further with the identification number for each site to which it was shipped. The report must be signed by the generator or an authorized representative to certify that the report is correct.

4. Personnel Training

A record must be maintained at the facility engaged in hazardous waste activities for each position related to management of the waste. The name of the current person filling each of these positions must be recorded. A written job description must be available for each position, describing the skills, education, and other qualifications needed for each job and the actual duties involved for each position.

The training requirements for haz-mat personnel managing waste under 49 CFR, Subpart H, are similar to the requirements for haz-mat personnel covered under 29 CFR 1910.120 and 40 CFR Subpart 265. Where they overlap, the training requirements for each of the three standards can be used rather than being duplicated for an employee. The owners and operators of a facility engaged in a hazardous waste management program must ensure that the personnel identified as working in the facility are properly trained so that the program will be conducted safely, with minimal risk to the environment and the general public. This can be done by the haz-mat employers providing classroom instruction and effective on-the-job training themselves or by providing access to public or commercial training programs. The person or persons responsible for the training must themselves be knowledgeable concerning hazardous waste management procedures. The training should be tailored to the specific duties of each person working in the facility. The employees must receive training in the facility's contingency plan, including initiation of an alarm and evacuation procedures, and appropriate responses to a fire, explosion, or an environmental contamination incident. They should know the procedures to safely secure the facility in an emergency.

A written description of the initial and maintenance training programs for each person or position must be available. Persons assigned to duties in a hazardous waste management program must successfully complete the training programs before they are allowed to work unsupervised and within 90 days after employment or assignment to a new task involving haz-mat duties. Each employee shall take part in a review of the initial training appropriate to their duties every 2 years. Records of the training for each continuing employee must be kept until closure of the facility, and for employees who leave the facility records must be kept for at least 3 years after the employee has left.

5. Preparedness and Prevention

Every facility involved with hazardous waste must be operated in such a manner as to minimize the chances of any incident, such as a fire, explosion, and release of a toxic substance into the environment, or which could endanger human life. It is essential to plan for such emergencies even though it is intended and hoped that there will never be a need for an emergency response.

Communications are a vital part of any emergency system. Provision must be made for an emergency alarm in the facility. Ready access must be available to either a telephone or two-way radio system to summon emergency help from appropriate agencies, e.g., fire department, police, rescue squads, and emergency response teams. Whenever hazardous material is being handled, access to communication equipment is especially important and must be available. Another occasion when communication capability is especially critical is when an employee is alone in a facility while operations are being conducted, although such a situation is undesirable and should be avoided if at all possible. Local hospitals should be made aware of the potential for exposures of personnel to hazardous materials and the types of injuries that could result from explosions, spills, or airborne releases of hazardous or toxic materials. Since incidents of this type are rare, it is recommended that local emergency rescue squads and emergency room personnel be alerted as to the possible situations which they might face.

Other emergency equipment that must be present are portable fire extinguishers (of different fire suppression classes, if needed), spill control equipment, and decontamination equipment. Caution needs to be used by personnel in using the portable fire extinguishers since fires involving chemicals are very likely to generate toxic fumes. Unless it is clear that the fire can be handled safely, basically this means small fires not yet involving chemicals, personnel should be advised to evacuate to a safe location and to notify appropriate emergency responders. Other desirable emergency equipment would be positive air-supplied breathing apparatus, fire and chemical protective clothing, and turnout clothing. Access to any area of the facility must be available in an emergency, so adequate aisle space must be provided. All emergency equipment must be tested and maintained as needed to ensure that it is operational in case of an emergency.

Participation with a local area emergency plan, such as required under Title III SARA, is essential. Arrangements should be made with local fire departments and police departments to familiarize them with the facility, its operations, and access to the facilities for emergency vehicles. Agreements should be established with state and local emergency response agencies, as well as commercial firms that might be needed to assist in an emergency. Other internal groups, such as physical plant or building maintenance groups, and internal security personnel should be similarly involved. As noted above, working agreements with nearby medical facilities should be established, and they should be made aware of the type of exposures or medical injuries which they might be asked to handle because of operations within the facility. Periodic drills would be highly desirable.

It is the responsibility of the owners and operators of an organization engaging in hazardous waste operations to make these emergency arrangements. If an agency refuses to accommodate the facility in establishing an appropriate arrangement, this refusal must be documented.

6. Contingency Plan

The previous section described some emergency provisions that must be made. However, a legal requirement of every hazardous waste facility is that a written contingency plan must be adopted. Copies of this plan must be maintained at the facility and provided to any group that may be called upon to assist in an emergency incident, including police, fire department, emergency response teams, and hospitals. This plan must be designed to reduce the adverse effects of any emergency incident due to the operations of the facility. The plan must be immediately activated upon any such emergency. There are a number of essential components of the plan.

1. The contingency plan must describe the actions facility personnel will take in case of emergencies such as fires, explosions, or releases of hazardous substances into the environment.
2. It must describe the agreements or arrangements made with the emergency response groups to coordinate the activities of these agencies in case of an emergency.
3. A person must be identified as the primary emergency coordinator. Others who are qualified to act as emergency coordinator are to be listed in the order they would be expected to act as alternates. The addresses and home and business, phone numbers of these individuals must be included on the list. At all times, at least one of these persons must be at the facility or on call in case of an emergency.
4. A current list of all the emergency equipment at the facility, and the specific location of each item, must be included. Each item must be identified by a brief physical description and its capabilities.
5. The plan must include an evacuation plan for facility personnel, identifying the signal(s) used to initiate the evacuation and primary and alternate escape routes.

7. Emergency Procedures

If an emergency appears likely or actually occurs, the individual acting as the emergency coordinator must immediately initiate an emergency response, including the following actions:

1. Activate the alarm system to alert facility personnel.
2. If necessary, notify emergency response groups to secure their aid.
3. If there has been a release of hazardous substances, a fire, or an explosion, the emergency coordinator must identify the nature of the material released, the exact source of the release, and the actual extent of the release.
4. The emergency coordinator must also immediately assess the possible impact of the release on human health or the environment away from the facility, from both direct causes and secondary effects, and include in the latter any adverse effects such as water runoff from efforts to control the incident.

5. If it is determined that adverse effects on humans or the environment could occur and evacuation of areas near the facility is needed, the emergency coordinator must immediately notify appropriate local authorities and remain available to help in making decisions as to which areas require evacuation.

6. In the event of an emergency of possible harm to humans and the environment away from the facility the emergency coordinator must also report the incident to the government coordinator at the scene or directly to the National Response Center (24-hour telephone number at time of writing, 1-800-424-8802). The report must include:

   a. Name, address and telephone number of the owner or operator of the facility.
   b. Name, address, and telephone number of the facility.
   c. Date, time, and character of incident (fire, explosion, spill, etc.).
   d. Name and quantity of materials involved, to the extent known.
   e. Extent of injuries, if any.
   f. Assessment of potential hazards to humans or the environment outside the facility where applicable.

7. The emergency coordinator must take all reasonable steps to contain the scope of the incident. If operations have not already ceased, shutting down operations must be considered. Waste containers and other hazardous materials not already involved in the incident should be moved to safer locations if feasible without risk to operating personnel. The shutdown of operations must be done to ensure there will not be any untoward equipment failures that could cause additional problems.

8. After the emergency, all recovered waste and contaminated soil, runoff water, etc., must be properly handled according to the provisions applicable to generators of hazardous waste.

9. All emergency equipment listed in the contingency plan must be cleaned and fit for its intended use before operations are resumed.

10. The operator or owner must notify the regional administrator and appropriate state and local authorities that the facility is again in compliance with the required standards before operations are resumed.

11. The operating records of the facility must reflect the time, date, and cogent details of any incident that requires implementation of the contingency plan. A report must be made to the regional administrator within 15 days after the incident. Besides the information listed under item 6 above, the report must include:

   a. An assessment of the actual or potential damages to human health or the environment because of the incident, where applicable.
   b. The estimated quantity and disposition of recovered material that resulted from the incident.

REFERENCES

1. DOT, 49 CFR Part 172, Subpart B
2. OSHA, 29 CFR Part 1910.120
3. EPA, 40 CFR Part 265

B. Practical Hazardous Waste Management Program

The sections.1-7, just preceding, briefly covered basic portions of the RCRA, OSHA regulations, and the DOT regulations affecting hazardous materials and hazardous waste. Hazardous waste management programs for generators must comply with applicable portions of these acts, including requirements that were not provided directly or paraphrased in those sections. It is incumbent upon generator waste management personnel to become familiar with all of the requirements, including changes as they occur.

1. Internal Waste Management Organization

In most organizations, it would be impractical for each individual laboratory, department, or division to establish a separate program, even if the regulatory organization in the state in which the facility was located was willing to deal separately with hundreds of different generator facilities at a single institution or research establishment. Thus, in most academic institutions and at many corporate research facilities, the responsibility for the general management of the hazardous waste program has been assigned to a centralized unit. This does not mean that the individual laboratory and laboratory personnel do not play a key role in an effective hazardous waste management program. Since they are the generators of the wastes, they can control, to a major degree, the amount of hazardous waste generated, and they are the primary source for identifying waste generated in their facility, especially that which is not in the original containers.

The hazardous waste organization thus starts with the laboratory itself. It is at this level that the waste is generated, collected, identified, and sometimes processed for reuse or rendered harmless. Each individual in the laboratory need not know all of the requirements under RCRA, but each person must be trained in the procedures by which the waste generated in the laboratory is to be handled before removal. Each worker must know what precautions to take to ensure that the waste will be acceptable for disposal, what information is required for each waste generated, and what internal records must accompany waste offered for disposal.

Unless the facility is so small that outside waste disposal firms come directly to the individual laboratories to package and remove the material, an internal waste management unit normally will be established within the organization to pick up the waste, classify it (according to the RCRA and DOT criteria), store it temporarily (up to 90 days), prepare appropriate documentation for each container, package it (in some cases), and arrange for the material classified as excess hazardous waste to be taken from the facility, according to the RCRA and DOT requirements.

It will be the responsibility of this internal agency to keep up with the current regulations, make all of the required reports, maintain all of the required records, and handle the waste itself. It will also probably be called upon to help develop and publish procedures for the institution's researchers to use in their part of the waste management program.

It is also likely that the same internal agency is the one assigned the responsibility of preparing and carrying out the required emergency contingency plan for chemical emergencies. This program must be integrated with the disaster emergency planning for the entire organization. As a technically trained group, some emergency responsibilities outside the waste program may also be assigned to waste management personnel.

2. Reduction of Hazardous Waste Volume

An effective hazardous waste management program cannot concern itself solely with disposal of unneeded materials. This is far too expensive and contradicts the RCRA concept. Chemicals are a resource that should be conserved, and where practical and legal, it is desirable for surplus chemicals to be put to a beneficial use rather than buried in a landfill, burned, or otherwise rendered unusable. However, as will be discussed later, there currently are limitations on how much effort an organization can devote to salvaging or treating a chemical waste without becoming a treatment and disposal facility.

One of the current requirements on generators is that they must document that they are making an effort to reduce the amount of waste that they generate. There are several effective ways in which laboratory operators themselves can aid in this effort. Among the more successful are:

1. *Planning of experiments.* Anticipation of waste generation should be a part of each operation. The research should be designed to reduce or eliminate the amount of dangerous substances generated as a byproduct of the research to the fullest extent possible, either as an ultimate or intermediary waste. Alternative reagents which would result in waste that is less dangerous or would reduce the volume of hazardous waste should be evaluated. The experiment should be planned to include rendering harmless any excess material generated in the research as a part of and a logical continuation of the work.

2. *Purchasing in smaller quantities.* It has been noted earlier that a substantial premium per unit chemical purchased is paid by buying chemicals in smaller containers rather than in larger ones. The cost of eight 1-pint bottles of a given material, purchased singly, is virtually certain to be much greater than the same material in a 1-gallon container, especially if one adds in the surcharge for buying the material in glass bottles covered with a protective plastic safety film. However, the total cost to the organization, if the cost of disposal of unused material is included, quite possibly will favor the smaller container. If individual containers are packed in a drum properly with an absorbent material, one typically can get only the equivalent of about 15 gallons of waste in the drum. If the original cost of the material is paid for by the laboratory facility while waste costs are paid for out of the organization's total budget, it is difficult to persuade the laboratory to pay the additional costs. It may require an explicit organizational policy to mandate purchasing of all chemicals in smaller sizes, unless the laboratory can document that their rate of use is sufficient to ensure the use of all the material in a short period, so that the material will not become degraded or become suspected of possible contamination and therefore unusable. If all materials in an institution or organization are purchased and distributed centrally this becomes an easier policy to enforce. This centralization also allows bulk purchasing of cases of smaller containers, perhaps making up for the original differential in cost. Proposals to charge for waste are usually self-defeating. Either the costs of collecting the charges in time and money exceed the funds collected, or some individuals attempt to evade the charges by disposing of their waste improperly down the sink.

3. *Using smaller quantities in experiments.* New technology has made it possible to do much chemical research with much smaller amounts of materials than was possible only a few years ago. More sensitive analytical instruments, balances, instruments operating on new principles, or more versatile or reconfigured instruments originally based on older principles often make it possible to perform an experiment with very small amounts of the reagents involved. For example, interfacing a small computer to an experiment may allow the simultaneous recording of many variables, with automatic control of the experiment to continuously modify the experiment, so that one run may replace a dozen or more. Not only will this reduce the volume of waste generated, but the experiments should become inherently safer. At least one institution made such a major commitment to the use of micro-scale techniques that they surplused most of their standard size glassware.

4. *Redistribution of used chemicals.* Excess chemicals are not necessarily waste chemicals. A specific part of any chemical waste program should be a classification and evaluation step, before which a chemical is only "surplus," not "waste." Clearly many materials are obviously clearly unusable and must be classified as waste and, if they meet any of the RCRA hazardous waste criteria, classified as hazardous waste or acutely hazardous waste if they fall in that category. However, many surplus chemicals are still usable for research or, if not, often are adequate for instructional laboratories. Unopened containers are the most likely to be attractive to potential users, but even partially used containers of reasonably fresh materials, which do not normally degrade by contact with air or moisture,

are often desirable to someone. A list of these materials should be maintained and circulated frequently to potential "customers" within the same organizations. Sometimes these materials are offered at a lower price than fresh material, but in most cases, charging will reduce the amount of surplus redistributed and may wind up increasing the cost to the organization because of the cost of disposing of the surplus material. There are "waste exchanges" to which lists of unwanted materials may be submitted which then circulate the information to organizations that might be interested. An individual taking advantage of materials from these exchange services must ensure that the material is really usable so that they will not incur expensive disposal costs.

Another category of materials that can often be reused are contaminated solvents. Many materials, such as xylene, methanol, and ethanol are heavily used materials, especially in the biological sciences, and frequently are available in the waste in sufficient volume to make it practical to reclaim by distillation. An automatic spinning band still can easily handle 20 to 30 liters per day. There will still be residues that must be treated as waste, but the volume will be far less than the original amount. The wastes must be treated carefully to ensure that two materials that form azeotropic mixtures are not mixed; otherwise simple distillation will not produce pure solvents. Prohibitions on disposal of liquids in landfills make this method of waste reduction even more attractive. Other materials such as mercury and silver from photographic processes are also readily recovered, and it is possible for the value of the reclaimed silver, for example, to more than pay for the effort involved.

5. *Management by tracking materials that degrade and become dangerous with time.* There are many chemicals that deteriorate over time, usually as they react with air or moisture. Ethers are a well-known example of this class of materials. The contents of partially empty containers of ethers, kept in storage are especially likely to form peroxides which are heat, shock, and friction sensitive so that an attempt to use them could result in a fatal explosion. Chemical waste vendors will not collect these potentially dangerous containers and remove them from the facility as part of a normal waste shipment. There are firms that specialize in disposing of such dangerous materials, but their services are extremely expensive. An effective program should be established for materials falling in this category for tracking them, beginning by dating each container when it is acquired, setting a target date for beginning checking the containers for dangerous degradation products, and establishing a firm date when they must be removed from the facility. Other materials that should be included in a tracking program should be pyrophoric materials, highly water-reactive materials or the converse, materials which become dangerous as they dry out, such as perchloric acid, and picric acid. Computer software now available makes it possible to identify with a bar code every container received so that all containers can be tracked. This software, coupled with a central receiving point for chemicals, should make it possible to virtually eliminate outdated containers as a problem.

6. *Elimination of unknowns.* There are two different ways in which unknowns are generated in a laboratory. One way is for information on the original labels on commercial containers to become lost either by becoming defaced or by disappearing (sometime this last is done deliberately). The other is when inadequate information is placed on labels to begin with, such as when a laboratory worker labels a container "Solution A" and then leaves the facility, so that no useful records are available. Unknown chemicals are among the most troublesome of all materials to handle as waste. There is a certain amount of risk in opening an unknown container, although the type of operations conducted in the laboratory in which it is found may provide enough information to reduce or eliminate the concern about shock or friction causing an explosion, or the possibility of a reaction with air or moisture in the air when the container is opened. Even after opening the container, the characteristics of the material may be sufficiently ambiguous to make identification very

difficult, especially if it is not originally a commercial product but has been generated in the laboratory. Internal characterization by organization analysis of these unknowns is one way of reducing the costs of their disposal.

The solution to the degraded or destroyed label problem is relatively straightforward. If the label on a container is damaged, then it should be laboratory policy for a new label to be affixed immediately by the first individual to be aware of the problem. Under the OSHA hazard communications standard and its secondary container requirements, every laboratory should have an ample supply of generic labels available for use. These generic labels should have space on them for the common name of the chemical, its CAS number, and basic hazard data, such as the NFPA 704 system color-coded numbering system or diamond.

The use of computer software for tracking chemical containers mentioned in item 5 immediately preceding would make it possible to eliminate truly old containers from accumulating if the organization were to adopt a policy that all chemical containers greater than 5 years old (with some justifiable exceptions) were to be discarded. The failure to properly label containers in the first place is a more difficult management problem, since there is typically only one person who knows what the container contents really are. In academic laboratories, the turnover in graduate students and postdoctoral research associates (and to some extent, faculty as well) is high. A policy should be adopted and enforced to have every person leaving the facility identify the contents of each container for which the individual has been responsible. Withholding graduation or requiring a financial deposit have been suggested as means of enforcing such a policy. Often, however, there would be few difficulties if the laboratory director or supervisor simply met with the individual shortly before the scheduled departure and went over the questionable containers. Unfortunately, some laboratory directors do not wish to take the time. There also have been cases where the individual in charge has compounded the problem by having the unknowns surreptitiously carried to an ordinary trash container outside the laboratory. Where this occurs and those responsible can be identified, they should be subject to severe punishment by the organization's authorities, because, in addition to the inappropriateness of such behavior, the organization can be held responsible for the consequences of such actions.

7. *Placing severe restrictions on "free" research materials or chemicals accepted by the research personnel.* It is frequently simpler, for example, for a supplier of agricultural chemicals to provide a researcher with a drum of one of their products (which may be a "numbered" experimental material) than to weigh out the amount that the person actually needs. A policy should be introduced that a supplier of a "free" material must either take back any excess material or guarantee to pay for the disposal cost of the excess. Institution of such a policy at the author's institution substantially reduced the amount of hazardous agricultural waste chemicals that required disposal. Other disciplines depend less on free materials, but the problem does exist elsewhere and should be restricted in the same manner. There are firms which attempt to "dump" unusable surplus materials on unsuspecting individuals.

8. *Miscellaneous.* There are a number of waste reduction techniques that are appropriate to the laboratory. These same procedures done by the internal waste disposal agency, could result in the institution being considered as a treatment and disposal facility if the procedures were employed after collection from the laboratory. For example, a container of ether, if there is only a small quantity remaining in the container should be placed in a suitable fume hood and allowed to evaporate. This eliminates the material as a solvent waste, as well as any chance of dangerous peroxides forming. Small quantities of other solvents can be dealt with in a similar fashion. Small quantities of acids and bases can be used in elementary neutralization processes. There are other purification,

reclamation, and neutralization techniques suggested in the literature for specific materials. Chemical procedures should be planned so that the end product consisting of any unused or generated materials not useful as a product of the procedure are rendered safe for ordinary disposal.

Since a nearly empty bottle requires as much space as a full one when packed in a drum for disposal, containers should be filled in the laboratory with either the same waste material or, less desirably, with compatible materials before being transferred from the laboratory to the waste collection group. In the latter case, a complete, accurate record of the contents of the container must be maintained and placed on the waste label affixed to the container when the transfer takes place.

A modest amount of material can be safely disposed of into the sanitary system if the material, primarily solvents, are miscible in water. However, there may be local restrictions that preclude this type of disposal, and if encouraged, some materials may be disposed of in this fashion for which the procedure is not appropriate. Unless done carefully under the direct supervision of a qualified laboratory employee, preferably a senior person who normally exercises managerial authority, disposal of materials into the sanitary system should be discouraged. The amount of money saved over several years of doing so safely would probably be minor compared to the cost of a single incident that subsequently causes a problem.

An excellent reference for means to render chemicals in the laboratory less hazardous is *Prudent Practices for Disposal of Chemicals from Laboratories,* published by the National Academy Press in 1983. Although the regulatory material included in the book is outdated, the chemistry remains valid. Some of this same material can be found in the newer *Prudent Practices in the Laboratory Handling and Disposal of Chemicals*.

3. Waste Collection

The RCRA regulations allow temporary storage of hazardous waste at or near the point of generation, which in the present document, means in or near a laboratory, but the storage area must be under the control and supervision of the generator. Independently operating laboratories cannot share the storage area. The amount is limited to 55 gallons of hazardous waste material or 1 quart of acutely hazardous material. If the procedure is to pick up the waste directly from the laboratory, then the only requirement will be to provide the proper documentation to the collection firm. Normally, the storage in the laboratory will only be a stage in the internal handling of the waste, with, typically, the waste being picked up and transferred to a central location under the control and supervision of the waste management group, where most of the waste management activities take place.

There are advantages if the determination of whether the surplus chemical meets the RCRA criteria for hazardous waste is delayed until after the material is taken to the central facility. If the excess chemical is defined at the laboratory as hazardous waste, transportation of the material may be a problem since one of the restrictions on transportation of hazardous waste, without meeting the requirements of a hazardous waste transporter, is that hazardous waste cannot be transported *along* a public street, although the waste can be moved *across* the same street. This problem does not arise for many corporate facilities since internal roads at corporate sites often are not accessible by the public. Most academic campuses, however, are open to the public and few streets in the vicinity of academic research buildings are restricted to the public. Sometimes, the strict interpretation of this requirement may be avoided by using small vehicles that can move on sidewalks and across limited access parking lots and other open spaces.

The delay of classification until the material is taken to a central location is itself a dubious circumvention of the standard if there is no real effort to classify the material brought to the facility as usable materials, harmless waste and hazardous waste at the facility. Only modest amounts of material should be transported at any time across a crowded campus to limit the consequences of an accident. Except in unusual circumstances, it would be desirable to limit the size of individual containers to 5-gallon pails for liquids and the equivalent for solids, which would be approximately 50 to 75 pounds, again as a safety measure. Containers of these sizes can be handled by hand, while 55-gallon drums would require equipment to lift in the event they were spilled in an accident. No chemicals should be moved in damaged containers or ones not sealed tightly. The containers should be securely loaded on the vehicle being used for transportation, so that they will not shift while the vehicle is in motion. Larger amounts in 55-gallon drums can be accumulated in secondary storage areas to avoid moving them through public areas. Waste management regulatory agencies must be notified of the locations of these satellite accumulation areas and the facility emergency contingency plan must apply to these areas.

Two separate trips should be made if two or more highly reactive materials or incompatible materials are to be transported. DOT regulations list which materials cannot be transported in the same vehicle or which can be transported simultaneously only under certain restrictions to protect against interactions of the materials.

It would be a good idea to have emergency supplies on the vehicle used for transportation of the chemicals, including: respiratory protection, gloves, chemical splash goggles, chemically protective coveralls, two class ABC fire extinguishers, a bag or two of an absorbent material, a package or two of Plug-n-Dike™ for plugging holes in a leaking container, a coil of rope and at least three lightweight standards (for establishing a temporary barrier), and a two-way radio or cellular telephone. All of this emergency gear should be where it is likely to be accessible in case of an accident. The individuals who pick up the material must be trained in how to respond in the event of an emergency. In the event of a fire, the same caution invoked earlier should be given. Personnel should only attempt to put the fire out if they can do so safely without exposing themselves to toxic fumes. If in a public area, immediate steps should be taken to cause the affected area to be evacuated.

4. Packaging

An upper limit to the size of the waste container packages was recommended in the previous section. Most surplus chemicals are not in 5-gallon metal cans, but are typically in smaller glass bottles of various English or metric sizes. The caps on containers should be required to be screw type, which are neither cracked nor corroded. Bottles with other types of stoppers, such as rubber, cork, or glass, may come open during transportation, allowing the contents to spill and cause a dangerous incident. The outside of any container offered for transportation should be clean. Small vials containing liquids may be placed in a larger container for transportation if packed carefully to avoid mixing incompatible materials.

Container sizes for flammable solvents, if the material has been transferred from the original container, cannot exceed the permissible sizes allowed by OSHA for flammable liquids of the appropriate class. The quality of the containers must be comparable to those used for the commercial versions of the chemicals. Chemical wastes should not be placed in polyethylene milk jugs or the equivalent. These containers are not sufficiently sturdy. Heavier duty polyethylene containers can be used if appropriate for the chemical.

Some solid materials normally come in bags. The original bags should be placed within one or two additional bags. One may suffice if the original bag is intact, while two should be used if the original bag is damaged and leaking. The external bag(s) preferably should be clear so the contents can be seen, especially if the original bag bears a legend identifying the contents. The thickness of the outer bag(s) should be no less than 6 mils. As noted earlier, compatible chemicals can be combined in a single container.

5. Characterization of the Waste

A typical waste label that might be used at any facility to identify and document waste internally was shown in Figure 4.13 earlier. A form such as this should be attached to each container of surplus chemical before it is removed from the laboratory and should remain with the container until final disposition of the material. A copy should be retained by the waste facility as part of their records. There are two sections to this form, one identifying the source of the material and the other characterizing the material itself. The latter information is to be used by the personnel responsible for classifying the material, and if necessary to complete the preparations for its disposal as a hazardous waste. Note that the material is not classified as a hazardous waste, allowing it to be evaluated, perhaps as a usable or recyclable material.

As can be noted, the information pertinent to the source is sufficiently complete to trace the material to its origin. A very important piece of information is the name of the person originating the waste, both as a printed name and as a full signature. The signature is important since it represents an acknowledgment by the individual signing the form that the remainder of the information is correct and eliminates confusion that might arise from a set of initials. If problems develop with the material, the person signing the form can be looked to for further information.

Providing information identifying the contents is the responsibility of the generator, not the waste management personnel. They can provide a reference or telephone numbers, but it is not their responsibility to perform this task for the generator. They are justified in refusing to take containers in which the contents have not been properly identified. All constituents of the material in the container must be listed by their proper chemical names. Abbreviations and formulas may cause confusion and should not be used as the sole identification. Similarly, trade names may be helpful, but again should not be used as a substitute for the correct chemical name. One useful additional identifier would be a CAS number, which, if not conveniently available to the researcher elsewhere, can be found for most common chemicals in virtually any chemical catalog. The term "inert ingredients" should not be used. All materials should be completely identified on the label.

Waste liquids containing acidic or basic materials must show the approximate pH of the liquid. If the liquid is not homogeneous but is stratified into layers, the pH of each layer must be provided.

Any material that is equivalent to a commercial product and is uncontaminated should be specifically marked as such as a candidate for the redistribution program.

The approximate percentage of each constituent should be provided, and the sum of these percentages must add up to 100%. If the container is one to which materials have been added from time to time, the log of these additions can be used to estimate the percentages of the various materials in the container.

It is advisable for waste management personnel to at least spot-check materials that they have collected from the various laboratories. A simple gas chromatograph analysis can be done on most organic solvents in about 15 minutes if the machine has the proper column and has been calibrated with appropriate standards. The pH of aqueous solutions can also be easily checked. If there is any doubt about the identity of most solid materials, the physical characteristics may be enough to resolve any questions and at least a rough qualitative identification can be made for most commonly used materials. Unknowns, of course, are a different story, and the analysis of these materials should be the responsibility of the originating laboratory, if possible. There are times when this is impossible, but these should be rare. The cost of analysis of unknowns by commercial disposal firms is high. In an academic institution with a graduate program in chemistry, unknowns can be characterized much less expensively by supporting a graduate student to do this task.

6. Packing of Waste for Shipment

The manner of packing the waste in preparation for transportation to a disposal facility depends upon the eventual means of disposal. The normal means of packaging bulk liquids is in 55-gallon steel drums, either DOT 1A1 closed-head or DOT 1A2 open-head steel drums, which are used for packing smaller containers. These are intended to be used only once, i.e., single-trip containers (STC), and are identified as such by stamping them with this legend. Normally containers are placed in the drum with enough filler material to completely fill the empty space. If the waste is a liquid, the filler material should be sufficient to absorb the liquid should the containers break. The space occupied by the awkward shape of most chemical containers, plus the need for enough inexpensive filler to completely absorb the liquid, limits the total volume of the bottles which can be placed in a 55-gallon drum to between 12 and 17 gallons. Smaller containers of incompatible chemicals, i.e., ones which can react vigorously together, cannot be placed in the same container. Smaller drums or pails also can be used if they meet DOT specifications for the materials placed in them. Occasionally there is a need to place a drum inside a larger one if there is a potential for risk of a leak. This is called an overpack and is relatively expensive. Very few liquids now are destined to be placed in a landfill, but these types of containers are used for transporting liquids to other types of disposal facilities or for recycling.

Some firms that accept materials for incineration will accept waste in metal drums, if they remove the waste containers from the drums at the disposal site. However, most firms accepting waste for incineration prefer the waste to be placed in combustible 1G2 fiber or 1H1 or 1H2 plastic drums that can be placed directly in the incinerator. These combustible fiber drums cannot be placed in landfills.

The drum types mentioned in the material above are the types meeting the newer international requirements.

7. Restrictions on Wastes

Some materials are forbidden to be transported as waste by DOT restrictions. Among these are:

1. Reactive wastes that can explode or release toxic vapors, gases, or fumes at standard temperature or pressure
2. Reactive materials that, when mixed with water or are exposed to moisture, can explode or generate toxic vapors, fumes, or gases
3. Materials that are shock sensitive
4. Materials that will explode, etc., if heated while confined
5. Class A or class B explosives.

These restrictions are imposed to ensure that the containers will not explode or evolve toxic gases, vapors, or fumes during transportation while loaded on a waste carrier and endanger the public. If a container of material were to explode on a truck loaded with other wastes, a major catastrophe could result.

The Hazardous and Solid Waste Amendments prohibit land disposal of some wastes unless the wastes are treated or it can be shown that there will be "no migration as long as the waste remains hazardous." Dioxin- and solvent-contaminated soils, and also dilute waste waters contaminated with solvents, were originally to be excluded, but there was not sufficient capacity to handle these otherwise; therefore, an extension was granted to continue placing these in landfills.

A more important restriction affecting laboratories went into effect on November 8, 1986, when it became illegal to place in a landfill wastes containing more than 1% of many solvents. These are listed in 40 CFR 268.30 and 268.31:

F001-The following spent halogenated solvents used in degreasing: tetrachloroethylene, trichloroethylene, methylene chloride, 1,1,1-trichloromethane, carbon tetrachloride, and chlorinated fluorocarbons, all spent solvent mixtures/blends used in decreasing containing before use, a total of 10% or more (by volume) of one or more of the above halogenated solvents or those solvents listed in F002, F004, and F005, and still bottoms from the recovery of these spent solvent mixtures.

F002 - The following spent halogenated solvents: tetrachloroethylene, methylene chloride, trichloroethylene, 1,1,1-trichloromethane, chlorobenzene, 1,1,2-trichloro-1,2,2-trifluoroethane, ortho-dichlorobenzene, and trichlorofluoromethane; orthodichlorobenzene and trichlorofluoromethane; all spent solvent mixtures/blends containing, before use, a total of 10% or more (by volume) of one or more of the above halogenated solvents or those solvents listed in F00l, F004, F005, and still bottoms from the recovery of these spent solvents and spent solvent mixtures.

F003 - The following spent nonhalogenated solvents: xylene, acetone, ethyl acetate, ethyl benzene, ethyl ether, methyl isobutyl ketone, n-butyl alcohol, cyclohexanone, and methanol; all spent solvent mixtures/blends containing, solely the above spent nonhalogenated solvents; and all spent solvent mixture/blends containing before use, one or more of the above nonhalogenated solvents, and a total of 10% or more (by volume) of one or more of those solvents listed in F00l, F002, F004, and F005; and still bottoms from the recovery of these spent solvents and spent solvent mixtures.

F004 - The following spent nonhalogenated solvents, creosols and cresylic acids and nitrobenzene, all spent solvent mixtures/blends containing, before use, a total of 10% or more (by volume) of one or more of the above nonhalogenated solvents or those solvents listed in F00l, F002, and F005; and still bottoms from the recovery of these spent solvents and spent solvent mixtures.

F005 - The following spent nonhalogenated, solvents; toluene, methyl ethyl ketone, carbon disulfide, isobutanol, and pyridine; all spent solvent mixture/blends containing, before use, a total of 10% or more (by volume) of one or more of the above nonhalogenated solvents or those solvent listed in F00l, F002, F004; and still bottoms from the recovery of spent solvents and solvent mixtures.

F020 - Wastes (except wastewater and spent carbon from hydrogen chloride purification) from the production or manufacturing use (as a reactant, chemical intermediate, or component in a formulating process) of tri- or tetrachlorophenol, or of intermediates used to produce their pesticide derivatives. (This listing does not include wastes from the production of hexachlorophene from highly purified 2,4,5-trichlorophenol.)

F021 - Wastes (except wastewater and spent carbon from hydrogen chloride purification) from the manufacturing use (as a reactant, chemical intermediate, or component in a formulating process) of pentachlorophenol, or of intermediates used to produce its derivatives.

F022 - Wastes (except wastewater and spent carbon from hydrogen chloride purification) from the manufacturing use (as a reactant, chemical intermediate, or component in a formulating process) of tetra-, penta-, or hexachlorobenzenes under alkaline conditions.

F023 - Wastes (except wastewater and spent carbon from hydrogen chloride purification) from the production of materials on equipment previously used for the production or manufacturing use (as a reactant, chemical intermediate, or component in a formulating process) of tri- and tetrachlorophenols. (This listing does not include wastes from equipment used only for the production or use of hexachlorophene from highly purified 2,4,5-trichlorophenol.)

F026 - Wastes (except wastewater and spent carbon from hydrogen chloride purification) from the production of materials on equipment previously used for the manufacturing use (as a reactant, chemical intermediate, or component in a formulating process) of tetra-, penta-, or hexachlorophene under alkaline conditions.

F027 - Discarded, unused formulations containing tri-, tetra-, or pentachlorophenol or discarded, unused formulations containing compounds derived from these chlorophenols. (This listing does not include formulations containing hexachlorophene synthesized from prepurified 2,4,5-trichiorophenol as the sole component.)

Many other materials have been placed on the restricted lists since this first list. These are listed in 40 CFR 268.32 through 40 CFR 268.39. Among the more critical in the context of laboratory wastes are (1) liquid hazardous wastes having a pH ≤ 2, (2) liquid hazardous wastes containing PCBs at concentrations ≥ 50 ppm, (3) liquid hazardous wastes that are primarily water and halogenated compounds at concentrations ≥ 1000 mg/L and $\leq 10,000$ mg/L, (4) mixed radioactive/hazardous wastes, and (5) RCRA hazardous wastes that contain naturally occurring radioactive materials.

As time passes, it is becoming clearer that there will be more limitations on the materials to be placed in landfills. Each of the sections listed in the previous paragraph have been amended, some as recently as 1998 and additional materials have been included in the individual wastes listed at the beginning of this section. Clearly, there will be additional restrictions on the construction and management of landfills, which will make them more expensive to use.* These costs will necessarily be passed on for payment to the generator. Obviously, it is going to become more important to the academic institution or corporation to restrain the amount of waste being produced. Disposal costs already are not a negligible item for a facility that is actively pursuing a waste program fully complying with the standards.

8. Shipping Waste

Once the waste has been collected and classified as hazardous, it must be disposed of by shipping it off-site for all those who are only classified as generators and do not have a permit for treatment, storage, disposal, or recycling facilities. A substantial amount of money can be saved if packing and preparation of the manifest are done by the internal waste management personnel rather than the disposal firm. This requires the management personnel to be thoroughly familiar with the current requirements for generators initiating shipments of hazardous waste equivalent to those given in Section X.A.2. Some hazardous waste disposal firms will not allow this, feeling that they cannot be sure that the waste is packaged properly. Others will accept waste packaged by the generator but will spot-check individual drums or will check the pH, for example, of some individual containers to confirm that the materials are packed properly, that the drums contain materials that agree with the accompanying lists, and that the chemicals are identified properly. It is essential that the generators subscribe to a reference service or have access to an on-line computer resource that will provide them with a current set of regulations for both the EPA requirements and DOT. The latter, especially change too frequently for the preparers of the shipments to be sure otherwise that they are in compliance, and, in any event, they need the references for the chemical identification numbers, proper shipping names, packing groups, labeling and placarding required, and any restrictions on shipping. In that vein, even the material presented here may have been partially superceded between the time of writing and the time read by the user of this handbook.

They will also need to conform to the specific requirements of the firm taking the waste. Some firms are essentially pure brokers, having no facilities of their own for the ultimate disposal of the waste. These firms sometimes collect wastes from several small generators at a central facility, group those materials that will be land-filled, incinerated, or used in a fuel-blending cement kiln facility, and then take the materials to firms that provide these services. If the size of the shipment is large enough, they may take it directly from the generator to the

---

* All landfill costs are increasing. Disposal of radioactive waste has risen by at least a factor of 4 during the past 5 years in the author's area, while disposal of ordinary solid wastes has gone up by a factor of even more. Due to landfill regulations, the cost of constructing and managing landfills has dramatically increased. As a result, many landfills are being closed. Similar increases in costs of landfill burials of hazardous waste or use of alternative technology are to be expected.

appropriate disposal site. In a case such as this, the generator will in effect have to conform to both the broker's requirements and those of the second firm, although it may make little difference to the generator.

Some firms offer a complete range of services for disposing of prepared waste. They offer transportation and either landfill disposal or incineration facilities. In principle at least, these firms should provide disposal at a lower cost per unit volume since they do not have to share the profits with an intermediary firm.

As noted earlier, there are materials that transporters are forbidden to carry because of the risk of an accident. There are firms that specialize in handling these dangerous materials, which may require special carriers that allow them to be moved safely or offer on-site destruction of materials such as shock-sensitive chemicals, explosives, highly reactive chemicals, gas cylinders that cannot be returned to the manufacturer or those that pose other special hazards. Some firms bring special equipment to a facility that allows them to take the materials, such as very old ethers containing large quantities of peroxides and other comparably dangerous materials, directly from the laboratory in cases of extreme risk. Local destruction will normally require permits and an isolated area in which to perform the task. Usually these will be the responsibility of the generator, and sometimes both may be difficult to obtain. Disposal of these items will be expensive.

The generator must accept one last responsibility that is not explicitly stated by the standards, and that is that the disposal firm must be capable of handling the waste properly. It would be assumed that the possession of an EPA permit and an Identification Number and being financially secure and able to provide a certificate of insurance would suffice to guarantee proper handling of the waste. However, there have been numerous fines, and closures of facilities operated by even major firms. It could be cost effective for a knowledgeable representative of the waste management group, or a colleague in the area whose judgment is known to be reliable, to visit a facility before selecting a firm. At the author's institution, such a practice has resulted in negative selection decisions in several cases. The facilities whose services were declined for this reason eventually were required by regulatory agencies to take corrective actions and in some cases ceased operations. Other protective steps a generator may take are (1) to require the disposal firm to provide a list of all citations for the past 5 years, (2) an insurance certificate or bond, and (3) very thorough checking of references. There are many cases in which the costs of remedying problems at disposal sites eventually have had to be borne by the generators that shipped waste to the sites.

9. Landfill Disposal

Until a few years ago landfill disposal has been the favored method of disposing of laboratory wastes. Because of potential problems such as damage to the incinerators and control of toxic emissions and hazardous constituents in the residual ash due to the variety of materials in mixed laboratory wastes, many incineration facilities did not wish to accept this category of materials, although some would accept properly segregated materials. In addition, or partially because of the operational problems, incineration was somewhat more expensive. However, because of the growing number of restrictions on materials that can be put into landfills (see Section X.B.7 of this chapter) and the increasing costs of constructing and managing landfills, incineration is becoming more popular and more cost effective.

Landfills are essentially elongated rectangular trenches cut into the earth in which the drums of hazardous waste are placed. Landfills are preferably sited in areas in which the water table is low compared to the depth of the trench and where the underlying earth consists of deep beds of clay with very low permeation properties. Clay layers underlying some of the more heavily used existing hazardous waste landfills are several hundred feet thick. Any material leaking (the current restrictions on liquid wastes make this eventuality unlikely) into these landfills from the waste containers should be contained much as water is in a bathtub. The containment is enhanced by current requirements to also use manufactured liners for the trench. Although steel drums would not deteriorate as rapidly as a combustible fiber drum, they still have a short lifetime compared to that of many of the materials contained within the drums. Orderly placement of the drums and having the drums completely filled is essential to keep the surface from sagging as the drums deteriorate. To prevent leakage from these sites, they should also be located in areas in which surface flooding is highly unlikely. All other things being equal, it would be desirable to have them in areas of low precipitation, but if the surface is properly configured, capped properly, and maintained, reasonable amounts of precipitation should not be a problem.

Low population density, the absence of commercial deposits of valuable minerals, and no historic or scenic attractions in the immediate vicinity also are desirable. The availability of good roads or, in some cases, rail access for bringing waste to the sit, is essential. Few people want a landfill near them so it is essential that the landfill be well managed, with trash dispersal being kept to an absolute minimum and a day's trash covered with soil at the end of each day's accumulation.

The typical size of these facilities is not very large, ranging from a few tens to a few hundreds of acres, and if managed properly, they are not an environmental problem to the surrounding area. Many existing landfills were not designed or managed properly and are among the SUPERFUND sites that must be cleaned up. Permitted landfills today must be lined with synthetic materials and have both a leachate collection system and a groundwater monitoring system. Because there are relatively few landfills that meet all of the essential criteria, waste disposal firms often must transport waste for long distances to them.

A fundamental concern about the land filling of hazardous waste is the fact that many materials retain their hazardous properties for very long periods. Although currently these sites can be managed properly, it is possible to imagine future circumstances that could arise in which good management would no longer be possible and even where records would be lost identifying the material buried at the site. Individuals raising these issues admit that these scenarios are remote at present, but the possibility of a negative impact on future generations has been raised as an issue weighing against the use of landfills for disposal of hazardous chemicals. They point out the historic short duration of social institutions (the United States is one of the longest lived continuous governments). The history of mismanagement by some collectors of hazardous waste has also contributed to an emotional reaction against landfills, so that siting additional landfills (or other hazardous waste operations) has become increasingly more difficult because of public concerns. Instead of being the method of choice, use of landfills is now often the last resort.

10. Incineration of Hazardous Waste

The two primary options of waste disposal off-site which have been available to generators of hazardous chemical waste have been landfills and incineration. It is unusual for a laboratory facility or even a facility with many laboratories at the same site to generate a sufficient waste stream that is consistent enough to be usefully reprocessed or recycled by a waste processor, although occasionally, as noted earlier, some operations may generate enough clean, nonhalogenated organic solvents to be of interest as fuel or to be recycled. Land filling has been the predominant disposal mechanism in the past, but incineration is now the preferred method by regulatory agencies.

Incineration has many major advantages. Many materials that are currently prohibited from being placed in landfills can, in principle, be destroyed by incineration. Regulatory requirements for most materials are that 99.99% be destroyed, although for some materials such as dioxin, the required destruction efficiency is 99.9999%. It is possible for most organic material to meet the RCRA requirements. The result is a much smaller volume of solid waste that may or may not have to be handled as a hazardous waste depending upon the materials in the original waste stream. Therefore, the size of the long-term management problem, if not totally eliminated, is at least greatly diminished. As a result, the residual liability for the generator for waste that has been incinerated may be eliminated.

There are some disadvantages. The emission of toxic materials, such as NO_x, SO_x, HCl, and some particulates containing metals from the stack must be controlled within newly stringent standards established by the EPA. This can be expensive for some waste streams. There are products of incomplete combustion (PICs), which may or may not be a problem. If an incinerator is managed properly, the PICs do not appear to be a problem in the large incinerators currently being used for waste disposal. The ash residue may contain concentrated quantities of heavy metals that must be disposed of as hazardous waste in a landfill. The combustion chamber itself may suffer physical damage or corrosion, and maintenance may be a problem. On the whole, however, especially as EPA standards on alternatives have become more restrictive, the advantages still may outweigh the disadvantages as long as the facilities are operated properly. However, as with landfills, construction of an incinerator is almost certain to meet with local opposition (and often by non-local environmental activists) so residents must be convinced that the unit is essential and that it will be well managed.

There are two types of incinerators used predominantly for incineration of hazardous waste. The first type is the rotary kiln, which operates at temperatures of approximately 2300°F. More hazardous wastes are burned in this type of kiln than any other and it is the one for which most tests on destruction efficiencies have been done. A second type is the cement kiln, which is now being used more often. The temperature of the cement kiln reaches 3000°F. The hazardous waste is blended with the fuel and used in the making of cement. By modifying the fuel blend, it is possible to obtain efficient destruction, even for chlorinated solvents.

The following article by Lawrence G. Doucet on incinerators provides a review of the factors that must be considered in establishing a local incineration program for an organizations hazardous waste.*

REFERENCES

1. Environmental Protection Agency, Hazardous Waste Regulations, 40 CFR Parts 260 -268, Washington, D.C.
2. Environmental Protection Agency, Designation, Reportable Quantities ("RQs"), and Notification, 40 CFR Part 302, Washington, D.C.
3. Environmental Protection Agency, Worker Protection in Hazardous Waste Operations, 40 CFR Part 311, Washington, D.C.
4. Environmental Protection Agency, Emergency Planning and Notification, 40 CFR Part 355, Washington, D.C.
5. Environmental Protection Agency, Hazardous Chemical Reporting: Community Right-to-Know, 40 CFR Part 370, Washington, D.C.
6. Occupational Safety and Health Administration, Employee Emergency Plans and Fire Prevention Plans, 29 CFR 1910.38, Washington, D.C.
7. Occupational Safety and Health Administration, Hazardous Waste Operations and Emergency Response, 29 CFR 1910.120, Washington, D.C.
8. Occupational Safety and Health Administration, Hazard Communication, 29 CFR 1910.1200, Washington, D.C.
9. Department of Transportation, Hazardous Material Regulations 49 CFR Parts 17 1-173, 177, Washington, D.C.

* This article was written prior to the newly restrictive rules on incinerators used for medical and infectious wastes. However, the information is still valid and useful for the selection and management of incinerators in general for waste disposal of other types. A discussion of the regulatory requirements for incinerators used for hospital and infectious wastes is given in the section immediately following this article.

10. Department of Transportation/United Nations Performance-Orientated Packaging Standards HM-81, 49 CFR Parts 172, 173, 177, 178., Washington, D.C.

All of the preceding regulatory references can be readily viewed at the online Internet site. http://www.access.gpo.gov/nara/cfr/cfr-retrieve.html#page1

C. State-of-the-Art Hospital and Institutional Waste Incineration, Selection, Procurement, and Operations[*]

1. Introduction

On-site incineration is becoming an increasingly important alternative for the treatment and disposal of institutional waste. Incineration reduces the weight and volume of most institutional solid waste by upwards of 90 to 95%, sterilizes pathogenic waste, detoxifies chemical waste, converts obnoxious waste such as animal carcasses into innocuous ash, and also provides heat recovery benefits. At most institutions, these factors provide a substantial reduction in off-site disposal costs such that on-site incineration is highly cost effective. Many systems have payback periods of less than 1 year. In addition, on-site incineration reduces potential exposures and liabilities associated with illegal or improper waste disposal activities.

Clearly, the most important factor currently affecting the importance of on-site incineration for healthcare organizations and research institutions across the country relates to infectious waste management and disposal. First of all, recent legislation and guidelines have dramatically increased the quantities of institutional waste to be disposed as "potentially infectious." For many institutions, particularly hospitals, incineration is the only viable technology for processing the increased, voluminous quantities of waste. Second, about half of the states and several major cities currently mandate that infectious waste be treated on-site, restrict its off-site transport, and/or prohibit it from being landfilled. Many additional states are planning similar, restrictive legislative measures within the next few years.

Off-site disposal difficulties and limitations probably contribute the greatest incentives for many health care and other institutions to consider or select on-site incineration as the preferred infectious waste treatment method. It has become increasingly difficult, if not impossible, to locate reliable, dependable infectious waste disposal service contractors. Many institutions able to obtain such services are literally required to transport their infectious waste across the country to disposal facilities. Furthermore, such services are typically very costly, if not prohibitive. Off-site disposal contractors are typically charging from about $0.30 to about $0.80 per pound of infectious waste, and some are charging as much as $1.50 per pound. For many hospitals and other institutions, this equates to hundreds of thousands of dollars per year, and several are paying in excess of one million dollars per year.

2. Incineration Technologies

Before the early 1960's, institutional incineration systems were almost exclusively multiple-chamber types, designed and constructed according to Incinerator Institute of America (IIA) *Incinerator Standards.* Since these systems operated with high excess air levels, most required scrubbers in order to comply with air pollution control standards. Multiple-chamber type systems are occasionally installed at modern facilities because they represent proven technology. However, the most widely and extensively used incineration technology over the past 20 years is "controlled-air" incineration. This has also been called "starved-air" incineration, "two-stage" incineration, "modular" combustion, and "pyrolytic" combustion. More than 7000 controlled-air type systems have been installed by approximately 2 dozen manufacturers over the past 2 decades.

---

[*] This article written by Lawrence G. Doucet, P.E., D.E.E.

Controlled-air incineration is generally the least costly solid waste incineration technology - a factor that has undoubtedly influenced its popularity. Most systems are offered as low cost, "pre-engineered" and prefabricated units. Costly air pollution control equipment is seldom required, except for compliance with some of the more current, highly stringent emission regulations, and overall operating and maintenance costs are usually less than for other comparable incineration technologies.

The first controlled air incinerators were installed in the late 1950s, and the first U.S. controlled-air incinerator company was formed in 1964. The controlled-air incineration industry grew very slowly at first. The technology received little recognition because it was considered unproven and radically different from the established and widely accepted IIA *Incinerator Standards.*

Approximately every 5 years the controlled-air incineration industry has gone through periods of rapid growth. In the late 1960s, this was attributable to the Clean Air Acts, in the early and late 1970s to the Arab oil embargoes, in the early 1980s to the enactment of hazardous waste regulations, and, recently to the enactment of infectious waste disposal regulations. Dozens of "new" vendors and equipment suppliers appeared on the scene during each of these growth periods. However, increased competition and rapid changes in the technology and market structure forced most of the smaller and less progressive companies to close. Generally, the controlled-air incineration industry has been in a state of almost constant development and change, with frequent turnovers, mergers, and company failures.

Today there are approximately a dozen listed "manufacturers"of controlled air incinerators. However, only about half of these have established successful track records with demonstrated capabilities and qualifications for providing first-quality installations. In fact, some of the "manufacturers" listed in the catalogs have yet to install their first system, and a few are no more than brokers who buy and install incinerator equipment manufactured by other firms.

Controlled-air incineration is basically a two-stage combustion process. Waste is fed into the first stage, or primary chamber, and burned with less than stoichiometric air. Primary chamber combustion reactions and turbulent velocities are maintained at very low levels to minimize particulate entrainment and carryover. This starved air burning condition destroys most of the volatiles in the waste materials through partial pyrolysis. Resultant smoke and pyrolytic products, along with products of combustion, pass to the second stage, or secondary chamber. Here, additional air is injected to complete combustion, which can occur either spontaneously or through the addition of auxiliary fuel. Primary and secondary combustion air systems are usually automatically regulated, or controlled, to maintain optimum burning conditions despite varying waste loading rates, composition, and characteristics.

Rotary kiln incineration systems have been widely promoted within the past few years. A rotary kiln basically features a cylindrical, refractory-lined combustion chamber which rotates slowly on a slightly inclined, horizontal axis. Kiln rotation provides excellent mixing, or turbulence, of the solid waste fed at one end — with high-quality ashes discharged at the opposite end. However, in general, rotary kiln systems have relatively high costs and maintenance requirements, and they usually require size reduction, or shredding, in most institutional waste applications. There are only a handful of rotary kiln applications in hospitals and other institutions in the U.S. and Canada.

"Innovative" incineration technologies also frequently appear on the scene. Some such systems are no more than reincarnations of older "failures," and others feature unusual applications and combinations of ideas and equipment. Probably the best advice when evaluating or considering an innovative system is to first investigate whether or not any similar successful installations have been operating for a reasonable period of time. Remember, so-called innovative systems should still be designed and constructed consistent with sound, proven principles and criteria.

Table 4.22 Classifications of Waste

Classification of Wastes			Approximate Composition % by Weight	Moisture Content %	Combustible Solids %	Value/lb of Refuse as Fired
Type	Description	Principal Components				
0	Trash	Highly combustible waste, paper, wood, cardboard cartons, treated papers, plastic or rubber scraps, from commercial and industrial sources	Trash 100%	10%	0%	6500
1	Rubbish	Combustible waste, paper, cartons, rags, wood scraps, combustible floor sweepings, domestic, from commercial, and industrial sources	Rubbish 80% Garbage 20%	25%	10%	6500
2	Refuse	Rubbish and garbage, residential sources	Rubbish 50% Garbage 50%	50%	7%	4300
3	Garbage	Animal and vegetable wastes, from restaurants, hotels, markets, institutional, commercial, and club sources	Garbage 65% Rubbish 35%	70%	5%	2500
4	Animal Solids and Organic Wastes	Carcasses, organs, solid organic wastes, hospital, laboratory, abattoirs, animal pounds, and similar sources	100% animal and human tissues	85%	5%	1000

3. Sizing and Rating

Classifications systems have been developed for commonly encountered waste compositions. These systems identify "average" characteristics of waste mixtures, including such properties as ash content, moisture, and heating value. The classification system published in the IIA *Incinerator Standards* is the most widely recognized and is almost always used by the incinerator manufacturers to rate their equipment. In this system, shown in Table 4.22, wastes are classified into seven types. Types 0 through 4 are mixtures of typical, general waste materials, and Types 5 and 6 (not listed in table) are industrial wastes requiring special analysis.

4. Primary Combustion Chambers
a. Heat Release Rates

Incinerator capacities are commonly rated as pounds of specific waste types, usually Types 0 through 4, that can be burned per hour. Incinerators generally have a different capacity rating for each type. For example, an incinerator rated for 1000 pounds per hour of Type I waste may only be rated for about 750 pounds per hour of Type 0 waste or about 500 pounds per hour of Type 4 waste. Such rating variations exist because primary chamber volumes are sized on the basis of internal heat release rates, or heat concentrations. Typical design heat release values range from about 15,000 to 25,000 BTU per cubic foot of volume per hour. In order to maintain design heat release rates, waste burning capacities vary inversely with the waste heating values (BTU/pound). As heating values increase, less waste can be loaded.

Since Type 3 waste, food scraps, and Type 4 (pathological) waste have heat contents of only 3500 and 1000 Btu per pound, respectively, it might be assumed that even higher capacity ratings could be obtained for these waste types. However, this is not the case. The auxiliary fuel inputs required to vaporize and superheat the high moisture contents of Types 3 and 4 wastes limit effective incineration capacities.

In essence, primary chamber heat release criterion establishes primary chamber volume for a specific waste type and charging rate. Heat release values are simply determined by multiplying burning rate (pound per hour) by heating value (BTU per pound) and dividing by primary volume (cubic foot).

b. Burning Rates

The primary chamber burning rate generally establishes the burning surface, or hearth area, in the primary chamber. It indicates the maximum pounds of waste that should be burned per square foot of projected surface area per hour (pounds per square foot per hour). Recommended maximum burning rates for various waste types are based upon empirical data, and are published in the IIA Incinerator Standards. Table 4.23 tabulates these criteria.

The figures in the table are calculated as follows: maximum burning rate (pounds per square foot per hour, or lbs/fl^2/hr) for Types 1, 2, and 3 wastes using factors as noted in the formula:

$$BR = \text{Factor for type waste} \times \text{log of capacity/hr}$$
$$BR = \text{Maximum burning rate lbs/ft}^2/\text{hr}$$

For example, for an incinerator capacity of 100 lbs/hr; for Type 1 waste:

$$BR = 13 \text{ (factor for Type 1 waste)} \times \log 100 \text{ (capacity/hr.)}$$

$$= 13 \times 2 = 26 \text{ lbs/ft}^2/\text{hr}$$

c. Secondary Combustion Chambers

Secondary chambers are generally sized and designed to provide sufficient time, temperature, and turbulence for complete destruction of combustibles in the flue gases from the primary chamber. Unless specified otherwise, secondary chamber design parameters are usually manufacturer specific. Typical parameters include:

- Flue gas retention times ranging from 0.25 seconds to at least 2.0 seconds
- Combustion temperatures ranging from 1400°F to as high as 2200°F
- Turbulent mixing of flue gases and secondary combustion air through the use of high velocity tangential air injectors, internal air injectors, abrupt changes in gas flow directions, or refractory orifices, baffles, internal injectors, and checker work in the gas flow passages.

Retention times, temperatures, and turbulence are interdependent. For example, secondary chambers that are specially designed for maximum turbulence but that have relatively short retention times may perform as well as other designs with longer retention

Table 4.23 Maximum Burning Rate (lbs/sq. ft/hr) of Various Type Wastes

Capacity (lbs/h)	Logarithm	Type of Waste			
		1 (Factor 13)	2 (Factor 10)	3 (Factor 8)	4 (No factor)
100	2.00	26	20	16	10
200	2.30	30	23	18	12[a]
300	2.48	32	25	20	14[a]
400	2.60	34	26	21	15[a]
500	2.70	35	27	22	16[a]
600	2.78	36	28	22	17[a]
700	2.85	37	28	23	18[a]
800	2.90	38	29	23	18[a]
900	2.95	38	30	24	18[a]
1000	3.00	39	30	24	18[a]

[a] The maximum burning rate in lb/ft²/hr for Type 4 waste depends to a great extent on the size of the largest animal to be incinerated. Therefore, whenever the largest animal to be incinerated exceeds 1/3 the hourly capacity of the incinerator, use a rating of 10 lblft²/hr for the design of the incinerator.

times but less effective turbulence. On many applications, increased operating temperatures may allow for decreased retention times, or vice versa, without significantly affecting performance. Regulatory standards and guidelines often dictate secondary chamber retention time and temperature requirements.

Flue gas retention time(s) is determined by dividing secondary chamber volume (cubic foot) by the volumetric flue gas flow rate (ft³/s) quantities and operating temperatures. They can be calculated or measured. However, during normal incinerator operations, flue gas flow rates vary widely and frequently.

d. Shapes and Configurations

Primary and secondary chamber shapes and configurations are generally not critical as long as heat release rates, retention time, and air distribution requirements are satisfactory. Chamber geometry is most affected by the fabrication and transport considerations of the equipment manufacturers. Although some primary and secondary chambers are rectangular or box-like, most are cylindrical.

Controlled air incinerators with a capacity of less than about 500 pounds per hour are usually vertically oriented, with primary and secondary chambers integral, or combined, within a single casing. Larger capacity controlled air incinerators are usually horizontally oriented and have non-integral, or separated, primary and secondary chambers. A few controlled air incinerator manufacturers offer systems with a third stage, or tertiary chamber, following the second stage. One manufacturer offers a fourth stage, termed a "reburn tunnel," which is primarily used to condition flue gases upstream of a heat recovery boiler. Most manufacturer "variations" are attempts to improve efficiency and performance. However, some of these may be no more than "gimmicks" that offer no advantages or improvements over standard, conventional systems. Adherence to proper design fundamentals, coupled with good operations, are the overall keys to the success of any system. Acceptance of unproven variations or design deviations is usually risky.

5. Selection and Design Factors

Highly accurate waste characterization and quantification data are not always required for selecting and designing incineration systems; however, vague or incomplete data can be very misleading and result in serious problems.

Waste characterization involves identification of individual waste constituents, relevant physical and chemical properties, and presence of any hazardous materials. A number of terms commonly used to characterize waste can be very misleading when used in specifications. As examples, vague terms such as "general waste," "trash," "biological waste," "infectious waste," and "solid waste" provide little information about the waste materials. An incineration system designed for waste simply specified as "general" waste would probably be inadequate if waste contained high concentrations of plastics. Likewise, the term "pathological" waste is frequently, but incorrectly, used to include an assortment of materials, including not only animal carcasses but also cage waste, laboratory vials, and biomedical waste materials of all types. "Pathological" incinerators are usually specifically designed for burning animal carcasses, tissues, and similar types of organic wastes. Unless the presence of other materials is clearly specified, resultant burning capacities may be inadequate for waste streams to be incinerated.

Waste characterization can range from simple approximations to complex and costly sampling and analytical programs. As discussed, the most frequently used approximation method is to categorize "average" waste mixtures into the five IIA classes, Types 0 through 4.

The popularity of this waste classification system is enhanced by the fact that most of the incinerator manufacturers rate their equipment in terms of these waste types. However, it should be noted that actual "average" waste mixtures rarely have the exact characteristics delineated for any of these indicated waste "types."

The other end of the characterization spectrum involves sampling and analysis of specific "representative" waste samples or items in order to determine "exact" heating values, moisture content, ash content, and the like. This approach is not generally recommended because it is too costly and provides no significant benefit over other acceptable approximation methods.

Virtually all components found in typical institutional type solid waste have been sufficiently well characterized in various engineering textbooks, handbooks, and other technical publications. Table 4.24 shows the various BTU values of materials commonly encountered in incinerator designs. The values given are approximate and may vary based on their exact characteristic or moisture content. In many cases, a reasonably accurate compositional analysis of the waste stream, in conjunction with such published data and information, could provide reliable and useful characterization data.

A key factor is that incineration systems must be designed to handle the entire range of the waste stream properties and characteristics, not just the "averages." System capacity and performance may be inadequate if the waste data does not indicate such ranges.

a. Capacity Determination

One of the primary criteria for selecting incineration system capacity is the quantity of waste to be incinerated. Such data should include not only average waste generation rates, but also peak rates and fluctuation cycles. The most accurate method of determining such data is a comprehensive weighing program over a period of about 2 weeks. However, the most common procedure has been to estimate waste quantities from the number and volume of waste containers hauled off-site to land disposal. Large errors have resulted from such estimates

Table 4.24   Waste Data Chart

Material	BTU Value per lb as Fired	Weight (lb/ft³) (Loose)	Weight (lb/ft³)	Content by Weight (%)	
				Ash	Moisture
Type 0 waste	8,500	8-10		5	10
Type 1 waste	6,500	8-10		10	25
Type 2 waste	4,300	15-20		7	50
Type 3 waste	2,500	30-35		5	70
Type 4 waste	1,000	45-55		5	85
Acetic acid	6280		65.8	0.5	0
Animal fats	17,000	50-60		0	0
Benzene	18,210		55	0.5	0
Brown paper	7,250	7		1	6
Butyl sole composition	10,900	25		30	1
Carbon	14,093		138	0	0
Citrus rinds	1,700	40		0.75	75
Coated milk cartons	11,330	5		1	3.5
Coffee grounds	10,000	25-30		2	20
Corn cobs	8,000	10-15		3	5
Corrugated paper	7,040	7		5	5
Cotton seed hulls	8,600	25-30		2	20
Ethyl alcohol	13,325		49.3	0	0
Hydrogen	61,000		0.0053	0	0
Kerosene	18,900		50	0.5	0
Latex	10,000	45	45	0	0
Linoleum scrap	11,000	70-100		20-30	1
Magazines	5,250	35-50		22.5	5
Methyl alcohol	10,250		49.6	0	0
Naphtha	15,000		41.6	0	0
Newspaper	7,975	7		1.5	6
Plastic-coated paper	7,340	7		2.6	5
Polyethylene	20,000	40-60	60	0	0
Polyurethane (foamed)	13,000	2	2	0	0
Rags (linen or cotton)	7,200	10-15		2	5
Rags (silk or wool)	8,400-8,900	10-15		2	5
Rubber waste	9,000-11,000	62-1 25		20-30	0
Shoe leather	7,240	20		21	7.5
Tar or asphalt	17,000	60		1	0
Tar paper 1/3 tar, 213 paper	11000	10-20		2	1
Toluene	18,440		52	0.5	0
Turpentine	17,000		53.6	0	0
1/3 wax, 2/3 paper	11,500	7-10		3	1
Wax paraffin	18,621		54-57	0	0
Wood bark	8,000-9,000		2-20	3	10
Wood bark (fir)	9,500		12-20	3	10
Wood sawdust	7,800-8,500		10-12	3	10
Wood sawdust (pine)	9,600		10-12	3	10

because of failures to account for container compaction densities or from faulty assumptions that the waste containers were always fully loaded.

Three major variables affect the selection of incineration system capacity or hourly burning rate: waste generation rates; waste types, form, and sizes; and operating hours. The quantity of waste to be incinerated is usually the primary basis for selecting system capacities. When waste generation rates are grossly underestimated, incineration capacity may be inadequate for the planned, or available, periods of operation. In such cases, the tendency is to overload the system, and operational problems ensue. On the other hand, incineration systems must be operated near their rated or design capacities for good performance; an oversized system must be operated less hours per day than may have been anticipated.

Such reduced operating hours could cause difficult problems with waste handling operations, particularly if waste storage areas are marginal. Furthermore, if waste heat recovery is necessary to justify system economics, insufficient waste quantities could be a serious problem.

Since incinerators are primarily sized according to heat release rates, waste heating value is a fundamental determinant of capacity. However, the physical form or consistency of waste may have a more significant impact on burning capacities. For example, densely packed papers, books, catalogs, and the like may have an effective incinerability factor of only about 20% compared to burning loosely packed paper. Likewise, animal bedding or cage waste typically has high ash-formation tendencies that may reduce burning rates by as much as 50%. Furthermore, highly volatile wastes, such as plastics and containers of flammable solvents, may require burning rate reductions of as much as 65% to prevent smoking problems.

The physical size of individual waste items and containers is also an important factor in the selection of incineration capacity. One rule-of-thumb is that an average incinerator waste load, or largest item, should weigh approximately 10% of rated, hourly system capacity. On this basis, a minimum 300 pounds per hour incinerator would be required for, say, Type I waste packaged in up to about 30-pound containers or bags. This capacity would be required regardless of the total daily quantity of Type I waste requiring incineration.

A typical daily operating cycle for a controlled air incinerator without automatic ash removal is as follows:

Operating Steps	Typical Durations
Clean-out	15-30 minutes
Preheat	15-60 minutes
Waste loading	12-14 hours
Burndown	2-4 hours
Cool-down	5-8 hours

It is important to note that waste loading for systems with manual clean-out is typically limited to a maximum of 12 to 14 hours per operating day.

b. Burning Rate vs. Charging Rate

When evaluating incineration equipment, it is important to distinguish between the terms "burning or combustion rate" and "charging or loading rate." Manufacturers may rate their equipment or submit proposals using either term. "Burning rate" refers to the amount of waste that can be burned or consumed per hour, while "charging rate" is the amount of waste that can be loaded into the incinerator per hour. For systems operating less than 24 hours per day, "charging rates" typically exceed "burning rates" by as much as 20%. Obviously, failure to recognize this difference could lead to selecting a system of inadequate capacity.

6. Incinerator System Auxiliaries

The incinerator proper is only a single component in a typical incineration "system." Other components or subsystems which require equal attention in the design and procurement process, include:

- Waste handling and loading systems
- Burners and blowers
- Residue or ash removal and handling systems
- Waste heat recovery boiler systems
- Emissions control systems
- Breeching, stacks and dampers
- Controls and instrumentation

Features of the major subsystems follow:

Incinerators with capacities less than about 200 pounds per hour are usually available only with manual loading capabilities. Manual loading entails charging waste directly into the primary chamber without the aid of a mechanical system. Units with capacities in the 200 to 500 pounds per hour range are usually available with mechanical loaders as a special option. Mechanical loaders are standardly available for most incinerators with capacities of more than about 500 pounds per hour.

The primary advantage of mechanical loaders is that they provide personnel and fire safety by preventing heat, flames, and combustion products from escaping the incinerator. In addition, mechanical loading systems prevent or limit ambient air infiltration into the incinerator. In most cases, air infiltration affects combustion conditions and, if excessive, substantially lowers furnace temperatures and causes smoking at the stack and into charging room areas. Infiltration also increases auxiliary fuel usage and usually accelerates refractory deterioration. Several states have recently enacted regulations requiring mechanical loading on *all* institutional waste incinerators.

Mechanical loaders enable incinerators to be charged with relatively small batches of waste at regulated time intervals. Such charging is desirable because it provides relatively stabilized combustion conditions and approximates steady-state operations. Limiting waste batch sizes and loading cycles also helps protect against overcharging and resultant operating problems.

The development of safe, reliable mechanical loaders has been a major step toward modernizing institutional waste incineration technology. The earliest incinerators were restricted to manual charging, which limited their capacities and applications. Of the loader designs currently available, most manufacturers use the hopper/ram system. With this system, waste is loaded into a charging hopper, a hopper cover closes, a primary chamber fire door opens and a charging ram then pushes the waste into the primary chamber. Hopper/ram systems are available with charging hopper volumes ranging from several cubic feet to nearly 10 cubic yards. The selection of proper hopper volume is a function of waste type, waste container size, method of loading the hopper, and incinerator capacity. An undersized hopper could result in spillage during waste loadings, an inability to handle bulky waste items, such as empty boxes, or the inability to charge the incinerator at rated capacity. On the other hand, an oversized hopper could result in frequent incinerator overcharging and associated operational problems.

A few manufacturers have recently developed mechanical loader systems capable of accepting as much as an hour's worth of waste loading at one time. These systems use internal rams to charge the primary chamber at intervals, as well as to prevent hopper bridging. Although these systems have had reportedly good success, they are still generally in the developmental stage.

One particular rotary kiln manufacturer uses an integral shredder at the bottom of the waste feed hopper. This system is termed an "auger feeder." It basically serves to process waste into a size that is compatible with the kiln dimensions.

With small capacity incinerators, less than about 500 pounds per hour, waste is usually loaded manually, bag by bag, into the charging hopper. Larger capacity systems frequently employ waste handling devices such as conveyors, cart dumpers, and sometimes skid-steer tractors to charge waste into the hopper. Pneumatic waste transport systems have been used to feed incinerator loading hoppers at a few institutions, but these have had limited success.

A cart-dumper loader basically combines a standard hopper/ram system with a device for lifting and dumping waste carts into the loading hopper. Several manufacturers offer these as integrated units. Cart dumpers also can be procured separately from several suppliers and adapted or retrofitted to almost any hopper/ram system. Cart-dumper loader systems have become increasingly popular because using standard, conventional waste carts for incinerator loading reduces extra waste handling efforts and often eliminates the need for intermediate storage containers and additional waste handling equipment.

Most modem hopper/ram assemblies are equipped with a water system to quench the face of the charging ram face after each loading cycle. This prevents the ram face from overheating due to constant, direct exposures to high furnace temperatures during waste injection. Without such cooling, plastic waste bags or similar materials could melt and adhere to the hot ram face. If these items did not drop from the ram during its stoking cycle, they could ignite and be carried back into the charging hopper, where they could ignite other waste remaining in the hopper or new waste loaded into the hopper. For additional protection against such possible occurrences, loading systems can also be equipped with hopper flame scanners and alarms, hopper fire spray systems, and/or an emergency switch to override the normal charging cycle timers and cause immediate injection of hopper contents into the incinerator.

a. Residue Removal and Handling Systems

Residue, or ash, removal has always been a particular problem for institutional type incineration systems. Most small capacity incinerators (less than about 500 pounds per hour) and most controlled-air units designed and installed before the mid-1970s must be cleaned manually. Operators must rake and shovel ashes from the primary chamber into outside containers. Small capacity units can be cleaned from the outside, but large capacity units often require operators to enter the primary chamber to clean ashes. The practice of manual clean-out is especially objectionable from many aspects, including:

- Difficult labor requirements
- Hazards to operating personnel because of exposures to hot furnace walls, pockets of glowing ashes, flaming materials, airborne dusts, and noxious gases
- Daily cool-down and start-up cycling requirements, which substantially increase auxiliary fuel usage and reduce available charging time
- Detrimental effects of thermal cycling on furnace refractories
- Severe aesthetic, environmental, and fire safety problems when handling hot, unquenched ashes outside the incinerator
- Possible regulatory restrictions

In multiple-chamber incinerators, automatic ash removal systems usually feature mechanical grates or stokers. In rotary kiln systems, ash removal is accomplished via the kiln rotation. However, automatic, continuous ash removal has historically been difficult to achieve in controlled air systems which have conventionally featured stationary, or fixed, hearths.

Early attempts at automatic ash removal in controlled air incinerators employed a "bomb bay" door concept. With these systems, the bottom of the primary chamber would swing open to drop ashes into a container or vehicle located below. Serious operating problems led to the discontinuance of these systems. More recent automatic ash removal systems use rams or plungers to "push" a mass of residue through the primary chamber and out a discharge door on a batch basis. Most of these systems have had only limited success.

Controlled-air incinerator automatic ash removal systems that have shown the most promise use the waste charging ram of the hopper/ram system to force waste and ash residues through the primary chamber to an internal discharge or drop chute for removal. Although charging rams usually extend no more than about 12 to 18 inches into the furnace during loading, this is sufficient to move materials across the primary chamber via the repetitive, positive-displacement actions of the rams. With proper design and operations, the waste should be fully reduced to ash by the time it reaches the drop chute. For incinerators with capacities greater than about 800 to 1000 pounds per hour, internal transfer rams are usually provided to help convey ashes through the furnace to the drop chute. Transfer rams are necessary because the ash displacement capabilities of charging rams are typically limited to a maximum length of about 8 feet. Primary chambers longer than about 16 feet usually have two or more sets of internal transfer rams.

The most innovative residue removal system uses a "pulse hearth" to transfer ashes through the incinerator. The entire floor of the primary chamber is suspended on cables and pulses intermittently via sets of end-mounted air cushions. The pulsations cause ash movement across the chamber and toward the drop chute.

After the ashes drop from the primary chamber through the discharge chute, there are two basic methods, other than manual, for collecting and transporting them from the incinerator. The first is a semiautomatic system using ash collection carts positioned within an air-sealed enclosure beneath the drop chute. A door or seal gate at the bottom of the chute opens cyclically to drop ashes into ash carts. Falling ashes are sprayed with water for dust suppression and a minor quenching. Because of weight considerations, ash cart volumes are usually limited to about 1 cubic yard.

Loaded ash carts are manually removed from the ash drop enclosure and replaced with empty carts. After the removed carts are stored on-site long enough for the hot ashes to cool, they are either emptied into a larger container for off-site disposal or are brought directly to the landfill and dumped. Adequate design and proper care are needed when dumping ashes into larger on-site containers to avoid severe dusting problems. In addition, some ashes could still be hot and may tend to ignite when exposed to ambient air during the dumping operations.

The second method of ash removal is a fully automatic system using a water quench trough and ash conveyor that continuously and automatically transports wet ashes from the quench trough to a container or vehicle. With these systems, the discharge chute terminates below water level in a quench trough in order to maintain a constant air seal on the primary chamber. Most manufacturers use drag, or flight, type conveyors but a few offer "backhoe" or "scoop" type designs to batch grab ashes from the quench trough. The important factor is that the selected ash conveyor system be of proven design and of heavy-duty construction for the severe services of ash handling.

b. Waste Heat Recovery

In most incineration systems, heat recovery is accomplished by drawing the flue gases through a waste heat boiler to generate steam or hot water. Most manufacturers use conventional firetube type boilers for reasons of simplicity and low costs. Both single and multipass firetube boilers have been used successfully at many installations. Several facilities incorporate supplemental fuel-fired waste heat boilers so that steam can be generated when the incinerator is not operating. Also, automatic soot blowing systems are being installed on an increasing number of firetube boilers, in order to increase on-line time and recovery efficiencies.

One manufacturer uses single-drum, watertube type waste heat boilers on incineration systems. Watertube boilers are also used by other manufacturers in installations where high steam pressures and flow rates are required. Another manufacturer offers heat recovery systems with water wall or radiant sections in the primary chamber. These water wall sections, which are usually installed in series with a convective type waste heat boiler, can increase overall heat recovery efficiencies by as much as 10 to 15%.

Many incinerator manufacturers typically "claim" system heat recovery efficiencies for their equipment ranging from 60% to as high as 80%. However, studies and EPA-sponsored testing programs have shown that realistic heat recovery efficiencies are typically on the order of 50% to 60%. The amount of energy or steam that can be recovered is basically a function of flue gas mass flow rates and inlet and outlet temperatures. Depending on boiler type and design, gas inlet temperatures are usually limited to a maximum of 2200°F. Outlet temperatures are limited to the dewpoint temperature of the flue gases in order to prevent condensation and corrosion of heat exchanger surfaces. Depending upon flue gas constituents, incinerator dewpoint temperatures are usually on the order of 400°F.

For estimating purposes, about 3 to 4 pounds of steam can be recovered for each pound of typical institutional type solid waste incinerated. However, the economic feasibility of providing a waste heat recovery system usually depends upon the ability to use the recovered energy. If only half of recovered steam can be used because of low seasonal steam demands, heat recovery may not be cost effective.

Some controlled air incinerator manufacturers offer air preheating, or "economizer packages," with their units. These primarily consist of metal jacketing around sections of the primary or secondary chambers. Combustion air is heated by as much as several hundred degrees when pulled through the shrouds by combustion air blowers. This preheating can reduce auxiliary fuel usage by as much as 10 to 15%. In addition, the shrouding on some systems also helps limit incinerator skin temperatures to within OSHA limits.

For safety and normal plant shutdown, waste heat boilers are equipped with systems to divert flue gases away from the boiler and directly to a stack. One such system comprises an abort, or dump, stack upstream of the incinerator and stack. Another system includes a bypass breeching connection between the incinerator and stack. Modern, well-designed bypass systems are equipped with isolation dampers either in the dump stack or in the bypass breeching section. In systems without isolation dampers, either hot flue gases can bypass the boiler or ambient air can dilute gases to the boiler. Because of these factors, boiler isolation dampers may improve overall heat recovery efficiencies by at least 5%.

c. Chemical Waste Incineration

An increasing number of institutions are disposing of chemical waste in their incineration systems. Incinerated chemicals are usually flammable waste solvents that are burned as fuels with solid waste. A simple method of firing solvents has been to inject them through an atomizer nozzle into the flame of an auxiliary fuel burner. Larger capacity and better designed systems use special, packaged burners to fire waste solvents. Such burners are either dedicated exclusively for waste solvent firing or have capabilities for switching to fuel oil firing when waste solvents are not available. Waste solvent firing, is usually limited to the primary chamber in order to assist in the burning of solid wastes and to maximize retention time by fully utilizing secondary chamber volumes. Injectors and burners must be located and positioned so as not to impinge on furnace walls or other burners. Such impingement results in poor combustion and often causes emission problems.

Chemical waste incineration systems must also include properly designed chemical waste handling systems. These include a receiving and unloading station, a storage tank, a pump set to feed the injector or burner, appropriate diking and spill protection, and monitoring and safety protection devices. Most of these components must be enclosed within a separate, fire-rated room that is specially ventilated and equipped with explosion-proof electrical fixtures.

When transporting, storing, and burning chemical waste, local, state and federal hazardous waste regulations must be followed. If the incinerated waste is regulated as a "hazardous waste," very costly trial burn testing, (Part B) permitting, and monitoring equipment are required. In addition, obtaining the permits could delay starting a new facility by as much as 12 to 18 months. Incinerators burning chemical solvents which are only hazardous due to "ignitability" are not likely to be considered hazardous waste incinerators, and the costly and lengthy hazardous waste incinerator permitting process is avoided. However, the storage and handling of these solvents will likely require a hazardous waste (Part B) permit.

At many institutions, bottles and vials of chemical wastes are often mixed with solid waste for incineration. If the quantities or concentrations of such containers and chemicals are very small with respect to the solid waste, incinerator operations may be unaffected. However, whenever solid waste loads are mixed with excessive concentrations of chemical containers, serious operating problems are likely, including rapid, uncontrolled combustion and volatilization resulting in heavy smoke emissions and potentially damaging temperature excursions. In addition, glass vials and containers tend to melt and form slag that can damage refractory materials and plug air supply ports.

d. Emission Control Systems

In general, only controlled air incinerators are capable of meeting the stringent emission standard of 0.08 grains of particulate per dry standard cubic foot of flue gas (gr/DSCF), corrected to 12% carbon dioxide, without emission control equipment. However, no incineration systems can meet the emission limits being recently enacted by many states which require compliance with best available control technology (BACT) levels. The BACT particulate level identified by many of the states is 0.015 gr/DSCF, corrected to 12% carbon dioxide. However, this is a controversial level which is being challenged by some in that it is only applicable to municipal waste incineration technology. Compliance with a 0.015 level will likely require a very high pressure drop, energy intensive, venturi scrubber system. Although "dry scrubbers," which comprise alkaline injection into the flue gas stream upstream of a baghouse filter, may also achieve a 0.015 level, as of this writing this technology has yet to be demonstrated on an institutional waste incineration system.

Most institutional solid waste streams, particularly hospitals, include significant concentrations of polyvinyl chloride (PVC) plastics. Upon combustion, PVC plastics break down and form hydrogen chloride (HCl) gas. The condensation of HCl gases results in the formation of highly corrosive hydrochloric acid. Therefore, flue gas handling systems, and particularly waste heat boilers, must be designed and operated above the dewpoint of the flue gases. Protection of scrubbing systems typically includes the provision of an acid neutralization system on the scrubber water circuitry and the use of acid-resistant components and materials.

Some states have identified BACT for HCl emissions as either 90% removal efficiency or 30 to 50 ppm, by volume, in the exhaust gases. For most well-designed wet scrubbers, 99% removal efficiencies are readily achievable. With respect to minimizing emissions of PICs, such as carbon monoxide and even dioxins and furans, the keys are proper furnace sizing, good combustion controls designed to accommodate varying waste compositions and charging rates, good operations and proper care, and adjustment of system components. Inadequacies in any of these could result in objectionable emissions.

e. Success Data

Incineration is considered proven technology in that a great many systems readily comply with stringent environmental regulations and performance requirements. Properly designed and

Table 4.25 Incineration System Performance Problems

Major Performance Difficulties	Examples
Objectionable stack emissions	Out of compliance with air pollution regulations
	Visible emissions
	Odors
	Hydrochloric acid gas (HCl) deposition and deterioration
	Entrapment of stack emissions into building air intakes
Inadequate capacity	Cannot accept "standard" size waste containers
	Low hourly charging rates
	Low daily burning rates (throughput)
Poor burnout	Low waste volume reduction
	Recognizable waste items in ash residue
	High ash residue carbon content (combustibles)
Excessive repairs and downtime	Frequent breakdowns and component failures .
	High maintenance and repair costs
	Low system reliability
Unacceptable working environment	High dusting conditions and fugitive emissions
	Excessive waste spillage
	Excessive heat radiation and exposed hot surfaces
	Blowback of smoke and combustion products from the incinerator
System inefficiencies	Excessive auxiliary fuel usage
	Low steam recovery rates
	Excessive operating labor costs

operated incineration systems provide "good" performance if they satisfy specific user objectives in terms of burning capacity; throughput, burnout, or destruction; environmental integrity; and on-line reliability. However, many incineration systems of both newer and older designs perform poorly. Performance problems range from minor nuisances to major disabilities, and needed corrective measures range from simple adjustments to major modifications or even total abandonment. Furthermore, performance problems occur as frequently and as extensively in small, dedicated systems as in large, complex facilities. The most common incineration system performance problems are shown in Table 4.25.

7. Incineration Performance and Procurement

It has been estimated that roughly 25% of incineration systems installed within the last 10 years either do not operate properly or do not satisfy user performance objectives. A 1981 University of Maryland survey of medical and academic institutions incinerating low-level radioactive wastes indicated that only about 50% of the institutions surveyed (23 total) "reported no problems," and about 47% of the institutions (20 total) reported problems ranging from mechanical difficulties to combustion difficulties. A survey conducted by the U.S. Army Corps of Engineers Research Laboratory in 1985 at 52 incineration facilities reported that 17% of the users were "very pleased with their systems," 71% were "generally satisfied with the performance of their systems" (but indicated that minor changes were needed to reduce maintenance and improve efficiency), and 12% were "not happy with their systems" (reporting severe problems). Results of this Army survey are summarized in Table 4.26.

a. Fundamental Reasons for Poor Performance

Underlying causes or reasons for poor incineration system performance are not always obvious. When performance difficulties are encountered, a typical reaction is often to "blame"

Table 4.26 Twenty Common Problems Found in Small Waste-To-Energy Plants[a]

Problems	Installations Reporting (%)
Castable refractory	71
Underfire air ports	35
Tipping floor	29
Warping	29
Charging ram	25
Fire tubes	25
Air pollution	23
Ash conveyer	23
Not on-line	21
Controls	19
Inadequate waste supply	19
Water tubes	17
Internal ram	15
Low steam demand	13
Induced draft fans	12
Feed hopper	10
High pH quench water	8
Stack damper	4
Charging grates	2
Front-end loaders	2
Consensus:	
17% very pleased	
71% generally satisfied - minor improvements needed	
12% not happy	

[a] Results of 1983 survey of 52 heat recovery incineration systems (5-50 TPD) conducted by U.S. Army Construction Engineering Research Laboratory.

the incinerator contractor for furnishing "inferior" equipment. While this may be the case on some installations, there are other possible reasons which are more common and sometimes more serious. Generally incineration system performance problems can be related to deficiencies or inadequacies in any or all of three areas:

1. Selection and/or design - before procurement
2. Fabrication and/or installation - during installation
3. Operation and/or maintenance - after acceptance

Examples of deficiencies in these three areas are as follows.

i. System Selection and/or Design Deficiencies

Deficiencies in this area are usually the result of basing incineration system election and design decisions on incorrect or inadequate waste data, as well as failures to address specific, unique facility requirements. The resultant consequences are that system performance objectives and design criteria are also inadequate. An example of this is the procurement of an incineration system of inadequate capacity because of underestimated waste generation rates. Not so obvious examples include the relationships between operating problems and inadequate waste characterization data. Since incinerators are designed and controlled to process specific average waste compositions, vague identification of waste types or wide variances between actual waste parameters and "selected" design parameters often result in poor system performance. Significant deviations in parameters such as heating values, moisture, volatility, density, and physical form could necessitate a capacity reduction of as much as two thirds in order to avoid objectionable stack emissions, unacceptable ash quality, and other related problems. Table 4.27 indicates examples of improper waste characterization affecting incineration capacity. The establishment of good performance objectives based upon sound data and evaluations is only the initial step toward procuring a successful installation. The next step would be to assure that system design criteria and associated contract documents are adequate to satisfy the performance objectives. A prime example of design inadequacies is the failure to relate incinerator furnace volumes to any specific criteria such as acceptable heat release rates. Another example is the specification of auxiliary components, such as waste loaders and ash removal systems, that are not suitable for the required operating schedules or rigors.

ii. Fabrication and/or Installation Deficiencies

Deficiencies in this area relate to inferior workmanship and/or materials in either the fabrication or installation of the system. The extent and severity of such deficiencies are largely dependent upon the qualifications and experience of the incinerator contractor. Unqualified incineration system contractors may be incapable or disinterested in providing a system in compliance with specified criteria. This could be either because of general inexperience in the field of incineration or because of a disregard of criteria that is different from their "standard" way of doing business or furnishing equipment.

It is typical for even the most experienced and qualified incineration system contractors to deviate to some extent from design documents or criteria. This is largely because there are no such things as "standard" or "universal" incineration systems or "typical" applications or facilities. Unless design documents are exclusively and entirely based upon and awarded to a specific, preselected incinerator manufacturer, different manufacturers usually propose various substitutions and alternate methodologies when bidding a project. The key to evaluating such proposed variations is to assess whether they comply with fundamental design and construction criteria and whether they reflect proven design and application. On the other hand, allowing such variations without proper assessment could have unfortunate consequences.

The number and severity of fabrication and installation deficiencies are also directly related to quality control efforts during construction phases of a project. For example, a review of contractor submittals, or shop drawings, usually helps assure compliance with contract documents *before* equipment is delivered to the job site. Site inspections during installation work may detect deficiencies in design or workmanship before they lead to operational problems and performance difficulties. In addition, specific operating and performance testing as a prerequisite to final acceptance is a key element in assuring that a system is installed properly.

Table 4.28 lists some of the most common reasons for deficiencies in the fabrication and installation of incineration systems.

iii. Operational and/or Maintenance Deficiencies

Deficiencies in this area are basically "self-inflicted" in that they usually result from owner or user omissions or negligence, and related problems occur *after* a system has been successfully tested and officially accepted.

Successful performance of even the best designed, most sophisticated, and highest quality incineration systems is ultimately contingent upon the abilities, training, and dedication of the operators. The employment of unqualified, uncaring, poorly trained, and unsupervised operators is one of the most positive ways of debilitating system performance in the shortest time.

Table. 4.27 Waste Characterization Data Deficiencies Necessitating System Capacity Reductions[a]

Actual Waste Characterizations

(Deviations front Selected "Design" Values)	Typical Examples	Basic Reasons for Reduced Capacities
Heating values (BTI/lb) excessive	Greater concentrations of paper and plastic components (or less moisture) than originally identified and specified	Incinerator volumetric heat release rates (BTU/ft³/h) exceed design limits[b]
Moisture concentrations excessive	Greater concentrations of high water content wastes, such as animal carcasses or food scraps (garbage), than originally identified and specified	Increased auxiliary fuel firing rates and additionaltime required for water evaporation and superheating
Volatiles excessive	Greater concentrations of plastic (such as polyethylene and polystyrene) or flammable solvents than originally identified and specified	Rapid (nearly instantaneous) releases of combustibles (volatiles) in large quantities along with excessively high temperature surges
Densities excessive	Computer printout, compacted waste, books, pamphlets, and blocks of paper	Difficulties in heat and flames penetrating and burning through dense layers of waste
High ash formation tendencies	Animal bedding or cage wastes—wood chips, shavings, or sawdust	Ash layer formation on surface of waste pile insulates bulk of waste from heat, flames, and combustion air

[a] Failure to reduce capacities, or hourly waste loading rates, to accommodate indicated deviations would likely result in other more serious operational problems.

[b] Based upon accepted, empirical values, primary chamber heat release rates should be in the range of 15,OO& to 20,000 BTU/ft³/h.

Table 4.28 Common Reasons for Fabrication and Installation Deficiencies

Incineration equipment vendor (manufacturer) unqualified

Equipment installation contractor (GC) unqualified

Inadequate instructions (and supervision) from the manufacturer for system installation by the GC

No clear lines of system performance responsibility between the manufacturer and the GC

Failure to review manufacturer's shop drawings, catalog cuts, and materials and construction data to assure compliance with contract (design) documents

Inadequate quality control during and following construction to assure compliance with design (contract) documents

Payment schedules inadequately related to system performance milestones

Final acceptance testing not required for demonstrating system performance in accordance with contract requirements

Table 4.29 Common Reasons for Operational and Maintenance Deficiencies

Unqualified operators

Negligent, irresponsible, and/or uncaring operators

Inadequate operator training programs

Inadequate operating and maintenance manuals

No record keeping or operating logs to monitor and verify performance

Inadequate operator supervision

Lack of periodic inspections, adjustments, and preventive maintenance

Extending equipment usage when repairs and maintenance work are needed

Incineration systems are normally subject to severe operating conditions, and they require frequent adjustments and routine preventive maintenance in order to maintain good performance. Failures to budget for and provide such adjustments and maintenance on a regular basis lead to increasingly bad performance and accelerated equipment deterioration. Also operating incineration equipment until it "breaks down" usually results in extensive, costly repair work and substantially reduced reliability.

Table 4.29 lists some of the most common operational and maintenance deficiencies which could result in poor incineration system performance.

The above problems are usually interrelated, and they usually occur in combination. They occur as frequently and as extensively in small, dedicated facilities as in large, complex facilities. They may range in severity from objectionable nuisances to major disabilities. Also, required corrective measures may range from minor adjustments to major modifications or even total abandonment.

Selection and design deficiencies are probably the most common as well as the most serious causes of problem incineration systems. Reputable incinerator contractors usually make every effort to satisfy specified design and construction criteria and meet their contractual obligations. Operating and maintenance deficiencies can usually be corrected. However, once a system has been installed and started, very little can be done to compensate for fundamental design inadequacies. Major, costly modifications and revisions to performance objectives are usually required.

The relatively frequent occurrence of design deficient systems may largely be attributable to a general misconception of the incineration industry as a whole. Incinerators are often promoted as standard, off-the-shelf equipment that can be ordered directly from catalogs, shipped to almost any job site and literally "plugged in." This impression has been enhanced by many of the incinerator vendors in a highly competitive market. Exaggerations, half-truths, and sometimes false claims are widespread relative to equipment performance capabilities. In addition, attractive, impressive brochures often suggest that implementation of an incineration system is simpler than it really is.

Incineration systems are normally subject to extremely severe operating conditions. These include very high and widely fluctuating temperatures, thermal shock, from wet materials, slagging residues which clinker and spall furnace materials, explosions from items such as aerosol cans, corrosive attacks from acid gases and chemicals, and mechanical abrasion from the movement of waste materials and from operating tools. These conditions are compounded by the complexity of the incineration process. Combustion processes are complicated in themselves, but in incineration this complexity is magnified by frequent, unpredictable, and often tremendous variations in waste composition and feed rates. To properly manage such severe and complex operating conditions, incineration systems require well-trained, dedicated operating personnel, frequent and thorough inspections, maintenance and repair, and administrative and supervisory personnel attuned to these requirements.

At many facilities, the practice is to operate the incineration system continuously until it breaks down because of equipment failures. This type of operation accelerates both bad performance and equipment deterioration rates. Repairs done after such breakdowns are usually far more extensive and costly than those performed during routine, preventive maintenance procedures. Also, items which are typically capable of lasting many years can fail in a fraction of that time if interrelated components are permitted to fail completely.

8. Key Step

A first step in procuring a good incineration system is to view the incineration "industry" in a proper perspective. There are four basic principles to bear in mind:

1. Incineration technology is not an "exact" science - It is still more of an art than a science, and there are no shortcuts, simplistic methods, or textbook formulas for success.

2. There is no "universal" incinerator - No design is universally suited for all applications. Incinerators must be specifically selected, designed, and built to meet the needs of each facility on an individual basis. Manufacturers catalogs identify typical models and sizes, but these are rarely adequate for most facilities without special provisions or modifications.

3. There is no "typical" incinerator application - Even institutions of similar type, size, and activities have wide differences in waste types and quantities, waste management practices, disposal costs, space availability, and regulatory requirements. Each application has unique incineration system requirements that must be identified and accommodated on an individual basis.

4. Incinerator manufacturers are not "equal" - There are wide differences in the capabilities and qualifications of the incinerator equipment manufacturers. Likewise, there are wide differences in the various systems and equipment which are offered by different manufacturers.

9. Recommended Procurement Steps

Table 4.30 outlines six steps, recommended for implementing an incineration system project. Each is considered equally important towards minimizing or eliminating the deficiencies discussed above and for increasing the likelihood of obtaining a successful installation. Performance difficulties on *most* problem incineration systems can usually be traced to a disregard or lack of attention to details in the first two steps; namely (1) evaluations and selections and (2) design documents. For example, many facilities have been procured strictly on the basis of "purchase orders" containing generalized requirements such as:

Table 4.30 Recommended Incineration System Implementation Steps

Evaluation and selections
  Collect and consolidate waste, facility, cost, and regulatory data
  Identify and evaluate options and alternatives
  Select system and components
Design (contract) documents
  Define wastes to be incinerated—avoid generalities and ambiguous terms
  Specify performance requirements
  Specify *full* work scope
  Specify minimum design and construction criteria
Contractor selection
  Solicit bids on quality and completeness—not strictly least cost
  Evaluate and negotiate proposed substitutions and deviations
  Negotiate payment terms
  Consider performance bonding
Construction and equipment installation
  Establish lines of responsibility
  Require shop drawing approvals
  Provide inspections during construction and installation
Startup and final acceptance
  Require punch-out system for contract compliance
  Require comprehensive testing: system operation, compliance with performance requirements and emissions
  Obtain operator training
After final acceptance
  Employ qualified and trained operators
  Maintain operator supervision
  Monitor and record system operations
  Provide regular inspections and adjustments
  Implement preventive maintenance and prompt repair; consider service contract

"Furnish an incineration system to burn
_____ lb/hr of institutional waste
in compliance with applicable regulations."

Obviously, the chances for success are marginal for any incineration system procured on the basis of such specifications.

On many projects, incinerator contractor evaluation and selection under step 3 involve no more than a solicitation of prices from a random listing of vendors with the award of a contract to that firm proposing a system for the "least cost." There are two basic problems with this approach. First, the selected incinerator contractors are assumed to have equivalent capabilities and qualifications. Second, "least cost" acceptance assumes that the equipment offered by each of the contractors is equivalent or identical. A comparative "value" assessment of proposals usually results in the procurement of a superior quality system for a negligible price difference. It is not uncommon to see cost proposals "low" by no more than 10%, but the equipment offered of only half the quality of the competition.

Again, although incineration is considered a proven technology, in many ways it is still more of an art than a science. There are no textbook formulas or shortcut methods for selecting and implementing a successful system, and there are no guarantees that a system will not have difficulties and problems. However, the probabilities of procuring a successful, cost effective system increase proportionally with attention to details and utilization of proven techniques, methodologies, and experience.

REFERENCES

1. **Bleckman, J., O'Reilly, L., and Welty, C.,** Incineration for Heat Recovery and Hazardous Waste Management, American Hospital Association, Chicago, 1983.

2. **Roegly, W.J., Jr.,** *Solid Waste Utilization—Incineration with Heat Recovery,* prepared for Argonne National Laboratory under Contract W-3 1-109-Eng-38 with the U.S. Department of Energy Publication No. ANL/CES/TE 78-3, Washington, D.C., 1978.

3. **Cooley, L.R., McCampbell, M.R., and Thompson, J.D.,** *Current Practice of Incineration of Low-Level Institutional Radioactive Waste,* prepared for U.S. Department of Energy, Publication No. EGG-2076, Washington, D.C., 1981.

4. **Doucet, L.G.,** Waste handling systems and equipment, *NFPA Fire Protection Handbook,* Chapter 14, Section 12, National Fire Protection Association, Quincy, MA, 1985.

5. **Doucet, L.G.,** Incineration: State-of-the-Art Design, Procurement and Operational Considerations, *Technical Document No. 055872,* American Society for Hospital Engineering, Environmental Management File, Chicago, 1988.

6. **Doucet, L.G. and Knoll, W.G., Jr.,** The craft of specifying solid waste systems, *Actual Specif. Eng.,* 107, May 1974.

7. **Doucet, L.G.,** Institutional waste incineration problems and solutions, *Proceedings of Incineration of Low Level and Mixed Wastes Conference,* St. Charles, IL, April 1987.

8. **Ducey, R.A., Joncich, D.M., Griggs, K.L., and Sias, S.R.,** Heat Recovery Incineration: A Summary of Operational Experience, Technical Report No. CERL SRE-85/06, prepared for U.S. Army Construction Engineering Research Laboratory, Champaign, IL, 1985.

9. **English, J.A., II,** Design aspects of a low emission, two-stage incinerator, *Proceedings 1974 National Incinerator Conference,* ASME, New York, 1974.

10. Environmental Protection Agency, Small Modular Incinerator Systems with Heat Recovery: A Technical, Environmental and Economic Evaluation, prepared by Systems Technology Corporation, Environmental Protection Agency, Publ. SW-177c, Washington, D.C., 1979.

11. **Hathaway, S.A.,** Application of the Packaged Controlled Air-Heat Recovery Incinerator of Army Fixed Facilities and Installations, Technical Report No. CERL-TR-E- 151, prepared for U.S. Army Construction Engineering Research Laboratory, Champaign, IL, 1979.

12. *Incinerator Standards,* Incinerator Institute of America, New York, 1968.

13. **Martin, A.E.,** Small-Scale Resource Recovery Systems, Noyes Data Corporation, Park Ridge, MN, 1982.

14. **McColgan, I.J.,** Air Pollution Emissions and Control Technology: Packaged Incinerators, Economic and Technical Review, EPS-3-AP-77-3, Canadian Environmental Protection Service, Burlington, Ontario, 1977.

15. **McRee, R.E.,** Controlled-air incinerators for hazardous waste application theory and practice, *Proc. APCD Int. Workshop Ser. on Hazardous Waste,* New York, April 1, 1985.

16. **McRee, R.E.,** Waste heat recovery from packaged incinerators, *Proc. ASME Incinerator Division Conf,* Arlington, VA, January 25, 1985.

17. **Theoclitus, G., Liu, H., and Dervay, J.R., II,** Concepts and behavior of the controlled air incinerator, *Proc. 1972 National Incinerator Conf,* ASME, New York, 1972.

C. Hospital, Medical and Infectious Waste

1. Incineration in Compliance With the Clean Air Act

Since the previous article by Lawrence Doucet was written, the Environmental Protection Agency (EPA) has passed, on September 15, 1997, new regulations and guidelines governing incinerators burning hospital/medical wastes (HMI). As noted earlier, this does not change the information presented in the earlier article on the selection, and operations of incinerators in general, but it does impose such severe restrictions on incinerators used for the disposal of medical and infectious wastes that a large majority of the existing approximately 2,400 incinerators used for that purpose will probably cease operations, at least for that portion of the time previously devoted to this application. This article deal with the operation of existing hospital and medical waste incinerators under the new rules and guidelines, and the restrictions placed on new units.

The definition of hospital/infectious medical wastes is given by EPA in 40 CFR Part 60.51c as follows. Note that the definition is important in that it also defines what are not infectious wastes.

"Medical/Infectious Waste means any waste which is generated in the diagnosis, treatment, or immunization of human beings or animals, in research pertaining thereto, or in the production or testing of biologicals, including cultures and stocks, human pathological wastes, human blood, and blood products, sharps and glassware, animal waste, isolation wastes, and unused sharps. A waste does not meet this definition if it is: (1) a hazardous waste listed under 40 CFR Part 261 (e.g., certain unused chemicals and pharmaceuticals; (2) household waste as defined in 40 CFR Section 261.4(b)(1); (3) ash from the incineration of medical waste; (4)human remains intended for interment or cremation; or (5) domestic sewage materials identified in 40 CR Part 261.4(a)(1)."

The new standards do not explicitly regulate medical and infectious waste but do so indirectly for those disposed of using incineration as a means of disposal. These rules were required under Section 129 of the Clean Air Act Amendments of 1990 to reduce air pollution from incinerators. The rules require states to develop plans to reduce pollution from existing MWIs, built on or before June 20, 1996, and from new units, built after that date. If a State already had rules in place, then those rules remain in effect but may have to be amended to meet the new Federal regulations. The State rules (and those for some local agencies) can be more stringent than the Federal Rules but must be as least as strong. The time for development of the State Plans has now passed as this is written, so all State Plans presumably have been completed and submitted for EPA approval.

The State plans require compliance by the operators either within one year after EPA has approved the State's Plan or within three years if the State has developed a schedule of steps which will bring the State Plan into compliance during that interval. The schedule must be verifiable. Regardless of the route a State takes, all existing Hospital/Medical Infectious Waste Incinerators (HMIWI) must be in compliance with the EPA regulations within five years after the final rule was published.

The federal standard sets numerical limits for a number of pollutants from HMIWI so that they will meet the limits required by the Clean Air Act. These pollutants are: particulate matter (PM), carbon monoxide (CO), dioxin/furan (CDD/CDF), hydrochloric acid (HCl), sulfur dioxide (SO_2, nitrogen oxides (NO_x), lead (Pb), cadmium (CD), and mercury (Hg). Opacity of the emissions is regulated as well. It also sets out a number of other requirements in addition to the emission limits.

Performance testing
Ongoing parameter testing
Inspections
Operator training
Waste management plans
Reporting and record keeping
Title V permit.

a. Incinerator Classes

There are four classes of incinerators established under the new rules, although one is relatively uncommon, small rural incinerators that are located more than 50 miles from a

Table 4.31 Emission Limits for Existing HMWI (corrected to 7% O$_2$)

Pollutant	Small Rural	Small	Medium	Large
PM, mg/dscm	197	115	69	34
CO, ppmv	40	40	40	40
CDD/CDF,	800 total,	125total,	125 total,	125 total,
HCl, ppmv	3,100	100 or 93 %[b]	100 or 93 %[b]	100 or 93 %[b]
SO$_2$, ppmv	55	55	55	55
NO$_x$, ppmv	250	250	250	250
Pb, mg/dscm	10	1.2 or 70%[b]	1.2 or 70%[b]	1.2 or 70%[b]
Cd, mg/dscm	4	0.16 or 65%[b]	0.16 or 65%[b]	0.16 or 65%[b]
Hg, mg/ dscm	7.5	0.55 or 85%[b]	0.55 or 85%[b]	0.55 or 85%[b]

[a] TEQ is toxic equivalency quantity determined by using international equivalency factors.
[b] Percent reduction across control device.
dscm = dry cubic meters at standard conditions
ppmv = parts per million by volume

standard metropolitan area, and burns less than 2000 pounds of waste per week. The emission limits are less restrictive for this latter class of incinerator. The other three classes are (1) "small," which have a burning capacity of 200 pounds per hour or less; (2) "medium," with a burning capacity of 200 - 500 ponds per hour and (3) "large," that have a capacity greater than 500 pounds per hour. Although these incinerators may have the capacity to burn the amounts stated, they can actually operate at the rate for a lower category and be classified at that level. The emission limits are given for all three classes and small rural units for existing and new incinerators in Table 4.31 and 4.32 respectively.

b. Other Regulatory Requirements for New and Existing Units

The three larger classes of existing HMW incinerator units require initial stack testing for all of the pollutants (including opacity) previously listed. Small rural units are only required to test for particulate matter, carbon monoxide, dioxin/furans, mercury and opacity. The tests are to be by specific EPA approved procedures. After the initial emission tests, the larger HMIWI units must test stack emissions for particulate matter, carbon monoxide, and hydrogen chloride annually, although passing the tests for three consecutive years allows the facility to forego testing for the next two years. Failure in any year would require the facility to pass for three consecutive years again. Opacity tests must be made annually. Small rural units must only perform annual equipment inspections after the initial stack tests. Table 4.34 provides data on the parameters to be monitored and the frequency of monitoring required for existing and new facilities respectively.

Operating parameters for the incinerators are to be established during the initial performance testing. This must be done within 180 days of initial startup. Subsequently,

Table 4.32 Emission Limits for New and Modified HMWI

Pollutant	Small	Medium	Large
PM, mg/dscm	115	69	34
CO, ppmv	40	40	40
CDD/CDF	125	25	25
HCl, ppmv	15 or 99 %[b]	15 or 99 %[b]	15 or 99 %[b]
SO$_2$, ppmv	55	55	55
NO$_x$, ppmv	250	250	250
Pb, mg/dscm	1.2 or 70% reduction	0.07 or 98% reduction	0.07or 98% reduction
Cd, mg/dscm	0.16 or 65% reduction	0.16 or 65% reduction	0.16 or 65% reduction
Hg, mg/ dscm	0.55 or 85%b reduction	0.55 or 85%b reduction	0.55 or 85% reduction
Opacity	10%	10%	10%

they must monitor charge rate, secondary chamber temperatures and bypass stack temperatures. Units using dry scrubbers and wet scrubbers must monitor several parameters associated with these accessory units.

Training requirements in the EPA emission guidelines require that all operators pass a course meeting the requirements spelled out in the emission guidelines. Information on site operations is to be developed and updated annually. Operators must review the site operations data and the updates.

It is desirable to minimize the amount of waste processed in the incinerator. The facility is required to develop a waste management plan to further this concept, specifically to determine what parts of the waste stream would be suitable for recycling.

The records accumulated in meeting these regulatory requirements must be retained for at least five years. However, it would be desirable to maintain them for a longer period in order to substantiate compliance should this be questioned. Records of the initial performance tests, later performance tests and inspection reports must be submitted to the regulatory agency responsible for their facility. If anything exceeds permissible limits, these data must be submitted on a semi-annual basis. The latter reports should also include information on efforts made to return the operations to acceptable levels.

Because of the costly measures needed to comply with the clean air act, it is anticipated that relatively few new medical and infectious waste incinerators will be constructed, less than 70 and perhaps as few as 10 by the year 2002. There is some relief that would allow some incinerators originally intended to be used forinfectious waste disposal to be employed to a lesser extent. They can be used for "pathological waste", *viz.,* animal and human remains, in some states. If the incinerator burns no more than 10% hospital and medical wastes, the incinerator may qualify as a "co-fired" combustor and would not be regulated as a HIMWI. An incinerator regulated under the Solid Waste Disposal Act or a

Table 4.33 Operating Parameters To Be Monitored

Operating Parameters	Data Measurement	Data Recording	Type of Control System Affected
Maximum charge rate	Continuous	Once per hour	All
Maximum fabric filter inlet temperature	Continuous	Once per minute	Dry scrubber followed by fabric filter
Maximum flue gas temperature	Continuous	Once per minute	Dry scrubber followed by fabric filter; wet scrubber
Minimum secondary chamber temperature	Continuous	Once per minute	All
Minimum dioxin/furan sorbent flow rate	Hourly	Once per hour	Dry scrubber followed by fabric filter
Minimum HCl sorbent flow rate	Hourly	Once per hour	Dry scrubber followed by fabric filter
Minimum mercury sorbent flow rate	Hourly	Once per hour	Dry scrubber followed by fabric filter
Minimum pressure drop across the wet scrubber or minimum horsepower or amperage to wet scrubber	Continuous	Once per minute	Wet scrubber
Minimum scrubber liquor flow rate	Continuous	Once per minute	Wet scrubber
Minimum scrubber liquor pH	Continuous	Once per minute	Wet scrubber

municipal incinerator may be able to accommodate the medical wastes. One should seek approval for using these alternatives with either the state or local regulatory body.

2. Alternative Technologies

Incinerators are still being sold but, as noted above, very few will be specifically designated for disposal of infectious wastes . Instead a variety of new technologies are now available as alternatives. Among these are the following in no particular order:

1. Autoclaving (not new but more advanced systems allow bulk treatment of substantial quantities)
2. Pyrolysis
3. Circulating fluidized-bed coal technologies
4. Thermal oxidation
5. Electron beam
6. Microwave incineration
7. Chemical disinfection
8. Gas/vapor sterilization
9. Biodegradation
10. Steam sterilization

Following are a few comments about these technologies. Additional information can be readily obtained in the Internet references, the last 14 of which are selected home pages for representative firms supplying some of the technologies.

i. Autoclaving

The primary advance in autoclaving to allow bulk processing of hospital waste is the use of pre-shredding and automatic feeding of the materials into the autoclaves, which otherwise acts in the same fashion as a typical laboratory autoclave. The shredding has the further advantage of making the end product unrecognizable. A major disadvantage is that there is no volume reduction and the final product takes up valuable space in rapidly filling up landfills.

ii. Pyrolysis

This process heats the waste in an enclosed chamber to a very high temperature (approximately 1200°C or 2200°F). This high temperature completely destroys any organisms and reduces the volume of organic materials by 97-98%. Pollutants are destroyed within the machine so release of air pollutants is negligible.

iii. Fluidized-bed Coal Technology

Circulating fluidized-bed burning of coal was originally developed to eliminate the need for expensive pollution controls in coal boilers. Approximately 10-12% hospital waste is burned with the coal. The same properties which make the coal burning less polluting reduce the generation of pollutants regulated under the EPA emission guidelines to well below the regulatory limits. This is not yet, at the time of writing, widely available commercially but is being used very successfully at one hospital. The residue is simply combined with the coal ash.

iv. Electron Beam Technology

Electron beam sterilization of medical waste is an effective means to accomplish this purpose. Recent improvements have made the process economically competitive to be applied to infectious waste. It can be applied to all types of infectious wastes, including liquids.

v. Microwave Technology

Microwave technology is used in a number of variations to treat infectious wastes. It is relatively expensive and not recommended for treating pathological or animal wastes. However, NASA has developed a small unit which would be suitable for processing small quantities (about 2 kg). There are companies which provide commercial units suitable for processing up to about 500 pounds per hour.

vi. Chemical Treatment

A variety of chemicals are used to kill the infectious organisms after the waste is ground. The grinding allows a volume (not weight) reduction of around 75%. One study shows that the treatment may not always completely be successful but commercial firms claim complete success. The chemicals used are disposed of into the sanitary sewer system.

vii. Gas/vapor Sterilization

This is similar to chemical treatment except it uses gases such as ethylene oxide or formaldehyde as sterilizing agents. Because of the negative properties of these agents, the EPA does not recommend this technique.

viii. Biodegradation

The use of enzymes to digest medical waste after it has been ground up has been employed. The resultant waste product is then suitable for disposal in a solid waste landfill.

ix. Steam Sterilization

In this process, the bulk waste is exposed to superheated steam (about 300°F) and then soaked in the hot water for a minute or more. Then, after grinding and cooking for a period, it is again exposed to the superheated steam for several minutes. After cooling, the water is drained off and the solids are disposed of as a solid waste. Again a substantial volume reduction can be obtained although a minimal weight reduction.

A list of more than 150 commercial medical infection waste disposal firms in this country can be found in the first Internet reference at the end of this article.

3. Storage and Transportation of Medical and Infectious Waste

The storage of infectious waste has been covered to a major extent in the previous article on the Bloodborne Pathogen OSHA standard, but there are a few aspects that were not covered. As in that standard, sharps especially need to be placed in sturdy, leak tight, puncture proof and clearly marked containers. Other medical waste should be placed in containers that will permit those handling it to do so safely and facilitate storage, handling, and eventual transportation. The containers should be stored in separately segregated areas in either double red bags or otherwise clearly marked containers.

The area in which infectious waste is stored should be well defined and secure. It should be easily decontaminated should a spill occur. It would be desirable for it to be close to the point at which it would be removed from the facility, typically by an outside transporter firm. Because much of hospital infectious waste is represented by tissue and other organic materials, or items contaminated by organic materials, it is likely that the waste will become putrescent if stored at room temperature. The storage space should incorporate refrigeration facilities. In some cases, all of these requirements can be met by the transporting firm providing refrigerated trailers on-site where the infectious waste can be accumulated until enough is ready for transportation.

Selection of a firm to transport the infectious waste off-site to its eventual disposal site must be done very carefully. The transporter should, of course, have all of the required permits, licenses, and liability insurance. The contract should not have a hold harmless clause for the generator unless the limits of this clause are very carefully delineated to distinguish unambiguously between generator and transporter errors. References should be checked thoroughly and any prior history of violations made part of the contract so that they can be reviewed. Cost is not always a valid criteria so the purchasing department for the generator must be willing to allow the generator to not select the lowest bidder for legitimate reasons.

REFERENCES

1. Standards for Performance for New Stationary Sources and Emission Guidelines for Existing Sources: Hospital/Medical/Infectious Waste Incinerators, *Federal Register*, Sept. 15, 1997, Vol. 62, No. 178, p. 48347.
2. Standards for Performance for New Stationary sources, Title 40 CFR Part 60.

INTERNET REFERENCES

1. http://ourworld.compuserve.com/homepages/ihewan/Nw6.htm
2. Hospital/Medical/ Infectious Waste Incineration in Region Five, http://www.epa.gov/reg5oair/mwi.htm
3. Hospital/Medical/ Infectious Waste Incineration, http://www.epa.gov/reg5oair/lim.htm
4. Medical Facility Waste, Air Pollution Regulations and Guidelines, http://sbeap.niar.twsu.edu/docs/medwaste/
5. Management of Medical/Infectious Waste, http://www. Afcee.brooks.af.mil/pro_act/fact/Oct98b.htm
6. **Huff, M.J. and Defur, P.,** Alternative Technology Descriptions, http://www..sustain.org/hcwh/hcwhmanual/hthburnalternatives.html
7. EcoRecycle Reports: http://www.ecorecycle.vic.gov.au/document/disposal/sdi-6.htm
8. http://ctoserver.arc.nasagov/TechOpps/micro.html
9. http://www.doe.gov/doe/whatsnew/pressrel/pr97035.html
10. http://www.ace-energy.com.au/waste/control/system.htm
11. http://www.jyd-1500.com/techspec.htm
12. http://www.nmwrc.com/auto-cla.html
13. http://www.med-dispose.com/pyrolysis.html
14. http://www.biosterile.com/medwaste.htm
15. http://www.etcusa.com/medwaste.htm
16. http://www.mediwaste.com/structure/SERV_INFO_Microwave.htm
17. http://www.nce-turboclean.com/page5.html
18. http://www.redbag.com_/aghtmlcode/ssm_150/ag_whatis.html
19. http://skychiefs.com/hcp/april5.98/long/L2enviroTech.html

XI. LABORATORY CLOSEOUT PROCEDURES

Occasionally laboratories cease operations and totally close down, due to the laboratory director retiring, changing jobs or locations, grant terminating, or for other reasons. In these circumstances, there are almost always large quantities of surplus chemicals for which disposal must be arranged. Prior planning and a cooperative effort involving laboratory personnel and the waste management group will ensure that there will not be substantial quantities of chemicals left behind which cannot be identified. It should be mandatory, wherever possible, that at least 30 days notice, and preferably more, should be given to the organization's hazardous waste group by the persons responsible for the laboratory, to allow the chemical identification procedure to be done carefully and thoroughly.

Most of the chemicals in the laboratory should be in the original containers and should pose no difficulty in identification. Many of these will still be useful and should be either directly distributed by the laboratory personnel to others who might use them, or should be transferred to the organization's redistribution program. Unfortunately, some will not be useful and must be placed in the category of material which will require disposal. The worst situation for those materials which are readily identified, will be those that have become unstable and require special measures to move and handle safely. Often, stuck at the back of cabinets and long forgotten, will be ancient bottles of ether, perchloric acid, picric acid, etc. However, even in this difficult situation, at least the material is known and is subject to standard operating procedures.

Possibly the greatest difficulty usually will be a substantial number of containers which often clutter up a facility that are not properly labeled, if at all. It will require the assistance of the laboratory personnel to help identify the substances in the containers that do not have labels or labels which have no significance to those who have not been working in the laboratory with those specific substances. However, with the aid of the individuals currently working in the laboratory, it should be possible to at least identify the contents of the containers for which they are directly responsible. If the labels bear initials so that it can be determined who owned or prepared the material, then from laboratory notebooks, reports, theses or dissertations, or even some familiarity with what the individual had been working on, it may be possible to determine what the contents are likely to be, which will greatly facilitate identification, if tests are required. In some cases, it may be possible to make a reasonable estimate from other containers in the vicinity, if they are obviously grouped by category, for example. In extreme cases, it may be cost efficient to ask a former worker to return to help with the task.

There will almost always be some containers left over which cannot be identified. Some may be dangerous while others may not. If materials are routinely used in the laboratory which degrade to dangerous materials, then these unknown containers must be treated with special care. In some cases, it may be necessary to call in a firm which specializes in handling dangerous materials, rather than take the risk of handling the material at all. When the materials normally generated in the laboratory are not shock, heat, or friction sensitive and procedures have not changed in character for several years, it is probably permissible to handle these unknown materials with only normal care. If the laboratory personnel were no longer available, it might not be possible to make this distinction.

There are occasions when former laboratory personnel may not choose to be cooperative in assisting with the identification process. Perhaps the reasons for the facility being closed were not agreeable to the individuals involved. Perhaps they are fearful that they may be held financially responsible in some degree. There are any number of scenarios where this could occur and it has occurred on a number of occasions in the author's experience. If they cannot be persuaded to help with the process, then, unless the situation is sufficiently important or costly to justify taking legal action, there is little that one can do except try all of the alternatives mentioned above.

Even if there are still unknowns left over after the best efforts of the laboratory personnel and waste group working together, there will certainly be far fewer than if the latter group had to do all the work themselves, or if the evaluation of the containers had to be done hastily because of inadequate warning. Most individuals will be surprised at the effort and time required to close out an average laboratory. It will rarely be done in a single day. Providing adequate warning and lead time will allow waste personnel to schedule their own duties to take maximum advantage of their own time and those of the laboratory personnel. The reduction in the amount of waste generated and the reduction in the number of containers requiring analysis will amply repay the effort.

Chapter 5

NONCHEMICAL LABORATORIES

I. INTRODUCTION

The emphasis in the previous chapters has been on laboratories in which the primary concerns were due to the use of chemicals, although in order to not completely avoid a topic unnecessarily, some of the problems arising in other types of operations have been addressed. For example, in the latter part of the last chapter some serious issues involving problems with specific contagious diseases were discussed due to the increasing importance these diseases have in our society and in our laboratories, as a continuation of the topic of health effects. The problems of dealing with infectious waste were considered at some length as well as chemical wastes as part of the larger problem of dealing with hazardous wastes. In this chapter laboratory operations which involve special problems in other classes of laboratories will be presented in greater detail. However, in responding to these special problems, one should be careful not to neglect the safety measures associated with those hazards already covered.

II. RADIOISOTOPE LABORATORIES

Exposure of individuals to ionizing radiation is a major concern in laboratories using radiation as a research tool or in which radiation is a byproduct of the research. Although there are many types of research facilities in which ionizing radiation is generated by the equipment, e.g., accelerator laboratories, X-ray facilities, and laboratories using electron microscopes, the most common research application in which ionizing radiation is a matter of concern is the use of unstable forms of the common elements which emit radiation. A very brief discussion of some atomic and nuclear terms will be given next, with apologies for those not requiring this introduction to the subject.

A. Brief Summary of Atomic and Nuclear Concepts

An atom of an element can be simply described as consisting of a positively charged nucleus and a cloud of negatively charged electrons around it. The electron cloud defines the chemical properties of the atom, which have been the subject up until now, while the processes primarily within the nucleus give rise to the nuclear concerns which will be addressed next. Although the nucleus is very complex, for the present purposes an atom of a given element may be considered to have a fixed number of positive protons in the nucleus, equal in number to the number of electrons around the neutral atom, but can differ in the number of neutral neutrons, the different forms being called isotopes. It is the property of the unstable forms, or radioisotopes, to emit radiation which makes them useful, since their chemical properties are essentially identical to the stable form of the element (where a stable form exists; for elements with atomic numbers greater than that of bismuth, there are no completely stable forms). The radiation which the radioisotopes emit allows them to be distinguished from the stable forms of the element in an experiment. There are three types of radiation normally emitted by various radio isotopes, alpha (α) particles, electrons (β), and gamma rays (γ). The properties of these radiations will be discussed later. A fourth type of radiation, neutrons, may be emitted under special circumstances, by a small number of radioactive materials. Most laboratories will not use materials emitting neutrons. The properties of these radiations will be discussed in more detail later.

B. Radiation Concerns

The radiation which makes radioisotopes useful also makes their use a matter of concern to the users and the general public. Exposure to high levels of radiation is known to cause health problems; at very high levels, death can follow rapidly. At lower, but still substantial levels, other health effects are known to occur, some of which, including cancer, can be delayed for many years. At very low levels, knowledge of the potential health effects is much more uncertain. The generally accepted practice currently is to extrapolate statistically known effects on individuals exposed to higher levels to large groups of persons exposed to low levels of radiation in a linear fashion. The concept is similar to the use of higher concentrations of chemicals using a limited number of animals in health studies of chemical effects, instead of more normal concentrations in a very large number of test animals. There are some who question the validity of this assumption in both cases, but it is a conservative assumption and, in the absence of confirmed data, is a generally safe course of action to follow. However, the practice may have led to a misleading impression of the risks of many materials. When a scientist makes the statement that he does not know whether a given material is harmful or not, he is often simply stating in a very honest way that the data do not clearly show whether, at low levels of use or exposure, a harmful effect will result. It does not necessarily imply, as many assume, that there is a lack of research in discovering possible harmful effects. In many cases, major efforts have been made to unambiguously resolve the issue, as in the case of radiation, and the data do not support a definite answer. There are levels of radiation below which no harmful effects can be detected directly. In the case of radiation, there is even a substantial body of experimental data (to which proponents of a concept called "hormesis" call attention) that supports possibly positive effects of radiation at very low levels. This position is, of course, very controversial. However, in chemical areas there are many examples of chemicals essential to health in our diets in trace amounts that are poisonous at higher levels. It is not the intent of this section to attempt to resolve the issue of the effects of low-level radiation, but to emphasize that there are concerns by many employees and the general public. It may well be that, by being very careful not to go beyond known information, scientists have actually contributed to these concerns. Another way of looking at the issue, and certainly a more comforting way, is that many unsuccessful attempts have been made to demonstrate negative effects at low levels. Radiation levels which normally accompany the use of radioisotopes are deliberately kept low, and the perception of risk by untrained individuals may be overstated. However, a linear dose-effect relation is the accepted basis for regulatory requirements at this time, and until better data are available scientists using radioactive substances must conform to the standards. Users owe it to themselves and the public to use the materials in ways known to be safe. However, as a general concept, it would be wise for scientists, when speaking to persons not trained in their field, to be sure that when they say they do not know of possible harmful effects of any material, that this statement is understood to be an informed uncertainty where this is the case, as opposed to being based on a lack of effort.

It is unfortunate that there is so much concern about radiation since there are many beneficial effects, but because of the dramatization of the concerns, many individuals fear radiation out of all proportion to any known risks. In an opinion poll in which members of the general public were asked to rank the relative risks of each of a number of hazards, nuclear radiation was ranked highest but in reality, was the *least* dangerous of all the other risks based on known data with all the others being much more likely to cause death or injury than radiation. Individuals who urgently have needed X-rays, the application of diagnostic use of radioactive materials, or radiation therapy have declined to have them because of this heightened fear. Used properly, radiation is an extremely valuable research tool and has many beneficial aspects. Used improperly, it can be dangerous, but so can many other things in the laboratory, many very much more so.

C. Natural Radioactivity

A common misconception is that radiation is an artificial phenomenon. Many of the most commonly used radioisotopes have been created artificially, but there are abundant sources of natural radiation which emit radiation of exactly the same three types as do artificially created radioisotopes. Elements such as potassium and carbon, which are major constituents of our body have radioactive isotopes. Many other elements, such as the rare earths, have radioactive versions. Every isotope of elements with atomic numbers (i.e., the number of protons in the nucleus of the element, or the number of electrons around the nucleus in a neutral atom) above 83 is unstable and these elements are common in the soils and rocks which make up the outer crust of the earth. There are areas in the world in which the natural levels of radiation could significantly exceed that permitted for the general public resulting from the operation of any licensed facility using radioisotopes. Radiation constantly bombards us from space due to cosmic rays. Persons who frequently take long airplane flights receive a significantly increased amount of radiation over a period of time compared to persons who fly rarely or not at all. Arguments that these natural forms of radiation are acceptable because they are natural has absolutely no basis in fact. As was mentioned earlier, there are only a modest number of varieties of radiation, and these are produced by both natural and artificially produced radioactive materials. Similarly, there are only a few ways in which radiation may interact with matter, and they also are the same for all sources of radiation.

The acknowledgment of natural sources of radiation is not intended to minimize concerns about radiation, even the natural forms, but to point out that if there are concerns about low levels of radiation, then these natural levels must be considered as well as the artificial sources. One of the naturally occurring radioactive materials, radon, has been receiving much attention and may be a significant hazard, perhaps contributing to an increase of 1 to 5% of the number of lung cancer deaths each year. This estimate, as in most cases dealing with attribution of specific effects of low levels of radiation, is supported by some and disputed by others. Note, however, that even in this case at least 95 to 99% of the lung cancer deaths are attributable to other causes. Radon as an issue will be discussed in a separate section later in this chapter. An isotope of potassium, an essential element nutritionally and present in substantial amounts in citrus fruits and bananas, for example, emits significant amounts of very penetrating radiation.

There are various estimates of the average source of radiation exposure for most individuals. An article by Komarov,[1] who is associated with the World Health Organization, provides the following data about sources of radiation: 37% from cosmic rays and the terrestrial environment, 28% from building materials in the home, 16% from food and water, 12% from medical usage (primarily X-rays), perhaps 4% from daily color television viewing, 2% from long-distance airplane flights, and 0.6% (under normal operating conditions) from living near a nuclear power plant. Note that the medical exposure to radiation is 20 times larger than from nuclear power plants even for those living near one. The Komarov article was written before the Chernobyl incident, but even this outstanding example of poor management is not sufficient to change the general picture. Unlike the Chernobyl reactor, commercial nuclear power plants in the United States are protected by very strong confinement enclosures to prevent unscheduled releases. In the case of the Three-Mile Island incident, in which the reactor core melted down, the confinement enclosure performed as designed and minimal amounts of radioactive material were released. As the news media reported some time after the initial furor, "the biggest danger from Three-Mile Island was psychological fear," to which the media contributed significantly by exaggerated news reports of the potential dangers.

In summary, radiation is a valuable research tool. In order to prevent raising public concerns and perhaps lead to further restrictions on its use, scientists need to be scrupulously careful to conform to accepted standards governing releases or over exposures. Fortunately, for common uses of radiation in research laboratories, this goal is easily achieved with reasonable care.

D. Basic Concepts

Each scientific discipline has its own special terms and basic concepts on which it is founded. This section is, of course, not necessary for most scientists who routinely work with radiation, but it may be useful for establishing a framework within which to define some needed terms. As scientists work with accelerators of higher and higher energies, the concept of matter is at once growing more complex and simpler; more complex in that more entities are known to make up matter, but simpler in that theorists working with the data generated by these gigantic machines are developing a coherent concept unifying all of the information. For the purposes of this discussion, a relatively simple picture of the atom will suffice, as noted earlier.

1. The Atom and Types of Decay

In the simple model of the atom employed here, as briefly described earlier in this chapter, the atom can be thought of as consisting of a very small dense nucleus, containing positively charged particles called protons and neutral particles called neutrons, surrounded by a cloud of negatively charged electrons. The number of protons and the number of electrons are equal for a neutral atom, but the number of neutrons can vary substantially, resulting in different forms of an element called, as already noted, isotopes of the element. Some elements have only one stable isotope, although tin has ten. There are unstable isotopes, logically called radioisotopes, in which, over a statistically consistent time, a transition of some type occurs within the nucleus. Different types of transitions lead to different types of emitted radiation. Hydrogen, for example, has two stable forms and one unstable one, in which a transition occurs to allow an electron to be generated and emitted from the nucleus, producing a stable isotope of helium. Prior to the transition, the electron did not exist independently in the nucleus. A neutron is converted to a proton in the process, and the electron is created by a transformation of energy into matter. This process is called beta decay. No element with more than 83 protons in the nucleus has a completely stable nucleus, although some undergo transitions (including by processes other than beta decay) extremely slowly.

In some cases, the mass energy of the nucleus favors emission of a positive electron (positron) instead of a normal electron which has a negative charge. This is called positive beta decay or positron decay. Here a proton is converted into a neutron. A competitive process to positive beta decay is electron capture (ϵ) in which an electron from the electron cloud around the nucleus is captured by the nucleus, a proton being converted into an neutron in the process. In the latter process, X-rays are emitted as the electrons rearrange themselves to fill the vacancy in the electron cloud. However, following positron emission, the positive

Table 5.1 Properties of Radioactive Emissions

Type	Mass (amu)	Charge (Electron Units)	Range of Energy
Alpha (a)	4	+2	4-6 MeV
Beta (B)	1/1 840	+1	eVs-4 MeV
Gamma	0	0	eVs-4 MeV
X-Rays	0	0	eVs-100 KeV

An amu is the mass of a single nucleon based on the 1/12th the mass of a carbon-12 nucleus.

electron eventually interacts with a normal electron in the surrounding medium, and the two vanish or annihilate each other in a flash of energy. The amount of energy is equal to the energy of conversion of the two electron masses according to $E = mc^2$. This amounts to, in electron volts, 1.02 million electron volts, or 1.02 MeV. In order to conserve momentum, two photons or gamma rays of 0.511 MeV each are emitted 180° apart in the process.

In many case, the internal transitions accompanying adjustments in the nucleus results in the emission of electromagnetic energy, or gamma rays. These can be in the original or parent nucleus, in which case they are called internal transitions, and the semi-stable states leading to these transitions are called metastable states. More often, the gamma-emitting transitions occur in the daughter nucleus after another type of decay such as beta decay (metastable states can exist in the daughter nucleus also). The gamma emission distribution can be very complex. In some instances, the internal transition energy is directly transferred to one of the electrons close to the nucleus in a process called internal conversion, and the electron is emitted from the atom. In this last case, energy from transitions in the orbital electron cloud is also emitted as X-rays.

Finally, the most massive entity normally emitted as radiation is the alpha (a) particle which consists of a bare (no electrons), small nucleus having two protons and two neutrons. The nucleons making up an alpha particle are very strongly bound together, and unlike electrons, the alpha particle appears to exist in the parent nucleus as a cohesive unit prior to the decay in our simple model. This process is somewhat more rare than β or γ decay.

The processes briefly described above are the key decay processes in terms of safety in the use of radioisotopes. There is another very important aspect of the decay processes, and that is the energy of the emitted radiation. The electrons emitted in beta decay can have energies ranging from a few eV to between 3 and 4 MeV. There is an unusual feature of the beta decay process in that the betas are not emitted monoenergetically from the nucleus as might be expected, and as does occur for alpha and gamma decay. The most probable energy of the betas in a decay process is approximately one third of the maximum energy beta emitted in the process. The reason is that, in addition to a beta being emitted, another particle, called a neutrino, of either zero mass or very close to it, is emitted simultaneously and shares the transitional energy, with varying amounts going to the two entities. The neutrino does not play a role in radiation safety as it interacts virtually negligible with matter, although its existence is very important for many other reasons. Gammas can have a similar range of energies to that of electrons, but the energies of the gammas are discrete instead of a distribution.

Alpha particles have a relatively high energy, normally ranging from 4 to 6 MeV. The decay of alphas with lower energies is so slow that it occurs very rarely while with an energy just a little higher, the nucleus decays very rapidly. The high energy, accompanying the high mass and the double positive charge, make the alpha particle a particularly dangerous type of radiation, if it is emitted in the proximity of tissue which can be injured. This last is an important safety qualification as will be seen later. Table 5.1 summarizes the properties of the types of radiation.

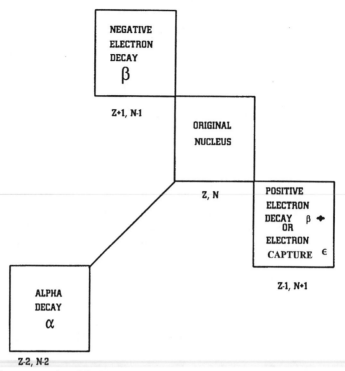

Figure 5.1 Schematic representation of the decay process.

Graphically the decay process can be depicted as shown in Figure 5.1, where N = the neutron number and Z = the nuclear charge. The box with N,Z is the parent nucleus and the others are the possible daughters for the processes shown.

2. The Fission Process

A major omission deliberately not mentioned in the preceding Section is not involved in most laboratories using radioisotopes. However, without this process many of the commonly used radioisotopes would not be available, since they are obtained from reprocessing spent fuel and recovery of the remnants left over after the fission process. The process of fission describes the process by which a few very heavy atoms decay by splitting into two major components and a few neutrons, accompanied by the release of large amounts of energy, ~200 MeV. The process can be spontaneous for some very heavy elements, e.g., Californium-252 but also can be initiated by exposing specific heavy nuclei to neutrons. There are no common radioisotopes that normally emit neutrons, but there are several interactions in which a neutron is generated. Among these are several reactions in which a gamma ray interacts with beryllium to yield neutrons, so that a portable source of neutrons can be created. There are many other ways to generate neutrons but there is no need to describe these in this book. However, if a source of neutrons, n, is available and is used to bombard an isotope of uranium, ^{235}U, the following reaction can occur.

$$n + {}^{235}U \rightarrow X + Y + \approx 2.5n + energy \qquad (1)$$

Here, X and Y are two major atomic fragments or isotopes resulting from the fission process. On average about 2.5 neutrons are emitted in the reaction plus energy. The process is enhanced if the initiating neutrons are slowed down until they are in or near thermal equilibrium with their surroundings. X and Y themselves will typically decay after the original fission event, a few by emitting additional neutrons, as well as betas and gammas. As noted earlier, about 200 MeV of energy are released in the process, much of it as kinetic energy shared by the particles. Some of these fission fragments are long lived, and can be chemically separated to provide radioisotopes of use in the laboratory. These fission fragment derived radioisotopes are the major source of the byproduct radioisotopes regulated by the Nuclear Regulatory Commission (NRC). The fission reaction can, under appropriate circumstances, be self-sustaining in a chain reaction. In some configurations, the chain reaction is extremely rapid, and an atomic bomb is the result. However, by using the neutrons emitted by the fission fragments (called delayed neutrons), the process can be controlled safely in a reactor. Over a period of time, the fission products build up in the uranium fuel eventually can be recovered when the fuel element is reprocessed.

Additional radioactive materials or radioisotopes are made by the following reaction:

$$n + {}^{A}X \rightarrow {}^{(A+1)}y* + a \qquad (2)$$

The asterisk indicates that the product nucleus, Y may be unstable and will undergo one (or more) of the modes of decay discussed previously. The 'a' indicates that there may be a particle directly resulting from the reaction. In many cases, the source of neutrons for radioisotopes created by this reaction is a nuclear reactor so these radioactive materials also are "byproduct materials," and are regulated by the Nuclear Regulatory Commission or State surrogates.

Plutonium is made in nuclear reactors by the above reaction where ${}^{238}U$ is the target nucleus. Although there are other reactions using different combinations of particles in Equation 2, in most cases these require energetic bombarding particles generated in accelerators. Also, since there are no common radioisotopes that generate neutrons, there is essentially no probability that other materials in laboratories will be made radioactive by exposure to radiation from byproduct materials.

Materials which will undergo fission and can be used to sustain a chain reaction are, in the nomenclature of the NRC, "special" nuclear materials. These include the isotopes of uranium with mass numbers 233 and 235, materials enriched in these isotopes, or the artificially made element, plutonium. Materials which have uranium or thorium, which also has a fissionable isotope, in them to the extent of 0.05% are called source materials.

3. Radioactive Decay

An important relationship concerning the actual decay of a given nucleus is that it is purely statistical, dependent only upon the decay constant for a given material, i.e., the activity A, is directly proportional to the number, N, of unstable atoms present:

$$\text{Activity} = A = dN/dt = C N \qquad (3)$$

This can be reformulated to give the number of radioactive atoms N at a time t in terms of the number originally present.

$$N(t) = N_0 e^{\lambda t} \qquad (4)$$

where $\lambda = \ln2/\tau$.

Table 5.2 Typical Decay of a Group of 1000 Radioactive Atoms

Number	Time (t)	Number	Time (t)
1000	0	14	6
502	1	7	7
249	2	4	8
125	3	1	9
63	4	1	10
31	5	0	11

Equation 4 shows that during any interval, $t = \tau$, theoretically half of the unstable nuclei at the beginning of the interval will decay. In practice, *approximately* half will decay in a half-life, τ. This is illustrated in Table 5.2.

The data in this table illustrate clearly that when small numbers are involved, the statistical variations cause the decrease to fluctuate around a decay of about one half of the remaining atoms during each successive half-life, but obviously between 3 and 4 half-lives in this table, it would have been impossible to go down by precisely half. The table also illustrates a fairly often used rule-of-thumb: after radioactive waste has been allowed to decay by 10 half-lives, the activity has often decayed sufficiently to allow safe disposal. This, of course, depends upon the initial activity.

The daughter nucleus formed after a decay can also decay as can the second daughter, and so forth. However, eventually a nucleus will be reached which will be stable. This is, in fact, what occurs starting with the most massive natural elements, uranium and thorium. All of their isotopes are unstable, and each of their daughters decays until eventually stable iso-topes of lead are reached. The existence of all of the elements above atomic number 83 owe their existence to the most massive members of these chains that have very long half-lives that are comparable to the age of the earth, so a significant fraction remains of that initially present.

4. Units of Activity

The units of activity are dimensionally the number of decays or nuclear disintegrations per unit time. Until fairly recently, the standard unit to measure practical amounts of activity was the curie (Ci), which was defined to be 3.7×10^{10} disintegrations per second (dps). Other units derived from this were the millicurie (mCi) or 3.7×10^7 dps, the microcurie (μCi) or 3.7×10^4 dps, the nanocurie (nCi) or 37 dps and the picocurie (pCi) or 0.037 dps. Many health physicists prefer to use disintegrations per minute (dpm), and the NRC also prefers the data logged in laboratory surveys to be expressed in dpm. The curie was originally supposed to equal the amount of activity of 1 g of radium. This unit, and the derivative units, are still the ones most widely used daily in this country; however, an international system of units, or SI system, has been established (and is used in scientific articles). In this system, one disintegration per second is defined as a becquerel (Bq). Larger units, which are multiples of 10^3, 10^6, 10^9, and 10^{12}, are indicated by the prefixes kilo, mega, giga, and tera, respectively. In most laboratories that use radioisotopes as tracers, the quantities used are typically about 10^4 to 10^8 dps. There are other uses of radioisotopes (e.g., therapeutic use of radiation) which use much larger amounts.

5. Interaction of Radiation with Matter
a. Alphas

As an alpha particle passes through matter, its electric field interacts primarily with the electrons surrounding the atoms. Because it is a massive particle, it moves comparatively slowly and spends a relatively significant amount of time passing each atom. Therefore, the alpha particle has a good opportunity to transfer energy to the electrons by either removing them from the atom (ionizing them) or raising them or exciting them to higher energy states. Because it is so much more massive than the electrons around the atoms, it moves in short, straight tracks through matter and causes a substantial amount of ionization per unit distance. An alpha particle is said to have a high linear energy transfer (LET). A typical alpha particle has a range of only about 0.04 mm in tissue or about 3 cm in air. The thickness of the skin is about 0.07 mm so that a typical alpha particle will not penetrate the skin. However, if a material that emits alphas is ingested, inhaled, or, in an accident, becomes imbedded in an open skin wound, so that it lodges in a sensitive area or organ, the alpha radiation can cause severe local damage. Since many heavier radioactive materials emit alpha radiation, this often makes them more dangerous than materials that emit other types of radiation, especially if they are chemically likely to simulate an element retained by the body in a sensitive organ. If they are not near a sensitive area, they may cause local damage to nearby tissue, but this may not cause appreciable damage to the organism as a whole.

b. Betas

Beta particles are energetic electrons. They have a single negative or positive charge and are the same mass as the electrons around the atoms in the material through which they are moving. Normally, they also are considerably less energetic than an alpha particle. They typically may move about two orders of magnitude more rapidly than alpha particles. They still interact with matter by ionization and excitation of the electrons in matter, but the rate of interaction per unit distance traveled in matter is much less. Typically, beta radiation, on the order of 1 MeV, can penetrate perhaps 0.5 cm deep into tissue, or about 4 meters of air, although this is strongly dependent upon the energy of the beta. Low-energy betas, such as from ^{14}C, would penetrate only about 0.02 cm in tissue or about 16 cm in air. Therefore, only those organs lying close to the surface of the body can be injured by external beta irradiation and then only by the more energetic beta emitters. Radioactive materials emitting betas taken into the body can affect tissues further away than those that emit alphas, but the LET is much less.

There is a secondary source of radiation from beta emitters. As the electrons pass through matter, they cause electromagnetic radiation called "bremsstrahlung," or braking radiation to be emitted as their paths are deflected by passing through matter. The energy that appears as bremsstrahlung is approximately ZE/3000 (where Z is the atomic charge number of the absorbing medium and E is the β energy in MeV.) This is not a problem with alpha particles since their paths through matter are essentially straight. Bremsstrahlung radiation can have important implications for certain energetic beta emitters such as ^{32}P. Protective shielding for energetic beta emitters should be made of plastic or other low-Z material instead of a high-Z material such as lead. Because of the silicon in glass, even keeping ^{32}P in a glass container can substantially increase the radiation dose to the hands while handling the material in the container as compared to the exposure that would result were bremsstrahlung not a factor.

c. Gammas

Since gamma rays are electromagnetic waves, they are not charged and do not have any mass, they interact differently with matter than do alpha and beta particles, although the net effect is usually still ionization of an orbital electron. They interact with the electrons in matter by three different mechanisms. In order to provide an understanding of the safety implications, a brief elaboration of these mechanisms follows. Until one of the three processes takes place, the gamma ray can continue to penetrate matter without hindrance.

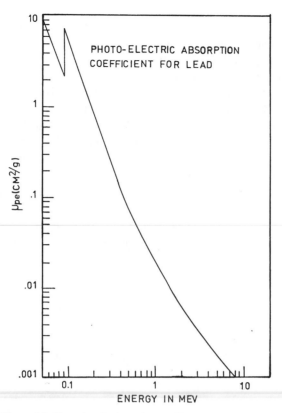

Figure 5.2 Photoelectric absorption coefficient for lead.

i. Photoelectric Effect

The gamma ray interacts with electrons around an atom. Electrons with which they interact normally are completely removed from the atom, i.e., the atom is ionized. In a photoelectric effect process, all the energy of the gamma ray is transferred to the electron and the gamma ray no longer exists. The photoelectric effect mechanism is dependent upon the gamma energy as shown in Figure 5.2. For low energy photons, the photoelectric effect depends upon the atomic number of the absorber, approximately as Z^4. As energies increase, the importance of the atomic number of the absorber decreases. The scattered electron interacts as would a beta.

ii. Compton Effect

The gamma ray can also scatter from an electron, transferring part of its energy to the electron and thus becoming scattered as a lower energy gamma. There is an upper limit to the amount of energy that can be transferred to the electron by this mechanism, so that in every scattering event, a gamma ray remains after the interaction. Dependent upon the energy transferred, the residual gamma can be scattered in any direction, relative to the original direction, up to 180°. This has important implications on shielding, since gammas can be scattered by the shielding itself, or by other nearby materials into areas shielded by a direct beam. Equation 5 gives the energy of the scattered gamma as a function of the angle of scattering. The interaction with matter is considerably less dependent upon the energy of the gamma. This is shown in Figure 5.3. Compton scattering is the primary mechanism of interaction for low atomic number elements, and decreases in relative importance as the atomic number increases.

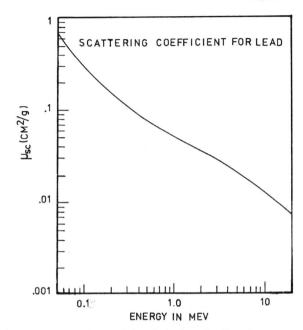

Figure 5.3 Scattering coefficient for lead as a function of energy.

$$E = \frac{E}{1 + \left(E / m_e C^2\right)\left(1 - \cos\theta\right)} \tag{5}$$

iii. Pair Production

If the energy of the gamma is greater than the energy needed to create an electron-positron pair, 1.02 MeV, then the gamma can interact with the absorbing medium to create the pair of electrons, an electron and a positron. The probability of this process increases as the energy increases. The excess energy over 1.02 MeV is shared by the two particles. The increases with the atomic number of the absorber, approximately proportionally to $Z^2 + Z$.

Gamma rays can penetrate deeply into matter, in theory infinitely, since unless the gamma interacts with an atom, it will go on unimpeded, just as in theory a rifle bullet fired into a forest can continue indefinitely unless it hits a tree (assuming no loss of energy for the bullet due to air friction). The intensity I of the original radiation at a depth x in an absorbing medium compared to the intensity of the radiation at the surface I_o is:

$$I = I_0 e^{-\mu x} \tag{6}$$

This equation is literally true if only gammas of the original energy are considered. If Compton scattering and the pair-production process are included, the decrease in the total number of gammas is less than that given by Equation 6, because of the scattered gammas from the Compton process, and the contribution of the annihilation gammas as the positron eventually is destroyed by interacting with a normal electron. The actual increase in the radiation levels is dependent on the gamma energy and the geometry of the scattering material.

If the total effect of all three mechanisms is considered at low and high energies, higher Z absorbers interact with gammas more strongly. However, between about 1 and 3 MeV there

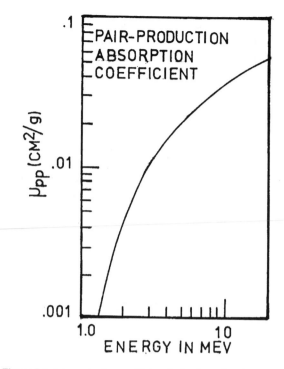

Figure 5.4 Pair-production coefficient for lead as a function of energy.

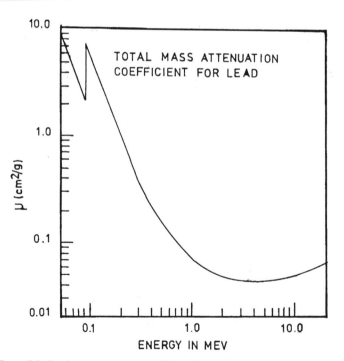

Figure 5.5 Total mass attenuation coefficient for γ absorption in lead as a function of energy.

is a relatively minor difference in the total absorption coefficient as a function of atomic number as shown in Figure 5.5. The discontinuities at lower energies are due to enhanced probability of interactions with electrons at the ionization thresholds.

As can be noted, all of the mechanisms by which a gamma interacts with matter (except the very small number of instances in which the gamma ray interacts with a nucleus) result in the energy being transferred to an electron, so a gamma is considered to have the same low LET characteristics as do betas. At very low energies the linear energy transfer characteristics of electrons increase some. However, unless a beta emitter is taken into the body, most internal organs will not be affected by beta radiation, while gammas can penetrate deeply into the body and injure very sensitive organs such as the blood-forming tissues. Thus, of the three types of radiation, gamma rays are usually considered the most dangerous for external exposures.

iv. Neutrons

As mentioned earlier, neutron radiation is rarely encountered in most laboratories that use radioisotopes in research programs. However, it is useful to understand the difference in the mechanisms by which a neutron interacts with matter compared to those involving other types of radiation since neutron radiation *may* make the matter with which it interacts radioactive. The neutron has no charge, but it does have about one fourth of the mass of an alpha particle, so that it does have an appreciable mass compared to the atoms with which it interacts.

An equation similar to Equation 6 gives the number of neutrons N, with an initial energy E, of an original number N_0 penetrating to a depth x in matter. Note that in both Equations 6 and 7, the units of x are usually converted into mg/cm^2 for the commonly tabulated values of μ and σ.

$$N = N_0 e^{-\sigma x} \qquad (7)$$

A neutron does not, as do electrons, alpha particles, and gammas, interact with the orbital electrons, but instead interacts directly with the nucleus. The neutron is not repelled by the positive charge on the nucleus as is the alpha because it has no charge. It is either scattered (elastically or inelastically) or captured by the nucleus. In a typical capture process usually several "capture" gammas with a total energy of about 8 MeV are emitted (the energies are somewhat less for some lighter nuclei). Thus, by this mechanism alone, the neutron could be considered more harmful than other radiations. Further, the nucleus in which it is captured may have been made radioactive, and the charge on the nucleus could change, so that the atom would no longer be chemically equivalent to its original form. In any event, the energy transferred to the participants in the interaction normally would be more than sufficient to break the chemical bonds.

Scattering events also typically would transfer enough energy to break the chemical bonds as long as the initial energy of the neutron is sufficiently high. As with Compton scattered gammas, the scattered neutrons can be scattered into virtually any direction, so that the equivalent of Equation 7 for neutrons of all energies would, as for gammas, have to be modified to include a buildup factor.

No figure showing the systematics of the reaction mechanisms will be given here because the relationships are extremely complex, varying widely not only between elements, but between isotopes of the same element. In addition, the interaction probabilities can vary extremely rapidly as a function of energy, becoming very high at certain "resonant" energies and far less only a few electron volts above or below the resonances. However, a few generalizations are possible. The probability of the capture process, excluding resonance effects, typically increases as the energy of the neutrons become lower, and for specific isotopes of certain elements, such as cadmium, gadolinium, samarium, and xenon, is extremely high at energies equivalent to thermal equilibrium (about 0.025 eV for room temperature matter). Energy can be lost rapidly by neutrons in scattering with low-Z materials, such as hydrogen, deuterium (^{2}H), helium, and carbon. Interposing a layer of water, paraffin, or graphite only a few inches thick, backed up by a thin layer (about 1/32 inch) of cadmium, in a beam of fast neutrons makes an effective shield for a beam of neutrons. Paraffin wax, in which boric acid has been mixed also makes an effective and cheap neutron shield (^{10}B has quite a respectable capture cross-section at thermal neutron energies).

Overall, the estimate of the danger of neutrons interacting with matter is estimated to be about ten times that of a gamma or electron, although this varies depending upon the energy of the neutrons, thermal neutrons are about two times as effective in causing atoms in tissue to be ionized, for example, as are betas and gammas while neutrons of 1 to 2 MeV energy are about 11 times more damaging.

6. Units of Exposure and Dose

There are two important concepts in measuring the relative impact of radiation on matter: one is the intensity of the radiation field, which represents a potential exposure problem, and the other is the actual energy deposited in matter, or the dose. Further, as far as human safety is concerned, the amount of energy absorbed in human tissue is more important than that absorbed in other types of matter. Each of these quantities have been assigned specific units in which they are measured.

The original unit of measuring radiation intensity was the roentgen, defined as the amount of X-ray radiation that would cause an ionization of 2.58×10^{-4} coulombs per kilogram of dry air at standard temperature and pressure. As noted, the dose or energy deposited in matter is more important, so another unit was subsequently defined, the rad, which was defined as the deposition of 0.01 joules per kilogram of matter. An exposure to 1 roentgen would result in an absorbed dose of 0.87 rads in air. A third unit, the rem, was subsequently defined which measured the equivalent dose, allowing for the relative effectiveness of the various types of radiation in causing biological damage. This originally was allowed for by multiplying the absorbed dose in rads by a relative biological effectiveness factor (RBE), to obtain a dose equivalent for tissue for the different varieties of radiation. Later, it was decided to restrict the term RBE to research applications and an equivalent multiplier called the quality factor, Q, was substituted. For practical purposes, RBE and Q factors are equivalent, although the latter is the one now commonly used.

The terms rads and rems are still used by most American health physicists in their daily work, and the current NRC regulations use these terms, as they do the curie and its derivative units. However, there are internationally accepted SI units for dose and also for activity. The equivalent units are:

1 Gray (Gy) = 1 joule/kilogram = 100 rad = absorbed dose
1 Sievert (Sv) = 1 Gray x Q x N = dose equivalent (N is a possible modifying factor, assigned a value of 1 at this time)
1 Sievert = 100 rem = dose equivalent

The quality factors for the various types of radiation are listed in Table 5.3. For neutrons of specific energies, the quality factor can be found in 10 CFR 20, Table 1004.

This concludes this very brief discussion of some basic terms and concepts in radiation physics that will be employed in the next few sections. Many important points and significant features have been omitted that would be of importance primarily to professional health

Table 5.3 Quality Factors
Type of Radiation

Type of Radiation	
Betas, gammas, X-rays	1
Alphas	20
Thermal neutrons	2
Fast neutrons (~ I MeV)	20
Neutrons (unspecified energy)	10

physicists, but are of less importance to those individuals that use radiation as a research tool to serve their more direct interests. The role of the Nuclear Regulatory Commission will be heavily stressed because the NRC very strictly regulates all aspects of radiation involving special nuclear materials and byproduct materials for safety and security through its licensing and oversight functions.

E. Licensing

This section will be restricted to a discussion of licensing of radioisotopes or byproduct materials, rather than other types of applications such as a research reactor. It has been some time since any new applications for construction of a nuclear power plant in the United States has been approved, and the number of operating non-governmental research reactors has been diminishing. Several of these research facilities are either in the process of terminating their license or going into an inactive status. At least some research reactors have closed rather than renew their license, as they must do periodically, because of excessive costs needed to meet the concerns of the public. The other major type of facility involved with radiation, laboratories using X-ray units, are usually regulated by state agencies, although the federal Food and Drug Administration sets standards for the construction of the machines and their applications. X-ray facilities will be discussed in a separate section.

Radioactive materials fall into two classes as far as regulation is concerned. Radioactive materials "yielded in or made radioactive by exposure to the radiation incident to the process of producing or utilizing special nuclear material" are regulated by the NRC, or by equivalent regulations in states with whom the NRC has entered into an agreement allowing for the states to act as the regulatory agency within their borders. Radioactive materials that are naturally radioactive or produced by means such as a cyclotron are regulated by the states in most cases.

The licensing of byproduct material is regulated under 10 CFR Part 30 or 33. Licenses are issued to "persons," a term which may refer to an individual but may also mean organizations, groups of persons, associations, etc. It is possible for individuals within an organization to have separate licenses, although it is more likely that instead of several individuals having separate licenses, an institution will apply for and be granted a license covering the entire organization, if they can show that they have established an appropriate internal organization so that they can ensure the NRC that the individual users will conform to the terms of the license and regulations governing the use of radioactive materials. This second class of licenses is denoted as a byproduct license of broad scope. There are different types of broad licenses, A, B, and C. A type A license is the least restriction and allows users to use radioisotopes as allowed in 10 CFR 30.100 Schedule A. A type B license is for users of larger quantities of various radioisotopes, on the order of curies or more, and a type C license is for users of smaller quantities. Schedule A in 10 CFR, Part 33.100 defines the quantities applicable to each of the last two licenses.

For most types of licenses of interest to research laboratories, the NRC has delegated licensing authority to five regional offices in Pennsylvania, Georgia, Illinois, Texas, and California. The current addresses of these regional offices can be obtained by writing to:

> Director
> Office of Nuclear Material Safety and Safeguards
> U.S. Nuclear Regulatory Commission
> Washington, D.C. 20555

Not all uses of radioisotopes require securing a license. There are many commercial products, such as watch dials and other self-luminous applications, and some types of smoke detectors that contain very small quantities of radioactive materials which the owner obviously does not require a license to possess. However, those who take advantage of exemptions must use no more than the "exempt quantities" listed in Schedule B, 10 CFR Section 30.71. In Table 5.4, the units are in microcuries. To convert to becquerels, multiply the number given in microcuries by 37,000.

Any byproduct material that is not listed in Table 5.4, other than alpha-emitting byproduct material, has an exempt quantity of 0.1 μCi or 3700 Bq.

Most users of radioisotopes would find it necessary to use more than the exempt quantities in Table 5.4 and should apply for a license. This is done through NRC Form 113, which can be obtained from the NRC office in the local region. If the activity planned has the potential for affecting the quality of the environment, the NRC will weigh the benefits against the potential environmental effects in deciding whether to issue the license. For most research-related uses of radioisotopes, environmental considerations will not usually apply, although where the isotopes will be used in the field, outside of a typical laboratory, the conditions and restrictions on their use to ensure that there will be no meaningful release into the environment will need to be fully included in the application.

There are three basic conditions that the NRC expects the applicant to meet in their application. In this context, "applicant" is used in the same sense as the word "person," which can be an individual or an organization, as noted earlier.

1. The purpose of the application is for a use authorized by the Act. Legitimate basic and applied research programs in the physical and life sciences, medicine, and engineering are acceptable programs.
2. The applicant's proposed equipment and facilities are satisfactory in terms of protecting the health of the employees and the general public, and being able to minimize the risk of danger to persons and property. The laboratories in which the radioisotopes are to be used need to be in good repair and contain equipment suitable for use with radioisotopes. Depending upon the level of radioactivity to be used and the scale of the work program, this may mandate the availability of hoods designed for radioisotope use. It could require specific areas designated and restricted for isotope use only, or the level of use and the amounts of activity may make it feasible to perform the research on an open bench in a laboratory. In any event, it must be shown in the application that the level of facilities and equipment must be adequate for the proposed uses of radiation.
3. The applicant must be suitably trained and experienced so as to be qualified to use the material for the purpose requested in a way that will protect the health of individuals and minimize danger to life and property The experience and training must be documented in the application.

Following are the specific NRC requirements for approval of a Type A, Broad License. The requirements for Type B and C licenses are a bit less stringent. Most major users will find that complying with the terms of Type A licenses to be most appropriate.

"An application for a Type A specific license of broad scope will be approved if:
(a) The applicant satisfies the general requirements specified in Sec. 30.33;
(b) The applicant has engaged in a reasonable number of activities involving the use of byproduct material; and
(c) The applicant has established administrative controls and provisions relating to organization and management, procedures, record keeping, material control, and accounting and management review that are necessary to assure safe operations, including:
  (1) The establishment of a radiation safety committee composed of such persons as a radiological safety officer, a representative of management, and persons trained and experienced in the safe use of radioactive materials;
  (2) The appointment of a radiological safety officer who is qualified by training and experience in radiation protection, and who is available for advice and assistance on radiological safety matters; and
  (3) The establishment of appropriate administrative procedures to assure:
      (i) Control of procurement and use of byproduct material;
      (ii) Completion of safety evaluations of proposed uses of byproduct material which take into consideration such matters as the adequacy of facilities and equipment, training and experience of the user, and the operating or handling procedures; and
      (iii) Review, approval, and recording by the radiation safety committee of safety evaluations of proposed uses prepared in accordance with paragraph (c)(3) (ii) of this section prior to use of the byproduct material."

Under item(3)(iii) the Radiation safety Committee also approves individual users if radioisotopes under the Broad License. In effect, they act as a local NRC governing use of the radioisotopes.

Before granting the license, the NRC may require additional information, or may require the application to be amended. The license is issued to a specific licensee and cannot be transferred without specific written approval of the NRC. The radioisotopes identified in the license can be used only for the purposes authorized under the license, at the locations specified in the license. If the licensee wishes to change the isotopes permitted to be used, to significantly modify the program in which they are used, or to change the locations where they are to be used, the license must be amended. This typically takes a substantial length of time, 1 to 3 months or even more not being unusual. Consequently, most substantial users of radioisotopes usually do apply for a "broad" license under 10 CFR Part 33.

Under the terms of a broad license, the application usually covers a request to use radioisotopes with atomic numbers from 3 to 83, with individual limits on the quantities held of specific isotopes, and an overall limit of the total quantity of all isotopes held at once. In addition, there should be specific identification of sealed sources held separately by the applicant on the license.

The license will be granted for a specific period, and the ending date will be written into the license. If the licensee wishes to renew the license as the end of the license period approaches, the applicant must be sure to submit a renewal request at least 30 days before the expiration date of the license. If this deadline is met, the original license will remain in force until the NRC acts on the request. This may take some time. During unusual periods when the NRC was under heavy work loads, it has taken over a year for action to take place.

Table 5.4 Exempt Quantities

Byproduct Material	μCi	Byproduct Material	μCi
Antimony 122	100	Indium 114m	10
Antimony	10	Indium 115m	100
Antimony 125	10	Indium 115	10
Arsenic 73	100	Iodine 125	1
Arsenic 74	10	Iodine 126	1
Arsenic 76	10	Iodine 129	0.1
Arsenic 77	100	Iodine 131	1
Barium 131	10	Iodine 132	10
Barium 133	10	Iodine 133	1
Barium 140	10	Iodine 134	10
Bismuth 210	1	Iodine 135	10
Bromine 82	10	Iridium 192	10
Cadmium 109	10	Iridium 194	100
Cadmium 115m	10	Iron 55	100
Cadmium 115	100	Iron 59	10
Calcium 45	10	Krypton 85	100
Calcium 47	10	Krypton 87	10
Carbon 14	100	Lanthanum 140	10
Cerium 141	100	Lutetium 177	100
Cerium 143	100	Manganese 52	10
Cerium 144	1	Manganese 54	10
Cesium 131	1,000	Manganese 56	10
Cesium 134m	100	Mercury 197m	100
Cesium 134	1	Mercury 197	100
Cesium 135	10	Mercury 203	10
Cesium 136	10	Molybdenum 99	100
Cesium 137	10	Neodymium 147	100
Chlorine 36	10	Neodymium 149	100
Chlorine 38	10	Nickel 59	100
Chromium 51	1,000	Nickel 63	10
Cobalt 58m	10	Nickel 65	100
Cobalt 58	10	Niobium 93m	10
Cobalt 60	1	Niobium 95	10
Copper 64	100	Niobium 97	10
Dysprosium 165	10	Osmium 185	10
Dysprosium 166	100	Osmium 191m	100
Erbium 169	100	Osmium 191	100
Erbium 171	100	Osmium 193	100
Europium 152 (9.2 hr)	100	Palladium 103	100
Europium 152 (13 yr)	1	Palladium 109	100
Europium 154	1	Phosphorus 32	10
Europium 155	10	Platinum 191	100
Fluorine 18	1000	Platinum 193m	100
Gadolinium 153	10	Platinum 193	100
Gadolinium 159	100	Platinum 197m	100
Gallium 72	10	Platinum 197	100
Germanium 71	100	Polonium 210	0.1
Gold 198	100	Potassium 42	10
Gold 199	100	Praseodymium 142	100
Hafnium 181	10	Praseodymium 143	100
Holmium 166	100	Promethium 147	10
Hydrogen 3	1000	Promethium 149	10
Indium 113m	100	Rhenium 186	100

Table 5.4 Exempt Quantities, Continued

Byproduct Material	μCi	Byproduct Material	μCi
Rhenium 188	100	Tellurium 127m	10
Rhodium 103m	100	Tellurium 127	100
Rhodium 105	100	Tellurium 129m	10
Rubidium 86	10	Tellurium 129	100
Rubidium 87	10	Tellurium 131m	10
Ruthenium 97	100	Tellurium 132	10
Ruthenium 103	10	Terbium 160	10
Ruthenium 105	10	Thallium 200	100
Ruthenium 106	1	Thallium 201	100
Samarium 151	10	Thallium 202	100
Samarium 153	100	Thallium 204	10
Scandium 46	10	Thulium 170	10
Scandium 47	100	Thulium 171	10
Scandium 48	10	Tin 113	10
Selenium 75	10	Tin 125	10
Silicon 31	100	Tungsten 181	10
Silver 105	10	Tungsten 185	10
Silver 110m	1	Tungsten 187	100
Silver 111	100	Vanadium 48	10
Sodium 24	10	Xenon 131m	1000
Strontium 85	10	Xenon 133	100
Strontium 89	1	Xenon 135	100
Strontium 90	0.1	Ytterbium 175	100
Strontium 91	10	Yttrium 90	10
Strontium 92	10	Yttrium 91	10
Sulfur 35	100	Yttrium 92	100
Tantalum 182	10	Yttrium 93	100
Technetium 96	10	Zinc 65	10
Technetium 97m	100	Zinc 69m	100
Technetium 97	100	Zinc 69	1000
Technetium 99m	100	Zirconium 93	10
Technetium 99	10	Zirconium 95	10

While the license is in effect, the NRC has the right to make inspections of the facility, the byproduct material, and the areas where the byproduct material is in use or stored. These inspections have to be at reasonable hours, but they are almost always unannounced. The inspector also will normally ask to see records of such items as surveys, personnel exposure records, transfers and receipts of radioactive materials, waste disposal records, instrument calibrations, radiation safety committee minutes, documentation of any committee actions, and any other records relevant to compliance with the terms of the license and compliance with other parts of 10 CFR, such as 19 and 20. Failure to be in compliance can result in citations of various levels or of financial penalties. Enforcement will be discussed further later. The NRC can require tests to be done to show that the facility is being operated properly, such as asking for tests of the instruments used in monitoring the radiation levels, or ask the licensee to show that security of radioactive material in the laboratory areas is not compromised.

Under Section 30.51, records of all transfers, receipts, and disposal of radioactive materials normally must be kept for at least 2 years after transfer or disposal of a radioactive material, or in some cases until the NRC authorizes the termination of the need to keep the records. There are other record keeping requirements in other parts of Title 10.

If for any reason there is a desire to terminate the license on or before the expiration date, Under 10 CFR 30.36 there are procedures that must be followed.

1. Terminate the use of byproduct material.
2. Remove radioactive contamination to the extent practicable (normally to background level, see 5 below).
3. Properly dispose of byproduct material.
4. Submit a completed Form NRC-314.
5. Submit a radiation survey documenting the absence of radioactive contamination, or the levels of residual contamination. In the latter case, an effort will be required to eliminate the contamination.
   a. The instruments used for the survey must be specified and then certified to be properly calibrated and tested.
   b. The radiation levels in the survey must be reported as follows:
      (1) Beta and gamma levels in microrads per hour at 1 cm from the surface, and gamma levels at 1 meter from the surface
      (2) Levels of activity in microcuries per 100 cm^2 of fixed and removable surface contamination
      (3) Microcuries per rub in any water
      (4) Picocuries per gram in contaminated solids and soils
6. If the facility is found to be uncontaminated, the licensee shall certify that no detectable radioactive contamination has been found. If the information provided is found to be sufficient, the NRC will notify the licensee that the license is terminated.
7. If the facility is contaminated, the NRC may require an independent survey acceptable to the NRC. The license will continue after the normal termination date. However, the use of byproduct materials will be restricted to the decontamination program and related activities. The licensee must submit a decontamination plan for the facility. They must continue to control entry into restricted areas until they are suitable for unrestricted use, and the licensee is notified in writing that the license is terminated.

In principle, the NRC has the right to modify, suspend, or revoke a license for a facility that is being operated improperly or if the facility were to submit false information to the NRC. If the failure to comply with the requirements of the license and other requirements for safely operating a facility can be shown to be willful or if the public interest, health, or safety can be shown to demand it, the modification, suspension, or revocation can be done without institution of proceedings which would allow the licensee an opportunity to demonstrate or achieve compliance.

Normally, an inspection will be followed up with a written report by the inspector in which any compliance problems will be identified. These may be minimal, serious (which would require immediate abatement), or graduated steps between these two extremes. The facility can (1) appeal the findings and attempt to show that they were complying with the regulations or that the violation was less serious than the citation described or (2) accept the findings. Unless the facility can show compliance, it must show how they will bring the facility into compliance within a reasonable period.

In recent years, there have been increasing numbers of occasions when the NRC has imposed substantial financial penalties on research facilities, including academic institutions, as they are entitled to do under Section 30.63, for violations that are sufficiently severe. Further, a few years ago, one city filed 179 criminal charges against a major university and several of its faculty members for failure to comply with radiation safety standards. Many individual violations were relatively minor, but apparently the city attorney thought he had a substantial case for a pattern of failure to comply with the terms of the license and the regulations.

The use of radioisotopes in research is continuing to increase, while the public concern about the safety of radiation continues unabated. It behooves all licensees to follow all regulations scrupulously, not only to ensure safety, but also to avoid aggravating the concerns of the public unnecessarily

1. Radiation Safety Committees

The primary function of the radiation safety committee (RSC), which is required under 10 CFR 33, is to monitor the performance of the users of ionizing radiation in a facility. It is, as noted earlier, a local surrogate of the NRC or the equivalent state agency in an agreement state. Usually, it is the ultimate local authority in radiation matters. In this one area at least, it is assigned more authority than the usual senior administrative officials. It is an operational committee, charged with an important managerial role in the use of ionizing radiation within the organization, not in directly managing the research program but assuring that the research is carried out safely. Due to this power, the NRC holds the committee responsible for compliance and will cite the committee and the parent organization for failure to provide appropriate oversight if the radiation users or radiation safety personnel under its supervision fail to ensure compliance with the regulations.

In addition to the responsibility of the RSC to ensure compliance with the provisions of the byproduct license and the other regulatory requirements of Title 10 CFR, it also must establish internal policies and procedures to guide those wishing to use radiation and to provide the internal operational structure in which this is done. The committee has other duties as well, which will be discussed after the makeup of the committee is considered.

The membership of an RSC should be carefully selected. It would be highly desirable to select much of the membership from among the active users of radiation within the organization and across the major areas or disciplines represented among the users. Each prospective member should be scrutinized very carefully. A RSC must enforce regulations set by one of the strongest regulatory agencies, and it must be fully willing to accept the delegated authority. Individuals on the committee must be willing, if necessary to establish policies that many users may feel are too restrictive. As active users themselves, they have a better chance of achieving compliance if the other users realize that the members of the RSC have accepted imposition of these same policies on their own activities. The members of the committee should have a reputation for objectivity, fairness, and professional credibility. A prima donna has no place on such a committee.

As professional scientists in their own right, the committee members will also understand the impact of a given procedure or policy on laboratory operations, and can often find legitimate ways to develop effective policies and procedures that are less burdensome on the users to carry out than would otherwise be the case.

The radiation safety officer (RSO) of the organization must be part of the RSC, and is a person who must maintain a current awareness of the rules and regulations required by the NRC and of radiation safety principles. This individual will serve to carry out the policies of the committee, and should be the individual to do the direct day-to-day monitoring of the operations of the laboratories using radioactive material. The decisions of the committee can be burdensome not only on the users, but, without input of the RSO, can be equally burdensome for this individual to carry out.

The working relationship between the RSO and the RSC is extremely important. No committee can effectively administer a program of any size on a daily basis. It must delegate some of its authority to a person, such as the RSO, or to an RSO through an alternate agency such as a Safety and Health Department charged with the daily administration of the area of responsibility assigned to an operational committee. However, especially when the RSO is a dynamic, effective person there is a tendency to defer to this person and to abrogate some of the committee's oversight responsibility. Both the RSC and the RSO should guard against this possibility. The RSO should have a voice and an influential one in the committee's deliberations, but should not be allowed to dictate policies independently.

The membership need not be limited to the persons already defined. A relatively new NRC requirement is that a senior management representative must be an *ex officio* member of the committee, and no official meeting can be held should the senior management represent-ative not be present. This individual cannot veto the actions of the committee, but by being present guarantees that higher management is aware of the actions of the committee. The head of the health and safety department, if different from the radiation safety officer, may be a member because this individual would bring in a wider perspective than would the RSO alone on the implications of some issues brought before the committee. Some large organi-zations may wish to have a representative of the organization's legal department as a member. Some may wish to have a representative of the public relations area as a member, especially if the facility is in an area in which there has been vigorous public opposition to the use of ionizing radiation. Some may wish to include a layperson, if not as a voting member, then perhaps as an observer, but the number of non-technical persons should not exceed those with sufficient technical expertise to fully understand the safety issues. The membership should not become too large, however, so that it will be practical to set up meetings without too much concern for having a quorum. Committees that are too large also tend to be less efficient, because of the time required for all the members to participate in discussions. On the other hand, each major scientific discipline using radiation should be represented. A reasonable size might be between 9 and 15 members, with a quorum established at between 5 and 8 members.

It is essential for the chair of the committee to be someone with prior experience with radiation, but it is also highly desirable if the chair is an individual with administrative cre-dentials. Such a person will normally ensure that committee meetings will be conducted efficiently, but if the administrative experience is at a level carrying budgetary and personnel responsibilities, the chair will bring still another dimension to the committee. Some actions of the committee may carry cost or manpower implications; an individual with managerial experience will recognize and perhaps have a feel for the feasibility of accommodating these requirements.

Besides the monitoring of existing programs, establishing policies, and providing guidance to radiation safety personnel, there are at least four other important functions that the committee must perform. The first of these is to perform the same function as the NRC in authorizing new participants to use radiation or radioactive materials. Basically the same information that the NRC requires for new applicants for a license should be required when a new internal facility is involved. The adequacy of the facility, the purpose of the program for which the use of radiation is involved, and the qualifications of the users should all be reviewed. At academic institutions especially there is a considerable turnover in users, represented by graduate students, postdoctoral research associates, and even faculty. Often individuals come from other facilities where internal practices may differ from local practices. To ensure that all users are familiar with not only the basic principles of radiation safety but also with local internal procedures, a simple written test, administered as part of the authorization procedure, is an effective and efficient means of documenting that the prospective users have familiarized themselves with the information. To avoid setting standards on who should take the test, it should be administered universally. Some faculty or researchers may object, but it serves an important legal point. A passed quiz shows unequivocally that the individual is familiar with the risks and the requirements associated with the use of radiation at the facility. An argument frequently put forward by those who object is that they are aware of the properties of the materials with which they are working, and this is undoubtedly true. A rebuttal argument, however, is that they probably are not as aware of the details of the NRC regulations with which they must comply and on the compliance with which they, and the organization, will be judged by an NRC inspector.

An internal authorization should be issued to an individual. Others may be added to the authorization, but one person should be designated as the local, ultimately responsible person, responsible for compliance with applicable safety and legal standards related to the use of radiation under the authorization. In laboratories that involve multiple users, it may be necessary to formally identify a senior authorized user so as to provide the additional authority to this individual.

The second additional function is to carefully review research or "new experiments," substantially different in the application of radioactive materials or radiation envisioned from work previously performed under the license. This role is easy to play when it is part of a new request for an authorization, but when an ongoing operation initiates a new direction in their program, it will be necessary for the committee to make it clear that the user must address the question to himself, "Is this application covered under the scope of work previously reviewed by the committee in my application?" If the answer is no, or "possibly not," then the responsible individual should ask for a review by the committee. The need to do this must be explicitly included in the internal policies administered by the committee. The RSC then must consider the proposed program in the same context as the institution's application to the NRC. Is the purpose of the work an approved purpose? This question must be answered positively in the context of the NRC facility license. Are the facilities adequate to allow the work to be done safely? Are the persons qualified because of training or experience to carry out the proposed research program safely? Incidentally, it is *not* within the purview of the committee's responsibility to judge the validity or worth of the research program, but only if the proposed research can be done safely according to radiation safety and health standards. Of course, obviously frivolous research is unacceptable for approval.

In the past, the use of proven research technology was sufficient to approve most research and routine experiments did not receive the scrutiny that new experiments did. In the last few years, the NRC has required the investigators to formally review even standard procedures for possible hazards, to establish procedures to prevent these potential hazards from occurring, and to develop a response protocol. Worst case failure modes must be reviewed. It is enlightening to see the results of these analyses. It is frequently found that there is far more potential for failure than most would anticipate. The committee must review and approve of these hazard analyses.

The fourth function not previously discussed is the role of the RSC as a disciplinary body. Occasions will arise when individual users will be found to not be in full compliance with acceptable standards. Often this will be done by the RSO in his periodic inspections, but many will be reported by the users themselves. The NRC will expect these situations to be evaluated and appropriate actions taken, which can include disciplinary measures. Not all violations are equally serious. Categories of violations should be established by the RSC to guide the RSO and the users. A single instance of faulty record keeping is not as serious as poor control over byproduct material usage, for example. Allowing material to be lost or radioactive material to escape into the environment is a serious violation. If the loss or discharge is due to an unforeseen accident, it is less serious than if the cause is negligence.

However, a continuing pattern of minor violations may show carelessness or lack of concern about compliance with the standards that could eventually lead to a more serious incident. Such a continuing pattern should be considered as a serious violation. Possible penalties should be listed in the internal policies and guidelines issued by the RSC.

Every case in which noncompliance is discovered or reported should be carefully investigated by the RSO and a report made to the RSC. The person responsible for the non-compliance and the responsible individual from the facility (if not the same person) should be invited to meet with the committee and present their sides of the issue if they wish. Few will normally contest minor citations, but most will contest a citation that could affect their ability to use radioactive materials. After hearing both sides, the committee should take appropriate disciplinary measures. Issues are rarely black or white, and the penalties or corrective actions should be adapted to the circumstances. An initial minor violation, for example, might elicit no more than a cautionary letter from the committee. However, a series of minor violations within a short interval probably should result in a mandatory cessation of usage of radioisotopes until the user can show the willingness and capability to comply with acceptable practice. A proven serious accidental violation should result in an immediate cessation of operations until procedures can be adopted to prevent future recurrence of the problem. A serious violation due to willful noncompliance should result in a mandatory cessation of the use of radioisotopes for a substantial length of time or even permanently. The elimination of the right to use radioisotopes is a very serious penalty, since the user's research program may depend upon this capability. In an academic institution, even a relatively short hiatus in a research program could result in the loss of a research grant or failure to get tenure. As a result, a permanent or extended loss of the right to use radioactive material should not be imposed lightly, but if the user shows by action and attitude that future violations are likely to occur, the committee may have no practical alternative except to do so to protect the rest of the organization's users from the loss of the institution's license or the imposition of a substantial fine by the NRC. The committee must be willing to accept the responsibility, unpleasant though it may be. If the users believe that the RSC is willing to be reasonable, but firm and fair, they will be more likely to comply with the required procedures.

The role of the RSO could be construed as the enforcement arm of the RSC and the radiation safety office (which may be part of a larger organization). If this were the case, the RSO would be the equivalent of an NRC inspector. However, as with many other persons working in safety and health programs who have enforcement duties, their primary function is service to the users. In later sections the other duties will make this clear.

F. Radiation Protection, Discussion, and Definitions

Many basic terms were defined in some detail in the several subsections of Section 5.1. However, several additional concepts will be introduced in the following sections, and some additional terms need to be defined.

The original definitions of dose units primarily were employed for external exposure by users of radioactive materials and other applications of ionizing radiation. Concerns relating to internal exposure generally were covered by establishing maximum permissible concentrations of radionuclides in air and water, in terms of the workplace and in terms of the public, in Appendix B to 10 CFR 20, Tables I and II, respectively. Protection was provided by considering the amount of radiation given to the most critical organ by the intake of specific radioisotopes and their physical or chemical form. The exposure limits to the whole body were established by the organs assigned the lowest dose limits, the bone marrow, gonads, and lens of the eye. The reasons for establishing these organs as the most sensitive to radiation were concerns about leukemia, hereditary effects, and cataracts, respectively.

The amounts of a radionuclide in organs, or the "burden," were calculated based on a constant exposure rate, maintained for a period sufficiently long so that an equilibrium would be established between the intake of the material and the effective elimination rate. The effective elimination rate, or effective half-life, is a combination of the radioactive half-life and the biological half-life based on the rate at which the material would be eliminated from the body. The relationship is given by the following equation:

$$\frac{1}{T_{eff}} = \frac{1}{T_\tau} + \frac{1}{T_b}$$

(8)

where T_{eff} = Effective half-life

T_τ = Radiological half-life

T_b = Biological half-life

The maximum permissible concentrations (MPC) were those that correspond to an organ burden that would cause the annual dose limits to be attained. Control measures, therefore, were designed to maintain the concentrations below the MPCs.

The current 10 CFR Part 20 that went into effect January 1, 1994, sets standards for protection for users of radioisotopes and requires combining internal and external exposures. The revision contains several other changes in Part 20 based on many of the recommendations contained in the International Commission on Radiological Protection (ICRP) Publications 26, 30, and 32. The remainder of this section and some succeeding sections will discuss the current regulations and will draw attention to significant changes to those areas in use for many years.

1. Selected Definitions

In order to discuss the current Part 20, some terms introduced into it from the ICRP 26 and 30 are needed.

The first six main definitions are related to dose terms:

1. Dose equivalent means the product of absorbed dose, quality factor, and all other necessary modifying factors at the location of interest in tissue.
2. External dose means that portion of the dose received from radiation sources from outside the body.
   a. Deep dose equivalent (H_d) applies to the external whole-body exposure and is taken as the dose equivalent at a tissue depth of 1 cm.
   b. Eye dose equivalent (He) applies to the external exposure of the lens of the eye and is taken as the dose equivalent at a tissue depth of 0.3 cm.
   c. Shallow dose equivalent (Ha) applies to the external exposure of the skin or an extremity and is taken as the dose equivalent at a tissue depth of 0.007 cm., averaged over an area of one square centimeter.
3. Internal dose is that portion of the dose equivalent received from radioactive material taken into the body.
   a. Committed dose equivalent $(H_{T,50})$ means the dose equivalent to organs or tissues of reference (T) that will be received from an intake of radioactive material by an individual during the 50-year period following the intake.
   b. Effective dose equivalent (H_E) is the sum of the products of the dose equivalent (H_T) to the organ or tissue (T) and the weighting factors (w_T) applicable to each of

Table 5.5 Definitions: Weighting Risk Coefficient

Organ or Tissue	Weighting Factor (W_T)	Risk Coefficient (per rem)	Probability (per rem)
Gonads	0.25	4×10^{-5}	1 in 25,000
Breast	0.15	2.5×10^{-5}	1 in 40,000
Red bone marrow	0.12	2×10^{-5}	1 in 50,000
Lung	0.12	2×10^{-5}	1 in 50,000
Thyroid	0.03	5×10^{-6}	1 in 200,000
Bone surfaces	0.03	5×10^{-6}	1 in 200,000
Any remaining organs or tissues receiving the highest dose at a relative sensitivity of 0.06 each	0.30	5×10^{-5}	1 in 20,000
Total	1.0	1.65×10^{-4}	1 in 6,000

the body organs or tissues that are irradiated.

$$H_E = \sum w_T H_T \tag{9}$$

   c. Committed effective dose equivalent is the sum of the products of the weighting factors applicable to each of the body organs or tissues that are irradiated and the committed dose equivalent to those organs or tissues.

$$H_{E,50} = \sum w_T H_{T,50} \tag{10}$$

   d. Collective effective dose equivalent is the sum of the products of the individual weighting dose equivalents received by a specified population from exposure to a specified source of radiation.

4. Weighting factor w_T, for an organ or tissue (T) is the proportion of the risk of stochastic effects resulting from the irradiation of that organ or tissue to the total risk of stochastic effects when the whole body is irradiated uniformly. See Table 5.6.

5. Occupational dose means the dose received by an individual in the course of employment in which the individuals assigned duties involve exposure to radiation or to radioactive material from licensed or unlicenced sources of radiation whether in the possession of the licensee or other person. It does not include doses received from background radiation, from any medical administration the individual has received, from exposure to individuals administered radioactive materials and released in accordance with 10CFR part 35.75, from voluntary participation in medical research programs, or as a member of the public.

6. Public dose is an exposure of a member of the public to radiation or to the release of radioactive material, or to another source, either in a licensee's controlled area or in unrestricted areas. This does not include background radiation or any kind of medically related exposures.

The next five definitions relate to dose control factors:

7. ALARA is an acronym that stands for "as low as reasonably achievable." It is a policy that means making every reasonable effort to maintaining exposures to radiation as far below the dose limits as is practical consistent with the purpose for which the licensed activity is undertaken, taking into account the state of technology, the economics of

improvements in relation to benefits to the public health and safety, and other societal and socioeconomic considerations, and in relation to utilization of nuclear energy and licensed materials in the public interest.

8. Annual limit of intake (ALI) means the derived limit for the amount of radioactive material taken into the body of an adult worker by inhalation or ingestion in a year. It is the smaller of (1) the value of the intake of a given radionuclide in 1 year by the reference man that would result in a committed dose equivalent of 5 rems (0.05 Sv); (2) a committed dose equivalent of 50 rem (0.5 Sv) to any individual organ or tissue.

9. Derived air concentration (DAC) means the concentration of a given radionuclide in air which, if breathed by the reference man for a working year of 2,000 hours under conditions of light activity (corresponding to an inhalation rate of 1.2 m^3 air per hour), results in an inhalation of 1 ALI. These are comparable to the MPCs in the older Part 20.

10. Dose limits means the permissible upper bounds of radiation doses. These are usually set for a calendar year. They apply to the dose equivalent received during the set interval, the committed effective dose equivalent resulting from the intake of radioactive material during the interval or the effective dose equivalent received in 1 year The external dose and the internal dose must be combined so as not to exceed the permissible limits. The following equation can be used to compute the relative amounts of each, for the annual intake I_J of nuclide J:

$$\frac{H}{5} + \frac{\sum_J I_J}{(ALI)_J} \leq 1 \tag{11}$$

Two terms are used to describe two different classes of effects of radiation.

11. Stochastic effects refers to health effects that occur randomly, so that the probability (generally assumed to be linear, without a threshold) of an effect occurring, such as the induction of cancer is a function of the dose rather than the severity of the effect.

12. Non-stochastic effects are health effects for which the severity depends upon the dose, and for which there is probably a threshold.

With these definitions in mind, the following section presents selected parts of the current 10 CFR Part 20, which went into effect on January 1, 1994. Holders of NRC licenses must have formal programs to ensure that all programs covered by the regulations and all individuals working with radioactive materials now comply with the revised standard.

2. Selected Radiation Protection Standards from 10 CFR Part 20

The following section contains the sense of the Part 20 regulations, paraphrased in some instances for clarity and brevity. A few comments are added in some sections, where they were felt to be helpful. Most of the areas covered by the standard will be gone into in more detail in later sections describing practical implementation of a program complying with the standard.

a. Occupational Limits for Adult Employees

The dose limits are the levels not to be exceeded, but they should not be considered as a goal. The revised standard requires the licensee to use, to the extent practicable, procedures and engineering controls based upon sound radiation protection principles to achieve occupational doses and doses to members of the public that are ALARA.

i. Whole body, head, trunk, arm above elbow, and leg above knee

The more limiting of:

- 5 rems per year (0.05 Sv per year)—includes summation of (external) deep dose equivalent and (internal) committed effective dose equivalent, or
- The sum of the deep-dose equivalent (the dose equivalent at a depth of 1 cm in tissue) and the committed dose equivalent to any individual organ or tissue other than the lens of the eye being equal to 50 rems (0.5 Sv).

Without any intake of radioactive materials, the limits above correspond to the external exposure limits of the previous standard.

ii. Eyes, Skin, Arms Below Elbows, Legs Below Knees
- 15 rems per year (0.15 Sv year) to lens of eye, and
- A shallow dose equivalent (at a tissue depth of 0.007 cm over a 1 cm^2 area) of 50 rems (0.5 Sv) to the skin and any extremity.

The assigned deep-dose equivalent and shallow-dose equivalent must be for the part of the body receiving the highest exposures. These data can be inferred from surveys or other measurements if direct data are not available.

The DAC and ALI values are given in Table 1, Appendix B, of Part 20 and can be used in the determination of an individual's internal dose and to demonstrate compliance with required occupational dose limits. Table 2, Appendix B lists the maximum permissible concentrations in effluents and Table 3 lists the maximum concentrations permitted for release into sewers.

The limits specified are for an individual. The total exposures from work for different employers cannot exceed these limits.

Significant changes are that the previous standard did not require addition of external and internal doses; the basic interval over which radiation is measured has been extended from a quarter to a year; the 5(N- 18) cumulative dose limit has been deleted; the higher limits to eye exposure and the extremities reflects more information on the sensitivity to radiation for these areas. In addition, the revised standard contains a new provision for planned special exposures for limited higher doses to an individual in exceptional cases when other alternatives are unavailable or impractical. The licensee must specifically authorize the planned special exposure in writing before the event. The individual who will receive the exposure must be informed in advance of (1) the purpose of the operation, (2) the estimated doses and potential risks involved in the operation, and (3) measures to be taken to comply with the principles of ALARA, taking into account other risks that might be involved. The licensee must also determine in advance the cumulative lifetime dose of the individual participating in the exercise. With these steps completed, the participating individual must not be caused to receive a dose from all planned special exposures and all other doses in excess of

- The numerical values of the allowable dose limits in any year, and
- Five times the annual dose limits during the individual's lifetime.

The licensee must retain records of all planned special exposures. In addition, the licensee must provide exposed individuals with a report of the dose received and submit a report to the regional director of the NRC within 30 days of the event.

It is not necessary in every case to sum the external and internal doses if the licensee can show that the internal dose does not contribute significantly. If, for example, the only intake of radioactivity is by inhalation, the total effective dose equivalent is not exceeded if the deep-dose equivalent divided by the total effective dose equivalent, plus an estimate of the internal dose as determined by one of three procedures stipulated in the regulation does not exceed 1, the internal dose need not be added to the external dose. Similarly, unless the amount of radioactivity ingested is more than 10% of the applicable ALI, it need not be included in the total dose equivalent. Most laboratories using radioactive materials at reasonable levels under normal conditions will find that they need only consider external exposures, just as they once did.

Only individuals likely to receive within 1 year more than 10% of the allowable dose limits are required to be monitored by the licensee. However, unless the dose is monitored, it is difficult to establish with certainty that an active user of radioisotopes may not have exceeded the 10% limit. Many licensees do monitor most users of radioactive materials by providing personnel dosimeters to measure external exposures, excluding those who only work with weak beta emitters. In order to monitor internal exposures, the licensee can perform measurements of (1) concentrations of radioactive materials in the air in the workplace, (2) quantities of radionuclides in the body, (3) quantities of radionuclides excreted from the body, or (4) combinations of these measurements.

The personnel monitors normally used for individual monitoring must be processed by a dosimetry processor holding personnel dosimetry accreditation from the National Voluntary Laboratory Accreditation Program (NVLAP) of the National Institute of Standards and Technology for the type of radiation or radiations included in the NVLAP program that most closely approximates the exposure characteristics for the individual wearing the dosimeter.

b. Occupational Limits for Minors (Under 18) and to an Embryo/Fetus

The annual occupational dose limits for minors are 10% of the limits for adult workers as listed in 10 CFR 20.1201.

The limits to a fetus are based on an exposure over the entire 9 months of the pregnancy of a declared pregnant woman. If the woman chooses not to declare her pregnancy to her employer, the licensee may be in a legally difficult position. Anti-discrimination laws prevent an employer from discriminating against a woman, but the normal work regimen may not be compatible with the NRC regulation. If the woman were to make an issue of a job change based on an obvious but undeclared pregnancy, there could be problems. It will be assumed here that nearly all women in the workplace would wish to limit the exposure to their unborn child as soon as possible and declare their pregnancy when they are sure. This could still cause up to an approximately 2-month period of higher exposure than desirable due to uncertainty in the early stages of pregnancy. Following is the text of Section 20.1208which governs this situation.

(a) The licensee shall ensure that the dose to an embryo/fetus during the entire pregnancy, due to occupational exposure of a declared pregnant woman, does not exceed 0.5 rem (5 mSv). (For record keeping requirements, see Sec. 20.2106.)

(b) The licensee shall make efforts to avoid substantial variation above a uniform monthly exposure rate to a declared pregnant woman so as to satisfy the limit in paragraph (a) of this section.

(c) The dose to an embryo/fetus shall be taken as the sum of—
  (1) The deep-dose equivalent to the declared pregnant woman; and
  (2) The dose to the embryo/fetus from radionuclides in the embryo/fetus and radionuclides in the declared pregnant woman.

(d) If the dose to the embryo/fetus is found to have exceeded 0.5 rem (5 mSv), or is within 0.05 rem (0.5 mSv) of this dose, by the time the woman declares the pregnancy to the licensee, the licensee shall be deemed to be in compliance with paragraph (a) of this section if the additional dose to the embryo/fetus does not exceed 0.05 rem (0.5 mSv) during the remainder of the pregnancy.

c. Dose Limits for Individual Members of the Public

The licensee must conduct operations in such a way as to ensure (1) that the total effective dose to individual members of the public from the licensed operation does not exceed 100 mrem (1 mSv) in 1 year, exclusive of the dose contribution from the licensee's disposal of radioactive materials into sanitary sewerage, and (2) the dose in any unrestricted area from external sources does not exceed 2 mrem in any 1 hour.

A licensee or license applicant may apply for prior NRC authorization to operate up to an annual dose limit for an individual member of the public of 0.5 rem (5 mSv). The licensee or license applicant shall include the following information in this application:(1) demonstration of the need for and the expected duration of operations in excess of the limit in paragraph (a) of this section; (2) the licensee's program to assess and control dose within the 0.5 rem (5 mSv) annual limit; and (3) the procedures to be followed to maintain the dose as low as is reasonably achievable. The licensee must make surveys, measurements, and/or calculations to prove that the limits would not be exceeded for an individual likely to receive the highest dose from the licensed operations.

There are also EPA regulations governing releases of radioactive materials into the environment that will be treated separately.

d. Surveys and Monitoring

Monitoring requirements have been covered in the section on occupational exposures for adult employees. In addition to personnel monitoring, surveys are required. The surveys can be done with portable instruments or samples can be obtained and tested in laboratory instruments. These surveys are used to evaluate (1) the extent of radiation levels, (2) concentrations or quantities of radioactive materials, and (3) the potential radiological hazards that could be present. The equipment used for the surveys must be calibrated periodically for the radiation measured. Normally this is done at least annually. A description of personnel monitoring and survey devices will be found later in this chapter, and recommended survey procedures.

i.   Controlled Areas, Restricted Areas, Radiation Areas, High Radiation Areas, and Very High Radiation Areas

Most laboratories using radioisotopes do not use sufficient amounts of radioactive materials so that they evoke the NRC requirements for areas defined by these labels. Access to a laboratory may be controlled for a number of other reasons — secure use of biological pathogens, toxic chemicals, or explosives — but the level of radiation and use of radioactive materials in most research laboratories is normally quite modest. Exceptions would be reactor facilities, medical labs for therapeutic radiation, fuel fabrication facilities, etc. These facilities normally will have their own radiation safety specialist who would be thoroughly familiar with these requirements, so they will not be covered further here. If information is needed, it can be found in 10CFR Part 20.1003.

ii. Storage and Control of Licensed Material

A matter of serious concern to the NRC is security of licensed materials. Although not explicitly stated in the regulations, there have been interpretations by individuals in the NRC that since the regulation specifies licensed materials, exempt quantities of materials may be construed as exempt from security requirements, but this interpretation is not necessarily firm. It would be advisable to treat exempt amounts of licensed materials as not exempt.

- A licensee must assure that licensed materials stored in controlled or unrestricted areas are secure from unauthorized removal or access.
- A licensee must control and maintain *constant* surveillance of licensed materials that are in a controlled or unrestricted area and that are not in storage.

The word "constant" was italicized in the last requirement for emphasis. It is not permissible to leave material <u>alone and unsecured</u> for brief visits to the soft drink machine, to step outside to smoke, or even to go to the bathroom.

iii. Posting of Areas or Rooms in which Licensed Material is Used or Stored

The licensee must post each area or room in which an amount of licensed material more than 10 times the quantity specified in Appendix C to 10 CFR Part 20 is used or stored with a conspicuous sign or signs bearing the radiation symbol and the words **CAUTION, RADIOACTIVE MATERIALS** or **DANGER, RADIOACTIVE MATERIALS.**

There are some exceptions to these posting requirements, but it is normally appropriate to post. There are additional posting requirements specified in 10 CFR Part 19 that will be discussed later.

iv. Labeling Containers

Every container of licensed material must bear a clear, durable label bearing either **CAUTION** or **DANGER, RADIOACTIVE MATERIAL.** The label must also provide additional information, such as the radionuclide in the container, the amount present, the date on which the activity was estimated, radiation levels, and the kinds of materials in order for individuals to determine the level of precautions needed to minimize exposures. The labels on empty, uncontaminated containers intended for disposal must be removed or defaced before disposal. There are exceptions to these labeling requirements, but most containers in use in a facility should be labeled.

v. Procedures for Ordering, Receiving, and Opening Packages

A supplier of radioactive materials must ensure that the facility ordering the material has a valid radioactive materials license before filling an order. This is accomplished by the vendor being provided with a copy of the license. Although not required, it is desirable if all radioactive materials are ordered through the radiation safety office and delivery specified to that office. With all materials being ordered through one point of sale and receipt, it is sometimes possible to arrange discounts for bulk purchases, even if individual items are ordered separately.

The regulations for receiving packages containing radioactive materials are covered by Part 20.1906. Briefly, these are: Licensees expecting to receive packages containing more than a type A quantity (defined in 10 CFR Part 71 and Appendix A to Part 71) must make arrangements to receive the package upon delivery or receive notification of its arrival by a carrier, and arrange to take possession of it expeditiously. In most facilities the responsibility for receiving radioactive deliveries for the entire organization is delegated to the radiation safety office. The package must be monitored for surface contamination and external radiation levels. The limits for loose contamination set by 49 CFR Part 173.443 are set for a swipe of 300 cm^2. The limits for beta, gamma, and low toxicity alpha emitter contamination are 22 dpm/cm^2. For all other alpha emitters, the loose surface contamination limits are 10% of these. The radiation level due to the contents of the package are set by both 10 CFR Part 20.71.47 and 49 CFR Part 173.441 to not exceed 200 mrem/hr (2mSv/hour) at any point on the surface. There are also limits set for the transporter. Levels in excess of these limits would require immediate notification of the NRC. If the levels are higher than anticipated or the package is damaged, the package should be carefully opened, as a minimum safety precaution, in a fume hood designated for use with radioactive materials. A glove box would be better.

Many carriers deliver packages outside normal working hours, and some organizations will receive them at the portions of their facility that function 24 hours per day, such as a security office. If the packages are received during normal working hours, required monitoring must be done no later than 3 hours after receipt. If packages are received outside normal working hours, monitoring must be done no later than 3 hours from the beginning of the next working day.

vi. Disposal of Radioactive Waste

The activity in radioactive waste from radioisotopes is included in the possession limits for a licensed facility until disposed of in an authorized manner. Therefore, records of materials transferred into waste must be maintained as with any other transfer by the individual user. At some point, the waste from individual laboratories will be combined and disposed of by one of the means approved by the NRC. The disposal options available to a generator of radioactive waste are

- Transfer to an authorized recipient according to the regulations in 10 CFR Parts 20 (Appendices E - G), 30, 40, 60, 61, 70, or 72
- Decay in storage
- By release of effluents into the sanitary system subject to specific limitations(20.2003)
- By other approved technologies by persons authorized to use the technologies

A person can be specifically licensed to receive radioactive waste for disposal from a facility for:

- Treatment prior to enclosure (this often means solidification of liquid wastes or compaction of dry wastes)
- Treatment or disposal by incineration
- Decay in storage
- Disposal at a land disposal facility licensed under 10 CFR Part 61
- Disposal at a geologic facility under 10 CER Part 60

The limitations on release into the sanitary system are

- The material must be readily soluble in water or is biological material readily dispersible in water.
- No more of a specific radioisotope can be released into the sanitary system in one month than an amount divided by the total volume of water released into the system in one month that does not exceed the concentration listed in Table 3 of Appendix B to Part 20.
- The total quantity of licensed and unlicenced material released into the sanitary system in one year does not exceed 5 Ci (185 GBq) of ^{3}H, 1 Ci (37 GBq) of ^{14}C, and one Ci of all other radioactive materials combined.
- Excreta containing radioactive material from persons undergoing medical diagnosis or therapy is not subject to the disposal limits.

The licensee can dispose of the following licensed materials as if they were not radioactive.

- 0.05 µCi (1.85 kBq) or less of ^{3}H or ^{14}C per gram of medium used for liquid scintillation counting. (Note that some liquid scintillation fluids are EPA-regulated RCRA wastes and must be treated as hazardous wastes. Alternative liquid scintillation fluids are available which are not hazardous in this context.)
- 0.05 µCi (1.85 kBq) or less of ^{3}H or ^{14}C per gram of animal tissue averaged over the weight of the entire animal.

Animal tissue containing radioactive material in the amounts specified cannot be disposed of in a way that would allow its use as either animal or human food. The licensee may use incineration to dispose of radioactive materials subject to the limitations of this section.

Records must be maintained by the licensee of disposal of radioactive wastes. Material transferred to a disposal firm is subject to requirements very similar to those covering chemical wastes. The manifest has specific requirements for the description of the physical and chemical form of the radioactive materials and the radiation characteristics of the wastes. There are specialists in transporting radioactive waste who normally serve as the intermediary or broker for the generator and disposal facility.

vii. Records of Survey, Calibration, and Personnel Monitoring Data

Records of facility surveys, package monitoring data, and equipment calibrations must be retained for 3 years after they are recorded. Where required to sum internal and external doses, data required to determine by measurement or calculation the doses to individuals or radioactive effluents released to the environment must be kept until the termination of the license requiring the record.

Personnel dose records, including internal and external doses where required, for all personnel for whom monitoring is required must be kept until the NRC terminates the license for which the record was required.

viii. Loss or Theft of Material

Losses of significant amounts of radioactive materials must be reported to the NRC both by telephone and by a written report. If the amount that is not accounted for is equal to or greater than 1000 times the quantity specified in Appendix C to 10 CFR Part 20, and it appears that an exposure could affect persons in unrestricted areas, the loss shall be reported by telephone as soon it is known to the licensee. Within 30 days after lost, stolen, or missing material is discovered, the NRC must be notified by telephone of any amount greater than 10 times the quantity specified in Appendix C that is still missing at that time. Most facilities will make their report to the NRC Operations Center.

After making the telephone report, the licensee must, within 30 days, make a written report to the NRC providing details of the loss, including (1) the amount, kind, chemical, and physical form of the material, (2) the circumstances of the loss, (3) the disposition or probable disposition of the material, (4) exposures to individuals, how the exposures occurred, and the probable dose equivalent to persons in unrestricted areas, (5) steps taken or to be taken to recover the material, and (6) procedures taken or to be taken to ensure against a recurrence of the loss. If at a later time the licensee obtains additional information, it must be reported to the NRC within 30 days. Names of individuals who may have received exposure due to the lost material are to be stated in a separate and detachable portion of the report.

e. Notification of Incidents

Immediate notification is required for any event due to licensed materials that would cause any of the following conditions: (1) an individual to receive a total effective dose of 25 rem

(0.25 Sv) or more, (2) an eye dose of 75 rems (0.75 Sv) or more, (3) a shallow dose equivalent to the skin or extremities of 250 rads (2.5 Gy) or more, (4) a release of material into the environment such that an individual, if present for 24 hours, could have received an intake of five times the occupational annual limit.

Notification is required within 24 hours of the discovery of an event involving loss of control of licensed radioactive material that may have caused or may cause any of the following conditions: (1) an individual to receive within a 24-hour period an effective dose exceeding the annual limits for an effective dose equivalent, and eye dose equivalent, a shallow dose equivalent to the skin or extremities, or an intake dose in excess of one occupational annual limit on intake. The times given are calendar days not working days.

For most facilities, the reports must be made by telephone to the NRC Operations Center *and* to the administrator of their regional office by telegram, mailgram, or facsimile.

i.   Reports of Exposures, Radiation Levels, and Concentrations of Radioactive Materials Exceeding the Limits

A written report must be submitted within 30 days after learning of the event for any of the following conditions: (1) an event requiring immediate or 24-hour notification, (2) doses in excess of the occupational dose limits for adults, the occupational dose limits for a minor, the limits for a fetus, or the limits for an individual member of the public, (3) any applicable limit in the license, or (4) the ALARA constraints for air emissions, (5) levels of radiation or concentrations of radioactive materials in a restricted area in excess of any applicable license limits or more than 10 times any applicable license limits in an unrestricted area. An individual need not be exposed in the last two conditions. The reports must contain the following information: (1) the extent of exposure of individuals, including estimates of each exposed individual's dose, (2) the levels of radiation and concentrations of radioactive materials involved, (3) the cause of the incident, and (4) corrective steps taken or planned to ensure against a recurrence, including the schedule for achieving conformance with applicable regulations.

The reports must include in a separate, detachable part the exposed individual's name, Social Security Number, and birth date. The reports are to be sent to the U.S. NRC Document Control Desk, Washington, D.C., 20555, with a copy to the regional administrator.

3.   EPA, National Emission Standards for Hazardous Air Pollutants, Radionuclides

The EPA enacted a rule limiting emissions of radionuclides into the atmosphere for several classes of source facilities, among them facilities licensed by the NRC. The rule became effective on December 15, 1989. However, shortly after that the rule was stayed for NRC facilities and federal facilities not owned by the Department of Energy. On January 28, 1994, NRC license holders were informed that the rule was in effect, and unless licensees could show that they were exempt, a report would have to be filed by March 31, 1994, and annually thereafter. The primary concerns were with the potential for increased risk of cancer due to inhalation of radionuclides. For laboratory facilities using radioactive materials, fume hood exhausts normally would be the source of emissions. The groups of concern would be the receptors (homes, offices, schools, resident facilities, businesses nearest the emission source). Another source of concern would be farms, including meat, vegetable, and dairy farms where the radionuclides could enter the food chain. The EPA provided several alternative means of determining whether a facility is exempt, in compliance, or needs to come into compliance. In their compliance guide, they stated that if a facility uses more than about six nuclides and has multiple release points, the licensee would be best served by using a computer program called COMPLY, available from the EPA. The results depend upon many input parameters, but once these have been determined, the program will quickly compute the results. There has since been a shift in regulatory responsibility to the

NRC and the EPA standards no longer apply to NRC licensed facilities except through the NRC. This went into effect on January 9, 1997. The basis for compliance is that the effective-dose equivalent to an individual is less than 10 mrem per year.

G. Radioisotope Facilities and Practices

The following sections describe facilities and practices that comply with the regulations outlined in Sections II.D to II.D.3 of this chapter.

1. Radiation Working Areas

The areas in which radiation is used should meet good laboratory standards for design, construction, equipment, and ventilation, as described in Chapter 3. The International Atomic Energy Agency (IAEA) has defined three classes of laboratories suitable to work with radionuclides. Their class II facility is essentially equivalent to a good quality level 2 facility as described in that chapter while their class I facility would be similar to the level 3 or 4 laboratory, depending upon the degree of risk, and would be especially equipped to safely handle even high levels of radioactive materials.

One additional feature which may need to be included in the design of a laboratory using radioactive materials is provision for using a HEPA (high efficiency particulate aerosol) filter, or an activated charcoal filter for nonparticulate materials, on the exhaust of any fume hood in which substantial levels of radioactive materials are used in order to comply with the constraints on release of airborne radionuclides described in the preceding section. A problem with this requirement is the possibility of the filter becoming rapidly "loaded up" by the chemicals used in the hood. A velocity monitor should be mandatory on any fume hood, but especially one equipped with a HEPA filter, to provide a warning should the face velocity fall below 100 fpm (assuming that is the standard established for a working face velocity). The definition of "substantial" will depend upon the type of radioactive material used. The levels of activity from an unfiltered hood exhaust should not exceed the levels permitted in Appendix B, Table II (10 CFR Part 20) under the worst possible circumstances, such as a spill of an entire container of a radioisotope in the hood. The volume of material, the physical properties of the material, and the rate at which air is pulled through the system should allow the maximum concentration in the exhausted air to be computed.

Most laboratories using radionuclides are not required to be limited access facilities. However, any area in which radioactive material is used or kept should be identified with a standard radiation sign such as shown in Figure 5.6. Unless the amount of material or exposure levels in the facility trigger a more explicit warning, the legend should say only **CAUTION, RADIOACTIVE MATERIAL.** Individuals working with radioactive materials must be trained in safe procedures in working with radiation under the rules and regulations described in Section II.D.2. Individuals working in the same area who do not use radioactive materials need not be as thoroughly trained, but should be sufficiently informed so that they understand the reasons for the care with which the others using the facility handle the radioactive materials, so they will not inadvertently become exposed to radiation levels in excess of those permitted for members of the public, and so that their actions or inactions do not cause an incident involving radiation.

If licensed materials are stored in an unrestricted area, the materials must be securely locked to prevent their removal from the area. Many materials used in the life sciences must be kept either in refrigerators or freezers, which should be purchased with locks or equipped with padlocks afterward. If the radioactive materials in an unrestricted area are not in storage, they must be under the constant surveillance and immediate control of the licensee.

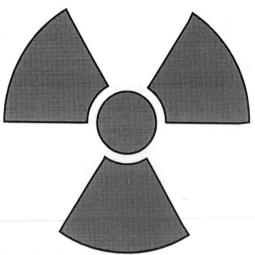

Figure 5.6   Standard radiation symbol.

As noted in Section II.D.2, this precludes leaving them unattended for any reason for even brief periods.

Some laboratory facilities, or portions of them, may need to be made into "restricted areas" because of the type of activities conducted within them or to ensure that members of the public will not be exposed to radiation in excess of that allowed by Part 20 for occupational exposures and members of the public. Access to a restricted area is not prohibited to a member of the public, but it must be controlled to provide the proper assurance of protection to them. A sign should be posted at the entrance to a restricted area stating

RESTRICTED AREA,
ACCESS LIMITED TO AUTHORIZED PERSONNEL ONLY

The entrance always should be locked when the area is unoccupied and at such times that unauthorized individuals could enter the area under circumstances that the occupants would be unaware of or unable to control their entrance.

Within a restricted area, there may be specifically defined areas where the levels of radiation may be significantly above the levels that would be acceptable for personnel to work on a normal 40 hours-per-week schedule. Depending upon the level, these areas would be designated as Radiation Areas, High Radiation Areas, and Very High Radiation Areas. Additional restrictions would be imposed on entering these areas.

2. External Radiation Exposure Areas

A Radiation Area within a restricted area is an area accessible to personnel where radiation exists (according to the legal definition) arising in whole or in part from licensed material at such levels that a major portion of the body could receive a dose of 5 mrem (0.05 mSv) in 1 hour, or in excess of 100 mrem (1 mSv) in 5 consecutive days. A Radiation Area must be conspicuously posted with one or more signs bearing the radiation symbol and the words

CAUTION
(or DANGER)
RADIATION AREA

Similarly, a High Radiation Area is an area within a restricted area accessible to personnel where there exists radiation, arising in whole or in part from licensed material at such levels that a major portion of the body could receive a dose of 100 mrem (1 mSv) in 1 hour. A High Radiation Area must be posted with one or more signs carrying the radiation symbol and the legend

CAUTION
(or DANGER)
HIGH RADIATION AREA

In addition to the warning signs, additional measures must be taken to prevent individuals from accidentally entering a High Radiation Area. These measures would include one or more of the following.

1. An automatic device must reduce the level to the 100 mrem per hour level upon entry of a person into the area.
2. The area must be equipped with an automatic visual or audible alarm to warn the individual and the licensee, or a supervisor, of the entry into the area.
3. The area must be kept locked except at such time that entry into the area is required and positive control is maintained over entry to the area at such times.
4. In place of the three alternatives just listed, the licensee provides continuous direct or electronic surveillance that is capable of preventing unauthorized entry.

None of the control measures that might be adopted can be configured to restrict individuals from leaving the high radiation area.

Part 20 also defines a Very High Radiation area, where even higher levels of activity may be encountered, i.e., where levels in excess of 500 rads per hour (5 grays per hour) could be received at 1 meter from a sealed source that is used for irradiation of materials. Note that a one hour exposure to levels of this magnitude would exceed the LD_{50} for individuals not receiving prompt medical attention. Instead of the word **DANGER** on the cautionary sign, the words **GRAVE DANGER** are to be used. Holders of the license may apply to the NRC for additional control measures.

3. Areas with Possible Internal Exposures

The previous limits within a restricted area were based on external exposures to a major portion of the whole body. The most probable means of radioactive materials entering the body to cause an internal exposure is through inhalation.[*] Therefore, Section 20.1202 -1204 establishes requirements for spaces in which airborne radioactivity is present. Any area defined as an airborne radiation area is one in which airborne radioactive material in excess of the derived air concentrations (DACs) listed in Appendix B to Part 20 exists or, which averaged over the number of hours present in the area in a week a person is in the area without respiratory protection, the person could exceed 0.6% of the annual limit on intake (ALI) or 12 DAC-hours. Each area meeting or exceeding the required limits must be conspicuously posted with one or more signs with the radiation symbol and with the legend

---

[*] Radiation can also enter the body by ingestion, by absorption through the skin, or through a break in the skin, such as a cut. However, the last of these is more likely to be due to unusual circumstances rather than due to the presence of a continuing source.

CAUTION
(or DANGER)
AIRBORNE RADIOACTIVITY AREA

Normally, individuals without respiratory protection should not work in such areas. If feasible, engineering practices should be used to eliminate the need for individual respiratory protective devices. The NRC in 10 CFR Part20, Subpart H, section 1703-1704, defines the procedures to be followed should a respiratory protection program be required. A discussion on the usage of respiratory protection will also be found in Chapter 6, Section I.

In most research laboratory situations, it would be unusual to find an airborne radioactivity area on other than a short-term basis. If there is any possibility of an approach to the limits while working on an open bench, the use of radioactive materials should be restricted to a hood, or in a glove box or hot cell. In a facility with a broad license, it is recommended that at least some of the more active individuals using radioisotopes should be included in a bioassay program for the same reason that others wear a personal monitoring device, to ensure and document that no one is receiving an internal dose over the limits.

H. Material Control Procedures

The amount of materials that a licensee may possess at any time is established by the terms of its byproduct license. As long as the total holdings are maintained within those limits, licensees may order radioactive materials, maintain them in storage, use them, transfer them to other licensees, and dispose of waste amounts as long as they comply with all of the applicable regulations. Accurate records must be maintained for individual laboratories and for a facility as a whole where the facility has a broad license. For short-lived isotopes, there will be a continuing reduction of the amount on hand of a given material due to decay alone, while the distribution of the remainder of the material in use, storage, or transferred to waste will fluctuate continuously.

Written records should be maintained of the amounts involved in each of these processes, and it should be possible to account for nearly all of the material from the time it is received until disposed of as waste. Some uncertainty is certain to be introduced during actual experimentation, especially if there is a gaseous metabolic or combustion product released through the exhaust from a hood. There also will be small quantities that will be retained on the interior of vessels containing radioactive liquids that will escape into the sanitary system when the container is washed. However, it should be possible to estimate these types of losses with reasonable accuracy.

Facility records of receipt, transfer, use, and disposal should be maintained at each individual laboratory authorized to use radioisotopes. This is the responsibility of the senior individual in charge of the facility to whom has been assigned the internal equivalence of a radioisotope license. A technician may perform the actual record keeping, but the ultimate responsibility for the radioactive material in the facility belongs to the principal authorized user. Except for the removal of material from the storage container for use within the laboratory, the organization's radiation safety officer (RSO) should be involved in all of the transactions involving radioactive materials and should be able to detect any anomalies or disparities. It is the RSO's responsibility to maintain records for the overall inventory of the radioisotopes within the organization, and these records can be used to compare with or audit the records for each internal facility. Laboratory managers may be assured that an NRC inspector will check the records of some laboratories at a facility during each unannounced inspection, so records must be able to pass these inspections at any time.

It has already been noted that materials in unrestricted areas must be kept in secure storage when the user is not present. This requirement for security is applicable whenever the facility is left vacant and open, even for short periods for going to the restroom, the stockroom, checking mail, etc. If multiple keys are available to a facility with some being held by non-licensed personnel, it is not an acceptable alternative to lock the doors to the room while absent, if unsecured radioactive materials are accessible. In restricted areas, which always should be locked when they are not occupied, or when access of unauthorized persons can be controlled, radioactive material can, in principle, be kept in storage areas that are not necessarily locked. However, if substantial amounts of radioactive materials are to be kept on hand, an effective key-control program must be in place or other security measures taken to protect against loss of the material. Whenever radioactive material is removed from or returned to storage, the type of material transferred, the amount, the time, and the person doing the transfer should be entered into a log.

Perhaps the most common type of loss is due to waste material being disposed of as ordinary trash. Up to this stage, the radioactive material is normally maintained in containers labeled as containing radioactive materials as required by Part 20, but contaminated trash may appear the same as ordinary waste to a custodian. This can be due to laboratory workers inadvertently putting the material in, with, or near ordinary trash. Under such circumstances, a custodian is liable to take the material away due to failure to recognize the material as being different from other waste, even if the radioactive waste containers may have "radioactive waste" written on them, may be marked with the radiation symbol, or may be in distinctive colors. Unfortunately, an increasing percentage of the population is becoming functionally illiterate (a recent estimate is as high as 30%), individuals may be color-blind, and if the symbol were to be accidentally concealed, the custodian may very conscientiously remove the material as ordinary trash unless additional precautions are taken. These errors can be reduced if the internal locations of radioactive waste (also broken glass, waste, or surplus chemicals) and ordinary solid waste are well separated, with radioactive waste being placed in distinctively shaped containers, used for no other purpose. These additional measures have proven helpful, if all personnel have been fully informed and cooperate. It would be desirable for new custodial personnel to receive special training, not only to reduce the possibility of loss of material, but also to ensure that they are not unduly concerned about servicing a laboratory in which radioactive materials are used.

If it is suspected or known that radioactive material has been lost from a laboratory, either by direct knowledge of the disappearance or as inferred by an examination of the records, the organization's RSO must be notified immediately, and the NRC as well (by telephone) if the amount lost is greater than 1000 times the quantity specified in Appendix C to Part 20 under such circumstances that it appears to the licensee that an exposure could occur to persons in unrestricted areas. If the loss is smaller but still 10 times greater than that specified in Appendix C, then the NRC must be notified within 30 days if the material is still missing. If, after investigation the loss or possible theft is confirmed, it is always desirable to notify the NRC. It is required that this be done for significant losses, according to the requirements of Section 20.2201 as described in Section I.D.2. of this chapter. The situation should be thoroughly investigated, independent of the size of the loss, to attempt to recover the material or to find out its actual fate. If the loss appears to be deliberate, it may be necessary to solicit police assistance. Although the actual monetary loss may be small, individuals who are not knowledgeable about the potential hazards have suffered severe injuries from the possession of radioactive materials. If the loss is a strong-sealed source used for irradiation, it is very important, as will be discussed in a later section, that the material be recovered quickly.

It is rare for radioactive material to be taken deliberately; one of the most common loss mechanisms is loss as trash, as already discussed. If it is a reportable incident, the NRC must be informed immediately by telephone or by a written report filed within 30 days of the specifics as to what was lost and the circumstances concerning the disappearance of the material. They will wish to know, to the best of the licensee's knowledge, what happened to the material and the possible risks to individuals in unrestricted areas. The steps taken to recover the material will need to be in the report and probably most important, the steps the facility intends to take to ensure that a similar incident will not recur.

Even if the amounts lost are minimal and pose no danger to the public (typically the amounts at any given time in the waste in a laboratory are in this category), the RSO and the person in charge of the laboratory in which the loss occurred should be required to make a full report to the organization's RSC. The RSC should consider whether disciplinary mea-sures are needed. It should be determined if the loss was due to a failure by the laboratory to follow established policies or if the internal policies are sufficient to prevent future incidents of the same type. If the policies need to be modified, the committee should develop new requirements and inform all users within the organization. A mechanism that works well to reduce such occurrences is to send a safety note or memorandum to all authorized users, informing them of the incident (with the identity of the laboratory involved not identified) to avoid unnecessarily embarrassing them and require that all users acknowledge reading the safety note in a log.

In a well-run organization with sound policies guiding the use of radiation, incidents involving loss of radioactive material should occur very rarely. In any active research institution, especially in academic ones where graduate students and even faculty change frequently, some mistakes will occur. A fairly common defense in instances in which radioactive material is lost is that the amounts lost were small and no one will be harmed. The RSC, however, needs to determine if internal procedures or enforcement of procedures is the basic problem, not the quantity lost. Individuals who have made a mistake should not be treated harshly if they recognize that a mistake has been made and learn from it. On the other hand, the RSC should take very firm steps to correct the situation if a facility shows a continuing pattern of laxity in following sound procedures or, far worse, deliberate neglect of safety and regulatory policies.

1. Ordering and Receipt of Materials

A very sound procedure to follow in ordering and receiving radioactive materials is for all orders and receipt of radioactive materials to be handled by the radiation safety office. This has several advantages to the organization in terms of record maintenance, as discussed in the previous section, and to the individual facility since it provides a parallel set of records. Some suppliers of radioactive materials will provide a discount for volume purchases if all the orders go through a common ordering center, even if the materials are for different users. Virtually the only problems involve the occasional user who wishes to have a custom compound prepared with the radioisotope in a specific location within the compound. They may wish to discuss their requirements by telephone with the vendor. Even such telephone orders can be handled through the radiation safety office satisfactorily if the vendor, the user, and the radiation safety office responsible for ordering work together. Since many biological materials deteriorate with time and at normal temperatures, these are sent packed in dry ice to keep them cool. If the radiation safety office is aware of the anticipated delivery date, the staff can ensure that it will be delivered promptly.

A major safety function of the radiation safety office is to receive all packages of radio-isotopes and process them according to the requirements of Section 20.1906. Unfortunately, there are occasional errors in shipping and packages can be damaged. On at least two occasions at the author's institution, packages have been received which contained substan-tially larger amounts of material, with consequently higher radiation levels than should have been the case. Most packages of radioactive materials are shipped in what are called type A packages, which are only required to withstand normal transportation conditions without a loss of their contents. The amounts of the various nuclides that can be shipped in a type A package are given in 10 CFR 71, Appendix A. It is the responsibility of the distributor to properly package their shipments to conform to these limits.

The receiver of packages of radioactive materials must be prepared to receive them when delivered, or pick them up at the carrier's location at the time of arrival or promptly upon notification of the arrival of the material. Most packages are probably delivered to the purchaser's location by the many freight and parcel delivery services that are available. Some of these deliver 24 hours per day. Except in the largest organizations that conduct research on a 24-hour basis, however, most radiation safety groups only work a normal daytime schedule. At a university, for example, the only support operation likely to be functioning on a 24-hour basis is the security or police unit, who would not normally be knowledgeable about how to do the required checks on packages received. It is desirable for packages to be placed in a secure area until they can be inspected, so one alternative would be to place a *few* permanently mounted, lockable, and reasonably shielded boxes in the security area in which packages could be placed until the following morning.

Packages are to be checked for external radiation levels, including loose surface contamination and direct radiation, within 3 hours of receipt during the day or within 3 hours after the beginning of the next working day if received after normal working hours. Levels that would require immediate notification of the NRC would be 200 mrem per hour (2 mSv per hour) on the surface or 10 mrem per hour (0.1 mSv per hour) 3 feet from the surface of the package. A limit of 22 dpm/cm^2 is set as well for loose contamination on the surface.

The radiation safety officer normally should not open individual packages (although some do so) unless there is reason to suspect that the material contained within the package is not consistent with the packing list, but rather should deliver the material to the user or have the users pick up their packages. The users should exercise due care in opening packages to reduce exposures to the persons performing this task. The precautions that need to be taken will be dependent upon the nature of the material, the anticipated radiation levels, or the potential for loose material to become airborne should the package not be tightly sealed. If there is any risk of personnel becoming contaminated while opening a package, it should be done in a radioisotope hood, with the package placed in a pan large enough to retain any spilled material, and with the pan sitting on a layer of plastic-backed absorbent paper. The employee should be double gloved and should be wearing a lab coat. For radioactive materials with high specific activity or with a small ALI, it could be desirable to open suspect packages in a glove box providing complete containment. If loose contamination were found on the surface of the package, a HEPA filtered respirator would be strongly recommended for persons handling it besides the other protective clothing. The need for shielding would depend upon the radiation characteristics of the material.

Any package found to be damaged such that more than the allowable limits of contamination is released must be reported at once to the NRC by telephone and telegraph mail-gram (or facsimile) and to the final delivering carrier. It is recommended to report any significant problems with the package, whatever the amounts of radioactive materials.

I. Operations

"Operations" is an all-inclusive term encompassing the program in which radioactive materials are used, and the support programs necessary to allow the materials to be used safely and in compliance with all regulations. Many of the latter requirements have been discussed in some detail, including, in some instances, the means by which compliance or safety can be enhanced, in preceding sections.

Although a direct relationship between low level exposures and adverse biological effects has not been shown conclusively, it is assumed, as a conservative premise, that many effects that are known to be caused by high doses will occur on a statistical basis to some of a large population exposed to lower levels. Stochastic effects that are based on probabilities, such as induction of cancer or genetic damage, are assumed to have no threshold, while non-stochastic effects, such as the causing of cataracts by radiation, are assumed to be related to the intensity of the exposure and are assumed to have a threshold. The stochastic effects are often the ones that limit the permissible exposures in using radioisotopes.

There is a wide variation in the susceptibility of individuals to radiation as with most other agents that can cause harm to humans. Some individuals are much more resistant than others, while there are individuals that are hypersensitive. There are many factors that can affect the damage done to an individual. Some of these may be sex linked and some dependent upon the age of the individual. Some may depend on exposure to other agents (synergistic effects), and others will depend upon the health of the individual at the time of an exposure. This section will not attempt to pursue these modifying factors but will assume, as did the sections in which chemical agents were discussed, that the workers radiation are normal, healthy individuals in their average working span, unless otherwise specified.

The following sections will be concerned with means to reduce exposures while using radioisotopes.

1. Reduction of Exposures, ALARA Program

The goal of an ALARA program is to make personnel exposures to radiation as low as reasonably achievable in the context of reasonable cost, technical practicality, and cost-benefit considerations. In such a program, each aspect of an operation is evaluated to decide if there are alternatives or means by which the procedure can be modified to reduce exposures.

One of the first considerations is to select the radionuclide that is to be used in the research. The half-life, the type of radiation emitted, the energies of the emissions, and the chemical properties of the radioisotopes are factors to be evaluated.

Control of the external hazard is based on manipulation of three primary variables, time, distance, and shielding.

Control of hazards for internal exposures is more complex. The properties of the materials must be considered, but the experimental techniques to be used, the design and construction of the facility the way it is equipped, and the use of personal protective equipment are all factors that must be considered in establishing an effective ALARA program.

a. Selection of Radioisotopes

To select the appropriate radioisotope to use in a given research program, several factors must be considered. If the choice is strictly based on relative safety involving nuclear properties, a beta emitter generally would be preferable to a gamma emitter, since gammas typically are much more penetrating than betas. Although alphas are much less penetrating than betas, an alpha emitter normally would not be selected over a beta emitter because, if ingested, among other reasons, the damage caused by an alpha as it moves through matter is so much greater than for a beta. However, the primary reason is that alpha emitters are generally not found among the elements usefully employed in biological research. For the same type of emitter, the isotope with the lower energy emission should normally be chosen. When half-lives are considered, a half-life as short as practicable would be preferable, so that the problem of disposal of the nuclear waste could be solved by simply allowing the radiation in the waste products to decay rather than having to ship them away for burial. A second consideration is that if the materials may be ingested by humans or used as a diagnostic aid, the total dose to an individual could be much less if the isotope used has a short half-life.

The desired chemical properties and costs may dictate the isotope chosen. Some isotopes are much more expensive than others, so for economic reasons the less expensive choice might be preferable. A custom-prepared, labeled compound is usually much more expensive

Table 5.6 Properties of Some Selected Nuclides

Nuclide	Half-Life	Beta(s) (MeV)	Gamma(s) (MeV)
^{45}Ca	163 days	0.257	0.01 24
^{109}Cd	453 days	ϵ	0.088
^{36}Cl	3.0×10^5 years	0.709	—
^{60}Co	5.27 years	0.318	1.173, 1.332
^{51}Cr	27.7 days	ϵ	0.320
^{137}Cs	30.2 years	0.512, 1.173	0.6616
^{59}Fe	44.6 days	0.467, 0.273	1.099, 1.292
^{203}Hg	46.6 days	0.212	0.279
^{125}I	59.7 days	ϵ	0.0355
^{131}I	8.04 days	0.606, others	0.364, others
^{54}Mn	312.5 days	ϵ	0.835
^{22}Na	2.6 years	ϵ, β^+ 0.545	1.27, 0.511 (annihilation gammas)
^{63}Ni	100 years	0.0659	—
^{65}Zn	243.8 days	ϵ, β^+ 0.325	1.116, 0.511 (annihilation gammas)

than a commercially available labeled compound, so that unless the safer isotope is available in a previously prepared compound, economic reasons would probably indicate using what is readily available. Chemical properties may dictate the choice of materials because of the risk to personnel ingesting them. An examination of Appendix B, 10 CFR Part 20, shows wide variation in the ALIs or DACs, even among those that have similar chemical properties. A suitable isotope with the greatest ALI or DAC should be selected.

Carbon-14 and tritium are two of the most frequently used radioisotopes since they can readily be incorporated into organic compounds. Unfortunately, ^{14}C has a very long half-life, approximately 5730 years, while tritium has a much shorter one, 12.3 years. Even the latter is long enough that it represents a disposal problem. Otherwise, these two isotopes have excellent radiological properties. The beta emitted by carbon has an energy of only 0.156 MeV so that its range in tissue is less than 0.3 mm and about 10 inches in air. The beta emitted by tritium is much weaker (0.0186 MeV) so that its range in tissue is only about 0.006 mm and about one fifth of an inch in air. Clearly, both isotopes would pose little risk as far as external exposures are concerned, although both can represent an internal problem. Tritium, for example, can easily become part of a water molecule that would be treated biologically by the body virtually the same as would any other water molecule. Other popular beta-emitting isotopes are ^{35}S, which emits a 0.167 MeV beta and has an 87.2-day half-life, and ^{32}P, which emits a 1.71 MeV beta and has a 14.3-day half-life. As noted in an earlier section, the high-energy beta from ^{32}P can cause a substantial radiation exposure problem due to emission of bremsstrahlung radiation, if the betas are allowed to interact with materials of significant atomic numbers. Another radioisotope of phosphorus can be used, ^{33}P, which has a longer half-life, 25.2 days, but a much lower energy beta, 0.248 MeV.

Some of the other more commonly used radioactive isotopes used in research are listed in Table 5.6.

b. Shielding

Shielding is usually the first protective measure that comes to mind when considering radiation protection. In reviewing properties of radioactive particles, the ranges or penetrating capacity of the various types were briefly discussed to provide some measure of understanding of their characteristics. These properties will now be considered to decide how best to provide shielding. The discussion will be limited to shielding against beta and gamma radiation.

i. Betas

The range of betas in any material can be calculated by the following two equations:

For energies $0.01 \leq$ Energy (MeV) ≤ 2.5

$$R = 412E^{(1.265-0.0954 \ln E)} \tag{12}$$

and for energies $E \geq 2.5$ MeV

$$R = 530E - 106 \tag{13}$$

The range, R, in these equations is given in mg/cm^2. This can be converted into centimeters in a given material by dividing the ranges given by these equations by the density of the material in mg/cm^3. The range R can be considered as the area density of a material, in contrast to the usual volume density. The range of the 1.71 MeV beta from ^{32}P in mg/cm^2 calculated from Equation 12 is 785 mg/cm^2. Tissue and many plastic materials have a density near 1, so this beta would have a range of about 0.8 cm, or about 0.3 inches in such materials. In lead the range would be 11.34 times less or 0.7 mm (0.027 inches). However, it was mentioned earlier that bremsstrahlung would be a major problem for such an energetic beta emitter if a high atomic number absorber were used. Therefore, a low Z material should be used for shielding of pure beta emitters. A piece of plastic 3/8 of an inch thick would shield against nearly all common betas used as radioisotopes in research, since there are few commercially available beta emitters that emit more energetic betas (quite a few neutron-activated nuclides emit higher energy betas, but these are not often used as radioisotope tracer materials). Research personnel who work with ^{32}P can reduce hand exposures from bremsstrahlung by slipping thick-walled plastic tubing over test tubes and other containers.

ii. Gammas

Equation 6, repeated below as Equation 14, can be used to calculate the attenuation of a monoenergetic narrow beam of gamma rays by a shielding material.

$$I = I_0 e^{-\mu x} \tag{14}$$

This equation, as noted earlier, will not take into account gammas scattered into the shielded area by Compton interactions by the shield, and by the walls (and floor and ceiling, if the gammas are not collimated so as to avoid these surfaces). Figure 5.7 shows a near optimal shielding geometry in which a shield just subtends a sufficient solid angle to completely block the direct rays from the source. This geometry reduces in-scatter. In Figure 5.8, a more typical geometry is shown where a worker is using a source in a fume hood. In this geometry the shielding would not be quite as effective, but with the low activities usually employed, the doses from scattered radiation would tend to be very small.

For work with gamma emitters and low energy betas, lead is a good shield. It is dense, 11.4 g/cm^3, and has a high atomic number, 82. It is relatively cheap and is more effective than any other material that is comparably priced. Iron or steel is often used as well although its lower density and atomic number requires more shielding. One advantage of using steel rods is that one can quickly fabricate a custom shield for a given configuration. The mass attenuation coefficient for lead is shown in Figure 5.9. The coefficients read from the graph can be converted to linear coefficients by multiplying them by the density expressed in g/cm^3.

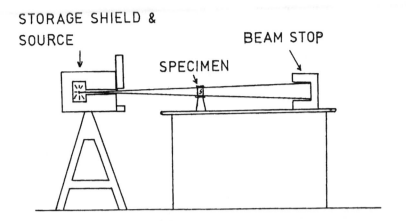

Figure 5.7 Shielding arrangement designed to minimize scatter radiation from shielding.

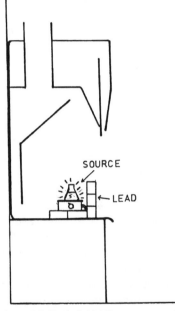

Figure 5.8 Typical shielding arrangement for hood operations using radio active materials.

If the absorption coefficients for lead at 0.5, 1.0, and 1.5 MeV are used to compute the thickness of lead required to attenuate a beam of gammas by factors of 2 and 10, the half-value thicknesses would be 0.4, 1.1, and 1.5 cm., respectively, and for the tenth-value thicknesses, 1.25, 3.5, and 5.0cm for the three energies. Thus, a lead brick 2 inches thick would reduce the intensity of the gammas from most of the common radioisotopes used in research by a factor of 10 or more, which is often enough for the average source strengths used in research. Shields should, however, be built of at least two layers of bricks so that the bricks can overlap to prevent radiation streaming in a direct path through a seam.

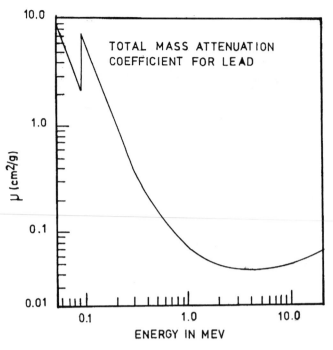

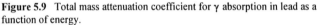

Figure 5.9   Total mass attenuation coefficient for γ absorption in lead as a function of energy.

Occasionally, individuals will set up a shield to protect themselves and forget that radiation from an open source extends over 4π steradians. Levels through the top, bottom, sides, and rear of the work area, and perhaps in the next room, may be excessive.

Extremely high-level gamma radiation operations are conducted in a special shielded enclosure called a "hot cell." Usually the shielding in such units is provided by making the enclosure of high density cast concrete. The thickness needed to reduce the radiation to levels acceptable for occupational exposures would depend upon the strength of the source. Often, the top of a hot cell is covered as well as the sides to avoid "sky-shine" radiation, i.e., radiation scattered from the air and ceiling due to radiation from the source within the cell. The researcher would perform the operations required in the cell using mechanically or electrically coupled manipulators. Vision typically would be through a thick leaded glass window, often doped with a material such as cerium to reduce the tendency of glass to discolor upon exposure to radiation, or by the use of closed-circuit television.

c. Distance

Distance is an effective means of reducing the exposure to radiation. The radiation level from a point source decreases proportionally to I/R^2, while from a point near an extended source, such as a wall, the level falls off in a more complicated manner, since the distance from each point on the wall to the measuring location will vary and the contribution of each point must be considered. There are many excellent references in which specific geometrical sources are discussed, but, unfortunately, most "real world" situations often do not lend themselves to simple mathematical treatment. If the distance to the source is large compared to the size of the source, then the approximation of a point source is reasonably accurate.

Many radionuclides emit more than one gamma and, for many, the gamma spectrum is extremely complex. For each original nuclear disintegration, which defines the number of curies or becquerels represented by the source, a specific gamma will occur a certain fraction,

f_i, of the time. If a gamma is emitted for every original nuclear decay then f_1 for that gamma would be 1. With this concept in mind, there are two simple expressions that can be used to calculate the dose at a distance R from a point source. The difference in the two expressions is simply a matter of units.

To compute the dose, D, in μSv per hour, the expression is

$$D = \frac{\sum_i Mf_i E_i}{6R^2} \tag{15}$$

Here, M is the source strength in MBq, E_i is the energy of the ith gamma in MeV and R is the distance in meters.

The same expression, expressed in the more traditional British units, is

$$D = 6\frac{\sum_i Cf_i E_i}{R^2} \tag{16}$$

Here, C is the source strength in Curies, E is the energy in Mev, R is in feet, and D is given in rads per hour.

It should always be kept in mind that even from a point source, both expressions give only the dose due to the *direct* radiation from the source. They do not include the radiation scattered from objects in the room, the walls, floors, or ceiling or the workbench on which the source is sitting. It also should be kept in mind that the radiation dose can be reduced further by placing a shield in the path of the radiation. If this is done, the levels calculated by either Equations 15 or 16 would be reduced by the factor $e^{-\mu x}$ where μ would be the linear attenuation coefficient for the shield material at the energy, E_i, of the gamma. The equation would have to be calculated for each gamma and the contribution of all the gammas would have to be summed. The result would be the *minimum* dose that would be received from the source. The actual dose would be higher due to in-scatter.

Because the exposure rate does go down rapidly with distance, sources should normally be handled with tongs rather than directly. Whenever a source is to be transferred, only an individual trained to handle the equipment and the source should do the transfer with perhaps an assistant nearby and the transfer should be done as expeditiously as possible.

When thinking of distance as a protective measure, the mental image is of a source with the user standing back from it. However, one way of thinking of contamination on the skin is of there being virtually a zero distance separating the irradiating material and the tissue, so that even relatively small amounts of radioactive material could eventually cause a significant local exposure, even if the material did not penetrate or permeate through the skin. To prevent contamination of the skin, gloves and protective covering, such as a laboratory coat, should always be worn when working with material that is not a single solid piece, such as a liquid or a powder that could be spilled or otherwise contaminate the worker and the experimental area.

d. Time

Time is perhaps the most easily achieved means of reducing the dose received. If the radiation level remains constant, reducing the exposure time reduces the dose received by the same proportion that the time is shortened. If the radioactivity of the material in use decays during the interval, the reduction in exposure would be even greater. This should be kept in mind when developing experimental procedures. Personnel should leave sources within their containers as much of the time as possible while working.

Another way in which time can be used effectively to at least reduce the level of exposure to a single individual is to share the exposure time among more individuals. Statistically, for those events that depend on the probability of an event being caused by exposure to individuals, i.e., stochastic events, little has been gained if the exposure to the population of the entire group of persons is considered, but it does reduce the probability of a problem developing for a specific individual and reduces the non-stochastic effects that depend upon the dose received by the person.

e. Quantity

Modern techniques often allow work with much smaller amounts of material than was possible only a few years ago. When working with radioactive materials, full advantage should be taken of any efficient and effective procedure for using less of the radioactive substance, not only at any given time, but over the course of the research. A carefully constructed written research plan and hazard analysis, in which the amounts of materials needed for each procedure are calculated, will help the minimization process.

f. Example of Time and Distance

Small sealed sources are frequently used in laboratories. If a source is dropped from a storage container, an employee may pick it up in his hands to return it to the storage container, handling it in direct contact with the hands for perhaps 1 minute. Even this short interval with weak sources can produce significant doses to the hands. Data have been published showing that at or very near the surface of a steel-clad sealed gamma source, electrons from the source cladding caused by interactions of the gammas with the material can contribute between 25 and 45% of the dose to the hands. The primary exposure to the hands still is due to direct radiation from the source, which is high because of the small distance between the source and the tissue. The tissue of the hands is estimated to absorb about 5 to 10% of the gamma energy per cm. With these data in mind, if a 1 Ci source of ^{60}Co is 1/4 of an inch in diameter, and is clad with 1/32 of an inch of 304 stainless steel, the approximate surface dose rate per minute for the source would be 2075 rads per minute. In less than 2 1/4 seconds, the hand of a person picking up the source would receive, locally, the entire year's permissible exposure of 75 rads. The rate falls off rapidly with depth in the tissue; at 1 cm deep, the rate is estimated to be 114 rads per minute and at 3 cm, 16 rads per minute. A reduction of any one of the three contributing effects, the quantity of radioactive material in the source, the time it is handled, or increasing the distance from the source will diminish the harm done to the individual.

Although this may be dismissed as unlikely, there have been a number of cases in which persons unfamiliar with what they were handling have picked up sources and placed them in their pockets. Often, the persons have been custodians or other support personnel. In some cases, many persons outside laboratory facilities have been exposed to high levels of radiation, because pellets of material from irradiation sources that were no longer being used were disposed of improperly and wound up in trash. Deaths have resulted from such instances. Not all high exposure incidents have been among the untrained and poorly educated. There have been instances of laboratory personnel who have had substantial exposures because of failure to exercise sufficient care in working with radioactive materials.

In several cases, the workers were exposed to substantial radiation levels due to exposed sources of which they were unaware because an interlock had failed. Safety features that depend upon a single micro-switch do not provide adequate protection. Some persons have died and a number have been seriously injured from unsuspected exposed sources. It should be noted that the exposures in these cases have been in the several hundreds of rads (or rems) range or greater, and sometimes localized exposures have been in the many tens of thousands of rad range, with the person surviving, but with extremely severe damage over the parts of the body exposed to these extraordinarily high levels.

In summary, to minimize external exposure:

1. Work with the safest isotope appropriate to achieve the desired results.
2. Use the smallest amount of radioactive material possible, consistent with the requirements of the experimental program.
3. Reduce the time exposed to the radiation.
4. Use shielding wherever possible.
5. Take advantage of distance as much as possible. Use extension tools to avoid direct handling of the material.
6. Make sure that any protective devices are fully functional. A checklist that must be followed before each use is a very inexpensive safety device.
7. Monitor the radiation field from the radioactive material with an appropriate instrument.

g. Internal Dose Limiting

There are, as with chemicals, only four basic means for radiation to enter the body: inhalation, ingestion, absorption through the skin, or through a break in the skin. The basic means of control of internal exposure is to restrict the entry into the body by any of these routes. However, other factors will affect the consequences of radioactive material that has succeeded in bypassing the defenses to prevent entry into the body. The form of the material is important. Is it in a soluble or insoluble form? Insoluble particles carried into the lungs following inhalation may remain there for long intervals. Soluble materials may be absorbed into body fluids. If the material is incorporated in a form in which it is likely to be metabolized, it is more likely to reach a critical organ if ingested than otherwise. The size of the particle is critical, if it is an airborne contaminant, in determining in what part of the respiratory system the material will be deposited. Part of the material that is not exhaled may be swallowed.

The chemical properties of the material are important in determining what organs are likely to be involved. In some cases, such as tritium, the tissue of the entire body is likely to be involved, since often the chemical form of released tritium will be as HTO, and the behavior within the body will be the same as that of ordinary water. Other materials, such as strontium, are bone seekers if in soluble form, while iodine in soluble form will most likely go to the thyroid.

i. Entry Through the Skin

Entry through the skin, either per cutaneously or through a break, is normally the least likely to occur and is the most easily prevented. Usually, entry would follow an inadvertent spill in which material reached the skin by the worker handling contaminated objects without using proper equipment, or by failure of protective items of apparel. However, in an exposure of an otherwise unprotected worker to water vapor containing an appreciable amount of HTO, a significant fraction of the total exposure could be through skin absorption (normally assumed to be one third of the total, the remainder being due to inhalation).

Prevention of contamination of items likely to be handled is the first step in prevention of skin exposure. For example, the work surface should be covered with a layer of plastic-backed absorbent paper to ensure that the permanent work surface is not contaminated by spilled materials, by aerosols generated in many different laboratory procedures, or even by vapors that might diffuse from a container, since vapors of most chemicals are heavier than air. Aerosols are perhaps the most likely source of local contamination as the droplets settle.

Aerosols can come from opening a centrifuge in which a tube has been broken, use of a blender or sonicator, the last drop falling from the end of a pipette, or even from dragging a wire across the surface of an uneven agar surface in a petri dish. The most obvious solution to most aerosol problems is doing such procedures in a fume hood, but some procedures still would contaminate the hands and arms. Frequent surveys for contamination of work surface and hands should help avoid contamination from aerosols. Setting up the apparatus in a shallow container, such as an inexpensive plastic or aluminum pan that can be obtained from a department store, large enough to contain all the spilled liquid in use at the time, is a simple way to confine the consequences of an accident to items that are disposable rather than ones that would have to be decontaminated. Covering some items of equipment with the plastic-backed absorbent paper, or with aluminum foil for warm surfaces, may prove profitable for some levels of work using materials of high toxicity or amounts of radioactive materials in excess of the typical laboratory application, or in procedures in which the possibility of contamination is higher than normal. Frequent checks with survey instruments of the apparatus in use and the immediate work area near the apparatus or smear or wipe tests of work surfaces and equipment being handled are good means of limiting contamination before it becomes widespread. Checks of the survey instrument itself and such areas as faucet handles on sinks used in the work are often overlooked in contamination evaluations, but if the hands are contaminated, any items that may have been handled need to be checked.

Protective gloves and laboratory coats or coveralls are the first level of protective gear that should be adopted in laboratory operations involving the use of radioisotopes. Wraparound lab coats may be a better choice than those that button up the front to avoid material passing through the front opening. The gloves should be chosen to be resistant to the chemicals of the materials in use. The dexterity allowed by the gloves also is a factor. Where the possibility of damage to the gloves is significant or the properties of the radioactive materials are especially dangerous if contact occurs, wearing two pairs of gloves should be considered. Many facilities make this a standard requirement instead of optional. Gloves used for protection while working with radioactive material should be discarded as radioactive waste after use. The cuffs on gloves should be long enough to allow sealing (with duct tape or the equivalent) to the sleeves of a garment worn to provide body protection if needed.

Any cuts or abrasions on the hands or other exposed skin areas on the forearms should be covered with a waterproof bandage while actively working with radioactive materials.

Cotton laboratory coats provide reasonable protection for the body of the worker in most laboratory uses of radioisotopes. However, if they do become contaminated, they must be washed to decontaminate them. Laundering facilities appropriate for cleaning of radioactive materials may not be available and consideration of disposable protective clothing made of materials such as Tyvek™ may be desirable. These garments are not washable, but if a thorough survey shows that they are free of contamination and if they are handled with reasonable care, they can be worn several times before being discarded. However, they are inexpensive enough to allow disposal whenever necessary if they do become contaminated. Disposable garments should have both welded and sewn seams to provide assurance that the seams will not split during use (wearing a garment a size or two larger than is actually needed also helps avoid this). Body contamination surveys should be made of the outer protective garment before removal and of the street clothes afterward. Checking the outerwear is important to find out if the procedures used generate contamination and to avoid contamination from a discarded garment. If the protective clothing is found to be contaminated while following routine procedures, the procedures need to be revised to eliminate the problem. The clothing is a final barrier, not the first line of defense.

Hands should be carefully washed at the conclusion of any operation in which radioactive materials are handled in procedures which offer any opportunity for contamination. If the skin has become contaminated, the affected area should be washed with tepid water and soap. The use of a soft brush aids in the removal of material on the surface. Harsh or abrasive soap should not be used. After washing for a few minutes, the contaminated area should be dried and checked for contamination. If the area is not free of radioactivity, rewashing can be tried. A mild detergent also can be tried, but repeated or prolonged application of detergents to an area may damage the skin and increase the likelihood that surface contamination will penetrate it. Organic solvents, acid or alkaline solutions should not be used since they will increase the chances of skin penetration. Difficult-to-clean areas should be checked with extra care.

Any contamination that cannot be removed by the above procedures should be reviewed by a radiation safety specialist. Further measures that could cause abrasion or injury to the skin should be done only under the advice and supervision of a physician.

ii. Ingestion

The inadvertent ingestion of radioactive material is often the consequence of transferral of radioactive material from the hands to the mouth while eating or drinking. No ingesting of food or drink, chewing gum, smoking, using smokeless tobacco products, or application of cosmetics should be allowed in the active work area of a laboratory in which radioactive materials are used. Before leaving the laboratory after working with such materials, as noted in the previous section, the hands should always be carefully washed and surveyed to ensure that no material is on the hands. If gloves and protective outerwear have been worn, the gloves should be discarded into a plastic container or double plastic bag, and the outerwear left at the entrance to the work area after ensuring that it is not contaminated. Washing the hands should take place after handling the clothing. It should be possible to effectively prevent ingestion of radioactive materials using reasonable care to prevent transfer of any activity to food and drink, or any other item that might be put in the mouth such as a cigarette or gum.

iii. Inhalation of Radioactive Materials

Airborne radioactivity is the most likely means by which radioactive materials can enter the body, resulting in an internal exposure. Most mechanisms by which either skin contamination or ingestion occurs involve contact of the user with the active material in a fixed location. However, airborne materials are not constrained to a given space, and unless the work done is confined to a glove box, they can fill the entire volume of the laboratory, surround the occupants of the room, and contaminate the air they breathe. The radioactive material also can be discharged from the laboratory by the building exhaust system or through the fume hoods. It is this last possibility that was the concern addressed by the EPA restrictions on air emissions, responsibility for which is now that of the NRC.

The size of particles deposited in the deep respiratory tract reaches a maximum at 1 to 2 microns (μ). Unfortunately, particles in this size range tend to remain suspended in air for relatively long periods of time. The gravitational settling rate for a 1 μ particle with a density of one is about 0.0035 cm per second. It would require about 12 hours for such a particle to settle about 5 feet under the influence of gravity, if allowed to do so without being disturbed. Thus, the particles of airborne radioactive contamination which have the greatest potential for deposit in the deep respiratory tract are among those that would remain airborne long enough for a worker to have an opportunity to inhale them.

It has been estimated that about 25% of all soluble particles inhaled are exhaled; about the same fraction are dissolved and absorbed into the body fluids. The remaining 50% are estimated to be deposited in the upper respiratory tract or swallowed within 24 hours after intake. For insoluble particles, about seven-eighths are exhaled or deposited in either the upper or deep respiratory tract, but swallowed within 24 hours after intake. However, the remaining one eighth is assumed (lacking specific biological data appropriate for the material in question) to be deposited and retained in the deep respiratory tract for 120 days. After being swallowed, the radioactive particles may either descend through the gastrointestinal system and be excreted, or may pass into the body. Varying portions may be excreted through the kidneys.

Engineering controls are the preferred means of keeping the DACs below the limits in 10 CFR Part 20, Appendix B. Areas in which radioactive materials are used should be ventilated with unrecirculated air. If the workplace is designed properly so that air movement is away from the workers breathing zone, work in which airborne releases are possible are done in an effective fume hood, and the quantities of activity are typical of most research programs that do not require access control, personnel exposures should be minimal. An air exchange of a minimum 6 to up to 12 air changes per hour should be sufficient to provide the needed environmental control for normal operations. The preferred hood type would be a good quality standard or bypass hood rather than an add-air type, since the latter type, with two airstreams and a more complicated design, has been found on occasion to be more prone to spillage if not used and maintained properly. An airflow velocity through the face of the hood of approximately 125 fpm (38 rn/min) would be a good design target. The hood should be equipped with an alarm to warn if the airflow through the face were to fall below 100 fpm (30.5 m/min). Making sure that any source of radioactive gaseous or vapor effluent is at least 8 inches (20 cm) from the entrance to the hood and keeping the sash down approximately halfway will further significantly improve the ability of a hood to capture and confine airborne hazardous materials.

If for a valid reason, full control of the levels of airborne contaminants cannot be maintained within the legal limits, it is permissible to use respiratory protection under some conditions. These are found in the current Part 20, Subpart H in Section 20.1703 and in Appendix A to Part 20. A key provision in 20.1703(a)(3) is for the licensee to have a formal written respiratory protection program.

There are several different types of respirators that provide various levels of protection. The various types and their characteristics will be discussed briefly below and more extensively in Chapter 6. However, their function is to reduce the level of radioactive material in the air being breathed by the user as far below the acceptable levels as is reasonably achievable.

Respirators fall into two primary classes, those in which air is supplied and those in which the air is purified by being passed through filters. The most common type of nonair-supplied units used for radioactive materials are HEPA filter-equipped respirators since most of the problem materials involve particulates. If solvents or other chemicals are involved, appropriate additional filters would be necessary. Supplied-air units are used where greater protection factors are needed. Supplied-air respirators themselves fall into several different classes. Units in which air is supplied on demand, and which do not provide a positive pressure inside the face mask with respect to the ambient air are the least effective because contaminants can enter should the seal between the mask and the face fail. These are rated to provide a protection factor of 5. Self-contained units in which air is supplied continuously or in which the air is always at positive pressure are rated highest and provide protection factors up to 10,000 although the usual rating is somewhat less. There are two other types, one in which the air is recirculated internally and purified chemically but is always at a positive pressure with respect to the outside, and another in which air is supplied to a hood that does not fit snugly to the face, but has a skirt that comes down over the neck and shoulders. These latter two types provide actual protection factors of 5000 and up to 1000, respectively, but are rated more conservatively. Self-contained units have a severe limitation in that they provide air for only a limited period. The most popular SCUBA type, with a 30-minute air tank, may provide air for only 15 minutes if the wearer is under heavy physical stress. Larger versions of the air-recirculating and air-purifying types can provide air for up to 4 hours, but even these are not suitable for continuous wearing.

Respirators that purify the air by filtering require a snug fit to the face to ensure that they are effective. Unfortunately, as a person works, the contact may be temporarily broken and the protection provided by the mask diminished. Facial hair will prevent a good seal, so that bearded workers cannot be allowed to use respirators requiring a good facial seal for protection. Some facial types are very hard to fit successfully although there are increasingly more models available which allow different types of features to be fitted. More models are also becoming available with a mask made of hypoallergenic materials that permit more persons to wear them without suffering irritation.

A major requirement is that the individual be physically able to wear a respirator. It takes some effort to breathe through a cartridge respirator, and a person with emphysema or some other breathing impairment would not be able to wear such a unit for extended periods. Passing a physical examination that includes a pulmonary function test is required of employees who would be expected to wear a respirator in doing some tasks assigned to them. The employer is required to have a comprehensive respirator program if advantage is taken of this method of achieving compliance with the permissible air limits of airborne concentrations of radioactive materials for the employees.

Although respiratory protection does offer an alternative to engineering controls to provide suitable breathing air quality, the use of respiratory protection is more commonly used for providing suitable air for temporary or emergency work situations.

In summary, to reduce internal exposure:

1. Do a hazard analysis on all procedures, including routine ones, for failure modes where releases of materials, contamination, or exposures could occur.
2. Use the isotope that would be the least dangerous if taken into the body.
3. Use the smallest amount of radioactive material possible consistent with the requirements of the experimental program.
4. Reduce the time in which an airbome exposure could occur.
5. Do the hands-on work in a radiological fume hood for most radioisotope research, or in a glove box if the possibility of airbome exposure mandates it because of the type, the amount, or the form of the material.
6. Take measures to reduce contamination with which it would be possible to come into contact, e.g., using absorbent plastic-backed paper to line the work surface, procedures to reduce the production of aerosols, and a tray underneath the experimental apparatus to catch any spills.
7. Wear gloves (two pairs when working with material in solvents that could cause a glove material to soften or weaken, or where a single pair might be breached due to abrasion or puncture).
8. Use protective outerwear that can be discarded if contaminated.
9. Do not eat, drink, chew gum, apply cosmetics, smoke, or do anything else that could lead to ingestion of radioactive materials while in the area in which radioactive materials are used.
10. Wash your hands after any use of radioactive material, before leaving the work area, and before eating, drinking, smoking, or any other activity in which radioactive material could be ingested.
11. Supplement the protection provided by the fume hood with the use of respiratory protection, if needed, to achieve compliance with airbome concentration limits.
12. Cover any cut with a waterproof bandage. If a minor cut occurs while working, wash the wound out, allowing it to bleed freely for a while. Then check for contamination.

If uncontaminated, cover with a waterproof bandage. If still contaminated, immediately contact radiation safety personnel for assistance.

13. After work is completed, survey the work area, any equipment handled, clothing, and any other potentially contaminated items or surfaces. Clean up any contamination found.

iv. Limitation of Dose to the Fetus

The revised Part 20.1208 places the responsibility of reducing the exposure to a fetus/embryo on both the employer and the women carrying the fetus/embryo. The woman's formal responsibility is met by the woman *declaring,* i.e., notifying her employee that she is pregnant. It is then the responsibility of the employer to limit the total effective dose (including external and internal exposure, if applicable) to the fetus/embryo to 0.5 mrem (5 mSv) during the pregnancy.

It is generally accepted that rapidly dividing cells are unusually radiosensitive, which, of course, is the situation that exists as the embryo develops. The early stages of fetal development, especially between the 10th and the 17th week, appear to be a critical time during which the fetus is especially susceptible to radiation damage. Unfortunately, during these first few critical months of a pregnancy, some women might not realize that they are pregnant and should begin to take precautions to limit the exposure to the developing fetus. However, if pregnancy is suspected, to avoid any unusually high exposures at potentially critical stages of development, the recommendation is that the exposure be spread uniformly throughout the prenatal period, or an average of 0.054 rem (0.54 mSv) per month.

Because of the uncertainty in the early critical stages of pregnancy of whether an individual is pregnant or not, a restriction on all fertile women to an annual dose of 0.5 rem (5 mSv) might be considered desirable. Since only a limited number of women of fertile age are pregnant at any given time, this could cause discrimination against hiring women and unnecessarily affect their employment opportunities. Radiation workers, on average, receive a dose equivalent of less than 0.5 rem per year, so that it does not appear justifiable to impose such a limit on fertile women as a class. It would appear reasonable for women to continue to exercise reasonable care in reducing the radiation exposure of any unborn child that they might be unknowingly carrying, especially if they are actively trying to have a child.

Part 20, Section 20.1208 sets limits to the occupational exposure of a woman who has voluntarily *declared* or informed her employer that she is pregnant. These limits include both external and internal exposures. Many materials do pass from the women to the unborn child through the placenta. Sometimes the fetus receives a disproportionate share. In addition, the fetus may be exposed to radiation from materials in the organs of the woman's body. The rule requires that the licensee ensure that the embryo/fetus, in such a case, does not receive an effective dose more than 0.5 rem (0.5 mSv) during the entire pregnancy. The effective dose is the sum of the deep dose equivalent to the declared pregnant woman and the dose to the fetus from radionuclides in the fetus and radionuclides in the declared pregnant woman. As noted in Section II.D.2 of this chapter, because of the uncertainty in establishing the actuality of the pregnancy, some women may not inform the licensee that they are pregnant until the fetus has already received a dose of 0.5 rem (0.005 Sv) or more. In such a case the licensee is enjoined from allowing the fetus to receive more than 0.05 rem (0.5 mSv) during the remainder of the pregnancy.

Compliance with the limitations on exposure to the embryo/fetus should not involve an economic penalty on the woman, nor should it jeopardize the woman's job security or employment opportunities. Additional guidance on compliance with the regulation on fetal exposure can be found in the latest revision of the NRC Regulatory Guide 8.13, now denoted as Regulatory guide DG-8014.

v. Personnel Monitoring

Determination of whether or not individuals receive doses over the permissible levels requires that a personnel monitoring program in which the doses to individuals are measured be in effect. Until the revision to Part 20, programs generally measured and maintained records for external doses only, since the exposures to airborne concentrations are limited to a relatively small number of specialized facilities. Part 20.1204 specifies the conditions under which it is necessary to include the internal doses in the total committed effective dose for an individual. Most programs will still find it necessary to monitor only external exposures.

Personnel monitoring requirements are given in Section 20.1501. The licensee must provide personal monitoring devices and require their use by employees who may receive an occupational dose in excess of 10% of the limits in 1 year specified in Section 20.1201, or of any individual who may enter a high radiation area. Personnel monitoring records must be kept according to Section 20.2 106 and shall be retained until the NRC terminates the license requiring the records.

Part 19 provides in Section 19.13 that workers or former workers can have access to their radiation monitoring records. Individual workers may request an annual report of their radiation exposure. Former workers may request their exposure history from an employer, and this must be provided within 30 days or, if the records have not been completed for the current quarter by the time the request is made, within 30 days after the data are available. An employee who is terminating employment can also request and receive his monitoring records. These data must provide the dates and locations at which the exposures occurred. There also are additional technical requirements for the content and form of the data provided.

Maintenance of these records by each licensee is important since each employer must attempt to determine a worker's prior exposure record. Under Section 20.2104, for records that are not available, the current employee must assume a prior exposure of 1.25 rem (12.5 mSv) per calendar quarter for each quarter the employee could have received an occupational exposure. The individual would not be available for planned special exposures if prior records were not available. These restrictions could restrict the employee's usefulness to the current employer.

h. Methods of Monitoring Personnel Exposures

There are several types of devices or dosimeters that can be used to implement a personnel radiation monitoring program. These devices are worn by the worker so they represent the actual radiation received by the wearer at the location on the body where they are worn. Since the most radiologically sensitive organs are the blood-forming tissues, which in an adult are located in the sternum, the genitals, and, to a lesser extent, the eyes, the most logical place to wear a dosimeter usually is on the outside of a shirt pocket. Working at a laboratory bench or in front of a hood in which the radioactive material is located, the distance to the source will probably be least in this location, and so the reading will most likely be higher than the average over the body. This is a conservative procedure. The most common monitoring devices are film badges, thermoluminescent dosimeter badges, and direct or indirect reading pocket ionization chambers (the latter are becoming much less common). Except for the latter type of dosimeters and dosimeters used to measure the dose to the extremities, Section 20.1501(c) requires that personnel dosimeters used by licensees to comply with the monitoring requirements must be processed and evaluated by a dosimetry processor.

1. Holding current personnel dosimetry accreditation from the National Voluntary Accreditation Program (NVLAP) of the National Bureau of Standards and Technology

2. Approved in this accreditation process for the type of radiation or radiations included in the NVLAP program that most closely approximates the type of radiations for which the individual wearing the dosimeter is monitored.

To accurately reflect the dose received by the body, a dosimeter ideally should have an energy response that would be the same as the tissue of the body. None of the dosimeters mentioned above fully meet this requirement, although the use of various filters in badges allows correction factors to be applied to the results. Each of the three types of units mentioned above will be discussed in the paragraphs below.

There are other dosimetry materials that are used in certain circumstances, for example, at very high exposure levels. There also are devices that electronically integrate the output of an electronic radiation detector and provide a digital reading of the accumulated dose, serving the same purpose as a simpler pocket ionization chamber, but at a considerably higher cost.

Versions of each of the following devices can be used to measure exposure to neutrons. However, since the use of neutrons outside reactor facilities or some types of accelerator facilities is uncommon, these will not be discussed here.

i. Film Badges

The sensitivity of film to radiation was the means by which Roentgen discovered the existence of penetrating radiation, and it still serves as one of the more commonly used materials to measure radiation exposures. The radiation level to which the film has been exposed is proportional to the blackening of the film. For gammas below about 200 KeV, the energy response of the film depends very strongly on the energy increasing rapidly as the gamma (or X-ray) energy decreases. In a typical case, the sensitivity of the film peaks at about 40 KeV falling off below this energy because of the attenuation of the soft gammas by the cover on the film to protect it from light. The sensitivity curve can be modified by using an appropriate filter, as shown in Figure 5.10. With filters, the sensitivity of a dosimeter using film can be made to be uniform to within about ±20% over an energy range of about 0.12 to 10 MeV. For intermediate energy gammas, film is usable (with care) as a dosimeter for exposures of about 10 mrad (0.1 mSv) to about 1800 rad (18 Sv). Film covered only by a light-tight cover also can be used to measure the dose from a beta emitter if the energy of the beta is above about 0.4 MeV for exposure levels of about 50 mrad to about 1000 rad (0.5 mSv to 10 Sv). There is no point in providing badges to individuals who work with low-energy pure beta emitters such as tritium, ^{14}C, and ^{35}S since those betas would not pass through the light-tight cover of the film.

Film badges can be fairly sophisticated in their design. Even the simplest will have at least one film packet (or part of a larger piece of film) covered only by its light cover, to allow detection of betas and soft gammas or X-rays.

Another area of the film covered only by the plastic of which the badge holder is typically made will help distinguish between the betas and less energetic gammas. Usually one or more metal filters of different atomic numbers will also be used that will change the response characteristic of the film, as well as help define the energies of the incident radiation. Usually the film has two emulsion layers, one of which is "faster," i.e., more sensitive than the other, to provide a wider response to different levels of radiation. Additional filters can be placed within the badge to allow it to detect and distinguish between thermal and fast neutrons. A diffuse source of radiation (such as scattered radiation) can be distinguished from a point source since the former will give diffuse edges to the images while the latter causes the edges of the filter images to be more sharply defined. The property of distinguishing between point sources and diffuse sources is especially important for X-ray users since it allows one to tell if the exposure is from the primary beam or from scattered radiation.

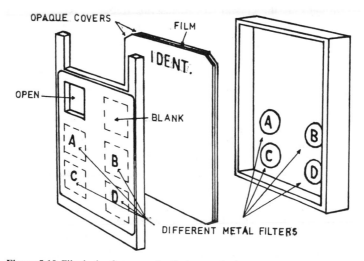

Figure 5.10 Film badge for personal radiation monitoring. The various areas shown on the front of the badge provide for information to be obtained about the radiation reaching the film. The "open" area may be covered with a a very thin mylar film which will allow betas to be detected in that area.

There are factors other than the varying energy response of the film that cause the reproducibility of their readings to vary. Each batch of film is a little different, the processing is subject to variation, and the temperature in the environment in which it is worn will affect the results. The average accuracy of film badge dosimeters to doses of 100 mrad (1 mSv) or more is about ±25%. Values below about 50 mrad (0.5 mSv) are often relatively meaningless. The requirement that dosimeter evaluation be done by processors belonging to NVLAP is a measure taken to control as many of the variable factors as possible. Additional variations are introduced by the way they are handled and worn. They should be worn on the body of the worker, on the outside of the clothes, since clothes would modify the radiation received. The wearer should be careful to orient the badge properly since the various sections of the badge depend upon radiation being incident upon the front of the badge. Obviously, the badges should never be placed in the employee's trouser pocket. Not only will the orientation be incorrect, but the badge will be shielded by coins, keys, and any other objects that might be present.

Although film badges have some disadvantages - specifically relatively poor accuracy and, a nonuniform energy response for low energy gammas- they have a major advantage in that they represent a long-lasting record. If the original developed film is stored properly, it can be reevaluated later if there is reason to suspect the original record is incorrect.

ii Thermoluminescent Dosimeters (TLDs)

The two most serious problems with a filmbadge—nonuniform energy response and reproducibility—are eliminated by a TLD dosimeter. Since a Lithium Fluoride (LiF) TLD with an effective atomic number of 8.1 is similar to tissue which has an effective atomic number of 7.4, the LiF energy response is about the same as tissue. The response of a LiF TLD is almost energy independent over an energy range of 0.1 to about 3 MeV. The accuracy of a TLD can be better than 10% over an exposure range of 10 mrad to 1000 rad (0.1 mSv to 10 Sv). The upper limit for a quantitative response is on the order of 100 times as great.

TLD materials are crystals in which radiation excites the atoms of the material, leaving them in long-lived metastable energy levels, locking the excitation energy into the crystal. Heating the crystals releases the energy in the form of light. If the heating process takes place at a uniform, controlled rate, the crystal will release the trapped energy in a reproducible "glow" or light curve as the temperature of the crystal rises. The area under the light curve generated in this manner is proportional to the quantity of the original radiation. The TLD material is often in the form of a small chip, although one vendor of TLD badges uses Teflon impregnated with the LiF instead. Obviously, the light seen by the reader from the detector will be affected by the condition of their surfaces, but if handled with care, a TLD dosimeter can be recycled many times.

As noted above, the energy response and reproducibility of a LiF TLD dosimeter is much better than that of a film badge. Badges can be configured as with film badges to provide dose information on betas, X-rays, soft gammas, and more energetic gammas. However, unlike film, the process of reading the dosimeter to obtain the exposure information will destroy the information stored in the crystal. They also cannot readily distinguish between a point and diffuse source. TLD dosimeters have gained a large share of the market in recent years because of their overall advantages, despite the deficiencies just noted.

Ionization Dosimeters: Both of the two previous types of dosimeters require processing in order to be read. The two most common types of ionization dosimeters can be read immediately, although the indirect type requires an auxiliary reading device to do so. Ionization dosimeters to be used as personnel dosimeters are typically about the size of a pen and are usually called pocket dosimeters.

The indirect reading type is simply a high-quality cylindrical capacitor in which the outer cylinder, made of an electrically conducting plastic or one having a conducting surface, forms one side or plate of the capacitor and a wire along the axis of the cylinder forms the other side. The capacitor is charged to a predetermined voltage, and radiation impinging on the walls of the dosimeter and the air in the cavity acts to discharge it, lowering the potential difference between the two electrodes. The potential difference is then read by an electrostatic voltmeter, calibrated so that the scale reads the amount of radiation to which the dosimeter and the wearer were exposed. The process of reading this type of ionization chamber discharges them so they must be recharged to be useful again.

The direct reading type also is a simple device. A fine, gold-plated quartz fiber, acting as a gold-leaf electroscope, is charged to a predetermined potential. As the air surrounding the fiber is exposed to radiation, the "electroscope" discharges, and the quartz fiber moves toward the discharged position. The motion of the fiber, superimposed on a scale, is viewed through a simple microscope incorporated in the device. The scale is calibrated so that the change in position can be read as the radiation to which the dosimeter has been exposed.

Both of these two types of dosimeters are primarily intended to read gammas and provide a response which is within about 15% of the actual exposure over gamma energies of about 0.04 to 2 MeV. Both can be configured to read over a wide range of exposures. Most of those used in the laboratories are usually set up to read from 0 to 200 mrem. The readings of the units are conservative since charge leakage, or discharging due to factors other than radiation, will cause the readings to be too high.

Pencil dosimeters are especially useful for measurements of short term exposures into areas where radiation is present and where an immediate information on the level of dose received is desired. They are also useful for issuing to visitors to radiation controlled areas to assure the visitors of the amount or lack thereof of radiation they have received.

i. Bioassays

The dosimeters discussed in the previous section measure the external fields of radiation to which the wearer is exposed. They do not measure the intake of any materials that will contribute to internal exposures. Methods of measuring airborne concentrations of radionuclides will be discussed in the following section, as part of the general subject of area radiation surveying. Measurements of the actual uptake of radioactive materials by the body is done by doing bioassays on individuals. There are routine laboratory uses of radioactive materials that are biologically active, such as the use of ^{125}I in certain laboratory analytical tests, which call for tests of specific organs to see if any of the material has been taken into the body. On other occasions, bioassays are used to determine if there has been any intake of material after a spill or other types of accidents in which there has been an unplanned release. Where there is active use of radioisotopes by a number of workers, it is also desirable to perform routine bioassays on selected employees as a precaution even if no intake is suspected.

There are two distinct aspects to doing a bioassay: (1) measurement of an activity related to the original intake, and (2) inferring the exposure from the measured activity. Of the two, measurement of an activity associated with the intake is the easier.

The basic assumption on which many bioassay procedures are based is that the activity in the excreta, e.g., the urine or feces, will be related to the amount of activity taken into the body. Urine is the usual choice for monitoring, especially if the original material is soluble, while for particulates that are less soluble, both the feces and urine might be used. The radioactivity from ^{40}K (0.012% isotopic abundance in natural potassium, 1.28×10^9 year half-life, 1.31 MeV β, 1.461 MeV γ) in the potassium in urine typically results in about 45 dps/L as an interfering activity when trying to measure the activity due to an intake of another radioactive material. Chemical removal of the potassium activity is possible and usually desirable to ease the measurement of the intruding isotope. Where the activity to be measured is due to a pure beta emitter and bremstrahlung radiation does not accompany the beta emission, determining the activity in the urine and feces is almost the only way to obtain a measurement.

The interpretation of the data obtained from urine and fecal measurements in terms of the activity in the body and the original amount taken in is complicated. Soluble materials containing radioisotopes that are distributed throughout the body lend themselves to the easiest analysis. Soluble materials which concentrate themselves in an organ represent a more complicated situation since they first must build up in the organ from the body fluids and then leave the organ to enter the excreta. Those that are absorbed in the bone have very long clearance times. Unfortunately, there are many factors that affect the metabolic processing of the materials and affect the interpretation of the results in terms of the actual internal exposure.

Urine analysis can be done for gamma emitters as well, but in cases in which the isotope is strongly concentrated in a single organ, a direct measurement of the activity in the organ using a scintillation counter* (or perhaps a Germanium detector) may be feasible. It also is possible to obtain a relatively accurate measurement of the amount in the organ by using a known amount of the isotope (or an isotope with a comparable gamma spectrum) in a model (or phantom) which simulates the physical characteristics of a person, as far as absorption and scattering of radiation from the organ are concerned. This is commonly done for iodine uptake measurements in the thyroid. The use of radioactive iodine is significant, so this technique is widely used.

An extension of the concept of measuring the activity in a single organ is to use a "whole-body-counter" in a scan of the activity throughout the body, which provides a measurement

---

* Scintillation counters and germanium detectors are two types of sophisticated nuclear radiation detectors which depend upon the interaction of radiation with solid materials, such as NaI (Tl) crystals or germanium solid state diodes. They have the ability to distinguish the energies of the radiation and have a higher sensitivity to radiation than the materials discussed heretofore. They also require sophisticated instrumentation associated with their use and are expensive.

of the entire body burden of gamma-emitting radioisotopes. These are large, expensive devices and are not often found in a typical radiation safety operation, although there are a number of locations where access to one is available in emergencies.

If it is suspected that a radioactive material has been inhaled, measuring the activity from nose swipes or in a tissue after the nose is blown can be used to confirm the inhalation. Measurement of radon concentrations in the breath is a technique that can be used if materials are inhaled which have radon as a decay product. Carbon-14 can be detected in exhaled breath as well.

j. Radiation Surveys

Measurements of the radiation levels in an area are a vital aspect of radiation safety. Every laboratory using radioisotopes should have ready access to the means to measure the ambient activity in the facility. In areas where there are fixed sources of substantial radiation such as a ^{60}Co irradiator, some types of accelerators, or a nuclear reactor, fixed radiation monitoring systems usually are installed. These permanently installed systems need to be supplemented by portable instruments to be used to detect radiation levels in localized areas that might not be "seen" by the fixed system, or arising from movable sources.

There are several types of instruments that can be used for radiation monitoring, each with its own characteristics, advantages, and limitations. There have been serious radiation exposures when individuals used an instrument that was not appropriate for the radiation field involved. This section will be limited to the type of radiation detection instruments that are commonly used for radiation surveys rather than those used for laboratory research. The instruments will primarily be those intended to measure external exposure rates.

A steady exposure to radiation that would result in a dose of approximately 100 mrem (1 mSv) per week or 2.5 mrem per hour (25 pSv per hour) for a 40-hour work week would allow a worker to meet the annual limits of occupational exposure for the whole body of 5 rem. The hands are often placed in radiation fields that would be at much higher levels than this, although they would rarely remain exposed to higher levels for an entire work week. The current limit of 75 rem per year would translate to about 1.5 rem (15 mSv) per week or approximately 37 mrem per hour (0.37 mSv per hour). A survey instrument for a laboratory using byproduct materials should be able to read low levels that are a small percentage of the average permissible whole-body exposure rate. Title 10 CFR Section 35.72 stipulates that a survey instrument should be able to read a minimum of 0.1 mrem per hour, and should, at a minimum, be able to read levels at or above the equivalent rate for extremities. The high end of the instrument's range probably should be selected to be somewhat greater than this guide would suggest, especially for spills in which an amount of material larger than that typically used in a single procedure may be involved. The maximum amounts of radioactive material that could be in the laboratory at a given time could be used to determine the high range of a survey instrument that might be selected. Survey instruments are required by 10 CFR Section 35.120 to be calibrated on all scales up to 1000 mrem per hour if they have that range.

Most laboratory survey instruments use gas-filled chambers as their detectors. In this type of instrument, as discussed earlier, the passage of radioactive emissions (α, β, γ) through matter creates a large number of ion pairs that can be translated into an electrical signal used to detect the radiation. There are several factors that affect the ability of an instrument to detect an interaction with the gas in the chamber as a discrete event or not. The combination of these factors is called the resolving time, with a shorter resolving time allowing more events per unit time to be detected individually than a longer resolving time. Other factors would affect the sensitivity of the detector and its ability to distinguish between emissions of different types or energies or normally used to simply detect the presence or absence of radiation, without regard to the energy or type. Most, however, do have the ability, by mechanical means, to distinguish between types of radiation (taking readings with and without a cap over the detector will allow the instrument to distinguish between betas and gammas). Since the radioisotope in use is usually known, and laboratories often use only one type of material, or at most a limited number, the person doing the survey rarely will need an instrument that can measure energies or varieties of radiation for simple surveys. When an instrument with more sophisticated capabilities is needed, the radiation safety office normally can be counted upon to provide one. For more information on the properties of gas filled counters, the reader is referred to any number of references on nuclear instrumentation.

Although there are many commercial varieties of instruments available with various advanced and useful features, they usually are only elaborations, with more versatile and sophisticated circuitry providing wider operating ranges, of the simple Geiger counters and ionization chambers that have been available for decades. The basic instruments are adequate to perform routine surveys, but higher range instruments should be available when needed.

All survey instruments kept in the laboratory should be calibrated at least once per year on all scales. If facilities are not available to do this locally, there are many commercial firms that will provide the service.

i. Geiger-Mailer Counter

The G-M counter, or Geiger counter as it is usually called, is the least sophisticated type of instrument that might be employed as a laboratory survey instrument. Its simplicity allows it to be built inexpensively as a rugged and, if used properly, relatively foolproof device. However, many knowledgeable and competent scientists have managed to use one incorrectly.

A simple Geiger counter in its basic form often has a cylindrical probe approximately 1 inch in diameter and about 6 inches long. The outer cylinder forms the cathode of the detector (the electrode toward which the positive ions move from the ion pairs created by the passage of a particle or ray into the gas-filled cylinder). A center wire forms the anode toward which the negative ions (electrons) portion of the ion pairs move. The center wire is connected to the external circuitry. One end of the cylinder usually has a thin "window" with an aerial density of about 1 to 2 mg/cm^2, through which betas of approximately 30 KeV and greater can pass. In order to protect the fragile window, it usually is covered with a removable cap. Alpha particles also can be detected with a thin window counter such as this, if the counter is held very close to the source of the alphas. A 4.5 MeV alpha has a range in air of about 3 cm, but it loses energy as it passes through the air, so that at a distance of a little less than 2 cm the alpha particle would not retain sufficient energy to penetrate even a 1 mg/cm^2 window. The geometry just described is not ideal for surveying a surface contaminated with emitters of alphas and betas where the detector has to be held very close to the contamination, but many thousands of such instruments have been constructed and used. There are different versions of the GM counter which offer a large surface area (pancake probes) so that one can survey a larger area more efficiently. Since gammas can penetrate the cylindrical walls of the counter as well as the thin end window, the geometry of a typical Geiger counter is not as disadvantageous for them.

When a charged particle, gamma or X-ray enters the active volume of a Geiger counter, not only does it create ion pairs, but the potential between the cathode and anode is set sufficiently high so that these ion pairs create an "avalanche" of charge along the entire length of the central anode so that a substantial space charge is developed around and along the center wire. The negative ions are electrons in the cloud and move very swiftly to the anode while the positive ions, being atoms and hence far more massive, move much more slowly toward the cathode, represented by the outer wall of the tube. While the positive ions are moving, the detector tube is not able to initiate another pulse, or at least not one as large as the first. The minimum time that must elapse before another event can be detected is the resolving time. For most Geiger counters, the resolving time is about 100 to 200 microseconds. This seems short, but it places a severe limitation on the number of events per second that can be detected. For example, let us calculate what a counter with a 200 micro second resolving time would actually appear to read if it is operated so that it should have a true count rate of 10,000 counts per second, if there were no problems with resolving time.

The equation that relates the true count rate C, the observed count rate c, and the resolving time t, is given by Equation 17.

$$C = \frac{c}{(1 - c\tau)} \tag{17}$$

If the assumed numbers are inserted in this equation, the observed count rate is 3333 cps, or a discrepancy of two thirds. Obviously, a Geiger counter cannot be used at a high interaction rate to obtain an accurate exposure rate. If the number of particles or rays becomes too high, the counter will saturate so that it will appear that no pulses are occurring and the apparent count rate will become zero. Under these conditions, the instrument would be worse than useless since a zero reading might be construed by the user as the true one, and on that basis enter a dangerously high radiation field. This has happened and serious radiation injuries have occurred as a result.

Because of the inability to function properly at high count rates, simple Geiger counters are not usable in high fields, where the immediate danger to real injury would exist. A Geiger counter can be calibrated against a known source of a known specific activity in terms of mrad (mGy) or mrem (mSv). A basic Geiger counter usually has a maximum scale of 20 to 50 mrem per hour (0.2 to 0.5 mSv per hour). If the activity being used is small enough, this may be adequate, although instruments that can read levels 10 to 20 times as high, but still have a scale on which the maximum reading is 1 to 2 mrem per hour (10 to 20 μSv) would be a more versatile general purpose instrument. Ionization chambers are available that meet this criterion, as discussed in the next section.

ii. Ionization Chambers

The basic geometry is similar to that of the Geiger tube except that the diameter of the chamber usually is substantially larger so that the volume of the gas in which the radiation may interact is much larger. An ion chamber is operated at a high enough potential between the electrodes so that the ion pairs created by the impinging radiation do not recombine, but not so high that the moving ions create any significant number of secondary ions, i.e., there is no avalanche. The negative and positive ions are collected on the anode and cathode, respectively, and can be detected as a weak current. A more intense (stronger) source will create more ion pairs and would produce a stronger current. Actually the current is not measured, but a large resistor (10^{10} ohms or larger) is connected from the anode to ground so that a voltage is developed across the resistor. A sensitive and stable electrometer circuit is used to measure this voltage. The larger the resistor, the more sensitive the counter. However, using larger resistors results in a penalty due to the time it requires for the current through the resistor to reach equilibrium (the time constant of the response is determined by the load resistor and the capacitance of the chamber). On a very sensitive scale, it could take a minute or more for a steady-state voltage to be attained. An ion chamber requires more care than does a Geiger counter. Accuracy requires that the resistance values not change and dirt and moisture can cause the effective resistance of the resistor to decrease, which would cause the apparent voltage measured for a given radiation field to decrease.

The advantage of the ion chamber is not that it can read very low fields, but that it can be designed, if the necessity arises or exists routinely, to read higher levels that represent a danger to the workers. At the higher levels, the slow response time observed at lower levels should pose no problems since the resistor and, therefore, the time constant, can be substantially lower and still provide an adequate voltage that can be measured.

As with the simple Geiger counter, the end window of the counter chamber can be made thin so that the unit can be capable of counting alphas, betas, and gammas. If only one survey instrument can be afforded by a facility, an ion chamber with an upper range of 10 R per hour (0.1 Sv per hour) might be a good choice. Typically however, one finds the less expensive and more rugged Geiger counter in most laboratories.

iii. Other Concepts

There are units designed to operate in what is called the proportional region of the confined gas, in which there is some multiplication of the charge deposited by the impinging radiation, but which permit much shorter resolving times so that count rate losses are much less severe. There are other arrangements of the electrodes, so that the detector presents a large sensitive area to the surface being surveyed and a shallow depth for the sensitive volume, instead of a long cylindrical geometry This usually will result in increased sensitivity, especially for alphas and betas that have short ranges in matter. It also allows "seeing" a larger surface area and facilitates a survey of an extended area, such as a work surface. If energy discrimination and sensitivity are needed, portable scintillation counters are available, incorporating NaI(Tl) crystals. If there is a possibility of personnel contamination, counters can be set up at the points of egress which can be used to automatically check the hands and feet of the employees as they leave the facility. "Friskers" as they are often called, which use a pancake-shaped detector, are available to rapidly scan a person's body to detect contamination on areas other than the hands and feet. There are specialized counters for virtually any application.

iv. General Surveys

Maps of the laboratory area should be prepared and, periodically, a general contamination survey of the area should be made using a general purpose survey meter, especially if the isotope in use emits gammas or strong beta emitters such as ^{32}P. The frequency of these general surveys should depend upon the level of use of material in the facility. If the use is very heavy, a daily survey of the immediate work area and a general survey of the entire laboratory at an interval of once a week might be necessary while very light use might make once a quarter adequate. The records of these surveys should be maintained in a permanent log and audited periodically by the institution's or company's radiation safety officer. They should also be retained for the NRC, should they be needed, as they are likely to be if the laboratory happens to be chosen for evaluation during an unannounced inspection by an NRC representative. More detailed surveys of the immediate work area and certain critical areas should be carried out as part of contamination control, as discussed in the following sections.

k. Measurement of Airborne Activities

The collection of samples for testing airborne radioactive materials follows essentially the same procedures as described in Section VIII.B of Chapter 4 for airborne monitoring of chemical contaminants. The basic procedure is to use a pump to circulate a known quantity of air through a collection device with known characteristics for capturing contaminants. If there are individuals working in an area where it is possible that airborne radioactivity may be a problem, it is recommended that routine periodic sampling of the air for activity be done and that the activities to which individuals may be exposed be directly and specifically monitored. Sampling can be done for an entire work day or sometimes, during times of potential peak exposures, by using small, lightweight pumps and collectors worn on the person near the breathing zone. If significant levels are found, a program to control the airborne contamination should be promptly initiated.

If the suspected contaminants are particulates, membrane filters are frequently used which have collection efficiencies of 100% for particles most likely to enter the deep respiratory tract (>0.3 to around 2 μ). A membrane filter traps particles on the surface, which is desirable for beta and alpha emitters, since energy losses in the filter material could be important if the particles were carried into the filter. The concentration of a specific airborne contaminant in air is derived by measuring the activity of the collected sample, correcting for counting efficiencies, and dividing by the volume of air circulated through the filter.

Natural airborne activity will contribute to most airborne measurements of activity. This arises primarily from the decay products of the various radon isotopes. Radon is a gaseous radioactive element arising from the decay of thorium and radium. The two principal radon isotopes of concern are ^{220}Rn (55 second half-life) and ^{222}Rn (3.82 day half-life). The first of these has such a short half-life that it will tend to decay before the gas diffuses very widely; therefore, the daughter products, which are not gaseous, will remain in the same vicinity as the ^{224}Ra, the parent nuclide for ^{220}Rn. Radon-222, with its much longer half-life, will diffuse away from the original location of its parent. The short-lived daughter products of ^{222}Rn, with half-lives ranging from 164 picoseconds for ^{214}Po to 26.8 minutes for ^{214}Pb, tend to become ionized and attach themselves to dust particles.

The activity in the particulates in the air is typically at an approximate state of equilibrium, since additional radon gas is diffusing into the air constantly from local sources. However, the activity due to the daughter products of ^{222}Rn on the particulates that have been collected on the filter is not being replenished and will decay until a much longer half-life daughter is formed, ^{210}Pb (half-life 22.3 years), which will have a much lower specific activity. A 4 or 5-hour wait will usually see most of the activity from the daughter products of ^{222}Rn gone. Although not as significant originally, the activity from the daughter products of ^{220}Rn, such as that of ^{212}Pb and its daughter products, may dominate the natural activity contribution after 4 or 5 hours. The half-life of ^{212}Pb is longer (10.6 hours) than that of any of the following daughters (or of the preceding ones) until a completely stable isotope is reached, so the activity in the collected sample due to the daughters of ^{220}Rn will have an equilibrium half-life of 10.6 hours. The activity of most radioisotopes used in the laboratory is typically much longer than this, so the activity of the airborne research isotope can readily be mathematically separated from the background natural activity by counting the collected sample several times over the next 2 or 3 days. If t_c is the half-life of the contaminant and t_n is the half-life of the natural background radiation in the air sample, the activity due to the contaminant in the collected sample can be obtained using any two of the delayed counts and by using the expression

$$C = \frac{C_2 - C_1 e^{-0.693t/t_n}}{e^{-0.693t/t_c} - e^{-0.693t/t_n}} \tag{18}$$

Here, t is the time interval between the two counts employed in the calculation.

If several counts are taken, Equation 18 can be used for several pairs to make sure that only two half-lives are contributing to the activity. Within statistical counting variations, the results should be the same for each pair. If they change systematically as the time from the beginning of the initial count changes, then additional components are contributing to the activity.

If the airborne contaminant is contained in a solvent aerosol or is gaseous instead of being a particulate, the basic procedures for sampling for nonparticulate materials in Section VIII.B, Chapter 4 can be used. The counting procedures may differ. Instead of using a standard counter, such as a proportional counter (for betas and gammas) or a solid scintillation detector (for gammas), a liquid scintillation detector might be used instead. The latter type of counter is particularly effective for betas, as its efficiency can approach 100%. If the contaminant is a gamma emitter, a more sophisticated counting system using either a NaI(Tl) or a germanium detector and a dedicated multichannel analyzer or computer can differentiate the energies of the gammas. These latter types of counting systems, with either a "hardwired" or software-based analytical program, cannot only positively identify the isotopes contributing to the energy, but can automatically determine the activity in the sample. These sophisticated analytical instruments are expensive compared to the type of instruments previously discussed, so they normally are found only in laboratories heavily engaged in work with gamma emitters or in a radiation safety facility.

l. Fixed and Loose Surface Contamination

Contamination is the presence of undesirable radioactivity. It may be either fixed or loose. Usually the acceptable limits for these in laboratories are set by the organization, except for certain items such as packages received in the laboratory for which the NRC and DOT establishes acceptable standards, at levels that should be sufficient to assure the occupational safety of personnel and members of the public. Although locally acceptable levels are generally set by the licensee, 10 CFR Section 35.70 stipulates that in medical facilities involving humans, the licensee must have instruments capable of measuring radiation levels as low as 0.1 mrem per hour or detect contamination of 2000 dpm with a wipe test. Although not explicitly stated, it could be inferred that these would be acceptable limits for surface contamination by the NRC in these facilities.

i. Fixed Contamination

Fixed contamination, by definition, is unlikely to either become airborne so that it may be inhaled or adhere to the hands upon being touched, so it is also not likely to become ingested. Material that has soaked into a porous surface would be a typical fixed source of radioactive contamination. The hazard is an external one to the body. Under the circum-stances, an acceptable level could be defined by the acceptable levels for external exposures. In an unrestricted area, a level of 0.25 mrem per hour (2.5 picoSv per hour) would barely meet this criterion. A level substantially below this level would be preferable, such as 50 picorem per hour (0.5 pSv per hour). The NRC Regulatory Guide No. 1.86, for the purposes of releasing an area to unrestricted use, uses a value of 5 μR per hour (0.05 pSv per hour) above background at a distance of 1 meter from the source. This is roughly one third of the average background in most parts of the country. Such a level would be for a permanent cessation of operations, and is probably lower than is necessary while the facility is licensed. While normal operations are being conducted and radiation safety survey and monitoring programs are actively done, a level five times as high would correspond to a conservative value of 10% of the permissible level for unrestricted areas. For betas, a value of 50 μR per hour (0.5 μSv per hour) for a reading close to the surface would not appear unreasonable for an unrestricted area. Any alpha level that is detectable in an unrestricted area would suggest a possible dispersal of some of the more toxic radionuclides in the area, which should not be allowed to occur, because of a research program. A nondetectable level of alpha contamination for fixed material in an unrestricted area would be desirable. However, levels equivalent to about 0.3 nanoCi or 10 Bq of loose contamination per 100 cm^2 are considered acceptable by some agencies.

For a restricted area, a level of fixed contamination well below the occupational level of

5 rem per year (50 mSv per year) or about 2.5 mrem per hour (25 µSv per hour) would be appropriate. Some organizations have set the level at 1 mrem per hour (10 pSv per hour). Another option might be to set it below a level that would require personnel monitoring, currently the equivalent of about 0.25 mrem per hour (2.5 µSv per hour).

ii. Loose Surface Contamination

The limits for loose surface contamination should be related to the amount of airborne contamination that would result if the material were to be disturbed so it could enter the body by inhalation or indirectly by ingestion through handling objects on which the material rested. The limiting factor is internal exposure. An examination of 10 CFR Part 20 readily reveals that the permissible DACs for the various isotopes vary by more than seven orders of magnitude, although the majority cluster around intermediate levels between the extremes. For the isotopes most commonly used in the laboratory, the maximum permissible concentrations still differ by almost three orders of magnitude. This disparity makes it difficult to establish a single limit for loose surface contamination.

As noted above, a level of removable alpha contamination between nondetectable and about 0.3 nanoCi (10 Bq) per 100 cm^2 is considered acceptable by many different groups for an unrestricted area. In restricted areas some organizations still do not accept any stray alpha contamination, except in the immediate work area, such as the inside of a hood. However, a level of about 0.8 nanoCi or 30 Bq per 100 cm^2 is considered acceptable by others.

Loose surface contamination is usually measured by rubbing or swiping an area of about 100 cm^2 with a filter paper, and then counting the activity that adheres to the paper. An estimate must be made of the fraction of the loose material that is removed from the surface and remains on the paper, to make a judgment of the amount of surface contamination.

For beta and gamma contamination, otherwise unspecified, levels of 0.1 to 1 nanoCi per 100 cm^2 are considered acceptable by various organizations for unrestricted areas. For restricted areas, these organizations usually make the acceptable levels ten times higher, i.e., 1.0 to 10 nanoCi per 100 cm^2. The United Kingdom and some others attempted to roughly take into account the differences in the MPCs and DACs by grouping the nuclides into toxicity classifications and recommending acceptable levels for each group. For the most toxic radionuclides (which includes the alpha emitters), an acceptable level of loose contamination in restricted areas was set at 0.8 nanoCi (30 Bq) per 100 cm^2. The levels for medium toxicity radioisotopes were increased by a factor of 10, while the levels for lesser toxicity materials were increased by another factor of 10 to 100. In late 1999, the Health Physics society and the American National Standards Institute published a standard, ANSI/HPS N13.12-1999 that provides similar recommendations for groups of radioactive contaminants.

All the numbers are for general area contamination. For personal clothing they should be at least ten times less and for the skin of the body, the levels should be either zero or as close as can be obtained by decontamination efforts that do not cause damage to the tissue.

In removing loose contamination, precautions must be taken to prevent additional areas from becoming contaminated and materials used in the decontamination process need to be treated as contaminated waste.

iii. Frequency of Surveys for Contamination

In laboratories where unsealed radioactivity is in use, a contamination survey of the work areas directly involved in the operations should be made at the end of each workday, or at the conclusion of that day's operations involving radioactive materials. Surveys of the workers' hands and clothing also should be checked at the end of each day's operations, or more frequently if they leave the work area or take a break to get something to eat or drink. At the end of any week that radioactive materials have been used, a contamination survey of the entire facility should be made. Particular attention should be paid to the work areas, items that might have been handled during the day, including controls of equipment, sink faucets, and the survey meter, radioactive material storage area, radioactive waste storage area, and the floor at the points of egress from the facility. The latter data are intended to provide assurance that no radioactive materials have been transported from the room on the soles of the employees' shoes.

iv. Accidents and Decontamination

Contamination may arise from poor laboratory practices, in which case the organization's radiation safety program must be prepared to take whatever steps are necessary to ensure that these practices are corrected. No such situation should be allowed to persist for long in a well-managed organization. In a properly run laboratory, accidents or spills should be the major source of contamination of the facility. Even such incidents should be very infrequent in a well-run facility.

The scale of an accident will determine the response to the incident. If, for example, the accident is a spill in a hood, if the base of the hood has been lined with absorbent paper, and the apparatus set up in a pan or tray that is sufficient to contain all the spilled fluids, there may be little or no contamination or dispersal of airborne activity within the room because the vapors will be discharged in the hood exhaust. The individual working at the hood, if not personally contaminated, should notify the laboratory supervisor and the material removed from the hood and placed in a suitable container. The container can be sealed and placed in the area reserved for radioactive waste. Any additional protection required for cleaning up, such as the use of two pairs of gloves, respirators, and coveralls, should be decided upon before beginning the remedial work. Every accidental spill should be reported to the radiation safety office. The radiation safety officer may wish to help or at least monitor the response, even for a minor spill, to ensure that all personnel are protected. After the clean up is completed, the entire area involved in the incident should be surveyed as well as all personnel involved in the original incident and the clean up. Controls of instruments that may have been handled should not be overlooked. The problem of personal contamination is covered below.

Larger accidents obviously pose more risk that individuals may have become directly contaminated or may have inhaled or ingested radioactive materials. Other individuals in the laboratory other than the ones directly using the material may have been exposed as well. A contingency plan should have been developed that should be put into effect when a radioactive emergency occurs. The following two scenarios illustrate some of the more common problems that should be considered in preplanning for an emergency.

v. A Spill Directly upon a Person

The individual should stay where he or she is or, at most, move a few feet away to avoid standing in spilled material and immediately remove all garments that are or may be contaminated by the spilled material. The spread of contamination should be minimized. The garments should be placed in a pail or other liquid-tight container, preferably one that can be sealed. The pail can be placed in a double garbage bag and the bags closed with a twist if a covered container is not available. After removing all contaminated clothing, the individual involved in the spill or an assistant should wash all contaminated areas of the body, being especially careful to wash areas where tissue creases or folds. If the spilled material is volatile, the contaminated individual and those assisting him should don appropriate respirators as soon as possible. If it is volatile, the most likely respirator needed would be one equipped with an organic material cartridge, unless the activity were in a different type of liquid. Individuals who assist the contaminated person should also thoroughly wash their hands. Potentially contaminated persons should be checked for contamination. Persons not immediately involved in the incident and not needed should immediately leave the area until the potential for aerosol dispersal is determined. However, those leaving the area should be checked as they leave to confirm that they are not carrying contamination with them. It would be prudent to take wipes of their nostrils to further confirm that they did not inhale any material.

Radiation safety personnel and laboratory employees should jointly decide how to clean up the spill after the immediate contamination to personnel problem has been resolved, and determine a schedule that will allow operations to resume. An investigation should take place for any incident of this size or greater, The Radiation Safety Committee should review the circumstances, the immediate actions taken, and decontamination program to ensure that all actions taken were consistent with maximum protection of personnel and compliance with regulatory standards.

vi. A Large Incident in Which Significant Airborne Activity is Generated

In any incident in which there is a likelihood of airborne contamination being generated, the most conservative approach is to immediately evacuate all personnel. This is based on the assumption that the situation will not rapidly worsen if it is left alone. An example of a situation that could worsen would be a small fire that involves small quantities of radioactive material and that could spread to involve not only larger quantities of radioactive materials, but other dangerous materials as well. If the fire appears clearly controllable with portable fire extinguishers (which should be conveniently available), one or two persons might choose to remain to attempt to rapidly put out the fire, but all other occupants should immediately evacuate. The individuals staying behind should leave as soon as possible. Note that this approach will depend upon factors such as the toxicity of the materials, the possibility of inhalation of airborne material, and the amounts involved, as well as the possibility of controlling the fire. If staying to fight the fire only to reduce loss of property would endanger a person's life, the recommendation is to evacuate all personnel immediately. Often, immediate reactions may have unforseen consequences so that, if the situation is contained and it is safe to allow the area to simply be left alone temporarily, this is recommended while thoughtful response plans are developed.

In a situation where the only problem is the generation of airborne radioactivity, individuals should leave the space in which the airborne activity is present, closing the doors behind them and turning off any ventilation in the area, if the controls are in the room. If the ventilation controls are in a switch room or breaker room elsewhere, they can be turned off from this location. If the accident occurs in or near a fume hood that would vent the material from the laboratory, it may be appropriate to allow the hood to remain on to reduce the airborne concentration of activity within the room; however, this would depend upon the radioactive toxicity of the material involved. If it is one of the more toxic materials, it would usually be preferable to avoid dispersal into a public area even if the levels are low, because of the possible concerns of exposure to persons in these areas. Even if the potential for exposure is minuscule, many persons will become very alarmed. The laboratory represents a controlled space that can be decontaminated and, in general, the best approach is to restrict the contaminated area as much as possible.

Individuals within the laboratory should, as in any other emergency, evacuate the area following routes that would reduce their risks. Any location within the facility should have two evacuation routes so that one, at least, should be relatively safe. The storage location of portable radiation survey instruments should be near the entrance to the facility or along the evacuation route deemed most likely to be safe in a radiation-related emergency.

Once the occupants have left the immediate area of danger and it has been isolated, the next steps to take in the emergency should be considered. The first order of business is for an individual (preferably one with the least potential of having experienced personal exposure in the incident) to notify the radiation safety department and other responsible authorities to request assistance. The police or security staff may be needed to control access to the facility. Concurrently, others should begin to evaluate any personal radiation exposure that may have taken place, by using the laboratory's portable instrumentation to check each individual. Assuming it is safe to remain, no one should leave the immediate area until they have been thoroughly checked for contamination, for their own safety and to prevent spread of contamination. In the case of airborne contamination, it would be essential to check the accessible portions of the respiratory tract for traces of inhalation. Nasal swipes should be taken. Personnel might blow up balloons, if available, for later checking. All individuals who are knowledgeable about the incident should remain to provide information to those persons arriving to take charge of the emergency response.

Once radiation safety personnel, security staff, and other persons such as the building authority have arrived and have been made thoroughly aware of the circumstances of the accident, a plan can be developed to respond to the situation. Other temporary steps to isolate the problem can be taken, such as taping around the edges of the doors and defining a secure area with barricades to prevent persons from entering or approaching the area and interfering with the response program.

Groups involved in planning the response should include the participation of laboratory personnel, radiation safety specialists, managerial personnel, media relations staff, and other support groups as dictated by the nature of the incident.

An operations center should be established outside the immediate area of the incident, and all operations and information releases should be managed from this center to ensure that all actions to control and remedy the situation are properly coordinated. Every incident is different and the correctional plan also would be different. However, in general, the approach is to establish a control point at the boundaries of the affected area through which all personnel, supplies, and waste would enter and leave the area. All persons and material leaving this area must be checked for contamination and all individuals entering would be checked to see that they are suitably protected. The normal program is to start at the edge of the contaminated area and work to reduce the area involved until all contamination has been removed or reduced to an acceptable level.

Assistance should be sought immediately for individuals who might have received internal and external contamination. Contact with a regional radioactive incident response center is advisable, since few local medical centers have sufficient training, experience, or facilities to respond to an internal radiation exposure incident. If the telephone number of the appropriate center is not immediately available, the regional NRC office, which must have been notified according to Part 20.2202, can provide it. The regional center may recommend transfer of the affected personnel to a facility in their area. Any information that could possibly aid medical personnel in their evaluation and treatment is invaluable. The response center may be able to suggest short-term measures to reduce the amount of radioactive material taken up by the body.

vii. Decontamination Techniques

Whether contamination has been caused by routine operations or by a major or a minor spill, the problem of decontamination arises. The cost and feasibility of decontamination of individual pieces of equipment and the added exposure risk to personnel must be weighed against the cost of replacement equipment and the cost of disposal of any material considered as radioactive waste. The latter costs are not negligible.

The facility itself, including major items of equipment such as hoods, work benches, and valuable specialized items of laboratory apparatus, are generally worth substantial decontamination efforts since they are expensive to replace. At current average building costs of $150 to $300 per square foot for laboratory space, a facility is not lightly abandoned or left idle for extended periods. If the facility has been originally built to simplify decontamination of spills of all types, and the major items of equipment specified for ease of maintenance, the decontamination process may be straightforward. Personnel, dressed in coated Tyvek™ coveralls, shoe covers, and head covers and wearing gloves and respirators, can clean most such areas by washing the surfaces with detergent and water since the surfaces should have been selected to resist absorption of water and chemicals. The equipment in the room should have been selected so that it would have few seams or cracks in which contamination can become lodged.

Unfortunately, many laboratory facilities using radiation have been located in older buildings with vinyl tile, unsealed concrete, or even wooden floors in which radioactive contamination can become fixed, and equipment has many seams and cracks that lend themselves to becoming contaminated. If surfaces are porous or if water would collect in the cracks, the use of detergent and water may not be desirable. In addition, materials on surfaces tend to adhere more strongly to them as time passes, due to moisture, oils, etc., in the air, so if it is not desirable to use a cleansing solution to remove the material, it would be best to begin decontamination as soon as practicable while material is still relatively loose. If a very limited area is involved, loose particulates often may be picked up by pressing the sticky side of masking or duct tape against the contaminated surface. If a larger area is contaminated, much of the loose material can be picked up by using a HEPA filter-equipped vacuum cleaner. The filter in a vacuum cleaner of this type is capable of removing 99.97% of all particles of 0.3 microns or greater from the air moving through it. Brushing lightly with a soft brush can break loose additional material that can be vacuumed with the special vacuum. To keep airborne material from spreading, the entrance to the area being cleaned can be isolated with a temporary "airlock" made of 6-mil polyethylene plastic that will block any air moving to and from the isolated area, but will allow passage of workers in and out of the room. If needed, the plastic enclosure can include a change area or even a temporary shower. Airborne material also can be continuously removed from the area by placing a movable, HEPA filter-equipped, air circulator within the space being cleaned. The circulator should be sized to pass the air in the space through the filter frequently. A 2000-cfm unit would pass the air (if thoroughly mixed) in one of the standard laboratory modules described in Chapter 3 through the filter about every 2 minutes. Local asbestos abatement firms use these routinely and one can possibly be obtained from them in an emergency.

If contamination has permeated deeply into the work and floor surfaces and into seams and cracks in the floor, it may be simpler to remove some of the surface and to remove and replace floor tile than to attempt cleaning. If the floor tile contains asbestos, as older 9 inch tiles very often do, the removal would have to be done according to OSHA and EPA restrictions on asbestos removal.

Every area that has been decontaminated must be thoroughly surveyed by the organization's radiation safety office and certified as meeting the limits for surface contamination established by the organization before release for use.

Pieces of electronic equipment are perhaps the most difficult items to clean since they frequently are equipped with fans that draw the air through them. Besides normal lubricants in them, over time they all accumulate at least some dust and grime to which the contamination is likely to adhere, even in rooms in which the air is carefully filtered. Normally, washing the interior components is impractical because of the potential damage to the components. It may be possible to use a very fine nozzle on a HEPA vacuum to clean the bulk of the removable dust from inside the instrument, and the remainder may be loosened and removed with the careful use of small swabs and solvent, followed by another vacuuming. The decision to make the effort will depend on the cost of the equipment, and the difficulty of the decontamination effort. A factor that might influence the decision to make the effort would be the possibility of a successful damage claim on the organization's insurer. The disposal cost of the equipment should be considered as well as the replacement cost. Disposal of any kind of hazardous waste has become increasingly expensive.

Small tools and glassware with hard, nonporous surfaces may be cleaned by standard techniques. Washing with a detergent and water may be sufficient. More resistant contaminating material in glassware can be cleaned with any of the standard solutions used in laboratories for cleaning glassware, such as a chromic acid solution. Contaminated metal tools that resist cleaning with detergent and water, accompanied by brushing, can be washed with a dilute solution of nitric acid. Sulfuric acid can be used on stainless steel tools. Again, the choice must be made among the relative cost of the tool, the cost of disposal, and the effort required to decontaminate it. Tools that have grease in and on them and hidden crevices in which radioactivity may become embedded may be too difficult to decontaminate to make the effort cost-effective.

If disposable protective clothing is used throughout the decontamination process the problem of laundering contaminated garments can be minimized. However, any items that need to be cleaned should be checked by the radiation safety office before sending them to a commercial laundry. If the quantity is not large, manual washing of some items might be possible, taking care not to get skin contaminated. Where more clothes are involved, purchasing a washing machine for the job might be cost effective. A simple, inexpensive model would normally be sufficient and probably can eliminate or reduce the contamination to an acceptable level unless the clothes are heavily contaminated in a form physically difficult to remove. If the clothes are contaminated with significant amounts of radioactive materials, any wastewater should be captured and retained until certified by the radiation safety officer as suitable to be disposed of in the sanitary system.

m. Radioactive Waste Disposal[*]

In past years, almost all off-site low-level non-federal radioactive waste disposal was sent to three commercially operated waste disposal burial grounds located in South Carolina, Nevada, and the State of Washington. The one in Nevada is now closed but a new commercially operated one is now being operated by Envirocare in Utah. The Hanford, Washington site only accepts waste from the Northwest compact. There have been other commercial facilities but they were closed for various environmental problems. The choice of the location to which generators of radioactive waste sent their waste was dictated by a number of operating restrictions for the three facilities that remained operating. As a result, many institutions found themselves sending much of their laboratory-type radioactive waste from the east coast and elsewhere to the Washington facility.

Because it was deemed inappropriate by Congress for one region to be responsible for handling another region's wastes, they dictated that a system of regional compacts was to be established, and each region would be required, when the system became fully operational, to take care of its own radioactive waste. Forty-two states now belong to compacts. Some regional compacts are fully operational while others are not. At this time, it is still possible for a generator in one region to send waste to another area, but at some point in the future, this will

---

[*] A good summary of the status of radioactive waste disposal in the United States can be found in an article by John R. Vincenti, *Low-Level Radioactive Waste Disposal in the United States* in the Supplement to the Health Physics Journal, Vol. 76 , No. 5, May 1999. It is noted in this article that the volume of this type of waste has decreased from 3.8 million ft^3 in 1990 to 0.32 million ft^3 in 1997.

no longer be allowed. The date at which this will occur is not wholly certain. Several states are developing regional sites at this time, but in the meantime, have secured the required permission from the compact in which they are located to send their wastes to either South Carolina or to Utah. Note that South Carolina has withdrawn from the Southeast compact of which it was once a member and now will accept waste from any other state except North Carolina, which should have been the next host state for the Southeast compact but has been slow to fulfill this role. The one in Utah can only accept low level waste while the other two can accept higher category wastes, but since this corresponds to the largest physical volume of radioactive waste although not the largest activity amount, many states now utilize this facility. The Utah site is also licensed to accept mixed waste. Its costs are competitive as well.

Although the nation's research and medical facilities do not produce the volume of waste generated by the electric utilities, they are much more vulnerable to the lack of a disposal facility for their waste because they do not have adequate facilities in which to store low-level waste on a long-term basis. It is essential for many research applications to use certain isotopes, which, because of their radiation characteristics, must go to a long-term disposal facility. The inability to use these isotopes would seriously hamper many research programs. These research applications often are in the life sciences and involve such critical areas as cancer, the search for cures to other diseases, recombinant DNA, and other comparably important areas of basic and applied research.

The actual amount of waste shipped off-site at many research facilities has decreased or stabilized as more facilities use waste reduction techniques to reduce the amount of radioactive waste that must be shipped away.

Decay-in-storage is the simplest method available for the reduction in the amount of waste that must be disposed of in a burial facility. Many of the most commonly used radio-isotopes have sufficiently short-half lives so that the radioactive content will decay to a "safe" level within a reasonably short time. For example, all of the following have half-lives of less than 65 days (the current limit for decay in storage), so that any material in the waste will go through at least 10 half-lives in less than 2 years, which is generally accepted as a reasonable storage period: ^{131}I ^{32}P, ^{47}Ca, ^{33}P, ^{51}Cr, ^{59}Fe, ^{203}Hg, and ^{125}I. Also the combination ^{99}Mo $\rightarrow$ ^{99m}Tc which has important nuclear medicine applications, has a short effective half-life so that any waste from its use also can be allowed to decay in storage. The initial activity of any radioactive material when placed in storage will determine if ten half-lives is sufficient to reduce the level to a point at which an unshielded reading with a survey meter on its most sensitive scale shows no reading over background. If this condition is not met, then additional decay time must be allowed before the material is disposed of as ordinary waste.

Another significant means of reducing the amount of waste is to take advantage of the provisions of Section 20.2003 to dispose of some of the radioactive waste into the sanitary system. However, Section 20.2003 prohibits the disposal of licensed material into the sanitary system unless it is readily soluble or dispersible in water. Some concern has been expressed about the dispersible materials and this may be modified. Many fluids used in liquid scintillation fluids do not meet these criteria, but there are now commercially available alternative scintillation liquids that do. In 1990, 71% of the "mixed waste", i.e. materials with both radioactive and chemically hazardous properties, were from liquid scintillation materials. The use of these newer scintillation fluids allows licensed or unlicensed material to be put into the sanitary system as long as the amounts are within the limits specified in 20.2003(a). Another restriction limits the yearly amounts to 5 Curies of ^{3}H or 1 Curie of ^{14}C and a total of 1 Curie of all the other radio-isotopes combined. Since ^{14}C and ^{3}H, two of the most commonly used radioactive research materials, have long half-lives and advantage cannot be taken of the decay-in-storage technique, the capability of using the sanitary system is of significant help in reducing the waste volume requiring commercial burial. Records of the amounts disposed of in this method must be maintained, and also the amounts disposed of by other means.

Very low specific activity (0.05 μCi or less per gram) ^{3}H or ^{14}C in scintillation fluid or in animal tissue (averaged over the weight of the entire animal) can be disposed of without consideration of the radioactivity, but is normally incinerated. The scintillation fluids can be incinerated only if they do not meet the EPA ignitability criterion. Some generators have incinerators licensed for disposal of radioactive waste that are used for this purpose.

Some dry solid wastes can be compacted to reduce their volume. Some generators have developed centralized local facilities to do this, while others place these materials into separate containers for the waste disposal firm to carry out this waste volume reduction technique.

These methods, plus encouraging the users to choose a material that can be allowed to decay in storage, can be employed to substantially reduce the amount and expense of radioactive waste disposal.

To take advantage of these waste and cost-reduction measures, users must cooperate to segregate the waste into appropriate categories: (1) aqueous, (2) nonaqueous, (3) aqueous containing ^{3}H and ^{14}C from other aqueous, (4) vials of nonmiscible scintillation fluids containing less than 0.05 μCi/g of ^{3}H and ^{14}C from other scintillation fluids and EPA compliant scintillation fluids from others, (5) dry solid wastes, and (6) animal carcasses separated into those containing more than 0.05 μCi/d per g of tissue of ^{3}H and ^{14}C from those that contain less than that amount. The generators of the waste must label each container of waste with the amount of each radioactive isotope contained in it. This information is used by either the radiation safety office to prepare the waste for shipping and to prepare the required manifests and forms as specified in Section 20.2006, or by a waste disposal firm who may do this service as well as transporting the waste to a burial ground. Each container or package of waste shipped must be classified and labeled according to 10 CFR Part 61, Section 60.55- 57.

Almost all waste is originally packed into 55-gallon steel drums that meet DOT specifications, segregated by types. Generally landfills will not take liquid wastes, even if they have been absorbed in an appropriate absorber. They must be solidified before being offered for burial. Some organizations have a commercial service do this procedure. The choice in most cases of whether to do a given operation locally or have it done by an outside contractor is usually based on economic factors. Nonaqueous waste is usually treated as RCRA waste as well.

Radioactive waste allowed to decay in storage may be disposed of according to the following schedules:

- Dry solids are taken to an ordinary landfill. Any labels on any container showing that the contents are radioactive must be removed or defaced.
- Nonhazardous, water-miscible waste is put into the sanitary system.
- Chemically hazardous waste is disposed of through the hazardous waste disposal program for that type of materials. Labels on a container showing that the contents are radioactive must be removed or defaced.
- Carcasses are normally incinerated but state regulations may affect this method, if the tissue is infectious, it must be disposed of appropriately to that classification.

Proper disposal of all radioactive waste is a major responsibility of the organization's radiation safety office. Proper records must be maintained of all transfers of radioactivity from the laboratory into the waste stream.

n. Individual Rights and Responsibilities

Individuals who are employed in licensed programs in which they work with byproduct materials (or those employed in several other activities regulated under Title 10 Chapter I, of the Code of Federal Regulations) have their rights and responsibilities spelled out in the regulations, primarily in Parts 19 and 21. Part 21 requires that the responsible parties in any licensed activity report to the NRC any safety deficiency, improper operations or defective equipment that could pose a "substantial safety hazard...to the extent that there could be a major reduction in the degree of protection provided to public health and safety..." However, any individual can make such a report. If a responsible party fails to do so, substantial civil penalties can be imposed on the offending party.

Part 19 is the section of the regulations that defines the rights of the employees and those of the NRC, and also the responsibilities of the licensee to make these rights available. The major provisions of Part 19 are summarized below:

i.   Information Requirements

The licensee must post several items of information relating to the operations under the license or post information describing the documents and where they may be examined. The latter choice is the one most often exercised. The documents involved are:

- The regulations in 10 CFR Part 19
- The regulations in 10 CFR Part 20
- The license, license conditions, and documents incorporated into the license by reference
- The operating procedures applicable to the licensed activities
- Form NRC-3, "Notice to Employees"

The following items must be posted, without the option of simply informing the employees of their existence and where they may be examined.

- Any violations of radiological working conditions
- Proposed civil penalties or orders
- Any response of the licensee

A critical section of Part 19 is 19.12, "Instructions to Workers," is given in its entirety below. It is brief, but is functionally equivalent to the OSHA Hazard Communication Standard for employees engaged in activities regulated by the NRC.

"All individuals working in or frequenting any portion of a restricted area shall be kept informed of the storage, transfer, or use of radioactive materials or of radiation in such portions of the restricted areas; shall be instructed in the health protection problems associated with exposure to such radioactive materials or radiation, in precautions or procedures to reduce exposure, and in the purposes and functions of protective devices employed; shall be instructed in, and instructed to observe, to the extent within the worker's control, the applicable provisions of Commission regulations and licenses for the protection of personnel from exposure to radiation or radioactive materials occurring in such areas; shall be instructed of their responsibility to report promptly to the licensee any condition which may lead to or cause a violation of Commission regulations and licenses or unnecessary exposure to radiation or to radioactive material; shall be instructed in the appropriate response to warnings made in case of any unusual occurrence or malfunction that may involve exposure to radiation or radioactive material; and shall be advised as to the radioactive exposure reports that workers may request pursuant to §19.13. The extent of these instructions shall be commensurate with potential radiological health protection problems in the restricted areas."

Implicit in the above instructions is that the employees must comply with the regulations and with procedures adopted to protect not only themselves but others, including the public outside the areas in which the licensed activities take place. However, it is not sufficient for an employer simply to place such responsibility on the employee. The employer must see that the employees comply with all applicable standards. There have been several instances in which heavy fines have been imposed on corporations and academic institutions that have failed to enforce compliance.

ii. Monitoring Data

Individual workers have the right to information concerning their radiation exposures, and directly related supporting data, under several provisions of Section 19.13:

- The worker can request an annual report.
- A former worker can request a report of his exposure records. The report must be provided within a 30-day interval or within 30 days after the exposure of the individual has been determined, whichever is later.
- An employee terminating work involving exposure to radiation either for the licensee or while working for another person can request the information. An estimate, clearly labeled as such, may be provided if the final data are not yet available at the time of the request.
- If the licensee is required to make a report to the NRC of an individual's exposure data, a report must be made to the individual no later than the report made to the NRC.

iii. Inspection Rights

The NRC has the right to make unannounced inspections of a licensed facility, including the "...materials, activities, premises, and records..." During an inspection, the inspectors have the right to consult privately with the workers. However, at other times, the licensee or a representative of the licensee can accompany the inspector.

The workers may have an official representative who may accompany the inspector during the inspection of physical working conditions. Normally, this is an individual who is also a worker engaged in licensed activities under the control of the licensee, and therefore could be expected to be familiar with the instructions provided to the employees by the employer. However, if agreeable to the employer and to the employees, an outside person can be allowed to accompany the inspector during the inspection of the physical working conditions. Different employee representatives can accompany the inspector during different parts of the inspection, but no more than one at any given time. If a person deliberately interferes with an inspector while conducting a reasonable inspection, he can refuse to let that person continue to accompany him. Certain areas, such as classified areas, or areas where proprietary information is involved, would require that only individuals normally having access to these areas or information can accompany the inspector.

During an inspection, an employee can bring privately to the attention of the inspector any matter relating to radiological safety pertaining to the licensee's operations that he wishes, either orally or in writing.

An employee or an employee representative can report a violation of the terms of the license or regulations covering the radiological activities engaged in by the employee. This should be done in writing to the commission or to the director of the commission's regional office and should provide pertinent details of the alleged violations. The identity of the employee making the report will not appear in any public report except under demonstrated good cause.

If the director of the regional office feels that there is valid cause to believe that a problem exists as described by the complainant, he will initiate an inspection as soon as practical. If an inspection does take place, the inspector will not necessarily confine himself to the original problem. The director may decide that an inspection is not justified. However, a complainant may request an informal conference to press his case or a licensee also may request one, although in the latter case the complainant has the right to not allow his identity to be made known.

The licensee is specifically prohibited against any act of discrimination against an employee who makes a complaint or asks for an inspection.

iv. Penalties

Violations of Part 19, as with other regulatory sections, can be classed as *de minimis* or higher. If the violations are sufficiently serious, the commission can obtain a court order prohibiting a violation or one can be obtained to impose a fine or a license may be revoked. If a violation is willful, in principle the person responsible can be considered guilty of a crime and be subject to a fine and/or imprisonment. Of course, none of the penalties can be invoked without the licensee having ample opportunity to respond to the charges and to propose a plan of correction.

REFERENCES

1.   **Komarov, E.I.,** Radiation in daily life, *Int. Civ Defense,* No. 295, 3.
2.   New A-bomb studies alter radiation estimates, *Science,* Vol. 212, May 22, 1981.
3.   Ionizing Radiation: Sources and Biological Effects, United Nations Scientific Committee on the Effects of Atomic Radiation (UNSCEAR), 1982 Report to the General Assembly, with Annexes, United Nations, NY, 1982.
4.   **Fabrikant, J.I.,** Carcinogenic effects of low-level radiation, *Health Phys. Soc. Newsl.,* X(10), October, 1982
5.   Radiation Exposure of the Population of the United States, NCRP Report No. 93, National Council on Radiation Protection, Bethesda, MD, 1987.
6.   **Luckey, T.D.,** Physiological benefits from low levels of ionizing radiation, *Health Phys.,*43(6), 771, 1982.
7.   Special issue on radiation hormesis, *Health Phys.,* 52(5), May, 1987.
8.   **Cohen, B.L.,** Alternatives to the BEIR relative risk model for explaining atomic-bomb survivor cancer mortality, *Health Phys.,* 52(1), 55, 1987.
9.   Location and Design Criteria for Area Radiation Monitoring Systems for Light Water Nuclear Reactors, ANSI/ANS-HPSSC-6. 8.1, American National Standards Institute, New York, 1981.
10. *Performance Requirements for Pocket-Sized Alarm Dosimeters and Alarm Rate Meters,* ANSI Ni 3.27, American National Standards Institute, New York, 1981.
11. **Hodges, H.D., Gibbs, W.D., Morris, A.C., Jr., and Coffey, W.C., II,** An improved high-level whole-body counter, *J. Nucl. Med, 15(7),* 610, 1974.
12. **Graham, C.L.,** A survey-instrument design for accurate B dosimetry, *Health Phys.,*52(4), 485, 1987.
13. **Kathren, R.L.,** *Radiation Protection, Medical Physics Handbook 16,* Adam Hilger, International Publishers Services, Accord, MA, 1985.
14. Standards for Protection Against Radiation, 10 CFR Part 20, Jan 1, 1998.
15. Standards for Protection Against Radiation, 10 CFR Part 20, *Fed Reg.,* 56 (97), 1991.
16. Rules of General Applicability to Domestic Licensing of Byproduct Material, 10 CFR Part 30, Jan 1, 1998.
17. Licensing Requirements for Land Disposal of Radioactive Waste, 10 CFR Part 61, *Fed Reg.,* 47(248), 57446, 1982.

18. Decontamination Limits, Regulatory Guide 1.86, Nuclear Regulatory Commission, Washington, D.C., 1974.

19. Decontamination Limits, ANSI/ANS-N13. 12, American National Standards Institute, New York.

20. Instruction Concerning Prenatal Radiation Exposure, Draft Regulatory Guide DG-8014, Nuclear Regulatory Commission, Oct 1992.

21. Information Relevant to Ensuring that Occupational Exposures at Medical Institutions Will Be As Low As Reasonably Achievable, Regulatory Guide 8.18, Nuclear Regulatory Commission, 1982.

22. Instructions Concerning Risks from Occupational Radiation Exposure, Draft Regulatory Guide DG-8012, Nuclear Regulatory Commission, February 1996.

23. National Emission Standards for Hazardous Air Pollutants: Radionuclides, 40 CFR Part 61, *Fed Reg.,* 54(240), 51654, 1989, Revised January 9, 1997.

24. A Guide for Determining Compliance with the Clean Air Act Standards for Radionuclide Emissions from NRC-Licensed and Non-DOE Federal Facilities, Revision 2, EPA 520/1 -89-002, Environmental Protection Agency, Washington, D.C., 1989.

25. User's Guide for the Comply Code, 520/1-89-003, Revision 2, Environmental Protection Agency Washington, D.C., 1989.

26. **Steere, N.V.,** Ed., Safe handling of radioisotopes, in *CRC Handbook of Laboratory Safely,* CRC Press, Cleveland, OH, 1971, 427.

27. **Strom, D.J.,** The four principles of external radiation protection: time, distance, shielding, and decay, *Health Phys.,* 54(3), 353, 1988.

28. **Saenger, E.L.,** Acute local irradiation injury, in *Proc. REACTS Int Conf Medical Basis for Radiation Accident Preparedness,* Hübner, K.F., and Fry, S.A., Eds., Elsevier/NorthHolland, 1980.

29. **Martin, A. and Harbison, S.A.,** *An Introduction to Radiation Protection,* 3rd. ed., Chapman and Hall, New York, 1986.

30. **Morgan, K.Z. and Turner, J.E.,** *Principles of Radiation Protection,* John Wiley & Sons, New York, 1967.

31. **Pelà, C.A., Ghilardi, A.J.P., Netto, and Ghilardi, T.,** Long-term stability of electret dosimeters, *Health Phys.,* 54(6), 669, 1988.

32. **Cember H.,** *Introduction to Health Physics,* Pergamon Press, Oxford, 1969.

33. *Radiological Health Handbook,* U.S. Department of Health, Education and Welfare, Washington, D.C., 1970.

34. Air Sampling in the Workplace, Nuclear Regulatory Commission, DG-8003, September, 1991.

35. Monitoring Criteria and Methods to Calculate Occupational Exposure, Nuclear Regulatory Commission, DG-8010, July, 1992.

36. Tests and Calibration of Protective Instrumentation, Nuclear Regulatory Commission, OP-032-5, September, 1984.

37. Specific Domestic Licenses of Broad Scope for Byproduct Material, Nuclear Regulatory Commission, January, 1998.

38. Packaging and Transportation of Radioactive Materials, Nuclear Regulatory Commission, 10 CFR part 71, January 1998.

39. Notices, Instructions and Reports to Workers, Inspections and Investigations, Nuclear Regulatory Commission, 10 CFR Part 19, July, 1998.

40. Contamination Control, Department of Transportation, 49 CFR, Part 173.443, October, 1998.

INTERNET REFERENCES

1. Code of Federal Register, http://www.gpo.gov/nara/cfr-retreive.html#page1

2. Regulatory Guides, http://www.johnglenn.com/Radiationsaety/regulato.htm

3. Envirocare homepage, http://www.envirocareutah.com

J. Radon

The problem of radon is not necessarily connected with laboratory operations involving radiation other than its being a contributor to background for counters used in these facilities and representing a problem when taking an air sample. However, many laboratories are built of concrete blocks and have concrete floors, drains that open to storm sewers, and other means for radon to enter a facility. The sand and aggregate used for the concrete in both the blocks and cement slabs may contain substantial amounts of natural radioactivity. Currently, a considerable amount of concern is being raised about the possibility of radon-induced lung cancer from this source. Since a typical research worker may spend an average of 8 or more hours per day in a research facility, it may be useful to provide a brief section on radon. A discussion of the physics of radon activity was provided in Section II.G.1.j. Figure 5.11 provides some additional information on the decay products of ^{222}Rn, the major environ-mental problem.

A study by the National Research Council has resulted in an estimate of the risk of lung cancer that is intermediate between those of earlier estimates. The report also suggests that the risk to smokers is ten or more times greater than that to nonsmokers. Although compiled by an eminent group of scientists using a more sophisticated set of assumptions than in the past, other well-known scientists have questioned some of the assumptions made and some features of the model. The results are given below, but before presenting them, a brief discussion of some of the terms is needed.

Risk estimates are made in terms of working-level-months (WLM). A WLM is defined as "an exposure to a concentration of any combination of the short-lived radon daughters in one liter of air that results in the ultimate release of 1.3×10^5 MeV of potential alpha energy for a working month of 170 hours." This amount of radiation is approximately equivalent to the radiation of a radon daughter in equilibrium with 100 pCi of radon. A level of 4 pCi/L of radon is equal to about 0.02 working levels. Over an entire working year, equal to 2000 hours, the equivalent exposure would be equal to about 0.25 WLM.

The BEIR IV report estimates that the excess lifetime risk of death from lung cancer from this source is 350 deaths per million person WLM. Based on this model and the assumed levels of exposures throughout the U.S., this is equivalent to an estimated 5000 to 20,000 lung cancer deaths due to radon per year in the U.S. Recently, the National Cancer Institute has published a newer study of eight epidemiologic studies of the cancer risk from radon. They found results that were not inconsistent from earlier studies. The following brief Table 5.7 of the radon associated cancer risks for non-smokers vs. smokers was taken from the first of the Internet references. It is clear that smoking has a dramatic effect on the rate of radon induced cancers. At the 4pCi/L level, the effect of smoking is to increase the number of lung cancer deaths by a factor of about 15, while at the 20 pCi/L level the ratio is close to 17. The fraction of persons smoking in the U.S. has declined to about 25% so that the effect of radon exposure estimates represent only about 5% of the total. In addition, since there are about 400,000 deaths attributable to lung cancer in the U.S. annually, even at the high end of the high end of the estimates, radon would be responsible for only 5% of the deaths and possibly only about 1%.

At the time of this writing, there are reputable scientists who both support and object to these findings. However, because there are public concerns about the potential of radon as a cause of lung cancer, many commercial services have been established which will provide devices that can be placed in the home for a period of time, and then can be sent to their laboratories for analysis of the ambient radon levels. These firms place advertisements in the media or often, in the "yellow pages," so that anyone wishing an analysis of their home or workplace can readily have it done at relatively reasonable costs. In some areas, state agencies will perform a survey upon request. Some localities now require a radon evaluation prior to the sale of a home. If one finds significant levels of 4 pCi/L or more in the home, there are steps that can be taken to reduce the levels. The major thrust of these measures is to block entry into the home of gaseous effluents from the soil, to which the radon particulates can become attached. Basements below grade are a particular concern due to the heavier than

NUCLIDE   PRIMARY RADIATIONS

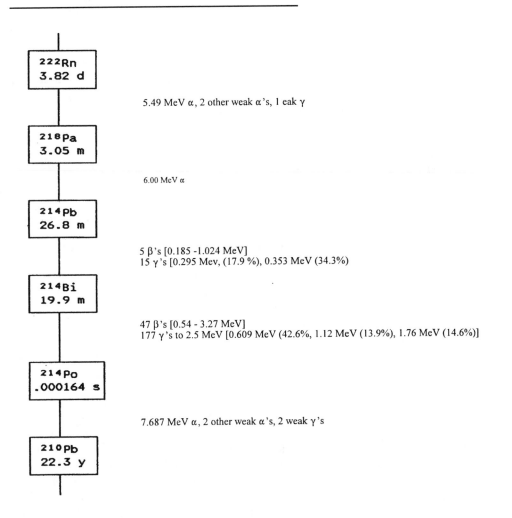

Figure 5.11 Primary decay path for ^{222}Rn

air nature of the particulates. There are links in the second Internet reference which discuss these measures at length.

REFERENCES

1. *Health Risks of Radon and Other Internally Deposited Alpha-Emitters; BEIR IV.* National Academy Press, Washington, D.C., 1988.
2. **Cohen, B.L. and Gromicko, N.,** Adequacy of time averaging with diffusion barrier charcoal adsorption collectors for ^{222}Rn measurements in homes, *Health Phys.,* 54, 195, February, 1988.

Table 5.7 Additional Lung Cancers per 1000 People

Exposure	Non-smokers	Smokers
4pCi/L	2	29
10pCi/L	4	71
20pCi/L	8	135

3. Radon in Buildings, NBS Special Publication 581, Colk, R. and McNall PB.,Jr., Eds., National Bureau of Standards, Gaithersburg, MD, 1980.
4. **Puskin, J.S. and Yang, Y.A., A** retrospective look at Rn-induced lung cancer mortality from the viewpoint of a relative risk model, *Health Phys.*, 54(6), 635, 1988.
5. **Cohen, B.L.,** Relationship between exposure to radon and various types of cancer, *Health Phys.*, 65(5), 529, 1993.
6. **Cole, L.A.,** *Element of Risk: The Politics of Radon,* AAAS Press, Washington, D.C., 1993.
7. **Lubin, J. and Boice, J, Jr.,** *Lung cancer risk from residential radon: Mata-analysis of eight epidemiological studies,* National Cancer Institute, 89;49-57, 1997.
8. National Radiological Protection Board., Risk of radiation-induced cancer at low doses and low dose rates for radiation purposes. Documents of the NRPB, 6(1), 1995.

INTERNET EFERENCES

1. http://www.oswego.edu/~gabel/g3rdn.html.
2. http://cancernet.nci.nih.gov.
3. Draft NCRP Scientific Committee, *Evaluation of the Linear Nonthreshold Dose-Response Model*, http://www.ncrp.com.

K. Acute Radiation Syndrome

It is improbable that an individual working with byproduct materials will ever be exposed to radiation levels that would cause him to be concerned about the immediate acute effects of radiation, leading to serious injury or death. However, there have been a small number of incidents in which massive exposures have occurred. These have normally involved large sources used for radiography, radiation therapy, sterilization of insects, machines producing intense beams of radiation, X-ray units, and critical reactor assemblies. Individuals have survived very high localized radiation doses, in the range of 10,000 rems or more, if the areas affected were relatively small and did not expose the more sensitive organs to a substantial dose. Following is a brief discussion of the immediate effects of a short-term exposure to high levels of radiation, summarized in Table 5.8. It should be noted that the levels that cause immediate harm to individuals are hundreds of times greater than that allowed individuals in normal usage spread over a period of months.

There is significant variation in the responses of individual persons to exposures of the whole body. The least amount of radiation for which clinical evidence is discernible in a "typical" person is approximately 25 rad (0.25 Gy). The effect will be noted as changes in the blood system. Above about 50 rad (0.5 Gy), most persons will show some effect in their blood system. There may be temporary sterility in men for exposures to the gonads at these levels. However, it is unlikely that an individual will personally feel any immediate discomfort at these relatively low levels.

As the exposure levels increase to approximately 100 rad (1 Gy), a small percentage of sensitive persons will begin to experience some of the symptoms of discomfort associated with acute exposures to substantial levels of radiation, nausea, fatigue, loss of appetite, sweating

Table 5.8 Dose-Effect Relationships Following Acute Whole-Body Irradiation (X-ray or gamma)

Whole-Body Dose (Gy)	Clinical or Laboratory Findings
05-0.25	Asymptomatic. Conventional blood studies are normal. Chromosome aberrations detectable.
0.50-0.75	Asymptomatic. Minor depressions of white cells and platelets detectable in a few persons, especially if baseline values established.
0.75-1.25	Minimal acute doses that produce prodromal symptoms (anorexia, nausea, vomiting, fatigue) in about 10% to 20% of persons within two days. Moderate depressions of white cells and platelets in some persons.
1.25-2.00	Symptomatic course with transient disability and clear hematologic changes in a majority of exposed persons. Lymphocyte depression of about 50% within 48 hours.
2.40-3.40	Serious, disabling illness in most persons, with about 50% mortality if untreated. Lymphocyte depression of about 75% within 48 hours.
5.00+	Accelerated version of acute radiation syndrome, with gastrointestinal complications within 2 weeks, bleeding and death in most exposed persons.
50.00+	Fulminating course, with cardiovascular gastrointestinal, and CNS complications resulting in death within 24 to 72 hours.

and a general feeling of malaise. As the exposure level increases further, the percentage of individuals exhibiting symptoms increases, until at the point a person experiences an exposure of 200 rad (2 Gy) most will be affected to some degree. At about this same level, some individuals who do not receive treatment may not survive. At an exposure between 300 and 450 rad (3 to 4.5 Gy), the survival rate without aid decreases to 50% by 30 days after the exposure. With proper treatment, survival is a distinct possibility up to about 500 rad (5 Gy,) above which, even with optimal treatment, survival chances decrease rapidly, until at about 800 rad (8 Gy), survival is highly unlikely.

Many individuals who have a finite chance of survival will begin to feel better after a day or two, but after a few weeks, their situation appears to worsen. They may have fevers, infections, loss of hair, severe lethargy, hemorrhaging, and other problems with their cardiovascular systems. During this 2- to 3-week interval, blood tests will reveal that significant changes are occurring in the blood system which reflect the damage to the blood-producing bone marrow. At about 200 rad (2 Gy) depression of the bone marrow function is apparent. At higher doses, complete ablation of the bone marrow occurs. Up to some exposure level, if the patient can be kept alive, the bone marrow will regenerate, but at a level in the area of 400 to 600 rad (4 to 6 Gy), this will become unlikely, and bone marrow transplants will be necessary for survival. There are problems with this technique, including the need for a close genetic match for the donor material and the possibility of infection. However, if the blood-forming function can be restored, and death does not occur due to secondary infections because of the lack of white blood cells, survival for 60 days or more is a promising sign of eventual survival.

As the level of radiation increases, the survival period decreases. At about 1000 rad (10 Gy), there will be serious damage to the gastrointestinal tract and survival may be a few weeks. At exposures of several thousand rad, the central nervous system will be strongly affected and survival will be measured in terms of hours. Much of the information above is derived from reference 3. An excellent current (1999) summary of the biological effects of radiation, including more on the acute radiation syndrome is found in the Internet reference to this section.

REFERENCES

1. **Upton, A.C. and Kimball, R.F.,** Radiation biology, acute radiation syndrome, in *Principles of Radiation Protection,* Morgan, K.Z. and Turner, J.E., Eds., John Wiley & Sons, New York, 1968, 427.
2. **Wald, N.,** Effects of radiation over-exposure in man, in *Principles of Radiation Protection,* Morgan, K.Z. and Turner, J.E., Eds., John Wiley & Sons, New York, 1968, 457.
3. **Voelz, G.L.,** Ionizing Radiation, Occupational Medicine, 3rd Ed. Zenz, C., Dickerson, O.B., and Horvath, E.P., Eds. St. Louis, MO, 1994.
4. **Andrews, G.A., Sitterson, B.W., Kretchnej, A.L., and Brucer, M.,** Criticality accident at the Y-12 Plant, Proc. Scientific Meeting on the Diagnosis and Treatment of Acute Radiation Injury, World Health Organization, Geneva, 17 October, 1960.
5. **Andrews, G.A., Hübner, K.E., and Fry, S.A.,** Report of 21-year medical follow-up of survivors of the Oak Ridge Y-12 accident, in *The Medical Basis for Radiation Accident Preparedness,* Hübner, K.F. and Fry, S.A., Eds., 59, Elsevier/North-Holland, New York, 1980.
6. Radiation accident grips Goiania, *Science,* 238, 1028, 1987.
7. **Schauer, D.A., Coursey, B.M., Dick, C.E., MacLaughlin, W.L., Puhl, J.M., Desrosiers, M.E, and Jacobson, A.D.,** A radiation accident at an industrial accelerator facility, *Health Phys.,* 65(2), 1993.
8. **Levin, S.G., Young, R.W., and Stohier, R.L.,** Estimation of median human lethal radiation dose computed from data of reinforced concrete structures in Nagasaki, Japan, *Health Phys.,* 63(5), 522, 1992.

INTERNET REFERENCE

1. GMA-12, Guide to Ionizing Radiation Exposures for the Occupational Physician, Atomic Energy Control Board of Canada, November 1998: http://198.103.98.211/reports/adv_comm/gma12_e/intro.htm.

III. X-RAY FACILITIES

There are many radiation safety issues associated with the use of X-rays in the laboratory. Much of the health physics information in standard texts, as applied to X-ray systems, discusses radiation safety in terms of a determination of the amount of shielding necessary in medical X-ray installations to protect persons in adjacent areas from the effects of radiation.

The radiation sources are the primary beams scattering from the patient or target and leakage from the radiation source. Except in specialized texts, there is little discussion of operational personnel protection. Although there are many research applications in which X-ray machines are used in a manner equivalent to medical usage, even here there are many applications in which there are significant differences in safety considerations. For example, in veterinary medicine, X-rays of large animals are not only likely to cover a larger area and involve more scattering mass, but it is far more likely that a holder (or more than one) will be needed than for most human patients. Besides briefly covering shielding requirements, the sections that follow will emphasize the exposures that the researcher receives, the doses to the patient, and the means by which these exposures can be reduced. Shielding design cannot be neglected, but will often be decided by the architect and his consultant before the research scientist becomes involved. Often, the research scientist will have relatively little control over his physical environment, especially those who begin working in existing facilities. Further, most architects will seek out a consultant to assist them should a need arise to create an X-ray facility in an existing structure. Most X-ray users' expertise and interests are primarily in applications, not facility design other than those features that enhance its usefulness to them. However, the scientist will exercise essentially complete control over the operations of the X-ray facility and should be held responsible for establishing operating procedures that maximize radiation protection for everyone.

Many individuals associate anything to do with radiation with the NRC, but it does not regulate the use of X-ray machines nor establish exposure levels for individuals working in these facilities. The Food and Drug Administration in Title 21 of CFR provides standards for X-ray devices, but for the most part the responsibility for regulating the use of equipment that emits X-rays is left up to the states. The degree of regulation provided varies among the various states. The National Committee on Radiation Protection and Measurements (NCRP) has recommended limits of exposure that are similar to those required by the NRC, basically 100 mrem per week (1 mSv per week) as an occupational limit and 10 mrem per week (0.1 mSv per week) for the public.

A. Generation of X-Rays

The general form of a device in which X-rays are generated is one in which electrons emitted from a cathode are accelerated through a voltage, V, and focused on a target, or anode, that has a high atomic number. The primary mechanism by which the X-rays are produced is bremstrahlung, as the electrons slow within the target. A substantial amount of heat is generated in the process, since only about 1% or less of the energy of the electron beam is converted to X-rays. In order to have a high atomic number for the maximally efficient conversion of energy into X-rays, the target is typically made of tungsten, which also has a high melting point. The target is usually embedded in an efficient heat conductor, usually copper, serving as the anode. Some designs add fins to conduct heat away from the anode, some cause the anode to rotate so the beam will not be continuously focused on the same spot, and other designs circulate cooling liquids through the anode.

The maximum value of the energy of the X-rays generated in an X-ray tube is determined by the accelerating potential, usually denoted by kVp. However, the energy spectrum of the X-rays is a continuum extending downward from this maximum energy. Thus, in any X-ray spectrum there are many "soft" X-rays that are often not of any significant value, although if the normal operating voltage is low, part of the wall of the tube may be made of a thin piece of beryllium to allow more of the lower energy components to escape from the tube. Usually, these unwanted X-rays are eliminated by interposing aluminum filters (or for higher energy machines, aluminum and copper are used to extend the energy of the filtered component to higher values) in the beam. These energy-modifying filters are typically 0.5 to 3 mm thick, depending upon the kVp of the X-ray machine.

B. Types of Machines

As noted above, X-ray machines for medical applications represent only one type of device that emits X-rays used in the laboratory. There are open and closed beam analytical instruments, used for X-ray diffraction work, cabinet systems, electron microscopes, and microprobes, among others. Each of these devices can, under appropriate circumstances, pose radiation exposure problems. Besides installing these in properly designed facilities, operating procedures must be established to limit occupational exposures, as well as those of the patients (for medical machines) and members of the public. Users of these machines should have access to a copy of the publication, *A Guide to Radiation Protection in the Use of X-Ray Optics Equipment,*[2] which provides much practical guidance to assist X-ray equipment users to do so safely.

1. Diagnostic Machines

There are three sources of radiation for a diagnostic X-ray machine (and for an open-beam analytical machine): direct radiation from the primary beam, scattered radiation from the target, and leakage from the protective tube housing. If the orientation of the machine can change, it may be necessary to design all the walls of the structure in which the machine is located to provide adequate shielding for the primary beam. If the machine has a permanent fixed orientation, one wall can be designed to protect against the primary beam, but the other walls, and possibly the floors and ceiling, will only have to be designed to reduce the intensity of the scattered and leakage radiation to an acceptable value. An acceptable value for the radiation level in a given space will depend upon whether the area is a controlled access area or not, and whether the exposures of the personnel working within the space are monitored. Maximum permissible exposure levels, P, are 100 mrem per week (1 mSv per week) in a controlled area, and 10 mrem per week (0.1 mSv per week) in other areas outside the controlled area. The primary shielding barrier thickness required can be obtained from the characteristics of the machine, the manner in which it is used, and the occupancy level of the spaces on the other side of the barrier.

a. Primary Beam Shielding

The information required to design an adequate shield for the primary beam is (1) the peak voltage (kVp) at which the machine may be operated, (2) the maximum current, I, of the electron beam used to excite the X-rays, usually expressed in milliamperes (mA), (3) the amount of time per week the unit is in use, usually expressed in minutes, and the fraction of time the machine is in use that it is aimed at the wall for which shielding is being calculated, U, and (4) a factor, T, which gives the occupancy level of the adjacent space. The required thicknesses of shielding can be computed by using this information and experimental data for the effectiveness of shielding of different materials for broad beam X-ray irradiation at different kVp's. The compiled data is given in terms of a factor.

For the primary beam, K can be calculated from:

$$K = \frac{Pd^2}{ItUT} \tag{19}$$

Here, d is the distance from the beam target in the X-ray tube to the desired point on the other side of the protective barrier. Usually the product, It, is replaced by the single factor W. All the factors on the right-hand side of the equation are normally known, so that K can be calculated. All that remains to obtain the required thickness is to find K, select the given kVp curve from a family of published curves for a given shielding material, and read the shielding thickness from the abscissa.

The factor T ranges from 1 for full occupancy (space occupied by full-time employees, children play areas, living quarters, employee restrooms, occupied space in immediately adjacent buildings, etc.) to 0.25 for areas with pedestrian traffic but unlikely to have long-term occupancy (corridors too narrow for desks, utility rooms, public parking lots, public rest rooms) to 1/16 for areas occupied occasionally (stairways, automatic elevators, closets, outside areas used by pedestrians or vehicular traffic). Older private facilities, such as physicians and dentists' offices were occasionally found, when testing programs were first initiated, to have substantial radiation levels in reception rooms and clerical areas.

b. Scattered Radiation

The corresponding equation for radiation scattered from the object being X-rayed is:

$$K_{eff} = \frac{400P(d_{sca})^2(d_{sec})^2}{aWTFf} \tag{20}$$

Here: K_{eff} = scattered radiation rate/workload in rads per milliampere per minute at 1 meter
a = scattered-to-incident ratio
d_{sec} = distance from a point to the scattering object in the X-ray beam in meters
F = area of beam impinging on scatterer (cm^2)
d_{sca} = distance from X-ray source to scatterer
f = factor that depends on the kVp and adjusts Equation 2 for the enhanced production of X-rays as the energy increases; below 500 kVp, f is taken to be 1

c. Leakage Radiation

A properly shielded diagnostic X-ray tube is limited to a maximum leakage of 100 mrem per hour (1 mSv per hour) at a distance of 1 meter. In order not to exceed the maximum weekly exposure at any distance, an attenuation of the leakage radiation is required as given by Equation 21:

$$B_x = \frac{600PId^2}{WT} \tag{21}$$

The thickness of shielding required to provide this degree of attenuation for a broad beam can be found by computing the number of half-value layers (thickness required to reduce the intensity of an incident beam by one half) that would be needed. To attenuate both the scattered and leakage radiation, the shielding required for each component is computed using Equations 20 and 21. If one of the answers is more than a tenth-value layer larger than the other, the thicker shield can be considered adequate. If not, then the thicker shield should be increased by one half-value layer.

Equation 21 also can be used for a therapeutic X-ray machine if the factor 600 in the numerator is changed to 60, reflecting the factor of 10 higher leakage allowed for this type of machine. Tables 5.9 and 5.10, adapted from NCRP 49, provide half- and tenth-value layers for lead and concrete for several peak voltages, and the scattered-to-incident ratio, a.

Table 5.9 Half-Value (HVL) and Tenth-Value (TVL) Layers

	Lead		Concrete	
Voltage (kVp)	HVL (mm)	TVL (mm)	HVL (cm)	TVL (cm)
50	0.06	0.17	0.43	1.5
70	0.17	0.52	0.84	2.8
100	0.27	0.88	1.6	5.3
125	0.28	0.93	2.0	6.6
150	0.30	0.99	2.24	7.4
200	0.52	1.7	2.5	8.4
250	0.88	2.9	2.8	9.4
300	1.47	4.8	3.1	10.4

Table 5.10 Scattered to Incident Exposure Ratio

(kV)	Scattering Angle (Relative to Incident Beam) Voltage					
	30	45	60	90	120	135
50	0.0005	0.0002	0.00025	0.00035	0.0008	0.0010
70	0.00065	0.00035	0.00035	0.0005	0.0010	0.0013
100	0.0015	0.0012	0.0012	0.0013	0.0020	0.0022
125	0.0018	0.0015	0.0015	0.0015	0.0023	0.0025
150	0.0020	0.0016	0.0016	0.0016	0.0024	0.0026
200	0.0024	0.0020	0.0019	0.0019	0.0027	0.0028
250	0.0025	0.0021	0.0019	0.0019	0.0027	0.0028
300	0.0026	0.0022	0.0020	0.0019	0.0026	0.0028

It will be noted that the minimum for the scattered ratio a in the energy range covered by Table 5.10 comes at an angle ranging from $45°$ to $90°$ with respect to the incident beam. At lower energies, the minimum is closer to $45°$, while at the upper end of the voltage range, the minimum shifts toward $90°$. Since the range of energies in Table 5.10 covers a large portion of the diagnostic voltages used, a person acting as a holder should position themselves accordingly and at as great a distance as feasible from the scatterer and the X-ray tube. A recent article concluded that the procedure described gives a value that will be too thick, and consequently more expensive than necessary and provides an alternate procedure which, however, is somewhat more complicated. Since the method in NCRP 49, and briefly outlined here, will result in a conservative shield, the additional thickness and cost may be justified since it does provide some margin for error and provides for lower exposures.

d. Exposure to Users

Many people are abnormally afraid of radiation because of a lack of knowledge or because of misinformation that they may have. A questionnaire given to second-year medical students before they had a course in biophysics or radiation applications in medical school, but after they had completed an undergraduate, presumably scientifically-oriented program, revealed a striking and troubling level of confusion about X-rays. If this is the situation that obtains for a well-educated group (the entrance requirements for medical schools are notoriously high), the lack of knowledge of the public is likely to be far worse. Four of the questions and the test results are given below.

1. Following the completion of an X-ray radiographic examination, objects within the room
    a. emit a large amount of radiation   29.3%
    b. emit a small amount of radiation   43.9%
    c. do not emit radiation   26.8%
2. Intravenous contrast materials used in angiograms and intravenous pyelograms are radioactive.
    a. true   37%
    b. false   61%
    c. don't know 19.5%
3. Gamma rays are more hazardous than X-rays.
    a. true   58.5%
    b. false   22.0%
    c. don't know 19.9%
4. Nuclear materials used in nuclear medicine are potentially explosive.
    a. true   24.4%
    b. false   68.3%
    c. don't know 7.3%

The intended correct answers are: 1. c, 2. b, 3. a, and 4. b. The only question about which there could be a quibble would be number 4 since some gammas are weaker than some X-rays although, in general, most gammas are stronger than diagnostic X-rays.

The normal practice in taking medical X-rays is for the patient to be situated with respect to the source and film cassette by the technician and, where possible, the patient is asked to "hold still" while the technician retreats to a shielded control room and activates the X-ray unit. The technician, under such circumstances, does not receive a dose other than that allowed by the scattered and leakage radiation levels behind the shield. Since the permissible limits are 100 mrem per week (1 mSv per week), the X-ray technician must be provided with a dosimeter capable of reading X-rays accurately. Both TLD and film badges are commercially available which are satisfactory for the purpose. The reproducibility and inherent radiological similarity to tissue of commonly used TLD dosimeters are desirable features for this type of dosimeter. The ability of film to distinguish between a diffuse extended source, as would be the case for the scattered radiation and that of a localized source, is an advantage of film dosimeters. The patient is exposed to not only the direct beam, but also the scattered and leakage radiation. However, the exposure of the patient will usually be a single episode, or at least a limited number of exposures, albeit at a much greater intensity, while the occupational exposures of the technician will continue over an extended period. When X-ray personnel are not wearing the dosimeter badges, they should be stored or kept in an area at background radiation levels, so that the dosimeter readings will reflect true occupational exposures.

The control room booth should have walls and a viewing window with sufficient shielding capability to ensure that the operator's exposure does not exceed the permissible limits for a controlled area. The dimensions of the booth must be of adequate height and width to ensure complete shielding while the beam is on, and the distance between the activation switch for the beam and the entrance to the door to the X-ray area should be sufficient to guarantee that an X-ray technician cannot operate the machine in an unshielded location. The entrances to an X-ray area should be posted with a sign bearing the words (or equivalent) **CAUTION: X-RAY EQUIPMENT.** It would also be desirable if a visual or audible warning were provided should anyone attempt to enter the beam area while the beam is on.

Another precaution that can be taken is to have the doors leading into the X-ray area other than from the control area interlocked so that they can only be opened from the outside when the beam is off or so that the X-ray machine will be turned off automatically upon the doors being opened.

Not all X-rays are taken under ideal circumstances. The patient may be in pain and find it difficult to stay in position. Children may be frightened or simply be too young to follow instructions. In certain fluoroscopic examinations, even healthy and cooperative patients may need assistance. Passive restraints should be used wherever possible, but on occasion patients must be held. This must not always be the responsibility of a single individual or a limited group of persons. Excessive personal doses are likely unless this duty is shared among several persons (note the comment in an earlier section, that sharing the dose, while to the benefit of an individual, may not be helpful for the exposure to the group as a whole). Since the X-ray beam is not on except for well-defined and determinable intervals, there should be no increase in the collective dose over that of the single individual by following this practice. No one who is not essential to the procedure should remain in the room while the X-ray is being taken, although relatives could restrain patients, especially children, who might need a familiar person present to avoid being frightened. Holders should remain as far from the primary beam as possible. As noted earlier, the intensity of the scattered radiation is a minimum between 45° to 90° with respect to the incident beam for most commonly used diagnostic X-ray voltages. Either a lead apron or a movable shield can substantially reduce the whole-body exposure. As noted in Table 5.9, less than 1 mm of lead will reduce the intensity of a broad beam of 150 kVp X-rays by a factor of 10. This amount of lead might be too heavy in an apron for comfort for some individuals, but even half a mm of lead will substantially reduce the level of radiation to the torso, where the critical organs are located.

If a lead apron is not worn, the dosimeter badge should be worn at the waist or above, so that the measured dose will be representative of a whole-body exposure. If it is worn on the upper part of the body, the dosimeter will not be partially shielded by the X-ray table in a vertical beam exposure when holding a patient lying down.

When a lead apron is worn, the dosimeter should be worn above the apron on the shirt collar or on the lapel of a laboratory coat so that it provides a reading of the dose to the exposed area above the apron. Especially in the early stages of her pregnancy, a pregnant woman should not act as a holder to avoid an exposure to the ferns. Should it be necessary for a pregnant woman to act as a holder, she should wear a second dosimeter under the lead apron near the fetus. If an employee is aware that she is pregnant, it is recommended that she "declare" (as under 10 CFR Part 20, although the NRC regulations do not apply to X-ray radiation) or inform her supervisor of her pregnancy. The supervisor, with the cooperation of the pregnant employee, should see that she limits the fetal exposure to 0.5 rem (5 rnSv) over the entire pregnancy. To confirm that this is so, a more frequent monitoring schedule should be established for this individual, unless work assignments can be made which preclude her exceeding the recommended levels.

Persons who assist with fluoroscopic procedures should wear ring badges and also body dosimeters. The beam stays on for longer periods than for a normal X-ray film exposure (sometimes longer than necessary when the radiologist is not conscientious about turning off the beam as promptly as possible at intervals during the procedure), and consequently the potential for exposure of the assistant is higher. Additional shielding to protect the viewer and any aides from scattered radiation should be provided as needed. If a direct viewing screen is employed, the entire primary beam must be intercepted by the screen, or the equipment should be interlocked so that it will not work unless it is. Units are now available so that the image is viewed on a high-resolution television screen instead of directly, which substantially reduces the amount of radiation exposure to the radiological physician.

Film and TLD dosimeters are normally read on a monthly or quarterly basis, instead of immediately, although the latter can be read in a few minutes if a calibrated and certified TLD dosimeter reader is available. If there is concern about potential exposures and an immediate reading is needed or desired, pocket dosimeters designed to be properly responsive to X-ray fields should be used to supplement the standard dosimeters, to provide an immediate exposure reading.

Many devices that decrease the dose to the patient, such as faster film or screen intensifiers, also serve to reduce the exposure to persons required to be in the room. Beam filters, usually of aluminum, are used to eliminate the nonuseful, soft X-rays from the X-ray beam. This reduces the skin exposure to the patient and the exposure to scattered radiation from this source to others. The recommended filter thicknesses are 0.5 mm aluminum (or equivalent) for machines operating at up to 50 kVp, 1.5 mm for operations between 50 and 70 kVp, 2 mm for 70 to 125 kVp, and 3 mm for 125 to 300 kVp. Collimators, which are used to define and limit the incident beam to only the area of clinical interest, also reduce the amount of scattering mass within the beam. The irradiated area is required by current standards to be defined by a light that is accurately coincident with the area exposed to the X-ray beam. The visible target will allow any required assistant to take a position that will be as far from the primary beam as possible and which will reduce the exposure to scattered radiation.

The genital areas of patients must be protected with a shield if the patient is of reproductive age, unless the shield would interfere with the diagnostic procedure. It would be desirable to know if a woman is pregnant before taking an X-ray of the pelvic area, and it is recommended that the physician ask a woman of a fertile age if there is a possibility that they may be pregnant. This is obviously a very sensitive topic and must be done extremely tactfully, and explain the possibility of enhanced damage to the fetus if exposed. However, failure to discuss the risk and warn the patient of possible problems with the development of the fetus could result in allegations of liability by the physician. A sign prominently posted in the X-ray facility such as **"IF YOU ARE PREGNANT IT IS IMPORTANT THAT YOU TELL THE TECHNICIAN"** could be helpful. In an emergency this concern should not delay taking an X-ray. Since the patient's body usually is exposed to scattered and leakage radiation from the areas being X-rayed, shields for other sensitive parts of the patient's body should be considered if a substantial series of "shots" are to be taken.

No more X-rays than necessary should be taken. No X-rays of humans should be taken for training, demonstration, or other purposes not directly related to the treatment of the patient.

One of the more disturbing problems of diagnostic X-rays is the frequent failure of either the physician or the radiologist to be responsive to the question by a patient of "how much radiation will be received." Too often the attitude appears to be that "it has been decided that the patient needs the X-ray and that is all that matters." There is substantial concern by many persons about the effects of radiation, and a substantial amount of misconceptions. This fear and confusion may inhibit the patient from having a needed X-ray. Not only should a reasonable answer be given but the significance of the answer should be explained, since it is unlikely that most persons will be knowledgeable about radiation safety terminology. The output from a typical X-ray machine is generally quite substantial. At 50 kVp, with no filtration and a thin beryllium window, the output of an X-ray tube is about 10 rad/min/mA at 1 meter or 100 mGy/min/mA at 1 meter. For a 100 kVp machine, with a 3 mm external aluminum filter, a typical exposure rate is 3 rad/min/mA at 1 meter or 30 mGy/minlmA at 1 meter. These rates, or preferably the actual rates for the machine being used, can be used to provide an estimate of the doses to the patient. The next section will describe the process of inspection and calibration of X-ray machines that should be followed. Since one of the

Table 5.11 Typical Form Used to Evaluate X-ray Units

TUBE REGISTRATION # _____

Machine

Registrant Name _____ Make _____ Purpose _____

Surveyor Name _____ Model _____ Serial # _____

Surveyor Signature _____ Max. kVp _____ Max. mA _____

Survey Date _____ Room # _____ Phase _____ 1 _____ 3

OK   N/A   N/S   S

___ ___ ___ ___ 1. Warning label not present. **NON-SERIOUS**

___ ___ ___ ___ 2. HVL at ____ kVp is ____ mm Al. Minimum is ____ mm Al. Deficiency of ≤ 0.2 mm, **NON-SERIOUS**, deficiency ≥ 0.2 mm, **SERIOUS**.

___ ___ ___ ___ 3. The length _____ width _____ misalignment between the X-ray and visual fields is ____ %. >2% to < 5% **NON-SERIOUS**, ≥ 5%, **SERIOUS**.

___ ___ ___ ___ 4. Misalignment between center of X-ray field and center of image receptor is ____% of the SID. >2% to < 5% **NON-SERIOUS**, ≥ 5%, **SERIOUS**.

___ ___ ___ ___ 5. SID not indicated to within 2% of SID. **NON-SERIOUS**.

___ ___ ___ ___ 6. The length ____ width ____ dimensions of the X-ray field not indicated to within 2% of the SID. **NON-SERIOUS**.

___ ___ ___ ___ 7. The length ____ of the X-ray field exceeds that of the image receptor by ____ % of the SID (special purpose systems only). >2% to < 5% **NON-SERIOUS**, ≥ 5% **SERIOUS**.

___ ___ ___ ___ 8. Radiographic control does not require constant operator pressure or does not terminate the exposure properly. **SERIOUS**.

___ ___ ___ ___ 9. Radiographic control switch not permanently located in an appropriate protected area or on a stretch cord of sufficient length as required. **SERIOUS**.

___ ___ ___ ___ 10. X-ray control not equipped with both visual and audible indication of X-ray production. **NON-SERIOUS**.

___ ___ ___ ___ 11. Timer reproducibility, coefficient of variation is ____ % at a technique setting of _____. > 10% to < 15%, **NON-SERIOUS**, ≥ 15% **SERIOUS**.

___ ___ ___ ___ 12. Exposure reproducibility: Coefficient of variation is ____ % at a technique setting of _____. > 10% to < 15%, **NON-SERIOUS**, ≥ 15% **SERIOUS**.

___ ___ ___ ___ 13. Timer accuracy is ± ____ % of the indicated time at ____ sec. > 10% to < 15%, **NON-SERIOUS**, ≥ 15% **SERIOUS**.

___ ___ ___ ___ 14. Standby radiation exposure is more than 2 mR/h Capacitor discharge systems only) **SERIOUS**.

___ ___ ___ ___ 15. mA Linearity: mR/mAs values at ____ Ma and ____ mA differ by more than 10% of their sum. **NON-SERIOUS**.

___ ___ ___ ___ 16. Positive beam limitation, if present, does not operate properly. **NON-SERIOUS**.

___ ___ ___ ___ 17. Light field illuminance is ____ ft. candles. Should be no less than 10 ft. candles at 100 cm. or at the max. SID whichever is less. <10 to 9 ft. candles: **NON-SERIOUS**, 9 ft. candles, **SERIOUS**.

___ ___ ___ ___ 18. kVp accuracy is ± _____ of indicated kVp at _____ kVp. > 10% to < 15%, **NON-SERIOUS**, ≥ 15% **SERIOUS**.

___ ___ ___ ___ 19. Exposure data: projection: _____ technique factors: _____ kVp ____ mA ____ MS _____ inches SID. Exposure results _____ mR.

___ ___ ___ ___ 20. Other/Remarks: _____

SID = source to image distance

results of the evaluations will be an exposure chart representative of various types of X-ray applications, the information is readily available if these evaluations are done. Techniques to reduce the exposures to the patient should be used. A normal "good" level of exposure to the patient for a chest X-ray is 10 mrad (1 mGy) or less.

The performance of a diagnostic X-ray machine should be checked at least annually with appropriate instruments by a qualified person. If the capability to do this is not available in-house, a qualified consultant should be hired to do the task. If any maintenance is done or if the machine is relocated, a survey should be undertaken for leakage radiation from the source. If the unit is moved to another facility, the exposure levels in the adjacent areas should be tested to ensure that the exposure levels are within the permissible limits for controlled and uncontrolled areas. Records of all maintenance, surveys, leakage checks, calibration, personnel monitoring, etc. should be maintained at the facility and at the radiation safety office. Because of the long latency period for cancer developing from radiation exposures, it would not be unreasonable to maintain personnel exposure records for up to 40 years.

2. X-Ray Quality Control in Medical X-Ray Laboratories[*]

This section is restricted to X-ray facilities, although another major use of radiation in larger nuclear facilities is the use of radioactive materials. The radioisotope usage is regulated by the NRC in 10 CFR Part 35, and by a medical industry group, the Joint Council on Accreditation of Healthcare Organizations (JCAHO). In terms of employees and the public, the same basic NRC restrictions on exposure generally must be followed. Virtually all hospitals use X-ray machines of various types.

Performance standards for X-ray equipment are developed by the Food and Drug Administration under 21 CFR Part 1020, Performance Standards for Ionizing Radiation Emitting Products. Regulation of the use of X-ray equipment is essentially left to the individual states and to the same health care industry group which monitors hospital quality control in many areas other than use of X-rays and radioisotopes, the Joint Council on Accreditation of Healthcare Organizations.

Inspections and tests of individual X-ray units and evaluation of general compliance with radiation safety rules and operating procedures is normally done by independent consultants, hired by the hospitals. The regulations of each state may vary, but the accrediting organization is a national organization, so their standards provide a baseline for guidance on the standards to be met during an inspection. Generally X-ray units are checked at least annually, but may be checked more often if needed. Each type of X-ray unit used for different applications has slightly different features that are checked. Fluoroscopic systems generally have much higher exposure rates than those used for general X-ray procedures, so they include more checks on scattered radiation. Machines used for soft tissue evaluation, e.g., mammography use lower accelerating voltages and a "softer" X-ray beam extending to lower energies, so the filters to remove the lowest energy X-rays are considerably thinner than for general application units. A typical inspection form used by a consultant inspector is shown in Table 5.11. The forms provide the range of acceptable values except for the filter thicknesses. For higher voltage exposures, the half-value filter thickness (HVL) may range from 2.3 to 3.2 mm of aluminum, while for low voltages, the HVL is much less, about 0.27 to 0.33 mm of aluminum.

---

[*] This section was coauthored by John Cure and A. Keith Furr. Mr. Cure is a professional consultant to hospitals for their radiation facilities.

Table 5.12 Entrance Skin Exposures

Projection	Patient Thickness (cm)	Grid	SID (in.)	ESE (mr) Film Speeds 300	ESE (mr) Film Speeds 400
Abdomen	23	yes	40	490	300
Lumbar spine	23	yes	40	570	330
Full spine	23	yes	72	260	145
Cervical spine	13	yes	40	135	95
Skull	15	yes	40	145	70
Chest	23	no	72	15	10
Chest	23	yes	72	25	15

MAMMOGRAPHY (CRANIOCAUDALVIEW)
Mean Glandular

Imaging System	Dose (mrad)
Screen film	
w/out grid	70
with grid	150
Xerox	
Positive mode	400
Negative mode	300

Computerized Tomography (CT Scans) Doses

Typical 3rd generation unit (tube and detector rotates)	4.4 Rads
Typical 4th generation unit (tube rotates, detectors stationary)	5.4 Rads

Fluoroscopic Exposure Rates

Examination	Dose Rate (Roentgens/min)
Routine typical patient	4.5
With barium	7.8
Maximum rate	8.1

SID = source to image distance.

The entrance skin exposures are the critical parameters in terms of safety to the patient. Generally, the higher kVp exposures reach the film, while depositing less energy in the exposures to the patients. As pointed out in the introduction to the JCAHO report, the largest contributor to total population exposure to radiation from man made sources is from diagnostic (dental and medical radiography). Clearly, X-rays need to be taken only when needed and under circumstances to reduce exposures to the patient, the medical staff, and others involved.

The fluoroscopic portion of Table 5.12 shows that the actual exposure depends upon the time the beam is on. If the beam is left on while adjustments are made to the patient's position, as possibly occurs during the barium enema used in the table, the exposures can be much higher than necessary. In one instance, a calculated dose of over 50 rem was delivered to the patient, while a similar test, done one year later with great care taken to turn off the beam between changes in position of the patient and acquiring the necessary data as quickly as possible gave a _measured_ dose to the patient of 3.4 rem, still a substantial amount. Note that the units are not mrem or mrads but are in effect rads, a thousand times larger, since there is only a modest difference in the energy deposited in tissue by rads and roentgens. Although not mandated by a hospital accrediting group and possibly not regulated by a state agency, it is recommended that a qualified X-ray consultant periodically inspect and evaluate the X-ray machines described in the following sections. It is also recommended that the need to minimize the exposure to the patient be emphasized in the training of X-ray technicians and nuclear medicine physicians. Reasonable care needs to be taken to limit exposures to employees as well as patients.

3. Open-Beam Analytical Machines

Open-beam analytical X-ray machines have many research applications that involve frequent manipulations of the equipment and samples. As a consequence, there are many opportunities for excessive occupational exposures. In crystallography studies, for example, a very high intensity beam is involved, although confined to small areas. In the following material, the words must and shall are used several times, These are simply ethical limits since the NRC does not regulate this type of radiation, nor in most states does a state agency. Should an injury occur, it is possible that OSHA could claim jurisdiction under the General Duty Clause with radiation being considered a recognized hazard.

If an individual were to expose the hands or any other portion of the body to the direct beam, very high localized exposures could occur in a few seconds. The occupational radiation levels considered acceptable for general whole body exposure are the same as for diagnostic machines, 100 mrem per week (1 mSv per week) in controlled areas. The maximum hourly levels are the levels that would be equivalent to this rate for a 40-hour week, 2.5 mr per hour (25 μSv per hour). Acceptable levels for hand exposures are 15 times greater than those for whole body exposures, 37.5 mrem per hour (37.5 μSv per hour). For exposures in uncontrolled areas, the levels should not exceed 2 mrem per hour (20 μSv per hour) or 100 mrem per week (1 mSv) on a short-term basis or 500 mrem (5 mSv) averaged over 1 year. Note that these levels are equivalent to the levels recommended by the NRC for exposure to radioactive materials in the 10 CFR Part 20 before January 1, 1994.

Laboratories that use open-beam analytical machines should possess a survey instrument capable of measuring the scattered radiation from the machine and associated apparatus around the experimental area. The scattered radiation from a crystal is nonuniform, which can cause narrowly defined "hot" spots if the radiation is allowed to leak from the apparatus into the laboratory area. A careful survey needs to be made whenever maintenance is done on any part of the system that could allow radiation to escape from the apparatus or when the system is reconfigured. If an alignment or maintenance procedure must be done with the beam on, surveys must be performed while this work is in progress, preferably by two persons, one of whom does the actual work while the other one monitors the radiation levels. Periodic surveys should also be made to ensure that no changes have been made or have occurred without everyone's knowledge which could affect the radiation field near the apparatus. A quarterly schedule is recommended while work is actively in progress. A survey should also be made any time anyone suspects that an abnormal condition could exist that might increase occupational exposures.

Personal dosimeters should be worn by individuals in the laboratory whenever the machine is operated, and finger badges should be worn whenever working directly on the apparatus, making adjustments, or doing any action that could cause the fingers or hand to receive an abnormal exposure. All badges, when not being worn by the individual, should be stored in a background level radiation area so that they will reflect the actual occupational exposure of the person to whom they have been assigned.

A standard operating procedure manual should definitely be prepared for systems as potentially dangerous as open beam analytical systems, and all persons must be made familiar with the procedures applicable to safe use of the equipment. The person in charge of the facility is responsible for ensuring that all personnel are informed of safe procedures and that it is required that all personnel must comply with the written safety policies. This should be done not only by instruction but by the laboratory director or supervisor setting an appropriate example as well. Every individual working in the facility should be trained in the basic principles of radiation safety, the permissible exposure levels for controlled and uncontrolled areas, what dosimeters to wear and when, what surveys are required and what records need to be maintained. All employees should also be informed of their rights to limit their exposures and of their rights to their exposure records.

Laboratories in which analytical X-ray machines are located should be posted with signs at all entrances, bearing the legend **CAUTION—APPARATUS CAPABLE OF PRO-DUCING HIGH-INTENSITY X-RAYS** or equivalent. Any time the machine is on, it should be attended or the doors to the experimental area should be locked. Keys should be restricted to those persons who work within the facility. It is especially critical to prevent access, or not to leave the apparatus unattended if the equipment is on and partially opened for alignment.

Measures must be taken to warn persons unfamiliar with the risks associated with the equipment in case they enter the facility while no one is present, even if the equipment is not on. The X-ray source housing of analytical X-ray equipment should have a conspicuous sign attached to it bearing the words **CAUTION—HIGH INTENSITY X-RAYS,** and a similar sign placed near any switch that could energize the beam, stating **CAUTION—THIS APPARATUS PRODUCES HIGH-INTENSITY X-RAYS WHEN THE SWITCH IS ENERGIZED**. In addition, a lighted sign stating **X-RAY SYSTEM ON**, or the equivalent, should be near the switch. The light should be incorporated in the activation circuit so that if the light fails, it will be impossible to activate the system. A key-operated switch would be an added safety feature.

It is highly desirable that an interlock system be incorporated into the system that will prevent any part of the body from being directly exposed to the primary beam, either by making it physically impossible to do so or by entry into the area causing the beam to be shut off. Where, for some compelling reason, this is not feasible, alternatives need to be in place to reduce the chance of an accidental overexposure. Conspicuous warning devices to alert persons working on or near the machine should be in place if the status of the apparatus is such that a danger could exist. These may take the form of lighted signs, for example, stating **CAUTION—SYSTEM IS ON** or **DANGER—SHUTTER OPEN** if a beam port is open. These warning devices should be fail safe. If they are not functional, the system should not be able to operate. Shutters on any unused ports should be firmly secured in the closed position. It would be desirable if the system could not operate with a port open, unless deliberately bypassed, in which case a fail safe warning sign would need to be used. Similarly, it should be impossible to open a shutter unless a collimator or an experimental device were connected to the port. This feature must be incorporated in commercial systems constructed after January 1, 1980. Beam catchers should be in place to intercept any beam beyond the point at which the beam has served its purpose.

Interlock systems should not depend solely upon the operation of a device such as a micro-switch. A micro-switch can operate for years without failing and then fail without warning. A more basic device, such as incorporating a male and female plug assembly into the shutter mechanism to complete a circuit, is virtually foolproof. A light in the circuit confirms that the circuit is complete. The completed circuit should always provide a warning and preferably be a condition of a safe system, i.e., if the light is on, the system is safe. If it is off, the activation circuit is interrupted by an unsafe condition or the light is burned out, so the system cannot operate in either case. In some cases, as with the open shutter above, a warning system needs to operate to warn of an unsafe condition. Here, the failure of the system to operate if the light is not on simply guarantees that the warning sign is on during the less safe condition.

Repair and alignment activities provide some of the best opportunities for accidental over-exposures. Every precaution should be taken to prevent these incidents. Several measures are listed below. Most of these should be made mandatory by the organization's written radiation safety policies.

1. Alignment procedures recommended by the manufacturer should be used. Departures from these procedures should require prior approval by radiation safety personnel.
2. Alignment procedures must be in writing, and the personnel doing them must have specific training on doing the procedures.
3. The radiation from an X-ray tube shall be confined by a suitable housing providing shielding during alignment and maintenance activities.
4. The main switch, instead of safety interlocks, must be used to turn the equipment off.
5. If the interlocks are bypassed to allow alignment or other work to be done, a sign must be placed on the equipment stating **DANGER—SAFETY INTERLOCKS BYPASSED.**
6. Personnel dosimeters and finger badges must be worn at all times the beam is on, by the person doing the work and the assistant (see item 11 below).
7. The smallest practical voltage and beam current that allow the alignment to be done shall be used.
8. Temporary shielding should be used to reduce the exposure to scattered radiation.
9. Long-handled tools and extension devices should be used to reduce the risk of direct exposure to the hands.
10. If the alignment and maintenance operations are such as to cause the radiation in an uncontrolled area to exceed the permissible levels for such an area, the area should be secured against entry by locks, surveillance, or both.
11. Alignment and repair operations with the beam on should not be done while working alone. A second person should be present as a safety observer, equipped with a survey meter to check radiation and prepared to immediately shut off the beam in case of a direct exposure of any part of the person working on the equipment.
12. The system should be checked to ensure all interlocks have been reconnected and to ensure that leakage radiation is at an acceptable level before operations resume.
13. A record of all repairs and the results of surveys taken subsequently to the work done should be recorded in the permanent log. The log can be used to identify recurring problem areas.

4. Closed-Beam Analytical Systems

Closed-beam diffraction cameras are used for purposes similar to those of open-beam systems. However, they are designed so that the X-ray tube, sample, detector, and diffracting crystal are enclosed in a chamber that should prevent entry into the system by any part of the body. However, some older systems allowed the hands to be put into the sample chamber without turning off the beam, although this was not supposed to be done. Instances of exposures of 50,000 to 100,000 rems (500 to 1000 Sv) to the hands are known to have occurred, resulting in severe damage to the hands. The ports on such units must now be equipped with safety interlocks by the manufacturer incorporating a warning light in series with the interlock, which guarantees that if the light is off, the beam is off. Any older units not so equipped should be retrofitted with this safety feature. Because there should be little or no leakage requirements with a properly operating closed beam analytical system, personnel need not be required to wear dosimeters.

5. Cabinet X-Ray Systems

Cabinet X-ray systems are enclosed systems that not only enclose the X-ray system but also the object to be irradiated. They are normally designed with a key-activated control so that X-rays cannot be generated when the key is removed. A conspicuous sign with the legend **CAUTION— X-RAYS PRODUCED WHEN ENERGIZED** must be placed near the controls used to activate the generation of X-rays. The doors to the cabinet are required to have two or more safety interlocks and each access panel is required to have at least one. If X-ray generation is interrupted through the functioning of a safety interlock, resetting the interlock shall not be sufficient to resume generation of X-rays. A separate control must be provided to reinitiate X-ray production.

The generation of X-rays must be indicated by two different indicators, one being a lighted warning light labeled **X-RAYS ON,** and the other can be the meter that reads the X-ray tube current.

The radiation at 5 cm from the surface of the cabinet must be no more than 0.5 mrem per hour (5 μSv per hour). The radiation limits in uncontrolled areas due to the operation of the machine are 2 mrem per hour (20 μSv per hour) or 100 mrem (1 mSv) in a 7-day period or 500 mrem per year (5 mSv per year). At least an initial radiation survey and an annual survey thereafter should be made to ensure conformity with these limits. The interlock systems should be tested periodically if the mode of operations does not automatically cause their function to be tested.

Personnel using a cabinet X-ray unit should be provided with a personal dosimeter that they should wear them whenever the machine is in operation.

6. Miscellaneous Systems

There are any number of systems that generate X-rays during normal operations, some as dangerous as or more so than the ones described above. Precautions appropriate to the system, including personnel dosimetry, surveys, safety interlocks, etc., must be taken to ensure the safety of all workers and those persons in uncontrolled areas nearby. An appropriate safety program must be worked out with the radiation safety department within the organization to ensure that the program not only meets all safety requirements recommended by the manufacturer but also is in conformity with regulatory requirements, including those policies established within the organization.

REFERENCES

1. **Marlin, E.B.M.,** *Guide to Safe Use of X-Ray Diffraction and Spectrometry Equipment,* Science Reviews, Leeds, England, 1983.

2. *A Guide to Radiation Protection in the Use of X-Ray Optics Equipment,* Science Reviews, Leeds, England, 1986.

3. **Simpkin, D.J.,** A general solution to the shielding of medical X and gamma rays by the **NRCP** report No. 49 methods, *Health Phys.,* 52, 431, 1987.

4. *Structural Shielding Design and Evaluation for Medical Use of X-Rays and Gamma Rays of Energies Up to*

10 MeV, NCRP Report No. 49, National Council on Radiation Protection and Measurements, Washington, D.C., 1976.

5. *X-Ray Equipment,* UL-187, Underwriters Laboratories, Chicago, IL, 1974.

6. Radiological Safety Standard for the Design of Radiographic and Fluoroscopic Industrial X-Ray Equipment, NBS Handbook 123, National Bureau of Standards, Washington, D.C., 1976.

7. *Radiation Safety for Diffraction and Fluorescence Analysis Equipment,* ANSI N43 .2, American National Standards Institute, New York, 1971.

8. Performance Standards for Ionizing Radiation Emitting Products, 21 CFR Chap I, 1020.30, 1988.

9. **Kaczmarek, R., Rednarek, D., and Wong, R.,** Misconceptions of medical students about radiological physics. *Health Phvs.,* 52. 106, 1987.

10. **Fleming, M.F. and Archer, M.E.,** Ionizing radiation, health hazards of medical use, *Consultant,* p. 167, January 1984.

11. **Lubenau, J.O., David, J.S., McDonald, D.J., and Gerusky, T.M.,** Analytical X-ray hazards: a continuing problem, *Health Phys.,* 16, 739, 1969.

12. **Weigensberg, I.J., Asbury, C.W., and Feldman, A.,** Injury due to accidental exposure to X-rays from an X-ray fluorescence spectrometer *Health Phys.,* 39, 237, 1980.

13. **Cember, H.,** *Introduction to Health Physics,* Pergamon Press, New York, 1969.

14. **Shapiro, J.,** *Radiation Protection, A Guide for Scientists and Physicians,* 3rd ed., Harvard Press, Cambridge, MA, 1990.

15. Average Patient Exposure Guides, CRCPD Pub. No. 92.4, Conf. of Radiation Control Program Directors, Frankfort, 1992.

16. Nationwide Evaluation of X-Ray Trends (NEXT), Conf. of Radiation Control Program Directors, 1974-89.

17. Accreditation Manual for Hospitals, Joint Commission on Accreditation of Healthcare Organizations, Chicago, 1982.

INTERNET REFERENCE

1. http://jacho.org/

IV. NONIONIZING RADIATION

In recent years, there has been substantial discussion about the effects of electromagnetic radiation on health for frequencies ranging from below ordinary 60 cycle power line fields to much higher frequencies in the megahertz (MHz) and Gigahertz (Ghz) ranges. The current OSHA guidelines are given in 29 CFR 1910.97. From 10 MHz to 100 GHz, the radiation protection guide provides that the incident radiation not exceed 10 mW/cm^2, as averaged over any possible 0.1-hour period. This refers to the power density. The guide provides that the energy density be no more than 1 mW/cm^2 for any 6-minute period.

There is a reference available on the Internet (Internet reference 1) which provides a link to the Lawrence Livermore National Laboratory Health and Safety Manual section on Non-Ionizing Radiation and Fields. The following information is taken from that fields of as much as 300 milligauss. Of course, these are usually short term exposures. Measurements made in an active laboratory computer center, which had a large number of computers, monitors, and printers gave values ranging from less than 5 mG to as high as 18 mG in one area near a cable bank. Note that these levels are significantly lower than those from the appliances listed above. The lower levels in the laboratory would have represented a typical occupational exposure for that location. The ANSI recommended exposure ELF limits are 2.5 mG but most of the other agencies cited allow higher limits. The second table, 5.14, provides recommended exposure limits for both occupational and public situations for the frequency range of DC to just over 4 MHz for a variety of exposure situations and durations.

Table 5.13 Bands of Radio-frequency and Subradio-frequency Fields and Radiation

Frequency	Wavelength	Name
>300 GHz	<1 mm	Infrared
300 GHz	1 mm	Extremely high frequency (EHF)
30 GHz	1 cm	Superhigh frequency (SHF)
3 GHz	10 cm	Ultra high frequency (UHF)
>300 MHz	<1 m	Microwaves
300 MHz	1 m	Very high frequency (VHF)
30 MHz	10 m	High frequency (HF)
3 MHz	100 m	Medium frequency (MF)
300 kHz	1 km	Low frequency (LF)
30 kHz	10 km	Very low frequency (VLF)
<3 kHz	>100 km	Subradio frequency
3 kHz	10 km	Voice frequency
300 Hz	1000 km	Extremely low frequency (ELF)
0 Hz	—	Static

Table 5.14 Exposure Guidelines for DC and AC Fields Below 3 kHz

Frequency	Exposure group	Exposure duration	Exposed part of body	Permissible exposure	
				Electric (kV/m)	Magnetic (mT)[a]
DC	Occupational	Shift	All	—	200
DC	Occupational	Ceiling	All	—	2000
DC	Occupational	Ceiling	Limbs	—	5000
DC	Public	24 hr/day	All	—	40[e]
50/60 Hz	Public	24 hr/day	All	5	0.1
50/60 Hz	Public	Few hrs/day[b]	All	10	1
1 .Hz—294 Hz	Occupational	Ceiling[c]	All	—	60/f[d]
1 Hz—294 Hz	Occupational	Ceiling[c]	Limbs	—	300/f[d]
DC—4.071 kHz	Occupational	Ceiling[c]	All	25	
100 Hz—4.071 kHz	Occupational	Ceiling[c]	All	2500/f[d]	
100 Hz—4.071 kHz	Occupational	Ceiling[c]	All	2500/f[d]	

a   1 mT = 10 G = 796 ($\approx$ 800 A/m).

b   Exposures to electric fields between 5 and 10 kV/m or magnetic fields between 0.1 and 1 mT should be limited to a few hrs/day, continuous exposures to electric fields >5 kV/m or magnetic fields >1 mT should not be allowed, and exposures to electric fields >10 kV/m or magnetic fields >1 mT should be limited to a few mins/day (electric field exposures >10 kV/m allowed if induced current density is ~2 mA/m^2).

c   Maximum exposure allowed at any time.

d   Frequency in units of Hz.

e   Exposures to higher fields in special facilities allowed if access controlled and occupational exposure limits are not exceeded.

Table 5.15 Controlled access exposure limits

Part A—Electromagnetic fields[a]

Frequency range (MHz)	E (Vim)	H (A/m)[c]	Power density, S [E,H] (mW/cm²)	Averaging time E², H², or S (min)
0.000294— 0.1	—	163	[100; 1,000,0001[b]	6
0.00407— 0.1	614	—	[100; 1,000,000][b]	6
0.1— 3	614	$16.3/f_m$	[100; 10 000/f_m^2][b]	6
3— 30	$1842/f_m$	$16.3/f_m$	[900/f_m^2; 10,000/f_m^2][b]	6
30— 100	61.4	$16.3/f_m$	[1.0; 10, 000/f_m^2][b]	6
100— 300	61.4	0.163	1.0	6
300—3000			$f_m/300$	6
3000—15000	—		10	6
15000—300000	—		10	616 000/f_m

Part B—Induced and contact radio-frequency

Frequency range (MHz)	Maximum current (mA)		
	Through both feet	Through each foot	Contact
0.003— 0.1	$2000f_m$	$1000f_m$	$1000f_m$
0.1—100	200	100	100

NOTE: f_m = frequency in units of MHz.

a   The exposure values in terms of electric and magnetic field strength are the values obtained by spatially averaging values over an area equivalent to the vertical cross section of the human body (projected area).

b   These plane-wave-equivalent, power-density values, although not appropriate for near-field conditions, are commonly used as a convenient comparison with MPEs at higher frequencies and are displayed on some instruments in use.

c   A/rn = Amp-turn/meter.

Table 5.16 Uncontrolled Access Exposure Limits

Part A—Electromagnetic fields[a]

Frequency range (MHz)	E (Vim)	H (A/m)[c]	Power density, S [E,H] (mW/cm²)	Averaging time E², H², or S (min)
0.000294— 0.1	—	163	[100; 1,000,0001[b]	6
0.00407— 0.1	614	—	[100; 1,000,000][b]	6
0.1— 3	614	$16.3/f_m$	[100; 10 000/f_m^2][b]	6
3— 30	$1842/f_m$	$16.3/f_m$	[900/f_m^2; 10,000/f_m^2][b]	6
30— 100	61.4	$16.3/f_m$	[1.0; 10, 000/f_m^2][b]	6
100— 300	61.4	0.163	1.0	6
300—3000			$f_m/300$	6
3000—15000	—		10	6
15000—300000	—		10	616 000/f_m

Part B—Induced and contact radio-frequency currents

Frequency range (MHz)	Maximum current (mA)		Contact
	Through both feet	Through each foot	
0.003— 0.1	2000f_m	1000f_m	1000f_m
0.1—100	200	100	100

Note: f_m = frequency in units of MHz.

a   The exposure values in terms of electric and magnetic field strength are the values obtained by spatially averaging values over an area equival the vertical cross section of the human body (projected area).

b   A/m = Amp-turn/meter.

c   These plane-wave-equivalent, power-density values, although not appropriate for near-field conditions, are commonly used as a convenient comparison with MPEs at higher frequency and are displayed on some instruments in use.

The third table, 5.15, provides guidelines in controlled access areas for a frequency range from just above the ELF band to the infra-red region and the fourth table, 5.16, does the same for uncontrolled access areas.

Note these tables provide only guidelines and do not have the force of regulatory standards. There is insufficient unanimity in the scientific community on the actual health effects of non-ionizing radiation to set firm standards.

REFERENCES

1. **Jammet, H.P.,** Guidelines on limits to radio-frequency electromagnetic fields in the frequency range from 100 kHz to 300 Ghz, *Health Phys.* 54, 115, 1988.

2. International Commission on Non-Ionizing Radiation Protection, Guidelines on Limits of Exposure to Static Magnetic Fields, *Health Phys.,* 66, 100, 1994.

3. American Conference of Governmental Industrial Hygienists, *Threshold Limit Values for Physical Agents in the Work Environment.* ACGIH, Cincinnati, OH (latest edition).

4. International Nonionizing Radiation Committee of the International Radiation Protection Association, "Interim Guidelines on Limits of Exposure to 50/60 Hz Electric and Magnetic Fields," *Health Phys.* 58, 113 (1990).

5. Institute of Electrical and Electronic Engineers, *American National Standard Safety Levels with Respect to Human Exposure to Radio Frequency Electromagnetic Fields, 3 kHz to 300 GHz,* IEEE, Piscataway, NJ, C95.l-1991 (1992).

6. Health & Safety Manual, Supp. 26.12 , *Nonionizing Radiation and Fields (Electromagnetic Fields and Radiation with frequencies below 300 Ghz),* Lawrence Livermore National Laboratory.

INTERNET REFERENCES

1. http://www.lessemf.com/standard.html
2. http://www.llnl.gov/ea_and_h/hsm/supplement_26.12/sup26-12.html

V. LASER LABORATORIES*

---

*   The material in this section for laser operations complies with Title 21 CFR Part 1040.1 for laser products, as well as ANSI Z-136 and the Threshold Limit Values (TLV©, a copyrighted trademark of the American Conference of Governmental Industrial Hygienists).

Table 5.17 Class I Accessible Emission Limits for Laser Radiation

Wavelength (nanometers)	Emission duration (seconds)	Class I—Acoessible emission limits		
		(value.)	(unit)	(quantity)[**]
>180 but <1400	$\leq 3.0 \times 10^4$	$2.4 \times 10^{-5} k_1 k_2$[*]	Joules(J)[*]	radiant energy
	$>3.0 \times 10^4$	$8.0 \times 10^{-10} k_1 k_2$[*]	Watts(W)	radiant power
>400 but <1400	$>1.0 \times 10^{-9}$ to 2.0×10^{-5}	$2.0 \times 10^{-7} k_1 k_2$	J	radiant energy
	$>2.0 \times 10^{-5}$ to 1.0×10^1	$7.0 \times 10^{-4} k_1 k_2 t^{0.75}$	J	radiant energy
	$>1.0 \times 10^1$ to 1.0×10^4	$3.9 \times 10^{-3} k_1 k_2$	J	radiant energy
	$>1.0 \times 10^4$	$3.9 \times 10^{-7} k_1 k_2$	W	radiant power
	and also			
	$>1.0 \times 10^{-9}$ to 1.0×10^1	$10 k_1 k_2 t^{1/3}$	$Jcm^{-2}sr^{-1}$	integrated
	$>1.0 \times 10^1$ to 1.0×10^4	$20 k_1 k_2$	$Jcm^{-2}sr^{-1}$	radiance
	$>1.0 \times 10^4$	$2.0 \times 10^{-3} k_1 k_2$	$Wcm^{-2}sr^{-1}$	integrated
>11400 but <2500	$>1.0 \times 10^{-9}$ to 1.0×10^{-7}	$7.9 \times 10^{-5} {}^3 k_1 k_2$	J	radiant energy
		$4.4 \times 10^{-3} k_1 k_2$	J	radiant energy
	$>1.0 \times 10^{-7}$ to 1.0×10^1	$7.9 \times 10^{-4} k_1 k_2$	W	radiant power
>2500 > but <1.0×10^6	$>1.0 \times 10^{-9}$ to 1.0×10^{-7}	$1.0 \times 10^{-2} k_1 k_2$	Jcm^{-2}	radiant exposure
	$>1.0 \times 10^{-7}$ to 1.0×10^1	$5.6 \times 10^{-1} k_1 k_2 t^{0.25}$	Jcm^{-2}	radiant exposure
	$>1.0 \times 10^1$	$1.0 \times 10^{-1} k_1 k_2 t$	Jcm^{-2}	radiant exposure

[14]Class 1 accessible emission limits for wavelengths equal to or greater than 180 nm but less than or equal to 400 nm shall not exceed the Class I accessible emission limits for the wavelengths greater than 1400 nm but lees than or equal to 1.0×10^6 nm with a k_1 and k_2 of 1.0 for comparable sampling intervals.

[**]Measurement parameters and test conditions shall be in accordance with paragraphs 21 CFR.

The risk to personnel from the use of a laser in the laboratory depends upon several factors. The first consideration is the classification of the laser itself, which is based on several parameters. Among these are (1) the frequency or frequencies of radiation emitted; (2) for a pulsed system, the pulse repetition frequency (PRF), the duration of each pulse, the maximum or peak power P in watts or maximum energy Q in joules per pulse, the average power output, and the emergent beam radiant exposure; (3) for a continuous wave (CW) machine, the average power output; and (4) for an extended-source laser, the radiance of the laser, and the maximum viewing angle subtended by the laser. Relatively few lasers used in the laboratory fall in this latter class, and the following discussion will be limited to lasers that are not considered extended source units.

All commercial units currently being built must have the classification identified on the unit. The classification must be according to that given in ANSI-Z 136.1 and in 21 CFR 1040. However, facilities that work with experimental units should determine the class in which their lasers fall. The specifications of noncommercial units should be compared with Tables 5.17, 5.18, 5.19 and 5.20 to determine the classification.

The least powerful class of laser is class I. In this class, the power and energy of the unit are such that the TLV© for direct viewing of the laser beam, if the entire beam passes through the limiting aperture of the eye, cannot be exceeded for the classification duration (the

Table 18A Class IIa Accessible Emission Limits for Laser Radiation
CLASS IIa ACCESSIBLE EMISSION LIMITS ARE IDENTICAL TO CLASS I ACCESSIBLE EMISSION LIMITS EXCEPT WITHIN THE FOLLOWING RANGE OF WAVELENGTHS AND EMISSION DURATIONS:

Wavelength (nanometers)	Emission duration (seconds)	Class IIa — Accessible emission limits		
		(value)	(unit)	(quantity)*
>1400 but <710	>1.0 K	3.9×10^{-6}	W	radiant power

TABLE 18B Class Ii Accessible Emission Limits for Laser Radiation
Class II Accessible Emission Limits Are Identical to Class I Accessible Emission Limits Except Within the Following Range of Wavelengths and Emission Durations:

Wavelength (nanometers)	Emission duration (seconds)	Class II — Accessible emission limits		
		(value)	(unit)	(quantity)*
>1400 but <710	$>2.5 \times 10^{-1}$	1×10^{-3}	W	radiant power

Table 19A ClassIIIa Accessible Emission Limits for Laser Radiation
Class IIIa Accessible Emission Limits Are Identical to Class I Accessible Emission Limits Except Within the Following Range of Wavelengths and Emission Durations:

Wavelength (nanometers)	Emission duration (seconds)	Class IIIa-.Accessible emission limits		
		(value)	(unit)	(quantity)*
> 400 but ≤710	$> 3.8 \times 10^{-4}$	5.0×10^{-3}	W	radiant power

Table 19B Class IIIb Accessible Emission Limits for Laser Radiation

Wavelength (nanometers)	Emission duration (seconds)	Class IIIb — Accessible emission limits		
		(value)	(unit)	(quantity)*
≥180 but <400	≤2.5×10^{-1}	$3.8 \times 10^{-4} k_1 k_2$	J	radiant energy
	$>2.5 \times 10^{-1}$	$1.5 \times 10^{-3}\ ^4 k_1 k_2$	W	radiant power
>400 but ≤1400	$>1.0 \times 10^{-9}$ to 2.5×10^{-1}	$10 k_1 k_2 t^{1/3}$	Jcm^{-2}	radiant exposure
	$>2.5 \times 10^{-1}$	5.0×10^{-1}	Jcm^{-2}	radiant exposure
>1400 but ≤1.0×10^6	1.0×10^{-9} to 1.0×10^1	10	Jcm^{-2}	radiant exposure
	$>1.0 \times 1.0 \times 10^1$	5.0×10^{-1}	W	radiant power

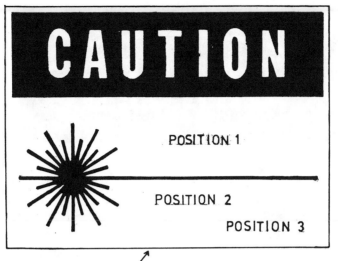

BLACK ON YELLOW BACKGROUND

Figure 5.12 Caution sign for low -to-moderate power lasers.

maximum duration of the exposure inherent in the design of the laser). In the spectral region of 400 to 1400 nanometers (1 nm $= 10^{-9}$ meters), and the limiting aperture of the eye is taken to be 7 mm. Class I lasers must emit levels below the accessible exposure limits (AEL) which are given in the first part of Table 5.17. Many of the recommended operating procedures are based on not exceeding the AELs for class I lasers.

These low power lasers require no control measures to prevent eye damage. Some class I lasers incorporate more powerful lasers, but are designated Class I because they are in an enclosure. If the more powerful enclosed laser is accessible, control measures appropriate to the higher class are required. Class II lasers are not considered hazardous for short-term viewing $\leq 1,000$ seconds and unintentional direct viewing of the laser beam, but should not be deliberately aimed at anyone's eyes. They are considered a chronic hazard if viewed for periods longer than 1,000 seconds. A class IIa laser must have a label affixed to the unit which bears the words **CLASS IIA LASER PRODUCT—AVOID LONG-TERM VIEWING OF DIRECT RADIATION.** A class II laser must have a label as shown in Figure 5.12 affixed to the unit in a conspicuous location, bearing the words **LASER RADIATION—DO NOT STARE INTO** BEAM in position 1 on the label. The words **CLASS II LASER PRODUCT** must appear in position 3. The labels of all class II, III, and IV units must give in position 2, in appropriate units, the maximum output of laser radiation, the pulse duration (when appropriate), and the laser medium or emitted wavelength(s).

Class III laser devices are hazardous to the eyes if the direct beam is looked at directly, or at specular reflections of the beam. Class IIIa units must have a label, as in Figure 5.13, affixed to the unit with the words **LASER—AVOID DIRECT EYE EXPOSURE** in position 1 and **CLASS IIIa LASER PRODUCT** in position 3. A Class IIIb unit would use a label, as in Figure 5.13, and have the words: **LASER RADIATION—AVOID DIRECT EXPOSURE TO THE BEAM** in position 1 and the words **CLASS IIIb LASER PRODUCT** in position 3. Mirror-like, smooth surfaces of any material that would reflect the beam as a beam should be eliminated as much as possible from the area in which the laser is situated. At a minimum, the user should take every precaution to avoid aiming the laser at such surfaces. It would be desirable to have the beam terminate on a surface that would only provide a diffuse reflection. It should, however, reflect well enough for a well-defined beam spot to be observable so that it is possible to visibly determine the point of contact. The laser

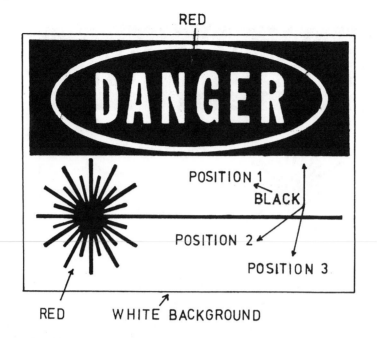

Figure 5.13 Danger sign for higher powered lasers.

should be set up in an area to which access can be controlled. When no responsible person is in the room, the room should be locked to prevent others from entering the room and changing the physical configuration, or accidentally exposing their eyes to the laser beam.

If there is any way in which it would be possible for either the direct or specularly reflected beam to enter the eye, appropriate eye protection must be worn in an area where a class IIIa laser is operated. Operation of class IIIb lasers generally should follow the practices for class IV lasers.

Class IV lasers must use a label as in Figure 5.13 and bear the words **LASER RADIA-TION—AVOID EYE OR SKIN EXPOSURE TO DIRECT OR SCATTERED RADIA-TION** in position 1 and the words **CLASS IV LASER PRODUCT** in position 3. Class IV lasers require many precautions to be used safely. The radiation from pulsed or CW units that lie in the visible and near-infrared can be focused by the eye and can cause damage to the retina from either direct and specular beams or diffuse reflections. The skin can also be injured. Pulsed units and CW units operating in the infrared and ultraviolet are a danger to the skin and the external portions of the eye and can provide enough energy to cause a fire in combustible materials.

Lasers of any class, except medical units and class II lasers that do not exceed the accessible exposure limits of class I lasers for any exposure duration of 1000 seconds or less, must have a label affixed near any aperture through which laser radiation is emitted in excess of the limits for class I lasers bearing the words: **AVOID EXPOSURE—LASER RADIATION EMITTED THROUGH THIS APERTURE.**

Table 5.20 provides the representative selected values of the wave-dependent factors k_1 and k_2. Table 5.21 provides the TLV$^©$'s for direct ocular exposure from a laser beam.

A. Protective Procedures for Class IIIb and Class IV Lasers

The primary means of protection is to physically prevent exposure. For laboratory workers, baffles may be used to physically intercept or terminate the primary beam and any reflected or secondary beams. Any windows in the facility should be covered during operations. Safety glasses are to be worn by workers while within the facility during operations. Interlocks of various sorts are another avenue of protection. Entrance to the facility by unauthorized personnel or unexpected entry by laboratory employees should be prevented by safety interlocks while the laser is in an operating condition, i.e., when it is on and in a condition to emit radiation. A warning light should be placed at the entrance. The interlocks should be capable of being bypassed to allow individuals to pass in and out of the controlled area in an emergency or to allow controlled access as needed to allow operations to be conducted. In the latter case, which might be considered a routine bypassing of the interlocks, the activation of the bypass should be limited to the person in charge of the operations at the time, who is aware of the condition of the system. In an emergency, there should be one or more readily available rapid shutoff switches to immediately disconnect power to the laser. One of these can be the required remote control unit stipulated in 21 CFR 1040.10. The access points to the controlled area should be so situated that there should is little or no likelihood that there would be dangerous levels of emitted radiation to persons entering or leaving through the portals or in areas beyond the entrances. Guests or visitors should be allowed in the controlled area during operations only under carefully controlled conditions, and with everyone in the area aware of their presence. Special care must be taken to provide such persons with protective eyewear or other protective gear as needed to ensure their safety. The applicable TLV® levels should not be exceeded for either employees or visitors.

It should be made impossible, by designing the firing circuit with sufficient fail safe safeguards, for a class IV pulsed laser to be discharged accidentally. A warning mechanism should be incorporated into the design to ensure that all persons in the room are aware that the discharge cycle has been initiated. If the laser unit and its power supply are more than 2 meters apart, both the laser and the power supply should have separate emission warning devices. Both the fail-safe system and the warning system should be designed so that no single component failure or a shorted or open circuit can disable the protective and warning features of the system. The system should not be capable of being operated if the redundancy of these circuits or of the emergency shut-off system has been compromised.

Very high-power infrared CW lasers, such as CO_2 units, represent not only a danger to operating personnel, they also constitute a fire hazard if their beams come into contact with combustible material. It would be desirable to have these units operated remotely or have them totally within an enclosure that affords good fire stopping capability. Asbestos should be avoided because it is a regulated carcinogen under current OSHA regulations, and under intense radiation could become friable. Commercial lasers incorporate many safety features. Following is a brief list of system safety features, besides the ones already mentioned, that either should be available in the laser as manufactured or should be incorporated in the system when set up for use. The items refer primarily to class IIIb and class IV systems, although they also apply to class II units, except those in which the acceptable emission limits of class I are not exceeded for any emission duration up to 1000 seconds. This list complies with the provisions of Title 21 CFR 1040 and ANSI Z136. When feasible, systems should be upgraded to the requirements of the latest editions of these two documents as they are revised.

1. All lasers are required to be in a protective housing, but safety interlocks should be provided on any portion that could be removed when the unit is operating, if the exposure limits for the class would be exceeded. Normally the enclosures should limit the radiation to no more than the limits for a class I laser. Some powerful laser systems should be required to be within enclosures, including the target or irradiation area, to protect personnel. Any portion of the housing that is not interlocked and could emit radiation over the accessible emission limits for each class must have a label attached to it bearing the words previously mentioned for position 1 for the basic identifying label for each class.

Table 5.20 Selected Numerical Solutions for k_1 and k_2

Wavelength (nanometers)	k_1	k_2				
		$t \leq 100$	$t = 300$	$t = 1000$	$1s3000$	$t \geq 10,000$
180	1.0					
300	1.0					
302	1.0					
303	1.32					
304	2.09					
306	3.31					
306	5.25					
307	8.32					
306	132					
309	20.9			1.0		
310	33.1					
311	52.8					
312	83.2					
313	132.0					
314	209.0					
315	330.0					
400	330.0					
401	1.0					
500	1.0					
600	1.0					
700	1.0					
710	1.06	1	1	1.1	3.3	11.0
720	1.09	1	1	2.!	6.3	21.0
730	1.14	1	1	7.1	9.3	31.0
740	1.20	1	1.2	4.1	12.0	41.0
750	1.25	1	1.5	5.0	15.0	50.0
780	1.31	1	1.8	6.0	18.0	60.0
770	1.37	1	2.1	7.0	21.0	70.0
780	1.43	1	2.4	8.0	24.0	80.0
790	1.50	1	2.7	9.0	27.0	90.0
800	1.56	1	3.0	10.0	30.0	100.0
850	1.95	1	3.0	10.0	30.0	100.0
900	2.44	1	3.0	10.0	30.0	100.0
960	3.05	1	3.0	10.0	30.0	100.0
1000	3.82	1	3.0	10.0	30.0	100.0
1060	4.78	1	3.0	10.0	30.0	100.0
1060	9.00	1	3.0	10.0	30.0	100.0
1100	5.00	1	3.0	10.0	30.0	100.0
1400	5.00	1	3.0	10.0	3~0	100.0
1500	1.0					
1540	100.0*			1.0		
1600	1.0					
1.0×10^6	10					

The factor $k_1 = 100.0$ when $t \leq 10^{-7}$ and $k_1 = 1$ when $t > 10^{-7}$. Note: The variable (t) is the magnitude of the sampling interval in units of seconds.

Table 5.21 Threshold Limit Values for Direct Ocular Exposures (Intrabeam Viewing) from a Laser Beam

Spectral Region	Wavelength (nm)	Exposure Time (t)(seconds)	TLV©	
UVC	200—280	10^{-9} — 3×10^4	3 mJ/cm²	
UVB	2 80—302	10^{-9} — 3×10^4	3 mJ/cm²	
	303		4 mJ/cm²	
	304		6 mJ/cm²	
	305		10 mJ/cm²	Not to
	306		16 mJ/cm²	exceed
	307		25 mJ/cm²	$0.56t^{1/4}$
	308		40 mJ/cm²	J/cm²
	309		63 mJ/cm²	≤ 10 S
	310		100 mJ/cm²	
	311		160 mJ/cm²	
	312		250 mJ/cm²	
	313		400 mJ/cm²	
	314		630 mJ/cm²	
UVA	315—400	10^{-9} — 10	$0.56t^{1/4}$ J/cm²	
		10—103	1.0 J/cm²	
		10^3 x10^4	1.0 mW/cm²	
Light	400—700	10^{-9} — 1.8 — 10^{-5}	5 x 10^{-7} J/cm²	
Light	400—700	1.8 x 10^{-5}— 10	$1.8(1/t^{0.75})$mJ/cm²	
	400—549	10—10^4	10 mJ/cm²	
	550—700	10 —T_1	$1.8(1/t^{0.75})$mJ/cm²	
	550—700	T_1 — 10^4	10 C_B mJ/cm²	
	400—700	10^4—3 x 10^4	$C_B\mu$W/cm²	
IR-A	700—1049	10^{-9} —1.8 x 10^{-5}	5 C_A 10^{-7}J/cm²	
	700—1049	1.8 x 10^{-5} —10^3	1.8	
$C_A(1/t^{0.75})$mJ/cm²	1050—1400	10^{-9}— 10^{-4}	5 x 10^{-6}J/cm²	
	1050—1400	10^3— 3 x 10^4	320 $C_A\mu$W/cm²	
IR-B and C	1.4μm — 10^3μm	10^{-9}— 10^{-7}	10^{-2} J/cm²	
		10^{-7}— 10	$0.56 t^{0.75}$J/cm²	
		10— 3 x10^4	0.1W/cm²	

Note: For C_A see Figure 5.14; $C_B = I$ for A = 400 to 549 nm; $C_B = 10^{[0.015(A-550)]}$ for A= 550 to 700 nm; $T_1 = 10$ s for A = 400 to 549 nm; $T_1 = 10 \times 10^{[0.02(A-550)]}$ for A = 550 to 700 nm.

2. Interlocks that are designed to prevent firing of a pulsed laser by turning off the power supply or interrupting the beam must not automatically allow the power supply to be reactivated when they have been reset, after serving their protective function.
3. An audible or visible warning device is needed if a required interlock is bypassed or defeated. The warning system should be fail safe, i.e., if it became inoperative the unit should be inoperable.
4. Class IV lasers must be key interlocked or activated, and it is recommended that class IIIb lasers have the same feature. The key must be removable and the unit must not be operable without the key in place and in the "on" position.
5. A portal, viewing window, or an attached optical device must be designed to prevent any exposure above the permissible TLV®.

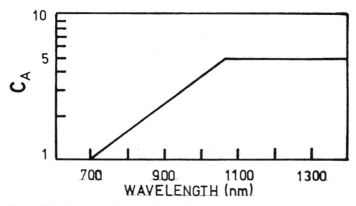

Figure 5.14   Correction factor, C_A for TLV's in Table 5.21.

6. Any class II (except units emitting less than class I limits for no more than 1000 seconds), III, or IV laser should be equipped with an appropriate safety device, such as a beam stop or optical attenuator, which will prevent emission of radiation in excess of those for a Class I unit in the controlled area.

7. The control units for laser units should be located in areas where the accessible emission limits for a Class I device are not exceeded.

8. Signs bearing essentially the same information as on the labels on the lasers in use within a controlled area should be posted at each entrance to the laser laboratory.

9. There is a possibility of radiation being emitted by other parts of the laser system, such as the power supply. This radiation associated with the operation of the laser system is called collateral radiation. The limits on this collateral radiation are (from Table VI, 21 CFR 1040.1):

   a. <u>Accessible emission limits</u> for collateral radiation having wavelengths greater than 180 nanometers but less than or equal to 1.0×10^6 nanometers are identical to the accessible emission limits of Class I laser radiation, as determined from Table 5.17 and Table IV in 21.CFR 1040.10 (Table 5.20 is derived from this latter Table):

      i. In the wavelength range of less than or equal to 400 nanometers, for all emission durations;

      ii. In the wavelength range of greater than 400 nanometers, for all emission durations less than or equal to 1×10^3 seconds and, when applicable under paragraph (f)(8) of 21 CFR 1040.10, for all emission durations.

   b. <u>Accessible emission limit</u> for collateral radiation within the x-ray range of wavelengths is 0.5 milliroentgen in an hour, averaged over a cross-section parallel to the external surface of the product, having an area of 10 square centimeters with no dimension greater than 5 centimeters.

B. Eye Protection

The shorter wavelength ultraviolet radiations, UV-C (100 to 280 nm) and UV-B (280 to 315 nm) are primarily absorbed within the conjunctiva and corneal portions of the eyes (see Figure 5.15), causing corneal inflammation. The ultraviolet frequencies just below the visible range, UV-A nm, (315 to 400) are absorbed largely within the lens of the eye. Although little UV-B radiation reaches the lens, UV-B is much more effective in causing cataracts to form than is UV-A. Radiations in the range of 400 to 1400 nm, which includes visible light and the near-infrared, are focused by the lens of the eye on the retina. The focusing properties of the lens may increase the energy per unit area for a point source by as much as 100,000 times.

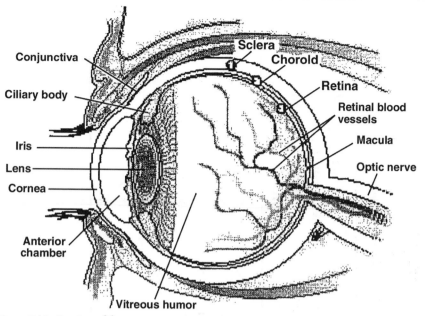

Figure 5.15. Structure of the eye.

The energy is primarily absorbed by the retinal pigmented epithelium and the choroid. At wavelengths longer than 1400 nm, the water in the eye tissues becomes opaque so that most of the energy is absorbed in the corneal region. The major mechanism for damage is thermal absorption, although there is some damage due to photochemical reactions.

Protective eyewear appropriate to the laser system in use should be worn if there is any eye hazard. The filter or filters in the protective goggles should be matched to the wavelength of the emissions of the laser. Since some lasers emit radiation at more than 1 wavelength, it may be necessary to have filters to cover each range of frequencies. It would be desirable to have the filters in the protective eyewear attenuate only narrow wavelength regions spanning those emitted by the laser, thereby allowing as much visible light through as possible to facilitate seeing by the wearer.

The term optical density is a convenient way to define the attenuation of incident radiation by a filter. A difference of one unit in the optical density of two filters corresponds to a transmission difference of a factor of ten. Thus, a filter that attenuates the incident radiation by a factor of 10 would have an optical density of 1, attenuation by a factor of 100 would mean an optical density of 2, and so forth. Since this is a logarithmic scale (to the base 10) the optical density of two filters stacked together is the sum of the optical densities of the two filters. Note, however, that this is true only if the two filters are for the same wavelength. If they are for different wavelengths and transmit essentially all of the radiation at the other's wavelength, the filtration of the stacked filter would be the same as that of the two filters considered individually.

When working with lasers of very high power or beam intensities, the absorption of energy in the filter can cause damage to the filters. For pulsed lasers, the threshold for damage to glass filters is approximately 10 to 100 joules/cm² and for plastic and dielectric coatings, also between 10 and 100 joules/cm². A continuous-wave laser operating at 10 watts or more can cause glass filters to fracture and can burn through plastic filters. If these numbers appear small, it should be considered that the power per square centimeter of a heating element on a range is approximately 15 to 20 watts. The filters should be inspected

Table 5.22 Selection Guide for Eye Protection for Direct Viewing of Laser Beams Emitting Radiation Between 400 and 1400 Nanometer Wavelengths

Q-Switched (1 ns—0.1 m)		Non-Q-Switched (0.4—10 ins)		Momentary 0.25—10 s		Continuous Long-Term Staring ≥3 h		Attenuation Factor	OD
Maximum Output Energy (J)	Maximum Beam Radiant (J/cm2)	Maximum Output Energy (J)	Maximum Power Radiant (J/cm2)	Maximum Beam Output (W)	Maximum Power Irradiance (W/cm2)	Maximum Beam Output (W)	Maximum Attenuation Irradiance (W/cm2)		
10	20	100	200	NR	NR	NR	NR	10^8	8
1	2	10	20	NR	NR	NR	NR	10^7	7
10^{-1}	2×10^{-1}	1	2	NR	NR	1	2	10^6	6
10^{-2}	2×10^{-2}	10^{-1}	2×10^{-1}	NR	NR	10^{-1}	2×10^{-1}	10^5	5
10^{-3}	2×10^{-3}	10^{-2}	2×10^{-2}	10	20	10^{-2}	2×10^{-2}	10^4	4
10^{-4}	2×10^{-4}	10^{-3}	2×10^{-3}	1	2	10^{-3}	2×10^{-3}	10^3	3
10^{-5}	2×10^{-5}	10^{-4}	2×10^{-4}	10^{-1}	2×10^{-1}	10^{-4}	2×10^{-4}	10^2	2
10^{-6}	2×10^{-6}	10^{-5}	2×10^{-5}	10^{-2}	2×10^{-2}	10^{-5}	2×10^{-5}	10^1	1

NR = Not recommended

routinely to be sure that they are not damaged to avoid damage to the eyes, including at this level, potential blindness.

In Table 5.22, there are two columns for each type of laser, one labeled "Maximum Output Power or Energy," which should be used for a focused beam, and the other labeled "Beam Radiant Exposure," which should be used for an unfocused beam that is larger than the pupil of the eye. Note that the pulsed laser outputs are defined in terms of energy (joules and joules/cm^2) while the CW laser outputs are defined in power (watts or watts/cm^2).

C. Medical Surveillance

All persons who routinely work with class III or class IV lasers should have a pre-employment medical examination. Others who occasionally work with lasers or in tasks for which it is conceivable that they could experience an eye exposure to the laser radiation should at least have an eye examination. The examination for those that normally work with lasers should include a medical history, stressing the visual and dermatologic systems. The examination should measure visual acuity, and special attention should be paid to those tissues most likely to be affected by the wavelengths emitted by the lasers that the employee will use. If employees suffer injuries or suspect that they have had a potential injury to the eye or skin, an examination of the potentially affected tissues should be done by a qualified physician, preferably an opthamologist, and the employees should be provided treatment as determined to be appropriate by the physician.

Although periodic examinations are not required by current guidelines, it would not be unreasonable to have a thorough eye examination on a 3- to 5-year schedule to determine if there are subtle changes occurring in the various systems of the eye.

REFERENCES

1. Performance Standards for Light-Emitting Products-Laser Products, 21 CFR Chap. I., Part 1040.10, Food and Drug Administration, 1998.
2. *A Guide for the Control of Laser Hazards,* American Conference of Governmental Industrial Hygienists, Cincinnati, OH, 1990.
3. American National Standard for the Safe Use of Lasers, ANSI Z-136.1, American National Standards Institute, New York, 1993.
4. *Guide for Selection of Laser Eye Protection, Pub 104, 4th ed.* Laser Institute of America, Orlando Florida
5. Threshold Limit Values, American Conference of Governmental Industrial Hygienists, Cincinnati, OH, latest revision.

INTERNET REFERENCES

1. http://www.hut.fi/~then/lights.html
2. http://www.retinoblastoma.com/guide/guide2.html

VI. MICROBIOLOGICAL AND BIOMEDICAL LABORATORIES

A. Introduction

Individuals who work in microbiological laboratories face a number of special problems when they work with organisms that are infectious to humans. There is evidence to show that biological laboratory personnel in such circumstances do have higher rates of incidence for a number of diseases associated with selected types of organisms than do comparable personnel working elsewhere, but there is no comprehensive system of data collection that defines the extent of the problem. However, virtually none of the primary laboratory infections which have occurred appear to have led to secondary infections for families, friends, or members of the general public. Only about 20% of the laboratory infections that have been reported have been attributed to specific incidents. The remainder have been assumed to be related to work practices within the laboratories, especially those which generate aerosols. In recent years, the fear which many laboratory workers feel about contracting the human immunodeficiency virus (HIV) virus, with the consequences which that implies, has emphasized the need for providing guidelines for safety in laboratories working with all types of infectious organisms. In 1988, the Centers for Disease Control (CDC) published its "Universal Precautions" due to the concern for transmittal of the HIV virus, but it has become the basis for handling blood and other body fluids that could contain contagious organisms. In 1991, OSHA incorporated the Universal Precautions in their standard for Occupational Exposure to Bloodborne Pathogens, covered in Chapter 4 of this book. The CDC has published comprehensive guidelines for Biosafety in Microbiological and Biomedical Laboratories, covering many other pathogenic organisms in addition to the HIV and hepatitis B (HBV) viruses which were emphasized in the OSHA standard. This section is based on the third edition of these guidelines, published in 1993.

Persons who work in microbiological laboratories in which the research involves organisms infectious to humans are usually well aware of the risks to themselves posed by the infectious agents. However, not everyone recognizes the many different operations and procedures which may result in significant quantities of aerosols being generated. Some of this awareness can be taught, some can be dictated by firm rules of laboratory practice, but some must be gained through experience. On the other hand, familiarity with procedures can occasionally lead to a casual attitude toward the risks associated with the research activities. Whether due to inexperience, too relaxed an attitude, or poor work performance, it only requires one mistake to cause a problem for everyone. Unlike radioisotopes, there is no simple means to detect contamination should it occur. There are methods for surveying for biological contamination, but they are much more complicated than those for radiation, for example, and do not provide instantaneous results, so that it is possible for an unsafe condition to persist for some time without the workers being aware of the situation. Although some diseases require a substantial exposure to the infectious organism, some can be transmitted with only a minimal exposure so that it does not require a major incident to create a problem. In addition, one can become as ill once the disease has been contracted whether it was due to an exposure to a few organisms or to a multitude, again unlike radiation where the risk is generally dose dependent.

B. Laboratory Safety and Hazard Communication Standards

Neither the OSHA Laboratory Safety nor the OSHA Hazard Communication standards explicitly cover exposures to contagious organisms. The OSHA bloodborne pathogen standard, by extension to sources of infection other than humans, comes closest to being an appropriate standard but as it is written, the emphasis is on human blood, tissues, and other bodily fluids. However, all of these standards contain the basic requirement that all employees must be made aware, by documented training programs, of the hazards associated with materials with which they work. The laboratory standard does not cover hazards associated with infectious organisms, but many of the procedures in microbiological laboratory involve chemicals covered by the standard. Individuals already working in the laboratory and all new employees, at the time of employment, must be instructed about the dangers associated with the specific chemical agents involved in their work and the safety procedures which have been adopted. Instruction on new materials brought into the laboratory must take place before employees begin using the material. As noted, the legal standard only applies to chemicals, but the concept is entirely appropriate to any microbiological hazards to which the workers may be exposed. If an employee is exposed to a known pathogenic organism due to a lack of information on the risks, means of preventing the exposure, and treatment options, it would be possible for citations to be issued under the general duty clause of the OSHA General Industrial standards. This is the approach by which the initial concerns about HIV and HBV were addressed, and has been considered for a response to the problems of exposure by health care and emergency personnel to resistant strains of tuberculosis.

C. Laboratory Director

The key person in any laboratory, but especially so in a microbiological laboratory, is the person in charge of the facility, the laboratory director. As noted in the previous section, there are guidelines and recommendations provided by many agencies, but there are relatively few mandatory governmental requirements for this class of laboratory. In the absence of a substantial body of formal regulations other than the internal policies which may have been set within the organization of which the laboratory is a part, the laboratory director is usually the individual that has been assigned the responsibility and the discretionary authority to set work practices. The attitude of this person will be reflected by others working in the facility. It is this individual who must see that laboratory workers are aware of the risks associated with their environment and set the work standards so that they are capable of performing all procedures safely and effectively. The laboratory authority must also determine if the available facilities are adequate for the research to be done, or to determine if special safety practices are needed. If the facilities are not adequate, then they must make the decision to either upgrade the facilities or modify the research so that it involves a lower level of risks. It is never appropriate to proceed if the facilities are not adequately supportive of the research program.

D. Miscellaneous Safety Practices

The microbiological laboratory shares many of the same safety needs as a standard chemical laboratory. Some of these are briefly summarized below. For a more detailed discussion of good general laboratory safety practices, refer to the comparable sections in the first four chapters.

1. Laboratory Line of Authority

The telephone numbers for the laboratory authorities and emergency phone numbers should be posted within the laboratory near the telephone, and outside the laboratory near the primary entrance. This list should include the overall building authority and an authorized alternate. In addition, a list of the significant hazards in the laboratory should be posted at the entrances to the area in the event that emergency personnel cannot reach knowledgeable persons, or if the time scale of the emergency requires immediate entrance. As a minimum, generally accepted signs such as the biohazard and radiation symbols or personal protective required should be posted at the entrance. Emergency workers always should be alerted to any risk to which they might be exposed. They are much less likely to have received training in infectious disease problems than in chemical emergencies. Exposure of emergency personnel to diseases has been addressed in the Ryan-White Act for Emergency Response Workers. This act *does* address potential infectious diseases other than HIV and HBV.

2. Spills and Emergencies Involving Chemically Dangerous Materials

Minor spills should be cleaned up immediately by laboratory personnel, providing that the material is not immediately dangerous to life and health (IDLH) and that equipment and supplies to do so are available. A readily accessible basic emergency spill kit should always be maintained within a facility. All personnel should be trained in where it is located, what are supplies are in it, and how to use those supplies.

Moderate spills of ordinarily dangerous materials may require technical guidance, supplies in excess of those normally kept on hand, or the assistance of the organizations safety or emergency response personnel.

For moderate to large spills of IDLH-level materials or for large spills of ordinarily dangerous materials, e.g., acids, etc., the following procedures are recommended:

1. Evacuate the area and initiate an evacuation of the building either personally or with the assistance of the building authority. The persons leaving the building should gather at a designated point, upwind from the building for further instructions.
2. Call the local emergency number (911, if available) and report the incident. The type of emergency must be clearly described so that the dispatcher can send the appropriate emergency responders. Do this from a nearby location but outside the affected area.
3. Those individuals directly involved in the incident are to remain available outside the building in a safe area to assist the emergency group unless they require immediate medical attention, which has first priority. In this case, their names should be taken and means of reaching them. After the initial notification of emergency personnel, the laboratory authority and the department head are to be notified.

The building should be evacuated and the fire department called for any fire other than a very small one, where there is confidence that it can be put out without risk of spreading or danger to anyone.

If there is an emergency involving an injury, be sure to alert the dispatcher of the circumstances of the injuries. If chemicals are involved, specific members of the local rescue squad having special training may need to respond. If there is exposure to the eye or skin, assist the injured person to an eyewash station, a deluge shower, or to a combination unit. In many states, the fire department is legally in charge of any hazardous material spill if called to the scene, but recognizing their limitations, they will often solicit advice from locally knowledgeable individuals or even defer to specialists from the organization involved.

3. Emergency Equipment for Non-biohazardous Spills

Everyone should be familiar with the location and use of all equipment in their laboratory area. As noted above, this includes means to initiate an evacuation (fire alarm pull stations, etc.), fire extinguishers, fire blankets, eyewash stations, deluge showers, first aid kits, spill kit materials, respiratory protective devices, and any other materials normally kept in the area for emergency response.

4. Evacuation

Everyone should be familiar with the primary and secondary evacuation routes from their area to the nearest exit, or an alternative one if the primary exit is blocked. Everyone should be told what method is used to signal a building evacuation, where to go, to check in with a responsible person, and not to reenter the building until an *official* clearance to do so is given by either the building authority or the senior emergency official responding to the emergency.

E. Accidents and Spills of Biohazardous Materials

Most public emergency agencies are not specifically equipped to handle biohazardous incidents so that laboratories handling organisms that are infectious to humans or organisms that could harm the environment should specifically plan for emergencies involving these materials. Concern for damage to the environment has caused an increasing number of emergency groups to obtain equipment that provides total personnel protection from airborne contaminants. The availability of these resources could make it possible to handle situations in which airborne dispersal of contagious organisms could be involved. The emergency respondents would depend greatly on the information available from laboratory personnel in determining the scope of their response. However, local emergency personnel could feel that the risks are unacceptable to them unless a person's life was in danger. This attitude should be understood by all concerned. No one should be asked to risk their lives unnecessarily.

Each facility should perform a hazard analysis of their operations to determine the worst-case scenarios involving all of their procedures, and to develop a plan to prevent these worst-case situations or to cope with them. The maximum credible adverse event for every procedure should be specifically addressed. For example, even a simple transfer of a container of active agents from one area of the room to another could result in dropping the container and releasing the organism to the room in the splashed material in the form of aerosols. Some of these, if small enough, could remain airborne for hours.

A minor spill would be one that remains contained within a biological safety cabinet, which provides personnel protection so no potentially infectious aerosols escape. It is assumed that no one is contaminated by direct contact with the spilled materials. Essentially, the procedures to be used in such a case should take advantage of the protection afforded by the cabinet, i.e., the cabinet should continue to operate. Eventual decontamination of the cabinet would be the measure to be taken to remedy the situation. In some minor cases, this can be done by using the routine procedures normally used for surface contamination. However, this may not decontaminate the fan, filters, and the airflow plenums which normally are not accessible. When there has been a substantial spill within the cabinet, the cabinet should be decontaminated by agents that could reach these concealed spaces. It could be necessary to use a strong agent such as formaldehyde gas. There are standard procedures for doing this available from a number of government agencies, such as NIH or the CDC. Individuals should be trained in how to perform this procedure safely and may wish to use it routinely on a number of occasions, such as before filter changes, before maintenance work, before relocating a cabinet, and upon instituting a different program in a biosafety cabinet.

A major spill would be one that is not contained within a biological safety cabinet. The response would depend upon the nature of the organism involved and the size of the spill, i.e., the probability of individuals being infected. In a later section, the design, special equipment, and practices will be described for laboratories for biosafety levels of 1 through 4. At the upper end of this scale, biosafety levels 3 and 4, there are legitimate concerns about the possibility of infections being carried out of the laboratory, so that the response would be different for these facilities than for the areas designed and operated to work with less dangerous organisms.

Following a spill of an infectious agent, one should immediately evacuate the room, breathing as little as possible of any aerosols. The door to the contaminated area should be closed and locked to avoid immediate airborne transfer of the material from the area in which the spill occurred, and to prevent unplanned entrance to the contaminated facility. Any outer garments that were contaminated, including laboratory coats, shoes, trousers, gloves, etc., should be removed and placed in sealed containers. These can be pails with covers, autoclavable bags, or, if no other alternatives are available, double plastic garbage bags. Everyone in the room should at least thoroughly wash their face and hands carefully with a disinfectant soap and, if possible, shower thoroughly. Once the area is isolated and the possibility of retaining infectious material on individuals minimized, the situation should be reviewed and a plan of correction determined by the responsible parties. Because of the possibility of liability, institutional and corporate management personnel as well as safety specialists should participate in the discussion. Any persons potentially infected should be referred to a physician as soon as possible. They should not participate in the clean up efforts and should be advised to take steps which would minimize potential spreading of infection until it is determined whether they were infected.

Any potentially contaminated materials should be autoclaved as soon as possible, or, depending on the value of the materials and the risk of infections to someone trying to sort the material, it may be decided that disposal as a hazardous waste is the proper action. This could still include autoclaving as an alternative but it could also include methods which would result in the total destruction of the contaminated items. Most organizations carry insurance to cover losses due to accidents. Some research items may represent irreproducible materials for which unusual efforts to recover them may be justified.

If the material is not highly infectious and the quantity is not great, it may be sufficient to have a volunteer reenter the area after aerosols have had an adequate time to settle, wearing a coverall covering the entire body, gloves, shoe covers, and a respirator, to decontaminate the immediate area of the spill with a suitable disinfectant and to wipe down nearby surfaces onto which some material may have splashed, or areas where the ventilation system in the laboratory may have carried aerosols. Once the preliminary clean up is completed, it may be sufficiently clean to warrant others entering the room wearing protective outer garments, gloves, and felt protective masks to proceed with a more thorough clean up. All disposable materials that are used in the clean up should go into hazardous waste containers, and materials which are to be retained, in autoclavable containers. It probably will be desirable to have tests made of the air to determine if there are still organisms present. All of these actions are not necessarily steps that should be taken in every case, but do represent a conservative approach which might be considered at the time of an incident depending upon the circumstances. One can more readily scale down after careful consideration than scale up a response in retrospect.

For an organism which is highly infectious to humans or for a massive spill, it may be necessary to seal off the area thoroughly, including shutting off ventilation ducts to and from the area to ensure that the organisms remain isolated. A specially designed decontamination program may be necessary. Consultation with specialists at the CDC and elsewhere may be desirable for advice and aid on the care of the exposed individuals and correction of the situation. Two things should be kept in mind: equipment can be replaced, people cannot. If the situation is not deteriorating, it is generally best to leave it alone while a plan of correction is developed which will provide maximum safety for everyone involved.

Custodial personnel should not be asked to assist in cleaning up a spill involving materials posing biohazards. They are not qualified to perform the work safely. Decontamination is a technical problem, not a custodial one.

F. Generation of Aerosols

It is generally conceded that aerosols are the primary means by which infectious diseases are contracted or spread in the microbiological laboratory although some cases are known to have occurred due to animal bites, needle sticks, and similar situations where direct contact can happen.

There are many opportunities for aerosols to be generated through normal laboratory procedures. Studies have been conducted of the average number of droplets created by many typical operations and some procedures are prolific generators of aerosols. Each droplet often contains several organisms. There are far more of these daily releases than there are accidents, and where the potential exists for many thousands of infectious organisms being released.

Only about 20% of all known laboratory infections have been traced to specific incidents. The majority of the remainder probably came from these ordinary, routine activities. In one comprehensive study, it was found that over 70% of the infections occurring in the laboratory were to scientific personnel, and that 98% of all laboratory-acquired infections were in institutions doing research or diagnostic work.

Some of the laboratory operations which release a substantial number of droplets are almost trivial in nature, such as breaking bubbles on the surface of a culture as it is stirred, streaking a rough agar plate with a loop, a drop falling off the end of a pipette, inserting a hot loop into a culture, pulling a stopper or a cotton plug from a bottle or flask, taking a culture sample from a vaccine bottle, opening and closing a petri dish in some applications, or opening a lyophilized culture, among many others. Most of these only take a few seconds and are often repeated many times daily. Other, more complicated procedures might be considered to be more likely to release organisms into the air, such as grinding tissue with a mortar and pestle, conducting an autopsy on a small animal, harvesting infected tissue from animals or eggs, intranasal inoculation of small animals, opening a blender too quickly, etc. Some incidents have resulted from failing to take into account the possibility that accidents can happen, such as a tube breaking in a centrifuge when not using a safety centrifuge cup. The possibility of aerosol production should always be considered while working with infectious organisms.

G. Infectious Waste

Any item that has been in contact with infectious organisms, or with materials such as blood, serums, excreta, tissue, etc., that may be infected must be considered infectious unless it has been rendered noninfectious. There are individuals who have excellent laboratory technique and take every precaution to avoid infection themselves and do not think of the possibility that someone who handles the materials which they discard may become infected from these materials. Some of the items used to carry out the procedures discussed in the previous paragraph are potentially infectious but are sometimes discarded as ordinary trash, in which case custodial workers can come into contact with them. Infected tissue has been found in clothing sent to a local laundry. Animal carcasses from animals that had rabies have been sent for disposal with no warnings about careful handling.

Waste from areas where the potential exists for coming into contact with infectious diseases should be treated as if it were hazardous and prepared so that it can be handled safely. It should be double bagged at least or put in a container which is not likely to break or rupture, then incinerated, steam sterilized, or perhaps chemically treated before disposal. Every organization which generates infectious waste should establish procedures to make sure that the waste is treated to render it harmless and that all personnel follow these procedures. One method suitable for a small generator of biological waste to safely collect waste is to double-bag it in heavy-duty plastic bags and freeze it until enough has been collected to run an incinerator economically (note that new EPA regulations under the clean air act are likely to have a severe impact on locally operated incinerators) or to justify the cost of a pickup by a biological waste disposal firm. There are now a number of commercial firms which offer services for disposing of dangerous biological waste just as there are firms for disposing of hazardous chemical wastes. Almost every state has now adopted regulations governing the handling and disposal of infectious waste and are required to have standards for incinerators used for hospital type medical wastes. The regulations are not equally stringent in every state, although they must be at least as stringent as the EPA rules require, so that laboratories generating infectious waste need to become familiar with the regulations that could affect their practices. There are many alternative technologies to steam sterilization and incineration (less likely under the new Clean Air Act rules) for processing infectious waste. These are discussed earlier in this book. Laboratories working with large animals have the most serious problems in dealing with routine infectious waste as existing autoclaving and incineration techniques are not as suitable for handling large animal carcasses and bedding as they are for processing blood, small tissue samples, "sharps," and other wastes characteristic of most biological facilities. The subject of disposing of medical/infectious biological waste was treated in some detail at the end of Chapter IV.

H. Laboratory Facilities — Design and Equipment

A microbiological laboratory designed for the level of activity expected to be housed within it is a major factor in protection of the employees. It must be designed properly to allow for safe working conditions. It will share many of the same basic features of a good chemistry laboratory. The ventilation should be a 100% fresh air system in most cases, with perhaps more stringent temperature and humidity controls than other types of laboratories. The temperature should be controlled over a relatively narrow range around 72 °F (22 °C) and the humidity should be maintained between 45 and 60%. There should be more than one means of egress in an emergency, although this is typically not a code requirement in most cases, since most laboratories in this class are usually occupied by only a small number of people. The interior layout should be conducive to free movement of personnel. Aisles should be sufficiently wide so that stools, chairs, or equipment placed temporarily in the aisles will not block them or constitute significant obstructions to travel. The floors, walls, and surfaces of the equipment should be of easily decontaminated material. The junctures of the floors with the walls and the equipment should be designed to be as seamless as possible to avoid cracks in which organic materials could collect and microorganisms thrive. There should be self-contained areas, either near the entrance to the laboratory or adjacent to it, where paperwork and records can be kept and processed, and where social conversation, studying, eating, etc., can be done safely outside the active work area. Adequate utilities should be provided. Proper experimental equipment, which will allow the laboratory operations to be done safely, should be provided and properly maintained. An eyewash fountain and deluge shower should be provided within the laboratory, and emergency equipment likely to be needed in the laboratory should be kept in a readily accessible place. A permanent shower and transition area between "clean" and "dirty" areas will be essential at some higher levels of risk. The laboratory should be at a negative pressure with respect to the corridor servicing it and the airflow within the laboratory should be away from the "social" area toward the work area.

Equipment is considered the primary barrier for protection of the employees. Items such as biosafety cabinets, safety centrifuges, enclosed containers, impervious work surfaces, autoclaves, foot-operated sinks, and other equipment specifically designed to prevent direct contact with infectious organisms or with aerosols must be available. Personal protective equipment can also be considered as an effective secondary barrier if engineering controls are not sufficient. These latter items can include, at minimum, a lab coat or wrap-around gown, possibly gloves, masks, or respirators, goggles, and head and foot covers.

I. Biosafety Levels

This section has been adapted with only minor changes, primarily syntax, from *Biosafety in Microbiological and Biomedical Laboratories* (BMBL) published by the U.S. Department of Health and Human Services. Some text has been added to extend the coverage to academic laboratories. The concept of protection in the following sections is based on a combination of three factors: a properly designed facility, appropriate equipment, and appropriate procedures. All three factors are the basis for the minimum level of needed protection not only for the workers but for others as well. One other factor must be added, and that is the attitude of everyone that safety is a very high priority that cannot be sacrificed to attain a scientific goal at a risk to personnel.

Microbiological laboratories are divided by the BMBL into four different classifications or levels, with level 1 intended for work with the lowest risk and level 4 designated for work with the highest risk. An ordinary laboratory would approximate a level 2 facility. Each level features a combination of design, standard and special laboratory practices and procedures, and standard and special equipment needed to allow the work appropriate for each level to

Table 5.23 Summary of Recommended Biosafety Levels(bsl) for Infectious Agents

BSL	Agents	Practices	Safety Equipment (Primary Barriers)	Facilities (Secondary Barriers)
1	Not known to cause disease in healthy adults	Standard Microbiological Practices	None required	Open bench top, sink required
2	Associated with human disease, hazard = auto-inoculation, ingestion, mucous membrane exposure	BSL-l practice plus: Limited access; Biohazard warning signs; 'Sharps' precautions;-Biosafety manual defining any needed waste decontamination or medical surveillance policies	Primmy barriers = Class I or II BSCs or other physical containment devices used for all manipulations of agents that cause splashes or aerosols of infectious materials; PPEs: laboratozy coats; gloves; face protection as needed	BSL-l plus: Autoclave available
3	Indigenous or exotic agents with potential for aerosol transmission low; disease may have serious or lethal consequences	BSL-2 practice plus: Controlled access; Decontamination of lab clothing before laundering; Baseline serum taken periodically	Primary barriers - Class I or II BCSs or other physical containment devices used for all manipulations of agents; PPE's -protective lab clothing; gloves, respiratory protection is needed	BSL-2 plus: Physical separation from access corridors; Self-closing, double door access; Exhausted air not recirculated; Negative airflow into laboratory
4	Dangerous/exotic agents which pose high risk of life-threatening disease, aerosol transmitted lab infections; or related with unknown risk of transmission	BSL-3 practices plus: Clothing change before entering; Shower on exit; All material decontaminated on exit from facility	Primary barriers All procedures conducted in Class Ill BSCs or Class I or II BSCs combination with full-body, air supplied, positive personnel suit	BSL-3 plus: Separate building or isolated zone; Dedicated supply/exhaust, vacuum, and decon systems ; Other requirements outlined in text.

be done safely. Refer to Table 5.23 for a summary of the recommended biosafety levels (BSL) for different levels of risk from infectious agents.

The following four sections describe each of these biosafety levels, after which several lists of organisms will be provided which are appropriate for each level.

1. Biosafety Level 1

The work at this level normally involves well-characterized agents not known to cause disease in healthy adult humans and of minimal risk to lab personnel and the environment. However, individuals who, because of poor health or who may be immunodeficient or immunosuppressed for any reason, may be at a higher risk, should inform the person responsible for the laboratory operation, to minimize the possibility of their acquiring an infection. The laboratory need not be separated from the remainder of the building. Work is generally conducted on open bench tops, using standard microbiological practices. No special containment equipment is required. The facilities and equipment are appropriate for undergraduate and secondary instruction and training in microbiological techniques. Laboratory personnel have been trained in the specific procedures used in the facility and are supervised by a person with general training in microbiology or a related science.

a. **Standard Microbiological Practices**
1. Access to the laboratory is limited or restricted at the discretion of the laboratory director when experiments or work with cultures and specimens are in progress.
2. Persons must wash their hands after they handle viable materials and animals, after removing gloves, and before leaving the laboratory.
3. Eating, drinking, smoking, handling contact lenses, and applying cosmetics are not permitted in the work areas where there is a reasonable likelihood of exposure to potentially infectious materials. Persons who wear contact lenses in laboratories should also wear goggles or a face shield. Food is to be stored outside the work area in cabinets or refrigerators designated and used for this purpose only.
4. Mouth pipetting is prohibited; mechanical pipetting devices are to be used.
5. All procedures are to be performed carefully to minimize the creation of splashes or aerosols.
6. Work surfaces in use are to be decontaminated at least once a day and after any spill of viable material.
7. All cultures, stocks, and other regulated wastes are decontaminated before disposal by an approved decontamination method, such as autoclaving. Materials to be decontaminated outside of the immediate laboratory are to be placed in a durable, leakproof container and closed for transport from the laboratory. Materials to be decontaminated off-site from the laboratory are to be packaged in accordance with applicable local, state, and federal regulations, before removal from the facility.
8. An insect and rodent control program is to be in effect.

b. **Special Practices**
None

c. **Safety Equipment (Primary Barriers)**
1. No special containment devices or equipment such as a biological safety cabinet are usually required for manipulations of agents assigned to biosafety level 1.
2. It is recommended that laboratory coats, gowns, or uniforms be worn to prevent contamination or soiling of street clothes.
3. Gloves should be worn if the skin of the hands is broken or if a rash exists.
4. Protective eyewear (chemical splash goggles and/or a face mask) should be worn for anticipated splashes of microorganisms or other hazardous materials to the face.

d. **Laboratory Facilities (Secondary Barriers)**
1. Each laboratory must contain a sink for hand washing.
2. The laboratory is designed so that it can be easily cleaned. Seamless or poured floor coverings are recommended. Epoxy paint is recommended for the walls. Rugs in laboratories are not appropriate and should not be used because proper decontamination of them after a spill is extremely difficult.
3. Laboratory furniture should be sturdy and incorporate as few seams and cracks as possible. Spaces between benches, cabinets, and equipment should be easily cleaned.
4. Bench tops must be impervious to water and resistant to acids, alkalies, organic solvents, and moderate heat.

5. Laboratory windows that can be opened must be equipped with fly screens.

2. Biosafety Level 2

This class of laboratory is suitable for work involving agents of moderate potential hazard to personnel and the environment. In addition to the characteristics of a biosafety level 1 facility: (1) laboratory personnel have specific training in handling pathogenic agents and are directed by competent scientists, (2) access to the laboratory is limited whenever work is being conducted, (3) extreme precautions (i.e., universal precautions) are taken with contaminated sharp items, and (4) some procedures in which infectious aerosols or splashes may be created or conducted will be performed in biological safety cabinets or other physical containment equipment. The laboratory facilities, equipment, and practices are suitable for clinical, diagnostic, and teaching programs involving the typical broad spectrum of indigenous moderate-risk agents present in the community. If the potential for aerosol production or splashes is low, work with these agents can be done safely on the open work bench.

a. Standard Microbiological Practices

1. Access to the laboratory is limited or restricted at the discretion of the laboratory director when experiments or work with cultures and specimens are in progress.
2. Persons must wash their hands after they handle viable materials and animals, after removing gloves, and before leaving the laboratory.
3. Eating, drinking, smoking, handling contact lenses, and applying cosmetics are not permitted in the work areas where there is a reasonable likelihood of exposure to potentially infectious materials. Persons who wear contact lenses in laboratories should also wear goggles or a face shield. Food is to be stored outside the work area in cabinets or refrigerators designated and used for this purpose only.
4. Mouth pipetting is prohibited; mechanical pipetting devices are to be used.
5. All procedures are to be performed carefully to minimize the creation of splashes or aerosols.
6. Work surfaces in use are to be decontaminated at least once a day and after any spill of viable material.
7. All cultures, stocks, and other regulated wastes are decontaminated before disposal by an approved decontamination method, such as autoclaving. Materials to be decontaminated outside of the immediate laboratory are to be placed in a durable, leakproof container and closed for transport from the laboratory. Materials to be decontaminated off-site from the laboratory are to be packaged in accordance with applicable local, state, and federal regulations, before removal from the facility.
8. An insect and rodent control program is to be in effect.

b. Special Practices

1. The laboratory director limits or restricts access to the laboratory when work with infectious agents is in progress. In general, persons who are at increased risk of acquiring infection or for whom infection might be unusually hazardous are not allowed in the laboratory or animal rooms. For example, persons who are immuno-compromised or immunosuppressed may be at risk of acquiring infections. The laboratory director has the final responsibility for assessing each circumstance and determining who may enter or work in the laboratory.
2. The laboratory director establishes policies and procedures whereby only persons who have been advised of the biohazards and meet any specific entry requirements (e.g.,

special training or immunization) may enter the laboratory or animal rooms.

3. When the infectious agent which is in use in the laboratory requires special provisions for entry (e.g., immunization) a hazard warning sign incorporating the universal biohazard symbol is to be posted on the door to the laboratory work area. The warning sign identifies the infectious agent, lists the name of the laboratory director or other responsible persons, and provides the special requirements for entering the laboratory.

4. Laboratory personnel are to receive appropriate immunizations or tests for the agents handled or potentially present in the laboratory (e.g., HBV vaccine or TB skin testing).

5. When appropriate, considering the agents handled, baseline serum samples for laboratory and other at-risk personnel are to be collected and stored. Additional serum samples may be taken periodically depending on the agents handled or the function of the facility.

6. A biosafety manual is to be prepared and adopted. Personnel are to be advised of special hazards and are required to read the manual and to follow instructions on practices and procedures.

7. Laboratory personnel are to receive appropriate training on the potential hazards associated with the work involved, the necessary precautions to prevent exposure, and the exposure evaluation procedures. Personnel receive annual updates or additional training as necessary for procedural or policy changes.

8. Laboratory personnel must always exercise a high degree of precaution with any contaminated sharp items, including needles and syringes, capillary tubes, and scalpels. Needles and syringes or other sharp instruments should be restricted in the laboratory, for use only when there is no alternative, such as parenteral injection, phlebotomy, or aspiration of fluids from laboratory animals and diaphragm bottles. Plasticware should be substituted for glassware whenever possible.

9. Only needle-locking syringes or disposable syringe-needle units (i.e., the needle is integral to the syringe) are to be used for the injection or aspiration of infectious fluids. Needles must not be bent, sheared, broken, recapped, removed from disposable syringes, or otherwise manipulated by hand before disposal; rather, they must be carefully placed in conveniently located, puncture-resistant containers used for sharps disposal. Nondisposable sharps must be placed in a hard-walled container for transport to a processing area for decontamination, preferably by autoclaving.

10. Syringes which resheathe the needle, needleless systems, and other safe devices should be used when appropriate.

11. Broken glassware must not be handled directly by hand, but must be removed by mechanical means such as a brush and dustpan, tongs, or forceps. Containers of contaminated needles, sharp equipment, and broken glass are to be decontaminated before disposal, according to any local, state, or federal regulations.

12. Cultures, tissues, or specimens of body fluids are to be placed in a container that prevents leakage during collection, handling, processing, storage transport, transport, or shipping.

13. Laboratory equipment and work surfaces should be decontaminated with an appropriate disinfectant on a routine basis, after work with infectious materials is finished, and especially after any overt spills, splashes, or other contamination by infectious materials. Contaminated equipment must be decontaminated before it is sent for repair or maintenance or packaged for transport in accordance with applicable local, state, or federal regulations before removal from the facility.

14. Spills and accidents which result in overt exposures to infectious materials are to be immediately reported to the laboratory director. Medical evaluation, surveillance, and treatment are to be provided as appropriate and written records maintained.

15. Animals not involved in the research being performed are not permitted in the laboratory.

c. Safety Equipment (Primary Barriers)

1. Properly maintained biological safety cabinets, preferably class II, or other appropriate personal protective or physical containment devices are to be used whenever:
   i. Procedures with a potential for creating infectious aerosols or splashes are conducted. These may include centrifuging, grinding, blending, vigorous shaking or mixing, sonic disruption, opening pressurized containers, inoculating animals intra-nasally, and harvesting infected tissues from animals or eggs.
   ii. High concentrations or large volumes of infectious agents are used. Such materials may be centrifuged in the open laboratory if sealed heads or centrifuge safety cups are used and if these rotors or safety cups are opened only in a biological safety cabinet.
2. Protective laboratory coats, gowns, smocks, or uniforms are to be worn while in the laboratory. This protective clothing is to be removed and left in the laboratory before leaving the laboratory for non-laboratory areas (e.g., cafeteria, library, administrative offices). All protective clothing is either disposed of in the laboratory or laundered by the institution; it should never be taken home by personnel.
3. Gloves are to be worn when handling infected animals and when hands may contact infectious materials, contaminated surfaces or equipment. Wearing two pairs of gloves may be appropriate; if a spill or spatter occurs, the hand will be protected after the contaminated glove is removed. Gloves are disposed of when contaminated, removed when work with infectious materials is completed, and are not to be worn outside the laboratory. Disposable gloves are not to be washed or reused.

d. Laboratory Facilities (Secondary Barriers)

1. Each laboratory is to contain a sink for hand washing.
2. The laboratory is designed so that it is easily cleaned. Seamless or poured floor coverings are recommended. Epoxy paint is recommended for the walls. Rugs in laboratories are not appropriate and should not be used because proper decontamination of them after a spill is extremely difficult.
3. Bench tops must be impervious to water and resistant to acids, alkalies, organic solvents, and moderate heat.
4. Laboratory furniture should be sturdy and incorporate as few seams and cracks as possible. Spaces between benches, cabinets, and equipment should be easily cleaned.
5. Laboratory windows that can be opened shall be equipped with fly screens.
6. A method for decontamination of infectious or regulated laboratory waste is available (e.g., autoclave, chemical disinfection, incinerator, or other approved decontamination system).
7. An eyewash facility (preferably one combined with a deluge shower) is to be readily available.

3. Biosafety Level 3

This level is applicable to clinical, diagnostic, teaching, and research facilities in which work is done with indigenous or exotic agents which may cause serious or potential lethal disease as a result of exposure by inhalation. All laboratory personnel are to be specifically trained in handling pathogenic and potentially lethal agents and are to be supervised by competent scientists who are experienced in working with these agents. All procedures involving manipulation of infectious agents are conducted within biological safety cabinets or other physical containment devices or by personnel wearing appropriate personal protective clothing or devices. The laboratory must be equipped with specific design features.

Many existing facilities may not have all the facility safeguards recommended for biosafety level 3 (e.g., access zone, sealed penetrations, and directional airflow, etc.). In these circumstances, acceptable safety may be achieved for routine or repetitive operations (e.g., diagnostic procedures involving the propagation of an agent for identification, typing, and susceptibility testing) in biosafety level 2 facilities. However, the recommended Standard Microbiological Practices, Special Practices, and Safety Equipment for Biosafety Level 3 must be rigorously followed. The decision to implement this modification of biosafety level 3 recommendations should be made only by the laboratory director.

a.   Standard Microbiological Practices

1. Access to the laboratory is limited or restricted at the discretion of the laboratory director when experiments or work with cultures and specimens are in progress.
2. Persons must wash their hands after they handle viable materials and animals, after removing gloves, and before leaving the laboratory.
3. Eating, drinking, smoking, handling contact lenses, and applying cosmetics are not permitted in the work areas where there is a reasonable likelihood of exposure to potentially infectious materials. Persons who wear contact lenses in laboratories should also wear goggles or a face shield. Food is to be stored outside the work area in cabinets or refrigerators designated and used for this purpose only.
4. Mouth pipetting is prohibited; mechanical pipetting devices are to be used.
5. All procedures are to be performed carefully to minimize the creation of splashes or aerosols.
6. Work surfaces in use are to be decontaminated at least once a day and after any spill of viable material.
7. All cultures, stocks, and other regulated wastes are decontaminated before disposal by an approved decontamination method, such as autoclaving. Materials to be decontaminated outside of the immediate laboratory are to be placed in a durable, leakproof container and closed for transport from the laboratory. Materials to be decontaminated off-site from the laboratory are to be packaged in accordance with applicable local, state, and federal regulations, before removal from the facility.
8. An insect and rodent control program is to be in effect.

b.   Special Practices

1. Laboratory doors are to be kept closed when experiments are in progress.
2. The laboratory director controls access to the laboratory and restricts access to persons whose presence is required for program or support purposes. For example, persons who are immunocompromised or immunosuppressed may be at risk of acquiring infections. Persons who are at increased risk of acquiring infection or for whom infection may be unusually hazardous are not to be allowed in the laboratory or animal rooms. The laboratory director has the final responsibility for assessing each circumstance and determining who may enter or work in the laboratory.
3. The laboratory director establishes policies and procedures whereby only persons who have been advised of the biohazards, who meet any specific entry requirements (e.g., special training or immunization), and who comply with all entry and exit procedures enter the laboratory or animal rooms.
4. When the infectious materials or infected animals are present in the laboratory or containment module, a hazard warning sign incorporating the universal biohazard symbol is to be posted on the door to the laboratory and animal room access doors. The warning sign identifies the infectious agent, lists the name and telephone number of the laboratory director or other responsible persons, and indicates any special requirements for entering the laboratory, such as the need for immunizations, respirators, or other personal

protective measures.

5. Laboratory personnel are to receive appropriate immunizations or tests for the agents handled or potentially present in the laboratory (e.g., HBV vaccine or TB skin testing).

6. Baseline serum samples are to be collected and stored for all laboratory and other at- risk personnel. Additional serum samples may be taken periodically depending on the agents handled or the function of the facility.

7. A biosafety manual is to be prepared and adopted. Personnel are to be advised of special hazards, and are required to read the manual and to follow instructions on practices and procedures.

8. Laboratory personnel are to receive appropriate training on the potential hazards associated with the work involved, the necessary precautions to prevent exposure, and the exposure evaluation procedures. Personnel are to receive annual updates or additional training as necessary for procedural or policy changes.

9. The laboratory director is responsible for ensuring that before working with organisms at biosafety level 3, all personnel must demonstrate proficiency in standard microbiological practices and techniques and in the practices and operations specific to the laboratory facility. This might include prior experience in handling human pathogens or cell cultures, or a specific training program provided by the laboratory director or other competent scientist proficient in safe microbiological practices and techniques.

10. Laboratory personnel must always exercise a high degree of precaution with any contaminated sharp items, including needles and syringes, capillary tubes, and scalpels. Needles and syringes or other sharp instruments should be restricted in the laboratory for use only when there is no alternative, such as parenteral injection, phlebotomy, or aspiration of fluids from laboratory animals and diaphragm bottles. Plasticware should be substituted for glassware whenever possible.

    i. Only needle-locking syringes or disposable syringe-needle units (i.e., needle is integral to the syringe) are to be used for the injection or aspiration of infectious fluids. Used disposable needles must not be bent, sheared, broken, recapped, removed from disposable syringes, or otherwise manipulated by hand before disposal; rather, they must be carefully placed in conveniently located puncture-resistant containers used for sharps disposal. Nondisposable sharps must be placed in a hard-walled container for transport to a processing area for decontamination, preferably by autoclaving.

    ii. Syringes which resheathe the needle, needleless systems, and other safe devices should be used when appropriate.

    iii. Broken glassware must not be handled directly by hand, but must be removed by mechanical means such as a brush and dustpan, tongs, or forceps. Containers of contaminated needles, sharp equipment, and broken glass are to be decontaminated before disposal, according to any local, state, or federal regulations.

11. All manipulations involving infectious materials are to be conducted in biological safety cabinets or other physical containment devices within the containment module. No work in open vessels is to be conducted on the open bench.

12. Laboratory equipment and work surfaces should be decontaminated with an appropri-ate disinfectant on a routine basis, after work with infectious materials, and especially after overt spills, splashes, or other contamination with infectious materials. Contaminated equipment must be decontaminated before it is sent for repair or main-tenance, or packaged for transport in accordance with applicable local, state, or federal regulations, before removal from the facility. Plastic-backed paper toweling used on nonperforated work surfaces within biological safety cabinets facilitates clean up.

13. Cultures, tissues, or specimens of body fluids are to be placed in a container that pre-vents leakage during collection, handling, processing, storage, transport, or shipping.

14. All potentially contaminated waste materials (e.g., gloves, lab coats, etc.) from labora-

tories or animal rooms are to be decontaminated before disposal or reuse.

15. Spills of infectious materials are to be decontaminated, contained, and cleaned up by appropriate professional staff or others properly trained and equipped to work with concentrated infectious materials.

16. Spills and accidents which result in overt or potential exposures to infectious materials are to be immediately reported to the laboratory director. Medical evaluation, surveillance, and treatment are to be provided as appropriate and written records maintained.

17. Animals and plants not related to the work being conducted are not permitted in the laboratory.

c. **Safety Equipment (Primary Barriers)**

1. Properly maintained biological safety cabinets (class II or III) are to be used for all manipulations of infectious materials.

2. For work done outside biological safety cabinets, appropriate combinations of personal protective equipment are to be used (e.g., special protective clothing, masks, gloves, face protection, or respirators), in combination with physical containment devices (e.g., centrifuge safety cups, sealed centrifuge rotors, or containment caging for animals).

3. This equipment must be used for manipulation of cultures and of those clinical or enviromnental materials which may be a source of infectious aerosols, the aerosol challenge of infected animals, harvesting of tissues or fluids from infected animals and embryonated eggs, and necropsy of infected animals.

4. Face protection (goggles and mask or face shield) is to be worn for manipulations of infectious materials outside of a biological safety cabinet.

5. Respiratory protection is worn when aerosols cannot be safely contained (i.e., outside of a biological safety cabinet), and in rooms containing infected animals.

6. Protective laboratory clothing such as solid-front or wrap-around gowns, scrub suits, or coveralls must be worn in the laboratory, but the same clothing is not to be worn outside.

7. Gloves must be worn when handling infected animals and when hands may contact infectious materials and contaminated surfaces or equipment. Disposable gloves should be discarded when contaminated and never washed for reuse.

d. **Laboratory Facilities (Secondary Barriers)**

1. The laboratory is to be separated from areas which are open to unrestricted traffic flow within the building. Passage through two sets of self-closing doors is the basic requirement when entry into the laboratory from access corridors or other contiguous areas is required. The doors should be separated by at least seven feet so that they cannot be opened simultaneously. A clothes change room (shower optional) may be included in the passage way.

2. Each laboratory must contain a sink for hand washing. The sink is to be foot, elbow, or automatically operated.

3. The interior surfaces of walls, floors, and ceilings are to be water resistant so that they can be easily cleaned. Penetrations in these surfaces are to be sealed or are to be capable of being sealed to facilitate decontamination.

4. Bench tops must be impervious to water and resistant to acids, alkalies, organic solvents, and moderate heat.

5. Laboratory furniture is to be sturdy, and spaces between benches, cabinets, and equipment must be easily cleaned.

6. Windows in the laboratory are to be closed and sealed.

7. A method of decontaminating all laboratory wastes must be available, preferably within the laboratory (i.e., autoclave, chemical disinfection, incineration, or other approved decontamination method).

8. A ducted exhaust air ventilation system is to be provided. The system is to be

designed so as to create directional airflow that draws air from "clean" areas into the laboratory toward "contaminated" areas. The exhaust air is not to be recirculated to any other area of the building, and is to be discharged to the outside with filtration and other treatment as needed. The outside exhaust must be dispersed away from occupied areas and air intakes. Laboratory personnel must verify that the direction of the airflow (into the laboratory) is proper (a static pressure gauge with an alarm between the laboratory and the adjacent spaces can ensure this).

9. The high-efficiency particulate air (HEPA)-filtered exhaust air from class II or class III biological safety cabinets is to be exhausted directly to the outside or through the building exhaust system. If the HEPA-filtered exhaust air from class II or III biological safety cabinets is to be discharged through the building exhaust air system, it is to be connected to this system in a manner (e.g., thimble unit connection) that avoids any interference with the air balance of the cabinets or building exhaust system. Exhaust air from class II biological safety cabinets may be recirculated within the laboratory if the cabinet is tested and certified at least every 12 months.

10. Continuous flow centrifuges or other equipment that may produce aerosols are to be contained in devices that exhaust air through HEPA filters before discharge into the laboratory.

11. Vacuum lines are to be protected with liquid disinfectant traps and HEPA filters, or their equivalent, which are routinely maintained and replaced as needed.

12. An eyewash facility (preferably one combined with a deluge shower) is to be readily available.

4. Biological Safety Level 4

Biosafety level 4 is required for work with dangerous and exotic agents which pose a high individual risk of aerosol-transmitted laboratory infections and life-threatening disease. These are expensive facilities and there are very few compared to Level 3 facilities. Agents that have a close or identical antigenic relationship to biosafety level 4 agents are to be handled at this level until sufficient data are available to either confirm continued work at this level or to work with them at a lower level. Members of the laboratory staff are to receive specific and thorough training in handling extremely hazardous infectious agents. They are to understand the primary and secondary containment functions of the standard and special practices, the containment equipment, and the laboratory design characteristics. They are to be supervised by competent scientists who are trained and experienced in working with these agents. Access to the laboratory is to be strictly controlled by the laboratory director. The facility is to be in either a separate building or in a controlled area within the building that is completely isolated from all other areas of the building. A specific facility operations manual is to be prepared and adopted. All laboratory personnel are to confirm that they have read this material.

Within the work areas of the facility, all activities are confined to Class III biosafety cabinets, or to class II biological safety cabinets used with one-piece positive pressure personnel suits ventilated by a life support system. The facility is to be designed to prevent discharge of microorganisms into the environment.

a. Standard Microbiological Practices

1. Access to the laboratory is limited or restricted at the discretion of the laboratory director when experiments or work with cultures and specimens are in progress.

2. Persons must wash their hands after they handle viable materials and animals, after removing gloves, and before leaving the laboratory.

3. Eating, drinking, smoking, handling contact lenses, and applying cosmetics are not

permitted in the work areas where there is a reasonable likelihood of exposure to potentially infectious materials. Persons who wear contact lenses in laboratories should also wear goggles or a face shield. Food is to be stored outside the work area in cabinets or refrigerators designated and used for this purpose only.

4. Mechanical or automatic pipetting devices must be used. Mouth pipetting is prohibited.

5. All procedures are to be performed carefully to minimize the creation of splashes or aerosols.

6. Work surfaces in use are to be decontaminated at least once a day and after any spill of viable material.

7. An insect and rodent control program is to be in effect.

b. **Special Practices**

1. Only persons whose presence in the facility or individual laboratory rooms is required for program or support purposes are authorized to enter. Persons who are immuno-compromised or immunosuppressed may be at risk of acquiring infections. Therefore, persons who may be at increased risk of acquiring infection or for whom infection may be unusually hazardous, such as children or pregnant women, are not allowed in the laboratory or animal rooms. The supervisor has the responsibility for assessing each circumstance and determining who may enter or work in the laboratory. Access to the facility is limited by means of secure, locked doors. Accessibility is managed by the laboratory director, biohazards control officer, or other person responsible for the physical security of the facility. All persons entering the facility are to be informed in advance of the potential biohazards within and instructed as to the appropriate safeguards for ensuring their safety. Authorized persons must comply with the instructions and all other applicable entry and exit procedures. A logbook, to be signed by all personnel, indicates the date and time of each entry and exit. Practical and effective emergency protocols for emergencies are to be established.

2. When infectious material or infected animals are present in the laboratory or animal rooms, hazard warning signs incorporating the universal biohazard symbol are to be prominently posted on all access doors. The sign is to identify the agent, list the name of the laboratory director or other responsible persons, and indicate any special requirements for entering the area, such as the need for immunizations or respirators.

3. The laboratory director is responsible for ensuring that, before working with or-ganisms at biosafety level 4, all personnel must demonstrate proficiency in standard microbiological practices and techniques, and in the special practices and operations specific to the laboratory facility. This might include prior experience in handling human pathogens or cell cultures, or a specific training program provided by the laboratory director or other competent scientist proficient in these unique safe micro-biological practices and techniques.

4. Laboratory personnel are to receive appropriate available immunizations for the agents handled or potentially present in the laboratory.

5. Baseline serum samples are to be collected and stored for all laboratory and other at-risk personnel. Additional serum samples may be collected periodically depending on the agents handled or the function of the laboratory. The decision to establish a serologic surveillance program takes into account the availability of methods for the assessment of antibody to the agents of concern. The program must make provision for the testing of serum samples at each collection interval and the communication of results to the participants.

6. A biosafety manual is to be prepared or adopted. Personnel are to be advised of special hazards and are required to read the manual and to follow instructions on practices and procedures.

7. Laboratory personnel are to receive appropriate training on the potential hazards associated with the work involved, the necessary precautions to prevent exposure, and the exposure evaluation procedures. Personnel are to receive annual updates or additional training as necessary for procedural or policy changes.

8. Personnel are to enter and leave the facility only through the clothing change and shower rooms, and are to shower each time they leave the facility. Personnel are to use the airlocks to enter or leave the laboratory only in an emergency.

9. Personal clothing is to be removed in the outer clothing change room and kept there. Complete laboratory clothing, including underclothes, pants, shirts or jump suits, shoes, and gloves, is to be provided and used by all personnel entering the facility. When leaving the laboratory and before proceeding into the shower area, personnel are to remove their laboratory clothing in the inner change room and store it in a locker or hamper in the inner change room.

10. Supplies and materials needed in the facility are to be brought in by way of the double-doored autoclave, fumigation chamber, or airlock which is appropriately decontaminated between each use. After the outer doors are secured, personnel within the facility are to retrieve the materials by opening the interior doors of the autoclave, fumigation chamber, or airlock. The interior doors are to be secured after the materials are brought into the facility.

11. Laboratory personnel must always exercise a high degree of precaution with any contaminated sharp items, including needles and syringes, capillary tubes, and scalpels. Needles and syringes or other sharp instruments should be restricted in the laboratory, for use only when there is no alternative, such as parenteral injection, phlebotomy, or aspiration of fluids from laboratory animals and diaphragm bottles. Plasticware should be substituted for glassware whenever possible.

12. Only needle-locking syringes or disposable syringe-needle units (i.e., needle is integral to the syringe) are to be used for the injection or aspiration of infectious fluids. Used disposable needles must not be bent, sheared, broken, recapped, removed from disposable syringes, or otherwise manipulated by hand before disposal; rather, they must be carefully placed in conveniently located puncture-resistant containers used for sharps disposal. Nondisposable sharps must be placed in a hard-walled container for transport to a processing area for decontamination, preferably by autoclaving.

13. Syringes which resheathe the needle, needleless systems, and other safe devices should be used when appropriate.

14. Broken glassware must not be handled directly by hand, but must be removed by mechanical means such as a brush and dustpan, tongs, or forceps. Containers of contaminated needles, sharp equipment, and broken glass are to be decontaminated before disposal, according to any local, state, or federal regulations.

15. Biological materials to be removed from the class III cabinet or from the biosafety level 4 laboratory in a viable or intact state are to be transferred to a nonbreakable, sealed primary container and then enclosed in a nonbreakable, sealed secondary container, which is to be removed from the facility through a disinfectant dunk tank, fumigation chamber, or an airlock designed for this purpose.

16. No materials, except for biological materials that are to remain in a viable or intact state, are to be removed from the maximum containment laboratory unless they have been autoclaved or decontaminated before they leave the facility. Equipment or materials which may be damaged by high temperatures or steam may be decontaminated by gaseous or vapor methods in an airlock or chamber designed for the purpose.

17. Laboratory equipment and work surfaces are to be decontaminated routinely after work with infectious materials is finished, and especially after overt spills, splashes, or other contamination with infectious materials. Contaminated equipment must be decontaminated before it is sent for repair or maintenance.

18. Spills of infectious materials are to be decontaminated, contained, and cleaned up by appropriate professional staff or others properly trained and equipped to work with concentrated infectious materials.

19. A system is to be established for proper reporting of laboratory accidents, exposures, and employee absenteeism, and for medical surveillance of potential laboratory-associated illnesses. Written records are to be prepared and maintained of the absentee records, exposures, and incidents. An essential adjunct to such a reporting-maintenance system is to have available a facility for the quarantine, isolation, and medical care of personnel with potential or known laboratory-associated illnesses.

20. Materials, such as plants, animals, and clothing, not related to the experiment being conducted, are not to be permitted in the facility.

c. Safety Equipment (Primary Barriers)

a. All procedures within the facility with agents assigned to biosafety level 4 are to be conducted in Class III biological safety cabinets or in Class II biological safety cabinets used in conjunction with one-piece positive pressure personnel suits ventilated by a life support system. Work with viral agents that require biosafety level 4 secondary containment capabilities can be conducted in class II biological safety cabinets within the facility without the one-piece positive pressure personnel suit if:

(1) the facility has been decontaminated, (2) no work is being conducted in the facility with other agents assigned to biosafety level 4, (3) all personnel are immunized against the specific agent being manipulated and demonstrate protective antibody levels, and (4) all other standard and special practices are followed.

2. All personnel entering the facility must don complete laboratory clothing, including undergarments, pants, shirts or jump suits, shoes, and gloves. All personnel protective equipment is to be removed in the change room before showering and leaving the laboratory.

d. Laboratory Facility (Secondary Barriers)

1. The biosafety level 4 facility consists of either a separate building or a clearly demarcated and isolated zone within a building. Outer and inner change rooms separated by a shower are to be provided for personnel entering or leaving the facility. A double-doored autoclave, fumigation chamber, or ventilated airlock is to be provided for passage of those materials, supplies, or equipment which are not to be brought into the facility through the change room.

2. Walls, floors, and ceilings of the facility are to be constructed to form a sealed internal shell which facilitates fumigation and is animal- and insect-proof. The internal surfaces of the shell are to be resistant to liquids and chemicals to facilitate cleaning and decontamination. All penetrations in these structures and surfaces are to be sealed. Any drains in the floor are to contain traps filled with a chemical disinfectant of demonstrated efficacy against the target agent. The drains are to be connected directly to the liquid waste decontamination system. Sewer vents and other ventilation lines are to contain HEPA filters.

3. Internal facility appurtenances, such as light fixtures and utility pipes, are to be installed in such a way as to minimize the horizontal surface area on which dust can settle.

4. Bench tops are to have seamless surfaces which are impervious to water and resistant to acids, alkalies, organic solvents, and moderate heat.

5. Laboratory furniture is to be of simple and sturdy construction and spaces between benches, cabinets, and equipment are to be easily cleaned.

6. A foot-, elbow-, or automatically-operated hand washing sink is to be provided near the door of each laboratory room in the facility.

7. Any installed central vacuum system is not to serve any areas outside the facility. In-line HEPA filters are to be placed as near as practicable to each use point or service connection. Filters are to be installed to permit in-place decontamination and replacement. Other liquid and gas services to the facility are to be protected by back-flow prevention devices.

8. If water fountains are provided, they are to be foot operated and are to be located in the facility corridors outside the laboratory. The water service to the fountains is not to be connected to the back-flow protected system supplying water to the laboratory areas.

9. Access doors to the laboratory are to be self-closing and lockable.

10. Any windows are to be sealed and breakage resistant.

11. A double-doored autoclave is to be provided for decontaminating materials passing out of the facility. The autoclave door that opens to the area external to the facility is to be sealed to the outer wall and is automatically controlled so that the outside door cannot be opened until after the "sterilization" cycle has been completed.

12. A pass-through dump tank, fumigation chamber, or an equivalent decontamination method is to be provided so that materials and equipment that cannot be decontaminated in the autoclave can be safely removed from the facility.

13. Liquid effluents from laboratory sinks, biological safety cabinets, floor drains (if used), and autoclave chambers are to be decontaminated by heat treatment before being discharged to the sanitary sewer. Effluents from showers and toilets may be discharged to the sanitary sewer without treatment. The process used for decontaminating liquid wastes must be validated physically and biologically by use of a constant recording temperature sensor in conjunction with an indicator microorganism having a defined heat susceptibility profile.

14. A dedicated nonrecirculating ventilation system is to be provided. The supply and exhaust components of the system are to be balanced to assure directional airflow from the area of least hazard to the areas of greatest potential hazard. The differential pressure/directional airflow between adjacent areas is to be monitored and alarmed to indicate malfunction of the system. The airflow in the supply and exhaust components is to be monitored and the components interlocked to assure inward (or zero) airflow is maintained.

15. The general room exhaust air from the facility in which the work is conducted in a class III cabinet system is to be treated by passage through HEPA filters prior to discharge to the outside. The air is to be discharged away from occupied spaces and air intakes. The HEPA filters are be located as near as practicable to the source in order to minimize the length of potentially contaminated ductwork. The HEPA filter housings are to be designed to allow for *in situ* decontamination of the filter prior to removal, or removal of the filter is to be in a sealed gas-tight primary container for subsequent decontamination and/or destruction by incineration. The design of the HEPA filter housing should facilitate validation of the filter installation. The use of precertified HEPA filters can be an advantage. The service life of the exhaust HEPA filters can be extended through adequate filtration of the supply air.

16. A specially designed suit area may be provided in the facility to provide personnel protection equivalent to that provided by class III cabinets. Personnel who enter this area are to wear a one-piece positive-pressure suit that is ventilated by a life support

system. The life support system is to include alarms and emergency backup breathing air tanks. Entry into this area is to be through an airlock fitted with airtight doors. A chemical shower is to be provided to decontaminate the surface of the suit before the worker leaves the area. The exhaust air from the suit area is to be filtered by two sets of HEPA filters installed in series. A duplicate filtration unit, exhaust fan, and an automatically starting emergency power source are to be provided. The air pressure within the suit area is to be lower than that of any adjacent area. Emergency lighting and communication systems are to be provided. All penetrations into the internal shell of the suit area are to be sealed. A double-doored autoclave is to be provided for decontaminating waste materials to be removed from the suit area.

17. The treated exhaust air from class II biological safety cabinets located in a facility in which workers wear a positive-pressure suit may be discharged into the animal room environment or to the outside through the facility air exhaust system. The biological safety cabinets are to be tested and certified at 12-month intervals. The exhaust air from class III biological safety cabinets is to be passaged through two sets of HEPA filters in series prior to discharge to the outside. If the treated exhaust air is discharged to the outside through the facility exhaust air system, it is to be connected to this system in such a manner as to avoid any interference with the air balance of the cabinets or the facility exhaust air system.

J. Vertabrate Animal Biosafety Level Criteria

Many biological laboratories use animals in their research, ranging from mice to large animals. The reference, *Biosafety in Microbiological and Biomedical Laboratories*, from which the information in the previous sections was taken also includes comparable material for such laboratories and their animal facilities. The following sections dealing with these facilities again were taken from this reference with only minor changes, primarily in syntax, where these changes were felt helpful to the reader.

If experimental animals are used, management must provide facilities and staff and establish practices which reasonably assure appropriate levels of environmental quality, safety, and care. Laboratory animal facilities in many ways are extensions of the laboratory. As a general principle, the biosafety level (facilities, practices, and operational requirements) recommended for working with infectious agents in vivo and in vitro are comparable. It is well to remember, however, that the animal room is not the laboratory, and can present some unique problems. In the laboratory, hazardous conditions are caused by personnel or the equipment that is being used. In the animal room the activities of the animals themselves can introduce new hazards. Animals may produce aerosols, and they may also infect and traumatize animal handlers by biting and scratching.

These recommendations presuppose that laboratory animal facilities, operational practices, and quality of animal care meet applicable standards and regulations and that appropriate species have been selected for animal experiments (e.g., *Guide for the Care and Use of Laboratory Animals,* HEW Publication No. (NIH) 86-23, and *Laboratory Animal Welfare Regulations* - 9 CFR, Subchapter A, Parts 1, 2, and 3).

Ideally, facilities for laboratory animals used for studies of infectious or noninfectious disease should be physically separate from other activities such as animal production and quarantine, clinical laboratories, and especially from facilities that provide patient care. Animal facilities should be designed and constructed to facilitate cleaning and housekeeping. Traffic flow that will minimize the risk of cross contamination should be considered in the plans. A "clean/dirty hall" layout is useful in achieving this. Floor drains should be installed in animal facilities only on the basis of clearly defined needs. If floor drains are installed, the drain trap should always contain water or a suitable disinfectant.

These recommendations describe four combinations of practices, safety equipment, and facilities for experiments on animals infected with agents which produce, or may produce, human infection. These four combinations provide increasing levels of protection to personnel and to the environment, and are recommended as minimal standards for activities involving infected laboratory animals. These four combinations, designated Animal Biosafety Levels ABSL 1-4, describe animal facilities and practices applicable to work on animals infected with agents assigned to corresponding Biosafety Levels BSL1-4.

As the reader will note, there are substantial similarities between ABSL level criteria and those of equivalent BSL, but there are significant differences as well.

1. Animal Biosafety Level 1
a. Standard Practices
1. Access to the animal facility is limited or restricted at the discretion of the laboratory or animal facility director.
2. Personnel are to wash their hands after handling cultures and animals, after removing gloves, and before leaving the animal facility.
3. Eating, drinking, smoking, handling contact lenses, applying cosmetics, and storing food for human use are not permitted in animal rooms. Persons who wear contact lenses in animal rooms should also wear goggles or a face shield.
4. All procedures are to be carefully performed to minimize the creation of aerosols.
5. Work surfaces are to be decontaminated after use or after any spill of viable materials.
6. Doors to animal rooms open inward, are self-closing, and are kept closed when experimental animals are present.
7. All wastes from the animal room are to be appropriately decontaminated, preferably by autoclaving, before disposal. Infected animal carcasses are to be incinerated after being transported from the animal room in leakproof, covered containers.
8. An insect and rodent control program is in effect.

b. Special Practices ABSL 1
1. The laboratory or animal facility director limits access to the animal room to personnel who have been advised of the potential hazard and who need to enter the room for program or service purposes when work is in progress. In general, persons who may be at increased risk of acquiring infection, or for whom infection might be unusually hazardous, are not allowed in the animal room.
2. The laboratory or animal facility director establishes policies and procedures whereby only persons who have been advised of the potential hazard and meet any specific requirements (e.g., immunization) may enter the animal room.
3. Bedding materials from animal cages are removed in such a manner as to minimize the creation of aerosols, and are disposed of in compliance with applicable institutional or local requirements.
4. Cages are washed manually or in a cage washer. Temperature of final rinse water in a mechanical washer should be 180°F.
5. The wearing of laboratory coats, gowns, or uniforms in the animal facility is recommended. It is further recommended that laboratory coats worn in the animal facility not be worn in other areas.
6. A biosafety manual is prepared or adopted. Personnel are advised of special hazards, and are required to read and to follow instructions on practices and procedures.

c. Safety Equipment (Primary Barriers) ABSL 1
Special containment equipment is not required for animals infected with agents assigned to Biosafety Level 1.

d. Animal Facilities (Secondary Barriers) ABSL 1

1. The animal facility is designed and constructed to facilitate cleaning and housekeeping.
2. A hand washing sink is available in the animal facility.
3. If the animal facility has windows that open, they are fitted with fly screens.
4. Exhaust air is discharged to the outside without being recirculated to other rooms, and it is recommended, but not required, that the direction of airflow in the animal facility is inward.

2. Animal Biosafety Level 2
a. Standard Practices

1. Access to the animal facility is limited or restricted at the discretion of the laboratory or animal facility director.
2. Personnel are to wash their hands after handling cultures and animals, after removing gloves, and before leaving the animal facility.
3. Eating, drinking, smoking, handling contact lenses, applying cosmetics, and storing food for human use are not permitted in animal rooms. Persons who wear contact lenses in animal rooms should also wear goggles or a face shield.
4. All procedures are to be carefully performed to minimize the creation of aerosols.
5. Work surfaces are to be decontaminated after use or after any spill of viable materials.
6. Doors to animal rooms open inward, are self-closing, and are kept closed when experimental animals are present.
7. All wastes from the animal room are to be appropriately decontaminated, preferably by autoclaving, before disposal. Infected animal carcasses are to be incinerated after being transported from the animal room in leakproof, covered containers.
8. An insect and rodent control program is in effect.

b. Special Practices ABSL 2

1. The laboratory or animal facility director limits access to the animal room to personnel who have been advised of the potential hazard and who need to enter the room for program or service purposes when work is in progress. In general, persons who may be at increased risk of acquiring infection, or for whom infection might be unusually hazardous, are not allowed in the animal room.
2. The laboratory or animal facility director establishes policies and procedures whereby only persons who have been advised of the potential hazard and meet any specific requirements (e.g., immunization) may enter the animal room.
3. When the infectious agent(s) in use in the animal room require(s) special entry provisions (e.g., the need for immunizations and respirators), a hazard warning sign incorporating the universal biohazard symbol, is to be posted on the access door to the animal room. The hazard warning sign identifies the infectious agent(s) in use, lists the name and telephone number of the animal facility supervisor or other responsible person(s), and indicates the special requirement(s) for entering the animal room.
4. Laboratory personnel receive appropriate immunizations or tests for the agents handled or potentially present in the laboratory (e.g., hepatitis B vaccine or TB skin testing).
5. When appropriate, considering the agents handled, baseline serum samples from animal care and other at-risk personnel are collected and stored. Additional serum samples may be collected periodically depending on the agents handled or the

function of the facility. The decision to establish a serologic surveillance program must take into account the availability of methods for the assessment of antibody to the agent(s) of concern. The program should provide for the testing of serum samples at each collection interval and the communication of results to the participants.

6. A biosafety manual is to be prepared and adopted. Personnel are to be advised of special hazards, and are required to read and to follow instructions on practices and procedures.

7. Laboratory personnel must receive appropriate training on the potential hazards associated with the work involved, the necessary precautions to prevent exposures, and the exposure evaluation procedures. Personnel are to receive annual updates, or additional training as necessary for procedural or policy changes.

8. A high degree of precaution must always be taken with any contaminated sharp items, including needles and syringes, slides, pipettes, capillary tubes, and scalpels. Needles and syringes or other sharp instruments are restricted in the animal facility for use only when there is no alternative, such as for parenteral injection, blood collection, or aspiration of fluids from laboratory animals and diaphragm bottles. Plasticware should be substituted for glassware whenever possible.

9. Only needle-locking syringes or disposable syringe-needle units (i.e., needle is integral to the syringe) are used for injection or aspiration of infectious materials. Used disposable needles must not be bent, sheared, broken, recapped, removed from disposable syringes, or otherwise manipulated by hand before disposal; rather, they must be carefully placed in conveniently located puncture-resistant containers used for sharps disposal. Non-disposable sharps must be placed in a hard-walled container for transport to a processing area for decontamination, preferably by autoclaving.

10. Syringes which re-sheathe the needle, needle-less systems, and other safe devices should be used when appropriate.

11. Broken glassware must not be handled directly by hand, but must be removed by mechanical means such as a brush and dustpan, tongs, or forceps. Containers of contaminated needles, sharp equipment, and broken glass should be decontaminated before disposal, according to any local, state, or federal regulations.

12. Cultures, tissues, or specimens of body fluids are to be placed in a container that prevents leakage during collection, handling, processing, storage, transport, or shipping.

13. Cages are appropriately decontaminated, preferably by autoclaving, before they are cleaned and washed. Equipment and work surfaces should be decontaminated with an appropriate disinfectant on a routine basis, after work with infectious materials is finished, and especially after overt spills, splashes, or other contamination by infectious materials. Contaminated equipment must be decontaminated according to any local, state, or federal regulations before it is sent for repair or maintenance or packaged for transport in accordance with applicable local, state, or federal regulations, before removal from the facility.

14. Spills and accidents which result in overt exposures to infectious materials are to be immediately reported to the laboratory director. Medical evaluation, surveillance, and treatment are provided as appropriate and written records are maintained.

15. Animals not involved in the work being performed are not permitted in the lab.

c. Safety Equipment (Primary Barriers) ABSL 2

1. Biological safety cabinets, other physical containment devices, and/or personal protective equipment (e.g., respirators, face shields) are used whenever procedures with a high potential for creating aerosols are conducted. These include necropsy of infected animals, harvesting of tissues or fluids from infected animals or eggs,

intranasal inoculation of animals, and manipulations of high concentrations or large volumes of infectious materials.

2. Appropriate face/eye and respiratory protection is worn by all personnel entering animal rooms housing nonhuman primates.
3. Laboratory coats, gowns, or uniforms are worn while in the animal room. This protective clothing is removed before leaving the animal facility.
4. Special care is taken to avoid skin contamination with infectious materials; gloves are worn when handling infected animals and when skin contact with infectious materials is unavoidable.

d. Animal Facilities (Secondary Barriers) ABSL 2

1. The animal facility is designed and constructed to facilitate cleaning and housekeeping.
2. A handwashing sink is available in the room where infected animals are housed.
3. If the animal facility has windows that open, they are fitted with fly screens.
4. If floor drains are provided, the drain traps are always filled with water or a suitable disinfectant.
5. Exhaust air is discharged to the outside without being recirculated to other rooms, and it is recommended, but not required, that the direction of airflow in the animal facility is inward.
6. An autoclave which can be used for decontaminating infectious laboratory waste is available in the building with the animal facility.

3. Animal Biosafety Level 3

a. Standard Practices

1. Access to the animal facility is limited or restricted at the discretion of the laboratory or animal facility director.
2. Personnel are to wash their hands after handling cultures and animals, after removing gloves, and before leaving the animal facility.
3. Eating, drinking, smoking, handling contact lenses, applying cosmetics, and storing food for human use are not permitted in animal rooms. Persons who wear contact lenses in animal rooms should also wear goggles or a face shield.
4. All procedures are to be carefully performed to minimize the creation of aerosols.
5. Work surfaces are to be decontaminated after use or after any spill of viable materials.
6. Doors to animal rooms open inward, are self-closing, and are kept closed when experimental animals are present.
7. All wastes from the animal room are to be appropriately decontaminated, preferably by autoclaving, before disposal. Infected animal carcasses are to be incinerated after being transported from the animal room in leakproof, covered containers.
8. An insect and rodent control program is in effect.

b. Special Practices ABSL 3

1. The laboratory director or other responsible person restricts access to the animal room to personnel who have been advised of the potential hazard and who need to enter the room for program or service purposes when infected animals are present. Persons who are at increased risk of acquiring infection, or for whom infection might be unusually hazardous, are not allowed in the animal room. Persons at increased risk may include children, pregnant women, and persons who are immunodeficient or immuno-suppressed. The supervisor has the final responsibility for assessing each circumstance and determining who may enter or work in the facility.

2. The laboratory director or other responsible person establishes policies and procedures whereby only persons who have been advised of the potential hazard and meet any specific requirements (e.g., for immunization) may enter the animal room.

3. When the infectious agent(s) in use in the animal room requires special entry provisions (e.g., the need for immunizations and respirators), a hazard warning sign incorporating the universal biohazard symbol is to be posted on the access door to the animal room. The hazard warning sign identifies the infectious agent(s) in use, lists the name and telephone number of the animal facility supervisor or other responsible person(s), and indicates the special requirement(s) for entering the animal room.

4. Laboratory personnel receive appropriate immunizations or tests for the agents handled or potentially present in the laboratory (e.g., hepatitis B vaccine or TB skin testing).

5. Baseline serum samples from all personnel working in the facility and other at-risk personnel should be collected and stored. Additional serum samples may be collected periodically and stored. The serum surveillance program must take into account the availability of methods for the assessment of antibody to the agent(s) of concern. The program should provide for the testing of serum samples at each collection interval and the communication of results to the participants.

6. A biosafety manual is prepared and adopted. Personnel are advised of special hazards, and are required to read and to follow instructions on practices and procedures.

7. Laboratory personnel receive appropriate training on the potential hazards associated with the work involved, the necessary precautions to prevent exposures, and the exposure evaluation procedures. Personnel receive annual updates, or additional training as necessary for procedural or policy changes.

8. A high degree of precaution must always be taken with any contaminated sharp items, including needles and syringes, slides, pipettes, capillary tubes, and scalpels. Needles and syringes or other sharp instruments are restricted in the laboratory for use only when there is no alternative, such as for parenteral injection, blood collection, or aspiration of fluids from laboratory animals and diaphragm bottles. Plasticware should be substituted for glassware whenever possible.

9. Only needle-locking syringes or disposable syringe-needle units (i.e., needle is integral to the syringe) are used for injection or aspiration of infectious materials. Used disposable needles must not be bent, sheared, broken, recapped, removed from disposable syringes, or otherwise manipulated by hand before disposal; rather, they must be carefully placed in conveniently located puncture-resistant containers used for sharps disposal. Non-disposable sharps must be placed in a hard-walled container for transport to a processing area for decontamination, preferably by autoclaving.

10. Syringes which re-sheathe the needle, needle-less systems, and other safe devices should be used when appropriate.

11. Broken glassware must not be handled directly by hand, but must be removed by mechanical means such as a brush and dustpan, tongs, or forceps. Containers of contaminated needles, sharp equipment, and broken glass should be decontaminated before disposal, according to any local, state, or federal regulations.

12. Cultures, tissues, or specimens of body fluids are to be placed in a container that prevents leakage during collection, handling, processing, storage, transport, or shipping.

13. Cages are autoclaved or thoroughly decontaminated before bedding is removed or before they are cleaned and washed. Equipment and work surfaces should be decontaminated with an appropriate disinfectant on a routine basis, after work with infectious materials is finished, and especially after overt spills, splashes, or other

contamination by infectious materials. Contaminated equipment must be decontaminated according to any local, state, or federal regulations before it is sent for repair or maintenance or packaged for transport in accordance with applicable local, state, or federal regulations, before removal from the facility.

14. Spills and accidents which result in overt exposures to infectious materials are immediately reported to the laboratory director. Medical evaluation, surveillance, and treatment are provided as appropriate and written records are maintained.

15. All wastes from the animal room are autoclaved before disposal. All animal carcasses are incinerated. Dead animals are transported from the animal room to the incinerator in leakproof covered containers.

16. Animals not involved in the work being performed are not permitted in the lab.

c. Safety Equipment (Primary Barriers) ABSL 3

1. Personal protective equipment is used for all activities involving manipulations of infectious materials or infected animals.

   i. Wrap-around or solid-front gowns or uniforms are worn by personnel entering the animal room. Front-button laboratory coats are unsuitable. Protective gowns should be appropriately contained until decontamination or disposal.

   ii. Personnel wear gloves when handling infected animals. Gloves are removed aseptically and autoclaved with other animal room wastes before disposal.

   iii. Appropriate face/eye and respiratory protection is worn by all personnel entering animal rooms housing nonhuman primates.

   iv. Boots, shoe covers, or other protective footwear, and disinfectant foot baths are available and used when indicated.

2. Physical containment devices and equipment appropriate for the animal species are used for all procedures and manipulations of infectious materials or infected animals.

3. The risk of infectious aerosols from infected animals or their bedding also can be reduced if animals are housed in partial containment caging systems, such as open cages placed in ventilated enclosures (e.g., laminar flow cabinets), solid wall and bottom cages covered with filter bonnets, or other equivalent primary containment systems.

d. Animal Facilities (Secondary Barriers) ABSL 3

1. The animal facility is designed and constructed to facilitate cleaning and housekeeping, and is separated from areas which are open to unrestricted personnel traffic within the building. Passage through two sets of doors is the basic requirement for entry into the animal room from access corridors or other contiguous areas. Physical separation of the animal room from access corridors or other activities may also be provided by a double-doored clothes change room (showers may be included), airlock, or other access facility which requires passage through two sets of doors before entering the animal room. The doors should be at least seven feet apart so that it is not possible to hold them both open at once.

2. The interior surfaces of walls, floors, and ceilings are water resistant so that they may be easily cleaned. Penetrations in these surfaces are sealed or capable of being sealed to facilitate fumigation or space decontamination.

3. A foot-, elbow-, or automatically operated hand-washing sink is provided in each animal room near the exit door.

4. If vacuum service (i.e., central or local) is provided, each service connection should be fitted with liquid disinfectant traps and a HEPA filter.

5. If floor drains are provided, they are protected with liquid traps that are always filled

with water or disinfectant.

6. Windows in the animal room are non-operating and sealed.
7. Animal room doors are self-closing and are kept closed when infected animals are present.
8. An autoclave for decontaminating wastes is available, preferably within the animal facility. Materials are transferred to the autoclave in a covered leak-proof container whose outer surface has been decontaminated.
9. A non-recirculating ventilation system is provided. The supply and exhaust components of the system are balanced to provide for directional flow of air into the animal room. The exhaust air is discharged directly to the outside and clear of occupied areas and air intakes. Exhaust air from the room can be discharged without filtration or other treatment. Personnel must periodically validate that proper directional airflow is maintained.
10. The HEPA filtered exhaust air from Class I or Class II biological safety cabinets or other primary containment devices is discharged directly to the outside or through the building exhaust system. Exhaust air from these primary containment devices may be recirculated within the animal room if the device is tested and certified at least every 12 months. If the HEPA filtered exhaust air from Class I or Class II biological safety cabinets is discharged to the outside through the building exhaust system, it is connected to this system in a manner (e.g., thimble unit connection) that avoids any interference with the performance of either the cabinet or building exhaust system.

4. Animal Biosafety Level 4
a. Standard Practices
1. Access to the animal facility is limited or restricted at the discretion of the laboratory or animal facility director.
2. Personnel wash their hands after handling cultures and animals, after removing gloves, and before leaving the animal facility.
3. Eating, drinking, smoking, handling contact lenses, applying cosmetics, and storing food for human use are not permitted in animal rooms. Persons who wear contact lenses in animal rooms should also wear goggles or a face shield.
4. All procedures are carefully performed to minimize the creation of aerosols.
5. Work surfaces are decontaminated after use or after any spill of viable materials.
6. Doors to animal rooms open inward, are self-closing, and are kept closed when experimental animals are present.
7. All wastes from the animal room are appropriately decontaminated, preferably by autoclaving, before disposal. Infected animal carcasses are incinerated after being transported from the animal room in leakproof, covered containers.
8. Cages are autoclaved before bedding is removed and before they are cleaned and washed. When feasible, disposable cages that do not require cleaning are recommended; however, these cages also autoclaved before disposal. Equipment and work surfaces should be decontaminated with an appropriate disinfectant on a routine basis, after work with infectious materials is finished, and especially after overt spills, splashes, or other contamination by infectious materials. Contaminated equipment must be decontaminated according to any local, state, or federal regulations before it is sent for repair or maintenance or packaged for transport in accordance with applicable local, state, or federal regulations, before removal from the facility.
9. An insect and rodent control program is in effect.

b. Special Practices ABSL 4
1. Only persons whose entry into the facility or individual animal room is required for

program or support purposes are authorized to enter. Persons who may be at increased risk of acquiring infection or for whom infection might be unusually hazardous are not allowed in the animal facility. Persons at increased risk may include children, pregnant women, and persons who are immunodeficient or immunosuppressed. The supervisor has the final responsibility for assessing each circumstance and determining who may enter or work in the facility. Access to the facility is limited by secure, locked doors. Accessibility is controlled by the animal facility supervisor, biohazards control officer, or other person responsible for the physical security of the facility. Before entering, persons are advised of the potential biohazards and instructed as to appropriate safe-guards. Personnel comply with the instructions and all other applicable entry and exit procedures. Practical and effective protocols for emergency situations are established.

2. Laboratory personnel receive appropriate immunizations or tests for the agents handled or potentially present in the laboratory (e.g., hepatitis B vaccine or TB skin testing).

3. Baseline serum samples are collected and stored for all laboratory and other at-risk personnel. Additional serum specimens may be collected periodically, depending on the agents handled or the function of the laboratory. The decision to establish a serologic surveillance program takes into account the availability of methods for the assessment of antibody to the agent(s) of concern. The program provides for the testing of serum samples at each collection interval and the communication of results to the participants.

4. A biosafety manual is prepared and adopted. Personnel are advised of special hazards, and are required to read and to follow instructions on practices and procedures.

5. When the infectious agent(s) in use in the animal room requires special entry provisions (e.g., the need for immunizations and respirators) a hazard warning sign, incorporating the universal biohazard symbol, is posted on the access door to the animal room. The hazard warning sign identifies the infectious agent(s) in use, lists the name and telephone number of the animal facility supervisor or other responsible person(s), and indicates the special requirement(s) for entering the animal room.

6. Laboratory personnel receive appropriate training on the potential hazards associated with the work involved, the necessary precautions to prevent exposures, and the exposure evaluation procedures. Personnel receive annual updates, or additional training as necessary for procedural or policy changes.

7. Hypodermic needles and syringes are used only for gavage, for parenteral injection, and aspiration of fluids from diaphragm bottles or well-restrained laboratory animals.

8. A high degree of precaution must always be taken with any contaminated sharp items, including needles and syringes, slides, pipettes, capillary tubes, and scalpels. Needles and syringes or other sharp instruments are restricted in the laboratory for use only when there is no alternative, such as for parenteral injection, blood collection, or aspiration of fluids from laboratory animals and diaphragm bottles. Plasticware is substituted for glassware whenever possible.

9. Only needle-locking syringes or disposable syringe-needle units (i.e., needle is integral to the syringe) are used for injection or aspiration of infectious materials. Used disposable needles are not bent, sheared, broken, recapped, removed from disposable syringes, or otherwise manipulated by hand before disposal; rather, they are carefully placed in conveniently located puncture-resistant containers used for sharps disposal. Non-disposable sharps are placed in a hard-walled container, preferably containing a suitable disinfectant, for transport to a processing area for decontamination, preferably by autoclaving.

10. Syringes which re-sheathe the needle, needle-less systems, and other safe devices

should be used when appropriate.

11. Broken glassware is not handled directly by hand, but is removed by mechanical means such as a brush and dustpan, tongs, or forceps. Containers of contaminated needles, sharp equipment, and broken glass are decontaminated before disposal, according to any local, state, or federal regulations.

12. Cultures, tissues, or specimens of body fluids are to be placed in a container that prevents leakage during collection, handling, processing, storage, transport, or shipping.

13. Spills and accidents which result in overt exposures to infectious materials are immediately reported to the laboratory director. Medical evaluation, surveillance, and treatment are provided as appropriate and written records are maintained.

14. Personnel enter and leave the facility only through the clothing change and shower rooms. Personnel shower each time they leave the facility. Head covers are provided to personnel who do not wash their hair during the exit shower. Except in an emergency, personnel do not enter or leave the facility through the airlocks.

15. Personal clothing is removed in the outer clothing change room and kept there. Complete laboratory clothing, including undergarments, pants and shirts or jumpsuits, shoes, and gloves, are provided and used by all personnel entering the facility. When exiting, personnel remove laboratory clothing in the inner change room before entering the shower area. Soiled clothing is autoclaved before laundering.

16. Supplies and materials are brought into the facility by way of a double-door autoclave, fumigation chamber, or airlock. After securing the outer doors, personnel inside the facility retrieve the materials by opening the interior door of the autoclave, fumigation chamber, or airlock. This inner door is secured after materials are brought into the facility. The autoclave fumigation chamber or airlock is decontaminated before the outer door is opened.

17. A system is established for the reporting of animal facility accidents and exposures, employee absenteeism, and for the medical surveillance of potential laboratory-associated illnesses. An essential adjunct to such a reporting-surveillance system is the availability of a facility for the quarantine, isolation, and medical care of persons with potential or known laboratory-associated illnesses.

18. Materials (e.g., plants, animals, clothing) not related to the experiment are not permitted in the facility.

c. Safety Equipment (Primary Barriers)

1. Laboratory animals, infected with agents assigned to Biosafety Level 4, are housed in a Class III biological safety cabinet or in a partial containment caging system (such as open cages placed in ventilated enclosures, solid wall and bottom cages covered with filter bonnets, or other equivalent primary containment systems), in specially designed areas in which all personnel are required to wear one-piece positive pressure suits ventilated with a life support system.

2. Animal work with viral agents that require Biosafety Level 4 secondary containment, and for which highly effective vaccines are available and used, may be conducted with partial containment cages and without the one-piece positive pressure personnel suit if: the facility has been decontaminated, no concurrent experiments are being done in the facility which require Biosafety Level 4 primary and secondary containment, and all other standard and special practices are followed.

d. Animal Facility (Secondary Barriers) ABSL 4

1. The animal rooms are located in a separate building or in a clearly demarcated and isolated zone within a building. Outer and inner change rooms, separated by a shower, are provided for personnel entering and leaving the facility. A double-doored

autoclave, fumigation chamber, or ventilated airlock is provided for passage of materials, supplies, or equipment which are not brought into the facility through the change room.

2. Walls, floors, and ceilings of the facility are constructed to form a sealed internal shell which facilitates decontamination and is animal and insect proof. The internal surfaces of this shell are resistant to liquids and chemicals, thus facilitating cleaning and decontamination of the area. All penetraions in these structures and surfaces are sealed.

3. Internal facility appurtenances, such as light fixtures, air ducts, and utility pipes, are arranged to minimize horizontal surface areas on which dust can settle.

4. A foot, elbow, or automatically operated hand washing sink is provided in each animal room near the exit door.

5. If there is a central vacuum system, it does not serve areas outside of the facility. The vacuum system has in-line HEPA filters placed as near as practicable to each use point or service connection. Filters are installed to permit in-place decontamination and replacement. Other liquid and gas services for the facility are protected by devices that prevent backflow.

6. External animal facility doors are self-closing and self locking.

7. Any windows must be resistant to breakage and sealed.

8. A double-doored autoclave is provided for decontaminating materials that leave the facility. The autoclave door which opens to the area external to the facility is automatically controlled so that it can only be opened after the autoclave "sterilization" cycle is completed.

9. A pass-through dunk tank, fumigation chamber, or an equivalent decontamination method is provided so that materials and equipment that cannot be decontaminated in the autoclave can be safely removed from the facility.

10. Liquid effluents from laboratory sinks, biological safety cabinets, floor drains (if used), and autoclave chambers are decontaminated by heat treatment before being discharged to the sanitary sewer. Effluents from showers and toilets may be discharged to the sanitary sewer without treatment. The process used for decontamination of liquid wastes must be validated physically and biologically by use of a constant recording temperature sensor in conjunction with an indicator microorganism having a defined heat susceptibility profile.

11. A dedicated non-recirculating ventilation system is provided. The supply and exhaust components of the system are balanced to assure directional airflow from the area of least hazard to the area(s) of greatest potential hazard. The differential pressure/directional airflow between adjacent areas is monitored and alarmed to indicate malfunction of the system. The airflow in the supply and exhaust components is monitored and the components interlocked to assure inward (or zero) airflow is maintained.

12. The general room exhaust air from a facility in which the work is conducted in a Class III cabinet system is treated by a passage through a HEPA filter(s) prior to discharge to the outside. The air is discharged away from occupied spaces and air intakes. The HEPA filter(s) are located as near as practicable to the source in order to minimize the length of potentially contaminated duct work. The HEPA filter housings are designed to allow for *in situ* decontamination of the filter prior to removal, or removal of the filter in a sealed gas-tight primary container for subsequent decontamination and/or destruction by incineration. The design of the HEPA filter housing should facilitate validation of the filter installation. The use of pre-certified HEPA filters can be an advantage. The service-life of the exhaust HEPA filters can be extended through

adequate filtration of the supply air.

13. The treated exhaust air from Class II biological safety cabinets located in a facility in which workers wear a positive pressure suit may be discharged into the animal room environment or to the outside through the facility air exhaust system. The biological safety cabinets are tested and certified at 9-month intervals. The air exhausted from Class III biological safety cabinets is passaged through two HEPA filter systems (in series) prior to discharge to the outside. If the treated exhaust is discharged to the outside through the facility exhaust system, it is connected to this system in a manner that avoids any interference with the air balance of the cabinets or the facility exhaust system.

14. A specially designed suit area may be provided in the facility. Personnel who enter this area wear a one-piece positive pressure suit that is ventilated by a life support system. The life support system is provided with alarms and emergency backup breathing air tanks. Entry to this area is through an airlock fitted with airtight doors. A chemical shower is provided to decontaminate the surface of the suit before the worker leaves the area. The exhaust air from the area in which the suit is used is filtered by two sets of HEPA filters installed in series. Duplicate filtration units and exhaust fans are provided. An automatically starting emergency power source is provided. The air pressure within the suit area is lower than that of any adjacent area. Emergency lighting and communication systems are provided. All penetrations into the inner shell of the suit area are sealed. A double-doored autoclave is provided for decontaminating waste materials to be removed from the suit area.

K. Recommended Biosafety Levels

The subsections of this part will contain, for a large number of potentially infectious agents, the biosafety level appropriate for typical laboratory-scale operations involving these agents. However, selection of an appropriate safety level for work with a specific agent or animal study depends upon a large number of factors. Some of the most important are (1) the virulence, pathogenicity, biological stability, route of spread, and communicability of the agent; (2) the nature or function of the laboratory, the procedures and manipulations involving the agent, and the endemicity of the agent; and (3) the availability of effective vaccines or therapeutic measures. The following sections will not present the rationale for the recommendations. For additional information, the reader should consult the original document, Reference 15, from which the lists in this section were taken. This information is also available from the Internet reference for this document.

The risk assessments and biosafety levels recommended presuppose a population of healthy, immunocompetent individuals. There are a number of parameters which influence this, including age, heredity, race, sex, pregnancy, predisposing diseases, surgery, and prior exposure to immunosuppressing agents.

Recommendations for the use of vaccines and toxoids are included where effective and safe versions are available. These may change over time and should be evaluated at the appropriate time. Appropriate precautions should be taken in the administration of live attenuated virus vaccines to individuals with altered immunocompetence. However, these specific recommendations should in no way preclude the routine use of such products as diphtheria-tetanus toxoids, polio virus vaccine, or influenza vaccine.

The basic biosafety level assigned to an agent is based on the activities typically associated with the growth and manipulation of quantities and concentrations of infectious agents required to accomplish identification or typing. If activities with clinical materials pose a lower risk to personnel than those activities associated with the manipulation of cultures, a lower biosafety level can be considered. On the other hand, if the activities involve large volumes ("production quantities") or highly concentrated preparations, manipulations which are likely to produce aerosols, or which are otherwise intrinsically hazardous, additional personnel precautions and increased levels of primary containment may be indicated. It may be possible to adapt biosafety levels up or down to compensate for the appropriate level of safety.

As noted in many places in the preceding material, it is the responsibility of the laboratory director to make these decisions. Risk assessment is ultimately a subjective process, but it is recommended that decisions should be biased toward more safety rather than less.

1. Agent Summaries

All of the recommendations in the following exclude working with production quantities or in situations where substantial amounts of aerosols may be generated. In such cases, a higher level of protection is recommended.

a. Parasitic Agents
- Cestode Parasites of Humans

Echinoccus granulosus	Level 2
Taenia solium (cysticercus cellulosae)	Level 2
Hymenolepsis nana	Level 2

- Nematode Parasites of Humans

Strongyloides spp.	Level 2
Ascaris spp.	Level 2

- Protozoal Parasites of Humans

Toxoplasma spp.	Level 2
Plasmodium spp. (including *P. cynomologi*)	Level 2
Trypanosoma spp.	Level 2
Leishmania spp.	Level 2
Sarcocystis spp.	Level 2
Coccidia spp.	Level 2
Entamoeba spp.	Level 2
Giardia spp.	Level 2
Cryptosporidia spp.	Level 2
Naegleria fowleri	Level 2
Toxoplasma cruzi	Level 2

- Trematode Parasites of Humans

Fasciola spp. (metacercaria)	Level 2
Schistosoma spp.	Level 2

b. Fungal Agents
- *Blastomyces dermatitidis* — Level 2
- *Coccidioides immitis*

Clinical specimens and animal tissue	Level 2
Animal studies when route of challenge is parenteral	Level 2
Sporulating mold form cultures, samples likely to contain infectious arthoconidia	Level 3

- *Cryptococcus neoformans* — Level 2
- *Histoplasma capsulatum*

Clinical specimens, animal tissues, and animal tissues when route of challenge is parenteral	Level 2
Processing of mold cultures, soil, etc., when likely to contain infectious conidia	Level 3

637

- *Sporothrix schenckii*   Level 2
- Pathogenic Members of the Genera *Epidermophyton, Microsporum,* and *Trichophyton*   Level 2
- Miscellaneous Molds, *Cladosporium (Xylohypha) trichoides, Cladosporium bantium, Penicillium marnefil, Exophiala (Wangiella) dermatitidis, Fonsecaea pedrosoi,* and *Dactylaria gallopava (Ochroconis gallopavum)*   Level 2

c. Bacterial Agents

- *Bacillus anthracis*   Level 2
  A licensed vaccine is available. Not normally recommended for ordinary use except for workers having frequent contact with clinical specimens or diagnostic specimens
- *Bordetella pertussis*   Level 2
- *Brucella (B. abortus, B. canis, B. melitensis, B. suis)*
  Activities with clinical materials of human or animal origin   Level 2
  Manipulation of cultures of pathogenic *Brucella* spp.   Level 3
- *Ampylobacter (C. jejuni/C. coli, C. fetus* ssp. *fetus)*   Level 2
- *Chlamydia psittaci, C. trachomatis*
  Diagnostic examination of tissues or cultures, necropsies, contact with clinical materials   Level 2
  Activities with concentrations of infectious materials or high potential for aerosol production   Level 3
- *Clostridium botulinum*
  All activities; a botulism toxoid is available from the CDC as an experimental new drug (ND)   Level 2
- *Clostridium tetani*   Level 2
  Administration of an adult diphtheria-tetanus toxoid at 10-year intervals is highly recommended   Level 2
- *Corynebacterium diphtheria*   Level 2
  Administration of an adult diphtheria-tetanus toxoid at 10-year intervals may be desirable
- *Francisella tularensis*
  Activities with clinical materials of human or animal origin   Level 2
  Manipulations of cultures and for experimental animal studies   Level 3
  An investigational live attenuated virus is available, and is recommended for those working with the agent or having potential contact in the laboratory or with infected animals
- *Leptospira* interrogans—all serovars
  All activities   Level 2
- *Legionella pneumophila;* other legionella-like agents
  All activities   Level 2
- *Mycobacterium leprae*
  All activities (special care with syringes)   Level 2
- *Mycobacterium* spp. other than *M. tuberculosis, M. bovis* or *M. leprae*
  Activities involving clinical materials and cultures   Level 2
- *Mycobacterium tuberculosis, M. bovis*
  Working with acid-fast smears or culturing sputa, other clinical specimens provided aerosol generating manipulations   Level 2

are done in class I or II biosafety cabinets; a few other
restricted exceptions

Propagation and manipulation of cultures and studies using nonhuman primates	Level 3

- *Neisseria gonorrhoeae*

All activities, with following exception:	Level 2
Activities generating aerosols or droplets	Level 3

- *Neisseria meningitidis*

All activities with following exception:	Level 2
Activities generating aerosol or droplets	Level 3

  Use of a licensed vaccine should be considered for work with
  high concentrations of infectious materials

- *Pseudomonas pseudomallei*

All activities with following exception:	Level 2
Activities with high potential for aerosol or droplet production	Level 3

- *Salmonella*—all serotypes except *typhi*

Clinical materials, cultures, and potentially contaminating agents	Level 2

- *Salmonella typhi*

All activities with following exception:	Level 2
Activities with high potential for aerosol or droplet production	Level 3

  A reasonably effective licensed vaccine is available and should be
  considered

- *Shigella* spp.

All activities	Level 2

- *Treponema pallidum*

All activities	Level 2

  Periodic serological monitoring should be considered for personnel

- Vibrionic enteritis (*Vibrio cholerae, V. parahaemolyticus*)   Level 2
  Short-term vaccine available but not recommended for routine
  use by laboratory personnel

- *Yersinia pestis*

All activities with following exception:	Level 2
Activities with a high potential for aerosol or droplet production and for work with antibiotic-resistant strains	Level 3

  Licensed inactivated vaccines available and use recommended

d. **Rickettisial Agents**

- *Coxiella burnetti*

Serological examinations, staining of impression smears	Level 2
Inoculation, incubation, etc., of eggs or tissue cultures, necropsies, manipulation of tissue cultures	Level 3

  New Q fever vaccine (IND) available from Fort Detrick, use
  should be limited to those at high risk, no demonstrated
  sensitivity to Q fever antigen

- *Rickettsia prowazekii, Rickettsia typhi (R. mooseri), Rickettsia tsutsugamushi, Rickettsia canada,* and spotted fever group agents *Rickettsia rickettsii, Rickettsia conorii, Rickettsia akari, Rickettsia australis,* and *Rickettsia siberica*

Nonpropagative laboratory procedures including serological and fluorescent antibody procedures, staining of impression smears	Level 2
All other manipulations of known or potentially infected materials	Level 3

Access to medical program and antibiotic therapy very important

e. Viral Agents

- Hepatitis A virus, Hepatitis E virus
  All activities   Level 2
  A licensed inactivated vaccine for hepatitis A is available in
  Europe, and as an investigational vaccine in the U.S.;
  it is recommended for laboratory personnel
- Hepatitis B virus, Hepatitis C virus (formerly known as
  nonA-nonB virus), Hepatitis D virus
  All activities with following exception: Level 2
  Activities with high potential for aerosol or droplet production   Level 3
  Licensed recombinant vaccines against hepatitis B are
  available and are highly recommended; if the requirements
  for the OSHA bloodborne pathogen standard are met, it is
  required that laboratory personnel be offered this vaccine
  and other measures.
- Herpes virus simiae (B-virus)
  All activities involving the use or manipulation of tissues,   Level 2
  body fluids, and primary tissue culture materials from
  macaques
  Activities involving the use or manipulation of any material   Level 3
  known to contain Herpes virus simiae.
- Human Herpes viruses
  All activities   Level 2
- Influenza
  All activities   Level 2
- Lymphocytic chloriomeningitis (LCM) virus
  All activities utilizing possibly infectious body fluids or   Level 2
  tissues and for tissue culture passage of mouse-brain
  passaged strains; manipulation of possibly infectious
  passage and clinical materials should be done in a
  biosafety cabinet
  Activities with high potential for aerosol or droplet production   Level 3
- Polio virus
  All activities   Level 2
  All laboratory personnel working with the agent must have
  documented polio vaccinations or demonstrated evidence of
  immunity to all three types of polio virus
- Poxviruses
  All activities   Level 2
  Persons working in or entering facilities where activities
  with vaccinia, monkeypox, or cowpox should have documented
  evidence of vaccination within the preceding 3 years
- Rabies virus
  All activities   Level 2
  Preexposure immunization for personnel working in facilities
  involved with diagnostic activities or research with rabies-
  infected materials; level 3 precautions should be used for activities
  with a high potential for droplet or aerosol production
- Retroviruses, including human and simian immunodeficiency
  viruses (HIV and SIV)

Follow "Universal Precautions" and other provisions of the OSHA
bloodborne pathogen standard; most activities   Level 2
Preparation of concentrated HIV or Sly   Level 3

- Transmissible spongiform encephalopathies (Creutzfeldt-Jakob
  and kura agents)
  All activities   Level 2
- Vesicular stomatitis virus (VSV)
  Activities utilizing laboratory-adapted strains of   Level 2
  demonstrated low virulence
  Activities involving the use or manipulation of infected   Level 3
  tissues and virulent isolates from infected livestock

f. Arboviruses Assigned to Biosafety Level 2

The classification of two of the following, marked with an asterisk, depend upon personnel being immunized. When performing some operations with highly infectious materials, some adaptation of biosafety level 3 conditions may be needed:

Acado	Batu	Catu	Flanders
Acara	Batama	Chaco	Fort Morgan
Aguacate	Bauline	Chagres	Frijoles
Alfuy	Bebaru	Chandipura	Gamboa
Almpiwar	Belmont	Changuinola	Gan Gan
Amapari	Benevides	Charleville	Gomoka
Ananindeua	Benfica	Chenuda	Gossas
Anhanga	Bertioga	Chilibre	Grand Arbaud
Anhembi	Bimiti	Chobar Gorge	Great Island
Anopheles A	Birao	Clo Mor	Guajara
Anopheles B	Bluetongue	Colorado Tick	Guam a
Apeu	Boracela	Fever	Guaratuba
Apoi	Botambi	Corriparta	Guaroa
Aride	Boteke	Cotia	Gumbo Limbo
Arkonam	Bouboui	Cowbone Ridge	Hart Park
Aroa	Bujaru	Csiro Village	Hazara
Aruac	Bunyamwera	Culaba	Huacho
Arumowot	Bunyip	D'aguilar	Hughes
Aura	Burg el Arab	Dakar Bat	Icoaraci
Avalon	Bushbush	Dengue- 1	Ieri
Abras	Bussuquara	Dengue- 2	Ilesha
Abu Hammad	Buttonwillow	Dengue- 3	Ilheus
Aabahoyo	Bwamba	Dengue- 4	Ingwavuma
Bagaza	Cacao	Dera Ghazi Khan	Inkoo
Bahig	Cache Valley	Eastern Equine	Ippy
Bakau	Caimito	Edge Hill	Irituia
Baku	California	Encephalomyelitis*	Isfahan
Bandia	Encephalomyelitis	Entebbe Bat	Itaporanga
Bangoran	Calovo	Epizootic	Itaqui
Bangui	Candiru	Hemorrhagic	Jamestown Canyon
Banzi	Cape Wrath	Disease	Japanaut
Bannah Forest	Capim	Erve	Jerry Slough
Barur	Caraparu	Eubenangee	Johnston Atoll
Batai	Carey Island	Eyach	Joinjakaka

Juan Diaz	Manawa	Pichinde	Tensaw
Jugra	Manzanilla	Pixuna	Tete
Jurona	Mapputta	Pongola	Tettnang
Jutiapa	Maprik	Ponteves	Thimiri
Kadam	Marco	Precarious	Thottapalayam
Kaeng Khoi	Marituba	Point	Tibrogargan
Kaikalur	Marrakai	Pretoria	Timbo
Kaisodi	Matariya	Prospect Hill	Timboteua
Kamese	Matruh	Puchong	Tindholmur
Kammavan Pettai	Matucare	Punta Salinas	Toscana
Kannaman Galam	Melao	Punta Toro	Toure
Kao Shuan	Mermet	Qalyub	Tribec
Karimabad	Minatitlan	Quaranfil	Triniti
Karshi	Minnal	Restan	Trivittatus
Kasba	Mirim	Rio Bravo	Trubanaman
Kemerovo	Mitchell River	Rio Grande	Tsuruse
Kem Canyon	Modoc	Ross River	Turlock
Ketapang	Moju	Royal Farm	Tyuleniy
Keterah	Mono Lake	Sabo	Uganda S
Keuraliba	Mont. Myotis	Saboya	Umatilla
Keystone	leukemia	Saint Floris	Umbre
Kismayo	Moriche	Sakhalin	Una
Klamath	Mosqueiro	Salehabad	Upola
Kokabera	Mossuril	San Angelo	Urucuri
Kolongo	Mount Elgon Bat	Sandfly f. (Naples)	Usutu
Koongol	Murutucu	Sandfly f. (Sicilian)	Uukuniemi
Kowanyama	Mykines	Saudjimba	Vellore
Kunjin	Navarro	Sathuperi	Venkatapuram
Kununurra	Nepuyo	Sawgrass	Vesicular Stomatitis
Kwatta	Ngaingan	Sebokele	Indiana
La Crosse	Nique	Seletar	Vesicular Stomatitis
La Joya	Nkolbisson	Sembalam	New Jersey
Lagos Bat	Nola	Serra do Navio	Wad Medani
Landjia	Ntaya	Shamonda	Wallal
Langat	Nugget	Shark River	Wanowrie
Lanjan	Nyamanini	Shuni	Warrego
Las Maloyas	Nyando	Silverwater	Western Equine
Latino	O'nyong-nyong	Simbu	Encephalomyelitis*
Le Dantec	Okhotskiy	Simian	Whataroa
Lebombo	Okola	Hemorrhagic	Witwatersrand
Lednice	Olifantsvlei	Fever	Wongal
Lipovnik	Oriboca	Tacajuma	Wongorr
Lokern	Ossa	Tacaribe	Wyeomyia
Lone Star	Pacora	Taggert	Yaquina Head
Lukuni	Pacui	Tahyna	Yata
M'Poko	Pahayokee	Tamiami	Yogue
Madrid	Palyam	Tanga	Zaliv Terpeniya
Maguari	Parana	Tanjong Rabok	Zegla
Mahogany	Pata	Tataguine	Zika
Hammock	Pathum Thani	Tembe	Zingilamo
Main Drain	Patois	Tembusu	Zirqa
Malakal	Phom-Penh Bat		

g. Vaccine Strains of Biosafety Level 3/4 Viruses which May be Handled at Biosafety Level 2

Virus	Vaccine Strain
Chikungunya	131/25
Junin	Candid #1
Rift Valley fever	Mp-12
Venezuelan equine encephalomyelitis	Tc-83
Yellow fever	17-D

h. Arboviruses and Certain Other Viruses Assigned to Biosafety Level 3 (on the basis of insufficient experience)

Adfelaide River	Estero Real	Meaban	Razdan
Agua Preta	Fomede	Mojui Dos Compos	Resistencia
Alenquer	Forecariah	Monte Dourado	Rochambeau
Almeirim	Fort Sherman	Munguba	Salanga
Altamira	Gabek Forest	Naranjal	San Juan
Andasibe	Gadgets Gully	Nariva	Santa Rosa
Antequeuera	Garba	Nasoule	Santarem
Araguari	Gordil	Ndelle	Saraca
Aransas Bay	Gray Lodge	New Minto	Saumarez Reef
Arbia	Gurupi	Ngari	Sedlec
Arboledas	Iaco	Ngoupe	Sena Madureira
Babanki Batken	Ibaraki	Nodamura	Sepik
Belem	Ife	Northway	Shokwe
Berrimali	Ingangapi	Odrenisrou	Slovaakia
Bimbo	Inini	Omo	Somone
Bobaya	Issyk-Kul	Oriximina	Spipur
Bobia	Itatuba	Quango	Tai
Bozo	Itimirim	Oubangui	Tamdy
Buenventura	Itupiranga	Oubi	Telok Forest
Cabassue[a,b]	Jacareacanga	Ourem	Termeil
Cacipacore	Jamanxi	Palestina	Thiafora
Calchaqui	Jari	Para	Tilligerry
Cananela	Kedougou	Paramushir	Tinaroo
Caninde	Khassan	Paroo River	Tiacotalpan
Chim	Kindia	Perinet	Tonate[a,b]
Coastal Plain	Kyzlagach	Petevo	Ttinga
Connecticut	Lake Clarendon	Picola	Xiburema
Corfu	Llano Seco	Playas	Yacaaba
Dabakala	Macaua	Pueblo Viejo	Yaounde
Douglas	Mapuera	Purus	Yoka
Ensenada	Mboke	Radi	Yug Bogkanovac

[a] It is recommended that work with this agent should be done only in biosafety level 3 facilities that provide for HEPA filtration of all exhaust air prior to discharge from the laboratory.

[b] A vaccine is available and is recommended for all persons working with this agent.

i. Arboviruses and Certain Other Viruses Assigned to Biosafety Level 3

Aino	Japanese	Nairobi Sheep	Seoul
Akabane	Encephalomyelitis	Disease	Spondweni
Bhanja	Jun[d]	Ndumu	St. Louis
Chikungunya[c,d]	Kairi	Negishi	Encephalomyelitis
Cocal	Kimberley	Oropouche[b]	ThogotoTocio
Dhori	Koutango	Orungo	Turuna
Dugbe	Louping III[a,b]	Peaton	Venezuelan Equine
Everglades [c,d]	Mayaro	Piry	Encephalitis[cd]
Flexal	Middleburg	Powassan	Vesicular Stomatitis
Germiston[c]	Mobala	Piumala	(Alagoas)
Getah	Mopeia[e]	Rift Valley[a,b,c,d]	Wesselsbron[a,c]
Hantaan	Mucambo[c,d]	Sagiyama	West Nile
Israel Turkey	Murray Valley	Sal Vieja	Yellow Fever[c,d]
Meningitis	Encephalomyelitis	San Perlita	Zinga[b]
		Semliki Forest	

[a] The importation, possession, or use of this agent is restricted by USDA regulation or administrative policy.

[b] Zinga virus is now recognized as being identical to Rift Valley Fever virus.
The Subcommittee on Arbovims Laboratory Safety recommends that work with this agent should only be done in a biosafety level 3 facility that provides for HEPA filtration of all exhaust air from the laboratory.

[c] A vaccine is available for this agent and is recommended for all persons working with this agent. This agent is presently being registered in the *Catalogue of Arboviruses.*

[d] A vaccine is available for this agent and is recommended for all persons working with this agent.

j. Arboviruses, Arenaviruses, and Filoviruses Assigned to Biosafety Level 4
Congo-Crimean hemorraghic fever

Tick-borne encephalitis virus complex (Absettarov, Hanzalova, Hypr., Kumlinge, Kyasanur Forest disease, Russian Omsk hemorrhagic fever, Spring-Summer encephalitis)
Marburg
Ebota
   Lassa
   Junin
   Machupo
   Guanarito

k. Restricted Animal Pathogens
Nonindigenous pathogens of domestic livestock and poultry may require special laboratory design, operation, and containment features not generally addressed in CDC guidelines. The importation, possession, or use of the following agents is prohibited or restricted by law or by U.S. Department of Agriculture regulations or administrative policies.

African Horse Sickness Virus	Bovine infectious petechial fever agent
African Swine Fever Virus	*Brucellosis melitensis*
Akabane Virus	Camelpox virus
Besnoitia besnoiti	*Cochliomyia hominivorax*
Borna disease virus	(screwworm)
Bovine spongiform encephalopathy	Ephemeral fever virus

Foot and mouth disease virus
Fowl plague virus (lethal avian influenza)
Hog cholera virus
Histoplasma (Zymonema) farciminosum
Louping III virus
Lumpy skin disease virus
Viral hemorrhagic disease of rabbits
Vesicular exanthema virus
Mycoplasma agalactiae
Mycoplasma mycoides
Nairobi Sheep Disease virus (Ganjam virus)
Newcastle disease virus (velogenic strains)
Peste des petits ruminants

Pseudomonas ruminantium (heartwater)
Rift Valley fever virus
Riderpest virus
Sheep and goat pox
Swine vesicular disease virus
Teschen disease virus
Theileria annulata
Theileria lawerencia
Theileria bovis
Theileria hirci
Trypanosoma evansi
Trypanosoma vivax
Wesselsbron disease virus

Additional information on the importation and interstate shipment of etiologic agents of human disease, diagnostic specimens, and other related materials may be obtained by contacting:

Centers for Disease Control and Prevention
Attention: Biosafety Branch
Office of Health and Safety,
Mail Stop F-05
1600 Clifton Road N.E.
Atlanta, GA 30333
Telephone: (404) 639-3883
Fax: (404) 639-2294

The importation, possession, use, or interstate shipment of animal pathogens other than those listed above may also be subject to regulations of the U.S. Department of Agriculture. Additional information may be obtained by writing to:

U.S. Department of Agriculture
Animal and Plant Health Inspection Service
Veterinary Services, Import-Export Products Staff
Room 756, Federal Building
6505 Belcrest Road
Hyattsville, MD 20782
Phone: (301) 436-7830 or (301) 436-8499, Fax:: (301) 436-8226

For general questions on biohazards and related topics, please contact the following resources.

Centers for Disease Control and Prevention
Attention: Biosafety Branch
Atlanta, GA 30333
Phone: (404) 639-3883

National Institutes of Health
Attention: Division of Safety
Bethesda, MD 20205
Phone: (301) 496-1357

National Animal Disease Center
U.S. Department of Agriculture
Ames, IA 50010
Phone:(515) 862-8258

The basic reference for all of the preceding sections is, of course, *Biosafety in Microbiological and Biomedical Laboratories, 3rd. Ed.* (Ref. 15 below). This reference includes 152 additional references from which the information in it was derived. The following brief list of references includes selected references from that longer list. In addition, the entire document is available on the Internet at the address cited in the Internet Reference below.

REFERENCES

1. **Favero, M.S.,** Biological hazards in the laboratory, in *Proceedings of Institute on Critical Issues in Health Laboratory Practice,* Richardson, J.W., Schoenfeld, E., Tullis, J.W., and Wagner, W.W., Eds., The DuPont Company, Wilmington, DE, 1986, 1.

2. **Pike, R.M.,** Laboratory-associated infections: incidence, fatalities, cases and prevention, *Annu. Rev .Microbiol.,* 33, 41, 1979.

3. **Pike, R.M.,** Past and present hazards of working with infectious agents, *Arch. Path. Lab. Med,* 102, 333, 1978.

4. **Pike, R.M.,** Laboratory-associated infections: summary and analysis of 3,921 cases, *Health Lab. Sci.,* 13, 105, 1976.

5. **Litsky, B.Y.,** Microbiology of sterilization, *AORNJ,* 26, 340, 1977.

6. Centers for Disease Control, Update: Universal Precautions for Prevention of Transmission of Human Immunodeficiency Virus, Hepatitis B Virus and other Bloodborne Pathogens in Healthcare settings, MMWR, 36 (25):1-7, 1988.

7. National Cancer Institute Safety Standards for Research Involving Oncogenic Viruses, U.S. Department of Health, Education and Welfare Publ. No. (NIH)75-790, October, 1974.

8. National Institutes of Health Biohazards Safety Guide, U.S. Department of Health, Education and Welfare, Public Health Service, National Institutes of Health, U.S. Government Printing Office, Stock No. 1740-003 83, 1974.

9. **Hellman, A., Oxman, M.N., and Pollack, R., Eds.,** *Biohazards in Biological Research,* Cold Spring Harbor Laboratory, Cold Spring Harbor, NY, 1974.

10. *Biosafety in the Laboratory, Prudent Practices for the Handling and Disposal of Infectious Materials,* National Research Council, National Academy Press, Washington, D.C., 1989.

11. **Bodily, J.L.,** General administration of the laboratory, in *Diagnostic Procedures for Bacterial, Mycotic and Parasitic Infections,* Bodily, H.L., Updyke, E.L., and Mason, J.O., Eds., American Public Health Association, New York, 1970, 11.

12. **Favero, M.S. and Bond, W.W.,** Sterization, Disinfection, and Antisepsis in the Hospital. In Lenette, E.H., Balows, A., Hausler, W.J. and Shadomy, H.J. Eds, *Manual of Clinical Microbiology, 4th ed, American Society for Microbiology* 183-200, 1991.

13. **Grist, N.R. and Emslie, J.A.N.,** Infections in British clinical laboratories, 1982-83, Clin. Pathol. 38:721-725, 1985.

14. **Grist, N.R. and Emslie, J.A.N.,** Infections in British clinical laboratories, 1984-85, Clin. Pathol. 40:826-829, 1987.

15. **Richmond, J.Y. and McKinney, R.W,** Eds., Biosafety in Microbiological and Biomedical Laboratories, 3rd ed., U.S. Department of Health and Human Services, Centers for Disease Control, and National Institutes of Health, HHS Publication No. (CDC) 93-8395, Washington, D.C., 1993.

16. **Darlow, H.M.,** Safety in the microbiological laboratory, in *Methods in Microbiology,* Norris, J.R. and Robbins, D.W., Eds., Academic Press, New York, 169, 1969.

17. **Collins, C.H., Hartley, E.G., and Pilsworth, R.,** *The Prevention of Laboratory Acquired Infection,* Public Health Laboratory Service, Monograph Series No. 6, 1974.

18. **Chatigny, M.A.,** Protection against infection in the microbiological laboratory: devices and procedures, in *Advances in Applied Microbiology* 3, Umbreit, W.W., Ed., Academic Press, New York, 131, 1961.

19. *International Catalog of Arboviruses Including Certain Other Viruses of Vertebrates*. The subcommittee on Arthropod-borne Viruses, 3rd ed., Karabatsos,N., Ed., American Society for Tropical Medicine and Hygiene, San Antonio, TX., 1985.
20. **Müller, H.E.,** Laboratory-acquired mycobacterial infection, *Lancet* 2:331, 1988.

INTERNET REFERENCE

1. http://ww.nih.gov/od/ors/ds/pubs/bmbl/contents.htm

VII. RECOMBINANT DNA LABORATORIES

Safety concerns, other than those already discussed for research in biological laboratories, for laboratories involved in recombinant DNA research are directed more toward the products of that research rather than the safety of the individuals involved. There has been considerable controversy about the possible release of genetically modified organisms into the environment. As a result of this concern, guidelines for the performance of recombinant DNA research have been developed.

The basic guidelines now in effect for recombinant DNA research appeared in the *Federal Register* on May 7, 1986 (51 CFR 16958). The following material represents selected portions taken directly from the most current version of the *NIH Guidelines* at the time of preparation of this section, the Spring of 1999. There will be a few additional comments and occasional bridging comments where intermediate material is deleted. Only those portions which apply to research organization personnel are included, although the deletions are not extensive. These are still evolving regulations, and any organization working in recombinant DNA research or planning to do so should conform to the latest revisions of the Guidelines. Changes have been published in the *Federal Register* on several occasions since the original Guidelines were published. These will be found in the references and the most recent developments can also be found at the Internet site listed in the references to this section. However, the basic Guidelines have not changed to a major degree since their original publication in 1986. The changes have primarily addressed large-scale uses of the technique, deliberate releases, gene therapy in humans, application to work outside the United States, and detailed changes in a fairly small number of sections for clarity. For most scientists working with the technique, or who wish to do so, additional guidance should be sought from their organization's Institutional Biosafety Committee or a local committee charged with the responsibility for reviewing an organization's recombinant DNA research.

The section heading will depart slightly from the practices used in the remainder of this book to parallel those in the *NIH Recombinant Guidelines* in order to facilitate comparison with later revisions in the published Guidelines.

A. Section I-B.   Definition of Recombinant DNA Molecules
The definition of recombinant DNA molecules has been changed slightly from that found in the original Guidelines.

In the context of these Guidelines, recombinant DNA molecules are defined as either (i) molecules which are constructed outside living cells by joining natural or synthetic DNA segments to DNA molecules that can replicate in a living cell, or (ii) DNA molecules that result from the replication of those described in (i) above.

Synthetic DNA segments likely to yield a potentially harmful polynucleotide or polypeptide (e.g., a toxin or a pharmacologically active agent) are considered as equivalent to their natural DNA counterpart. If the synthetic DNA segment is not expressed *in vivo* as a biologically active polynucleotide or polypeptide product, it is exempt from the Guidelines.

Genomic DNA of plants and bacteria that has acquired a transposable element, even if the latter was donated from a recombinant vector no longer present, is not subject to the NIH Guidelines unless the transposon itself contains recombinant DNA.

B. I-C. General Applicability

I-C-1. The *NIH Guidelines* are applicable to:

a. All recombinant DNA research within the United States (U.S.) or its territories that is within the category of research described in either (1) research that is conducted at or sponsored by an institution that receives any support for recombinant DNA research from NIH, including research performed directly by NIH. An individual who receives support for research involving recombinant DNA must be associated with or sponsored by an institution that assumes the responsibilities assigned in the *NIH Guidelines,* or as follows. (2) Research that involves testing in humans of materials containing recombinant DNA developed with NIH funds, if the institution that developed those materials sponsors or participates in those projects. Participation includes research collaboration or contractual agreements, not mere provision of research materials.

b. All recombinant DNA research performed abroad that is within the category of research described in either (1) research supported by NIH funds or (2) research that involves testing in humans of materials containing recombinant DNA developed with NIH funds, if the institution that developed those materials sponsors or participates in those projects. Participation includes research collaboration or contractual agreements, not mere provision of research materials. (3) If the host country has established rules for the conduct of recombinant DNA research, then the research must be in compliance with those rules. If the host country does not have such rules, the proposed research must be reviewed and approved by an NIH-approved Institutional Biosafety Committee or equivalent review body and accepted in writing by an appropriate national governmental authority of the host country. The safety practices that are employed abroad must be reasonably consistent with the *NIH Guidelines.*

C. l-D. Compliance with the NIH Guidelines

As a condition for NIH funding of recombinant DNA research, institutions shall ensure that such research conducted at or sponsored by the institution, irrespective of the source of funding, shall comply with the NIH Guidelines.

Information concerning noncompliance with the *NIH Guidelines* may be brought forward by any person. It should be delivered to both NIH/ORDA and the relevant institution. The institution, generally through the Institutional Biosafety Committee, shall take appropriate action. The institution shall forward a complete report of the incident recommending any further action to the Office of Recombinant DNA Activities, National Institutes of Health/MSC 7010, 6000 Executive Boulevard, Suite 302, Bethesda, Maryland 20892-7010, (301) 496-9838.

In cases where NIH proposes to suspend, limit, or terminate financial assistance because of noncompliance with the *NIH Guidelines*, applicable DHHS and Public Health Service procedures shall govern.

The policies on compliance are as follows: (1) All NIH funded projects involving recombinant DNA techniques must comply with the *NIH Guidelines.* Non-compliance may result in: (i) suspension, limitation, or termination of financial assistance for the noncompliant NIH-funded research project and of NIH funds for other recombinant DNA research at the institution, or (ii) a requirement for prior NIH approval of any or all recombinant DNA projects at the institution. (2) All non-NIH funded projects involving recombinant DNA techniques conducted at or sponsored by an institution that receives NIH funds for projects involving such techniques must comply with the *NIH Guidelines*. Noncompliance may result in: (i) suspension, limitation, or termination of NIH funds for recombinant DNA research at the institution, or (ii) a requirement for prior NIH approval of any or all recombinant DNA projects at the institution.

D. Section II, Safety Considerations
1. IIA-1. Risk Groups

Much of this will be duplicative of the material in the previous sections on Biosafety in Microbiological and Biological Laboratories but due to the concerns about transfer of genetically modified organisms from the facility into the wider biosphere, there will be some different slants from time to time. Risk assessment is ultimately a subjective process. The investigator must make an initial risk assessment based on the Risk Group (RG) of an agent. Agents are classified into four Risk Groups (RGs) according to their relative pathogenicity for healthy adult humans by the following criteria: (1) Risk Group 1 (RG1) agents are not associated with disease in healthy adult humans. (2) Risk Group 2 (RG2) agents are associated with human disease which is rarely serious and for which preventive or therapeutic interventions are *often* available. (3) Risk Group 3 (RG3) agents are associated with serious or lethal human disease for which preventive or therapeutic interventions *maybe* available. (4) Risk Group 4 (RG4) agents are likely to cause serious or lethal human disease for which preventive or therapeutic interventions are *not* usually available.

2. II-A-2. Risk Assessment

Classification of agents on the basis of hazard, is based on the potential effect of a biological agent on a healthy human adult and does not account for instances in which an individual may have increased susceptibility to such agents, e.g., preexisting diseases, medications, compromised immunity, pregnancy, or breast feeding (which may increase exposure of infants to some agents).

Personnel may need periodic medical surveillance to ascertain fitness to perform certain activities; they may also need to be offered prophylactic vaccines and boosters (see Section IV-B-1-f, *Responsibilities of the Institution, General Information*).

3. IIA-3. Risk Assessment Factors

In deciding on the appropriate containment for an experiment, the initial risk assessment should be followed by a thorough consideration of the agent itself and how it is to be manipulated. Factors to be considered in determining the level of containment include agent factors such as virulence, pathogenicity, infectious dose, environmental stability, route of spread, communicability, operations, quantity, availability of vaccine or treatment, and gene product effects such as toxicity, physiological activity, and allergenicity. Any strain that is known to be more hazardous than the parent (wild-type) strain should be considered for handling at a higher containment level. Certain attenuated strains or strains that have been demonstrated to have irreversibly lost known virulence factors may qualify for a reduction of the containment level compared to the Risk Group assigned to the parent strain.

A final assessment of risk based on these considerations is then used to set the appropriate containment conditions for the experiment. The containment level required may be equivalent to the Risk Group classification of the agent or it may be raised or lowered as a result of the above considerations. The Institutional Biosafety Committee must approve the risk assessment and the biosafety containment level for recombinant DNA experiments described in Sections III-A, Experiments that Require Institutional Biosafety Committee Approval, RAG Review, and NIH Director Approval Before Initiation, III-B, Experiments that Require

NIH/ORDA and Institutional Biosafety Committee Approval Before Initiation, III-C, Experiments that Require Institutional Biosafety Committee and Institutional Review Board Approvals and NIH/ORDA Registration Before Initiation and, III-D, Experiments that Require Institutional/Biosafety Committee Approval Before Initiation.

Careful consideration should be given to the types of manipulation planned for some higher Risk Group agents. For example, the RG2 dengue viruses may be cultured under the Biosafety Level (BL) 2 containment; however, when such agents are used for animal inoculation or transmission studies, a higher containment level is recommended. Similarly, RG3 agents such as Venezuelan equine encephalomyelitis and yellow fever viruses should be handled at a higher containment level for animal inoculation and transmission experiments.

Individuals working with human immunodeficiency virus (HIV), hepatitis B virus (HBV) or other bloodborne pathogens should consult *Occupational Exposure to Bloodborne Pathogens; Final Rule* (56 FR 64175-64182). BL2 containment is recommended for activities involving all blood-contaminated clinical specimens, body fluids, and tissues from all humans, or from HIV- or HBV-infected or inoculated laboratory animals. Activities such as the production of research-laboratory scale quantities of HIV or other bloodborne pathogens, manipulating concentrated virus preparations, or conducting procedures that may produce droplets or aerosols, are performed in a BL2 facility using the additional practices and containment equipment recommended for BL3. Activities involving industrial scale volumes or preparations of concentrated HIV are conducted in a BL3 facility, or BL3 Large Scale if appropriate, using BL3 practices and containment equipment.

Exotic plant pathogens and animal pathogens of domestic livestock and poultry are restricted and may require special laboratory design, operation and containment features not addressed in *Biosafety in Microbiological and Biomedical Laboratories.* For information regarding the importation, possession, or use of these agents see Sections V-G and V-H, Footnotes and References of Sections I, IV.

E. II-B. Containment

Effective biological safety programs have been operative in a variety of laboratories for many years. Considerable information already exists about the design of physical containment facilities and selection of laboratory procedures applicable to organisms carrying recombinant DNA (see material on Biosafety in Microbiological and Biomedical Laboratories in the preceding sections to this topic). The existing programs rely upon mechanisms that can be divided into two categories: (i) a set of standard practices that are generally used in microbiological laboratories; and (ii) special procedures, equipment, and laboratory installations that provide physical barriers that are applied in varying degrees according to the estimated biohazard.

Experiments involving recombinant DNA lend themselves to a third containment mechanism, namely, the application of highly specific biological barriers. Natural barriers exist that limit either: (i) the infectivity of a vector or vehicle (plasmid or virus) for specific hosts, or (ii) its dissemination and survival in the environment. Vectors, which provide the means for recombinant DNA and/or host cell replication, can be genetically designed to decrease, by many orders of magnitude, the probability of dissemination of recombinant DNA outside the laboratory (see Appendix I, *Biological Containment*).

Since these three means of containment are complementary, different levels of containment can be established that apply various combinations of the physical and biological barriers along with a constant use of standard practices. Categories of containment are considered separately in order that such combinations can be conveniently expressed in the *NIH Guidelines.*

Physical containment conditions within laboratories, described in Appendix G, *Physical Containment,* may not always be appropriate for all organisms because of their physical size, the number of organisms needed for an experiment, or the particular growth requirements of the organism. Likewise, biological containment for microorganisms described in Appendix I, *Biological Containment,* may not be appropriate for all organisms, particularly higher eukaryotic organisms. However, significant information exists about the design of research facilities and experimental procedures that are applicable to organisms containing recombinant DNA that is either integrated into the genome or into microorganisms associated with the higher organism as a symbiont, pathogen, or other relationship. This information describes facilities for physical containment of organisms used in non-traditional laboratory settings and special practices for limiting or excluding the unwanted establishment, transfer of genetic information, and dissemination of organisms beyond the intended location, based on both physical and biological containment principles. Research conducted in accordance with these conditions effectively confines the organism.

For research involving plants, four biosafety levels (BL1-P through BL4-P) are described in Appendix P, *Physical and Biological Containment for Recombinant DNA Research Involving Plants.* BL1-P is designed to provide a moderate level of containment for experiments for which there is convincing biological evidence that precludes the possibility of survival, transfer, or dissemination of recombinant DNA into the environment, or in which there is no recognizable and predictable risk to the environment in the event of accidental release. BL2-P is designed to provide a greater level of containment for experiments involving plants and certain associated organisms in which there is a recognized possibility of survival, transmission, or dissemination of recombinant DNA containing organisms, but the consequence of such an inadvertent release has a predictably minimal biological impact. BL3-P and BL4-P describe additional containment conditions for research with plants and certain pathogens and other organisms that require special containment because of their recognized potential for significant detrimental impact on managed or natural ecosystems. BL1-P relies upon accepted scientific practices for conducting research in most ordinary greenhouse or growth chamber facilities and incorporates accepted procedures for good pest control and cultural practices. BL1-P facilities and procedures provide a modified and protected environment for the propagation of plants and microorganisms associated with the plants, and a degree of containment that adequately controls the potential for release of biologically viable plants, plant parts, and microorganisms associated with them. BL2-P and BL3-P rely upon accepted scientific practices for conducting research in greenhouses with organisms infecting or infesting plants in a manner that minimizes or prevents inadvertent contamination of plants within or surrounding the greenhouse. BL4-P describes facilities and practices to provide containment of certain exotic plant pathogens.

For research involving animals, which are of a size or have growth requirements that preclude the use of conventional primary containment systems used for small laboratory animals, four biosafety levels (BL1-N through BL4-N) are described in Appendix Q, Physical and Biological Containment for Recombinant DNA Research Involving Animals. BL1-N describes containment for animals that have been modified by stable introduction of recombinant DNA, or DNA derived therefrom, into the germ-line (transgenic animals) and experiments involving viable recombinant DNA-modified microorganisms and is designed to eliminate the possibility of sexual transmission of the modified genome or transmission of recombinant DNA-derived viruses known to be transmitted from animal parent to offspring only by sexual reproduction. Procedures, practices, and facilities follow classical methods of avoiding genetic exchange between animals. BL2-N describes containment which is used for transgenic animals associated with recombinant DNA-derived organisms and is designed to eliminate the possibility of vertical or horizontal transmission. Procedures, practices, and facilities follow classical methods of avoiding genetic exchange between animals or controlling arthropod transmission. BL3-N and BL4-N describe higher levels of containment for research with certain transgenic animals involving agents which pose recognized hazard.

In constructing the *NIH Guidelines*, it was necessary to define boundary conditions for the different levels of physical and biological containment and for the classes of experiments to which they apply. These definitions do not take into account all existing and anticipated information on special procedures that will allow particular experiments to be conducted under different conditions than indicated here without affecting risk. Individual investigators and Institutional Biosafety Committees are urged to devise simple and more effective containment procedures and to submit recommended changes in the NIH Guidelines to permit the use of these procedures.

F. Section III Experiments Covered by the NIH Guidelines

This section describes six categories of experiments involving recombinant DNA: (i) those that require Institutional Biosafety Committee (IBC) approval, RAC review, and NIH Director approval before initiation (see Section III-A), (ii) those that require NIH/ORDA and Institutional Biosafety Committee approval before initiation (see Section III-B), (iii) those that require Institutional Biosafety Committee and Institutional Review Board approvals and NIH/ORDA registration before initiation (see Section III-C), (iv) those that require Institutional Biosafety Committee approval before initiation (see Section IIII-D), (v) those that require Institutional Biosafety Committee notification simultaneous with initiation (see Section III-E), and (vi) those that are exempt from the NIH Guidelines (see Section III-F).[*]

Any change in containment level, which is different from those specified in the NIH Guidelines, may not be initiated without the express approval of NIH/ORDA.

1. Section Ill-A.[*] Experiments that Require Institutional Biosafety Committee[**] Approval, RAC[***] Review, and NIH Director Approval Before Initiation

a. III-A-1, Major Actions Under the Guidelines

Experiments considered as Major Actions Under the NIH Guidelines cannot be initiated without submission of relevant information on the proposed experiment to the Office of Recombinant DNA Activities,[****] National Institutes of Health/MSC 7010,6000 Executive Boulevard, Suite 302, Bethesda, Maryland 20892-7010, (301)496-9838, the publication of the proposal in the *Federal Register* for 15 days of comment, review by RAC, and specific approval by NIH. The containment conditions or stipulation requirements for such experiments will be recommended by RAC and set by NIH at the time of approval. Such experiments require Institutional Biosafety Committee approval before initiation. Specific experiments already approved are included in Appendix D, *Major Actions Taken Under the NIH Guidelines,* which may be obtained from the address given above.

---

[*] If an experiment falls into Sections III-A , III-B, or III-C and one of the other sections, the rules pertaining to Sections III-A, III-B, or III-C shall be followed . If an experiment falls into Section III-F and into either Sections III-D or III-E as well, the experiment is considered exempt from the *NIH Guidelines.*

[**] An Institutional Biosafety Committee is an institutional committee that meets the membership criteria of the NIH Guidelines and reviews, approves, and oversees projects according to the NIH guidelines.

[***] The Recombinant DNA Advisory Committee is a public advisory committee that advises on recombinant DNA research.

[****] ORDA is the office within NIH for reviewing and coordinating activities involving *NIH Guidelines.*

b. Section III-A-1-a

The deliberate transfer of a drug resistance trait to microorganisms that are not known to acquire the trait naturally if such acquisition could compromise the use of the drug to control disease agents in humans, veterinary medicine, or agriculture, will be reviewed by RAG.

2. Section III-B. Experiments That Require NIH/ORDA and Institutional Biosafety Committee Approval Before Initiation

Experiments in this category cannot be initiated without submission of relevant information on the proposed experiment to NIH/ORDA. The containment conditions for such experiments will be determined by NIH/ORDA in consultation with adhoc experts. Such experiments require Institutional Biosafety Committee approval before initiation.

a. Section III-B-1. Experiments Involving the Cloning of Toxin Molecules with LD_{50} Less than 100 Nanograms per Kilogram Body Weight

Deliberate formation of recombinant DNA containing genes for the biosynthesis of toxin molecules lethal for vertebrates at an LD_{50} less than 100 nanograms per kilogram body weight (e.g., microbial toxins such as the botulinum toxins, tetanus toxin, diphtheria toxin, and *Shigela dysenteriae* neurotoxin). Specific approval has been given for the cloning in *Eacherichia* coli K-12 of DNA containing genes coding for the biosynthesis of toxic molecules which are lethal to vertebrates at 100 nanograms to 100 micrograms per kilogram body weight. Specific experiments already approved under this section may be obtained from the Office of Recombinant DNA Activities, National Institutes of Health/MSC 7010, 6000 Executive Boulevard, Suite 302, Bethesda, Maryland 20892-7010, (301) 496-9838.

3. Section III-C. Experiments that Require Institutional Biosafety Committee and Institutional Review Board Approvals and NIH/ORDA Registration Before Initiation

a. Section III-C-1. Experiments Involving the Deliberate Transfer of Recombinant DNA or DNA or RNA Derived from Recombinant DNA into One or More Human Subjects

Research proposals involving the deliberate transfer of recombinant DNA, or DNA or RNA derived from recombinant DNA, into human subjects (human gene transfer) will be considered through a review process involving both NIH/ORDA and RAC. Investigators shall submit relevant information on the proposed human gene transfer experiments to NIH/ORDA. Submission of human gene transfer protocols to NIH will be in the format described in Appendix M-l of the NIH Guidelines. Submission Requirements — Human Gene Transfer Experiments. Submission to NIH/ORDA shall be for registration purposes and will ensure continued public access to relevant human gene transfer information in compliance with the NIH Guidelines. Investigational New Drug (IND). Applications should be submitted to FDA in the format described in 21 CFR, Chapter I, Subchapter D, Part 312, Subpart B, Section 23, IND Content and Format.

Institutional Biosafety Committee approval must be obtained from each institution at which recombinant DNA material will be administered to human subjects (as opposed to each institution involved in the production of vectors for human application and each institution at which there is *ex vivo* transduction of recombinant DNA material into target cells for human application).

RAC prefers that submission to NIH/ORDA in accordance with Appendix M-l, *Submission Requirements — Human Gene Transfer Experiments,* contain no proprietary data or trade secrets, enabling all aspects of the review to be open to the public. Following receipt by NIH/ORDA, relevant information shall be entered into the NIH human gene transfer database for registration purposes. Summary information pertaining to the human gene transfer protocol will be forwarded to RAC members. NIH/ORDA summary information shall include comparisons to previously registered protocols. Specific items of similarity to previous experiments include (but are not limited to): (i) gene delivery vehicle, (ii) functional gene, (iii) marker gene, (iv) packaging cell (if applicable), (v) disease application, (vi) route of administration, and (vii) patient selection criteria.

RAC members shall notify NIH/ORDA within 15 working days if the protocol has been determined to represent novel characteristics requiring further public discussion.

Full RAC review of an individual human gene transfer experiment can be initiated by the NIH Director or recommended to the NIH Director by: (i) three or more RAC members, or (ii) other Federal agencies. An individual human gene transfer experiment that is recommended for full RAC review should represent novel characteristics deserving of public discussion. RAC recommendations on a specific human gene transfer experiment shall be forwarded to the NIH Director, the Principal Investigator, the sponsoring institution, and other DHHS components, as appropriate.[*]

4. Section III-D. Experiments that Require Institutional Biosafety Committee Approval Before Initiation

Prior to the initiation of an experiment that falls into this category, the Principal Investigator must submit a registration document to the Institutional Biosafety Committee which contains the following information: (i) the source(s) of DNA; (ii) the nature of the inserted DNA sequences; (iii) the host(s) and vector(s) to be used; (iv) if an attempt will be made to obtain expression of a foreign gene and, if so, indicate the protein that will be produced; and (v) the containment conditions that will be implemented as specified in the *NIH Guidelines*. For experiments in this category, the registration document shall be dated, signed by the Principal Investigator, and filed with the Institutional Biosafety Committee. The Institutional Biosafety Committee shall review and approve all experiments in this category prior to their initiation. Requests to decrease the level of containment specified for experiments in this category will be considered by NIH (see Section IV-C-1 -b-(2)-(c), *Minor Actions* of the NIH Guidelines).

a. Section III-D-1. Experiments Using Risk Group 2, Risk Group 3, Risk Group 4, or Restricted Agents as Host-Vector Systems

i. Section III-D-1-a. Experiments involving the introduction of recombinant DNA into Risk Group 2 agents will usually be conducted at Biosafety Level (BL) 2 containment. Experiments with such agents will usually be conducted with whole animals at BL2 or BL2-N (Animals) containment.

ii. Section III-D-1-b. Experiments involving the introduction of recombinant DNA into Risk Group 3 agents will usually be conducted at BL3 containment. Experiments with such agents will usually be conducted with whole animals at BL3 or BL3-N containment.

iii. Section III-D-1-c. Experiments involving the introduction of recombinant DNA into Risk Group 4 agents shall be conducted at BL4 containment. Experiments with such agents shall be conducted with whole animals at BL4 or BL4-N containment.

iv. Section III-D-1-d. Containment conditions for experiments involving the introduction of recombinant DNA into restricted agents shall be set on a case-by-case basis following

---

[*] For specific directives concerning the use of retroviral vectors for gene delivery, consult Appendix B-V-1, *Murine Retroviral Vectors* of the *NIH Guidelines*.

NIH/ORDA review. A U.S. Department of Agriculture permit is required for work with plant or animal pathogens. Experiments with such agents shall be conducted with whole animals at BL4 or BL4-N containment.

b. Section III-D-2. Experiments in Which DNA From Risk Group 2, Risk Group 3, Risk Group 4, or Restricted Agents is Cloned into Nonpathogenic Prokaryotic or Lower Eukaryotic Host-Vector Systems

i. Section III-D-2-a. Experiments in which DNA from Risk Group 2 or Risk Group 3 agents is transferred into nonpathogenic prokaryotes or lower eukaryotes may be performed under BL2 containment. Experiments in which DNA from Risk Group 4 agents is transferred into nonpathogenic prokaryotes or lower eukaryotes may be performed under BL2 containment after demonstration that only a totally and irreversibly defective fraction of the agent's genome is present in a given recombinant. In the absence of such a demonstration, BL4 containment shall be used. The Institutional Biosafety Committee may approve the specific lowering of containment for particular experiments to BL1. Many experiments in this category are exempt from the NIH Guidelines. Experiments involving the formation of recombinant DNA for certain genes coding for molecules toxic for vertebrates require NIH ORDA approval (see Section II I-B-1, *Experiments Involving the Cloning of Toxin Molecules with* LD_{50} *of Less than 100 Nanograms Per Kilogram Body Weight,* or shall be conducted under NIH specified conditions as described in Appendix F of the NIH Guidelines.

ii. Section III-D-2-b. Containment conditions for experiments in which DNA from restricted agents is transferred into nonpathogenic prokaryotes or lower eukaryotes shall be determined by NIH/ORDA following a case-by-case review.

c. Section III-D-3. Experiments Involving the Use of Infectious DNA or RNA Viruses or Defective DNA or RNA Viruses in the Presence of Helper Virus in Tissue Culture Systems[*]

i. Section III-D-3-a. Experiments involving the use of infectious or defective Risk Group 2 viruses in the presence of helper virus may be conducted at BL2.

ii. Section III-D-3-b. Experiments involving the use of infectious or defective Risk Group 3 viruses in the presence of helper virus may be conducted at BL3.

iii. Section III-D-3-c. Experiments involving the use of infectious or defective Risk Group 4 viruses in the presence of helper virus may be conducted at BL4.

iv. Section III-D-3-d. Experiments involving the use of infectious or defective restricted poxviruses in the presence of helper virus shall be determined on a case-by-case basis following NIH/ORDA review. A U.S. Department of Agriculture permit is required for work with plant or animal pathogens.

v. Section lII-D-3-e. Experiments involving the use of infectious or defective viruses in the presence of helper virus which are not covered in Sections III-D-3-a through III-D-3-d may be conducted at BL1.

---

[*]   Caution: Special care should be used in the evaluation of containment levels for experiments which are likely to either enhance the pathogenicity (e.g., insertion of a host oncogene) or to extend the host range (e.g., introduction of novel control elements) of viral vectors under conditions that permit a productive infection. In such cases, serious consideration should be given to increasing physical containment by at least one level.

d. Section III-D-4. Experiments Involving Whole Animals*

This section covers experiments involving whole animals in which the animal's genome has been altered by stable introduction of recombinant DNA, or DNA derived therefrom, into the germ-line (transgenic animals) and experiments involving viable recombinant DNA-modified microorganisms tested on whole animals. For the latter, other than viruses which are only vertically transmitted, the experiments may *not* be conducted at BL1-N containment. A minimum containment of BL2 or BL2-N is required.

i.   Section III-D-4-a. Recombinant DNA, or DNA or RNA molecules derived therefrom, from any source except for greater than two-thirds of eukaryotic viral genome may be transferred to any non-human vertebrate or any invertebrate organism and propagated under conditions of physical containment comparable to BL1 or BL1-N and appropriate to the organism under study. Animals that contain sequences from viral vectors, which do not lead to transmissible infection either directly or indirectly as a result of complementation or recombination in animals, may be propagated under conditions of physical containment comparable to BL1 or BL1-N and appropriate to the organism under study. Experiments involving the introduction of other sequences from eukaryotic viral genomes into animals are covered under Section III-D-4-b, *Experiments Involving Whole Animals.* For experiments involving recombinant DNA-modified Risk Groups 2, 3, 4, or restricted organisms, see Sections V-A, V-G, and V-L. It is important that the investigator demonstrate that the fraction of the viral genome being utilized does not lead to productive infection. A U.S. Department of Agriculture permit is required for work with plant or animal pathogens.

ii. Section III-D-4-b. For experiments involving recombinant DNA, or DNA or RNA derived therefrom, involving whole animals, including transgenic animals, and not covered by Sections III-D-1, the appropriate containment shall be determined by the Institutional Biosafety Committee.

iii.   Section III-D-4-c. Exceptions under Section **III-D-4**, *Experiments Involving Whole Animals.*

iv. Section III-D-4-c-(1). Experiments involving the generation of transgenic rodents that require BL1 containment are described under Section III-E-3, *Experiments Involving Transgenic Rodents.*

v.   Section III-D-4-c-(2). The purchase or transfer of transgenic rodents is exempt from the NIH Guidelines under Section III-F.

e. Section III-D-5. Experiments Involving Whole Plants**

Experiments to genetically engineer plants by recombinant DNA methods, to use such plants for other experimental purposes (e.g., response to stress), to propagate such plants, or to use

---

*   Caution - Special care should be used in the evaluation of containment conditions for some experiments with transgenic animals. For example, such experiments might lead to the creation of novel mechanisms or increased transmission of a recombinant pathogen or production of undesirable traits in the host animal. In such cases, serious consideration should be given to increasing the containment conditions.

**   For recombinant DNA experiments falling under Sections III-D-5-a through III-D-5-d, physical containment requirements may be reduced to the next lower level by appropriate biological containment practices, such as conducting experiments on a virus with an obligate insect vector in the absence of that vector or using a genetically attenuated strain.

plants together with microorganisms or insects containing recombinant DNA, may be conducted under the containment conditions described in Sections III-D-5-a through III-D-5-e. If experiments involving whole plants are not described in Section III-D-5 and do not fall under Sections III-A, III-B, III-D, or III-F, they are included in Section III-E.

i.   Section III-D-5-a. BL3-P (Plants) or BL2-P + biological containment is recommended for experiments involving most exotic infectious agents with recognized potential for serious detrimental impact on managed or natural ecosystems when recombinant DNA techniques are associated with whole plants.

ii.   Section III-D-5-b. BL3-P or BL2-P+ biological containment is recommended for experiments involving plants containing cloned genomes of readily transmissible exotic infectious agents with recognized potential for serious detrimental effects on managed or natural ecosystems in which there exists the possibility of reconstituting the complete and functional genome of the infectious agent by genomic complementation *in plants.*

iii. Section III-D-5-c. BL4-P containment is recommended for experiments with a small number of readily transmissible exotic infectious agents, such as the soybean rust fungus *(Phakospora pachyrhizi)* and maize streak or other viruses in the presence of their specific arthropod vectors, that have the potential of being serious pathogens of major U.S. crops.

iv.   Section III-D-5-d. BL3-P containment is recommended for experiments involving sequences encoding potent vertebrate toxins introduced into plants or associated organisms. Recombinant DNA containing genes for the biosynthesis of toxin molecules lethal for vertebrates at an LD_{50} of <100 nanograms per kilogram body weight fall under Section III-B-1, and require NIH/ORDA, and Institutional Biosafety Committee approval before initiation.

v.   Section III-D-5-e. BL3-P or BL2-P+ biological containment is recommended for experiments with microbial pathogens of insects or small animals associated with plants if the recombinant DNA-modified organism has a recognized potential for serious detrimental impact on managed or natural ecosystems.

f.   Section III-D-6. Experiments Involving More than 10 Liters of Culture
The appropriate containment will be decided by the Institutional Biosafety Committee. Where appropriate, Appendix K of the *NIH Guidelines.*

g.   Section III-E. Experiments that Require Institutional Biosafety Committee Notice Simultaneous with Initiation
Experiments not included in Sections III-A, III-B, III-C, III-D, III-F, and their sub-sections are considered in Section III-E. All such experiments may be conducted at BLI containment. For experiments in this category, a registration document shall be dated and signed by the investigator and filed with the local Institutional Biosafety Committee at the time the experiment is initiated. The Institutional Biosafety Committee reviews and approves all such proposals, but Institutional Biosafety Committee review and approval prior to initiation of the experiment is not required. For example, experiments in which all components derived from non-pathogenic prokaryotes and non-pathogenic lower eukaryotes fall under Section III-E and may be conducted at BL1 containment.

i.   Section III-E-1. Experiments Involving the Formation of Recombinant DNA Molecules Containing No More than Two-Thirds of the Genome of any Eukaryotic Virus

Recombinant DNA molecules containing no more than two-thirds of the genome of any eukaryotic virus (all viruses from a single Family being considered identical may be propagated and maintained in cells in tissue culture using BL1 containment. For such experiments, it must be demonstrated that the cells lack helper virus for the specific Families of defective viruses being used. If helper virus is present, procedures specified under Section III-D-3 should be used. The DNA may contain fragments of the genome of viruses from more than one Family but each fragment shall be less than two-thirds of a genome.

ii. Section III-E-2. Experiments Involving Whole Plants

This section covers experiments involving recombinant DNA-modified whole plants, and/or experiments involving recombinant DNA-modified organisms associated with whole plants, except those that fall under Section III-A, III-B, III-D, or III-F. It should be emphasized that knowledge of the organisms and judgment based on accepted scientific practices should be used in all cases in selecting the appropriate level of containment. For example, if the genetic modification has the objective of increasing pathogenicity or converting a non-pathogenic organism into a pathogen, then a higher level of containment may be appropriate depending on the organism, its mode of dissemination, and its target organisms. By contrast, a lower level of containment may be appropriate for small animals associated with many types of recombinant DNA-modified plants.

iii. Section III-E-2-a. BL1-P is recommended for all experiments with recombinant DNA-containing plants and plant-associated microorganisms not covered in Section III-E-2-b or other sections of the *NIH Guidelines.* Examples of such experiments are those involving recombinant DNA-modified plants that are not noxious weeds or that cannot interbreed with noxious weeds in the immediate geographic area, and experiments involving whole plants and recombinant DNA-modified non-exotic microorganisms that have no recognized potential for rapid and widespread dissemination or for serious detrimental impact on managed or natural ecosystems (e.g., *Rhizobium* spp. and *Agrobacterium* spp.).

iv. Section III-E-2-b. BL2-P or BL1-P + biological containment is recommended for the following experiments:

Section III-E-2-b-(1). Plants modified by recombinant DNA that are noxious weeds or can interbreed with noxious weeds in the immediate geographic area.

Section III-E-2-b-(2). Plants in which the introduced DNA represents the complete genome of a non-exotic infectious agent.

Section III-E-2-b-(3). Plants associated with recombinant DNA-modified non-exotic microorganisms that have a recognized potential for serious detrimental impact on managed or natural ecosystems.

Section III-E-2-b-(4). Plants associated with recombinant DNA-modified exotic microorganisms that have no recognized potential for serious natural ecosystems.

Section III-E-2-b-(5). Experiments with recombinant DNA-modified arthropods or small animals associated with plants, or with arthropods or small animals with recombinant DNA-modified microorganisms associated with them if the recombinant DNA-modified microorganisms have no recognized potential for serious detrimental impact on managed or natural ecosystems .

v. Section III-E-3. Experiments Involving Transgenic Rodents

This section covers experiments involving the generation of rodents in which the animal's genome has been altered by stable introduction of recombinant DNA, or DNA derived therefrom, into the germ-line (transgenic rodents). Only experiments that require BL1 containment are covered under this section; experiments that require BL2, BL3, or BL4 containment are covered under Section III-D-4.

h. Section III-F. Exempt Experiments

The following recombinant DNA molecules are exempt from the NIH Guidelines and registration with the Institutional Biosafety Committee is not required:

i. Section III-F-1. Those that are not in organisms or viruses.

ii. Section III-F-2. Those that consist entirely of DNA segments from a single nonchromosomal or viral DNA source, though one or more of the segments may be a synthetic equivalent.

iii. Section III-F-3. Those that consist entirely of DNA from a prokaryotic host including its indigenous plasmids or viruses when propagated only in that host (or a closely related strain of the same species), or when transferred to another host by well established physiological means.

iv. Section III-F-4. Those that consist entirely of DNA from an eukaryotic host including its chloroplasts, mitochondria, or plasmids (but excluding viruses) when propagated only in that host (or a closely related strain of the same species).

v. Section III-F-5. Those that consist entirely of DNA segments from different species that exchange DNA by known physiological processes, though one or more of the segments may be a synthetic equivalent. A list of such exchangers will be prepared and periodically revised by the NIH Director with advice of the RAC after appropriate notice and opportunity for public comment.

vi. Section III-F-6. Those that do not present a significant risk to health or the environment as determined by the NIH Director, with the advice of the RAG, and following appropriate notice and opportunity for public comment.

G. Section IV. Roles and Responsibilities

1. Section IV-A. Policy

The safe conduct of experiments involving recombinant DNA depends on the individual conducting such activities. The *NIH Guidelines* cannot anticipate every possible situation. Motivation and good judgment are the key essentials to protection of health and the environment. The *NIH Guidelines* are intended to assist the institution, Institutional Biosafety Committee, Biological Safety Officer, and the Principal Investigator in determining safeguards that should be implemented. The *NIH Guidelines* will never be complete or final since all conceivable experiments involving recombinant DNA cannot be foreseen. Therefore, it is the responsibility of the institution and those associated with it to adhere to the intent of the *NIH Guidelines* as well as to their specifics. Each institution (and the Institutional Biosafety Committee acting on its behalf) is responsible for ensuring that all recombinant DNA research conducted at or sponsored by that institution is conducted in compliance with the *NIH Guidelines*. General recognition of institutional authority and responsibility properly establishes accountability for safe conduct of the research at the local level. The following roles and responsibilities constitute an administrative framework in which safety is an essential and integral part of research involving recombinant DNA molecules. Further clarifications and interpretations of roles and responsibilities will be issued by NIH as necessary.

2. Section IV-B. Responsibilities of the Institution

a. Section IV-B-1. General Information

Each institution conducting or sponsoring recombinant DNA research which is covered by the *NIH Guidelines* is responsible for ensuring that the research is conducted in full conformity with the provisions of the NIH Guidelines. In order to fulfill this responsibility, the institution shall:

i. Section IV-B-1-a. Establish and implement policies that provide for the safe conduct of recombinant DNA research and that ensure compliance with the *NIH Guidelines*. As part of its general responsibilities for implementing the *NIH Guidelines*, the institution may establish additional procedures, as deemed necessary, to govern the institution and its components in the discharge of its responsibilities under the *NIH Guidelines*. Such procedures may include: (i) statements formulated by the institution for the general implementation of the *NIH Guidelines*, and (ii) any additional precautionary steps the institution deems appropriate.

ii. Section IV-B-1-b. Establish an Institutional Biosafety Committee that meets the requirements set forth in Section IV-B-2-a and carries out the functions detailed in Section IV-B-2-b.

iii. Section IV-B-1-c. Appoint a Biological Safety Officer (who is also a member of the Institutional Biosafety Committee) if the institution: (i) conducts recombinant DNA research at Biosafety Level (BL) 3 or BL4, or (ii) engages in large scale (greater than 10 liters) research. The Biological Safety Officer carries out the duties specified in Section IV-B-3.

iv. Section IV-B-1-d. Appoint at least one individual with expertise in plant, plant pathogen, or plant pest containment principles (who is a member of the Institutional Biosafety Committee) if the institution conducts recombinant DNA research that requires Institutional Biosafety Committee approval in accordance with Appendix P of the NIH guidelines, *Physical and Biological Containment for Recombinant DNA Research Involving Plants.*

v. Section IV-B-1-e. Appoint at least one individual with expertise in animal containment principles (who is a member of the Institutional Biosafety Committee) if the institution conducts recombinant DNA research that requires Institutional Biosafety Committee approval in accordance with Appendix *Q, Physical and Biological Containment for Recombinant DNA Research Involving Animals.*

vi. Section IV-B-1-f. Ensure that when the institution participates in or sponsors recombi-nant DNA research involving human subjects: (i) the Institutional Biosafety Committee has adequate expertise and training (using *ad hoc* consultants as deemed necessary), and (ii) all aspects of Appendix M, *Points to Consider in the Design and Submission of Protocols for the Transfer of Recombinant DNA Molecules into One or More Human Subjects (Points to Consider),* have been appropriately addressed by the Principal Investigator prior to submission to NIH/ORDA. Institutional Biosafety Committee approval must be obtained from each institution at which recombinant DNA material will be administered to human subjects (as opposed to each institution involved in the production of vectors for human application and each institution at which there is *ex vivo* transduction of recombinant DNA material into target cells for human application).

vii. Section IV-B-1-g. Assist and ensure compliance with the NIH Guidelines by Principal Investigators conducting research at the institution as specified in Section IV-B-4.

viii. Section IV-B-1-h. Ensure appropriate training for the Institutional Biosafety Committee Chair and members, Biological Safety Officer and other containment experts (when applicable), Principal Investigators, and laboratory staff regarding laboratory safety and implementation of the *NIH Guidelines*. The Institutional Biosafety Committee Chair is responsible for ensuring that Institutional Biosafety Committee members are appropriately trained. The Principal Investigator is responsible for ensuring that laboratory staff are appropriately trained. The institution is responsible for ensuring that the Principal Investigator has sufficient training; however, this responsibility may be delegated to the Institutional Biosafety Committee.

ix. Section IV-B-1-i. Determine the necessity for health surveillance of personnel involved in connection with individual recombinant DNA projects; and, if appropriate, conduct a health surveillance program for such projects. The institution shall establish and maintain a health surveillance program for personnel engaged in large scale research or production activities involving viable organisms containing recombinant DNA molecules which require BL3 containment at the laboratory scale. The institution shall establish and maintain a health surveillance program for personnel engaged in animal research involving viable recombinant DNA-containing microorganisms that require BL3 or greater containment in the laboratory. The *Laboratory Safety Monograph* discusses various components of such a program (e.g., records of agents handled, active investigation of relevant illnesses, and the maintenance of serial serum samples for monitoring serologic changes that may result from the employees' work experience). Certain medical conditions may place a laboratory worker at increased risk in any endeavor where infectious agents are handled. Examples cited in the *Laboratory Safety Monograph* include gastrointestinal disorders and treatment with steroids, immunosuppressive drugs, or antibiotics. Workers with such disorders or treatment should be evaluated to determine whether they should be engaged in research with potentially hazardous organisms during their treatment or illness. Copies of the *Laboratory Safety Monograph* are available from the Office of Recombinant DNA Activities, National Institutes of Health/MSC 7010, 6000 Executive Boulevard, Suite 302, Bethesda, Maryland 20892-7010, (301) 496-9838.

x. Section IV-B-1-j. Report any significant problems, violations of the NIH Guidelines, or any significant research-related accidents and illnesses to NIH/ORDA within thirty days, unless the institution determines that a report has already been filed by the Principal Investigator or Institutional Biosafety Committee. Reports shall be sent to the Office of Recombinant DNA Activities, National Institutes of Health/MSC 7010, 6000 Executive Boulevard, Suite 302, Bethesda, Maryland 20892-7010, (301) 496-9838.

3. Section IV-B-2. Institutional Biosafety Committee (IBC)

The institution shall establish an Institutional Biosafety Committee whose responsibilities need not be restricted to recombinant DNA. The Institutional Biosafety Committee shall meet the following requirements:

a. Section IV-B-2-a. Membership and Procedures

i. Section IV-B-2-a-(1). The Institutional Biosafety Committee must be comprised of no fewer than five members so selected that they collectively have experience and expertise in recombinant DNA technology and the capability to assess the safety of recombinant DNA research and to identify any potential risk to public health or the environment. In addition to the members identified in Section 1V-B-1, at least two members shall not be affiliated with the institution (apart from their membership on the Institutional Biosafety Committee) and who represent the interest of the surrounding community with respect to health and protection of the environment (e.g., officials of state or local public health or environmental protection agencies, members of other local governmental bodies, or persons active in medical, occupational health, or environmental concerns in the community).*

ii. Section IV-B-2-a-(2). In order to ensure the competence necessary to review and approve recombinant DNA activities, it is recommended that the Institutional Biosafety Committee: (i) include persons with expertise in recombinant DNA technology, biological safety, and physical containment; (ii) include or have available as consultants persons knowledgeable in institutional commitments and policies, applicable law, standards of professional conduct and practice, community attitudes, and the environment, and (iii) include at least one member representing the laboratory technical staff.

iii. Section IV-B-2-a-(3). The institution shall file an annual report with NIH/ORDA which includes: (i) a roster of all Institutional Biosafety Committee members clearly indicating the Chair, contact person, Biological Safety Officer (if applicable), plant expert (if applicable), animal expert (if applicable), human gene therapy expertise or *ad hoc* consultant (if applicable); and (ii) biographical sketches of all Institutional Biosafety Committee members (including community members).

iv. Section IV-B-2-a-(4). No member of an Institutional Biosafety Committee may be involved (except to provide information requested by the Institutional Biosafety Committee) in the review or approval of a project in which he/she has been or expects to be engaged or has a direct financial interest.

v. Section IV-B-2-a-(5). The institution, that is ultimately responsible for the effectiveness of the Institutional Biosafety Committee, may establish procedures that the Institutional Biosafety Committee shall follow in its initial and continuing review and approval of applications, proposals, and activities.

vi. Section IV-B-2-a-(6). When possible and consistent with protection of privacy and proprietary interests, the institution is encouraged to open its Institutional Biosafety Committee meetings to the public.

vii. Section IV-B-2-a-(7). Upon request, the institution shall make available to the public all Institutional Biosafety Committee meeting minutes and any documents submitted to or received from funding agencies which the latter are required to make available to the public.

---

* Individuals, corporations, and institutions not otherwise covered by the *NIH Guidelines*, are encouraged to adhere to the standards and procedures set forth in Sections I through IV. The policy and procedures for establishing an Institutional Biosafety Committee under *Voluntary Compliance,* are specified in Section IV-D-2.

If public comments are made on Institutional Biosafety Committee actions, the institution shall forward both the public comments and the Institutional Biosafety Committee's response to the Office of Recombinant DNA Activities, National Institutes of Health/MSC 7010, 6000 Executive Boulevard, Suite 302, Bethesda, Maryland 20892-7010, (301) 496-9838.

b.   **Section IV-B-2-b. Functions**[*]
On behalf of the institution, the Institutional Biosafety Committee is responsible for:

i.   **Section IV-B-2-b-(1).** Reviewing recombinant DNA research conducted at or sponsored by the institution for compliance with the *NIH Guidelines* and approving those research projects that are found to conform with the *NIH Guidelines*. This review shall include: (i) independent assessment of the containment levels required by the *NIH Guidelines* for the proposed research; (ii) assessment of the facilities, procedures, practices, and training and expertise of personnel involved in recombinant DNA research; and (iii) ensuring compliance with all surveillance, data reporting, and adverse event reporting requirements mandated by the *NIH Guidelines*.

ii.   **Section IV-B-2-b-(2).** Notifying the Principal Investigator of the results of the Institutional Biosafety Committee's review and approval.

iii.   **Section IV-B-2-b-(3).** Lowering containment levels for certain experiments as specified In Section III-D-2-a.

iv.   **Section IV-B-2-b-(4).** Setting containment levels as specified in Sections III-D-4-b and III-D-5.

v.   **Section IV-B-2-b-(5).** Periodically reviewing recombinant DNA research conducted at the institution to ensure compliance with the *NIH Guidelines*.

vi.   **Section IV-B-2-b-(6).** Adopting emergency plans covering accidental spills and person-nel contamination resulting from recombinant DNA research.

vii.   **Section IV-B-2-b-(7).** Reporting any significant problems with or violations of the NIH Guidelines and any significant research-related accidents or illnesses to the appropriate institutional official and NIH/ORDA within 30 days, unless the Institutional Biosafety Committee determines that a report has already been filed by the Principal Investigator. Reports to NIH/ORDA shall be sent to the Office of Recombinant DNA Activities, National Institutes of Health/MSC 7010, 6000 Executive Boulevard, Suite 302, Bethesda, Maryland 20892-7010, (301) 496-9838.

viii.   **Section IV-B-2-b-(8).** The Institutional Biosafety Committee may not authorize initia-tion of experiments which are not explicitly covered by the *NIH Guidelines* until NIH (with the advice of the RAC when required) establishes the containment requirement.

---

[*]   In the context of Section IV-B-2-b-(6) of this section, the *Laboratory Safety Monograph* describes basic elements for developing specific procedures dealing with major spills of potentially hazardous materials in the laboratory, including information and references about decontamination and emergency plans. The NIH and the Centers for Disease Control and Prevention are available to provide consultation and direct assistance, if necessary, as posted in the *Laboratory Safety Monograph*. The institution shall cooperate with the state and local public health departments by reporting any significant research-related illness or accident that may be hazardous to the public health.

ix. Section IV-B-2-b-(9). Performing such other functions as may be delegated to the Institutional Biosafety Committee under Section IV-B-2, Institutional Biosafety Committee.

4. Section IV-B-3. Biological Safety Officer (BSO)[*]

a. Section IV-B-3-a. The institution shall appoint a Biological Safety Officer if it engages in large scale research or production activities involving viable organisms containing recombinant DNA molecules.

b. Section IV-B-3-b. The institution shall appoint a Biological Safety Officer if it engages in recombinant DNA research at BL3 or BL4. The Biological Safety Officer shall be a member of the Institutional Biosafety Committee.

c. Section IV-B-3-c. The Biological Safety Officer's duties include, but are not be limited to:

i. Section IV-B-3-c-(1). Periodic inspections to ensure that laboratory standards are rigorously followed;

ii. Section IV-B-3-c-(2). Reporting to the Institutional Biosafety Committee and the institution any significant problems, violations of the *NIH Guidelines*, and any significant research-related accidents or illnesses of which the Biological Safety Officer becomes aware unless the Biological Safety Officer determines that a report has already been filed by the Principal Investigator;

iii. Section IV-B-3-c-(3). Developing emergency plans for handling accidental spills and personnel contamination and investigating laboratory accidents involving recombinant DNA research;

iv. Section IV-B-3-c-(4). Providing advice on laboratory security;

v. Section IV-B-3-c-(5). Providing technical advice to Principal Investigators and the Institutional Biosafety Committee on research safety procedures.

5. Section IV-B-7. Principal Investigator (P1)
On behalf of the institution, the Principal Investigator is responsible for full compliance with the *NIH Guidelines* in the conduct of recombinant DNA research.

a. Section IV-B-7-a. General Responsibilities
As part of this general responsibility, the Principal Investigator shall:

i. Section IV-B-7-a-(1). Initiate or modify no recombinant DNA research which requires Institutional Biosafety Committee approval prior to initiation until that research or the proposed modification thereof has been approved by the Institutional Biosafety Committee and has met all other requirements of the NIH Guidelines.

ii. Section IV-B-7-a-(2). Determine whether experiments are covered by Section III-E, and ensure that the appropriate procedures are followed;

---

[*] See the *Laboratory Safety Monograph for* additional information on the duties of the Biological Safety Officer.

iii. Section IV-B-7-a-(3). Report any significant problems, violations of the *NIH Guidelines*, or any significant research-related accidents and illnesses to the Biological Safety Officer (where applicable), Greenhouse/Animal Facility Director (where applicable), Institutional Biosafety Committee, NIH/ORDA, and other appropriate authorities (if applicable) within 30 days. Reports to NIH/ORDA shall be sent to the Office of Recombinant DNA Activities, National Institutes of Health/MSC 7010, 6000 Executive Boulevard, Suite 302, Bethesda, Maryland 20892-7010, (301) 496-9838;

iv. Section IV-B-7-a-(4). Report any new information bearing on the *NIH Guidelines* to the Institutional Biosafety Committee and to NIH/ORDA.

v. Section IV-B-7-a-(5). Be adequately trained in good microbiological techniques;

vi. Section IV-B-7-a-(6). Adhere to Institutional Biosafety Committee approved emergency plans for handling accidental spills and personnel contamination; and

vii.Section IV-B-7-a-(7). Comply with shipping requirements for recombinant DNA molecules (see Appendix H, *Shipment,* for shipping requirements and the *Laboratory Safety Monograph* for technical recommendations).

b. Section IV-B-7-b. Submissions by the Principal Investigator to NIH/ORDA
The Principal Investigator shall:

i.   Section IV-B-7-b-(1). Submit information to NIH/ORDA for certification of new host-vector systems; Section IV-B-7-b-(2). Petition NIH/ORDA, with notice to the Institutional Biosafety Committee, for proposed exemptions to the *NIH Guidelines*.

ii. Section IV-B-7-b-(3). Petition NIH/ORDA, with concurrence of the Institutional Biosafety Committee, for approval to conduct experiments specified in Sections III-A-1, and IIIB.

iii. Section IV-B-7-b-(4). Petition NIH/ORDA for determination of containment for experiments requiring case-by-case review and Section IV-B-7-b-(5). Petition NIH/ORDA for determination of containment for experiments not covered by the *NIH Guidelines*.

iv. Section IV-B-7-b-(6). Ensure that all aspects of Appendix M have been appropriately addressed prior to submission of human gene therapy experiments to NIH/ORDA.

c. Section IV-B-7-c. Submissions by the Principal Investigator to the Institutional Biosafety Committee
The Principal Investigator shall:

i. Section IV-B-7-c-(1). Make an initial determination of the required levels of physical and biological containment in accordance with the *NIH Guidelines*;

ii. Section IV-B-7-c-(2). Select appropriate microbiological practices and laboratory techniques to be used for the research;

iii. Section IV-B-7-c-(3). Submit the initial research protocol and any subsequent changes (e.g., changes in the source of DNA or host-vector system), if covered under Sections III-A, III-B, III-C, III-D, or III-E to the Institutional Biosafety Committee for review and approval or disapproval; and

iv. **Section IV-B-7-c-(4).** Remain in communication with the Institutional Biosafety Committee throughout the conduct of the project.

d. **Section IV-B-7-d. Responsibilities of the Principal Investigator Prior to Initiating Research**
The Principal Investigator shall:

i. **Section IV-B-7-d-(1).** Make available to all laboratory staff the protocols that describe the potential biohazards and the precautions to be taken;

ii. **Section IV-B-7-d-(2).** Instruct and train laboratory staff in: (i) the practices and techniques required to ensure safety, and (ii) the procedures for dealing with accidents; and

iii. **Section IV-B-7-d-(3).** Inform the laboratory staff of the reasons and provisions for any precautionary medical practices advised or requested (e.g., vaccinations or serum collection).

e. **Section IV-B-7-e. Responsibilities of the Principal Investigator During the Conduct of the Research**
The Principal Investigator shall:

i. **Section IV-B-7-e-(1).** Supervise the safety performance of the laboratory staff to ensure that the required safety practices and techniques are employed;

ii. **Section IV-B-7-e-(2).** Investigate and report any significant problems pertaining to the operation and implementation of containment practices and procedures in writing to the Biological Safety Officer (where applicable), Greenhouse/Animal Facility Director (where applicable), Institutional Biosafety Committee, NIH/ORDA, and other appropriate authorities (if applicable).

iii. **Section IV-B-7-e-(3).** Correct work errors and conditions that may result in the release of recombinant DNA materials; and

iv. **Section IV-B-7-e-(4).** Ensure the integrity of the physical containment (e.g., biological safety cabinets) and the biological containment (e.g., purity and genotypic and phenotypic characteristics).

v. **Section IV-B-7-e-(5).** Comply with reporting requirements for human gene transfer experiments conducted in compliance with the *NIH Guidelines*.

The sections of the *NIH Guidelines* that deal with the responsibilities of the NIH, RAC, and ORDA are omitted here. The areas of their responsibilities that affect the research facility have been covered indirectly in the preceding material.

H. Section IV-D. Voluntary Compliance
1. Section IV-D-1. Basic Policy - Voluntary Compliance
Individuals, corporations, and institutions not otherwise covered by the *NIH Guidelines*

are encouraged to follow the standards and procedures set forth in Sections I through IV. In order to simplify discussion, references hereafter to "institutions" are intended to encompass corporations and individuals who have no organizational affiliation. For purposes of complying with the *NIH Guidelines,* an individual intending to carry out research involving recombinant DNA is encouraged to affiliate with an institution that has an Institutional Biosafety Committee approved under the *NIH Guidelines.*

Since commercial organizations have special concerns, such as protection of proprietary data, some modifications and explanations of the procedures are provided in Sections IV-D-2 through IV-D-5-b in order to address these concerns.

2. Section IV-D-2. Institutional Biosafety Committee Approval-Voluntary Compliance

It should be emphasized that employment of an Institutional Biosafety Committee member solely for purposes of membership on the Institutional Biosafety Committee does not itself make the member an institutionally affiliated member. Except for the unaffiliated members, a member of an Institutional Biosafety Committee for an institution not otherwise covered by the *NIH Guidelines* may participate in the review and approval of a project in which the member has a direct financial interest so long as the member has not been, and does not expect to be, engaged in the project. Section IV-B-2-a-(4) is modified to that extent for purposes of these institutions.

3.   Section IV-D-3. Certification of Host-Vector Systems-Voluntary Compliance

A host-vector system may be proposed for certification by the NIH Director in accordance with the procedures set forth in Appendix I-II. In order to ensure protection for proprietary data, any public notice regarding a host-vector system which is designated by the institution as proprietary under Section IV-D will be issued only after consultation with the institution as to the content of the notice.

4.   Section IV-D-4. Requests for Exemptions and Approvals - Voluntary Compliance

Requests for exemptions or other approvals as required by the *NIH Guidelines* should be submitted based on the procedures set forth in Sections I through IV. In order to ensure protection for proprietary data, any public notice regarding a request for an exemption or other approval which is designated by the institution as proprietary under Section IV-D-5-a will be issued only after consultation with the institution as to the content of the notice.

5.   Section IV-D-5. Protection of Proprietary Data - Voluntary Compliance
a.   Section IV-D-5-a.   General

In general, the Freedom of Information Act requires Federal agencies to make their records available to the public upon request. However, this requirement does not apply to, among other things, "trade secrets and commercial or financial information that is obtained from a person and that is privileged or confidential." Under 18 U.S.C. 1905, it is a criminal offense for an officer or employee of the U.S. or any Federal department or agency to publish, divulge, disclose, or make known "in any manner or to any extent not authorized by law any information coming to him in the course of his employment or official duties or by reason of any examination or investigation made by, or return, report, or record made to or filed with, such department or agency or officer or employee thereof, which information concerns or relates to the trade secrets, (or) processes of any person, firm, partnership, corporation, or association." This provision applies to all employees of the Federal Government, including special Government employees. Members of RAC are "special Government employees."

In submitting to NIH for purposes of voluntary compliance with the *NIH Guidelines*, an institution may designate those items of information which the institution believes constitute trade secrets, privileged, confidential, commercial, or financial information. If NIH receives a request under the Freedom of Information Act for information so designated, NIH will promptly contact the institution to secure its views as to whether the information (or some portion) should be released. If NIH decides to release this information (or some portion) in response to a Freedom of Information request or otherwise, the institution will be advised and the actual release will be delayed in accordance with 45 Code of Federal Regulations, Section 5.65(d) and (e).

b. Section IV-D-5-b. Pre-submission Review

Any institution not otherwise covered by the NIH Guidelines, which is considering submission of data or information voluntarily to NIH, may request pre-submission review of the records involved to determine if NIH will make all or part of the records available upon request under the Freedom of Information Act.

A request for pre-submission review should be submitted to NIH/ORDA along with the records involved to the Office of Recombinant DNA Activities, National Institutes of Health/MSC 7010, 6000 Executive Boulevard, Suite 302, Bethesda, Maryland 20892-7010, (301) 496-9838. These records shall be clearly marked as being the property of the institution on loan to NIH solely for the purpose of making a determination under the Freedom of Information Act. NIH/ORDA will seek a determination from the responsible official under DHHS regulations (45 CFR Part 5) as to whether the records involved, (or some portion) will be made available to members of the public under the Freedom of Information Act. Pending such a determination, the records will be kept separate from NIH/ORDA files, will be considered records of the institution and not NIH/ORDA, and will not be received as part of NIH/ORDA files. No copies will be made of such records.

NIH/ORDA will inform the institution of the DHHS Freedom of Information Officers determination and follow the institution's instructions as to whether some or all of the records involved are to be returned to the institution or to become a part of NIH/ORDA files. If the institution instructs NIH/ORDA to return the records, no copies or summaries of the records will be made or retained by DHHS, NIH, or ORDA. The DHHS Freedom of Information Officer's determination will represent that official's judgment at the time of the determination as to whether the records involved (or some portion) would be exempt from disclosure under the Freedom of Information Act if at the time of the determination the records were in NIH/ORDA files and a request was received for such files under the Freedom of Information Act.

F. Appendices

1. Appendix A. Exemptions Under Section III-f-5-Sublists of Natural Exchangers

Certain specified recombinant DNA molecules that consist entirely of DNA segments from different species that exchange DNA by known physiological processes, though one or more of the segments may be a synthetic equivalent are exempt from these *NIH Guidelines* Institutional Biosafety Committee registration is not required for these exempt experiments. A list of such exchangers will be prepared and periodically revised by the NIH Director with advice from the RAC after appropriate notice and opportunity for public comment. For a list of natural exchangers that are exempt from the *NIH Guidelines*, see Appendices A-I through A-VI. Section III-F-5 describes recombinant DNA molecules that are: (1) composed entirely of DNA segments from one or more of the organisms within a sublist, and (2) to be propagated in any of the organisms within a sublist (see *Classification of Bergey's Manual of Determinative Bacteriology* 8th edition, R. E. Buchanan and N. E. Gibbons, editors, Williams and Wilkins Company; Baltimore, Maryland 1984). Although these experiments are exempt, it is recommended that they be performed at the appropriate biosafety level for the host or recombinant organism (see *Biosafety in Microbiological and Biomedical Laboratories,* 3rd edition, May 1993, U.S. DHHS, Public Health Service, Centers for Disease Control and Prevention, Atlanta, Georgia, and NIH Office of Biosafety, Bethesda, Maryland).

a. Appendix A-I. Sublist A
Genus *Escherichia*
Genus *Shigella*
Genus *Salmonella*- including *arizona*
Genus *Enterobacter*
Genus *Citrobacter*- including *levinea*
Genus *Klebsiella*- including *oxytoca*
Genus *Erwinia*
Pseudomonas aeruginosa, Pseudomonas putida, Pseudomonas fluorescens, and Pseudomonas mendocina
Serratia marcescens
Yersinia enterocolitica

b. Appendix A-II. Sublist B
Bacillus subtilis
Bacillus licheniformis
Bacillus pumilus
Bacillus globigii
Bacillus niger
Bacillus nato
Bacillus amyloliquefaciens
Bacillus aterrimus

c. Appendix A-III. Sublist C
Streptomyces aureofaciens
Streptomyces rimosus
Streptomyces coelicolor

d. Appendix A-IV. Sublist D
Streptomyces griseus
Streptomyces cyaneus
Streptomyces venezuelae

e. Appendix A-V. Sublist E
One way transfer of *Streptococcus mutans* or *Streptococcus lactis* DNA into *Streptococcus sanguis*

f. Appendix A-VI. Sublist F
Streptococcus sanguis
Streptococcus pneumoniae
Streptococcus faecalis
Streptococcus pyogenes
Streptococcus mutans

2. Appendix B. Classification of Human Etiologic Agents on the Basis of Hazard
This appendix includes those biological agents known to infect humans as well as selected

animal agents that may pose theoretical risks if inoculated into humans. Included are lists of representative genera and species known to be pathogenic; mutated, recombined, and non-pathogenic species and strains are not considered. Non-infectious life cycle stages of parasites are excluded.

This appendix reflects the current state of knowledge and should be considered a resource document. Included are the more commonly encountered agents and is not meant to be all inclusive. Information on agent risk assessment may be found in the *Agent Summary Statements of* the CDC/NIH publication, *Biosafety in Microbiological and Biomedical Laboratories.* Further guidance on agents not listed in Appendix B may be obtained through: Centers for Disease Control and Prevention, Biosafety Branch, Atlanta, Georgia 30333, Phone: (404) 639-3883, Fax: (404) 639-2294; National Institutes of Health, Division of Safety, Bethesda, Maryland 20892, Phone: (301) 496-1357; National Animal Disease Center, U.S. Department of Agriculture, Ames, Iowa 50010, Phone: (515) 862-8258.

A special committee of the American Society for Microbiology will conduct an annual review of this appendix and its recommendation for changes will be presented to the Recombinant DNA Advisory Committee as proposed amendments to the NIH Guidelines.

a. Appendix B: Table 1. Basis for the Classification of Biohazardous Agents by Risk Group (RG)

i. Risk Group 1 (RG1)
Agents that are not associated with disease in healthy adult humans.

ii. Risk Group 2 (RG2)
Agents that are associated with human disease which is rarely serious and for which preventive or therapeutic interventions are often available.

iii. Risk Group 3 (RG3)
Agents that are associated with serious or lethal human disease for which preventive or therapeutic interventions *may be* available (high individual risk but low community risk).

iv. Risk Group 4 (RG4)
Agents that are likely to cause serious or lethal human disease for which preventive or therapeutic interventions are *not* usually available (high individual risk and high community risk).

b. Appendix B-I. Risk Group 1 (RG1) Agents
RG1 agents are not associated with disease in healthy adult humans. Examples of RG1 agents include asporogenic *Bacillus subtilis* or *Bacillus licheniformis*, *Escherichia* coli-K 12, and adeno-associated virus types 1 through 4.

Those agents not listed in Risk Groups (RGs) 2, 3, and 4 are not automatically or implicitly classified in RG1; a risk assessment must be conducted based on the known and potential properties of the agents and their relationship to agents that are listed.

c. Appendix B-II. Risk Group 2 (RG2) Agents
RG2 agents are associated with human disease which is rarely serious and for which preventive or therapeutic interventions are *often* available.

d. Appendix B-II-A. Risk Group 2 (RG2) - Bacterial Agents Including Chlamydia
Acinetobacter baumannii (formerly *Acinetobacter calcoaceticus)*

Actinobacillus
Actinomyces pyogenes (formerly *Corynebacterium pyogenes)*
Aeromonas hydrophila
Amycolata autotrophica
Archanobacterium haemolyticum (formerly *Corynebacterium haemolyticum)*
Arizona hinshawii: all serotypes
Bacillus anthracis
Bartonella henselae, B. quintana, B. vinsonil
Bordetella including *B. pertussis*
Borrelia recurrentis, B. burgdorferi
Burkholderia (formerly *Pseudomonas* species) except those listed in Appendix B-Ill-A (RG3))

Campylobacter coli, C. fetus, C. jejuni
Chlamydia psittaci, C. trachomatis, G. pneumoniae
Clostridium botulinum, Cl. chauvoei, Cl. haemolyticum, Cl. histolyticum, Cl. novyi, Cl. septicum, Cl. tetani
Corynebacterium dipththeriae, C. pseudotuberculosis, C. renale
Dermatophilus congolensis
Edwardsiella tarda
Erysipelothrix rhusiopathiae
Escherichia coli; all enteropathogenic, enterotoxigenic, enteroinvasive and strains bearing K1 antigen, including *E.* coli O157:H7
Haemophllus ducreyi, H. influenzae
Helicobacter pylori
Kiebsiella- all species except *K oxytoca(RG1)*
Legionella including *L. pneumophila*
Leptospira interrogans: all serotypes
Listeria
Moraxella
Mycobacterium (except those listed in Appendix B-Ill-A (RG3)) including *M.* aviurn complex, *M. asiaticum, M. bovis* BCG vaccine strain, *M. chelonei, M. fortuitum, M. kansasii, M. leprae, M. malmoense, M. marinum, M. paratuberculosis, M. scrofulaceum, M. simiae, M. szulgai, M. ulcerans, M. xenopi.*
Mycoplasma, except *M. mycoides and M. agalactiae which* are restricted animal pathogens
Neisseria gonorrhoeae, N. meningitidis
Nocardia asteroides, N. brasiliensis, N. otitidiscaviarum, N. transvalensis
Rhodococcus equi
Salmonella including *S. arizonae, S. cholerasuis, S. enteritidis. S. gallinarum-pullorum, S. meleagridis, S. paratyphi;* A, B, C, *S. typhi, S. typhirnurium*
Shigella including *S. boydii; S. dysenteriae,* type 1, *S. flexneri, S. sonnei*
Sphaerophorus necrophorus
Staphylococcus aureus
Streptobacillus moniliformis
Streptococcus including *S. pneumoniae, S. pyogenes*
Treponema pallium, T. carateum
Vibrio cholerae, V. parahemolyticus, V. vulnificus
Yersinia enterocolitica

e. Appendix B-II-B. Risk Group 2 (RG2)- Fungal Agents
Blastomyces dermatitidis

Cladosporium bantianum, C. (Xylohypha) trichoides
Cyptococcus neoformans
Dactylaria galopava (Ochroconis gallopavum)
Epidermophyton
Exophiala (Wangiella) dermatitidis
Fonsecaea pedrosoi
Microsporum
Paracoccidioides braziliensis
Penicillium marneffei
Sporothrix schenckii
Trichophyton

f. Appendix B-II-C. Risk Group 2 (RG2)- Parasitic Agents

Ancylostoma human hookworms including *A. duodenale, A. ceylanicum*
Ascaris including *A. lumbricoides suum*
Babesa including *B. divergens, B. microti*
Brugia filaria worms including *B. malayi B. timori*
Coccidia
Cryptosporidium including *C. parvum*
Cysticercus cellulosae (hydatid cyst, larva of *T. solium)*
Echinococcus including *E. granulosis, E. multilocularis, E. vogeli*
Entamoeba histolytica
Enterobius
Fasciola including *F. gigantica, F. hepatica*
Giardia including *G. lamblia*
Heterophyes
Hymenolepis including *H. diminuta, H. nana*
Isospora
Leishmania including *L. braziliensis, L. donovani, L. ethiopia, L. major, L. mexicana, L. peruvania, L. tropica*
Loa loa filaria worms
Microsporidium
Naegleria fowleri
Necator human hookworms including *N. americanus*
Onchocerca filaria worms including, *0. volvulus*
Plasmodium including simian species, *P. cynomologi P. falciparum, P. malariae, P. ovale, P. vivax*
Sarcocystis including *S. suihorninis*
Schistosorna including *S. haematobium, S. intercalaturn, S. japonicum, S. mansoni, S. mekongi*
Strongyloides including S. *stercoralis*
Taenia solium
Toxocara including *T. canis*
Toxoplasma including *T.gondii*
Trichinella spiralis
Trypanosoma including *T. brucei brucei, T. brucei gambiense, T. brucei rhodesiense, T. cruzi*
Wuchereria bancrofti filaria worms

g. Appendix B-II-D. Risk Group 2 (RG2) - Viruses

i. Adenoviruses, human: all types

ii. Alphaviruses (Togaviruses): Group A Arboviruses
Eastern equine encephalomyelitis virus
Venezuelan equine encephalomyelitis vaccine strain TC-83
Western equine encephalomyelitis virus

iii. Arenaviruses
Lymphocytic choriomeningitis virus (non-neurotropic strains)
Tacaribe virus complex
Other viruses as listed in the reference source (see Section V-C, *Footnotes and References of Sections I through IV)*

iv. Bunyaviruses
Bunyamwera virus
Rift Valley fever virus vaccine strain MP-1 2
Other viruses as listed in the reference source (see Section V-C, *Footnotes and References of Sections IV*

v. Calciviruses

vi. Coronaviruses

vii. Flaviviruses (Togaviruses): Group B Arboviruses
Dengue virus serotypes 1, 2, 3, and 4
Yellow fever virus vaccine strain 17D
Other viruses as listed in the reference source (see Section V-C, *Footnotes and References of Sections IV*

viii. Hepatitis A, B, C, D, and E viruses

ix. Herpesviruses - except Herpesvirus simiae (Monkey B virus) (see Appendix B-IV-D, *Risk Group 4 (RG4) -Viral Agents)*
Cytomegalovirus
Epstein Barr virus
Herpes simplex types 1 and 2
Herpes zoster
Human herpes virus types 6 and 7

x. Orthomyxoviruses
Influenza viruses types A, B, and C
Other tick-borne orthomyxoviruses as listed in the reference source (see Section V-C, *Footnotes and References of Sections I through IV)*

xi. Papovaviruses
All human papilloma viruses

xii. Paramyxoviruses
Newcastle disease virus
Measles virus
Mumps virus

Parainfluenza viruses types 1, 2, 3, and 4
Respiratory syncytial virus

xiii. Parvoviruses
Human parvovirus (B19)

xiv. Picornaviruses
Coxsackie viruses types A and B
Echoviruses: all types
Polioviruses: all types, wild and attenuated
Rhinoviruses: all types

xv. Poxviruses: all types except Monkeypox virus (see Appendix B-III-D, and restricted poxviruses including Alastrim, Smallpox, and Whitepox (see Section V-L, *Footnotes and References of Sections I through* IV)

xvi. Reoviruses: all types including Coltivirus, human Rotavirus, and Orbivirus (Colorado tick fever virus)

xvii. Rhabdoviruses
Rabies virus: all strains
Vesicular stomatitis virus: laboratory adapted strains including VSV-lndiana, San Juan, and Glasgow

xvii. Togaviruses (see Alphaviruses and Flaviviruses)
Rubivirus (rubella)

h. **Appendix B-III. Risk Group 3 (RG3) Agents**
    RG3 agents are associated with serious or lethal human disease for which preventive or therapeutic interventions *may be* available.

i. **Appendix B-Ill-A. Risk Group 3 (RG3) - Bacterial Agents Including Rickettsia**
Bartonella
Brucella including *B. abortus, B. canis, B. suis*
Burkholderia (Pseudomonas) mallei, B. pseudomallei
Coxiella burnetii
Franciseila tularensis
Mycobacterium bovis (except BCG strain, see Appendix B-Il-A, *Bacterial Agents including Chiamydia), M. tuberculosis*
Pasteurella muitocida type B -"buffalo" and other virulent strains
Rickettsia akar, R. australls, R. canada, R. conorii R. prowazekii, R. rickettsii, R, siberica, R.   tsutsugamushi,
R. typhi (R. mooseni)
Yersinia pestis

j. **Appendix B-III-B. Risk Group 3 (RG3) - Fungal Agents**
Coccidioides immitis (sporulating cultures; contaminated soil)
Histoplasma capsulatum, H. capsulatum var.. duboisii

k. **Appendix B-III-C. Risk Group 4 (RG3) - Parasitic Agents**
None

l.   Appendix B-III-D. Risk Group 3 (RG3) - Viruses and Prions

i.   Alphaviruses (Togaviruses) - Group A Arboviruses

Semliki Forest virus

St. Louis encephalitis virus

Venezuelan equine encephalomyelitis virus (except the vaccine strain TC-83, see Appendix
   B-II-D (RG2))

Other viruses as listed in the reference source (see Section V-C)

ii.   Arenaviruses

Flexal

Lymphocytic choriomeningitis virus (LCM) (neurotropic strains)

iii.   Bunyaviruses

Hantaviruses including Hantaan virus

Rift Valley fever virus

iv.   Flaviviruses (Togaviruses) - Group B Arboviruses

Japanese encephalitis virus

Yellow fever virus

Other viruses as listed in the reference source (see Section V-C)

v.   Poxviruses

Monkeypox virus

vi.   Prions

Transmissible spongiform encephalopathies (TME) agents (Creutzfeldt-Jacob disease and
   kuru agents)(see Section V-C)

vii. Retroviruses

Human immunodeficiency virus (HIV) types 1 and 2

Human T cell lymphotropic virus (HTLV) types 1 and 2

Simian immunodeficiency virus (SIV)

viii. Rhabdoviruses

Vesicular stomatitis virus

m.   Appendix B-IV. Risk Group 4 (RG4) Agents

   RG4 agents are likely to cause serious or lethal human disease for which preventive or
therapeutic interventions are not usually available.

n.   Appendix B-IV-A. Risk Group 4 (RG4) - Bacterial Agents

None

o.   Appendix B-IV-B. Risk Group 4 (RG4) - Fungal Agents

None

p.   Appendix B-IV-C. Risk Group 4 (RG4) - Parasitic Agents

None

q. Appendix B-IV-D. Risk Group 4 (RG4) - Viral Agents

i. Arenaviruses
Guanarito virus
Lassa virus
Junin virus
Machupo virus
Sabia

ii. Bunyaviruses (Nairovirus)
Crimean-Congo hemorrhagic fever virus

iii. Filoviruses
Ebola virus
Marburg virus

iv. Flaviviruses (Togaviruses) - Group B Arboviruses
Tick-borne encephalitis virus complex including Absetterov, Central European encephalitis, Hanzalova, Hypr, Kumlinge, Kyasanur Forest disease, Omsk hemorrhagic fever, and Russian spring-summer encephalitis viruses

v. Herpesviruses (alpha)
Herpesvirus simiae (Herpes B or Monkey B virus)

vi. Paramyxoviruses
Equine morbillivirus

vii. Hemorrhagic fever agents and viruses as yet undefined

r. Appendix B-V. Animal Viral Etiologic Agents in Common Use
The following list of animal etiologic agents is appended to the list of human etiologic agents. None of these agents is associated with disease in healthy adult humans; they are commonly used in laboratory experimental work.

A containment level appropriate for RG1 human agents is recommended for their use. For agents that are infectious to human cells, e.g., amphotropic and xenotropic strains of murine leukemia virus, a containment level appropriate for RG2 human agents is recommended.

i. Baculoviruses

ii. Herpesviruses
Herpesvirus ateles
Herpesvirus saimiri
Marek's disease virus
Murine cytomegalovirus

iii. Papovaviruses
Bovine papilloma virus
Polyoma virus
Shope papilloma virus
Simian virus 40 (SV4O)

iv. Retroviruses
Avian leukosis virus
Avian sarcoma virus
Bovine leukemia virus
Feline leukemia virus
Feline sarcoma virus
Gibbon leukemia virus
Mason-Pfizer monkey virus
Mouse mammary tumor virus
Murine leukemia virus
Murine sarcoma virus
Rat leukemia virus

s.   Appendix B-V-1. Murine Retroviral Vectors
Murine retroviral vectors to be used for human transfer experiments (less than 10 liters) that contain less than 50% of their respective parental viral genome and that have been demonstrated to be free of detectable replication competent retrovirus can be maintained, handled, and administered, under BL1 containment.

3.   Appendix C. Exemptions Under Section III-F-6
Section III-F-6 states that exempt from these *NIH Guidelines* are "those that do not present a significant risk to health or the environment as determined by the NIH Director, with the advice of the RAG, and following appropriate notice and opportunity for public comment." See Appendix C, Exemptions under Sections III-F-6. The following classes of experiments are exempt under Section III-F-6:

a.   Appendix C-I. Recombinant DNA in Tissue Culture
Recombinant DNA molecules containing less than one-half of any eukaryotic viral genome (all viruses from a single family (see Appendix C-VI-D) being considered identical) (see Appendix C-VI-E) that are propagated and maintained in cells in tissue culture are exempt from these *NIH Guidelines* with the exceptions listed in Appendix C-I-A.

b.   Appendix C-I-A. Exceptions
The following categories are not exempt from the N/H Guidelines (i) experiments described in Section III-A which require Institutional Biosafety Committee approval, RAC review, and NIH Director approval before initiation, (ii) experiments described in Section III-B which require NIH/ORDA and Institutional Biosafety Committee approval before initiation, (iii) experiments involving DNA from Risk Groups 3, 4, or restricted organisms (see Appendix B and Sections V-G and V-L) or cells known to be infected with these agents, (iv) experiments involving the deliberate introduction of genes coding for the biosynthesis of molecules that are toxic for vertebrates (see Appendix F), and (v) whole plants regenerated from plant cells and tissue cultures are covered by the exemption provided they remain axenic cultures even though they differentiate into embryonic tissue and regenerate into plantlets.

c.   Appendix C-II. *Escherichia coli* K-12 Host-Vector Systems
Experiments which use *Escherichia coli* K-12 host-vector systems, with the exception of those experiments listed in Appendix C-II-A, are exempt from the *NIH Guidelines*

provided that: (i) the *Escherichia coli* host does not contain conjugation proficient plasmids or generalized transducing phages; or (ii) lambda or lambdoid or Ff bacteriophages or non-conjugative plasmids (see Appendix C-VI-B) shall be used as vectors. However, experiments involving the insertion into *Escherichia coli* K-12 of DNA from prokaryotes that exchange genetic information (see Appendix C-VI-C) with *Escherichia coli may* be performed with any *Escherichia coli* K-12 vector (e.g., conjugative plasmid). When a non-conjugative vector is used, the *Escherichia coli* K-12 host may contain conjugation-proficient plasmids either autonomous or integrated, or generalized transducing phages. For these exempt laboratory experiments, Biosafety Level (BL) 1 physical containment conditions are recommended. For large scale fermentation experiments, the appropriate physical containment conditions need be no greater than those for the host organism unmodified by recombinant DNA techniques; the Institutional Biosafety Committee can specify higher containment if deemed necessary.

d. Appendix C-II-A. Exceptions

The following categories are not exempt from the *NIH Guidellnes*: (i) experiments described in Section III-A which require Institutional Biosafety Committee approval, RAC review, and NIH Director approval before initiation, (ii) experiments described in Section III-B which require NIH/ORDA and Institutional Biosafety Committee approval before initiation, (iii) experiments involving DNA from Risk Groups 3, 4, or restricted organisms (see Appendix B), or cells known to be infected with these agents may be conducted under containment conditions specified in Section III-D-2 with prior Institutional Biosafety Committee review and approval, (iv) large scale experiments (e.g., more than 10 liters of culture), and (v) experiments involving the cloning of toxin molecule genes coding for the biosynthesis of molecules toxic for vertebrates (see Appendix F).

e. Appendix C-III. *Saccharomyces* Host-Vector Systems

Experiments involving *Saccharomyces cerevisiae* and *Saccharomyces uvarum* host-vector systems, with the exception of experiments listed in Appendix C-III-A, are exempt from the NIH Guidelines. For these exempt experiments, BL1 physical containment is recommended. For large scale fermentation experiments, the appropriate physical containment conditions need be no greater than those for the host organism unmodified by recombinant DNA techniques; the Institutional Biosafety Committee can specify higher containment if deemed necessary.

f. Appendix C-III-A. Exceptions

The following categories are not exempt from the *NIH Guidellnes*: (i) experiments described in Section III-A which require Institutional Biosafety Committee approval, RAC review, and NIH Director approval before initiation, (ii) experiments described in Section III-B which require NIH/ORDA and Institutional Biosafety Committee approval before initiation, (iii) experiments involving DNA from Risk Groups 3, 4, or restricted organisms (see Appendix B), or cells known to be infected with these agents may be conducted under containment conditions specified in Section III-D-2 with prior Institutional Biosafety Committee review and approval, (iv) large scale experiments (e.g., more than 10 liters of culture), and (v) experiments involving the deliberate cloning of genes coding for the biosynthesis of molecules toxic for vertebrates (see Appendix F).

g. Appendix C-IV. *Bacillus subtilis* or *Bacillus licheniformis* Host-Vector Systems

Any asporogenic *Bacillus subtilis* or asporogenic *Bacillus licheniformis* strain which does not revert to a spore-former with a frequency greater than 10^{-7} may be used for cloning DNA with the exception of those experiments listed in Appendix C-IV-A. Exceptions: For these exempt laboratory experiments, BL1 physical containment conditions are recommended. For large scale fermentation experiments, the appropriate physical containment conditions need be no greater than those for the host organism unmodified by recombinant DNA techniques; the Institutional Biosafety Committee can specify higher containment if it deems necessary.

h. Appendix C-IV-A. Exceptions

The following categories are not exempt from the NIH Guidelines: (i) experiments described in Section III-A which require Institutional Biosafety Committee approval, RAC review, and NIH Director approval before initiation, (ii) experiments described in Section Ill-B which require NIH/ORDA and Institutional Biosafety Committee approval before initiation, (iii) experiments involving DNA from Risk Groups 3, 4, or restricted organisms (see Appendix B), and Sections VG and V-L, or cells known to be infected with these agents may be conducted under containment conditions specified in Section III-D-2 with prior Institutional Biosafety Committee review and approval, (iv) large scale experiments (e.g., more than 10 liters of culture), and (v) experiments involving the deliberate cloning of genes coding for the biosynthesis of molecules toxic for vertebrates (see Appendix F).

i. Appendix C-V. Extrachromosomal Elements of Gram Positive Organisms

Recombinant DNA molecules derived entirely from extrachromosomal elements of the organisms listed below (including shuttle vectors constructed from vectors described in Appendix C), propagated and maintained in organisms listed below are exempt from these *NIH Guidelines.*

Bacillus amyloliquefaciens	*Staphylococcus aureus*
Bacillus amylosacchariticus	*Staphylococcus carnosus*
Bacillus anthracis	*Staphylococcus epidermidis*
Bacillus aterrimus Bacillus brevis	*Streptococcus agalactiae*
Bacillus careus	*Streptococcus anginosus*
Bacillus giobigii	*Streptococcus avium*
Bacillus licheniformis	*Streptococcus cremoris*
Bacillus megaterium	*Streptococcus dorans*
Bacillus natto	*Streptococcus equisimilis*
Bacillus niger	*Streptococcus faecalis*
Bacillus pumilus	*Streptococcus ferus*
Bacillus sphaericus	*Streptococcus lactis*
Bacillus stearothermophilis	*Streptococcus ferns*
Bacillus subtilis	*Streptococcus mitior*
Bacillus thuringiensis	*Streptococcus mutans*
Clostridium acetobutylicum	*Streptococcus pneumoniae*
Lactobacilius casei	*Streptococcus pyogenes*
Listeria grayi	*Streptococcus salivarious*
Listeria monocytogenes	*Streptococcus sanguis*
Listeria murrayi	*Streptococcus sobrinus*
Pediococcus acidilactici	*Streptococcus thermophylus*
Pediococcus damnosus	
Pediococcus pentosaceus	

The following categories are not exempt from the *NIH Guide/lines* (i) experiments described in Section III-A which require Institutional Biosafety Committee approval, RAC review, and NIH Director approval before initiation, (ii) experiments described in Section III-B which require NIH/ORDA and Institutional Biosafety Committee approval before initiation, (iii) experiments involving DNA from Risk Groups 3, 4, or restricted organisms (see Appendix B), and Sections VG and V-L, or cells known to be infected with these agents may be conducted under containment conditions specified in Section III-D-2 with prior Institutional Biosafety Committee review and approval, (iv) large scale experiments (e.g., more than 10 liters of culture), and (v) experiments involving the deliberate cloning of genes coding for the biosynthesis of molecules toxic for vertebrates (see Appendix F).

j. Appendix C-VI. The Purchase or Transfer of Transgenic Rodents
The purchase or transfer of transgenic rodents for experiments that require BL1 containment (See Appendix G-III-M), are exempt from the *NIH Guidelines*.

k. Appendix C-VII-A and A-1.
The NIH Director, with advice of the RAC, may revise the classification for the purposes of these *NIH Guidelines* (see Section IV-C-1-b-(2)-(b)). The list of organisms in each Risk Group is located in Appendix B.

l. Appendix C-VII-B.
A subset of non-conjugative plasmid vectors are poorly mobilizable (e.g., pBR322, pBR31 3). Where practical, these vectors should be employed.

m. Appendix C-VII-C.
Defined as observable under optimal laboratory conditions by transformation, transduction, phage infection, and/or conjugation with transfer of phage, plasmid, and/or chromosomal genetic information. Note that this definition of exchange may be less stringent than that applied to exempt organisms under Section III-F-5.

n. Appendix C-VII-D.
As classified in the *Third Report of the International Committee on Taxonomy of Viruses: Classification and Nomenclature of Viruses,* R. E. F. Matthews (Ed.), Intervirology 12 (129-296), 1979.

o. Appendix C-VII-E.
The total of all genomes within a Family shall not exceed one-half of the genome.

4. Appendix D. Major Actions Taken under the NIH Guidelines
As noted in the subsections of Section IV-C-1 -b-(1), the Director, NIH, may take certain actions with regard to the *NIH Guidelines* after the issues have been considered by the RAG. Some of the actions taken to date include the following: These are typically specific to a particular institution or experimenter and are omitted for that reason. There are 115 listed in the current *NIH Guidelines*.

5. Appendix E. Certified Host-vector Systems (See Appendix I, *Biological Containment*)

While many experiments using *Escherichia* coli K-12, *Saccharomyces cerevisiae,* and *Bacillus subtilis* are currently exempt from the NIH Guidelines under Section III-F, some derivatives of these host-vector systems were previously classified as Host-Vector 1 Systems or Host-Vector 2 Systems. A listing of those systems follows:

a.   Appendix E-I. *Bacillus subtilis*

b.   Appendix E-I-A. *Bacillus subtilis* **Host-Vector 1 Systems**
    The following plasmids are accepted as the vector components of certified *B. subtilis* systems: pUB110, pC194, pS194, pSA2100, pE194, pT127, pUB112, pC221, pC223, and pAB124. *B. subtilis* strains RUB331 and BGSC1S53 have been certified as the host component of Host-Vector 1 systems based on these plasmids.

c.   Appendix E-I-B. *Bacillus subtilis* **Host-Vector 2 Systems**
    The asporogenic mutant derivative of *Bacillus subtilis,* ASB 298, with the following plasmids as the vector component: pUB110, pC194, pS194, pSA2100, Pe194, pT127, pUB112, pC221, pC223, and pAB124.

d.   Appendix E-II. *Saccharomyces cerevisiae*

e.   Appendix E-II-A. *Saccharomyces cerevisiae* **Host-Vector 2 Systems**
    The following sterile strains of *Saccharomyces cerevisiae,* all of which have the ste-VC9 mutation, SHYl, SHY2, SHY3, and SHY4. The following plasmids are certified for use: YIpl, YEp2, YEp4, Ylp5, YEp6, YRp7, YEp20, YEp21, YEp24, YIp25, YIp26, YIp27, YIp28, YIp29, YIp3O, YIp3l, YIp32, and YIp33.

f.   Appendix E-III. *Escherichia coli*

g.   Appendix E-III-A. *Escherichia coli* **(EK2) Plasmid Systems**
    The *Escherichia coli* K-12 strain chi-1776. The following plasmids are certified for use: pSCl0l, pMB9, pBR313, pBR322, pDH24, pBR325, pBR327, pGL101, and pHB1. The following *Escherichia coli/S. cerevisiae* hybrid plasmids are certified as EK2 vectors when used in *Escherichia coli* chi-1776 or in the sterile yeast strains, SHYl, SHY2, SHY3, and SHY4: Ylpl, YEp2, YEp4, YIp5, YEp6, YRp7, YEp20, YEp21, YEP24, YIp25, YIp26, YIp27, YIp28, YIp29, YIp30, YIp3l, YIp32, and YIp33.

h.   Appendix E-III-B. *Escherichia coli* **(EK2) Bacteriophage Systems**
    The following are certified EK2 systems based on bacteriophage lambda:

Vector	**Host**
gt *WESB*	DP50*sup*F
gt ZJ*vir*B	*Escherichia coli* K-12
gtALO. B	DP50*sup*F
Charon 3A	DP50 or DP50supF
Charon 4A	DP50 or DP50supF
Charon 16A	DP50 or *DP50sup*F
Charon 21A	DP50*sup*F
Charon 23A	DP50 or DP50*sup*F
Charon 24A	DP50 or DP50*sup*F

Escherichia coli K-12 strains chi-2447 and chi-2281 are certified for use with lambda vectors that are certified for use with strain DP50 or DP50*sup*F provided that the su-strain not be used as a propagation host.

i. Appendix E-IV. *Neurospora crassa*

j. Appendix E-IV-A. *Neurospora crassa* Host-Vector 1 Systems

The following specified strains of *Neurospora crassa* which have been modified to prevent aerial dispersion:

In1 (inositolless) strains 37102, 37401, 46316, 64001, and 89601. Csp-1 strain UCLA37 and
   csp-2 strains FS 590, UCLA101 (these are conidial separation mutants).
Eas strain UCLA1 91 (an "easily wettable" mutant).

k. Appendix E-V. *Streptomyces*

l. Appendix E-V-A. *Streptomyces* Host-Vector 1 Systems

The following *Streptomyces* species: *Streptomyces coelicolor, S. lividans, S. parvulus,* and S. *griseus.* The following are accepted as vector components of certified *Streptomyces* Host-Vector 1 systems: *Streptomyces* plasmids SCP2, SLP1.2, pIJ101, actinophage phi C31, and their derivatives.

m. Appendix E-VI. *Pseudomonas putida*

n. Appendix E-VI-A. *Pseudomonas putida* Host-Vector 1 Systems

Pseudomonas putida strains KT2440 with plasmid vectors pKT262, pKT263, and pKT264.

6. Appendix F. Containment Conditions for Cloning of Genes Coding for the Biosynthesis of Molecules Toxic for Vertebrates

a. Appendix F-I. General Information

Appendix F specifies the containment to be used for the deliberate cloning of genes coding for the biosynthesis of molecules toxic for vertebrates. The cloning of genes coding for molecules toxic for vertebrates that have an LD_{50} of <100 nanograms per kilograms body weight (e.g., microbial toxins such as the botulinum toxins, tetanus toxin, diphtheria toxin, *Shigella dysenteriae neurotoxin*) are covered under Section III-B-1 and require Institutional Biosafety Committee and NIH/ORDA approval before initiation. No specific restrictions shall apply to the cloning of genes if the protein specified by the gene has an LD_{50} of 100 micrograms per kilograms of body weight. Experiments involving genes coding for toxin molecules with an LD_{50} of <100 micrograms per kilograms and >100 nanograms per kilograms body weight require Institutional Biosafety Committee approval and registration with NIH/ORDA prior to initiating the experiments. A list of toxin molecules classified as to LD_{50} is available from NIH/ORDA. Testing procedures for determining toxicity of toxin molecules not on the list are available from the Office of Recombinant DNA Activities, National Institutes of Health/MSC. The results of such tests shall be forwarded to NIH/ORDA, which will consult with *ad hoc* experts, prior to inclusion of the molecules on the list

b. Appendix F-II. Cloning of Toxin Molecule Genes in *Escherichia coli* K-12

c. Appendix F-II-A.

Cloning of genes coding for molecules toxic for vertebrates that have an LD_{50} of >100 nanograms per kilograms and <1000 nanograms per kilograms body weight (e.g., abrin, *Clostridium perfringens* epsilon toxin) may proceed under Biosafety Level (BL) 2 + EK2 or BL3 + EK1 containment conditions.

d. Appendix F-II-B.

Cloning of genes for the biosynthesis of molecules toxic for vertebrates that have an LD_{50} of >1 microgram per kilogram and <100 microgram per kilogram body weight may proceed under BL1 + EK1 containment conditions (e.g., *Staphylococcus aureus* alpha toxin, *Staphylococcus aureus* beta toxin, ricin, *Pseudomonas aeruginosa* exotoxin A, *Bordetella pertussis* toxin, the lethal factor of *Bacillus anthracis*, the *Pasteurella pestis* murine toxins, the oxygen-labile hemolysins such as streptolysin O, and certain neurotoxins present in snake venoms and other venoms).

e. Appendix F-II-C.

Some enterotoxins are substantially more toxic when administered enterally than parenterally. The following enterotoxins shall be subject to BL1 + EK1 containment conditions: cholera toxin, the heat labile toxins of *Escherichia coli, Kiebsiella,* and other related proteins that may be identified by neutralization with an antiserum monospecific for cholera toxin, and the heat stable toxins of *Escherichia coli* and of *Yersinia enterocolitica.*

f. Appendix F-III. Cloning of Toxic Molecule Genes in Organisms Other Than *Escherichia coli* K-12

Requests involving the cloning of genes coding for toxin molecules for vertebrates at an LD_{50} of <100 nanograms per kilogram body weight in host-vector systems other than *Escherichia* coli K-12 will be evaluated by NIH/ORDA in consultation with adhoc toxin experts (see Sections III-B-1).

g. Appendix F-IV. Specific Approvals

An updated list of experiments involving the deliberate formation of recombinant DNA containing genes coding for toxins lethal for vertebrates at an LD_{50} of <100 nanograms per kilogram body weight is available from the Office of Recombinant DNA Activities, National Institutes of Health/MSC.

7. Physical Containment

Appendix G specifies physical containment for standard laboratory experiments and defines Biosafety Level 1 through Biosafety Level 4. For large scale (over 10 liters) research or production, Appendix K *(Physical Containment for Large Scale Uses of Organisms Containing Recombinant DNA Mo/ecu/es)* supersedes Appendix G. Appendix K defines Good Large Scale Practice through Biosafety Level 3- Large Scale. For certain work with plants, Appendix P *(Physical and Biological Containment for Recombinant DNA Research Involving Plants)* supersedes Appendix G. Appendix P defines Biosafety Levels 1 through 4: Plants. For certain work with animals, Appendix *Q (Physical and Biological Containment for Recombinant DNA Research Involving Animals)* supersedes Appendix G. Appendix Q defines Biosafety Levels 1 through 4- Animals.

The first principle of containment is strict adherence to good microbiological practices such as described earlier in this Chapter, Section VI- Microbiological and Biomedical Laboratories. In most cases the material for recombinant DNA facilities is very similar, often requiring only the substitution of the words containing recombinant DNA material for containing viable infectious material. However, one should refer to the complete text of the NIH Guidelines for additional information, particularly for suggestions as to alternative methods to achieve the desired level of protection.

The purpose of physical containment is to confine organisms containing recombinant DNA molecules and thus to reduce the potential for exposure of the laboratory worker, persons outside of the laboratory, and the environment to organisms containing recombinant DNA molecules. The selection of alternative methods of primary containment is dependent, however, on the level of biological containment provided by the host-vector system used in the experiment. Consequently, all personnel directly or indirectly involved in experiments using recombinant DNA shall receive adequate instruction. At a minimum, these instructions include training in aseptic techniques and in the biology of the organisms used in the experiments so that the potential biohazards can be understood and appreciated.

8. Appendix H. Shipment*

Recombinant DNA molecules contained in an organism or in a viral genome shall be shipped under the applicable regulations of the U.S. Postal Service (39 Code of Federal Regulations, Part 3); the Public Health Service (42 Code of Federal Regulations, Part 72); the U.S. Department of Agriculture (9 Code of Federal Regulations, Subchapters D and E; 7 CFR, Part 340); and/or the U.S. Department of Transportation (49 Code of Federal Regulations, Parts 171 -179).

a. Appendix H-I.

Host organisms or viruses will be shipped as etiologic agents, regardless of whether they contain recombinant DNA, if they are regulated as human pathogens by the Public Health Service (42 Code of Federal Regulations, Part 72) or as animal pathogens or plant pests under the U.S. Department of Agriculture, Animal and Plant Health Inspection Service (Titles 9 and 7 Code of Federal Regulations, respectively).

b. Appendix H-II.

Host organisms and viruses will be shipped as etiologic agents if they contain recombinant DNA when: (i) the recombinant DNA includes the complete genome of a host organism or virus regulated as a human or animal pathogen or a plant pest; or (ii) the recombinant DNA codes for a toxin or other factor directly involved in eliciting human, animal, or plant disease or inhibiting plant growth, and is carried on an expression vector or within the host chromosome and/or when the host organism contains a conjugation proficient plasmid or a generalized transducing phage; or (iii) the recombinant DNA comes from a host organism or virus regulated as a human or animal pathogen or as a plant pest and has not been adequately characterized to demonstrate that it does not code for a factor involved in eliciting human, animal, or plant disease.

c. Appendix H-III. Additional Resources

For further information on shipping etiologic agents, contact: (i) The Centers for Disease Control and Prevention, ATTN: Biohazards Control Office, 1600 Clifton Road, Atlanta, Georgia 30333, (404) 639-3883, FTS 236-3883; (ii) The U.S. Department of Transportation, ATTN: Office of Hazardous Materials Transportation, 400 7th Street, S.W., Washington, D.C. 20590, (202) 366-4545; or (iii) U.S. Department of Agriculture, ATTN: Animal and Plant Health Inspection Service (APHIS), Veterinary Services, National Center for Import-Export, Products Program, 4700 River Road, Unit 40, Riverdale, Maryland 20737. Phone:

---

* A host-vector system may be proposed for certification by the NIH Director in accordance with the procedures set forth in Appendix I-II. In order to ensure protection for proprietary data, any public notice regarding a host-vector system which is designated by the institution as proprietary under Section IV-D will be issued only after consultation with the institution as to the content of the notice (see Section IV-D-3).

(301) 734-8499; Fax: (301) 734-8226.

9. Appendix I. Biological Containment (See Appendix E)

a. Appendix I-I. Levels of Biological Containment

In consideration of biological containment, the vector (plasmid, organelle, or virus) for the recombinant DNA and the host (bacterial, plant, or animal cell) in which the vector is propagated in the laboratory will be considered together. Any combination of vector and host which is to provide biological containment shall be chosen or constructed so that the following types of "escape" are minimized: (i) survival of the vector in its host outside the laboratory, and (ii) transmission of the vector from the propagation host to other non-laboratory hosts. The following levels of biological containment (host-vector systems) for prokaryotes are established. Appendices I-I-A through I-II-B describe levels of biological containment (host-vector systems) for prokaryotes. Specific criteria will depend on the organisms to be used.

b. Appendix I-I-A. Host-Vector 1 Systems

Host-Vector 1 systems provide a moderate level of containment. Specific Host-Vector 1 systems are:

i. Appendix I-I-A-1. *Escherichia coli* K- 12 Host-Vector 1 Systems (EK1)

The host is always *Escherichia coli* K-12 or a derivative thereof, and the vectors include non-conjugative plasmids (e.g., pSC101, ColEl, or derivatives thereof (see Appendices I-III-A through G) and variants of bacteriophage, such as lambda (see Appendices I-III-H through O.)The *Escherichia* coli K-12 hosts shall not contain conjugation-proficient plasmids, whether autonomous or integrated, or generalized transducing phages.

ii. Appendix l-l-A-2. Other Host-Vector 1 Systems

At a minimum, hosts and vectors shall be comparable in containment to *Escherichia coli* K-12 with a non-conjugative plasmid or bacteriophage vector. Appendix I-II describes the data to be considered and mechanism for approval of Host-Vector 1 systems.

c. Appendix I-I-B. Host-Vector 2 Systems

Host-Vector 2 Systems provide a high level of biological containment as demonstrated by data from suitable tests performed in the laboratory. Escape of the recombinant DNA either via survival of the organisms or via transmission of recombinant DNA to other organisms should be $<1/10^8$ under specified conditions. Specific Host-Vector 2 systems are:

i. Appendix I-I-B-1.

For *Escherichia coli* K-12 Host-Vector 2 systems (EK2) in which the vector is a plasmid, no more than $1/10^8$ host cells shall perpetuate a cloned DNA fragment under the specified non-permissive laboratory conditions designed to represent the natural environment, either by survival of the original host or as a consequence of transmission of the cloned DNA fragment.

ii. Appendix I-I-B-2.

For *Escherichia coli* K-12 Host-Vector 2 systems (EK2) in which the vector is a phage, no more than $1/10^8$ phage particles shall perpetuate a cloned DNA fragment under the specified non-permissive laboratory conditions designed to represent the natural environment, either as a prophage (in the inserted or plasmid form) in the laboratory host used for phage propagation, or survival in natural environments and transferring a cloned DNA fragment to other hosts (or their resident prophages).

d. Appendix I-II. Certification of Host-Vector Systems

i. Appendix I-II-A. Responsibility

Host-Vector 1 systems (other than *Escherichia coli* K-12) and Host-Vector 2 systems may not be designated as such until they have been certified by the NIH Director. Requests for certification of host-vector systems may be submitted to the Office of Recombinant DNA Activities, National Institutes of Health/MSC. Proposed host-vector systems will be reviewed by the RAC (see Section IV-C-1-b-(1)-(f)). Initial review will be based on the construction, properties, and testing of the proposed host-vector system by a subcommittee composed of one or more RAC members and/or ad hoc experts. The RAC will evaluate the subcommittee's report and any other available information at the next scheduled RAC meeting. The NIH Director is responsible for certification of host-vector systems, following advice of the RAC. Minor modifications to existing host-vector systems (i.e., those that are of minimal or no consequence to the properties relevant to containment) may be certified by the NIH Director without prior RAC review. Once a host-vector system has been certified by the NIH Director, a notice of certification will be sent by NIH/ORDA to the applicant and to the Institutional Biosafety Committee Chairs. A list of all currently certified host-vector systems is available from the Office of Recombinant DNA Activities, National Institutes of Health/MSC. The NIH Director may rescind the certification of a host-vector system. If certification is rescinded, NIH will instruct investigators to transfer cloned DNA into a different system or use the clones at a higher level of physical containment level, unless NIH determines that the already constructed clones incorporate adequate biological containment. Certification of an host-vector system does not extend to modifications of either the host or vector component of that system. Such modified systems shall be independently certified by the NIH Director. If modifications are minor, it may only be necessary for the investigator to submit data showing that the modifications have either improved or not impaired the major phenotypic traits on which the containment of the system depends. Substantial modifications to a certified host-vector system requires submission of complete testing data.

e. Appendix I-II-B. Data to be Submitted for Certification

i. Appendix I-II-B-1. Host-Vector 1 Systems Other than *Escherichia coli* K-12

The following types of data shall be submitted, modified as appropriate for the particular system under consideration: (i) a description of the organism and vector; the strain's natural habitat and growth requirements; its physiological properties, particularly those related to its reproduction, survival, and the mechanisms by which it exchanges genetic information; the range of organisms with which this organism normally exchanges genetic information and the type of information exchanged and any relevant information about its pathogenicity or toxicity; (ii) a description of the history of the particular strains and vectors to be used, including data on any mutations which render this organism less able to survive or transmit genetic information; and (iii) a general description of the range of experiments contemplated with emphasis on the need for developing such a Host-Vector 1 system.

ii. Appendix I-II-B-2. Host-Vector 2 Systems

Investigators planning to request Host-Vector 2 systems certification may obtain instructions from NIH/ORDA concerning data to be submitted. In general, the following types of data are required: (i) description of construction steps with indication of source, properties, and manner of introduction of genetic traits; (ii) quantitative data on the stability of genetic traits that contribute to the containment of the system; (iii) data on the survival of the host-vector system under non-permissive laboratory conditions designed to represent the relevant natural environment; (iv) data on transmissibility of the vector and/or a cloned DNA fragment under both permissive and non-permissive conditions; (v) data on all other properties of the system which affect containment and utility, including information on yields of phage or plasmid molecules, ease of DNA isolation, and ease of transfection or transformation; and (vi) in some cases, the investigator may be asked to submit data on survival and vector transmissibility from experiments in which the host-vector is fed to laboratory animals or one or more human subjects. Such *in vivo* data may be required to confirm the validity of predicting *in vivo* survival on the basis of *in vitro* experiments. Data shall be submitted 12 weeks prior to the RAC meeting at which such data will be considered by the Office of Recombinant DNA Activities, National Institutes of Health. Investigators are encouraged to publish their data on the construction, properties, and testing of proposed Host Vector 2 systems prior to consideration of the system by the RAC and its subcommittee. Specific instructions concerning the submission of data for proposed *Escherichia coli* K-12 Host-Vector 2 system (EK2) involving either plasmids or bacteriophage in *Escherichia coli K-12,* are available from the Office of Recombinant DNA Activities, National Institutes of Health/MSC.

10. Appendix K. Physical Containment for Large Scale Uses of Organisms Containing Recombinant DNA Molecules

Appendix K specifies physical containment guidelines for large scale (greater than 10 liters of culture) research or production involving viable organisms containing recombinant DNA molecules. It shall apply to large scale research or production activities as specified in Section III-D-6. It is important to note that this appendix addresses only the biological hazard associated with organisms containing recombinant DNA. Other hazards accompanying the large scale cultivation of such organisms (e.g., toxic properties of products; physical, mechanical, and chemical aspects of downstream processing) are not addressed and shall be considered separately, albeit in conjunction with this appendix.

All provisions shall apply to large scale research or production activities with the following modifications: (i) Appendix K shall supersede Appendix G, when quantities in excess of 10 liters of culture are involved in research or production. Appendix K-II applies to Good Large Scale Practice; (ii) the institution shall appoint a Biological Safety Officer if it engages in large scale research or production activities involving viable organisms containing recombinant DNA molecules, (iii) the institution shall establish and maintain a health surveillance program for personnel engaged in large scale research or production activities involving viable organisms containing recombinant DNA molecules which require Biosafety Level (BL) 3 containment at the laboratory scale. The program shall include: preassignment and periodic physical and medical examinations; collection, maintenance, and analysis of serum specimens for monitoring serologic changes that may result from the employee's work experience; and provisions for the investigation of any serious, unusual, or extended illnesses of employees to determine possible occupational origin.

a. Appendix K-I. Selection of Physical Containment Levels

The selection of the physical containment level required for recombinant DNA research or production involving more than 10 liters of culture is based on the containment guidelines established in Section III. For purposes of large scale research or production, four physical containment levels are established. The four levels set containment conditions at those appropriate for the degree of hazard to health or the environment posed by the organism, judged by experience with similar organisms unmodified by recombinant DNA techniques and consistent with Good Large Scale Practice. The four biosafety levels of large scale physical containment are referred to as Good Large Scale Practice, BL1-Large Scale, BL2-Large Scale, and BL3-Large Scale. Good Large Scale Practice is recommended for large scale research or production involving viable, non-pathogenic, and non-toxigenic recombinant strains derived from host organisms that have an extended history of safe large scale use. Good Large Scale Practice is recommended for organisms such as those included in Appendix C which have built-in environmental limitations that permit optimum growth in the large scale setting but limited survival without adverse consequences in the environment. BL1-Large Scale is recommended for large scale research or production of viable organisms containing recombinant DNA molecules that require BL1 containment at the laboratory scale and that do not qualify for Good Large Scale Practice. BL2-Large Scale is recommended for large scale research or production of viable organisms containing recombinant DNA molecules that require BL2 containment at the laboratory scale. BL3-Large Scale is recommended for large scale research or production of viable organisms containing recombinant DNA molecules that require BL3 containment at the laboratory scale. No provisions are made for large scale research or production of viable organisms containing recombinant DNA molecules that require BL4 containment at the laboratory scale. If necessary, these requirements will be established by NIH on an individual basis.

Only the general Good Large Scale Practice Guidelines will be given here. Those for individual BL1-3, Large Scale, are provided in the complete version of the *NIH Guidelines* readily available at the Internet reference to this topic.

b. Appendix K-II. Good Large Scale Practice (GLSP)
i. Appendix K-II-A. Institutional codes of practice shall be formulated and implemented to assure adequate control of health and safety matters.

ii. Appendix K-II-B. Written instructions and training of personnel shall be provided to assure that cultures of viable organisms containing recombinant DNA molecules are handled prudently and that the work place is kept clean and orderly.

iii. Appendix K-II-C. In the interest of good personal hygiene, facilities (e.g., hand washing sink, shower, changing room) and protective clothing (e.g., uniforms, laboratory coats) shall be provided that are appropriate for the risk of exposure to viable organisms containing recombinant DNA molecules. Eating, drinking, smoking, applying cosmetics, and mouth pipetting shall be prohibited in the work area.

iv. Appendix K-II-D. Cultures of viable organisms containing recombinant DNA molecules shall be handled in facilities intended to safeguard health during work with microorganisms that do not require containment.

v. Appendix K-II-E. Discharges containing viable recombinant organisms shall be handled in accordance with applicable governmental environmental regulations.

vi. Appendix K-II-F. Addition of materials to a system, sample collection, transfer of culture fluids within/between systems, and processing of culture fluids shall be conducted in a manner that maintains employee's exposure to viable organisms containing recombinant DNA molecules at a level that does not adversely affect the health and safety of employees.

vii. Appendix K-II-G. The facility's emergency response plan shall include provisions for handling spills.

11. **Appendix M. Points to Consider in the Design and Submission of Protocols for the Transfer of Recombinant DNA Molecules into One or More Human Subjects (Points to Consider)**

Appendix M applies to research conducted at or sponsored by an institution that receives any support for recombinant DNA research from NIH. Researchers not covered by the NIH Guidelines are encouraged to use Appendix M.

This appendix will be abridged, as was Appendix K, because it applies to a much smaller group of laboratories than most recombinant DNA operations. It is, however, an extremely sensitive area and anyone to whom it applies must adhere to all sections. In addition to this introductory portion, Appendix M-3 and M-4 will be included which apply to informed consent on the part of the participants. Any research involving human subjects must be cleared through the Institutions Institutional Review Board (IRB) according to the regulations found in Title 45 CFR 46 which provides for the protection of Human subjects. Preparation of an adequate Informed Consent form often appears to provide difficulties to a researcher who is more comfortable with the technical aspects of the research.

The acceptability of human somatic cell gene therapy has been addressed in several public documents as well as in numerous academic studies. In November 1982, the President's Commission for the Study of Ethical Problems in Medicine and Biomedical and Behavioral Research published a report, *Splicing Life*, which resulted from a two-year process of public deliberation and hearings. Upon release of that report, a U.S. House of Representatives subcommittee held three days of public hearings with witnesses from a wide range of fields from the biomedical and social sciences to theology, philosophy, and law. In December 1984, the Office of Technology Assessment released a background paper, *Human Gene Therapy*, which concluded that civic, religious, scientific, and medical groups have all accepted, in principle, the appropriateness of gene therapy of somatic cells in humans for specific genetic diseases. Somatic cell gene therapy is seen as an extension of present methods of therapy that might be preferable to other technologies. In light of this public support, RAC is prepared to consider proposals for somatic cell gene transfer.

RAC will not, at present, entertain proposals for germ line alterations but will consider proposals involving somatic cell gene transfer. The purpose of somatic cell gene therapy is to treat an individual patient, e.g., by inserting a properly functioning gene into the subject's somatic cells. Germ line alteration involves a specific attempt to introduce genetic changes into the germ (reproductive) cells of an individual, with the aim of changing the set of genes passed on to the individual's offspring.

Research proposals involving the deliberate transfer of recombinant DNA, or DNA or RNA derived from recombinant DNA, into human subjects (human gene transfer) will be considered through a review process involving both NIH/ORDA and RAC. Investigators shall submit their relevant information on the proposed human gene transfer experiments to NIH/ORDA. Submission of human gene transfer protocols to NIH will be in the format described in Appendix M-I.. Submission to NIH shall be for registration purposes and will ensure continued public access to relevant human gene transfer information conducted in compliance with the NIH Guidelines. Investigational New Drug (IND) applications should be submitted to FDA in the format described in 21 CFR, Chapter I, Subchapter D, Part 312,

Subpart B, Section 23.

Institutional Biosafety Committee approval must be obtained from each institution at which recombinant DNA material will be administered to human subjects (as opposed to each institution involved in the production of vectors for human application and each institution at which there is *ex vivo* transduction of recombinant DNA material into target cells for human application).

Factors that may contribute to public discussion of a human gene transfer experiment by RAC include: (i) new vectors/new gene delivery systems, (ii) new diseases, (iii) unique applications of gene transfer, and (iv) other issues considered to require further public discussion. Among the experiments that may be considered exempt from RAC discussion are those determined not to represent possible risk to human health or the environment. Full RAC review of an individual human gene transfer experiment can be initiated by the NIH Director or recommended to the NIH Director by: (i) three or more RAC members, or (ii) other Federal agencies. An individual human gene transfer experiment that is recommended for full RAC review should represent novel characteristics deserving of public discussion. If the Director, NIH, determines that an experiment will undergo full RAC discussion, NIH/ORDA will immediately notify the Principal Investigator. RAC members may forward individual requests for additional information relevant to a specific protocol through NIH/ORDA to the Principal Investigator. In making a determination whether an experiment is novel, and thus deserving of full RAC discussion, reviewers will examine the scientific rationale, scientific context (relative to other proposals reviewed by RAC), whether the preliminary *in vitro* and *in vivo* safety data were obtained in appropriate models and are sufficient, and whether questions related to relevant social and ethical issues have been resolved. RAC recommendations on a specific human gene transfer experiment shall be forwarded to the NIH Director, the Principal Investigator, the sponsoring institution, and other DHHS components, as appropriate. Relevant documentation will be included in the material for the RAC meeting at which the experiment is scheduled to be discussed. RAC meetings will be open to the public except where trade secrets and proprietary information are reviewed. RAC prefers that information provided in response to Appendix M contain no proprietary data or trade secrets, enabling all aspects of the review to be open to the public.

Any application submitted to NIH/ORDA shall not be designated as 'confidential' in its entirety. In the event that a sponsor determines that specific responses to one or more of the items described in Appendix M should be considered as proprietary or trade secret, each item should be clearly identified as such. The cover letter (attached to the submitted material) shall: (1) clearly indicate that select portions of the application contain information considered as proprietary or trade secret, (2) a brief explanation as to the reason that each of these items is determined proprietary or trade secret.

Public discussion of human gene transfer experiments (and access to relevant information) shall serve to inform the public about the technical aspects of the proposals, meaning and significance of the research, and significant safety, social, and ethical implications of the research. RAC discussion is intended to ensure safe and ethical conduct of gene therapy experiments and facilitate public understanding of this novel area of biomedical research.

In its evaluation of human gene transfer proposals, RAC will consider whether the design of such experiments offers adequate assurance that their consequences will not go beyond their purpose, which is the same as the traditional purpose of clinical investigation, namely, to protect the health and well being of human subjects being treated while at the same time gathering generalizable knowledge. Two possible undesirable consequences of the transfer of recombinant DNA would be unintentional: (i) vertical transmission of genetic changes from an individual to his/her offspring, or (ii) horizontal transmission of viral infection to other persons with whom the individual comes in contact. Accordingly, Appendices M-l through M-V request information that will enable RAC and NIH/ORDA to assess the possibility that the proposed experiment(s) will inadvertently affect reproductive cells or lead to infection of other people (e.g., medical personnel or relatives).

Appendix M will be considered for revisions as experience in evaluating proposals accumulates and as new scientific developments occur. This review will be carried out periodically as needed.

12. Appendix M-III. Informed Consent

In accordance with the Protection of Human Subjects (45 CFR Part 46), investigators should indicate how subjects will be informed about the proposed study and the manner in which their consent will be solicited. They should indicate how the Informed Consent document makes clear the special requirements of gene transfer research. If a proposal involves children, special attention should be paid to the Protection of Human Subjects (45 CFR Part 46), Subpart D, Additional Protections for Children Involved as Subjects in Research.

a. Appendix M-III-A. Communication About the Study to Potential Participants
i.   Appendix M-III-A-1.

Which members of the research group and/or institution will be responsible for contacting potential participants and for describing the study to them? What procedures will be used to avoid possible conflicts of interest if the investigator is also providing medical care to potential subjects?

ii. Appendix M-III-A-2.

How will the major points covered in Appendix M-ll, Description of Proposal, be disclosed to potential participants and/or their parents or guardians in language that is understandable to them?

iii. Appendix M-III-A-3.

What is the length of time that potential participants will have to make a decision about their participation in the study?

iv. Appendix M-III-A-4.

If the study involves pediatric or mentally handicapped subjects, how will the assent of each person be obtained?

v. Appendix M-III-B. Informed Consent

Investigators submitting human gene transfer proposals must include the Informed Consent document as approved by the local Institutional Review Board. A separate Informed Consent document should be used for the gene transfer portion of a research project when gene transfer is used as an adjunct in the study of another technique, e.g., when a gene is used as a "marker" or to enhance the power of immunotherapy for cancer.

Because of the relative novelty of the procedures that are used, the potentially irreversible consequences of the procedures performed, and the fact that many of the potential risks remain undefined, the Informed Consent document should include the following specific information in addition to any requirements of the DHHS regulations for the Protection of Human Subjects (45 CFR 46): indicate if each of the specified items appears in the Informed Consent document or, if not included in the Informed Consent document, how those items will be presented to potential subjects; include an explanation if any of the following items are omitted from the consent process or the Informed Consent document.

b. Appendix M-III-B-1. General Requirements of Human Subjects Research
i. Appendix M-III-B-1-a. Description/Purpose of the Study
The subjects should be provided with a detailed explanation in non-technical language of the purpose of the study and the procedures associated with the conduct of the proposed study, including a description of the gene transfer component.

ii. Appendix M-III-B-1-b. Alternatives
The Informed Consent document should indicate the availability of therapies and the possibility of other investigational interventions and approaches.

iii. Appendix M-III-B-1-c. Voluntary Participation
.The subjects should be informed that participation in the study is voluntary and that failure to participate in the study or withdrawal of consent will not result in any penalty or loss of benefits to which the subjects are otherwise entitled.

iv. Appendix M-III-B-1-d. Benefits
The subjects should be provided with an accurate description of the possible benefits (to themselves), if any, of participating in the proposed study. For studies that are not reasonably expected to provide a therapeutic benefit to subjects, the Informed Consent document should clearly state that no direct clinical benefit to subjects is expected to occur as a result of participation in the study, although knowledge may be gained that may benefit others.

v. Appendix M-III-B-e. Possible Risks, Discomforts, and Side Effects
There should be clear itemization in the Informed Consent document of types of adverse experiences, their relative severity, and their expected frequencies. For consistency, the following definitions are suggested: side effects that are listed as mild should be those which do not require a therapeutic intervention; moderate side effects require an intervention; and severe side effects are potentially fatal or life-threatening, disabling, or require prolonged hospitalization.

If verbal descriptors (e.g., "rare," "uncommon," or "frequent") are used to express quantitative information regarding risk, these terms should be explained.

The Informed Consent document should provide information regarding the approximate number of people who have previously received the genetic material under study. It is necessary to warn potential subjects that, for genetic materials previously used in relatively few or no humans, unforeseen risks are possible, including some that could be severe.

The Informed Consent document should indicate any possible adverse medical consequences that may occur if the subjects withdraw from the study once the study has started.

vi. Appendix M-III-B-1-f. Costs
The subjects should be provided with specific information about any financial costs associated with their participation in the protocol and in the long-term follow-up to the protocol that are not covered by the investigators or the institution involved.

Subjects should be provided an explanation about the extent to which they will be responsible for any costs for medical treatment required as a result of research-related injury.

c. Appendix M-III-B-2. Specific Requirements of Gene Transfer Research
i. Appendix M-III-B-2-a. Reproductive Considerations

To avoid the possibility that any of the reagents employed in the gene transfer research could cause harm to a fetus/child, subjects should be given information concerning possible risks and the need for contraception by males and females during the active phase of the study. The period of time for the use of contraception should be specified.

The inclusion of pregnant or lactating women should be addressed.

ii. Appendix M-III-B-2-b. Long-Term Follow-Up

To permit evaluation of long-term safety and efficacy of gene transfer, the prospective subjects should be informed that they are expected to cooperate in long-term follow-up that extends beyond the active phase of the study. The Informed Consent document should include a list of persons who can be contacted in the event that questions arise during the follow-up period. The investigator should request that subjects continue to provide a current address and telephone number.

The subjects should be informed that any significant findings resulting from the study will be made known in a timely manner to them and/or their parent or guardian including new information about the experimental procedure, the harms and benefits experienced by other individuals involved in the study, and any long-term effects that have been observed.

iii. Appendix M-III-B-2-c. Request for Autopsy

To obtain vital information about the safety and efficacy of gene transfer, subjects should be informed that at the time of death, no matter what the cause, permission for an autopsy will be requested of their families. Subjects should be asked to advise their families of the request and of its scientific and medical importance.

iv. Appendix M-III-B-2-d. Interest of the Media and Others in the Research

To alert subjects that others may have an interest in the innovative character of the protocol and in the status of the treated subjects, the subjects should be informed of the following: (i) that the institution and investigators will make efforts to provide protection from the media in an effort to protect the participants' privacy, and (ii) that representatives of applicable Federal agencies (e.g., the National Institutes of Health and the Food and Drug Administration), representatives of collaborating institutions, vector suppliers, etc., will have access to the subjects' medical records.

d. Appendix M-IV. Privacy and Confidentiality

Indicate what measures will be taken to protect the privacy of patients and their families as well as to maintain the confidentiality of research data.

i. Appendix M-IV-A.

What provisions will be made to honor the wishes of individual patients (and the parents or guardians of pediatric or mentally handicapped patients) as to whether, when, or how the identity of patients is publicly disclosed.

ii. Appendix M-IV-B

What provisions will be made to maintain the confidentiality of research data, at least in cases where data could be linked to individual patients?

13. Appendix P. Physical and Biological Containment for Recombinant DNA Research Involving Plants

Appendix P specifies physical and biological containment conditions and practices suitable to the greenhouse conduct of experiments involving recombinant DNA-containing plants, plant-associated microorganisms, and small animals. All provisions of the *NIH Guidelines* apply to plant research activities with the following modifications:

Appendix P shall supersede Appendix G when the research plants are of a size, number, or have growth requirements that preclude the use of containment conditions described in Appendix G. The plants covered in Appendix P include but are not limited to mosses, liverworts, macroscopic algae, and vascular plants including terrestrial crops, forest, and ornamental species.

Plant-associated microorganisms include viroids, virusoids, viruses, bacteria, fungi, protozoans, certain small algae, and microorganisms that have a benign or beneficial association with plants, such as certain *Rhizobium* species and microorganisms known to cause plant diseases. The appendix applies to microorganisms which are being modified with the objective of fostering an association with plants.

Plant-associated small animals include those arthropods that: (i) are in obligate association with plants, (ii) are plant pests, (iii) are plant pollinators, or (iv) transmit plant disease agents, as well as other small animals such as nematodes for which tests of biological properties necessitate the use of plants. Microorganisms associated with such small animals (e.g., pathogens or symbionts) are included.

The Institutional Biosafety Committee shall include at least one individual with expertise in plant, plant pathogen, or plant pest containment principles when experiments utilizing Appendix P require prior approval by the Institutional Biosafety Committee.

a. Appendix P-I. General Plant Biosafety Levels
i. Appendix P-I-A.

The principal purpose of plant containment is to avoid the unintentional transmission of a recombinant DNA-containing plant genome, including nuclear or organelle hereditary material or release of recombinant DNA-derived organisms associated with plants.

ii. Appendix P-I-B.

The containment principles are based on the recognition that the organisms that are used pose no health threat to humans or higher animals (unless deliberately modified for that purpose), and that the containment conditions minimize the possibility of an unanticipated deleterious effect on organisms and ecosystems outside of the experimental facility, e.g., the inadvertent spread of a serious pathogen from a greenhouse to a local agricultural crop or the unintentional introduction and establishment of an organism in a new ecosystem.

iii. Appendix P-I-C.

Four biosafety levels, referred to as Biosafety Level (BL) 1-Plants (P), BL2-P, BL3-P, and BL4-P, are established in Appendix P-II, *Physical Containment Levels.* The selection of containment levels required for research involving recombinant DNA molecules in plants or associated with plants is specified in Appendix P-III, *Biological Containment Practices.* These biosafety levels are described in Appendix P-II, *Physical Containment Levels.* This appendix describes greenhouse practices and special greenhouse facilities for physical containment.

iv. Appendix P-I-D.

BL1-P through BL4-P are designed to provide differential levels of biosafety for plants in the absence or presence of other experimental organisms that contain recombinant DNA. These biosafety levels, in conjunction with biological containment conditions described in Appendix P-III, *Biological Containment Practices,* provide flexible approaches to ensure the safe conduct of research.

v. Appendix P-I-E.

For experiments in which plants are grown at the BL1 through BL4 laboratory settings, containment practices shall be followed as described in Appendix G. These containment practices include the use of plant tissue culture rooms, growth chambers within laboratory facilities, or experiments performed on open benches. Additional biological containment practices should be added by the Greenhouse Director or Institutional Biosafety Committee as necessary if botanical reproductive structures are produced that have the potential of being released.

b. Biological Containment Practices

Appropriate selection of the following biological containment practices may be used to meet the containment requirements for a given organism. The present list is not exhaustive; there may be other ways of preventing effective dissemination that could possibly lead to the establishment of the organism or its genetic material in the environment resulting in deleterious consequences to managed or natural ecosystems.

i.   Appendix P-III-A. Biological Containment Practices (Plants)
ii. Appendix P-III-A-1.

Effective dissemination of plants by pollen or seed can be prevented by one or more of the following procedures: (i) cover the reproductive structures to prevent pollen dissemination at flowering and seed dissemination at maturity; (ii) remove reproductive structures by employing male sterile strains, or harvest the plant material prior to the reproductive stage; (iii) ensure that experimental plants flower at a time of year when cross-fertile plants are not flowering within the normal pollen dispersal range of the experimental plant; or (iv) ensure that cross-fertile plants are not growing within the known pollen dispersal range of the experimental plant.

iii. Appendix P-III-B. Biological Containment Practices (Microorganisms)

iv. Appendix P-III-B-1.

Effective dissemination of microorganisms beyond the confines of the greenhouse can be prevented by one or more of the following procedures: (i) confine all operations to injections of microorganisms or other biological procedures (including genetic manipulation) that limit replication or reproduction of viruses and microorganisms or sequences derived from microorganisms, and confine these injections to internal plant parts or adherent plant surfaces; (ii) ensure that organisms, which can serve as hosts or promote the transmission of the virus or microorganism, are not present within the farthest distance that the airborne virus or microorganism may be expected to be effectively disseminated; (iii) conduct experiments at a time of year when plants that can serve as hosts are either not growing or are not susceptible to productive infection; (iv) use viruses and other microorganisms or their genomes that have known arthropod or animal vectors, in the absence of such vectors; (v) use microorganisms that have an obligate association with the plant; or (vi) use microorganisms that are genetically disabled to minimize survival outside of the research facility and whose natural mode of transmission requires injury of the target organism, or assures that inadvertent release is unlikely to initiate productive infection of organisms outside of the experimental facility.

v. Appendix P-III-C. Biological Containment Practices (Macroorganisms)
vi. Appendix P-III-C-1.

Effective dissemination of arthropods and other small animals can be prevented by using one or more of the following procedures: (i) use non-flying, flight-impaired, or sterile arthropods; (ii) use nonmotile or sterile strains of small animals; (iii) conduct experiments at a time of year that precludes the survival of escaping organisms; (iv) use animals that have an obligate association with a plant that is not present within the dispersal range of the organism; or (v) prevent the escape of organisms present in run-off water by chemical treatment or evaporation of run-off water.

14. Appendix Q. Physical and Biological Containment for Recombinant DNA Research Involving Animals

Appendix Q specifies containment and confinement practices for research involving whole animals, both those in which the animal's genome has been altered by stable introduction of recombinant DNA, or DNA derived therefrom, into the germ-line (transgenic animals) and experiments involving viable recombinant DNA-modified microorganisms tested on whole animals. The appendix applies to animal research activities with the following modifications.

Appendix Q shall supersede Appendix G when research animals are of a size or have growth requirements that preclude the use of containment for laboratory animals. Some animals may require other types of containment. The animals covered in Appendix Q are those species normally categorized as animals including but not limited to cattle, swine, sheep, goats, horses, and poultry.

The Institutional Biosafety Committee shall include at least one scientist with expertise in animal containment principles when experiments utilizing Appendix Q require Institutional Biosafety Committee prior approval.

The institution shall establish and maintain a health surveillance program for personnel engaged in animal research involving viable recombinant DNA-containing microorganisms that require Biosafety Level (BL) 3 or greater containment in the laboratory.

a. Appendix Q-l. General Considerations
b. Appendix Q-I-A. Containment Levels

The containment levels required for research involving recombinant DNA associated with or in animals is based on classification of experiments in Section III. For the purpose of animal research, four levels of containment are established. These are referred to as BL1-Animals (N), BL2-N, BL3-N, and BL4-N and are described in the appendices of Appendix Q. The descriptions include: (i) standard practices for physical and biological containment, and (ii) animal facilities.

c. Appendix Q-l-B. Disposal of Animals (BL1 -N through BL4-N)
i. Appendix Q-I-B-1.

When an animal covered by Appendix Q containing recombinant DNA or a recombinant DNA-derived organism is euthanized or dies, the carcass shall be disposed of to avoid its use as food for human beings or animals unless food use is specifically authorized by an appropriate Federal agency.

ii. Appendix Q-l-B-2.

A permanent record shall be maintained of the experimental use and disposal of each animal or group of animals.

Since animals are mobile and could conceivably escape or intermingle with other animals, the provisions of Appendix Q contain many provisions to make sure the animals involved with recombinant DNA do not have opportunities to spread and transfer any modified genetic materials either directly or via experimental procedures. Those intending to use recombinant DNA procedures in work with animals should be thoroughly familiar with the provisions of appendix Q of the NIH Guidelines.

REFERENCES

1. National Institutes of Health guidelines for research involving recombinant DNA molecules, *Fed Reg., 51,* May 7, 1986, 16958.

2. Classification of Etiologic Agents on the Basis of Hazard, 4th Edition, U.S. Department of Health Education and Welfare, Center for Disease Control, Office of Biosafety, Atlanta, GA, July, 1974.

3. Laboratory Safety at the Center for Disease Control, Publ. No. CD.C. 75-8118, U.S. Department of Health, Education and Welfare, 1974.

4. National Cancer Institute Safety Standards for Research Involving Oncogenic Viruses, Pub. No. (NIH) 75-790, U.S. Department of Health, Education and Welfare, 1974.

5. National Institutes of Health Biohazards Safety Guide, Stock No. 1740-003 83, Public Health Service, National Institutes of Health, U.S. Department of Health, Education and Welfare, Washington, D.C., 1974.

6. **Hellman, A., Oxman, M.N., and Pollack, R.,** Eds., *Biohazards in Biological Research,* Cold Spring Harbor Laboratory, New York, 1974.

7. **Bodily, J.L.,** General administration of the laboratory, in *Diagnostic Procedures for Bacterial, Mycotic and Parasitic Infections,* Bodily, H.L., Updyke, E.L., and Mason, J.O., Eds., American Public Health Association, New York, 1970, 11.

8. **Darlow, H.M.,** Safety in the microbiological laboratory, in *Methods in Microbiology,* Norris, J.R. and Robbins, D.W, Eds., Academic Press, New York, 1969, 169.

9. **Collins, C.H., Hartley, E.G., and Pilsworth, R.,** The Prevention of Laboratory Acquired Infection,, Monogr. Ser. No. 6, Public Health Laboratory Service, 1974.

10. **Chatigny, M.A.,** Protection against infection in the microbiological laboratory: devices and procedures, in *Advances in Applied Microbiology,* Vol. 3, Umbreit, W.W., Ed., Academic Press, New York, 1961, 131.

11. Design Criteria for Viral Oncology Research Facilities, Publ. No. (NIH) 75-891, U.S. Department of Health, Education and Welfare, Washington, D.C., 1975.

12. **Kuehne, R.W.,** Biological containment facility for studying infectious disease, *Appl. Microbiol.,* 26, 239, 1973.

13. **Runkle, R.S. and Phillips, G.E.,** *Microbial Containment Control Facilities,* Van Nostrand Reinhold, New York, 1969.

14. **Chatigny, M.A. and Clinger, D.I.,** Contamination control in aerobiology, in *An Introduction to Experimental Aerobiology,* Dimmick, R.L. and Akers, A.B., Eds., John Wiley & Sons, New York, 1969, 194.

15. **Matthews, R.E.E,** Ed., Third report of the international committee on taxonomy of viruses: classification and nomenclature of viruses, *Intervirology,* 12, 129, 1979.

16. **Buchanan, R.E. and Gibbons, N.E.,** Eds., *Bergey's Manual of Determinative Bacteriology* 8th ed., Williams and Wilkins, Baltimore, MD, 1974.

17. **Richmond, J.Y. and McKinney, R.W., Eds.,** *Biosafety in Microbiological and Biomedical Laboratories,* 3rd ed., Publ. No. (CD.C.) 84-8395, U.S. Department of Health, and Human Services, National Institutes of Health, Centers for Disease Control, Washington, D.C., 1993.

18. Laboratory Safety Monograph — A Supplement to the NIH Guidelines for Recombinant DNA Research, Office of Recombinant DNA Activities, National Institutes of Health, Bethesda, MD.

19. **Hershfield, V., Boyer H.W., Yanofsky, C., Lovett, M.A., and Heliaski, D.R.,** Plasmid Col El as a molecular vehicle for cloning and amplification of DNA, *Proc.Natl. Acad. Sci.,* U.S.A., 71, 3455, 1974.

20. **Wensink, P.E., Finnegan, D.J., Donelson, J.E., and Hogness, D.S.,** A system for mapping DNA sequences in the chromosomes *of Drosophila melanogaster, Cell,* 3, 315, 1974.

21. **Tanaka, T. and Weisbium, B.,** Construction of a coli in E1-R factor composite plasmid in vitro: means for amplification of deoxyribonucleic acid, *J. Bacteriol.,* 121,354, 1975.

22. **Armstrong, K.A., Hershfield, V., and Helsinki, D.R.,** Gene cloning and containment properties of plasmid Col El and its derivatives, *Science,* 196, 172, 1977.

23. **Bolivar, F., Rodriguez, R.L., Batlach, M.C., and Boyer, H.W.,** Construction and characterization of new cloning vehicles. I. Ampicillin-resistant derivative of pMB9, *Gene,* 2, 75, 1977.

24. **Cohen, S.N., Chang, A.C.W., Boyer, H., and Helling, R.,** Construction of biologically functional plasmids in vitro, *Proc. Nati. Acad. Sci. U.S.A.,* 70, 3240, 1973.

25. **Bolivar, F., Rodriguez, R.L., Greene, R.J., Batlach, M.C., Reynekei H.L., Boyer, W., Crosa, J.H., and Falkow, S.,** Construction and characterization of new cloning vehicles. II. A multi-purpose cloning system, *Gene,* 2, 95, 1977.

26. **Thomas, M., Cameron, I.R., and Davis, R.W,** Viable molecular hybrids of bacteriophage lambda and eukaryotic DNA, *Proc. Natl. Acad. Sci. U.S.A.,* 71, 4579, 1974.

27. **Murray, N.E. and Murray, K.,** Manipulation of restriction targets in phage lambda to form receptor chromosomes for DNA fragments, *Nature,* 251, 476, 1974.

28. **Rambach, A. and Tiolais, P.,** Bacteriophage having ecor1 endonuclease sites only in the non-essential region of the genome, *Proc. Natl. Acad. Sci. U.S.A.,* 71, 3927, 1974.

29. **Blattner, E.R., Williams, G.G., Bleche, A.E., Denniston-Thompson, K., Faber, H.E., Furlong, L.A., Gunwald, D.J., Kiefer, D.O., Moore, D.D., Shumm, J.W., Sheldon, E.L., and Smithies,O,** Charon phages: safer derivatives of bacteriophage lambda for DNA cloning, *Science,* 196, 163, 1977.

30. **Donoghue, D.J. and Sharp, P.A.,** An improved lambda vector: construction of model recombinants coding for kanamycin resistance, *Gene,* 1, 209, 1977.

31. **Leder, P., Tiemeier, D., and Enquist, L.,** EK2 derivatives of bacteriophage lambda useful in the cloning of DNA from higher organisms: the gt WES System, *Science,* 196, 175, 1977.

32. **Skalka, A.,** Current Status of Coliphage EK2 Vectors, *Gene,* 3, 29, 1978.

33. **Szybalski, W., Skalka, A., Gottesman, S., Campbell, A., and Botstein, D.,** Standardized laboratory tests for EK2 certification, *Gene,* 3, 36, 1978.

INTERNET REFERENCE

1. Guidelines for Research Involving Recombinant DNA Molecules, May 1998, http://www.nih.gov/od/orda/guidelines.pdf/toc.htm

VIII. RESEARCH ANIMAL CARE AND HANDLING*

A. Introduction

Animal use in research, teaching, and testing has provided advances in health care and preventive medicine for both animals and humans. Experimental results are greatly dependent upon the humane care and treatment of animals used in research. There is a large body of laws, regulations, and guidelines governing the use of animals in research to assure humane animal care and use. These regulations should not be feared as inhibitory to scientific freedom. Rather, as Aristotle said, "Shall we not like the archer who has a mark to aim at, be more likely to hit upon that which is right?" Compliance with these laws and guidelines assures healthy, high-quality animal models for use in research, assuring consistency from laboratory to laboratory throughout the nation, thus enhancing experimental reliability. The use of high-quality healthy animals and experimental methodologies which seek to minimize or eliminate pain or discomfort to the animals should be incorporated because not only is it the law, but because it makes scientific sense and is the most humane and ethical thing to do. The following sections briefly describe the laws and regulations governing animal care and use, and programs for health maintenance of animals and research personnel who come into contact with those animals.

---

* This Section was written by David M. Moore, D.V.M.. It was written in 1994 so the reader should refer to current versions of any regulatory references.

B. Laws and Regulations Relating to Animal Care and Use

Two major types of regulatory activities impacting the use of animals at a research facility involve voluntary and involuntary regulations. Involuntary regulations are statutory in nature, are uncompromising, and include federal and state laws which dictate minimum standards for the acquisitions of animals, provision of veterinary and husbandry care, use, and disposition of laboratory animals, and personal and institutional compliance is mandatory. Voluntary regulations are those which a research facility imposes on itself above and beyond the minimum standards set forth by the government. Knowledge of and compliance with applicable institutional, state, and federal policies, regulations, and laws will assure humane care of animals, improve scientific reliability, and deflect criticism from the small segment of society which questions whether animals used in research are humanely treated.

1. Animal Welfare Act

The major federal law affecting and regulating use of animals in research is the Federal Animal Welfare Act (PL. 89-544, and its amendments, PL. 9 1-579, PL. 94-279, and PL. 99-198). Full text copies of the Act are published in the Code of Federal Regulations CFR Title 9, Animals and Animal Products, Subchapter A, Animal Welfare, Parts 1, 2, and 3, and copies of these laws and regulations can be obtained from the Director, Regulatory Enforcement and Animal Care, USDA, APHIS, Room 207, Federal Building, 6505 Belcrest Road, Hyattsville, MD 20782, or the U.S. Department of Agriculture, Animal and Plant Health Inspection Service, Veterinary Services, Import-Export Products Staff, Room 756, Federal Building, 6505 Belcrest Road, Hyattsville, MD 20782. They can also be obtained via the Internet.

The Act is administered and enforced by the United States Department of Agriculture Animal, Plant Health Inspection Service (USDA/APHIS). All research institutions using animals defined by the Act (dog, cat, nonhuman primate, guinea pig, hamster, or rabbit) must complete and submit VS Form 18-11 "Applications for Registration of a Research Facility" to the USDA-APHIS-VS Veterinarian-in-Charge for the state in which the facility is located (contact the national office listed above for the local address). The form asks for the location of animal holding facilities, and the species and numbers of animals to be used. Failure to register as a research facility using covered animals may result in fines or sanctions prohibiting future use of animals at that institution. Currently, rodent species, birds, farm animals, and exotic species are not considered "Animals" under the definition contained in the Act. These may be included in subsequent amendments, so it would be advisable to contact the veterinarian-in-charge for your state to determine if you must register. The Act addresses and sets minimum standards for the care and use of animals in research in the areas of facilities, construction, caging, and operations of the facility; animal health and husbandry, standards covering feeding, watering, sanitation, employee qualifications, separation of species, record keeping, and provision of adequate veterinary care; and transportation standards of animals to and from the facility. Please refer to a copy of the Act for detailed specifications. The size of a facility and its research program may determine whether a full-time veterinarian is required on the staff, or if the animal health care needs can be met with a part-time or consulting veterinarian with laboratory animal training or experience. The institution must file a "Program of Veterinary Care" with the USDA/APHIS-VS Veterinarian-in-Charge for that state, detailing programs of disease control and prevention, euthanasia methods, and use of appropriate anesthetic, analgesic, or tranquilizing drugs when necessary as determined by the institutional veterinarian.

Each registered research facility must submit VA 18-23 "Annual Report of Research Facility" with USDA/APHIS each year, listing the numbers of animals by species and by category of potential pain or discomfort used during that year, with the signatures of the attending institutional veterinarian and a designated senior administrative official at the institution. Federal penalties may be invoked against the signatories for falsification of information in the body of the report.

The Improved Standards for Laboratory Animals Act (P.L. 99-198), the most recent amendment to the Animal Welfare Act, sets even more important requirements. Each research facility must have an institutional animal care committee of not fewer than three members to be appointed by the chief executive officer of the facility, to include a doctor of veterinary medicine (usually the institutional veterinarian), another facility employee, and a nonemployee, community member who has no family member affiliated with the facility. The committee must inspect all animal study areas and animal facilities twice annually keeping all inspection reports on file for 3 years, and notifying the administrative representative of the facility of any deficiencies or deviations from the Act. Additionally, each research facility must establish a program for the training of scientists, animal technicians, and other personnel involved with animal care and treatment to include instruction on:

1. Humane practice of animal maintenance and experimentation
2. Research or testing methods that minimize or eliminate animal pain or distress
3. Utilization of the information service at the National Agricultural Library
4. Methods whereby deficiencies in animal care and treatment should be reported

This amendment also calls for establishing institutional standards for exercise of dogs, and provision for environmental enrichment for nonhuman primates.

2. The Good Laboratory Practices Act

The Good Laboratory Practices Act (22 December 1978 issue of the *Federal Register* 43 FR *59986-60025)* regulates "nonclinical laboratory studies that support applications for research or marketing permits for products regulated by the Food and Drug Administration, including food and color additives, animal food additives, human and animal drugs, medical devices for humane use, biological products, and electronic products."

Standards addressed by GLP regulations include: (i) compliance with the "NIH Guide for the Care and Use of Laboratory Animals" (to be discussed in the following pragraph); (ii) establishment of standard operating procedures (SOPs) for animal husbandry and experimental treatment; (iii) meticulous record keeping and documentation of activities; and the establishment of a functional quality assurance unit reporting to the highest administrative levels of the facility. Please consult this document for further details and applicability to your studies.

3. The Guide for the Care and Use of Laboratory Animals

The "Guide" (NIH Publications 85-23) was prepared by the Committee on Care and Use of Laboratory Animals of the Institute of Laboratory Animal Resources, National Research Council. The Guide presents recommendations and basic guidelines for: (i) appropriate cage and enclosure sizes for a variety of commonly used laboratory species; (ii) social environmental enrichment; (iii) appropriate environmental temperature and humidity ranges; (iv) ventilation of animal facilities, levels of illumination, noise; (v) separation of species, sanitation of caging and facilities; (vi) provision for quarantine, and provision of adequate veterinary care.

Whereas the standards in the Guide are more stringent than those in the Animal Welfare Act, they are simply recommendations and do not by themselves carry legal penalties. However, other government agencies use the Guide as a measure for animal care, and will terminate funding support for an Institution for noncompliance with the Guide. Single copies of the Guide can be obtained from the Animal Resources Program, Division of Research Resources, National Institutes of Health, Bethesda, Maryland 20205.

4. Public Health Service Policies

Institutions receiving federal grant support for animal research activities from the Public Health Service (including the National Institutes of Health) must file an Animal Welfare Assurance statement with the Office for Protection from Research Risks (OPRR), National Institutes of Health, 9000 Rockville Pike, Building 31, Room 4B09, Bethesda, Maryland 20892. The components of the Assurance are listed in the publications "Public Health Service Policy on Humane Care and Use of Laboratory Animals" which can be obtained from OPRR. The intent of this policy is to "require Institutions to establish and maintain proper measures to ensure the appropriate care and use of all animal involved in research, research training, and biological lasting activities."

The PHS "requires that institutions, in their Assurance Statement, use the Guide for the Care and Use of Laboratory Animals as the basis for developing and implementing an institutional program for activities involving animals." In contrast to the Animal Welfare Act, PHS mandates an institutional animal care and use committee with a minimum of five members (the membership credentials mirror those specified in the Act). Institutions with approved assurance statements must file an annual report with OPRR detailing any changes in the animal program, and listing the dates of the twice-yearly facility inspections by the committee. Failure to comply with the provision of this policy will result in non-funding or withdrawal of funding for ongoing activities.

As with the Animal Welfare Act, investigators submitting proposals to PHS agencies must submit a research protocol detailing animal use for review of and approval by the institutional animal care and use committee. This committee must approve the project before funding is released, and may require modification of the project if it is not in compliance with Animal Welfare Act standards or PHS policies. The institutional animal care and use committee may halt or terminate an ongoing project for noncompliance with federal laws and policies.

5. Voluntary Regulations

Institutions may develop their own internal policies regarding animal care and use, providing that they are equal to or more stringent than those contained in the Guide or the Animal Welfare Act. Internal policies might include SOPs for animal care and use, mechanisms for selection of and purchase from commercial animal vendors, quarantine policies, and human health monitoring programs.

C. Personnel

"The Guide for the Care and Use of Lab6ratory Animals" promotes Institutional personnel policies requiring the use of technicians qualified to provide proper, humane animal care and husbandry, and recommends that these individuals apply for and receive certification from the American Association for Laboratory Animal Science (AALAS, 70 Timbercreek Drive, Suite 5, Cordova, TN 38018). There are three levels of certification based upon educational background and training, and on-the-job experience dealing with laboratory animals: Assistant Technician, Technician, and Laboratory Animal Technologist. Most facilities require facility supervisors or managers to have the AALAS Technologist certification. In-house training of technicians using AALAS course materials will satisfy the training requirement set forth in the Improved Standards for Laboratory Animals Act. Training of the scientific staff in humane animal care would be best accomplished by the lab animal veterinarian or an AALAS technologist. The qualifications for full-time or consulting veterinarians to the research facility should include either specialty board certification by the American College of Laboratory Animal Medicine (ACLAM) or indications of postdoctoral training or experience with laboratory animals. The role of the veterinarian in assuring the provision of "adequate, veterinary care" (as referred to in the laws and policies section of this chapter) is described in a report by ACLAM on "Adequate Veterinary Care" issued in October 1966.

D. Animal Holding Facilities

Animal holding facilities should be designed and constructed following the recommendations of "The Guide for the Care and Use of Laboratory Animals," which also assures compliance with the Animal Welfare Act. Consult the references at the end of this section for further information on animal facility design and management.

Facilities should be designed and operated for the comfort of the animals and the convenience of the investigator. Another critical factor in facility design operations involves the prevention of transmission of latent diseases from animal to animal, or animals to humans. The first step involves purchase of animals from "clean" sources who have a documented animal health quality assurance program. Newly arrived animals should be held in a quarantine area in the facility to prevent potential contamination of existing research animal populations. The quarantine facility should be located in an area adjacent to the main colony, but with separate access to prevent cross-contamination of the colony as would be the case with common traffic flows.

The second step for maintaining clean animals involves control of the micro-environment through the use of appropriate housing units (i.e., micro-isolator caging), laminar airflow housing racks, or mass air displacement "clean" rooms. The items suggested above may be cost prohibitive for some facilities, and adequate care could involve simply following sanitation and hygiene recommendations in the "Guide." Air pressure in the room can be changed to protect either the animals or humans working in the facility. Making the room air pressure slightly positive with respect to the hallways will minimize the chances of entry of airborne disease agents into an animal room. However, if the research involves animals infected with animal or human pathogens, or they are treated with toxic or carcinogenic agents, then the room air pressure should be made slightly negative as compared to the adjacent hallways, to prevent contamination of animals in other nearby rooms or humans who use that hallway. Hessler and Moreland discuss the use of HEPA filtration in rooms using nonvolatile carcinogens.

E. Animal Care and Handling

The Improved Standards for Laboratory Animals Act requires that experimental procedures "ensure that animal pain and distress are minimized." A stressed animal is not a good experimental model, since its biochemical and physiological attributes are altered during stress. Minimizing animal stress can be easily accomplished by:

1. Purchasing animals free of latent and overt clinical diseases, and providing adequate veterinary care to maintain their health.
2. Familiarizing animals with experimental devices or rooms prior to the start of the experiment.
3. Limiting restraint to that which is necessary to accomplish the experimental goals, preconditioning the animals to the restraint apparatus, or using other nonrestraint alternatives.
4. Controlling or eliminating environmental stressors (i.e., inappropriate temperature, humidity, light, noise, aggressive cage mates, cage size).
5. Providing environmental enrichment programs and/or exercise for animals, especially dogs, cats, and nonhuman primates.
6. Selecting and using appropriate anesthetics, tranquilizers, sedatives, and analgesic drugs for procedures in which pain or distress are likely.
7. Allowing only skilled, trained individuals to perform surgery.

8. Providing training for technicians and investigators in humane animal care and use techniques.

F. Human Health Monitoring

It is imperative that a human health monitoring program be established for those individuals having limited or full-time contact with research animals. Caretakers and investigators can be exposed to hazardous aerosols, bites, scratches, bodily wastes and discharges, and fomites contaminated with zoonotic agents. A preemployment physical should be conducted to obtain baseline physical and historical data, and a serum sample should be drawn and frozen for future reference. Additional examinations should be scheduled periodically depending on the nature and risks in the work environment.

Training programs should be established to acquaint personnel with biologic (zoonotic), chemical, and physical hazards within the animal facility. Appropriate hygiene should be stressed, and protective clothing and equipment (gloves, protective outer garments, masks, respirators, face shields or eye protectors) should be made available, and their use made mandatory where appropriate in SOPs. Technicians should be aware of clinical signs of disease, notifying the facility veterinarian for confirmation, treatment, isolation of the animal(s), or euthanasia of the affected animal.

REFERENCES

1. **Clark, J.D.,** Regulation of animal use: voluntary and involuntary, *I. Vet Med Educ.,* 6(2), 86, 1979.
2. **McPherson, C.W.,** Legislation regulations pertaining to laboratory animals - United States, in *Handbook of Laboratory Animal Science,* Vol. I, Melby, E.C. and Altman, N.H., Eds., CRC Press, OH, Cleveland, 1974, 3.
3. **McPherson, C.W.,** Laws, regulations, and policies affecting the use of laboratory animals, in *Laboratory Animal Medicine,* Fox, J.G. et al., Eds., Academic Press, Orlando, FL, 1984, 19.
4. **Poiley, S.M.,** Housing requirements - general considerations, in *Handbook of Laboratory Animal Science,* Vol. I., Melby, E.C. and Altman, N.H., Eds., CRC Press, Cleveland, OH, 1974, 21.
5. **Hessler, J.R. and Moreland, A.F.,** Design and management of animal facilities in *Laboratory Animal Medicine,* Fox, J.G. et al., Eds., Academic Press, Orlando, FL, 1984, 505.
6. **Simmonds, R.C.,** The design of laboratory animal homes, in *Aeromedical Review,* Vol 2, USAF School of Aerospace Medicine, Brooks Air Force Base, San Antonio, TX, 1973.
7. **Runkle, R.S., Laboratory animal housing** - part II, *Am. Inst. Architect. J.,* 41, 77, 1964.
8. Comfortable Quarters for Laboratory Animals, Animal Welfare Institute, Washington, D.C., 1979.
9. **Richmond, J,Y. and McKinney, R.W.,** Biosafety in Microbiological and Biomedical Laboratories, 3rd ed., U.S. Department of Health and Human Services, HITS Publ. 93-8395, 1993.

INTERNET REFERENCE

1. Internet URL for laws cited: http://www.access.gpo.gov/nara/cfr/waisidx/9cfrv1_99.html

Chapter 6

PERSONAL PROTECTIVE EQUIPMENT

I. INTRODUCTION

This chapter is intended to be a reasonably extensive but not exhaustive treatment of personal protective equipment. It is designed to provide an overview of the topic in the context of normal laboratory usage. The concept of laboratories will be extended to include field studies, where workers are often exposed to hazardous materials such as agricultural chemicals so that the protective equipment discussion will also apply to them.

II. RESPIRATORY PROTECTION

Respiratory protective devices range all the way from the simple soft felt mask, which is often used to provide protection against nuisance levels of dusts and particulates, to the self-contained, positive-pressure, fully-enclosing suit, which, if properly matched to the antici-pated exposure, offers total body and respiratory protection from the toxic substance involved. As will be noted in a later section on gloves and protective clothing, there is no single material that will protect against all possible chemicals in the workplace so there is no "universal"material which will provide protection in all cases.

If a laboratory is properly designed and operated, with adequate ventilation and efficient fume hoods or other types of safety cabinets, additional respiratory protection normally will not be needed for most procedures. However, in the event of an accident or an unusual operation which cannot be performed in a hood, or when working with highly toxic substances, laboratory persons working with toxic materials should be included in a respiratory protection program, managed by the institution or corporation. There are a few situations which should require the availability of respiratory protection immediately at hand.

The selection of the proper respiratory protective device will depend upon a number of factors, the most important of which relate to the properties of the chemical or material for which protection is needed. What is the permissible exposure limit (PEL) or threshold level value (TLV) of the material? Is it immediately dangerous to life and health (IDLH)? Does it pose a skin absorption problem? What type of material is it, i.e., acid, solvent, dust, radioactive, carcinogen, asbestos, ammonia, etc? Will air or oxygen need to be supplied? What maximum levels of the material might be expected to be present? Does the material provide an adequate warning of its presence by an odor or by irritation of the respiratory system or the eyes, at a level well below the legal limits? The respiratory protection devices to be used must comply with the specifications of ANSI Z88.2, bear an appropriate NIOSH approval number, and provide the degree of protection needed under the existing working conditions. In addition to the OSHA respiratory protection standard briefly discussed in the following paragraph, OSHA also deals with respiratory protection in 29 CFR 120 and provides selection criteria for respirators in Appendix B to that standard.

The Occupational Safety and Health Administration (OSHA) issued a new version of the respirator standard, 29 CFR Part 1910.134 on April 8, 1998 which contains a substantial number of significant changes from the previous version. For example, it still requires that the individual who is to wear the respirator must be capable of wearing the respirator under working conditions, i.e., a person must be physically able to perform their work and safely use the equipment. A physician must determine if the respirator user's medical condition permits wearing a respirator but differs in detail how this is to be done. The following five paragraphs in this section provide some of the highlights covered by the standard but, because of the changes in the standard, major portions of the standard will be provided in a separate section.

As noted, an individual with poor respiratory function should not be asked or permitted to wear many types of respirators which require breathing through a protective filter or cartridge. The employee is to fill out a medical history questionnaire to help determine this and is evaluated by a physician. The medical status of the respirator user should be checked afterwards on a regular basis. Several medical conditions would preclude the wearing of a respirator. Among these are emphysema, asthma, reduced pulmonary function (variety of causes other than the preceding two), severe hypertension, coronary artery disease, cerebral blood vessel disease, epilepsy, claustrophobia (brought on by the wearing of the unit), or other relevant conditions as determined by the examining physician. Since the worker is being asked to wear a respirator to protect his health by reducing or eliminating inhalation of noxious vapors, it would appear logical that the employee should want to participate in a comprehensive medical surveillance program to check on the status of the individual's health.

The respirator must be properly fitted to the user, and the new standard details how this is to be done in detail. It is not possible to obtain a proper seal of the respirator to the user's face if there is facial hair where the respirator comes into contact with the face. Facial hair also may interfere with the operation of the exhalation valve. There are skin conditions which make it very difficult to obtain a proper seal, such as scarring due to acne, or injuries would also make it difficult to fit an individual. Facial structure also will have a bearing on the quality of fit. For example, a small or petite face is often difficult to fit, as is a narrow face with a prominent nasal bridge. In recent years, significant advances have been made in the design of half-face cartridge respirators, which are the most common type used, to improve the seal to the face. Different size units are now available. There are units which provide a second seal to the skin to aid in keeping toxic fumes from entering. The means of holding respirator units more securely in place have been improved. Hypoallergenic materials can be used in the construction of respirators to prevent the skin from becoming irritated when wearing a unit for an extended period is required.

Each person wearing a respirator must be individually fitted to ensure that the respirator is providing the needed protection to the wearer. The worker should be trained in the proper use of the respirator to enable it to fulfill its function and to maintain it in good working condition. Respirators kept for common use in a facility may be acceptable under certain conditions but it is most desirable that each individual should be issued a personal respirator, and held responsible for maintaining the unit in a clean, good operating condition.

The fitting program, at a minimum, must include a qualitative fit test administered by a knowledgeable person (usually someone from the safety or medical departments), using a particulate irritant, such as a nontoxic smoke generated by a smoke tube, or an organic solvent generating a distinctive odor, such as isoamyl acetate, to challenge the respirator. It would be preferable to perform a quantitative test, using known concentrations of an appropriate test material while the wearer performs simulated work movements. Even in qualitative tests, simulated work movements are helpful in detecting poor fits. For more toxic materials, a quantitative test should be done.

The training program should include, at a minimum: (a) how to care for the unit, including how to inspect it for proper functioning, as well as normal care; (b) how to put the respirator on and how to check to see that it is performing its function; (c) the function and limitations of the respirator; and (d) the health risks associated with either not using the protection or failure to use it properly. Refresher training should be given on an annual basis.

The following four sections describe the various types of respirators that are commonly used.

A. "Dust" Masks

The use of the term "dust" mask for the nonrigid soft felt mask is somewhat of a misnomer since, in modified forms, they can be used for other applications such as limited protection against paint fumes, moderate levels of organics, acid fumes, mercury, etc., although their biggest use is against nuisance dust.

These units are the simplest form of the air-purifying respirator. These respirators normally should not be employed for hazardous dusts, but are helpful for exposures to inert or nuisance dust levels below 15 mg/m^3. More elaborate versions of these felt masks include such features as exhalation valves, molded bridge pieces, and small sections of metal over the bridge of the nose, which can be bent to help the mask remain in contact with the nose and cheeks. Some have chemical absorbent incorporated in the mask to absorb some fumes and gases. In most cases these inexpensive masks are meant to be worn at relatively low levels of air pollutants (although the better versions which provide good facial contact can theoretically provide a protection factor of 10) and to be disposed of after a limited period of wear. For particulates, the felt mask tends to become more effective in removing particulates as they are worn in a contaminated atmosphere, since the space for air to pass through the filter becomes more limited as loading of the filter increases. This increases the difficulty in breathing as the filter offers more resistance to the flow of air. The use of an exhalation valve eases the outflow of air, but does not decrease the effort during the intake of air.

Filters used to absorb chemical gases or vapors do not use a mechanical action to trap the material but use an absorbent material (or in some cases, a chemical reactant) to prevent the material from passing through. When the absorbent is saturated or the reactant exhausted, the filter will no longer be effective. The relatively small amount of absorbent material incorporated in a simple felt mask limits their lifetime. Since they are usually discarded after use, they are typically intended to be used for about 8 hours or less.

Unless the contaminant in the air has an effective warning property such as a distinctive odor or acts as an irritant, respirators that only purify the air should not be worn for protection against such contaminants. The sensations experienced by an individual due to lack of oxygen do not constitute a sufficient warning signal for this hazard. At oxygen concentrations of 10 to 16%, it is possible to continue to function for short intervals but with significantly impaired judgment as the oxygen supply to the brain is decreased. At levels below 6%, death occurs in only a few minutes. It is quite likely that more persons have died wearing air-purifying respirators because of the failure to recognize the lack of oxygen in the air than have died from the direct effects of toxic materials. The new OSHA standard establishes lower limits on the acceptable oxygen concentration as a function of altitude. As noted earlier, no air-purifying respirators, even the more elaborate types, are approved for use in atmospheres which are IDLH.

B. Half-Face Cartridge Respirators

The half-face cartridge respirator is the type most frequently used, especially in atmospheres in which there is little or no problem of irritation or absorption of material through the skin. The facepiece of most of these units is molded of a flexible plastic or silicone rubber, which provides a seal to the face when properly adjusted. As noted earlier, facial hair between the mask and the face will prevent the seal from being effective, and it is not permitted for a person with a beard or extended sideburns in the area of the seal to be fitted with a respirator. Accommodation for individuals who wear glasses also must not break the seal to the face. The facepieces of most brands of these units are provided with receptacles for two sets of cartridges and/or filters. The respirators are certified as complete units, i.e., the facepiece equipped with specific filters. Cartridges from one vendor cannot be used on another manufacturer's facepiece. The major advantage of this type of unit is that by interchanging cartridges and filters, or by using one or more additional filters and cartridges in series, a single facepiece can be adapted to provide protection against a large variety of contaminants. However, some cartridge respirators are now being built with nonremovable and noninterchangeable cartridges. These are disposable units since the protective devices cannot be replaced. However, the capacity of the cartridges is considerably larger than the felt mask type, and often can be worn for several shifts if the levels of contaminants challenging the filter are not excessive. The models of these disposable cartridge respirators currently available are shaped somewhat differently than the usual half-face respirator and some individuals prefer them for this reason. The price is usually competitive with the replacement costs of cartridges for dual-cartridge half-face respirators. Since combinations of filters and cartridges cannot be modified, a large variety of models must be maintained in stock to fit a variety of exposure conditions.

The normal protection factor provided by a half-face respirator, which is accepted by OSHA and the Environmental Protection Agency (EPA), is either ten times the PEL or the cartridge limit, whichever is lower. In order to maintain the usefulness of the cartridge respirators, they must be properly maintained and stored. Fit tests should be repeated periodically to ensure that they still provide the required protection. Details on performing fit tests are to be found in the standard. Replacement parts for the exhalation valves should be maintained.

A problem alluded to in an earlier part of this section is the effort required to breathe through the filters. Although check valves can be designed so as to require little effort during the exhalation cycle, breathing air must pass through the cartridges on the intake cycle. Power-assisted breathing units are now available from a number of vendors which provide the flexibility of movement provided by the independent respirator, but remove much of the additional breathing effort incurred by wearing a respirator. In this system air is fed into the facepiece by a small battery-operated pump. The intake air is passed through cartridge filters located on the pump instead of on the facepiece. The pump is usually attached to the belt and, including the weight of the battery, is still light enough not to represent any significant problem. The batteries are usually designed to provide a nominal 8 hours of continuous operation, if they are properly maintained. Most of the batteries are nickel-cadmium which can lose capacity if they are not routinely taken through a complete discharge-charge cycle.

Since a power-assisted air-purifying (PAAP) respirator is a positive-pressure system, i.e., the air within the facepiece is at a higher pressure than the outside air, this type of unit intrinsically provides more protection than the ordinary half-face respirator. If provision is made for an "escape" mode of operation, i.e., the wearer can continue to breathe through the filters should the pump fail, and escape from the contaminated atmosphere. The ANSI Z88.2 standard (American National Standard Institute) would permit the use of this type of respirator in an IDLH atmosphere. Some of the early models of this type of unit had some problems with the seals on the pumps, but these problems have been corrected and they represent a desirable alternative if the wearer is to remain in a contaminated atmosphere for extended periods. There have been a few problems with the pumps overheating while being used at elevated temperatures.

Since a PAAP type of unit will permit entry into an IDLH atmosphere, assuming all other personal protective equipment is suitable, the user should be aware of a number of essential safety practices associated with this use. At least one standby person with the proper equipment for entry and rescue must be present outside the affected area. Communications must be continuously maintained between the worker within the area and the standby person. Persons within the area must be equipped with devices such as harnesses and safety lines to facilitate rescue operations should they be necessary.

In the best of circumstances, speaking is difficult while wearing a respirator. Throat microphones may be connected to amplifiers or radios so that the wearer can communicate with others without having to use the hands to activate a microphone.

If the possibility of chemical splashing exists, the eyes can be protected by wearing chemical splash goggles and/or face masks in addition to the half-face respirator. Goggles which provide ports or other means to allow air into the space behind the lens will not protect the eyes from vapors and gases in the air since the air behind the lens is room air. Some goggles and respirators are physically incompatible and, if both are needed, they must be selected taking compatibility into account.

The use of any type of respirator in almost any type of interior laboratory should be the exception rather than the rule, since these spaces should be engineered to be normally safe as far as atmospheric contaminants are concerned, if workers follow safe laboratory practices. However, field workers often must depend upon the proper selection of personal protective equipment to protect them from the hazards of exposure to contaminants. In many cases, the half-face respirator is the minimum acceptable respiratory protection. It is often worn under adverse personal comfort conditions as well. The weather may be very hot and the respirator may become very uncomfortable, so that the user may wish to forego wearing a unit and "take his chances." There is also the tendency for some workers to dismiss the need for the units and to make fun of the persons asking for a respirator or wearing one. Unfortunately, management sometimes shares this attitude. It is extremely important that responsible respiratory protection programs be made available, and the use of protective equipment required. This is especially true in academic institutions in which future managers of commercial farm operations and agricultural research operations are being trained. They should be taught by formal instruction and example to follow good safety practices, including some that might appear inconvenient at the time, such as wearing the right respiratory protection in contaminated environments. There are training courses taught for pesticide applicators, which stipulate that the applicator must wear a suitable "respirator" but fail to provide sufficient information on how to select a unit. Contaminated environments may not be immediately obvious. It is clear that when spraying operations are being conducted, respiratory protection is likely to be needed, but it is also likely to be needed when entering a treated field a day or two later, depending upon the rapidity of the biodegradation of the material used. Often, in agricultural research, experimental chemicals are used and the data to determine exposure problems may be incomplete or unavailable. An even more serious problem in encouraging field workers to wear personal protective equipment is the need to provide overall body protection for dispersal of pesticides and herbicides. This will be discussed further in a later section.

C. Full-Face Respirators

Full-face air-purifying respirators are similar in many respects to half-face respirators, with the obvious difference that the mask covers the upper part of the face, protecting the eyes. This has advantages and disadvantages. It is often easier to obtain a fit to the user than with a half-face unit. As a result, both the current ANSI Z88.2 standard and OSHA allow a higher protection factor (note that the present OSHA selection standards are based on the 1969 ANSI Z88.2 version instead of the current one), generally by a factor of 5 or less depending upon the contaminant. OSHA has announced that in the very near future from the time of this writing, it intends to accept the current standards for personal protection equipment (PPE).

A major difficulty with wearing a full-face respirator exists for persons who require prescription lenses for seeing. The temple piece extending back over the ears will interfere with the seal at those points. Some units are built to accommodate eyeglasses. In other cases, the wearer may decide to temporarily remove the temple pieces from the glasses and tape them to the bridge of the nose. This solution is acceptable for occasional, sporadic wearing but not for extended periods or use in which the facepiece is taken off and on frequently. Very recently "stick-on" prescription lenses have become available which could be used.

Another problem with a full-face respirator is fogging due to the warm, moist air exhaled in breathing. Full-face units are designed so that the incoming air flows across the lens of the unit. This feature, plus antifogging coatings on the lens, will normally prevent fogging at normal temperatures. However, at temperatures below freezing, fogging becomes an increasingly serious problem as the temperature decreases. Some full-face respirators include nose cups so that the warm air from the nose is directed through the exhalation valve and does not come into contact with the lens. These units should be able to go down to about - 32°C (- 25°F) and still allow adequate vision through the lens. At very low temperatures, the warm, moisture-laden air passing out through the exhalation valve may be a problem. The valve may stick open or closed because of ice, or the moisture may freeze and block the free flow of air through the valve.

The cost of full-face respirators is substantially more than the half-face units, typically by a factor of 4 or 5.

A variation of the full-face respirator is the PAAP unit discussed in the previous section. In one version, the only difference is that it supplies a full-face mask instead of a half-face mask. However, a useful variation is for the PAAP unit to supply air into the top of a hood which has a cape extending down to the shoulders. Some of these have a transparent section extending all the way around the upper part of the hood and provide an unusual degree of flexibility in vision and comfort, with a minimal loss in protection. This style accommodates both facial hair and glasses.

For other features of full-face respirators other than those covered here, the information on half-face units in Section B.2 of this chapter will apply.

D. Air-Supplied Respirators

An air-supplied respirator is intended to provide a source of breathing air to the user independent of the air in the surrounding space so that they can be used in oxygen-deficient atmospheres. However, depending upon the design they may or may not be approved for IDLH atmospheres.

There are two basic designs, one in which the supply of air is from a source outside the contaminated area, while in the other, the wearer of the respirator unit carries the air supply in a tank. There are also subdivisions within these two major types. A major subdivision, common to both, is whether the units operate as "demand" or "pressure demand" units. For the former, the demand valve permits the flow of air only during inhalation, and a negative pressure exists at that time within the facepiece, which may allow inward leakage from the contaminated atmosphere. The pressure-demand type maintains a positive pressure within the facepiece at all times and is unlikely to allow leakage of outside air into the respirator Pressure demand units are much more desirable and should be used in most applications. Demand units are not approved for IDLH atmospheres.

Air-supplied units are supplied with air through a hose from a source outside the contamination area, either from cylinders or air compressors, and must be of high purity. Cylinders may be used to supply oxygen instead of compressed air. Compressed air may possibly contain low concentrations of oil. Since contact of high-pressure oxygen with oil may result in a fire or explosion, it is not permissible to use oxygen with air-supplied units that have previously used compressed air. A compressor used to supply breathing air must be equipped with a high-temperature alarm or carbon-monoxide alarm, or both, if the compressor is lubricated with oil. The air provided by the compressor must be passed through an absorbent and filter to ensure that the air supplied is pure. A hose up to a maximum length of 300 feet long is permissible.

A major difficulty with the units supplied by air through a hose is the hose itself. The person wearing the unit is constrained to move only as permitted by the hose. The hose is subject to damage or kinking, and one must retrace one's steps when leaving the area. The major advantage is that the external source effectively provides an infinite supply of air, if an ample number of cylinders are available, or if the supply is from a compressor. An acceptable provision for escape from a contaminated area, should the pump fail or the hose fail or become constricted, would be an auxiliary tank of air to be carried by the user and connected to the respirator. This would also permit the wearer to disconnect from the supply hose and leave the contaminated area by perhaps a shorter and safer route in an emergency.

There are two different types of units in which the user carries his own air supply (in addition to the demand and pressure-demand versions). These self-contained breathing apparatus (SCBA) units have a basic limitation in that they provide only the limited amount of air that the user can carry. There are considerable differences in the useful life among different types of units within this class. Most units incorporate a full-facepiece, but other styles are available. Again, pressure-demand versions are the most desirable for most applications and, if equipped with escape provisions, are acceptable for IDLH atmospheres.

The basic type of SCBA unit is a tank of breathing air carried on the user's back and fed into the facepiece through appropriate regulators and valves. If the tank holds ordinary air a nominal "30 minute" tank may last only 15 minutes for a person engaged in strenuous activity. These units also suffer from being bulky and relatively heavy. Because of their limited life, often there is little time for productive labor while wearing one.

There are several commercial units which use pure compressed oxygen instead of compressed air. After reduction in pressure from that in the oxygen cylinder, the oxygen is used in a system in which the air exhaled by the wearer is passed through a chemical pack which removes the carbon dioxide from the exhaled air and returns the purified air to the system to supplement the oxygen from the portable canister. The system is often much smaller and lighter than the basic system incorporating an air tank and, by appropriate sizing of the oxygen tank and the air purifying chemical, can be designed to last a fixed amount of time. A typical system will last 1 hour, although there are systems which are designed to last considerably longer. One hour is usually long enough to allow a significant amount of productive work. Some persons do not like to use this type where a fire is involved because of the pure oxygen, but it would be difficult for the oxygen to come into contact with a flame so this should not be a major concern. The system is designed so that the "scrubber" chemical will outlast the oxygen supply so there is no danger that the purification process will cease too early (not recommended, but it is possible to remove a depleted oxygen cylinder and replace it while continuing work to gain additional time). A disadvantage of the scrubber units is that the purification action releases heat so that the air becomes warm and moist. This is a problem for some users.

Pressure-demand units which are self-contained and which use full-facepieces are approved for IDLH atmospheres.

Exposure to contaminants can damage respirator components, even after they have been "cleaned" and put in storage, due to permeation of chemicals into the materials of which they are made. An examination of the units should be made each time they are worn and a careful check made on a definite schedule. This is important for all types of units, but especially for those which are intended to be used in unusually hazardous applications. Records should be kept of all maintenance.

In addition to the following OSHA standard for general respirator protection usage, individuals working under circumstances where they could be exposed to tuberculosis, there is a specific respiratory protection standard applicable to them, 29 CFR 1910.139. The NRC also specifies respirator usage in 10 CFR Part 20.

E. OSHA 29 CFR 1910.134 (Slightly Abridged)*

(a)*Permissible practice.* (1) In the control of those occupational diseases caused by breathing air contaminated with harmful dusts, fogs, fumes, mists, gases, smokes, sprays, or vapors, the primary objective shall be to prevent atmospheric contamination. This shall be accomplished as far as feasible by accepted engineering control measures (for example, enclosure or confinement of the operation, general and local ventilation, and substitution of less toxic materials). When effective engineering controls are not feasible, or while they are being instituted, appropriate respirators shall be used pursuant to this section.

(2) Respirators shall be provided by the employer when such equipment is necessary to protect the health of the employee. The employer shall provide the respirators which are applicable and suitable for the purpose intended. The employer shall be responsible for the establishment and maintenance of a respiratory protection program which shall include the requirements outlined in paragraph (c) of this section.

(b) *Definitions.* The following definitions are important terms used in the respiratory protection standard in this section.

Air-purifying respirator means a respirator with an air-purifying filter cartridge, or canister that removes specific air contaminants by passing ambient air through the air-purifying element.

Atmosphere-supplying respirator means a respirator that supplies the respirator user with breathing air from a source independent of the ambient atmosphere, and includes supplied-air respirators (SARs) and self-contained breathing apparatus (SCBA) units.

Canister or cartridge means a container with a filter, sorbent, or catalyst, or combination of these items, which removes specific contaminants from the air passed through the container.

Demand respirator means an atmosphere-supplying respirator that admits breathing air to the facepiece only when a negative pressure is created inside the facepiece by inhalation.

Emergency situation means any occurrence such as, but not limited to, equipment failure, rupture of containers, or failure of control equipment that may or does result in an uncontrolled significant release of an airborne contaminant.

Employee exposure means exposure to a concentration of an airborne contaminant that would occur if the employee were not using respiratory protection.

End-of-service-life indicator (ESLI) means a system that warns the respirator user of the approach of the end of adequate respiratory protection, for example, that the sorbent is approaching saturation or is no longer effective.

Escape-only respirator means a respirator intended to be used only for emergency exit.

---

*   The formatting in this section will be that of the OSHA standard.

Filter or air purifying element means a component used in respirators to remove solid or liquid aerosols from the inspired air.

Filtering facepiece (dust mask) means a negative pressure particulate respirator with a filter as an integral part of the facepiece or with the entire face-piece composed of the filtering medium.

Fit factor means a quantitative estimate of the fit of a particular respirator to a specific individual, and typically estimates the ratio of the concentration of a substance in ambient air to its concentration inside the respirator when worn.

Fit test means the use of a protocol to qualitatively or quantitatively evaluate the fit of a respirator on an individual. (See also Qualitative fit test QLFT and Quantitative fit test QNFT.)

Helmet means a rigid respiratory inlet covering that also provides head protection against impact and penetration.

High efficiency particulate air (HEPA) filter means a filter that is at least 99.97% efficient in removing monodispersed particles of 0.3 micrometers (or greater) in diameter. The equivalent NIOSH 42 CFR 84 particulate filters are the N100, R100, and P100 filters.

Hood means a respiratory inlet covering that completely covers the head and neck and may also cover portions of the shoulders and torso.

Immediately dangerous to life or health (IDLH) means an atmosphere that poses an immediate threat to life, would cause irreversible adverse health effects, or would impair an individual's ability to escape from a dangerous atmosphere.

Loose-fitting facepiece means a respiratory inlet covering that is designed to form a partial seal with the face.

Negative pressure respirator (tight fitting) means a respirator in which the air pressure inside the facepiece is negative during inhalation with respect to the ambient air pressure outside the respirator.

Oxygen deficient atmosphere means an atmosphere with an oxygen content below 19.5% by volume.

Physician or other licensed health care professional (PLHCP) means an individual whose legally permitted scope of practice (*i.e..* license, registration, or certification) allows him or her to independently provide, or be delegated the responsibility to provide, some or all of the health care services required by paragraph (e) of this section.

Positive pressure respirator means a respirator in which the pressure inside the respiratory inlet covering exceeds the ambient air pressure outside the respirator.

Powered air-purifying respirator (PAPR) means an air-purifying respirator that uses a blower to force the ambient air through air-purifying elements to the inlet covering.

Pressure demand respirator means a positive pressure atmosphere-supplying respirator that admits breathing air to the facepiece when the positive pressure is reduced inside the facepiece by inhalation.

Qualitative fit test (QLFT) means a pass/fail fit test to assess the adequacy of respirator fit that relies on the individual's response to the test agent.

Quantitative fit test (QNFT) means an assessment of the adequacy of respirator fit by numerically measuring the amount of leakage into the respirator.

Respiratory inlet covering means that portion of a respirator that forms the protective barrier between the user's respiratory tract and an air-purifying device or breathing air source, or both. It may be a facepiece, helmet, hood, suit, or a mouthpiece respirator with nose clamp.

Self-contained breathing apparatus (SCBA) means an atmosphere-supplying respirator for which the breathing air source is designed to be carried by the user.

Service life means the period of time that a respirator, filter or sorbent, or other respiratory equipment provides adequate protection to the wearer.

Supplied-air respirator (SAR) or airline respirator means an atmosphere-supplying respirator for which the source of breathing air is not designed to be carried by the user.

Tight-fitting facepiece means a respiratory inlet covering that forms a complete seal with the face.

User seal check means an action conducted by the respirator user to determine If the respirator is properly seated to the face.

(c) *Respiratory protection program.* This paragraph requires the employer to develop and implement a written respiratory protection program with required worksite-specific procedures and elements for required respirator use. The program must be administered by a suitably trained program administrator. In addition, certain program elements may be required for voluntary use to prevent potential hazards associated with the use of the respirator. The Small Entity Compliance Guide contains criteria for the selection of a program administrator and a sample program that meets the requirements of this paragraph. Copies of the Small Entity Compliance Guide have been available since April 8, 1998 from the Occupational Safety and Health Administration's Office of Publications, Room N 3101, 200 Constitution Avenue, NW, Washington, D.C., 20210 (202-219-4667).

(1) In any workplace where respirators are necessary to protect the health of the employee or whenever respirators are required by the employer, the employer shall establish and implement a written respiratory protection program with worksite-specific procedures. The program shall be updated as necessary to reflect those changes in workplace conditions that affect respirator use. The employer shall include in the program the following provisions of this section, as applicable:

(i) Procedures for selecting respirators for use in the workplace;

(ii) Medical evaluations of employees required to use respirators;

(iii) Fit testing procedures for tight-fitting respirators;

(iv) Procedures for proper use of respirators in routine and reasonably foreseeable emergency situations;

(v) Procedures and schedules for cleaning, disinfecting, storing, inspecting, repairing, discarding. and otherwise maintaining respirators;

(vi) Procedures to ensure adequate air quality, quantity, and flow of breathing air for atmosphere-supplying respirators;

(vii) Training of employees in the respiratory hazards to which they are potentially exposed during routine and emergency situations;

(viii) Training of employees in the proper use of respirators, including putting on and removing them, any limitations on their use, and their maintenance; and

(ix) Procedures for regularly evaluating the effectiveness of the program.

(2) Where respirator use is not required:

(i) An employer may provide respirators at the request of employees or permit employees to use their own respirators, if the employer determines that such respirator use will not in itself create a hazard. If the employer determines that any voluntary respirator use is permissible, the employer shall provide the respirator users with the information contained in Appendix D to this section C "Information for Employees Using Respirators When Not Required Under the Standard;" and

(ii) In addition, the employer must establish and implement those elements of a written respiratory protection program necessary to ensure that any employee using a respirator voluntarily is medically able to use that respirator, and that the respirator is cleaned, stored, and maintained so that its use does not present a health hazard to the user. Exception: Employers are not required to include in a written respiratory protection program those employees whose only use of respirators involves the voluntary use of filtering facepieces (dust masks).

(3) The employer shall designate a program administrator who is qualified by appropriate training or experience that is commensurate with the complexity of the program to administer

or oversee the respiratory protection program and conduct the required evaluations of program effectiveness.

(4) The employer shall provide respirators, training, and medical evaluations at no cost to the employee.

(d) *Selection of respirators.* This paragraph requires the employer to evaluate respiratory hazard(s) in the workplace, identify relevant workplace and user factors, and base respirator selection on these factors. The paragraph also specifies appropriately protective respirators for use in IDLH atmospheres, and limits the selection and use of air-purifying respirators.

(1) *General requirements.* (i) The employer shall select and provide an appropriate respirator based on the respiratory hazard(s) to which the worker is exposed and workplace and user factors that affect respirator performance and reliability.

(ii) The employer shall select a NIOSH-certified respirator. The respirator shall be used in compliance with the conditions of its certification.

(iii) The employer shall identify and evaluate the respiratory hazard(s) in the workplace; this evaluation shall include a reasonable estimate of employee exposures to respiratory hazard(s) and an identification of the contaminant's chemical state and physical form. Where the employer cannot identify or reasonably estimate the employee exposure, the employer shall consider the atmosphere to be IDLH.

(iv) The employer shall select respirators from a sufficient number of respirator models and sizes so that the respirator is acceptable to, and correctly fits, the user.

(2) *Respirators for IDLH atmospheres.* (i) The employer shall provide the following respirators for employee use in IDLH atmospheres:

(A) A full facepiece pressure demand SCBA certified by NIOSH for a minimum service life of thirty minutes, or

(B) A combination full facepiece pressure demand supplied-air respirator (SAR) with auxiliary self-contained air supply.

(ii) Respirators provided only for escape from IDLH atmospheres shall be NIOSH-certified for escape from the atmosphere in which they will be used.

(iii) All oxygen-deficient atmospheres shall be considered IDLH. Exception: If the employer demonstrates that, under all foreseeable conditions, the oxygen concentration can be maintained within the ranges specified in Table 6.1 of this section (i.e., for the altitudes set out in the table), then any atmosphere-supplying respirator may be used.

(3) *Respirators for atmospheres that are not IDLH.* (i) The employer shall provide a respirator that is adequate to protect the health of the employee and ensure compliance with all other OSHA statutory and regulatory requirements, under routine and reasonably foreseeable emergency situations.

(ii) The respirator selected shall be appropriate for the chemical state and physical form of the contaminant.

(iii) For protection against gases and vapors, the employer shall provide: (A) An atmosphere-supplying respirator, or (B) An air-purifying respirator, provided that:

(1) The respirator is equipped with an end-of-service-life indicator (ESLI) certified by NIOSH for the contaminant; or

(2) If there is no ESLI appropriate for conditions in the employer's workplace, the employer implements a change schedule for canisters and cartridges that is based on objective information or data that will ensure that canisters and cartridges are changed before the end of their service life. The employer shall describe in the respirator program the information and data relied upon and the basis for the canister and cartridge change schedule and the basis for reliance on the data.

(iv) For protection against particulates, the employer shall provide:

(A) An atmosphere-supplying respirator; or

(B) An air-purifying respirator equipped with a filter certified by NIOSH under 30 CFR part

TABLE 6.1 Respirator Oxygen Limits vs. Altitude

Altitude(ft.)	Oxygen deficient atmospheres (% O_2) for which the employer may rely on atmosphere supplying respirators
Less than 3,001	16.0-19.5
3,001-4,000	16.4-19.5
4,001-5,000	17.1-19.5
5,001-6,000	17.8-19.5
6,001-7,000	18.5-19.5
7,001-8,000	19.3-19.5

Above 8,000 feet, the exception does not apply. Oxygen enriched breathing air must be supplied above 14,000 feet.

11 as a high efficiency particulate air (HEPA) filter, or an air-purifying respirator equipped with a filter certified for particulates by NIOSH under 42 CFR part 84; or

(C) For contaminants consisting primarily of particles with mass median aerodynamic diameters (MMAD) of at least 2 micrometers, an air-purifying respirator equipped with any filter certified for particulates by NIOSH.

(e) *Medical evaluation.* Using a respirator may place a physiological burden on employees that varies with the type of respirator worn, the job and workplace conditions in which the respirator is used, and the medical status of the employee. Accordingly, this paragraph specifies the minimum requirements for medical evaluation that employers must implement to determine the employee's ability to use a respirator.

(1) *General.* The employer shall provide a medical evaluation to determine the employee's ability to use a respirator before the employee is fit tested or required to use the respirator in the workplace. The employer may discontinue an employee's medical evaluations when the employee is no longer required to use a respirator.

(2) *Medical evaluation procedures.* (i) The employer shall identify a physician or other licensed health care professional (PLHCP) to perform medical evaluations using a medical questionnaire or an initial medical examination that obtains the same information as the medical questionnaire.

(ii) The medical evaluation shall obtain the information requested by the questionnaire in Sections 1 and 2, Part A of Appendix C of this section.

(3) *Follow-up medical examination.* (i) The employer shall ensure that a follow-up medical examination is provided for an employee who gives a positive response to any question among questions 1 through 8 In Section 2, Part A of Appendix C, or whose initial medical examination demonstrates the need for a follow-up medical examination.

(ii) The follow-up medical examination shall include any medical tests, consultations, or diagnostic procedures that the PLHCP deems necessary to make a final determination.

(4) Administration of the medical questionnaire and examinations: (1) The medical questionnaire and examinations shall be administered confidentially during the employee's normal working hours or at a time and place convenient to the employee. The medical questionnaire shall be administered in a manner that ensures that the employee understands its content.

(ii) The employer shall provide the employee with an opportunity to discuss the questionnaire and examination results with the PLHCP.

(5) *Supplemental information for the PLHCP.* (i) The following information must be provided to the PLHCP before the PLHCP makes a recommendation concerning an employee's ability to use a respirator:

(A) The type and weight of the respirator to be used by the employee;

(B) The duration and frequency of respirator use (including use for rescue and escape);

(C) The expected physical work effort;

(D) Additional protective clothing and equipment to be worn; and

(E) Temperature and humidity extremes that may be encountered.

(ii) Any supplemental information provided previously to the PLHCP regarding an employee need not be provided for a subsequent medical evaluation if the information and the PLHCP remain the same.

(iii) The employer shall provide the PLHCP with a copy of the written respiratory protection program and a copy of this section. NOTE TO PARAGRAPH (e)(5)(iii): When the employer replaces a PLHCP, the employer must ensure that the new PLHCP obtains this information, either by providing the documents directly to the PLHCP or having the documents transferred from the former PLHCP to the new PLHCP. However, OSHA does not expect employers to have employees medically reevaluated solely because a new PLHCP has been selected.

(6) *Medical determination.* In determining the employee's ability to use a respirator, the employer shall:

(i) Obtain a written recommendation regarding the employee's ability to use the respirator from the PLHCP. The recommendation shall provide only the following information:

(A) Any limitations on respirator use related to the medical condition of the employee, or relating to the workplace conditions in which the respirator will be used, including whether or not the employee is medically able to use the respirator;

(B) The need, if any, for follow-up medical evaluations; and

(C) A statement that the PLHCP has provided the employee with a copy of the PLHCP's written recommendation.

(ii) If the respirator is a negative pressure respirator and the PLHCP finds a medical condition that may place the employee's health at increased risk if the respirator is used, the employer shall provide a PAPR if the PLHCP's medical evaluation finds that the employee can use such a respirator; if a subsequent medical evaluation finds that the employee is medically able to use a negative pressure respirator, then the employer is no longer required to provide a PAPR.

(7) *Additional medical evaluations.* At a minimum, the employer shall provide additional medical evaluations that comply with the requirements of this section if:

(i) An employee reports medical signs or symptoms that are related to ability to use a respirator;

(ii) A PLHCP, supervisor, or the respirator program administrator informs the employer that an employee needs to be reevaluated;

(iii) Information from the respiratory protection program, including observations made during fit testing and program evaluation, indicates a need for employee reevaluation; or

(iv) A change occurs in workplace conditions (e.g., physical work effort, protective clothing, temperature) that may result in a substantial increase in the physiological burden placed on an employee.

(f) *Fit testing.* This paragraph requires that, before an employee may be required to use any respirator with a negative or positive pressure tight-fitting facepiece, the employee must be fit tested with the same make, model, style, and size of respirator that will be used. This paragraph specifies the kinds of fit tests allowed, the procedures for conducting them, and how the results of the fit tests must be used.

(1) The employer shall ensure that employees using a tight-fitting face-piece respirator pass

an appropriate qualitative fit test (QLFT) or quantitative fit test (QNFT) as stated in this paragraph.

(2) The employer shall ensure that an employee using a tight-fitting face-piece respirator is fit tested prior to initial use of the respirator, whenever a different respirator facepiece (size, style, model or make) is used, and at least annually thereafter.

(3) The employer shall conduct an additional fit test whenever the employee reports, or the employer, PLHCP, supervisor, or program administrator makes visual observations of, changes in the employee's physical condition that could affect respirator fit. Such conditions include, but are not limited to, facial scarring, dental changes, cosmetic surgery, or an obvious change in body weight.

(4) If after passing a QLFT or QNFT, the employee subsequently notifies the employer, program administrator, supervisor, or PLHCP that the fit of the respirator is unacceptable, the employee shall be given a reasonable opportunity to select a different respirator facepiece and to be retested.

(5) The fit test shall be administered using an OSHA-accepted QLFT or QNFT protocol. The OSHA-accepted QLFT and QNFT protocols and procedures are contained in Appendix A of this section.

(6) QLFT may only be used to fit test negative pressure air-purifying respirators that must achieve a fit factor of 100 or less.

(7) If the fit factor, as determined through an OSHA-accepted QNFT protocol, is equal to or greater than 100 for tight-fitting half facepleces, or equal to or greater than 500 for tight-fitting full facepieces, the QNFT has been passed with that respirator.

(8) Fit testing of tight-fitting atmosphere-supplying respirators and tight-fitting powered air-purifying respirators shall be accomplished by performing quantitative or qualitative fit testing in the negative pressure mode, regardless of the mode of operation (negative or positive pressure) that is used for respiratory protection.

(i) Qualitative fit testing of these respirators shall be accomplished by temporarily converting the respirator user's actual facepiece into a negative pressure respirator with appropriate filters, or by using an identical negative pressure air-purifying respirator facepiece with the same sealing surfaces as a surrogate for the atmosphere-supplying or powered air-purifying respirator facepiece.

(ii) Quantitative fit testing of these respirators shall be accomplished by modifying the facepiece to allow sampling inside the facepiece in the breathing zone of the user, midway between the nose and mouth. This requirement shall be accomplished by installing a permanent sampling probe onto a surrogate facepiece, or by using a sampling adapter designed to temporarily provide a means of sampling air from inside the facepiece.

(iii) Any modifications to the respirator facepiece for fit testing shall be completely removed, and the facepiece restored to NIOSH-approved configuration, before that facepiece can be used in the workplace.

(g) *Use of respirators.* This paragraph requires employers to establish and implement procedures for the proper use of respirators. These requirements include prohibiting conditions that may result in facepiece seal leakage, preventing employees from removing respirators in hazardous environments, taking actions to ensure continued effective respirator operation throughout the work shift, and establishing procedures for the use of respirators in IDLH atmospheres or in interior structural firefighting situations.

(1) *Facepiece seal protection.* (i) The employer shall not permit respirators with tight-fitting facepieces to be worn by employees who have:

(A) Facial hair that comes between the sealing surface of the facepiece and the face or that interferes with valve function; or

(B) Any condition that interferes with the face-to-facepiece seal or valve function.

(ii) If an employee wears corrective glasses or goggles or other personal protective equipment, the employer shall ensure that such equipment is worn in a manner that does not interfere with the seal of the facepiece to the face of the user.

(iii) For all tight-fitting respirators, the employer shall ensure that employees perform a user seal check each time they put on the respirator using the procedures in Appendix B-l or procedures recommended by the respirator manufacturer that the employer demonstrates are as effective as those in Appendix B-1 of this section.

(2) *Continuing respirator effectiveness.* (i) Appropriate surveillance shall be maintained of work area conditions and degree of employee exposure or stress. When there is a change in work area conditions or degree of employee exposure or stress that may affect respirator effectiveness, the employer shall reevaluate the continued effectiveness of the respirator.

(ii) The employer shall ensure that employees leave the respirator use area:

(A) To wash their faces and respirator facepieces as necessary to prevent eye or skin irritation associated with respirator use; or

(B) If they detect vapor or gas breakthrough, changes in breathing resistance, or leakage of the facepiece; or

(C) To replace the respirator or the filter, cartridge, or canister elements.

(iii) If the employee detects vapor or gas breakthrough, changes in breathing resistance, or leakage of the face-piece, the employer must replace or repair the respirator before allowing the employee to return to the work area.

(3) *Procedures for IDLH atmospheres.* For all IDLH atmospheres, the employer shall ensure that:

(i) One employee or, when needed, more than one employee is located outside the IDLH atmosphere;

(ii) Visual, voice, or signal line communication is maintained between the employee(s) in the IDLH atmosphere and the employee(s) located outside the IDLH atmosphere;

(iii) The employee(s) located outside the IDLH atmosphere are trained and equipped to provide effective emergency rescue;

(iv) The employer or designee is notified before the employee(s) located outside the IDLH atmosphere enter the IDLH atmosphere to provide emergency rescue;

(v) The employer or designee authorized to do so by the employer, once notified, provides necessary assistance appropriate to the situation;

(vi) Employee(s) located outside the IDLH atmospheres are equipped with:

(A) Pressure demand or other positive pressure SCBAs, or a pressure demand or other positive pressure supplied-air respirator with auxiliary SCBA; and either

(B) Appropriate retrieval equipment for removing the employee(s) who enter(s) these hazardous atmospheres where retrieval equipment would contribute to the rescue of the employee(s) and would not increase the overall risk resulting from entry; or

(C) Equivalent means for rescue where retrieval equipment is not required under paragraph (g) (3) (vi) (B).

(h) *Maintenance and care of respirators.* This paragraph requires the employer to provide for the cleaning and disinfecting, storage, inspection, and repair of respirators used by employees.

(1) *Cleaning and disinfecting.* The employer shall provide each respirator user with a respirator that is clean, sanitary, and in good working order. The employer shall ensure that respirators are cleaned and disinfected using the procedures in Appendix B-2 of this section, or procedures recommended by the respirator manufacturer, provided that such procedures are of equivalent effectiveness. The respirators shall be cleaned and disinfected at the following intervals:

(i) Respirators issued for the exclusive use of an employee shall be cleaned and disinfected as often as necessary to be maintained in a sanitary condition;

(ii) Respirators issued to more than one employee shall be cleaned and disinfected before being worn by different individuals;

(iii) Respirators maintained for emergency use shall be cleaned and disinfected after each use; and

(iv) Respirators used in fit testing and training shall be cleaned and disinfected after each use.

(2) *Storage.* The employer shall ensure that respirators are stored as follows:

(i) All respirators shall be stored to protect them from damage, contamination, dust, sunlight, extreme temperatures, excessive moisture, and damaging chemicals, and they shall be packed or stored to prevent deformation of the facepiece and exhalation valve.

(ii) In addition to the requirements of paragraph (h)(2)(i) of this section, emergency respirators shall be:

(A) Kept accessible to the work area;

(B) Stored in compartments or in covers that are clearly marked as containing emergency respirators; and

(C) Stored in accordance with any applicable manufacturer instructions.

(3) *Inspection.* (i) The employer shall ensure that respirators are inspected as follows:

(A) All respirators used in routine situations shall be inspected before each use and during cleaning;

(B) All respirators maintained for use in emergency situations shall be inspected at least monthly and in accordance with the manufacturer's recommendations, and shall be checked for proper function before and after each use; and

(C) Emergency escape-only respirators shall be inspected before being carried into the workplace for use.

(ii) The employer shall ensure that respirator inspections include the following:

(A) A check of respirator function, tightness of connections, and the condition of the various parts including, but not limited to, the facepiece, head straps, valves, connecting tube, and cartridges, canisters or filters; and

(B) A check of elastomeric parts for pliability and signs of deterioration.

(iii) In addition to the requirements of paragraphs (h)(3)(i) and (ii) of this section, self-contained breathing apparatus shall be inspected monthly. Air and oxygen cylinders shall be maintained in a fully charged state and shall be recharged when the pressure falls to 90% of the manufacturer's recommended pressure level. The employer shall determine that the regulator and warning devices function properly.

(iv) For respirators maintained for emergency use, the employer shall:

(A) Certify the respirator by documenting the date the inspection was performed, the name (or signature) of the person who made the inspection, the findings, required remedial action, and a serial number or other means of identifying the inspected respirator; and

(B) Provide this information on a tag or label that is attached to the storage compartment for the respirator, is kept with the respirator, or is included in inspection reports stored as paper or electronic files. This information shall be maintained until replaced following a subsequent certification.

(4) *Repairs.* The employer shall ensure that respirators that fail an inspection or are otherwise found to be defective are removed from service, and are discarded or repaired or adjusted in accordance with the following procedures:

(i) Repairs or adjustments to respirators are to be made only by persons appropriately trained to perform such operations and shall use only the respirator manufacturer's NIOSH-approved parts designed for the respirator;

(ii) Repairs shall be made according to the manufacturer's recommendations and specifications for the type and extent of repairs to be performed; and

(iii) Reducing and admission valves, regulators, and alarms shall be adjusted or repaired only by the manufacturer or a technician trained by the manufacturer.

(i) *Breathing air quality and use.* This paragraph requires the employer to provide employees using atmosphere-supplying respirators (supplied-air and SCBA) with breathing gases of high purity.

(1) The employer shall ensure that compressed air, compressed oxygen, liquid air, and liquid oxygen used for respiration accords with the following specifications:

(i) Compressed and liquid oxygen shall meet the United States Pharmacopoeia requirements for medical or breathing oxygen; and

(ii) Compressed breathing air shall meet at least the requirements for Grade D breathing air described in ANSI/Compressed Gas Association Commodity Specification for Air. G7.1-1989, to include:

(A) Oxygen content (volume) of 19.5 to23.5%;

(B) Hydrocarbon (condensed) content of 5 milligrams per cubic meter of air or less;

(C) Carbon monoxide (CO) content of 10 ppm or less;

(D) Carbon dioxide content of 1,000 ppm or less; and

(E) Lack of noticeable odor.

(2) The employer shall ensure that compressed oxygen is not used in atmosphere-supplying respirators that have previously used compressed air.

(3) The employer shall ensure that oxygen concentrations greater than 23.5% are used only in equipment designed for oxygen service or distribution.

(4) The employer shall ensure that cylinders used to supply breathing air to respirators meet the following requirements:

(i) Cylinders are tested and maintained as prescribed in the Shipping Container Specification Regulations of the Department of Transportation (49 CFR part 173 and part 178);

(ii) Cylinders of purchased breathing air have a certificate of analysis from the supplier that the breathing air meets the requirements for Grade D breathing air; and

(iii) The moisture content in the cylinder does not exceed a dew point of - 50°F (- 45.6°C) at 1 atmosphere pressure.

(5) The employer shall ensure that compressors used to supply breathing air to respirators are constructed and situated so as to:

(i) Prevent entry of contaminated air into the air-supply system;

(ii) Minimize moisture content so that the dew point at 1 atmosphere pressure is 10 degrees F (5.56°C) below the ambient temperature;

(iii) Have suitable in-line air-purifying sorbent beds and filters to further ensure breathing air quality. Sorbent beds and filters shall be maintained and replaced or refurbished periodically following the manufacturer's instructions.

(iv) Have a tag containing the most recent change date and the signature of the person authorized by the employer to perform the change. The tag shall be maintained at the compressor.

(6) For compressors that are not oil lubricated, the employer shall ensure that carbon monoxide levels in the breathing air do not exceed 10 ppm.

(7) For oil-lubricated compressors, the employer shall use a high-temperature or carbon monoxide alarm, or both, to monitor carbon monoxide levels. If only high-temperature alarms are used, the air supply shall be monitored at intervals sufficient to prevent carbon monoxide in the breathing air from exceeding 10 ppm.

(8) The employer shall ensure that breathing air couplings are incompatible with outlets for nonrespirable work site air or other gas systems. No asphyxiating substance shall be introduced into breathing air lines.

(9) The employer shall use breathing gas containers marked in accordance with the NIOSH respirator certification standard, 42 CFR part 84.

(j) *Identification of filters, cartridges, and canisters.* The employer shall ensure that all filters, cartridges, and canisters used in the workplace are labeled and color coded with the NIOSH approval label and that the label is not removed and remains legible.

(k) *Training and information.* This paragraph requires the employer to provide effective training to employees who are required to use respirators. The training must be comprehensive, understandable, and recur annually, and more often if necessary. This paragraph also requires the employer to provide the basic information on respirators in Appendix D of this section to employees who wear respirators when not required by this section or by the employer to do so.

(1) The employer shall ensure that each employee can demonstrate knowledge of at least the following:

(i) Why the respirator is necessary and how improper fit, usage, or maintenance can compromise the protective effect of the respirator;

(ii) What the limitations and capabilities of the respirator are;

(iii) How to use the respirator effectively in emergency situations, including situations in which the respirator malfunctions;

(iv) How to inspect, put on and remove, use, and check the seals of the respirator;

(v) What the procedures are for maintenance and storage of the respirator;

(vi) How to recognize medical signs and symptoms that may limit or prevent the effective use of respirators; and

(vii) The general requirements of this section.

(2) The training shall be conducted in a manner that is understandable to the employee.

(3) The employer shall provide the training prior to requiring the employee to use a respirator in the workplace.

(4) An employer who is able to demonstrate that a new employee has received training within the last 12 months that addresses the elements specified in paragraph (k)(1)(i) through (vii) is not required to repeat such training provided that, as required by paragraph (k)(1), the employee can demonstrate knowledge of those element(s). Previous training not repeated initially by the employer must be provided no later than 12 months from the date of the previous training.

(5) Retraining shall be administered annually, and when the following situations occur:

(i) Changes in the workplace or the type of respirator render previous training obsolete;

(ii) Inadequacies in the employee's knowledge or use of the respirator indicate that the employee has not retained the requisite understanding or skill; or

(iii) Any other situation arises in which retraining appears necessary to ensure safe respirator use.

(6) The basic advisory information on respirators, as presented in Appendix D of this section, shall be provided by the employer in any written or oral format, to employees who wear respirators when such use is not required by this section or by the employer.

(l) *Program evaluation.* This section requires the employer to conduct evaluations of the workplace to ensure that the written respiratory protection program is being properly implemented, and to consult employees to ensure that they are using the respirators properly.

(1) The employer shall conduct evaluations of the workplace as necessary to ensure that the provisions of the current written program are being effectively implemented and that it continues to be effective.

(2) The employer shall regularly consult employees required to use respirators to assess the employees' views on program effectiveness and to identify any problems. Any problems that are identified during this assessment shall be corrected. Factors to be assessed include, but are not limited to:

(i) Respirator fit (including the ability to use the respirator without interfering with effective workplace performance);

(ii) Appropriate respirator selection for the hazards to which the employee is exposed;

(iii) Proper respirator use under the workplace conditions the employee encounters; and

(iv) Proper respirator maintenance.

(m) *Record keeping.* This section requires the employer to establish and retain written information regarding medical evaluations, fit testing, and the respirator program. This information will facilitate employee involvement in the respirator program, assist the employer in auditing the adequacy of the program, and provide a record for compliance determinations by OSHA.

(1) *Medical evaluation.* Records of medical evaluations required by this section must be retained and made available in accordance with 29 CFR 1910.1020.

(2) *Fit testing.* (i) The employer shall establish a record of the qualitative and quantitative fit tests administered to an employee including:

(A) The name or identification of the employee tested;

(B) Type of fit test performed;

(C) Specific make, model, style, and size of respirator tested;

(D) Date of test; and

(E) The pass/fall results for QLFTs or the fit factor and strip chart recording or other recording of the test results for QNFTs.

(ii) Fit test records shall be retained for respirator users until the next fit test is administered.

(3) A written copy of the current respirator program shall be retained by the employer.

(4) Written materials required to be retained under this paragraph shall be made available upon request to affected employees and to the Assistant Secretary or designee for examination and copying.

(n) *Dates-* (l) Effective date. This section became effective April 8, 1998. The obligations imposed by this section commenced on the effective date unless otherwise noted in this paragraph. Compliance with obligations that did not commence on the effective date shall occur no later than the applicable start-up date.

(2) *Compliance dates.* All obligations of this section commenced on the effective date except as follows:

(i) The determination that respirator use is required [(paragraph (a)] shall have been completed no later than September 8, 1998.

(ii) Compliance with provisions of this section for all other provisions shall have been completed no later than October 5, 1998.

(3) The provisions of 29 CFR 1910.134 and 29 CFR 1926.103, contained in the 29 CFR parts 1900 to 1910.99 and the 29 CFR part 1926 editions, revised as of July 1, 1997, were in effect and enforceable until October 5, 1998, or during any administrative or judicial stay of the provisions of this section.

(4) *Existing respiratory protection programs.* If, in the 12 month period preceding April 8, 1998, the employer conducted annual respirator training, fit testing, respirator program evaluation, or medical evaluations, the employer may use the results of those activities to comply with the corresponding provisions of this section, providing that these activities were conducted in a manner that meets the requirements of this section.

(o) *Appendices.* (I) Compliance with Appendix A, Appendix B-1, Appendix B-2, and Appendix C of this section is mandatory.

(2) Appendix D of this section is non-mandatory and is not intended to create any additional obligations not otherwise imposed or to detract from any existing obligations.

Appendix A to §1910.134: Fit Testing Procedures (Mandatory)

Part I. OSHA-Accepted Fit Test Protocols

A. Fit Testing Procedures—General Requirements

The employer shall conduct fit testing using the following procedures. The requirements in this appendix apply to all OSHA-accepted fit test methods, both QLFT and QNFT.

1. The test subject shall be allowed to pick the most acceptable respirator from a sufficient number of respirator models and sizes so that the respirator is acceptable to and correctly fits the user.

2. Prior to the selection process, the test subject shall be shown how to put on a respirator, how it should be positioned on the face, how to set strap tension, and how to determine an acceptable fit. A mirror shall be available to assist the subject in evaluating the fit and positioning of the respirator. This instruction may not constitute the subject's formal training on respirator use because it is only a review.

3. The test subject shall be informed that he/she is being asked to select the respirator that provides the most acceptable fit. Each respirator represents a different size and shape, and if fitted and used properly, will provide adequate protection.

4. The test subject shall be instructed to hold each chosen facepiece up to the face and eliminate those that obviously do not give an acceptable fit.

5. The more acceptable facepieces are noted in case the one selected proves unacceptable; the most comfortable mask is donned and worn at least five minutes to assess comfort. Assistance in assessing comfort can be given by discussing the points in the following item A.6. If the test subject is not familiar with using a particular respirator, the test subject shall be directed to don the mask several times and to adjust the straps each time to become adept at setting proper tension on the straps.

6. Assessment of comfort shall include a review of the following points with the test subject and allowing the test subject adequate time to determine the comfort of the respirator:

(a) Position of the mask on the nose

(b) Room for eye protection

(c) Room to talk

(d) Position of mask on face and cheeks

7. The following criteria shall be used to help determine the adequacy of the respirator fit:

(a) Chin properly placed

(b) Adequate strap tension, not overly tightened

(c) Fit across nose bridge

(d) Respirator of proper size to span distance from nose to chin

(e) Tendency of respirator to slip

(f) Self-observation in mirror to evaluate fit and respirator position

8. The test subject shall conduct a user seal check, either the negative and positive pressure seal checks described in Appendix B-1 of this section or those recommended by the respirator manufacturer which provide equivalent protection to the procedures in Appendix B-1. Before conducting the negative and positive pressure checks, the subject shall be told to seat the mask on the face by moving the head from side-to-side and up and down slowly while taking in a few slow deep breaths. Another facepiece shall be selected and retested if the test subject fails the user seal check tests.

9. The test shall not be conducted if there is any hair growth between the skin and the facepiece sealing surface, such as stubble beard growth, beard, mustache, or sideburns which cross the respirator sealing surface. Any type of apparel that interferes with a satisfactory fit shall be altered or removed.

10. If a test subject exhibits difficulty in breathing during the tests, she or he shall be referred to a physician or other licensed health care professional, as appropriate, to determine whether the test subject can wear a respirator while performing her or his duties.

11. If the employee finds the fit of the respirator unacceptable, the test subject shall be given the opportunity to select a different respirator and to be retested.

12. Exercise regimen. Prior to the commencement of the fit test, the test subject shall be given

a description of the fit test and the test subject's responsibilities during the test procedure. The description of the process shall include a description of the test exercises that the subject will be performing. The respirator to be tested shall be worn for at least 5 minutes before the start of the fit test.

13. The fit test shall be performed while the test subject is wearing any applicable safety equipment that may be worn during actual respirator use which could interfere with respirator fit.

14. Test Exercises. (a) The following test exercises are to be performed for all fit testing methods prescribed in this appendix, except for the CNP method. A separate fit testing exercise regimen is contained in the CNP protocol. The test subject shall perform exercises. in the test environment, in the following manner:

(l) Normal breathing. In a normal standing position, without talking, the subject shall breathe normally.

(2) Deep breathing. In a normal standing position, the subject shall breathe slowly and deeply, taking caution so as not to hyperventilate.

(3) Turning head side to side. Standing in place, the subject shall slowly turn his/her head from side to side between the extreme positions on each side. The head shall be held at each extreme momentarily so the subject can inhale at each side.

(4) Moving head up and down. Standing in place, the subject shall slowly move his/her head up and down. The subject shall be instructed to inhale in the up position (i.e., when looking toward the ceiling).

(5) Talking. The subject shall talk slowly and loud enough so as to be heard clearly by the test conductor. The subject can read from a prepared text such as the Rainbow Passage, count backward from 100, or recite a memorized poem or song.

Rainbow Passage

When the sunlight strikes raindrops in the air, they act like a prism and form a rainbow. The rainbow is a division of white light into many beautiful colors. These take the shape of a long round arch, with its path high above, and its two ends apparently beyond the horizon. There is, according to legend, a boiling pot of gold at one end. People look, but no one ever finds it. When a man looks for something beyond reach, his friends say he is looking for the pot of gold at the end of the rainbow.

(6) Grimace. The test subject shall grimace by smiling or frowning. (This applies only to QNFT testing; it is not performed for QLFT.)

(7) Bending over. The test subject shall bend at the waist as if he/she were to touch his/her toes. Jogging in place shall be substituted for this exercise in those test environments such as shroud type QNFT or QLFT units that do not permit bending over at the waist.

(8) Normal breathing. Same as exercise (1).

(b) Each test exercise shall be performed for one minute except for the grimace exercise which shall be performed for 15 seconds. The test subject shall be questioned by the test conductor regarding the comfort of the respirator upon completion of the protocol. If it has become unacceptable, another model of respirator shall be tried. The respirator shall not be adjusted once the fit test exercises begin. Any adjustment voids the test, and the fit test must be repeated.

B. Qualitative Fit Test (QLFT) Protocols
1. General

(a) The employer shall ensure that persons administering QLFT are able to prepare test solutions, calibrate equipment, and perform tests properly, recognize invalid tests, and ensure that test equipment is in proper working order.

(b) The employer shall ensure that QLFT equipment is kept clean and well maintained so as to operate within the parameters for which it was designed.

2. Isoamyl Acetate Protocol

NOTE: This protocol is not appropriate to use for the fit testing of particulate respirators. If used to fit test particulate respirators, the respirator must be equipped with an organic vapor filter.

(a) Odor Threshold Screening

Odor threshold screening, performed without wearing a respirator, is intended to determine if the individual tested can detect the odor of isoamyl acetate at low levels.

(1) Three 1 liter glass jars with metal lids are required.

(2) Odor-free water (e.g., distilled or spring water) at approximately 25°C (77°F) shall be used for the solutions.

(3) The isoamyl acetate (IAA) (also known at isopentyl acetate) stock solution is prepared by adding 1 ml of pure IAA to 800 ml of odor-free water in a 1 liter jar, closing the lid, and shaking for 30 seconds. A new solution shall be prepared at least weekly.

(4) The screening test shall be conducted in a room separate from the room used for actual fit testing. The two rooms shall be well ventilated to prevent the odor of IAA from becoming evident in the general room air where testing takes place.

(5) The odor test solution is prepared in a second jar by placing 0.4 ml of the stock solution into 500 ml of odor-free water using a clean dropper or pipette. The solution shall be shaken for 30 seconds and allowed to stand for two to three minutes so that the IAA concentration above the liquid may reach equilibrium. This solution shall be used for only one day.

(6) A test blank shall be prepared in a third jar by adding 500 cc of odor-free water.

(7) The odor test and test blank jar lids shall be labeled (e.g., 1 and 2) for jar identification. Labels shall be placed on the lids so that they can be periodically peeled off and switched to maintain the integrity of the test.

(8) The following instruction shall be typed on a card and placed on the table in front of the two test jars (i.e., 1 and 2): "The purpose of this test is to determine if you can smell banana oil at a low concentration. The two bottles in front of you contain water. One of these bottles also contains a small amount of banana oil. Be sure the covers are on tight, then shake each bottle for two seconds. Unscrew the lid of each bottle, one at a time, and sniff at the mouth of the bottle. Indicate to the test conductor which bottle contains banana oil."

(9) The mixtures used in the IAA odor detection test shall be prepared in an area separate from where the test is performed, in order to prevent olfactory fatigue in the subject.

(10) If the test subject is unable to correctly identify the jar containing the odor test solution, the IAA qualitative fit test shall not be performed.

(11) If the test subject correctly identifies the jar containing the odor test solution, the test subject may proceed to respirator selection and fit testing.

(b) Isoamyl Acetate Fit Test

(1) The fit test chamber shall be a clear 55-gallon drum liner suspended inverted over a 2-foot diameter frame so that the top of the chamber is about 6 inches above the test subject's head. If no drum liner is available, a similar chamber shall be constructed using plastic sheeting. The inside top center of the chamber shall have a small hook attached.

(2) Each respirator used for the fitting and fit testing shall be equipped with organic vapor cartridges or offer protection against organic vapors.

(3) After selecting, donning, and properly adjusting a respirator, the test subject shall wear it to the fit testing room. This room shall be separate from the room used for odor threshold screening and respirator selection, and shall be well-ventilated, as by an exhaust fan or lab hood, to prevent general room contamination.

(4) A copy of the test exercises and any prepared text from which the subject is to read shall

be taped to the inside of the test chamber.

(5) Upon entering the test chamber, the test subject shall be given a 6-inch by 5-inch piece of paper towel, or other porous, absorbent, single-ply material, folded in half and wetted with 0.75 ml of pure IAA. The test subject shall hang the wet towel on the hook at the top of the chamber. An IAA test swab or ampule may be substituted for the IAA wetted paper towel provided it has been demonstrated that the alternative IAA source will generate an IAA test atmosphere with a concentration equivalent to that generated by the paper towel method.

(6) Allow two minutes for the IAA test concentration to stabilize before starting the fit test exercises. This would be an appropriate time to talk with the test subject; to explain the fit test, the importance of his/her cooperation, and the purpose for the test exercises; or to demonstrate some of the exercises.

(7) If at any time during the test, the subject detects the banana-like odor of IAA, the test is failed. The subject shall quickly exit from the test chamber and leave the test area to avoid olfactory fatigue.

(8) If the test is failed, the subject shall return to the selection room and remove the respirator. The test subject shall repeat the odor sensitivity test, select and put on another respirator, return to the test area and again begin the fit test procedure described in (b) (1) through (7) above. The process continues until a respirator that fits well has been found. Should the odor sensitivity test be failed, the subject shall wait at least 5 minutes before retesting. Odor sensitivity will usually have returned by this time.

(9) If the subject passes the test, the efficiency of the test procedure shall be demonstrated by having the subject break the respirator face seal and take a breath before exiting the chamber.

(10) When the test subject leaves the chamber, the subject shall remove the saturated towel and return it to the person conducting the test, so that there is no significant IAA concentration buildup in the chamber during subsequent tests. The used towels shall be kept in a self-sealing plastic bag to keep the test area from being contaminated.

3. Saccharin Solution Aerosol Protocol

The entire screening and testing procedure shall be explained to the test subject prior to the conduction of the screening test.

(a) Taste threshold screening. The saccharin taste threshold screening, performed without wearing a respirator, is intended to determine whether the individual being tested can detect the taste of saccharin.

(1) During threshold screening as well as during fit testing, subjects shall wear an enclo-sure about the head and shoulders that is approximately 12 inches in diameter by 14 inches tall with at least the front portion clear and that allows free movements of the head when a respirator is worn. An enclosure substantially similar to the 3M hood assembly, parts # FT 14 and # FT 15 combined, is adequate.

(2) The test enclosure shall have a 0.75-inch (1.9 cm) hole in front of the test subject's nose and mouth area to accommodate the nebulizer nozzle.

(3) The test subject shall don the test enclosure. Throughout the threshold screening test, the test subject shall breathe through his/her slightly open mouth with tongue extended. The subject is instructed to report when he/she detects a sweet taste.

(4) Using a DeVilbiss Model 40 Inhalation Medication Nebulizer or equivalent, the test conductor shall spray the threshold check solution into the enclosure. The nozzle is directed away from the nose and mouth of the person. This nebulizer shall be clearly marked to distinguish it from the fit test solution nebulizer.

(5) The threshold check solution is prepared by dissolving 0.83 grams of sodium saccharin USP in 100 ml of warm water. It can be prepared by putting 1 ml of the fit test solution (see (b)(5) below) in 100 ml of distilled water.

(6) To produce the aerosol, the nebulizer bulb is firmly squeezed so that it collapses completely, then released and allowed to fully expand.

(7) Ten squeezes are repeated rapidly and then the test subject is asked whether the saccharin

can be tasted. If the test subject reports tasting the sweet taste during the ten squeezes, the screening test is completed. The taste threshold is noted as ten regardless of the number of squeezes actually completed.

(8) If the first response is negative, ten more squeezes are repeated rapidly and the test subject is again asked whether the saccharin is tasted. If the test subject reports tasting the sweet taste during the second ten squeezes, the screening test is completed. The taste threshold is noted as twenty regardless of the number of squeezes actually completed.

(9) If the second response is negative, ten more squeezes are repeated rapidly and the test subject is again asked whether the saccharin is tasted. If the test subject reports tasting the sweet taste during the third set of ten squeezes, the screening test is completed. The taste threshold is noted as thirty regardless of the number of squeezes actually completed.

(10) The test conductor will take note of the number of squeezes required to solicit a taste response.

(11) If the saccharin is not tasted after 30 squeezes (step 10), the test subject is unable to taste saccharin and may not perform the saccharin fit test.

NOTE TO PARAGRAPH 3. (a): If the test subject eats or drinks something sweet before the screening test, he/she may be unable to taste the weak saccharin solution.

(12) If a taste response is elicited, the test subject shall be asked to take note of the taste for reference in the fit test.

(13) Correct use of the nebulizer means that approximately 1 ml of liquid is used at a time in the nebulizer body.

(14) The nebulizer shall be thoroughly rinsed in water, shaken dry, and refilled each morning and afternoon or at least every four hours.

(b) Saccharin solution aerosol fit test procedure.

(1) The test subject may not eat, drink (except plain water), smoke, or chew gum for 15 minutes before the test.

(2) The fit test uses the same enclosure described in 3. (a) above.

(3) The test subject shall don the enclosure while wearing the respirator selected in Section I.A. of this appendix. The respirator shall be properly adjusted and equipped with a particulate filter(s).

(4) A second DeVilbiss Model 40 Inhalation Medication Nebulizer or equivalent is used to spray the fit test solution into the enclosure. This nebulizer shall be clearly marked to distinguish it from the screening test solution nebulizer.

(5) The fit test solution is prepared by adding 83 grams of sodium saccharin to 100 ml of warm water.

(6) As before, the test subject shall breathe through the slightly open mouth with tongue extended, and report if he/she tastes the sweet taste of saccharin.

(7) The nebulizer is inserted into the hole in the front of the enclosure and an initial concentration of saccharin fit test solution is sprayed into the enclosure using the same number of squeezes (either 10, 20, or 30 squeezes) based on the number of squeezes required to elicit a taste response as noted during the screening test. A minimum of 10 squeezes is required.

(8) After generating the aerosol, the test subject shall be instructed to perform the exercises in section I.A.14. of this appendix.

(9) Every 30 seconds, the aerosol concentration shall be replenished using one half the original number of squeezes used initially (e.g., 5, 10, or 15).

(10) The test subject shall indicate to the test conductor if at any time during the fit test the taste of saccharin is detected. If the test subject does not report tasting the saccharin, the test is passed.

(11) If the taste of saccharin is detected, the fit is deemed unsatisfactory and the test is failed.

A different respirator shall be tried and the entire test procedure is repeated (taste threshold screening and fit testing).

(12) Since the nebulizer has a tendency to clog during use, the test operator must make periodic checks of the nebulizer to ensure that it is not clogged. If clogging is found at the end of the test session, the test is invalid.

4. Bitrex™ (Denatonium Benzoate) Solution Aerosol Qualitative Fit Test Protocol

The Bitrex™ (Denatonium benzoate) solution aerosol QLFT protocol uses the published saccharin test protocol because that protocol is widely accepted. Bitrex is routinely used as a taste aversion agent in household liquids which children should not be drinking and is endorsed by the American Medical Association, the National Safety Council, and the American Association of Poison Control Centers. The entire screening and testing procedure shall be explained to the test subject prior to the conduct of the screening test.

(a) Taste Threshold Screening.

The Bitrex™ taste threshold screening, performed without wearing a respirator, is intended to determine whether the individual being tested can detect the taste of Bitrex.™

(1) During threshold screening as well as during fit testing, subjects shall wear an enclosure about the head and shoulders that is approximately 12 inches (30.5 cm) in diameter by 14 inches (35.6 cm) tall. The front portion of the enclosure shall be clear from the respirator and allow free movement of the head when a respirator is worn. An enclosure substantially similar to the 3M hood assembly, parts # FT 14 and # FT 15 combined, is adequate.

(2) The test enclosure shall have a ¾ inch (1.9 cm) hole in front of the test subject's nose and mouth area to accommodate the nebulizer nozzle.

(3) The test subject shall don the test enclosure. Throughout the threshold screening test, the test subject shall breathe through his or her slightly open mouth with tongue extended. The subject is instructed to report when he/she detects a bitter taste.

(4) Using a DeVilbiss Model 40 Inhalation Medication Nebulizer or equivalent, the test conductor shall spray the Threshold Check Solution into the enclosure. This Nebulizer shall be clearly marked to distinguish it from the fit test solution nebulizer.

(5) The Threshold Check Solution is prepared by adding 13.5 milligrams of Bitrex™ to 100 ml of 5% salt (NaCl) solution in distilled water.

(6) To produce the aerosol, the nebulizer bulb is firmly squeezed so that the bulb collapses completely, and is then released and allowed to fully expand.

(7) An initial ten squeezes are repeated rapidly and then the test subject is asked whether the Bitrex™ can be tasted. If the test subject reports tasting the bitter taste during the ten squeezes, the screening test is completed. The taste threshold is noted as ten regardless of the number of squeezes actually completed.

(8) If the first response is negative, ten more squeezes are repeated rapidly and the test subject is again asked whether the Bitrex™ is tasted. If the test subject reports tasting the bitter taste during the second ten squeezes, the screening test is completed. The taste threshold is noted as twenty regardless of the number of squeezes actually completed.

(9) If the second response is negative, ten more squeezes are repeated rapidly and the test subject is again asked whether the Bitrex™ is tasted. If the test subject reports tasting the bitter taste during the third set of ten squeezes, the screening test is completed. The taste threshold is noted as thirty regardless of the number of squeezes actually completed.

(10) The test conductor will take note of the number of squeezes required to solicit a taste response.

(11) If the Bitrex™ is not tasted after 30 squeezes (step 10), the test subject is unable to taste Bitrex™ and may not perform the Bitrex™ fit test.

(12) If a taste response is elicited, the test subject shall be asked to take note of the taste for reference in the fit test.

(13) Correct use of the nebulizer means that approximately 1 ml of liquid is used at a time in

the nebulizer body.

(14) The nebulizer shall be thoroughly rinsed in water, shaken to dry, and refilled each morning and afternoon or at least every four hours.

(b) Bitrex™ Solution Aerosol Fit Test Procedure.

(1) The test subject may not eat, drink (except plain water), smoke, or chew gum for 15 minutes before the test.

(2) The fit test uses the same enclosure as that described in 4. (a) above.

(3) The test subject shall don the enclosure while wearing the respirator selected according to section 1.A. of this appendix. The respirator shall be properly adjusted and equipped with any type particulate filter(s).

(4) A second DeVilbiss Model 40 Inhalation Medication Nebulizer or equivalent is used to spray the fit test solution into the enclosure. This nebulizer shall be clearly marked to distinguish it from the screening test solution nebulizer.

(5) The fit test solution is prepared by adding 337.5 mg of Bitrex™ to 200 ml of a 5% salt (NaCl) solution in warm water.

(6) As before, the test subject shall breathe through his or her slightly open mouth with tongue extended, and be instructed to report if he/she tastes the bitter taste of Bitrex™

(7) The nebulizer is inserted into the hole in the front of the enclosure and an initial concentration of the fit test solution is sprayed into the enclosure using the same number of squeezes (either 10, 20, or 30 squeezes) based on the number of squeezes required to elicit a taste response as noted during the screening test.

(8) After generating the aerosol, the test subject shall be instructed to perform the exercises in Section 1.A.14. of this appendix.

(9) Every 30 seconds the aerosol concentration shall be replenished using one half the number of squeezes used initially (e.g., 5, 10, or 15).

(10) The test subject shall indicate to the test conductor if at any time during the fit test the taste of Bitrex™ is detected. If the test subject does not report tasting the Bitrex,™ the test is passed.

(11) If the taste of Bitrex™ is detected, the fit is deemed unsatisfactory and the test is failed. A different respirator shall be tried and the entire test procedure is repeated (taste threshold screening and fit testing).

5. Irritant Smoke (Stannic Chloride) Protocol

This qualitative fit test uses a person's response to the irritating chemicals released in the "smoke" produced by a stannic chloride ventilation smoke tube to detect leakage into the respirator.

(a) General Requirements and Precautions

(1) The respirator to be tested shall be equipped with high efficiency particulate air (HEPA) or P100 series filter(s).

(2) Only stannic chloride smoke tubes shall be used for this protocol.

(3) No form of test enclosure or hood for the test subject shall be used.

(4) The smoke can be irritating to the eyes, lungs, and nasal passages. The test conductor shall take precautions to minimize the test subject's exposure to irritant smoke. Sensitivity varies, and certain individuals may respond to a greater degree to irritant smoke. Care shall be taken when performing the sensitivity screening checks that determine whether the test subject can detect irritant smoke to use only the minimum amount of smoke necessary to elicit a response from the test subject.

(5) The fit test shall be performed in an area with adequate ventilation to prevent exposure of the person conducting the fit test or the build-up of irritant smoke in the general atmosphere.

(b) Sensitivity Screening Check. The person to be tested must demonstrate his or her ability to detect a weak concentration of the irritant smoke.

(1) The test operator shall break both ends of a ventilation smoke tube containing stannic chloride, and attach one end of the smoke tube to a low flow air pump set to deliver 200 milliliters per minute, or an aspirator squeeze bulb. The test operator shall cover the other end of the smoke tube with a short piece of tubing to prevent potential injury from the jagged end of the smoke tube.

(2) The test operator shall advise the test subject that the smoke can be irritating to the eyes, lungs, and nasal passages and instruct the subject to keep his/her eyes closed while the test is performed.

(3) The test subject shall be allowed to smell a weak concentration of the irritant smoke before the respirator is donned to become familiar with its irritating properties and to determine if he/she can detect the irritating properties of the smoke. The test operator shall carefully direct a small amount of the irritant smoke in the test subject's direction to determine that he/she can detect it.

(c) Irritant Smoke Fit Test Procedure

(1) The person being fit tested shall don the respirator without assistance and perform the required user seal check(s).

(2) The test subject shall be instructed to keep his/her eyes closed.

(3) The test operator shall direct the stream of irritant smoke from the smoke tube toward the face seal area of the test subject, using the low flow pump or the squeeze bulb. The test operator shall begin at least 12 inches from the facepiece and move the smoke stream around the whole perimeter of the mask. The operator shall gradually make two more passes around the perimeter of the mask, moving to within six inches of the respirator.

(4) If the person being tested has not had an involuntary response and/or detected the irritant smoke, proceed with the test exercises.

(5) The exercises identified in Section 1.A.14 of this appendix shall be performed by the test subject while the respirator seal is being continually challenged by the smoke, directed around the perimeter of the respirator at a distance of six inches.

(6) If the person being fit tested reports detecting the irritant smoke at any time, the test is failed. The person being retested must repeat the entire sensitivity check and fit test procedure.

(7) Each test subject passing the irritant smoke test without evidence of a response (involuntary cough, irritation) shall be given a second sensitivity screening check, with the smoke from the same smoke tube used during the fit test, once the respirator has been removed, to determine whether he/she still reacts to the smoke. Failure to evoke a response shall void the fit test.

(8) If a response is produced during this second sensitivity check, then the fit test is passed.

C. Quantitative Fit Test (QNFT) Protocols. Should quantitative fit testing of respirators be required or deemed desirable, please refer to the OSHA Standard text covering this topic.

Appendix B-1: User Seal Check Procedures (Mandatory)

The individual who uses a tight-fitting respirator is to perform a user seal check to ensure that an adequate seal is achieved each time the respirator is put on. Either the positive and negative pressure checks listed in this appendix, or the respirator manufacturer's recommended user seal check method shall be used. User seal checks are not substitutes for qualitative or quantitative fit tests.

I. Facepiece Positive and/or Negative Pressure Checks

A. *Positive pressure check.* Close off the exhalation valve and exhale gently into the facepiece. The face fit is considered satisfactory if a slight positive pressure can be built up inside the facepiece without any evidence of outward leakage of air at the seal. For most respirators this method of leak testing requires the wearer to first remove the exhalation valve cover before closing off the exhalation valve and then carefully replacing it after the test.

B. *Negative pressure check.* Close off the inlet opening of the canister or cartridge(s) by covering with the palm of the hand(s) or by replacing the filter seal(s), inhale gently so that the facepiece collapses slightly, and hold the breath for ten seconds. The design of the inlet opening of some cartridges cannot be effectively covered with the palm of the hand. The test can be performed by covering the inlet opening of the cartridge with a thin latex or nitrile glove. If the facepiece remains in its slightly collapsed condition and no inward leakage of air is detected, the tightness of the respirator is considered satisfactory.

II. Manufacturer's Recommended User Seal Check Procedures

The respirator manufacturer's recommended procedures for performing a user seal check may be used instead of the positive and/or negative pressure check procedures provided that the employer demonstrates that the manufacturer's procedures are equally effective.

Appendix B-2: Respirator Cleaning Procedures (Mandatory)

These procedures are provided for employer use when cleaning respirators. They are general in nature, and the employer as an alternative may use the cleaning recommendations provided by the manufacturer of the respirators used by their employees, provided such procedures are as effective as those listed here in Appendix B-2. Equivalent effectiveness simply means that the procedures used must accomplish the objectives set forth in Appendix B-2, i.e., must ensure that the respirator is properly cleaned and disinfected in a manner that prevents damage to the respirator and does not cause harm to the user.

I. Procedures for Cleaning Respirators

A. Remove filters, cartridges, or canisters. Disassemble facepieces by removing speaking diaphragms, demand and pressure-demand valve assemblies, hoses, or any components recommended by the manufacturer. Discard or repair any defective parts.

B. Wash components in warm, 43°C (110°F) maximum, water with a mild detergent or with a cleaner recommended by the manufacturer. A stiff bristle (not wire) brush may be used to facilitate the removal of dirt.

C. Rinse components thoroughly in clean warm, 43°C (110°F) maximum, preferably running water. Drain.

D. When the cleaner used does not contain a disinfecting agent, respirator components should be immersed for two minutes in one of the following:

1. Hypochlorite solution (50 ppm of chlorine) made by adding approximately one milliliter of laundry bleach to one liter of water at 43°C (110° F); or,

2. Aqueous solution of iodine (50 ppm iodine) made by adding approximately 0.8 milliliters of tincture of iodine (6 to 8 grams ammonium and/or potassium iodide/100 cc of 45% alcohol) to one liter of water at 43°C (110°F); or,

3. Other commercially available cleansers of equivalent disinfectant quality when used as directed, if their use is recommended or approved by the respirator manufacturer.

E. Rinse components thoroughly in clean, warm, 43°C (110°F) maximum, preferably running water. Drain. The importance of thorough rinsing cannot be overemphasized. Detergents or disinfectants that dry on facepieces may result in dermatitis. In addition, some disinfectants may cause deterioration of rubber or corrosion of metal parts if not completely removed.

F. Components should be hand-dried with a clean lint-free cloth or air-dried.

G. Reassemble facepiece, replacing filters, cartridges, and canisters where necessary.

H. Test the respirator to ensure that all components work properly.

Appendix C: OSHA Respirator Medical Evaluation Questionnaire (Mandatory)

To the employer: Answers to questions in Section 1. and to question 9 in Section 2 of Part A. do not require a medical examination.

To the employee:

Can you read (circle one): Yes/No

Your employer must allow you to answer this questionnaire during normal working hours, or at a time and place that is convenient to you. To maintain your confidentiality, your employer or supervisor must not look at or review your answers, and your employer must tell you how to deliver or send this questionnaire to the health care professional who will review it.

Part A. Section 1. (Mandatory) The following information must be provided by every employee who has been selected to use any type of respirator (please print).

1. Today's date: _____
2. Your name: _____
3. Your age (to nearest year): _____
4. Sex (circle one): Male/Female
5. Your height: _____ft. _____in.
6. Your weight: _____lbs.
7. Your job title: _____
8. A phone number where you can be reached by the health care professional who reviews this questionnaire (include the Area Code): _____
9. The best time to phone you at this number: _____
10. Has your employer told you how to contact the health care professional who will review this questionnaire (circle one): Yes/ No
11. Check the type of respirator you will use (you can check more than one category):
    a._____ N, R, or P disposable respirator (filter-mask, non-cartridge type only).
    b._____ Other type (for example, half- or full-facepiece type, powered-air purifying, supplied-air, self-contained breathing apparatus).
12. Have you worn a respirator (circle one): Yes/No
    If' "yes." what type(s):

Part A. Section 2. (Mandatory) Questions 1 through 9 below must be answered by every employee who has been selected to use any type of respirator (please circle "yes" or "no").

1. Do you *currently* smoke tobacco, or have you smoked tobacco in the last month: Yes/No
2. Have you *ever had* any of the following conditions?
   a. Seizures (fits): Yes/No
   b. Diabetes (sugar disease): Yes/No
   c. Allergic reactions that interfere with your breathing: Yes/No
   d. Claustrophobia (fear of closed-in places): Yes/No
   e. Trouble smelling odors: Yes/No
3. Have you *ever had* any of the following pulmonary or lung problems?
   a. Asbestosis: Yes/No
   b. Asthma: Yes/No
   c. Chronic bronchitis: Yes/No
   d. Emphysema: Yes/No
   e. Pneumonia: Yes/No
   f. Tuberculosis: Yes/No
   g. Silicosis: Yes/No
   h. Pneumothorax (collapsed lung): Yes/No
   i. Lung cancer: Yes/No
   j. Broken ribs: Yes/No

k. Any chest injuries or surgeries: Yes/No

l. Any other lung problem that you've been told about: Yes/No

4. Do you *currently* have any of the following symptoms of pulmonary or lung illness?

a. Shortness of breath: Yes/No

b. Shortness of breath when walking fast on level ground or walking up a slight hill or incline: Yes/No

c. Shortness of breath when walking with other people at an ordinary pace on level ground: Yes/No

d . Have to stop for breath when walking at your own pace on level ground: Yes/No

e. Shortness of breath when washing or dressing yourself: Yes/No

f. Shortness of breath that interferes with your job: Yes/No

g. Coughing that produces phlegm (thick sputum): Yes/No

h. Coughing that wakes you early in the morning: Yes/No

i. Coughing that occurs mostly when you are lying down: Yes/No

j. Coughing up blood in the last month: Yes/No

k. Wheezing: Yes/No

l. Wheezing that interferes with your job: Yes/No

m. Chest pain when you breathe deeply: Yes/No

n. Any other symptoms that you think may be related to lung problems: Yes/No

5. Have you *ever had* any of the following cardiovascular or heart problems?

a. Heart attack: Yes/No

b. Stroke: Yes/No

c. Angina: Yes/No

d. Heart failure: Yes/No

e. Swelling in your legs or feet (not caused by walking): Yes/No

f. Heart arrhythmia (heart beating irregularly):Yes/No

g. High blood pressure: Yes/No

h. Any other heart problem that you've been told about: Yes/No

6. Have you *ever had* any of the following cardiovascular or heart symptoms?

a. Frequent pain or tightness in your chest: Yes/No

b. Pain or tightness in your chest during physical activity: Yes/No

c. Pain or tightness in your chest that interferes with your job: Yes/No

d. In the past two years, have you noticed your heart skipping or missing a beat: Yes/No

e. Heartburn or indigestion that is not related to eating: Yes/No

f. Any other symptoms that you think may be related to heart or circulation problems: Yes/No

7. Do you *currently* take medication for any of the following problems?

a. Breathing or lung problems: Yes/No

b. Heart trouble: Yes/No

c. Blood pressure: Yes/No

d. Seizures (fits): Yes/No

8. If you've used a respirator, have you *ever had* any of the following problems? (If you've never used a respirator, check the following space and go to question 9:)

a. Eye irritation: Yes/No

b. Skin allergies or rashes: Yes/No

c. Anxiety: Yes/No

d. General weakness or fatigue: Yes/No

e. Any other problem that interferes with your use of a respirator: Yes/No

9. Would you like to talk to the health care professional who will review this questionnaire about your answers to this questionnaire: Yes/No

Questions 10 to 15 below must be answered by every employee who has been selected to use either a full-facepiece respirator or a self-contained breathing apparatus (SCBA). For employees who have been selected to use other types of respirators, answering these questions is voluntary.

10. Have you *ever lost* vision in either eye (temporarily or permanently): Yes/No

11. Do you *currently* have any of the following vision problems?

a. Wear contact lenses: Yes/No

b. Wear glasses: Yes/No

c. Color blind: Yes/No

d. Any other eye or vision problem: Yes/No

12. Have you *ever had* an injury to your ears, including a broken ear drum: Yes/No

13. Do you *currently* have any of the following hearing problems?

a. Difficulty hearing: Yes/No

b. Wear a hearing aid: Yes/No

c. Any other hearing or ear problem: Yes/ No

14. Have you *ever had* a back injury: Yes/No

15. Do you *currently* have any of the following musculoskeletal problems?

a. Weakness in any of your arms, hands, legs, or feet: Yes/No

b. Back pain: Yes/No

c. Difficulty fully moving your arms and legs: Yes/No

d. Pain or stiffness when you lean forward or backward at the waist: Yes/No

e. Difficulty fully moving your head up or down: Yes/No

f. Difficulty fully moving your head side to side: Yes/No

g. Difficulty bending at your knees: Yes/No

h. Difficulty squatting to the ground: Yes/ No

i. Climbing a flight of stairs or a ladder carrying more than 25 lbs: Yes/No

j. Any other muscle or skeletal problem that interferes with using a respirator: Yes/No

Part B. Any of the following questions, and other questions not listed, may be added to the questionnaire at the discretion of the health care professional who will review the questionnaire

1. In your present job, are you working at high altitudes (over 5,000 feet) or in a place that has lower than normal amounts of oxygen: Yes/No

   If 'yes', do you have feelings of dizziness. shortness of breath, pounding in your chest, or other symptoms when you're working under these conditions: Yes/No

2. At work or at home, have you ever been exposed to hazardous solvents, hazardous airborne chemicals (e.g., gases, fumes, or dust), or have you come into skin contact with hazardous chemicals: Yes/No

   If "yes," name the chemicals if you know them: ⎯⎯⎯⎯⎯⎯⎯⎯⎯⎯⎯⎯⎯⎯⎯⎯

───

3. Have you ever worked with any of the materials, or under any of the conditions, listed below:

a. Asbestos: Yes/No

b. Silica (e.g., in sandblasting): Yes/No

c. Tungsten/cobalt (e.g., grinding or welding this material): Yes/No

d. Beryllium: Yes/No

e. Aluminum: Yes/No

f. Coal (for example, mining): Yes/No

g. Iron: Yes/No

h. Tin: Yes/No

i. Dusty environments: Yes/No

j. Any other hazardous exposures: Yes/No

   If "yes," describe these exposures:

_____

_____

4. List any second jobs or side businesses you have: _____

_____

_____

5. List your previous occupations: _____

6. List your current and previous hobbies: _____

7. Have you been in the military services: Yes/No

   If "yes," were you exposed to biological or chemical agents (either in training or combat): Yes/No

8. Have you ever worked on a HAZMAT team: Yes/No

9. Other than medications for breathing and lung problems, heart trouble, blood pressure, and seizures mentioned earlier in this questionnaire, are you taking any other medications for any reason (including over-the-counter medications): Yes/ No

   If "yes," name the medications if you know them: _____

_____

10. Will you be using any of the following items with your respirator(s):

   a. HEPA Filters: Yes/No

   b. Canisters (for example, gas masks): Yes/ No

   c. Cartridges: Yes/No

11. How often are you expected to use the respirator(s) (circle "yes" or "no" for all answers that apply to you)::

   a. Escape only (no rescue): Yes/No

   b. Emergency rescue only: Yes/No

   c. Less than 5 hours *per week:* Yes/No

   d. Less than 2 hours *per day:* Yes/No

   e. 2 to 4 hours *per day:* Yes/No

   f. Over 4 hours *per day:* Yes/No

12. During the period you are using the respirator(s), is your work effort:

   a. *Light* (less than 200 kcal per hour): Yes/ No

   If "yes," how long does this period last during the average shift::

   hrs,_____ mins._____

   Examples of a light work effort are *sitting* while writing, typing, drafting, or performing light assembly work: or *standing* while operating a drill press (1-3 lbs.) or controlling machines,

   b. *Moderate* (200 to 350 kcal per hour): Yes/ No

   If "yes," how long does this period last during the average shift:

   hrs,_____mins._____

   Examples of moderate work effort are *sitting* while nailing or filing; *driving* a truck or bus in urban traffic; *standing* while drilling, nailing, performing assembly work, or transferring a moderate load (about 35 lbs.) at trunk level; *walking* on a level surface about 2 mph or down a 5-degree grade about 3 mph. or *pushing* a wheelbarrow with a heavy load (about 100 lbs.) on a level surface,

   c. *Heavy* (above 350 kcal per hour): Yes/No

   If "yes," how long does this period last during the average shift:

   hrs._____mins._____

   Examples of heavy work are *lifting* a heavy load (about 50 lbs.) from the floor to your waist or shoulder; *working* on a loading dock; *shoveling, standing* while bricklaying or chipping castings; *walking* up an 8-degree grade about 2 mph; *climbing* stairs with a heavy load (about 50 lbs.).

13. Will you be wearing protective clothing and/or equipment (other than the respirator) when you're using your respirator: Yes/No

   If "yes," describe this protective clothing and/or equipment:_____

14. Will you be working under hot conditions (temperature exceeding 77°F): Yes/No

15. Will you be working under humid conditions: Yes/No

16. Describe the work you'll be doing while you're using your respirator(s):_____

17. Describe any special or hazardous conditions you might encounter when you're using your respirator(s) (for example, confined spaces, life-threatening gases):_____

18. Provide the following information, if you know it, for each toxic substance that you'll be exposed to when you're using your respirator(s):

   Name of the first toxic substance: _____Estimated maximum exposure level per shift:_____ Duration of exposure per shift:_____ _____

   Name of the second toxic substance: _____Estimated maximum exposure level per shift:_____ Duration of exposure per shift:_____

   Name of the third toxic substance:_____ Estimated maximum exposure level per shift:_____ Duration of exposure per shift:_____

   The name of any other toxic substances that you'll be exposed to while using your respirator:

19. Describe any special responsibilities you'll have while using your respirator(s) that may affect the safety and well-being of others (for example, rescue, security):

Appendix D to 11910.134 (Mandatory) Information for Employees Using Respirators When Not Required under the Standard

Respirators are an effective method of protection against designated hazards when properly selected and worn. Respirator use is encouraged, even when exposures are below the exposure limit, to provide an additional level of comfort and protection for workers, However, if a respirator is used improperly or not kept clean, the respirator itself can become a hazard to the worker. Sometimes, workers may wear respirators to avoid exposures to hazards, even if the amount of hazardous substance does not exceed the limits set by OSHA standards. If your employer provides respirators for your voluntary use, or if you provide your own respirator, you need to take certain precautions to be sure that the respirator itself does not present a hazard.

You should do the following:

1. Read and heed all instructions provided by the manufacturer on use, maintenance, cleaning and care, and warnings regarding the respirators limitations.

2. Choose respirators certified for use to protect against the contaminant of concern. NIOSH, the National Institutes for Occupational Safety and Health of the U.S. Department of Health and Human Services, certifies respirators. A label or statement of certification should appear on the respirator or respirator packaging. It will tell you what the respirator is designed for and how much it will protect you.

3. Do not wear your respirator into atmospheres containing contaminants for which your respirator is not designed to protect against. For example, a respirator designed to filter dust particles will not protect you against gases, vapors, or very small solid particles of fumes or smoke.

4. Keep track of your respirator so that you do not mistakenly use someone else's respirator.

INTERNET REFERENCE

1.   http//www.access.gpo.gov/nara/cfr/waisidx/29cfr1910.html

III. EYE PROTECTION

The primary concerns for laboratory workers are impact and chemical splash protection. There are numerous commercial products available which meet the standards for eye protection in OSHA 29 CFR Part 1910.133. This standard is based on ANSI standard Z87.1-1968. Later revisions of this standard have been issued, and as other revisions are adopted, eye protection programs should incorporate more protective portions of these standards. Any protective devices purchased after July 5, 1994, must comply with the ANSI Z87.1-1989 standard or be demonstrated to be equally effective. Devices purchased earlier must comply with the 1968 version of the same standard. In the latter case, while not required, it is recommended that compliance with the later standard be followed.

Eye protection should be worn at all times while working with chemicals in the laboratory. In the author's experience, failure to wear goggles is one of the most common safety violations in the laboratory. All too frequently, one will observe a large sign on the entrance to a laboratory requiring the wearing of goggles and yet the occupants of the facility failing to wear them. A large percentage of the issues of *Chemical and Engineering News* contain letters reporting unexpected explosions in the laboratory, many of which could cause injury to the eye for persons not protected by goggles. Visitors should be provided with temporary protective goggles or, at least, protective glasses if they are allowed in any area in which the occupational use of eye protection is required.

A. Chemical Splash Goggles

There are dozens of brands of chemical splash goggles available, almost all of which meet the basic standards in ANSI Z87.1 for this type of eye protection. However, there are wide variations in the degree of acceptance of these goggles by the users. Chemical splash goggles should fit snugly and comfortably around the eyes. The goggles should "breathe," i.e., the wearer should not overheat under them and perspire. They should not fog. They should provide good peripheral vision. Preferably, they should be compatible with the wearing of respirators. It would be desirable (if not essential) if prescription glasses could be worn under the goggles, although some styles of glasses are too big for most chemical splash goggles. The goggles should be easy to clean. Goggles made of hypoallergenic materials are available for wearers irritated by specific materials.

Just because a pair of chemical splash goggles is provided with "ports" to allow air into the space behind the lens of the goggle and has an antifogging coating, it does not necessarily mean that all models with these features will be comparably effective. Ports on the side of the goggles appear to be less successful in most cases in eliminating fogging than openings around the edge of the lens, where motion of the head moves air directly across the lens. The latter configuration also appears to work well in removing heat. Not all antifogging coatings appear equally effective, nor in some cases does product control quality appear to be uniform, even within a single brand. In at least one case involving over 1000 pairs of goggles, the coating worked very well on about half of the goggles, while for the remainder, the lens fogged up within 15 minutes of the users donning them. Where ventilation is through openings around the edge of the lense, the design needs to be such that air enters through a circuitous route in order that no chemical could be splashed through the openings.

Because it is so important to ensure that personnel in laboratories wear eye protection under circumstances in which it is needed, it is desirable to test goggles under actual use conditions. Not only must chemical splash goggles meet the required physical specifications, but they must be sufficiently comfortable to be accepted by the users as well. A pair of goggles pushed up on the forehead or lying on the work bench does not afford eye protection. Price is not necessarily a valid guarantee of quality. A mid-price unit may perform as well or better than a higher price unit. Products change over time, and newer products are continually coming on the market. Prior to selecting a specific chemical splash goggle, it is recommended that it be tested under actual use conditions in comparison with a selection of other units which meet not only the OSHA and ANSI standards, but the criteria mentioned in the introductory paragraph to this section. A vendor's claim that a given product will perform as well as a tested unit needs to be verified. Making sure of the quality and efficacy of a goggle is especially important if buying large quantities to be used in instructional labora-tories in school or in a process involving a large number of employees.

If the probability of a vigorous reaction appears to be substantial, or the material involved in the work in progress is very corrosive to tissue, a face mask should be used to supplement the splash goggles and provide additional protection to the face and throat. Where there is a risk of a minor explosion, an explosion shield should be placed between the worker and the reaction vessel. A wraparound shield will provide protection to the sides as well as directly in front of the shield.

The most commonly used lens material in safety goggles and safety glasses is polycarbonate. Typically the material used in the goggles is 0.060 inches (1.52 mm) thick. It is also used in face masks and, in somewhat greater thicknesses, in explosion shields. This material is lightweight and tough and resists impacts and scratches. Models coated with silicone are resistant to a number of chemicals.

The close fit to the face provided by chemical splash goggles and the strap around the head provides good stability against lateral impacts which might knock ordinary safety spectacles off.

Many laboratory workers need to wear prescription glasses and not all goggles will accommodate the larger lenses and unusual frames favored by many wearers. The need to accommodate these individuals should be taken into account. The temple pieces of ordinary glasses will prevent a tight fit in the area where the temple pieces pass under the edge of the goggles, but the gap should be small and, in any event, will be further back on the head than the eye area.

B. Safety Spectacles

Safety spectacles (which resemble ordinary prescription glasses) that meet ANSI Z87.1 for impact protection offer very limited protection against chemicals. They do not fit tightly against the face and would not prevent chemicals from running down from the forehead into the eyes. They would provide some protection against flying glass in the event of a reaction vessel exploding. Side shields are used to protect the eyes from flying objects from the side. However, as noted in the previous section, chemical splash goggles are form fitting and are held on tightly, so that they resist lateral impacts considerably better than most safety spectacles.

Because they do not fit snugly against the face, safety spectacles do not have any more problems with heat and fogging than would ordinary glasses. In laboratory facilities which do not use chemicals, but do offer opportunities for mechanical injuries, safety glasses are acceptable. Many companies offer safety spectacles as prescription glasses and in attractive choices of frames. Some individuals who would resist wearing safety spectacles because they think that they must be unattractive can be provided with a choice of glasses which should prove satisfactory to almost any taste.

C. Contact Lenses

The suitability of wearing contact lenses in chemical laboratories has long been under discussion. In the event of a chemical accident to the eyes, there could be some protection but, on the other hand, the presence of the lens would be an impediment to prompt and thorough flushing of the eyes. The lens would have to be removed which might result in damage to the eye in itself. If, however, the wearer of contacts lens would conscientiously wear a good quality pair of goggles *at all times* during which the possibility of an incident might occur, there is probably little risk in wearing contact lens. Even in the latter case, where extremely corrosive vapors are likely to be involved, there is a possibility of capillary action causing these vapors to be drawn under the contact lens, and the wearer should exercise caution if there is any suspicion that such could be happening.

There are a few rare medical conditions where it is essential that contact lens be worn, to maintain the proper shape of the eye lens. Clearly, contact lens must be worn in such cases but with due precautions.

IV. MATERIALS FOR PROTECTIVE APPAREL

As Keith[5] stated, "How many times have we seen the phrase 'use appropriate protective materials'?" This is true in some other areas as well for specific types of protective gear. However, in the field of chemically protective materials, where personnel must depend on protective gear, it is essential that they select effective protective clothing and articles, such as gloves, which will provide protection to the wearer against contact with the hazardous substances which they use in their laboratories. An examination of the catalog description of many of the items of protective apparel shows that some of the advertisements at best provide only qualitative descriptions of the efficacy of the materials used in the products. However, some firms will provide technical supportive data upon request.*

A. Recommended Information Sources

Two publications (one of which is a two-component computerized book and expert selection system) provide a substantial amount of detailed information, not only by types of materials but by brand names, since not all versions of a given material have identical properties. These two publications are briefly described below. Both are recommended.**

The first component of the computerized system is a computerized reference book compiled by Kristan Forsberg:[8] *Chemical Permeation and Degradation Database and Selection Guide for Resistant Protective Materials.* According to the vendor, the database contains over 4200 permeation tests on more than 540 compounds and mixtures. Included are more than 6000 breakthrough times or permeation rates and a total of over 20,000 pieces of associated data, which includes information on the test material, manufacturer model number, thickness, comments, a safety guide number, and references. The stress is on gloves.

---

* NORTH HAND PROTECTION (Siebe North, Inc.) is a good example in their advertisements for their Silver Shield™ glove where they provide detailed breakthrough times and permeation rates for 49 chemicals for their gloves and give comparison data for several other materials. They claim, "It resists permeation and breakthrough by more toxic/hazardous chemicals than any other type of glove on the market today." Tests seem to support this statement; see Reference 10. Chemical companies also are improving in their information in their catalogs. Life-Net FAX Hotline will provide data at 1-800-447-2436 24 hours per day.

** Unfortunately, the first of these may be out of print but may still be available in research libraries.

The program is an outgrowth of "Guidelines for the Selection of Chemical Protective Clothing," which appeared in *Performance of Protective Clothing* edited by R.L. Barker and G.C. Coletta, American Society for Testing and Materials, 1986.

The second component of the computerized system is an "expert" system called GlovEs, which, given a set of initial conditions or parameters, screens the information in the data base and makes recommendations based on the needs as defined by the user of the system. The system allows considerable flexibility in seeking information and quickly provides the data in a useful form. The program runs on a standard IBM or compatible PC.

The second source of data is the third edition of *Guidelines for the Selection of Chemical Protective Clothing.* The work was sponsored by the U.S. EPA and the U.S. Coast Guard. The publication is organized in two volumes, the first of which contains an overview of the general topic of chemical protective clothing (CPC) and tables which can be used to properly select and use CPC. Twelve major clothing materials are evaluated in the context of about 500 different chemicals and permeation data, including 25 multiple-component organic solutions. Of particular interest are Appendices I and J, which define equipment needed to provide different levels of protection and procedures for using protective gear in decontamination efforts. The latter information is of value to organizations who may have to occasionally respond to emergencies involving chemical spills but, fortunately, not sufficiently frequently so that they find it easy to maintain their skills.

The second volume is a technical support base for Volume I. It provides the data on which the recommendations in Volume I are based and some of the theoretical material used in arriving at the recommendations.

B. Overview of Chemical Protective Clothing[*]

The purpose of chemical protective clothing (CPC) is to prevent chemicals from reaching the skin. The chemicals can do this in two ways: permeate the material of which the clothing is made, or enter through penetrations in the clothing. The two sources described in the previous section concentrate on the problem of permeation and breakthrough, although the second of the two, in the introductory portions, discusses the penetration issue.

Much of the data in the two resources is based on manufacturer's subjective evaluations, i.e., the material provided "excellent," "good," "fair," or "poor" protection, and much of these subjective evaluations were based on visible degradation of the product. However as noted in the *Guidelines,* "...it has been found that chemicals can permeate a material without there being any visible sign of problems." In addition, in the reported evaluations, the temperatures at which the tests were performed often are not given and the permeation rate has been found to depend significantly on the temperature. The thicknesses of the materials tested also are not given in many cases. More scientific means of comparative evaluations need to be adopted, and the techniques are available.

In some cases, the chemical to which the barrier material is exposed will simply diffuse through the barrier. In others, the chemical will react with the barrier material and degrade the performance of the barrier, for example, by changing the chemical properties of the material or by leaching out some of the components in the material. As far as penetration is concerned, it is obviously desirable for the material to be a poor absorber of the challenging chemical and the rate of permeation to be slow.

Chemicals, once they have begun to permeate a material will continue to do so, even after the challenge has been removed from the surface, because of the chemical already absorbed within the material. If the amount of material absorbed is large and the rate of permeation is relatively high, the chemical may eventually penetrate through the barrier material after

[*] Many of the concepts in this section are developed more fully in Reference 9.

the protective clothing has been removed and stored. The contaminant may cause exposure the next time it is used if it has not had an opportunity to diffuse away. If the clothing is carefully folded and placed in a container, the possibility of contaminant being trapped in the clothing is substantial. Another possibility is that the breakthrough will result in pinhole leaks and the item, which still afforded protection when removed, will no longer do so the next time it is used. If the amount absorbed is small and the permeation rate is slow, the diffusion back through the entering surface may result in a negligible amount penetrating to the interior surface.

Some articles of protective clothing, such as gloves, have no openings which could come into contact with chemicals. However, other items such as coats, jackets, and trousers usually have one or more edges which are intended to be opened and closed, and all of these garments and others have seams where sections of the materials are joined together.

Protective clothing is normally fabricated of sections of materials which are often welded together chemically or by heat. The resulting garments should have no means for chemicals to penetrate due to the fabrication process, but the seams do have a tendency to split under strain. Many have sewn seams for added strength. The sewn seams, unless sealed afterward, will leave pinholes through which chemicals can penetrate.

The normal openings, such as the fronts of coats or overalls, are fastened by zippers, pressure-locking lips, or buttons. All of these should be supplemented by inner and outer flaps if they are intended to eliminate the possibility of chemicals penetrating through these openings. For total encapsulating clothing, boots, hoods, and gloves may be integral parts of the suits with a minimal number of openings for which secure seals have to be provided. Where separate items are used, arrangements for seals at the neck, feet, and hands must be provided. Many of these totally encapsulating suits will be used at a slight positive pressure to keep vapors which might occur from penetrating through any small gaps in the protective clothing.

The visor is a key component of any chemical splash suit. Not only must the visor material withstand the effects of the chemicals and maintain good visibility, but a good seal between the visor and the suit is essential and the seal to the fabric portion of the suit should be checked frequently. Normally visors are made of materials such as polycarbonates, acrylics, fluorinated ethylene propylene (FEP), and clear polyvinyl chloride. The first two of these are subject to "crazing" upon contact with some chemicals. Covering these two types of material with a thin layer of FEP is sometimes done to protect them.

A major problem with a totally enclosing chemical protective suit is elimination of body heat. If the air supply is carried with the user (see the material on respiratory protection earlier in this chapter), a common practice is to wear a cooling vest which has pockets for ice cubes and cooling liquid. If the source of air is an external supply, a vortex tube cooling unit is often incorporated in the system. In order to protect the tubes, valves, etc. associated with the air supply system from the effects of chemicals, the air supply components are often worn inside the protective suit. High body temperature is an especially troublesome problem in extended field uses of protective clothing. Workers should take a break for at least 10 minutes of each hour during normal summer use, open the clothes in a safe location, and replenish the salt in the body with cool liquids containing a modest level of sodium. Heat exhaustion or heat stroke could result from the overheating due to wearing air-tight protective clothing, unless care is taken.

The second reference source mentioned earlier provides a wealth of information on the properties of protective materials for individual chemicals. Reference 10 reports on several later studies of many of the same materials challenged by many of the same chemicals. They generally support the findings in reference 9. Table 6.2 is adapted from this source for the resistance of several materials for various classes of chemicals.

The nomenclature used in the table is the following. RR, R, rr, and r represent various positive degrees of resistance while NN, N, nn, and n represent various degrees of poor resistance. Double characters indicate that the rating is based on test data, and single characters, on qualitative data. Upper case letters indicate a large body of consistent data, while lower case letters indicate either a small quantity of data or inconsistent information. Asterisks (**) mean that the material varied considerably in its resistance to chemicals within a class and data for specific chemicals should be used if available, or an alternative selected.

The column headings in Table 6.2 stand for the following materials:

Butyl—Butyl rubber
CPE—Chlorinated polyethylene
Viton/Neoprene—Layered material, first material on surface
Natural Rubber—same
Neoprene—same
Nitrile + PVC—Nitrile rubber + polyvinyl chloride
Nitrile—Nitrile rubber
PE—Polyethylene
PVA—Polyvinyl alcohol
PVC—Polyvinyl chloride
Viton—same
Butyl/Neoprene—Layered material, first material on surface

In addition to the chemical properties of materials, a number of physical properties of materials are also of importance in selecting an appropriate material. Table 6.3 is also adapted from Reference 9. The qualities listed in the table may also be affected by factors such as thickness, formulation, and whether or not there is a fabric backing to the material.

Much of the information used in the preparation of both resource books came from the use of the tested materials in the fabrication of gloves, and there are some differences in how gloves are made and how items of protective clothing are made. However, the basic information should be similar, if not identical.

The severity of the application will govern the choice of protective clothing in most cases. If the clothing is to be exposed to severe abuse and the environment is unusually hazardous, the choice should be the most durable and protective material available. In other cases, less protective and/or less durable units, or even disposable items, might prove entirely acceptable. Experience should assist in selecting brands. There have been instances where well-known brands have experienced runs of high levels of problems with seam failures in their moderately priced lines, while other brands of comparable price have not demonstrated these difficulties. Such problems should be documented and, unless the difficulties are resolved with the manufacturer, used in the selection process. User confidence in the protection offered by the protective items is too important to allow competitive cost to be the only, or even the primary factor, in selecting CPC items. One area of critical importance to avoid rupture of the clothing is to ensure that it is of ample size. Protective clothing is often worn over ordinary clothing so the protective apparel should afford ample room for these garments while bending, stretching, turning, etc. in the course of the job task.

An important area not touched upon as yet as far as protective clothes are concerned is an alternative to the asbestos gloves used in laboratories to handle hot objects. Asbestos gloves observed in the laboratory often are in poor condition, i.e., the material of which they are made has become very friable. Any asbestos items in poor condition, especially those employed as asbestos gloves are, should be discarded as hazardous material through the organization's hazardous waste program. It would be desirable to eliminate asbestos gloves entirely. Alternatives that have been found acceptable for many high-temperature laboratory applications are gloves made of materials such as Kevlar™, Nomex™, Zetex™, and fiberglass.

Table 6.2 Resistant Properties of Selected Materials by class

Chemical	Butyl	CPE	Viton/ Neoprene	Natural Rubber	Neoprene	Nitrile + PVC	Nitrile	PE	PVA	PVC	Viton	Butyl Neoprene
Acids, carboxylic, aliphatic												
Unsubstituted	R	r	r	**	rr	**	rr	NN	**	**	**	r
Polybasic					rr	rr	rr	rr	n	rr		
Aldehydes												
Aliphatic and alicyclic	RR	NN	r	**	NN	nn	NN	**	NN	NN	**	r
Aromatic and heterocyclic	rr		n	nn	nn	n	nn	NN	rr	N		r
Amides	rr			**	nn		nn	nn			nn	
Amines, aliphatic, alicyclic												
Primary	**	**	n	NN	**		rr		NN	**	**	
Secondary	**		n	NN	nn		**		**	NN	nn	n
Tertiary	**	**		**	rr	**	**		**	**	rr	
Polyamine	**			NN	rr	nn				NN	rr	
Cyanides					r	nn						
Esters, carboxylic												
Formates			n									
Acetates	**	**	n	NN	nn		NN	NN	**	NN	n	**
Higher monobasic	nn	nn	**	NN	nn		nn	NN	rr	NN		**
Polybasic			r	r	r		**			rr		r
Aromatic phthalate	rr		r	**	**	**	**			nn	rr	r
Ethers												
Aliphatic	**	rr	**	NN	**	**	**		**	**		**
Halogen compounds												
Aliphatic, unsubstituted	nn	nn	r	NN	NN	NN	NN	NN	**	NN	**	n
Aliphatic, substituted	**			NN	rr		nn		**	NN	rr	
Aromatic, unsubstituted	nn	nn	r	NN		n	nn	NN		N	rr	n
Polynuclear				NN	nn					n	rr	
Vinyl halides				NN						n	rr	

Table 6.2 Resistant Properties of Selected Materials by class (continued)

Chemical	Butyl	CPE	Viton/Neoprene	Natural Rubber	Neoprene	Nitrile + PVC	Nitrile	PE	PVA	PVC	Viton	Butyl Neoprene
Heterocyclic compounds												
Epoxy compounds	**			**	nn		nn	NN	**	nn	NN	
Furan derivativs	nn	nn	nn									
Hydrazines	**		n	**	**		**		nn	**	**	n
Hydrocarbons												
Aliphatic and alicyclic	N	r	r	NN	**	**	**	**	**	NN	RR	N
Aromatic	**	rr	r	NN	NN	NN	**	NN	**	NN	RR	r
Hydroxyl compounds aliphatic, and alicyclic												
Primary	RR	rr	rr	nn	**	nn	**	**	**	**	rr	**
Secondary	rr	rr	r	**	**	**	rr		rr	**	rr	r
Tertiary	r			**	rr	rr	rr			**		
Polycis	r		**	rr	rr	rr	rr			**		**
Aromatic	**	**	r	**	**	**	**	**	nn	**	rr	r
Inorganic acids	**	**	rr	**	**	**	**	**	nn	**	rr	r
Inorganic bases	r	r		RR	RR	**	RR	**	n	**	rr	**
Inorganic gases	**	r	n	n	r			**		**	**	**
Inorganic salts	r		n		r	r	r			R		
Isocyanates				NN	n				rr			
Ketones, aliphatic	**	NN	n	NN	NN	N	**	NN	**	NN	NN	**
Nitriles, aliphatic	rr			NN	**			NN	rr	NN	rr	
Nitro componds, unsubstituted	rr	r		NN	**		nn		**	**	**	**
Organophosphorus compounds			r									r
Peroxides				r								
Sulfur compounds												
Thios			**									n

Note: see text for explanation of abbreviations.

743

Table 6.3 Physical Characteristics of Chemical-Resistant Materials

Material	Abrasion Resistance	Cut Resistance	Flexibility	Heat Resistance	Ozone Resistance	Puncture Resistance	Tear Resistance	Relative Cost
Butyl rubber	F	G	G	E	E	G	G	High
Chlorinated Polyethylene (CPE)	E	G	G	G	E	G	G	Low
Natural rubber	E	E	E	F	P	E	E	Medium
Nitrile-butadiene rubber (NBR)	E	E	E	G	F	E	G	Medium
Neoprene	E	E	G	G	E	G	G	Medium
Nitrile rubber (nitrile)	E	E	E	G	F	E	G	Medium
Nitrile rubber + polyvinyl chloride (Nitrile + PVC)	G	G	G	F	E	G	G	Medium
Polyethylene	F	F	G	F	F	P	F	Low
Polyurethane	E	G	E	G	G	G	G	High
Polyvinyl alcohol (PVA)	F	F	P	G	E	F	G	Very High
Polyvinyl chloride (PVC)	G	P	F	P	E	G	G	Low
Styrene-butadiene rubber (SBR)	E	G	G	G	F	F	F	Low
Viton	G	G	G	G	E	G	G	Very High

Note E = excellent, G = Good, F = fair, p = poor

Table 6.4 Permissable Noise Exposure

Duration (Hours/Day)	Sound Level (dBA Slow Response)
8	90
6	92
4	95
3	97
2	100
1.5	102
1.0	105
0.5	110
0.25 or less	115

or combinations of these materials. Zetex ^{TM}, for example, is specifically advertised as a replacement for asbestos for high-temperature applications.

V. Hearing Protection

The noise levels in most laboratories are usually not excessive, but there are laboratory facilities in which noise can reach levels for which hearing protection should be provided or the employees required to be involved in a hearing conservation program. It would be preferable, of course, if the noise levels could be lowered rather than to depend upon personal protective devices.

OSHA has adopted a comprehensive hearing conservation program in 29 CFR Part 1910.95. Under this standard, any employee exposed to an 8-hour time weighted average of 85 dB, as measured on a properly calibrated sound level instrument on the slow response (A) scale, must be placed in an employer-run hearing conservation program. Among other requirements, an annual audiometric test is required. Loss of hearing as one grows older is normal, and there are diseases resulting in hearing losses which are not occupationally related. A properly administered audiometric test should, in many cases, be able to distinguish these two causes from a loss of hearing due to external factors. One of the difficulties in determining if the loss of hearing is occupationally related is that the individual's lifestyle during non-working hours can affect his hearing as well. Prolonged listening to loud music can cause problems, and so can frequent shooting of firearms for recreation. Workers' compensation claims are sometimes disallowed because the evidence is not clear that the hearing loss is occupationally related. Sound level measurements and sound dosimetry measurements in the workplace, as well as periodic audiometric tests, are all needed. At this time, the record of enforcement of the OSHA standard is not impressive, and the number of workers' compensation cases processed has been small. Hearing is vital, especially to a laboratory employee, since loss of hearing may make it impossible to communicate adequately or, in some cases, even perform some laboratory operations.

According to Table 6.4, at average noise levels over an 8-hour day in excess of 90 dB, unprotected workers must have their working hours reduced. The table provides that the work interval should be cut in half if the sound level goes up by 5 dB, implying that the sound level goes up by two when the measured level increases by 5 dB. Actually, the sound level increases by a factor of two for a measured increase of 3 dB, so that the sound level at 105 dB is 32 times that at 90 dB instead of 8, as could be inferred from the table.

There are many different types of hearing protection on the market. The simplest type is an earplug which is placed within the ear. Typically, these are soft foam which conform to the ear

Table 6.5 Laboratory First Aid Kits

Adhesive bandages, various sizes	Antiseptic wipes
Sterile pads, various sizes	Cold packs
Sterile sponges	Burn cream
Bulk gauze	Antiseptic cream
Eye pads	Absorbent cotton
Adhesive tape	Scissors
First aid booklet	Tweezers

canal and are effective in reducing noise levels. Many can be washed and reused, if desired. Some persons do not like to use this type of hearing protection because they do not like to keep the plugs in their ears and do not like to be continually taking them in and out as they go from noisy areas to quieter ones. There are many inexpensive earmuffs on the market which attenuate sounds by 20 to 30 dB. The attenuation of the hearing protection devices varies with frequency. The ear muffs can be taken on and off and kept available for reuse. These should be individually assigned or cleaned thoroughly between use if used by more than one person. Over a relatively short interval, they will be more economical than ear plugs, and often have a higher degree of acceptance by the users.

VI. FIRST AID KITS

A first aid kit for a laboratory or for most areas should be intended to provide immediate treatment for most minor injuries or burns, not serve as a substitute for the family medical cabinet. It also need not be a large unit. Most injuries within a laboratory are to individuals. If a major accident were to occur, it would be necessary to call in emergency medical personnel rather than to try local treatment. The contents of a first kit should include the supplies listed in Table 6.5. Other optional items can be added such as those mentioned in the following paragraph, but not medicines to be given internally.

Note that there are no tourniquets, aspirin (or other medicines to be taken internally), iodine, or merthiolate. Possible additions to the above list could be Ipecac, used to induce vomiting, and activated charcoal, to help absorb poisons internally, but if these are supplied, persons should receive specific training in how to use them properly. As noted in Chapter 1, it would be highly desirable if a number of persons in a laboratory facility would receive formal training in first aid and CPR. A specific person should be designated to be responsible for maintaining the supplies in the first aid kit.

VII. OTHER PERSONAL PROTECTIVE EQUIPMENT

There are several other items classified under OSHA personal protective equipment regulations. For example, in high voltage laboratories, there are a number of specific devices which are required to assure protection against electrocution, including hard hats and gloves specifically designed to withstand conduction of electricity. Where electrical protection is not needed, there are requirements for hand, foot, and head protection. Where individuals are involved in welding, regulations define what sort of goggles are needed. There are OSHA regulations for the types of equipment needed to protect divers. There are legitimate research areas where all of these types of protective equipment would be needed as well as many others, but it is unlikely that many of these areas would fall under the intended scope of this handbook. Should they do so, reference to the appropriate sections of OSHA would, of course, be dictated in addition to whatever specific training programs would be required to safely perform the research.

REFERENCES

1. General Industry Standards, 29 CFR Section 1910.134, Respiratory Protection, OSHA, Washington, D.C.
2. Practices for Respiratory Protection, ANSI Z88.2, American National Standards Institute, New York, 1980.
3. Practices for Occupational and Educational Eye and Face Protection, ANSI Z87. 1, American National Standards Institute, New York, 1989.
4. Recommendations for Prescription Ophthalmic Lenses, ANSI Z80. 1, American National Standards Institute, New York, 1989.
5. **Keith, L.H.,** Technology blends computers, books, for comparing protective materials, *Occup. Health Saf.,* 56(11), 74, 1987.
6. **Forsberg, K.,** *Chemical Permeation and Degradation Database and Selection Guide for Resistant Protective Materials,* Instant Reference Sources, Austin, TX, 1987.
7. **Forsberg, K.,** *Chemical Protective Clothing Permeation and Degradation Database,* Keith, L.H., Ed., Lewis Publishers, Boca Raton, FL, 1992.
8. **Schwope, A.D., Costas, P.P., Jackson, J.O., Stull, J.O., and Weitzman, D.J.,** Eds., *Guidelines for the Selection of Chemical Protective Clothing,* 3rd ed., Arthur D. Little, U.S. Environmental Protection Agency and U.S. Coast Guard, American Conference of Governmental Industrial Hygienists, Cincinnati, OH, 1987.
9. **Briarty J.P. and Henry, N.W.,** *Performance of Protective Clothing,* Vol. 4, American Society for Testing and Materials, West Conshohacken, PA, 1992.
10. ASTM F 1001, Standard Guide for Selection of Chemicals to Evaluate Protective Clothing Materials, American Society for Testing and Materials, West Conshohacken, PA, 1989.

INTERNET REFERENCE

1. http://www.access.gpo.gov/nara/cfr/wisidx/29cfr1910.html

Appendix: Laboratory Checklist

Each organization's chemical hygiene plan should incorporate a laboratory checklist so the laboratory manager and the other employees will have a benchmark as to what standards their facility should meet. The following is an example of a representative BASIC check list. Organizations will likely have specific variations on this checklist which will better meet their needs. Many will probably be more extensive and detailed.

A. General

1. Housekeeping satisfactory
2. Aisles not cluttered, paths of egress maintained free of obstructions
3. Hazard warning signs on outside at entrance(s)
4. Laboratory Authority List at entrances on outside
5. Work area separated from study/social areas
6. Laboratory Hygiene Plan written and available, including
   Written Standard Operating Procedures available
   Material Safety Data Sheets available
   Emergency response and evacuation plan
7. Food not stored in laboratory refrigerators
8. Equipment in good condition, preventive maintenance plan in place

B. Fire Safety

1. Flammables stored in flammable material storage cabinets
2. Class ABC fire extinguisher in laboratory, near door or on path of egress
3. Fire blankets available
4. Flammable material stocks maintained at minimal levels, within limits
5. Refrigerators used for flammables are flammable material storage units or explosion-proof
6. Two well-separated exits, doors swing outward for hazard class labs
7. Flammables not stored along path of egress

C. Chemical Handling

1. Chemicals stored according to compatibility
2. Ethers identified by date of receipt and latest date for disposal. Other chemicals which degrade and become unsafe also treated similarly.
3. Other chemicals with notably dangerous properties identified and stored safely
4. All containers properly labeled, including secondary containers
5. Quantities of chemicals not excessive
6. Chemicals stored at safe levels, in cabinets or on stable shelving; no chemicals on floor
7. Chemical waste properly labeled and segregated prior to disposal, removed frequently
8. Gas cylinders strapped firmly in place; cylinders not in use capped; oxidizing and reducing gases properly segregated
9. Perchloric acid quantities maintained at minimum levels; perchloric acid hoods available for hot perchloric acid applications
10. Apparatus marked with warning signs or protected by barriers if susceptible to damage
11. Work generating toxic and hazardous fumes done in hoods
12. Work capable of causing an explosion behind protective barriers; vacuum vessels taped; warning signs in place; employees made fully aware of risks, etc.

D. Ventilation

1. Ventilation 100% fresh air, 6 to 12 air changes per hour
2. Laboratory at negative pressure with respect to corridors
3. Hoods located in low traffic draft-free zones
4. Hoods capable of maintaining 100 fpm face velocity with sash fully open
5. Low-velocity warning alarm on hoods
6. Fume generating apparatus with hoods at least 20 cm from sash opening
7. Local exhaust units used where hoods not suitable
8. Hoods not used for storage of surplus materials
9. No modifications made to hoods that would reduce their effectiveness

E. Electrical

1. All electrical circuits three-wire
2. No circuits overloaded with extension cords or multiple connection
3. No extension cords used unsafely, cords protected or in raceways
4. Apparatus equipped with three-prong plugs or double-insulated
5. Motors are nonsparking
6. Heating apparatus equipped with redundant temperature controls
7. Adequate lighting, lights in hoods protected from vapors

8. Circuits, equipment provided with ground-fault interrupters as needed
9. Electrical equipment properly covered
10. Breaker panel accessible
11. Deluge shower located so water will not splash on electrical equipment or circuits
12. GFI devices in use where use is indicated for personnel safety

F. Safety Devices

1. Eyewash station available, checked at least semiannually
2. Deluge shower available, checked at least semiannually
3. First aid kit available and fully maintained
4. Personal protective equipment: goggles, face masks, gloves, aprons, respirators, explosion shields, escape breathing masks, available and used as needed
5. Evacuation routes marked
6. No smoking, other safety signs posted and observed
7. Chemical waste properly labeled and segregated prior to disposal, removed frequently
8. Biological wastes segregated, stored properly in clearly marked containers, removed frequently
9. Sharps containers available and used.
10. Radioactive wastes segregated, stored properly in clearly designated containers, removed frequently
11. Gas cylinders strapped firmly in place, cylinders not in use capped, oxidizing and reducing gases properly segregated

G. Records and Training

1. All required records properly maintained
2. All required training up-to-date

INDEX